8th Biennial Autonomous VTOL Technical Meeting & 6th Annual Electric VTOL Symposium 2019

Mesa, Arizona, USA
28 January – 1 February 2019

ISBN: 978-1-5108-8874-6

Printed from e-media with permission by:

Curran Associates, Inc.
57 Morehouse Lane
Red Hook, NY 12571

Some format issues inherent in the e-media version may also appear in this print version.

Printed by Curran Associates, Inc. (2019)

For permission requests, please contact Vertical Flight Society
at the address below.

Vertical Flight Society
2701 Prosperity Ave, Suite 210
Fairfax, VA 22031
USA

Phone: (703) 684-6777
Fax: (703) 739-9279

www.vtol.org

Additional copies of this publication are available from:

Curran Associates, Inc.
57 Morehouse Lane
Red Hook, NY 12571 USA
Phone: 845-758-0400
Fax: 845-758-2633
Email: curran@proceedings.com
Web: www.proceedings.com

Thank you to our sponsors and exhibitors

Platinum-Level Sponsors

BELL · BOEING · KAREM AIRCRAFT, INC. · NEXA Capital Partners, LLC · TERRAFUGIA

Gold-Level Sponsors

E|S|AERO · Gannett Fleming · Honeywell THE POWER OF CONNECTED · SentientScience · THALES · XTI AIRCRAFT

Exhibitors

CDI Continuum Dynamics, Inc · DSIAC Defense Systems Information Analysis Center · NEXA Capital Partners, LLC · NFT · Pi Innovo · SAE INTERNATIONAL

Vertical Flight Society

www.vtol.org

Monday, January 28, 2019: Registration Desk Hours 7:00 am–5:00pm *(Canyon Foyer)*		
8:00 am – 5:00 pm: **Short Course on Electric VTOL Design**, Dr. James Wang		
Tuesday, January 29, 2019: Registration Desk Hours 7:00 am–5:00pm *(Canyon Foyer)*		
Tuesday-Thursday: **Exhibits** *(Canyon AB Foyer)*		
Plenary Presentations *(Canyon Ballroom)*		
Needs for Future Autonomous VTOL Aircraft Moderator: Dr. Ram JanakiRam, The Boeing Co.		
8:00–8:10	**Welcome,** Dr. Ram JanakiRam, The Boeing Co.	
8:10–8:30	**Autonomy for Next Generation VTOL Systems,** Dr. Bill Lewis, US Army, Aviation Development Directorate	
8:30–9:00	**The Electric VTOL Revolution**, Mike Hirschberg, Vertical Flight Society	
9:00–9:30	**Army VTOL UAS Development Efforts,** LTC Mike Osmon, US Army, Aviation Development Directorate	
9:30–10:00	**Agility, Cyber and Emerging Technologies for Naval UAS,** Dr. Robert Ernst, NAVAIR PMA-266	
10:00–10:30 am Break Sponsored by Honeywell		
10:30–11:00	**Analyzing Electric VTOL for the Joint Force,** Lt Col Nathan P. Diller, US Air Force, J8 Joint Force Integration Cell, Joint Staff	
11:00–11:30	**Boeing's Autonomous Systems Portfolio,** Mark Jordan, Boeing Autonomous Systems	
11:30–12:00	**Needs for the Uber Elevate Ecosystem,** Mark Moore, Uber	
12:00–1:30 pm Lunch Sponsored by The Boeing Co.		
Track	Autonomous VTOL Technical Meeting *(Canyon Rooms)*	Electric VTOL Symposium *(Canyon Rooms)*
Session *Moderator*	**UAV Design, Analysis and Testing** *Dave Mittleider, Boeing*	**Progress in Electric VTOL** *Mike Hirschberg, Vertical Flight Society*
1:30–2:00	**Development of the Quad Tilt Prop UAV,** Seongwook Choi, Korea Aerospace Research Institute 1	**The XTI TriFan 600**, Bob Labelle, CEO, XTI Aircraft N/A
2:00–2:30	**Development of UAV-Scale Cyclocopter,** Adam Kellen, Joel White and Moble Benedict, Texas A&M University 8	**Terrafugia TF-2 Overview**, Carl Dietrich, Co-Founder/CTO, Terrafugia N/A
2:30–3:00	**Configuration Redesign & Prototype Flight Testing of an Unmanned Fixed-Wing eVTOL Aircraft with Under-Fuselage Hover Lift & Pusher Wingtip Propulsion System,** Philipp Stahl, Christian Rößler and Mirko Hornung, Technical University of Munich 19	**Karem Butterfly**, Ben Tigner, President, Karem Aircraft N/A
3:00–3:30	**Comparative eVTOL UAV Design Optimization,** Thomas Seren, Christian Rößler and Mirko Hornung, Technical University of Munich 29	**Electric VTOL on the Boeing Horizon**, Troy Rutherford, Senior Director for eVTOL, Boeing NeXt N/A
3:30–4:00 pm Break Sponsored by XTI Aircraft		
Session *Moderator*	**UAV Design, Analysis and Testing** (con't) *Dave Mittleider, Boeing*	**Challenges in Electric VTOL** *Jim Sherman, SAE International*
4:00–4:30	**Analysis Methods for Tilting Wing and Tailsitter E-VTOL Configurations,** Todd R. Quackenbush, Daniel A. Wachspress, Glen R. Whitehouse and Michael K. Yu, Continuum Dynamics, Inc. 42	**Distributed Electric Propulsion (DEP) Failure Modes,** Patrick Darmstadt, The Boeing Co. [technical paper] N/A
4:30–5:00	**NexGen UAS Take-off and Landing Methods and Sizing Studies Solution for UAS,** Robin Preator, The Boeing Co. 52	**Gust Rejection,** Nick Brake, Empirical Systems Aerospace N./A
5:00–5:30	**NexGen UAS in the Counter Integrated Air Defense Environment,** Scott Swinsick, Hendrik Gideonse, Keith Meeker and Mark Clemente, The Boeing Co. N/A	**VTOL Flight Control for Safety,** Justin Paines, Chief Test Pilot, Joby Aviation N/A
6:30–7:00 pm Reception Sponsored by Karem Aircraft *(Courtyard, weather permitting; Canyon rooms, inclement weather)*		

STEM Trailer Tour

Presented by Rick Gutridge, President, Technical Research & Engineering Co.

Thursday, January 31 @ 9:00 am–8:00 pm

Public viewing of a trailer for science, technology, engineering and math (STEM) education and a demonstration of its educational activities and simulators will be held outside the conference center entrance. Light refreshments will be available.

Note: Electric VTOL symposium presentations will vary in length from 20–30 minutes. Symposium panels will be primarily discussion.

Thank you to our host Vertical Flight Society **Arizona Chapter**

A special thanks to the following chapter officers for their tireless support of this event: ***Kendra Befort*** *(Deputy Technical Chair),* ***Christa Luna*** *(Administrative Chair),* ***Ram JanakiRam*** *(Technical Chair) and* ***Tonja Reinert*** *(Arizona Chapter Presiden*

Wednesday, January 30, 2019: Registration Desk Hours 7:00 am–5:00pm *(Canyon Foyer)*		
Track	Autonomous VTOL Technical Meeting *(Canyon rooms)*	Electric VTOL Symposium *(Canyon rooms)*
Session *Moderator*	**UAV Design, Analysis and Testing II** *Hieu Ngo, Boeing*	**Standards and Regulations Panel** *Tom Gunnarson, Kitty Hawk*
8:00–8:30	**Precision Docking for Multi-UAV Cooperative Transportation,** Kevin Webb, Jared Elinger and Jonathan Rogers, Georgia Institute of Technology 62	• Anna Dietrich, Co-Founder, Terrafugia • Ajay Sehgal, Chief Engineer, KBRWyle • Dr. Mark DeAngelo, Aerospace Standards Engineer, SAE International • Greg Bowles, VP, Global Innovation & Policy, GAMA
8:30–9:00	**New Architectures for VTOL All-Electric Craft Using Enhanced Specific Power Electric Motor,** Oved Zucker, POLARIX Corporation and Christopher Liu, University of California 69	
9:00–9:30	**Algorithmic Icing Detection for On-Demand Electric Aircraft,** Robert McKillip, Jr., Andrew Kaufman, Todd Quakenbush, Continuum Dynamics, Inc. 75	
9:30-10:00	**Fleetwide Integrated Prognostics Health Management for Emerging eVTOL Operations,** Andrew Vechart, Sentient Science and Jacek Kawecki, Uber 86	
10:00–10:30 am	**Break Sponsored by Empirical Systems Aerospace**	
Session *Moderator*	**UAV Design, Analysis and Testing (cont)** *Hieu Ngo, Boeing*	**System Safety Panel** *Dan Newman, The Boeing Co.*
10:30–11:00	**Multi Objective Optimization of Wiring Harnesses for Overall System Performance of Electric Fixed Wing VTOL UAVs,** Sebastian Oberschwendtner, Christian Rößler and Mirko Hornung, Technical University of Munich 95	• Jonathan Hartman, Disruptive Technologies Lead, Sikorsky, a Lockheed Martin Company
11:00–11:30	**Resolving the Conflict Between Ship Design and UAV Launch and Recovery Deck Limits,** Bernard Ferrier, Dr. Robert Ernst and Ajay Sehgal, Naval Air Systems Command 112	• Kyle Heironimus, Innovation Engineer, Bell
11:30–12:00	**Challenges for Vehicle and Occupant Safety in Autonomous Electric Vertical Take-off and Landing (eVTOL) Vehicle Crashworthiness,** Justin Littell, NASA Langley Research Center 127	• Andrea Signorini, Head of Safety & Airworthiness, Sistemi Dinamici, a Leonardo company
12:00–1:30 pm	**Lunch Sponsored by Terrafugia**	
Session *Moderator*	***UAV Aerodynamics and Noise*** *Robin Preator, Boeing*	**Vehicle Technologies** *Johnny Doo, International Vehicle Research*
1:30–2:00	**Aerodynamic Modelling of Coaxial Counter-Rotating UAV Propellers,** Moritz Thiele and Mirko Hornung, Technical University of Munich 144	• **Hybrid Electric Propulsion System Advantages & Disadvantages**, Eric Blumer, Honeywell N/A • **New Lithium Sulfur Battery Development**, Dr. Zachary Favors, Nextech Batteries N/A • **Fuel Cell Technology**, Dr. Rick Sickenberger, Teledyne Energy Systems N/A • **Demonstration of an Autonomous Optical Launch & Recovery System**, Avinash Gandhe, SSCI [technical paper] N/A • **A Multi-Physics Prognostics Framework for Urban Mobility Operations**, Dr. Amine Chigani, Sentient Science N/A
2:00–2:30	**Scaling Investigation of Electric Quasi-Quadrotor Hover Endurance,** Budhyant Venepalli and Dr. Stefan Dancila, University of Texas at Arlington 157	
2:30–3:00	**Empirical Study of Aerodynamic Interference for Mid-air Docking Strategies for Multirotors,** Karan Jain, Trey Fortmuller and Mark W. Mueller, University of California, Berkeley N/A	
3:00–3:30	**CFD-CSD Coupled Aeroelastic Analysis of a Highly Flexible Flapping Wing in Hover,** Xuan Yang and Moble Benedict, Texas A&M University 179	
3:30–4:00 pm	**Break Sponsored by Sentient Science**	
Session *Moderator*	***UAV Aerodynamics and Noise (cont)*** *Robin Preator, Boeing*	**Missions and Applications** *Jim Sherman, SAE International*
4:00–4:30	**Computational Aeroacoustics of Different Propeller Configurations for eVTOL Applications,** Bernadine Passe and James Baeder, University of Maryland (UMD) 195	• **eVTOL Roadmapping**, Mike Hirschberg for Michael Dudley, NASA Ames N/A • **Group 2 – Intracity Missions**, Candice Carter N/A • **Group 3 – Intercity Missions**, Seren Webber, University of Hawaii N/A • **Group 4 – Public Service Missions**, Johnny Doo, International Vehicle Research N/A
4:30–5:00	**Developing Noise Metrics for Urban VTOL Aircraft: A Progress Report,** David Josephson, Josephson Engineering 205	
5:00–5:30	**Aerodynamic and Mechanical Design of a Morphing Winglet for a Quadrotor Biplane Tail-sitter,** Peter Ryseck, Vikram Hrishikeshavan, Derrick Yeo and Inderjit Chopra, UMD 211	
5:30–7:00 pm	**Reception and Banquet Sponsored by Bell *(Canyon Ballrooms)*; Cash Bar** **Banquet Keynote: *"Autonomy at Bell," Michael Thacker, Executive VP for Technology & Innovation, Bell***	

Note: Electric VTOL symposium presentations will vary in length from 20–30 minutes. Symposium panels will be primarily discussion.

Want to learn more about in Electric VTOL? Check out the Vertical Flight Society's "Electric VTOL News" and subscribe to the free eVTOL eNewsletter or check out the VFS World eVTOL Aircraft Directory: **www.eVTOL.news**

Thursday, January 31, 2019: Registration Desk Hours 7:00 am–5:00 pm *(Canyon Foyer)*

Track	Autonomous VTOL Technical Meeting *(Canyon rooms)*	Electric VTOL Symposium *(Canyon rooms)*
Session *Moderator*	**UAV Controls, Flight Dynamics and Navigation I** *Jason Graham, Boeing*	**Infrastructure and Passenger Experience** Michael Dyment, NEXA Advisors
8:00–8:30	**Stabilization of a Quadrotor in Wind with Flow Sensing: Linear Modeling and Control for Attitude and Position Hold,** William Craig and J.T. Lewis, Derek A. Paley, University of Maryland 219	**Urban Air Mobility (UAM) — Economics and Global Markets,** Michael Dyment, NEXA Advisors N/A
8:30–9:00	**An Assessment of Heave Response Dynamics for Electrically Driven Rotors of Increasing Diameter,** Ariel Walter, Michael McKay, Robert Niemiec, Farhan Gandhi, Colin Hamilton, Rensselaer Polytechnic Institute 229	**Vertiport Integration Constraints**, Rex Alexander, 5-Alpha N/A
9:00–9:30	**Kinematic Velocity Commander for Obstacle Field Navigation,** Chad Goerzen, San Jose State Unvi., Marc Takahashi, Matt Whalley, US Army ADD; Maj. Mark Cleary, Jeffrey Cox, US Army AMRDEC 237	**Vertiport Design Possibilities**, Mark Pilwallis, Gannett Fleming N/A
9:30-10:00	**Position Control of an Autonomous Unmanned Aerial Vehicle (UAV) Based on Accelerometer Response for Indoor Navigation,** Sagufta Kapadia and Syed Misbahuddin, Vaughn College of Aeronautics and Technology N/A	**eVTOL Noise Considerations**, Dr. Ken Brentner, Penn State University N/A
10:00-10:30 am Break Sponsored by Thales		
Session *Moderator*	**UAV Controls, Flight Dynamics and Navigation I** (cont) *Jason Graham, Boeing*	**eVTOL Opportunities** *Mike Hirschberg, Vertical Flight Society*
10:30–11:00	**Development of a Reconfigurable Multicopter Flight Dynamics Model from Flight Data Using System Identification,** Mark J. S. Lopez, Mark B. Tischler, Ondrej Juhasz, Anthony Gong, Frank C. Sanders, US Army Aviation Development Directorate; San Jose State University 261	**A Review of Current Technology and Research in Urban On-Demand Air Mobility Applications**, Nicholas Polaczyk and Enzo Trombino, Iowa State University [technical paper] 333
11:00–11:30	**A Comprehensive Simulation Framework to Certify Autonomous eVTOL, Aircraft,** Kara Gremillion, ANSYS, Inc. 272	**A Venture Capitalist's Perspective on eVTOL**, Peter Shannon, Levitate Capital N/A
11:30–12:00	**Improvements Gust Tolerance of a Cyclocopter UAV using Onboard Flow Sensing,** Elena Shrestha, Derrick Yeo, Vikram Hrishikeshavan and Inderjit Chopra, University of Maryland 249	*Exhibitor Showcase:* • ***eVTOL Aircraft Analysis Tools*** *Continuum Dynamics, Dan Wachspress N/A* • ***DSIAC Overview:*** *Defense Systems Information Analysis Center, Daniel Whipp* N/A • ***Door-to-Door eVTOL****, NFT, Guy Kaplinsky* N/A • ***OpenECUs for eVTOL,*** *Pi Innovo, Dwight Hansell* N/A
12:00-1:30 pm Lunch Sponsored by NEXA		
Session *Moderator*	**UAV Controls, Flight Dynamics and Navigation II** *Jason Graham, Boeing*	**Autonomy and Air Operations Panel** Bruce Holmes, SmartSky Networks
1:00–1:30	**Reinforcement Learning Control for Quadrotors using Snapdragon Flight,** Abhishek Shastry, Eric Solomon, Vikram Hrishkeshavan and Inderjit Chopra, University of Maryland 280	• **Enabling Urban Air Mobility (UAM) and eVTOL Operations,** Steve Jacobson, CEO, Autonodyne N/A • **Enabling Operations in the U-Space,** Maurizio Manzo, UTM Product Manager, Leonardo Electronics Division N/A • **How UTM Will Enable the Future of All Autonomous Aircraft,** Jessie Mooberry, Airbus UTM N/A
1:30–2:00	**Horizon Estimation from Thermal Cameras for Unmanned Aerial Vehicles,** Keith Sponsler and Srikanth Saripalli, Texas A&M University 290	
2:00–2:30	**Amphibious Quadcopter with Smooth Powered Air-Water Transition,** Illya Semenov, Vikram Hrishikeshavan and Inderjit Chopra, University of Maryland 295	
2:30–3:00 pm Break Sponsored by Gannett Fleming		
Session *Moderator*	***eVTOL and VTOL Design*** *Hieu Ngo, Boeing*	**eVTOL for Good** Johnny Doo, International Vehicle Research
3:00–3:30	**Disruptive Advance in P.M. Motor Technology Provides Higher Top Speed and Extended Range,** Lawrence Zepp, Dura-Trac Motors, Inc. 306	• **Distributed On-Demand TVF Search & Rescue & Disaster Relief Network**, Johnny Doo, IVR N/A • **eVTOL for Search & Rescue**, Eddie Bennet, The Ripper Group, Australia N/A • **eVTOL applications for Fire Fighting**, Michael Ralston, Menlo Fire District, California N/A • **From Humanitarian Efforts to Easier Commutes**, Jessie Mooberry, Airbus UTM N/A • **A Unique eVTOL Concept and Applications**, Arnaud Didey, Neoptera N/A
3:30–4:00	**IGBT Robustness Capabilities and Setting Appropriate Protection Thresholds in High Power Propulsion Inverters,** Satyavrat Laud, Renesas Electronics America, Automotive Unit 122	
4:00–4:30	**Usability of a Steering Wheel Control Concept for a Highly Augmented Personal Helicopter,** Bianca Schuchardt, DLR Institute of Flight Systems, German Aerospace Center 317	
4:30–5:00	**Integrated Occupant Safety of Urban Air Transport Emergency Landing Applications,** Gerardo Olivares, F. Caralt, and A. Vina; National Institute for Aviation Research and BRS Aerospace 326	
5:00–5:30	**Complexities of Lightning Certification of VTOL Aircraft,** George Tian and Justin P. McKennon, NTS Lightning Technologies 312	**Electric VTOL Symposium Wrap-up,** Mike Hirschberg, VFS Executive Director **N/A**

Additional Paper:

Evaluation of Reconfigurable Rotorcraft Concepts as Viable Combined High Efficient Hover and Forward Flight Solutions for Unmanned Aerial Systems (UAS), David Adams, Daniel P. Schrage, Carter Crouch 165

Note: Electric VTOL symposium presentations will vary in length from 20–30 minutes. Symposium panels will be primarily discussion.

Development of the Quad Tilt Prop UAV

Seongwook Choi
Principal Researcher,
Korea Aerospace Research Institute,
Daejeon, Republic of Korea

ABSTRACT

The Quad Tilt Prop UAV that consists of a tandem wing and four tiltable and variable pitch propellers has been developed. The air vehicle was designed to meet two basic requirement: VTOL and high-speed capability. The unpowered and powered windtunnel tests for the full-scale air vehicle were conducted for constructing aerodynamic database and finding trim data for all flight conditions respectively. The tethered test was conducted to investigate hovering characteristics as well as to check system soundness. The rotary, transition, and high-speed flight test have been carried out and showed smooth and stable flight characteristics in all flight modes.

NOMENCLATURE

KARI Korea Aerospace Research Institute
LSWT Low Speed Wind Tunnel
Prop Propeller
QTP Quad Tilt Prop
SOC State of Charge
UAV Unmanned Aerial Vehicles
VTOL Vertical Take-off and Landing

INTRODUCTION

The electrically driven Quad Tilt Prop (QTP) UAV has been developed. The drivetrain consists of four tiltable nacelles that include variable pitch props and reduction mechanism to match rotational speeds of prop and electric motor. The QTP-UAV has been designed based on the requirement: gross weight 48 kg, payload 3 kg, maximum speed 160 km/h, and maximum endurance 30 minutes for Li-Ion battery powered and 2 hours for hybrid propulsion system respectively.

The air vehicle has a tandem wing structure with nacelles mounted at the tip of wing as shown in Figure 1. The prop radius was selected 0.55 m and the rotational speed was set 1800 rpm to have minimum required power in hovering condition. The 2.4:1 reduction system was adopted to match rotational speed of motor and prop. In the process of determining rotational speed of prop and configuration design, the in-house performance analysis code for a Quad Tilt Prop air vehicle has been used. The performance code is based on the blade element momentum theory and airframe aerodynamic database obtained from unpowered windtunnel test (Ref. 1).

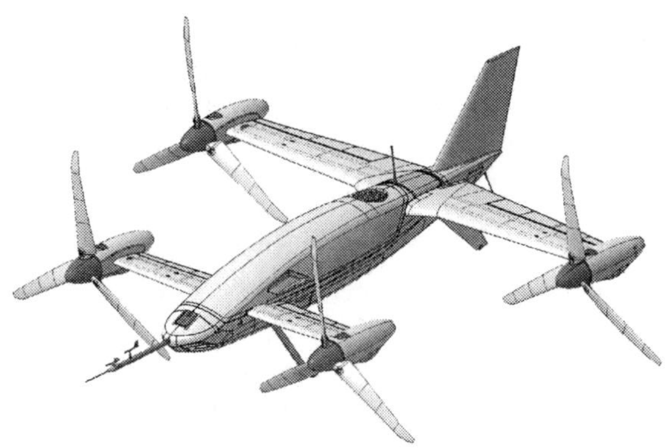

Fig. 1. The QTP-UAV in airplane mode.

Presented at the Autonomous VTOL Technical Meeting & eVTOL Symposium, Mesa, AZ, January 29-31, 2019.

The attitude of air vehicle can be controlled primarily using four variable pitch props in rotary mode and four flaps were used as an aileron and elevator in airplane mode. The differential collective pitch was used for directional control in airplane mode. The conversion corridor, which scheduling the relation between tilt angle and flight speed, has been estimated from performance analysis and windtunnel test. Table 1 shows basic specification and performance of QTP-UAV (Ref. 2).

Table 1. Specification of QTP-UAV.

Weight		Airframe	
Payload (kg)	3.0	Overall Length (m)	2.034
Airframe (kg)	8.0	Overall Height (m)	0.775
Prop System (kg)	6.0	Front Wing Span (m)	1.500
Motor + Reduction + Tilt (kg)	7.0	Front Wing Chord (m)	0.190
Avionics (kg)	2.0	Rear Wing Span (m)	2.200
Electric (W/H etc.) (kg)	3.0	Rear Wing Chord (m)	0.305
Battery (3.2kWh) (kg)	19.0	Rear Wing Area (m^2)	0.629
Gross Weight (kg)	48.0	Wing Loading (kg/m^2)	76.31
Prop		Performance	
No. of Props	4	Loitering Speed (km/h) @Tilt 30	123.0
No. of Blades	3	Max Speed (km/h) @ Tilt 0	170.0
Rotor Radius (m)	0.55	Hover Power (kW)	8.0
Rotor Disk Area (m^2)	3.80	Loitering Power (kW)	5.0
Disk Loading (kg/m^2)	12.6	Hover Endurance (min) @ Tilt 90	18.0
Rotor RPM (HL Mode)	1800	Max Endurance (min) @ Tilt 0	21.0
Rotor RPM (AP Mode)	1440	Max Endurance (min) @ Tilt 30	29.0

The windtunnel test has been conducted using the full-scale flight air vehicle in the open jet test section of KARI-LSWT. The basic aerodynamic database has been built from the result of unpowered windtunnel test and flight trim data was obtained from the powered windtunnel test including rotating props. The tethered test for investigating and tuning hovering characteristics has been performed. The flight test for rotary, transition, and airplane modes has been carried out and characteristics of each flight mode has been investigated.

WINDTUNNEL TEST

The windtunnel test has been conducted in 6 x 4 m open test section of KARI-LSWT using full-scale air vehicle for unpowered and powered configuration. The unpowered windtunnel test, which uses the air vehicle without props, was conducted for constructing aerodynamic database for performance analysis. And the powered test, which uses the air vehicle with rotating props as shown in Figure 2, has been conducted to obtain trim points that lifting 48 kgf and making drag and pitching moment zero by controlling collective pitches and flap angles for entire tilt and flight speed.

Fig. 2. Windtunnel test of QTP-UAV.

The windtunnel model is equipped with a Flight Control Computer (FCC) so that all of the control surfaces of the model including the rotational speed of the prop can be remotely controlled through wire communication from the Ground Control Station (GCS) in control room. A 15 kW external power supply was used to drive props of the windtunnel model. The wind speed was operated up to 40 m/s (144 km/h) in this test.

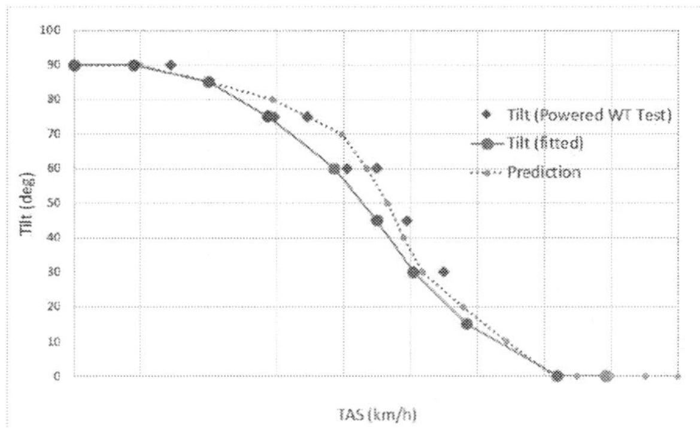

Fig. 3. Speed vs. tilt angle.

Figure 3 is the flight speed versus tilt angle and shows that the transition final speed at which the tilt angle is zero degrees is 144 km/h. For the transition test in the windtunnel, the tilt angle and wind speed were controlled so that the angle of attack maintains 4 degrees. It can be seen that the transition path by the performance analysis is predicted close to the path measured by the windtunnel test.

Figure 4 and 5 are changes of front and rear collective pitches to the speed increase. In the transition region up to 120 km/h, in order to trim, the rear collective pitch is maintained higher than that of front, and the front and rear collective pitches approach an angle as the wind speed increased over the speed. For the front collective pitch, the test and the predicted data are so close, but for the rear collective pitch, the predicted data is lower than the test data in the transition region. This indicates that the pitch-up moment of the air vehicle is much larger than expected in the transition region, and that the rear collective pitch was increased to counteract the pitching moment.

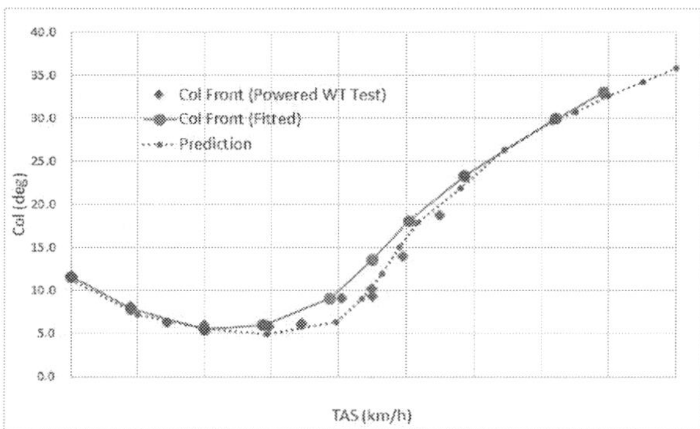

Fig. 4. Speed vs. front collective pitch.

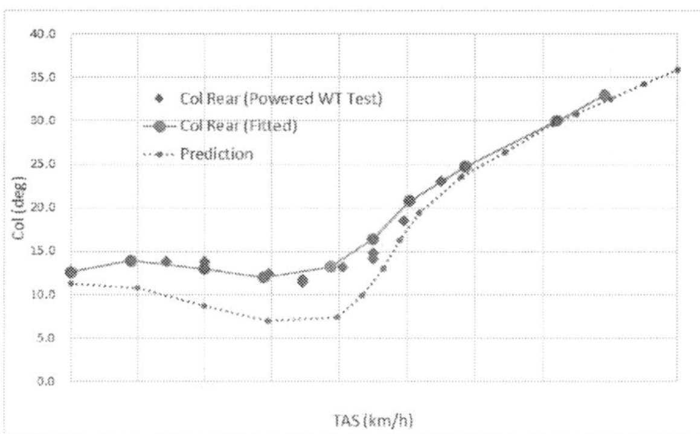

Fig. 5. Speed vs. rear collective pitch.

Figure 6 shows that 6.7 kW is required for hovering and that 4.3 kW, which is the minimum power in entire flight, is required at a speed of 70 km/h. From the tendency of the electric power curve, in order to reach the speed of 160 km/h, the required power expected to be about 6.5 kW which is close to hovering power.

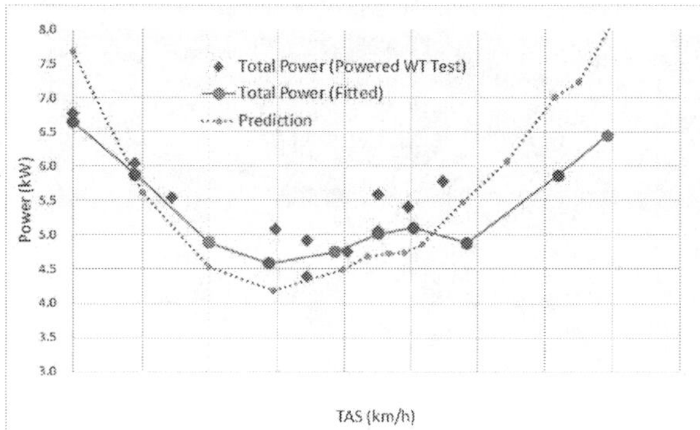

Fig. 6. Speed vs. electric power.

Figure 7 shows that the pitch link load measured with miniature load cell installed on the collective pitch linkage. It shows almost constant load 2.2 kgf up to 80 km/h, and then linearly increases to 5.4 kgf, which is within the rated load 6.74 kgf of the collective pitch actuator (Ref. 3), to the speed of 160 km/h.

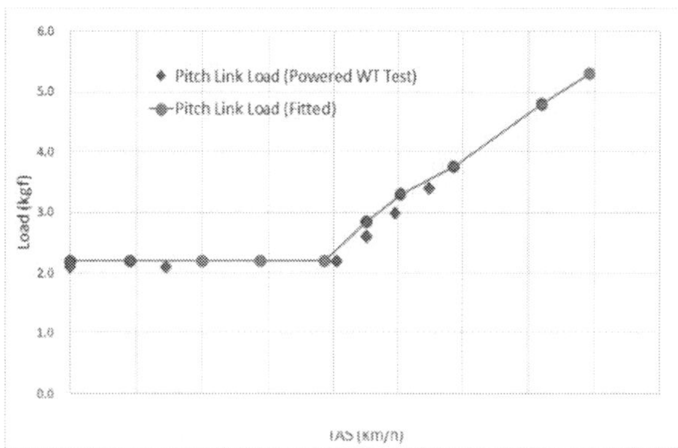

Fig. 7. Speed vs. pitch link load.

TETHERED TEST

The tethered test, as shown in Figure 8, was carried out by connecting the center of gravity of the air vehicle to a tower structure with a safety wire to investigate hovering characteristics as well as to check the overall soundness of the QTP-UAV system. Through the tethered hovering test, the rotational speed of props were changed to 1900 rpm instead of initial 1800 rpm to enhance hovering characteristics.

Fig. 8. Tethered test.

FLIGHT TEST

Transient Flight Test

The flight test of the battery-powered QTP-UAV was performed. The gross weight of air vehicle in flight was 44 kg and the battery has the capacity of 3.5 kWh. For ensuring stable flight in rotary and hovering mode, the SOC of the battery has to be maintained over 40 %, so the actual available battery capacity is about 55 %. The transition and airplane mode were tested through the point navigation flight with altitude 150 m and radius 600 m as shown in Figure 9 and 10.

The flight was carried out in the order of automatic takeoff, point navigation, increasing speed, deceleration, and landing. The transition final speed at which the tilt was zero degrees was 144 km/h, and reached 155 km/h through the acceleration. The maximum speed flight was taken 9.7 minutes from takeoff to landing. The battery SOC was 95.5 % at take-off and 75.8 % at landing then total consumption was 19.7 % resulting in consumption rate of 2.03 SOC/min.

Figure 11 shows that the power required was about 7 kW at 155 km/h, which is similar to the power required at takeoff and the minimum power is about 3 kW at the speed of 70 km/h.

Figure 12 compares the tilt angle and the collective pitch change of the front and rear props in flight test with the powered wind tunnel test results. The tilt angle variation to the flight speed is close to the windtunnel test results in the entire flight speed because the conversion corridor used in the control law was based on the results obtained from the powered windtunnel test.

Fig. 9. Rotary, conversion, and airplane mode flight.

The front and rear collective pitches are also close to the windtunnel test results. The collective pitch starts at about 10 degrees at takeoff, and as speed increases, the front collective pitch decreases to less than 5 degrees and rear collective pitch increases to about 14 degrees then the difference between two collective pitches maintained over 10 degrees. The two collective pitches approach to an angle from 80 km/h, and they reached 31 degrees when the speed increase to 155 km/h. It noted that the operating range of the front and rear collective pitches are zero to 40 degrees. The rotational speed of four props are maintained at 1900 rpm in the entire flight speed as shown in Figure 13.

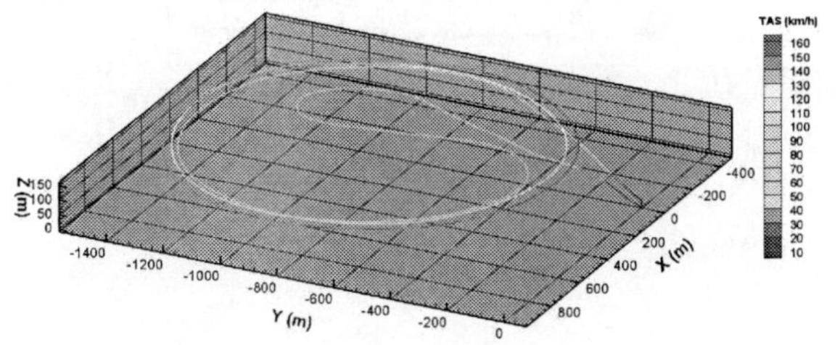

Fig. 10. Trajectory for transition flight.

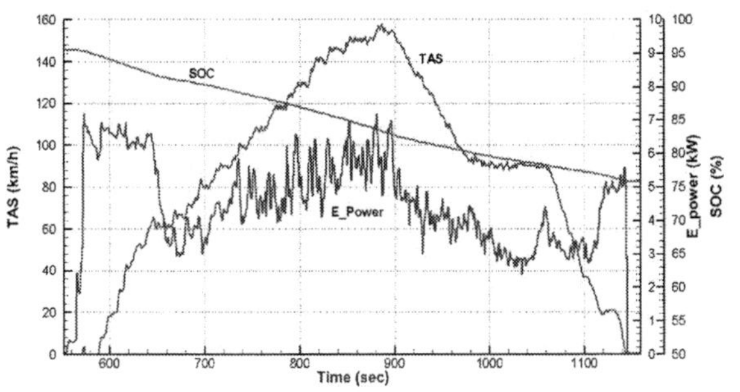

Fig. 11. Electric power and SOC.

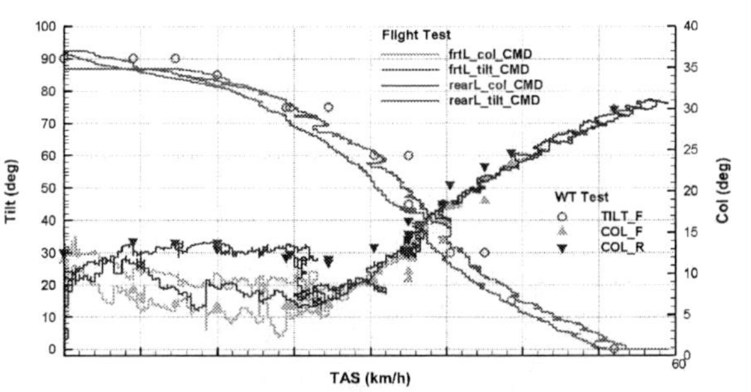

Fig. 12. Tilt angle and collective pitch.

Figure 14 shows the variation of electric power to the flight speed. It absorbs 6.5 kW of power while accelerating and ascending from take-off to the speed of 60 km/h. Then the speed and altitude were stabilized and the power decreased to 3.5 kW. The speed gradually increases and reaches 7 kW at 155 km/h. During deceleration and descending, the power reduced to less than 3 kW at the speed of 60 km/h.

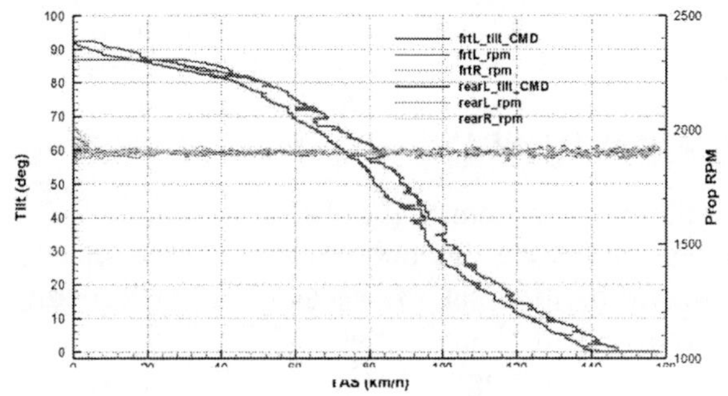

Fig. 13. Rotational speed of four props.

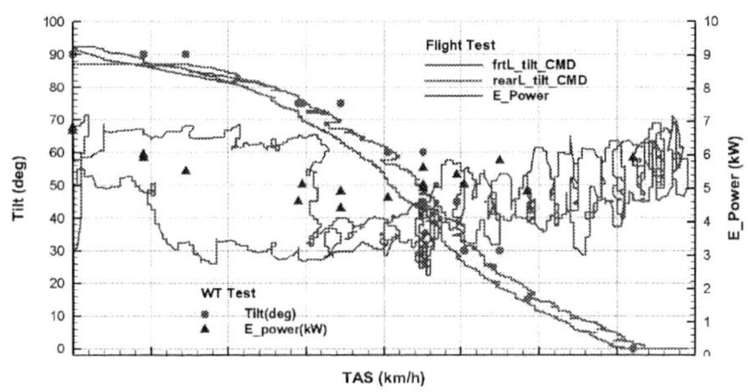

Fig. 14. Electric power.

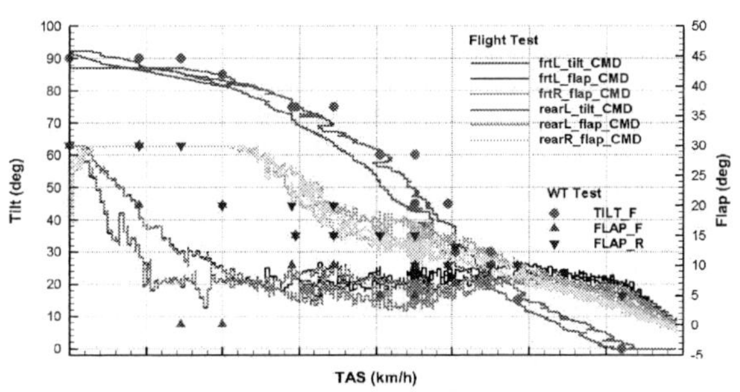

Fig. 15. Front and rear flap angle.

Figure 15 shows the variation of front and rear flap angles to the flight speed. The front flap and rear flap angles started with 30 degrees at speed zero. As the flight speed increases, the front flap decrease abruptly but the rear flap maintains with the initial angle. The difference between the front and rear flap angle was used as an auxiliary means to maintain the flight attitude against pitch-up moment occurring as the speed increased up to 40 km/h. The front and rear flap angles are getting closer to same angle above the flight speed of 100 km/h and approach near zero angles as speed increased to 155 km/h.

Endurance Flight Test

Through the transient flight test, it was found that the required power is minimized at a flight speed of 70 km/h. In order to measure the maximum endurance of the battery powered QTP-UAV, the point navigation flight of 70 km/h, altitude of 150 m and radius of 600 m was carried out as shown in Figure 16. From Figures 17 and 19, it can be seen that the average current is 80 Amperes and the electric power is 3.5 kW at the speed of 70 km/h. The total flight time from take-off to landing was 17.3 min and the SOC decreased from 95.4 % to 65.1 % consuming 29.5% as shown in Figure 19. In case of effective SOC of the battery is 55 %, the maximum flying time can be estimated about 32 minutes.

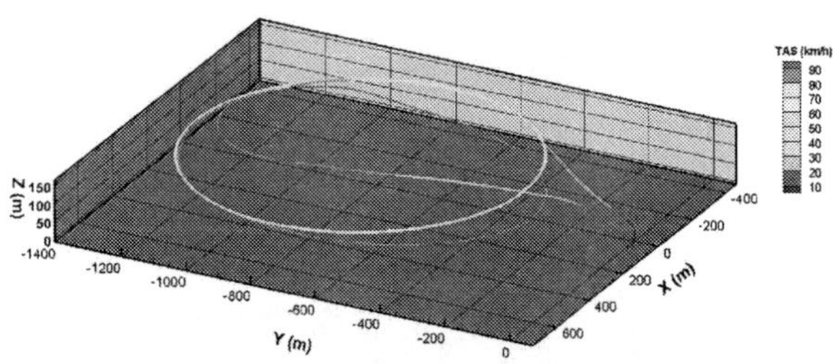

Fig. 16. Trajectory of endurance flight.

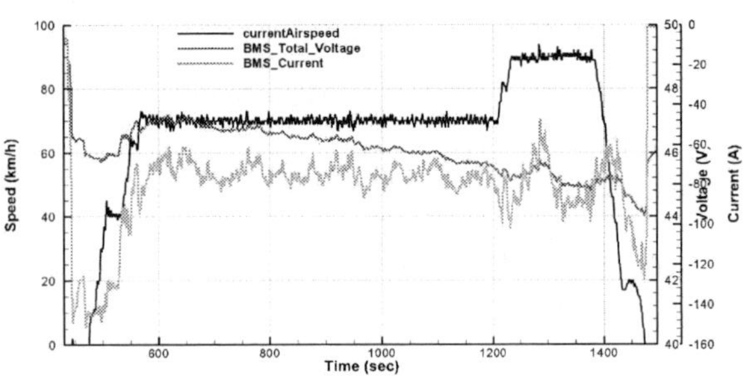

Fig. 17. Voltage and current of endurance flight.

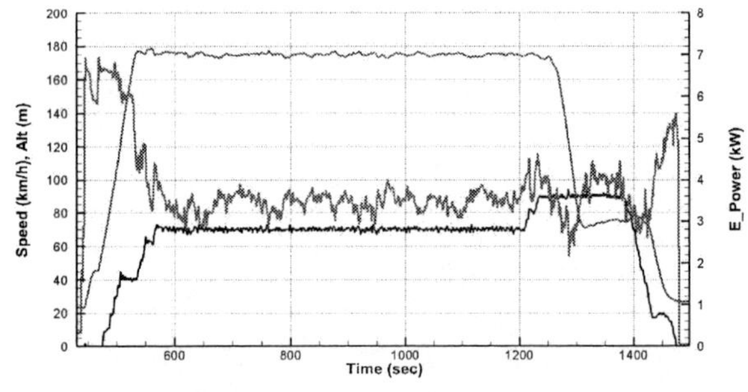

Fig. 18. Electric power of endurance flight.

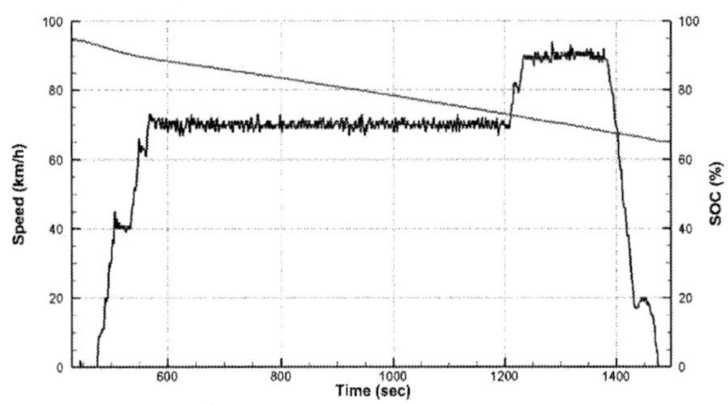

Fig. 19. SOC of endurance flight.

Maximum Speed Flight Test

In order to investigate the tilt speed and maximum speed of QTP-UAV, transition and accelerated flight tests were carried out by point navigation at an altitude of 250 m and a radius of 700 m as shown in Figure 20.

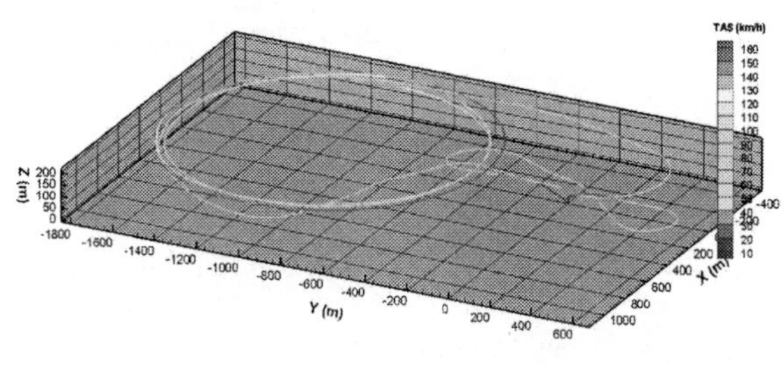

Fig. 20. Trajectory of maximum speed flight.

Figure 21 shows that the successive transitions were performed from 60 km/h to 145 km/h and it was taken 44 seconds, which corresponds to the preset speed rate 2 kph/sec in the control system. After transition, the speed was incrementally increased to 165 km/h. From this flight test, the maximum speed of the QTP-UAV can be estimated up to 170 km/h considering the electric power consumption, collective pitch limitation, and pitch link load of props.

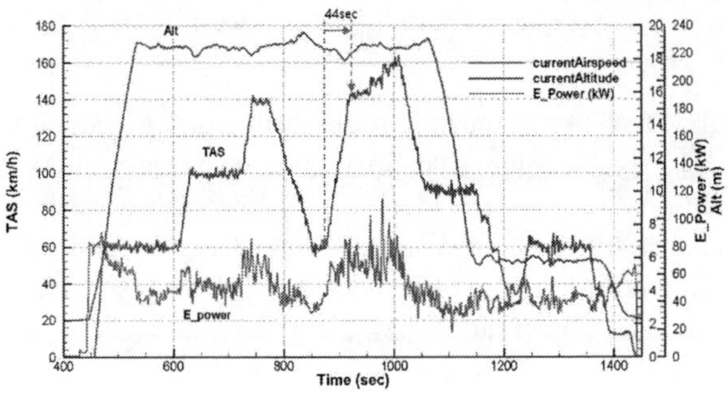

Fig. 21. Speed, Altitude, and electric power of maximum speed flight.

CONCLUSIONS

The electrically driven quad tilt prop UAV that consists of a tandem wing and four tiltable and variable pitch props has been developed. The unpowered and powered windtunnel test were conducted to build aerodynamic database and to obtain trim points for all flight speed. The powered windtunnel test alleviated many risks that would be encountered during the flight test. The tethered test was performed for stabilizing hovering mode and check overall soundness of the UAV system. The flight test for rotary, transition, and airplane mode were conducted through point navigation and shows very stable behavior for all flight modes and reached the maximum speed of 165 km/h. In order to increase the flight endurance of the QTP-UAV, a hybrid propulsion system has been developing and will be integrated into the air vehicle system.

ACKNOWLEDGMENTS

This study has been supported by the KARI internal program - Research for Next Generation Aircraft Technology Integration. The author would like to thank all the researchers who have worked on this flight test: Youngshin Kang, Yushin Kim, Sungho Chang, Am Cho, Jongmin Bae, and Myungjae Kim.

REFERENCES

[1]Choi, S., "Performance Analysis of QTP-UAV," KARI-AVSRT-ELN-2017-002, 2017.

[2]Choi, S., "Configuration Design of QTP-UAV," KARI-AVSRT-ELN-2017-001, 2017.

[3]Choi, S., et al., "Research for Next Generation Aircraft Technology Integration," KARI Internal Report, 2017. 12.

Development of a UAV-Scale Cyclocopter

Adam Kellen
Graduate Research Assistant

Joel White
Undergraduate Research Assistant

Moble Benedict
Assistant Professor

Advanced Vertical Flight Laboratory
Department of Aerospace Engineering
Texas A&M University
College Station, TX 77845

ABSTRACT

This paper discusses the detailed vehicle design, control strategy, gimbal testing and tethered flight testing of a 17-pound novel unmanned aerial vehicle (UAV) utilizing cycloidal rotors, known as a cyclocopter. Cycloidal rotors (or cyclorotors) are a horizontal-axis rotary-wing concept. The current vehicle configuration features two identical cyclorotors rotating in the opposite directions positioned at the front and rear of the vehicle. The cyclorotors utilized in the vehicle design were optimized through an extensive experimental parametric study. At an operating Reynolds number of 200,000, the optimal cycloidal rotor configuration had a chord-by-radius ratio of 0.66, 3 blades featuring a blade aspect ratio (span/chord) of 4 and a NACA 0020 airfoil, rotor aspect ratio (span/diameter) of 1.33 and pitch amplitude of 35 deg and obtained a figure of merit of 0.6. The final cyclorotor geometric parameters chosen for the UAV were a 6 in. chord, 18 in. span, and 18 in. diameter. The novel cyclorotor design features a revolutionary split-blade with independent collective and cyclic pitch control for the left and right blade halves. The independent pitch kinematics allows the vehicle to achieve both roll and yaw control while maintaining a compact and light-weight design. Pitch control was achieved using differential rotational speed between front and rear rotors. A state-of-the-art 1.7-g autopilot was integrated onto the vehicle for attitude stability through feedback control. Attitude stability of the vehicle was verified through a series of experiments on a three degree-of-freedom gimbal stand. The vehicle's ability to liftoff and remain stable in pitch and yaw was verified through tethered flight tests.

NOTATION

A	=	Rectangular projected area, ($b \times 2R$)
A_S	=	Blade swept area, $2\pi Rb$
b	=	Blade Span
c	=	Blade chord length
c/R	=	Chord by Radius ratio
C_T	=	Thrust coefficient, $T_{Res} / \rho A (\Omega R)^2$
DL	=	Disk Loading, T_{Res} / A
N_b	=	Number of Blades
P	=	Total aerodynamic power, $Q\,\Omega$
PL	=	Power Loading, T_{Res} / P, N W^{-1}
Q	=	Aerodynamic Torque
R	=	Radius
T_{Res}	=	Resultant thrust, $\sqrt{T_Y^2 + T_Z^2}$
T_z	=	Rotor vertical thrust
T_y	=	Rotor sideward thrust
ρ	=	Air density
σ	=	Rotor solidity, $N_b c / 2\pi R$
Φ	=	Thrust angle, $tan^{-1} T_y/T_z$
ψ	=	Azimuthal position of blade
Ω	=	Rotational speed of rotor

Presented at the Vertical Flight Society Autonomous VTOL Technical Meeting and Electric VTOL Symposium, Mesa, Arizona, UAS, Jan. 29-31, 2019.

INTRODUCTION

Interest in Unmanned Air Vehicles (UAVs) and their research has greatly expanded in both the civilian and military sectors. UAVs are capable of accomplishing tasks previously too difficult, costly, or unsafe to complete from the ground or via manned aircraft. Creating the next generation of vertical takeoff and landing (VTOL) capable UAVs requires revolutionary vehicle concepts. Traditionally these concepts rely on conventional rotors/propellers as their primary means of propulsion. Unfortunately, conventional rotors tend to suffer from low aerodynamic efficiency compared to full-scale helicopter rotors, high susceptibility to gust disturbances and high acoustic signatures [Ref. 1]. Cyclorotors are now being investigated as an alternative propulsion system for larger UAVs.

A cyclorotor is a propulsion system consisting of multiple blades rotating about a horizontal axis (Fig. 1). Each blade can rotate about its pitch axis with the span positioned parallel to the axis of rotation. To create a net vertical force, the blades are cyclically pitched in a 1/rev fashion such that the blades experience positive geometric angles of attack at both the top and bottom halves of its circular trajectory (Fig. 2). The

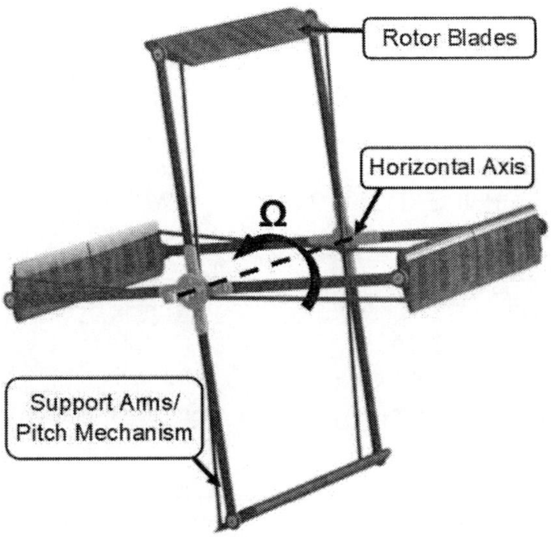

Fig. 1: Cyclorotor schematic.

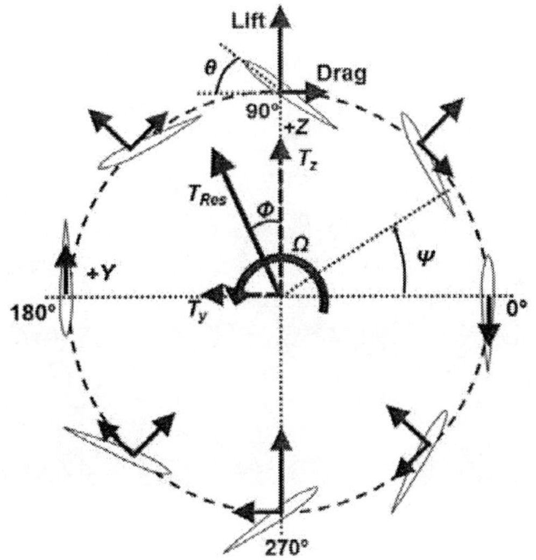

Fig. 2: Blade pitching kinematics and force vectors on a cyclorotor.

magnitude and direction of the net thrust vector can be changed by varying the amplitude and phase of the cyclic blade pitch.

The cyclorotor's unique design allows for several key advantages over conventional rotors. The ability to near instantaneously vector the thrust in any direction perpendicular to the axis of rotation provides increased maneuverability compared to a UAV utilizing conventional rotors. This increased maneuverability is particularly important for operating in space-constrained or gusty environments. The positioning of the cyclorotor's blades ensures the entire blade span operates at similar aerodynamic conditions simplifying blade optimization. Additionally, the blades positioning partnered with proper pitch scheduling can mitigate the effects of retreating blade stall providing efficient high-speed flight capabilities [Refs. 2-4]. Another important advantage of a cyclorotor is the reduced acoustic signature resulting from the reduced blade speed. The major drawback associated with cyclorotors is the large rotating structure which results in increased mechanical complexity and structural weight. An important focus during the vehicle's design and construction was a strong yet lightweight cyclorotor.

The first cyclocopter designs originated shortly after wright brother's first successful flight. One of the first designs was created in 1909 by E. P. Sverchkov in Russia as a rotating wing structure, but it failed to ever achieve flight [5]. Kirsten attempted again to create a cyclocopter in the 1920s, but it too failed to achieve liftoff [6-7]. Multiple other attempts were made in the 1920s and 30s to create stable full-scale cyclocopters but none of them were successful as flying vehicles. The primary challenges that could not be overcome were associated with the mechanical complexity of early designs and high inertial loading on the heavy rotor blades.

The development of advanced composite material and expanded interest in unmanned aerial vehicles led to the reemergence of interest in the study of cyclorotors in the late 1990s [Ref. 8]. Entering the early 21st century pioneering research on micro air vehicle (MAV) scale cyclorotors was conducted at the University of Maryland and Texas A&M University. Their research has resulted in the creation and optimization of cyclocopters between 29 and 800 grams (Fig. 3) [Ref. 9-10]. The research established a fundamental understanding of the aeromechanics and performance benefits of MAV-scale cyclorotors [Ref. 11-15]. Moving to larger scale cyclorotors, Seoul National University developed two UAV-scale cyclocopters [Ref. 16-17]. The 110 kg dual cyclorotor configuration demonstrated tethered flight while the 10 kg quad cyclorotor configuration has demonstrated hovering flight.

The work being presented now includes the development of a novel dual-rotor configuration for a cyclocopter. Prior dual cyclorotor configurations have featured a conventional tail rotor to counteract the primary cyclorotors torque and produce excess lift. The cyclocopter developed as part of the current study features two cyclorotors positioned at the front

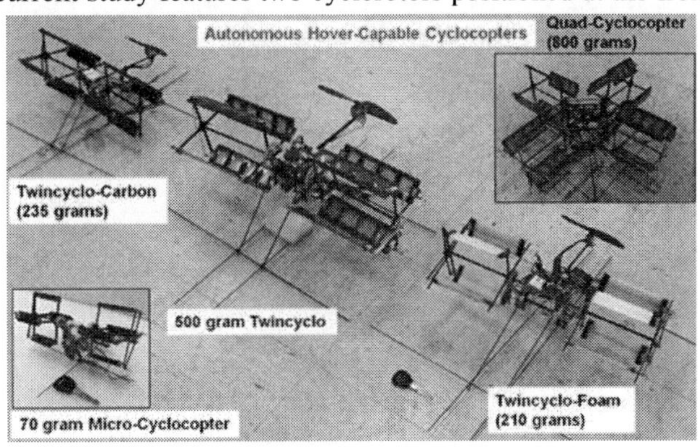

Fig. 3: Previous cyclocopter vehicles.

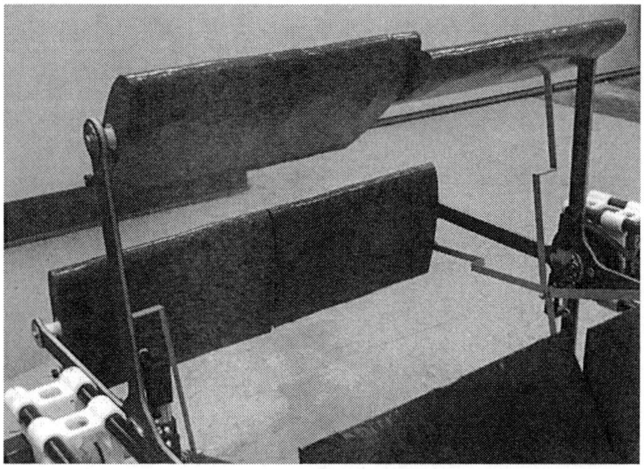

Fig. 4: Split blade design.

and rear of the vehicle. In order to achieve stable hovering flight, the vehicle's cyclorotors feature a novel split-blade design. The focus of the present paper is to discuss the design and development of the vehicle, the formulation and implementation of the control strategy for hover, and the initial flight testing.

VEHICLE DESIGN

The current vehicle is designed to demonstrate and take full advantage of the cycloidal rotors thrust vectoring and high-speed flight capability. The concept features two cyclorotors on the front and rear of the vehicle's fuselage. To maintain VTOL capabilities, the cyclorotors must be capable of producing differential forces capable of providing roll and yaw control authority. To generate these forces the cyclorotors feature a novel split-blade design, where the blade is split at mid-span location into left and right halves as shown in Fig. 4. The split-blade design allows independent pitch amplitude and phase control for each half enabling different thrust magnitude and direction for the left and right sides of a single cyclorotor for roll and yaw control. Pitch control could

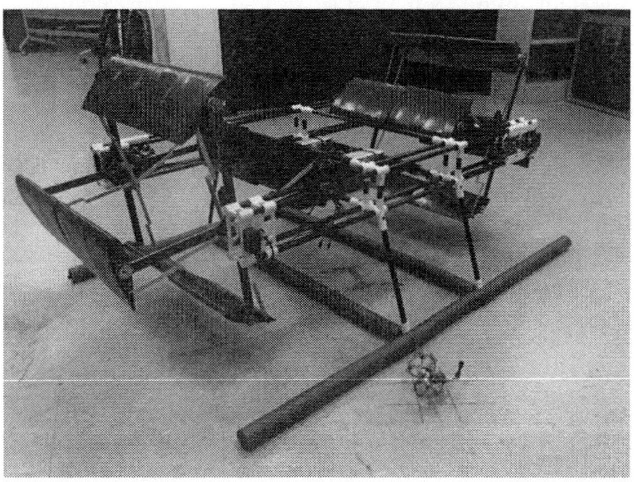

Fig. 5: Current UAV-scale cyclocopter pictured with 29g cyclocopter.

Table 1. Design specifications for cyclocopter.

Design Feature	UAV-scale Cyclocopter
Cyclorotor diameter	18 inches
Blade span	18 inches
Blade chord	6 inches
Blade airfoil section	NACA 0020
Nominal pitch amplitude	±35°
Electric motor	KDE10218XF-105
Transmission method	Timing Belt
Transmission ratio	4.17:1
Collective linear servo	Actuonix L12
Cyclic rotation servo	Turnigy TGY-6109MD
Autopilot	ELKA-R
Battery	2x 25c 6S 6000 mAh

be achieved by a differential pitch amplitude or rotational speed between front and rear rotors.

The cyclocopter developed for the current study is shown in Fig. 5. The vehicle measures roughly 2 feet in width, 5 feet in length, and 2.5 feet in height. The gross weight of the vehicle with batteries is 17 lbs. The fuselage of the vehicle was created using a system of ½" composite tubes and 3D printed joints to create a lightweight and stiff frame for the cyclorotors. A central composite plate provides a mounting place for the vehicle's electronics and potential payloads. The cyclorotors have an 18-inch diameter, 18 in span, and 6 in blade chord. The rotors are independently driven by a 105 kV outrunner KDE motor via a single stage timing belt transmission. The nominal operating rotor rotational speed is 900 RPM.

Pitch amplitude and phase (analogous to collective and cyclic) control is achieved via a four-bar pitch mechanism controlled by a linear servo and rotational servo. Power is provided by two 25S 6S LIPO 6000 mAh batteries connected in series to create a 12S battery. The separate batteries allows the weight to be distributed across the vehicle. Attitude stabilization is achieved using an ultralight inertial navigation system developed at the University of Maryland, ELKA-R [Ref. 18]. A summary of the design specifications is provided in Table 1 and a weight breakdown is provided in Table 2. The detailed design of the cyclorotor and its subassemblies is provided in the subsequent section.

Table 2. Weight breakdown for cyclocopter.

Component	Weight (lbs)	Total (%)
Fuselage	5.50	32.4
Cyclorotors (combined)	3.50	20.6
Pitch mechanism + servos	1.85	10.9
Motors + ESCs	2.65	15.6
Batteries	3.40	20.0
Wiring + Misc.	0.10	0.5
Total	17	100

CYCLOROTOR DESIGN

Like most VTOL aircraft, the design and development of the rotor system is the most crucial step in the design process. The present cyclorotor was designed based on the results from an extensive experimental hover performance study we conducted recently and summarized in Ref. 19.

The design of the cyclorotor can be broken into (1) obtaining optimal blade kinematics and rotor geometry, (2) pitch mechanism design, (3) blade structural design, and (4) the design of the remainder of the rotor structure and transmission.

Rotor Geometry

Prior to the development of the current cyclocopter, a systematic experimental parametric study was conducted on a UAV-scale cyclorotor by varying blade kinematics and rotor geometry to investigate the effect of rotor geometry on cyclorotor performance in terms of both thrust and aerodynamic efficiency [Ref. 19]. During the experimental study the two key design parameters were determined to be the chord-by-radius ratio and number of blades. Changing either of these parameters would change the rotor solidity (σ), which is defined as the ratio of total blade area to the swept area.

The importance of the chord-by-radius ratio is derived from the implication of chord-by-radius ratio on flow curvature effects (Fig. 6). The flow curvature effects are a result of the velocity distribution across the blade chord which results in a symmetric airfoil in curvilinear flow being aerodynamically similar to a cambered airfoil in rectilinear flow. The effect becomes more substantial as the chord-by-radius ratio is increased due to the increase in the velocity variation along the chord. The effects of the chord-by-radius ratio on aerodynamic performance is best visualized when examining the plot for thrust vs chord-by-radius ratio for a two-blade

6

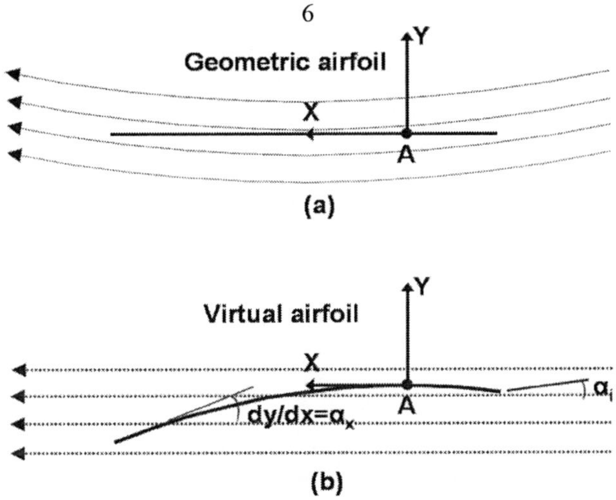

Fig. 6: Virtual camber in a curvilinear flow.

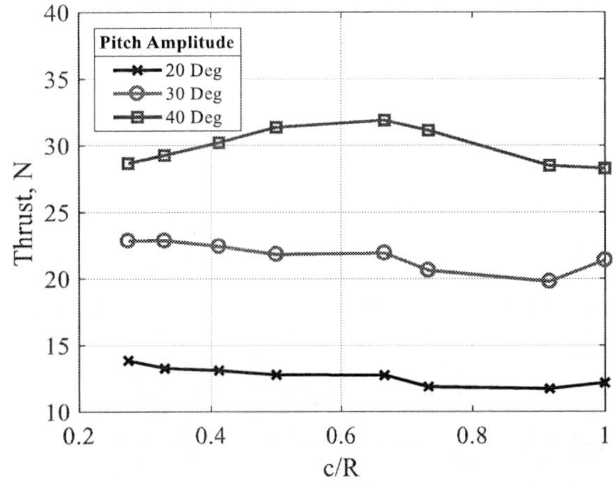

Fig. 7: Thrust vs. chord-by-radius ratio for fixed blade speed and varying pitch amplitude.

rotor operating at a fixed speed (Fig. 7). For the 40° pitch amplitude case the same 5.5 in. cyclorotor blade operating at a fixed speed is capable of producing as much as 20% additional thrust at a chord-by-radius ratio of 0.66 compared to lower or higher ratios. Coupling the thrust measurements with power measurements, the optimal chord-by-radius ratio for a two bladed rotor was found to 0.66.

During the experimental study, varying the number of blades was found to exhibit highly variable results depending on the chord-by-radius ratio. Selecting the optimal number of blades should instead be determined in conjunction with rotor solidity. Figure 8 presents thrust per blade area versus solidity at a 40° pitch amplitude for the different cases of increasing chord-by-radius ratio and increasing number of blades at three fixed chord-by-radius ratios. The results demonstrated maximizing rotor thrust per blade area requires selecting an optimized chord-by-radius ratio and minimizing the number of blades. To quantify the aerodynamically optimal design, Fig. 9 presents the figure of merit vs. rotor solidity. Based on these results, an optimal cyclorotor configuration was found

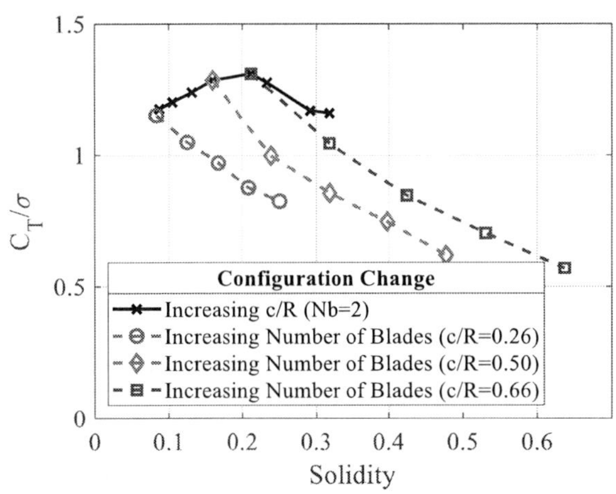

Fig. 8: C_T/σ vs. solidity for fixed blade speed and different methods to change solidity (PA=40°).

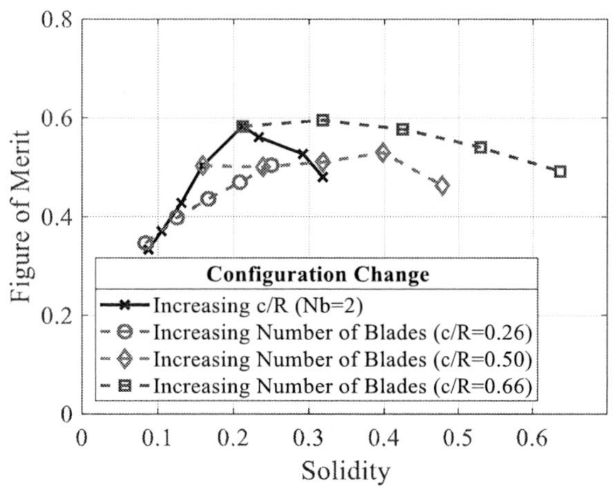

Fig. 9: *FM* vs. solidity for fixed blade speed and different methods to change solidity (PA=40°).

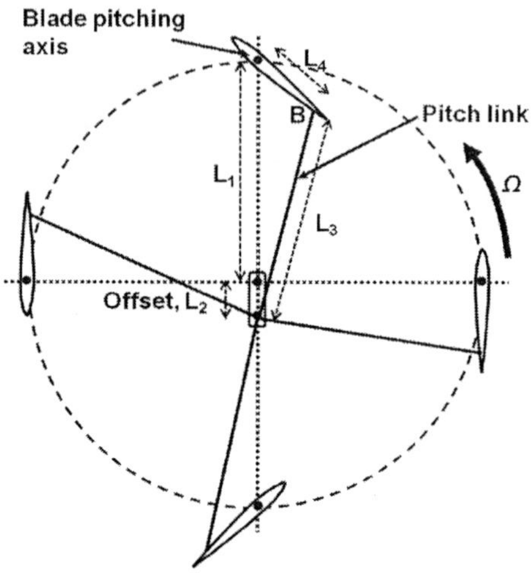

Fig. 10: Four bar pitch mechanism schematic.

to be a three bladed cyclorotor with a chord-by-radius ratio of 0.66.

Additional conclusions from the experimental study included the determination that the blade aspect ratio resulted in minimal change in cyclorotor performance and the optimal symmetric blade thickness was found to be 20% of chord (NACA 0020).

Transitioning to the cyclocopter UAV design, the first step was defining the desired rotor size. Considering the targeted vehicle weight of around 20 lbs and previous experimental studies, the rotor diameter was chosen to be 18 inches. Based on this diameter and the optimal chord-by-radius ratio of 0.66, the blade chord was defined to be 6 inches. The rotor span was chosen to be 18 inches due to the limited effect aspect ratio has on performance and the desire for a compact vehicle width and minimized rotor weight. The rotor was designed with four rotor blades utilizing NACA 0020 airfoils as opposed to the slightly more aerodynamically efficient three blade design. The reason for this was to increase maximum lift capability as well as to simplify rotor balancing and minimize vibrations in the system.

Blade Kinematics and Pitch Mechanism

Designing the cyclorotor with the proper blade pitching kinematics is equally important as choosing the proper rotor geometry. As part of the experimental parametric study the performance of symmetric and asymmetric pitching was examined [Ref. 19]. Results from the study showed that the optimal pitch kinematics varied significantly with the rotor geometry. Based on the experimental study the optimal pitch kinematics was found to be a symmetric sinusoidal motion with a maximum pitch amplitude centered around ±35°. Even though asymmetric pitching was found to be beneficial during previous studies [Ref. 11], its effects were found to be negligible or even detrimental on rotor performance for UAV-scale cyclorotors.

The pitch control mechanism developed for use on the cyclocopter must be capable of reproducing the optimal pitch kinematics while also providing a means for actively controlling both the phase and pitch amplitude. The pitch mechanism design was created using a four-bar linkage (Fig. 10). The radius or the rotor serves as linkage length L_1. An offset link serves as the primary control link (L_2). L_3 is the length of the pitch link and transfers the necessary loads to achieve cyclic pitching. The distance between the connection point on the blades and the support arms serves as L_4.

The mechanism implemented on the cyclocopter was designed to be strong enough to transfer all cyclic pitching loads while minimizing assembly size and weight (Fig. 11). Length L_1 was established via the rotor support and is equal

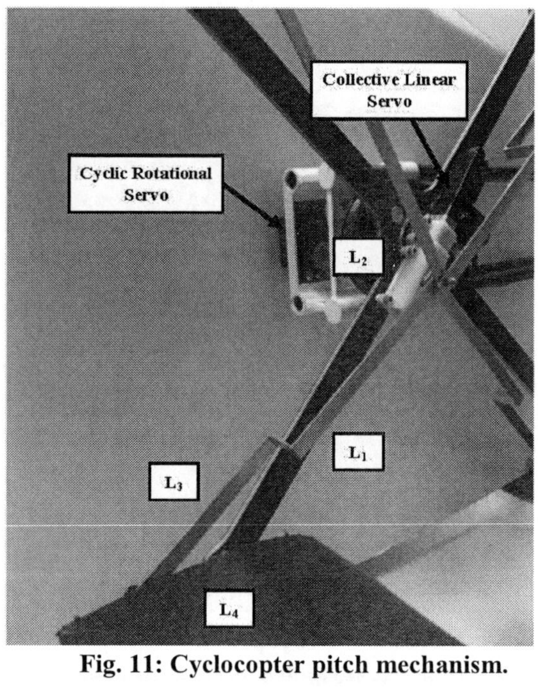

Fig. 11: Cyclocopter pitch mechanism.

to the rotor radius, 9 inches. Length L_4 is equal to the separation distance between the mounting of the rotor radius and pitch link mounting points, 2 inches. Length L_3, 9.133 in, is the primary pitch link and was designed to bend around the central collective linear servo. The link was machined out of a sheet of AL 7075 to maximize stiffness and eliminate stress resulting from bent metal manufacturing techniques.

The final length, L_2, was the most complicated component of the vehicles design. To achieve the desired performance the link assembly must be capable of rotating freely and changing length. The structure of the L_2 link included a machined aluminum bracket which threaded onto a machined threaded tube. The aluminum assembly serves as the rotational axis for the cyclorotor and is rotated using the rotational servo via a 2:3 gear ratio. The linear servo used to achieve collective control was mounted to the aluminum bracket via a secondary bracket machined from plastic. The presence of the linear servo in the rotational frame means power and signal must be fed through the aluminum bracket and tube via a micro-slip ring before being fed to the vehicle's autopilot. The final pitch mechanism design is a lightweight passively pitching system capable of independently achieving cyclic and collective control.

Blade Structural Design

The span-wise split blade design coupled with the high rotational speeds introduces several structural concerns to ensure durability and functionality. The loads on the blade are split between aerodynamic and centrifugal loads. The contribution of aerodynamic loads to blade deformation is minimal compared to the centrifugal loads. To that end, the blades are designed primarily with concern for the centrifugal loads (Fig. 12). The absence of one continuous surface on the blades drastically reduces the overall structural stiffness. The main spar (pitch axis) is the only continuous structural member and thus served as the primary load path to the side supports (Fig. 12d). The two blade halves are free to rotate along this pitch axis. The free rotation of each half of the blades introduces the possibility of pinching between the two blade halves. Along with these concerns, significant effort has been taken at the design stage to minimize blade weight in order to reduce the centrifugal loads.

The blade was designed to allow for maximum structural strength at the lowest possible weight (Fig. 12). To accomplish this, each half of the blade is created from three ribs cut from a carbon fiber and foam sandwich structure linked by a filler foam (Fig. 12a). The composite material used for the ribs provides a high strength to weight ratio with the only compromise being reduced compressive strength and a laminated structure. To ensure free rotation of each half of the blades in the split blade design a bearing was installed in each rib. Shear forces from the bearing mounted in the rib could delaminate the structure, drastically reducing the structural strength of the rib. A 1mm thick plastic ring was 3D printed and pressed into every rib to provide a mounting surface for the bearings, ensuring loads transferred to the entire rib structure. The filler foam is a rigid low-density Rohacell foam. The foam formed the rigid shape of each half blade and provided a constant surface for mounting the Monokote® cover. The foam provides minimal bending strength, so its weight was minimized through the use of the bare amount of material necessary to adhere the Monokote® material. To reduce weight the larger leading-edge foam pieces are connected using wooden dowel pins while the

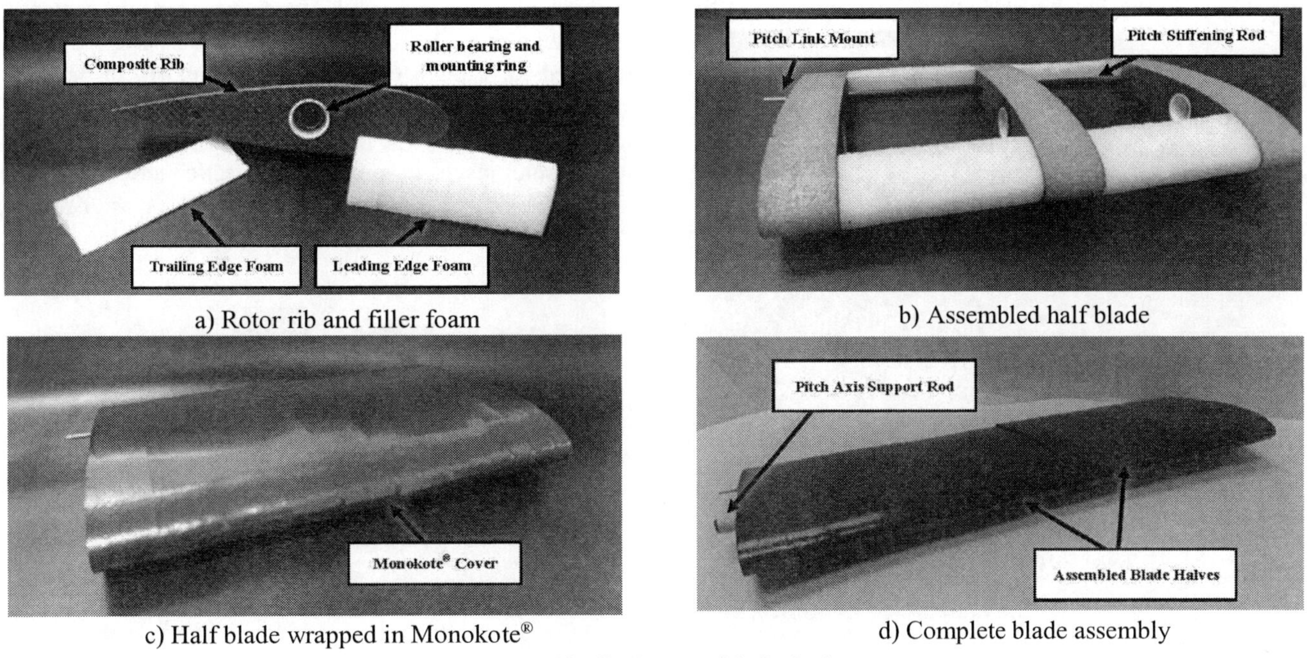

a) Rotor rib and filler foam

b) Assembled half blade

c) Half blade wrapped in Monokote®

d) Complete blade assembly

Fig. 12: Cyclorotor blade design.

trailing edge pieces are held in place with masking tape and minimal epoxy until the Monokote® is applied. The assembled blade halves include a 3/16" pultruded carbon fiber rod, that runs through the rear of the half blade (Fig. 12b). The rods provide additional structural stiffness as well as the mounting point for the rear pitch control rod. The Monokote® wrap creates a clean aerodynamic surface (Fig. 12c). The complete blade assembly features the continuous 3/8" pultruded carbon fiber rod along with ABS spacers to position the blades and prevent pinching between the blade halves (Fig. 12d). The total blade weight comes in at 0.25 lbs. Despite the added complexity of the split blade design the current blade weight represents one of the lowest ratios of blade weight to blade area achieved from any design to date.

Rotor Support Structure and Transmission

The rotor support structure must be designed to handle the large tension loads created by the centrifugal forces on the rotor blades. Despite the lightweight blade design 50 lbs of centrifugal force is created by each blade at the operating rotational rate of 900 RPM. To ensure the rotor is capable of handling the large loads with minimal deflection and to minimize rotor weight the supports were machined from an extremely strong laminate composite material (Fig. 13). The carbon fiber laminate utilized a birch core which provided sufficient compressive strength for handling loads while the carbon fiber provided the high tensile strength required to handle the centrifugal forces. The final design utilized arms machined from a 5/32" thick plate of the carbon fiber birch laminate. A central 5/8" radial bearing is pressed and epoxied into the support arms.

Cyclorotors are designed to operate at much lower RPMs than conventional rotors, but at a much higher torque. Current off-the-shelf electric motors are not designed to operate efficiently in these high-torque/low-RPM environments. As such, a transmission system was designed to be mounted directly to the rotor support arms and motor. A timing belt drive system was chosen for the design due to its reduced acoustic signature. The final design featured a 4.17:1 ratio between the pulleys. The driving motor was a 105 kV outrunner KDE motor controlled by a 75A KDE ESC (KDEXF-UAS75HVC). The rotor support structure and transmission system, combined, provided a lightweight design capable of supporting and driving the cyclorotor to necessary operating RPMs.

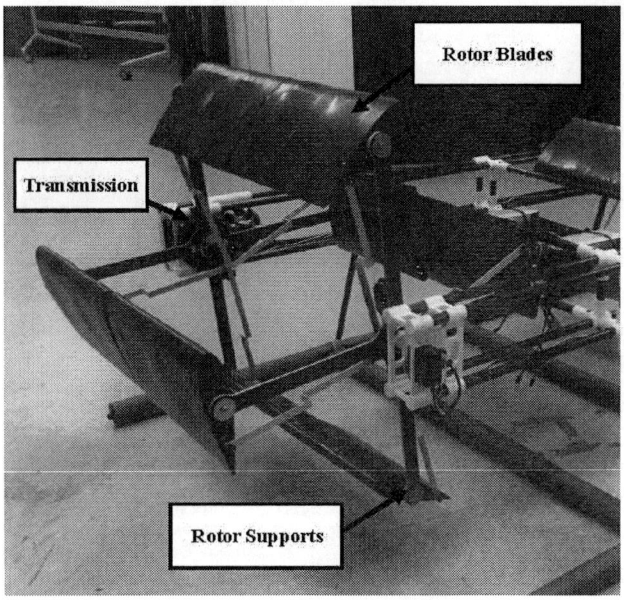

Fig. 13: Cyclorotor diagram.

ATTITUDE CONTROL STRATEGY

Achieving stable hovering flight for a novel vehicle concept requires the careful development of attitude control strategies. The current vehicle features 10 independent control mechanisms (2 rotor RPMs, 4 pitch amplitude controls, and 4 pitch phasing controls). The control redundancy present allows for several unique control strategies to be implemented.

The definition for the vehicles pitch, roll, and yaw degrees of freedom is provided in Fig. 14. The front and rear rotors rotate in opposite directions minimizing any inherent pitching moment. The present control strategy is the simplest possible with each control mechanism providing independent control of a degree of freedom. The control strategy is detailed in Fig. 15 where the thrust vectors are shown for each cyclorotor halve. Pitch control is achieved via differential rotor RPM (Fig. 15a), i.e. differential rotor thrust, between the front and rear cyclorotor. The direction of rotation of the cyclorotors must be opposite to mitigate inherent pitch moments, but whether both rotors rotate away from or towards the center of the vehicle must be carefully considered. The effect is visualized for a pitch up maneuver in Fig. 16. In the case where the rotors are rotating towards the center (Fig, 16a), the increase in rotor RPM required to increase the net thrust for the front rotor results in a pitch down reaction torque from the motors. The decrease in rotor RPM for the rear rotor results

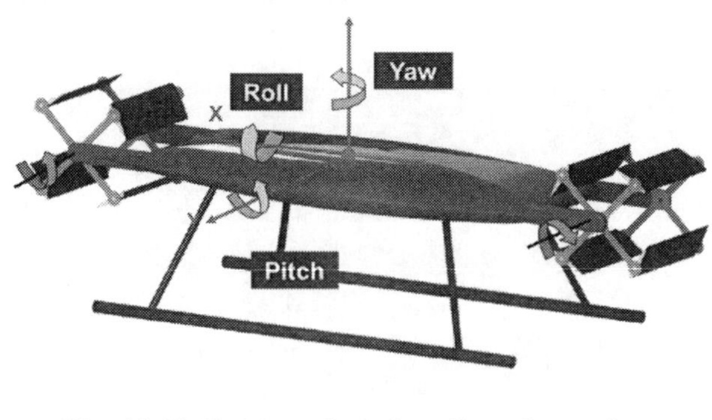

Fig. 14: Definition of pitch, roll, and yaw degrees of freedom.

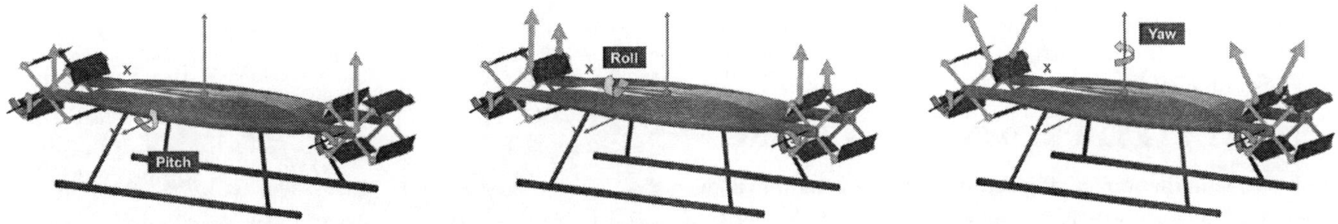

a) Pitch control b) Roll control c) Yaw control

Fig. 15: Attitude control strategy.

in a similar pitch down reaction torque. The resulting reaction torque would act against the torque resulting from the thrust differential. In fact, depending on the magnitude of the command input, the rapid response of the reaction torque could cause an initial pitch down followed by a pitch up. However, if the cyclorotors were rotated away from the center, as they are on the current vehicle, the reaction torque would be reversed causing the reaction torque and thrust differential to work together resulting in an increase in responsiveness (Fig. 16b). Roll control relies on differential pitch amplitudes, i.e. differential rotor thrust, between the left and right halves of the vehicle (Fig. 15b). Yaw control is obtained by vectoring the thrust in opposite directions between the right and left halves of the vehicle (Fig. 15c). The current control strategy minimizes coupling from commanded inputs. Vehicle trim is accomplished initially via geometric positioning of the blade-pitch control length. Additional vehicle trim is achieved during the flight-testing process.

Avionics and Feedback Control

A feedback control system is required to provide sufficient attitude damping and stiffness to achieve stable hover. A state-of-the-art inertial navigation system was incorporated into the vehicle to handle the onboard processing necessary. The navigation system known as ELKA-R (Embedded Lightweight Kinematic Autopilot-Revised) was developed at the University of Maryland [Ref. 18]. The board features a 32-bit low-power ARM Cortex-MR microprocessor and is capable of stabilization rates as high as 1000 Hz. 2.4 GHz XBEE and PPM receivers were installed to handle communication of control inputs from the remote as well handle communication of telemetry data to a computer via LabVIEW for active recording. The telemetry connection also provides a means to actively update the vehicle's feedback gains and trim inputs.

The on-board gyros measure the pitch (q), roll (p) and yaw (r) angular rates while the accelerometers record the tilt of the gravity vector in the body frame. The vehicle attitude can then be extracted by integrating the gyro measurements with time. However, it is known that this leads to drift in attitude measurements. Accelerometers on the other hand offer stable bias but are sensitive to vibrations and in general offer poor high frequency information. Therefore, a complementary filter was incorporated to extract the pitch and roll Euler angles using a high-pass filter for the gyros (4 Hz cut-off) and a low-pass filter for accelerometers (6 Hz cut-off). The rotational vibrations were filtered out since they are sufficiently higher than the body dynamics. On-board inner loop feedback was implemented using a proportional-derivative (PD) controller as shown in Fig. 17. The feedback

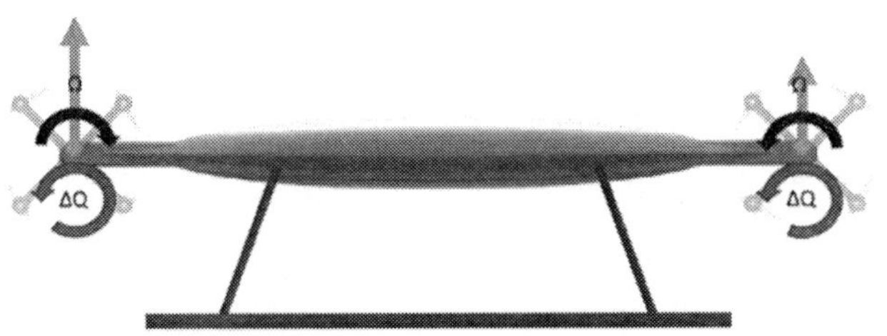

a) Rotors rotated towards the center

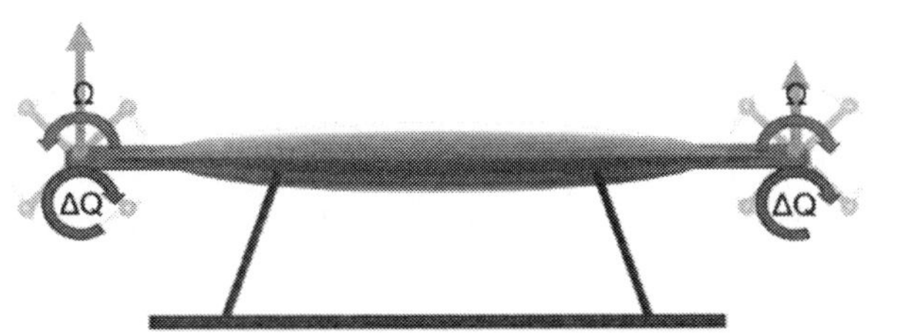

b) Rotors rotated away from the center

Fig. 16: Motor reaction torque's effect on pitch stability. Pitch up maneuver represented.

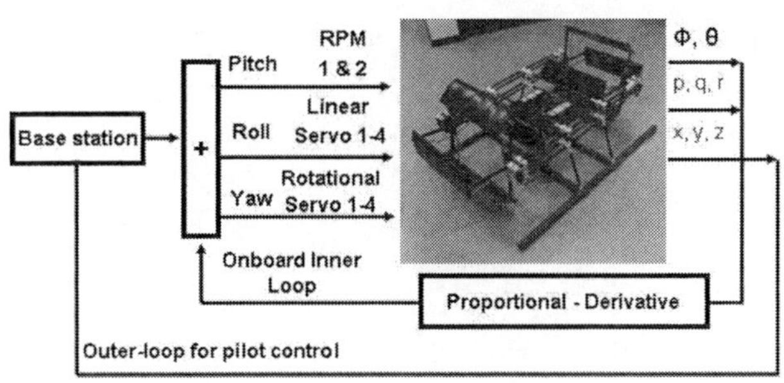

Fig. 17: Schematic showing the inner-loop feedback control system implemented.

states are the pitch and roll Euler angles θ and φ, and the attitude rates, p, q and r. The outer loop feedback capability was provided for translational positioning by a human pilot or for a position tracking system.

GIMBAL / TETHERED FLIGHT TESTING

Prior to the initiation of flight-testing gimbal testing and tethered flight testing was conducted as a means to verify proper operation of all mechanical and electrical systems. Gimbal testing involved mounting the vehicle through a spherical bearing allowing for three rotational degrees of freedom, namely, pitch, roll and yaw (Fig. 18). Conducting experiments on the gimbal test stand was a crucial step to flight testing as a means for troubleshooting with minimal risk to the vehicle. Systematic tests allowed for the determination of the gains necessary to achieve acceptable stiffness and damping to reject external disturbances with minimal oscillations.

Following gimbal tests, tethered flight tests were conducted to verify vehicle liftoff and control responsiveness (Fig. 19). Slight modifications to the feedback gains were determined necessary to account for the change in point of rotation (center of gravity vs center of spherical bearing) and removal of friction from the bearing. Unfortunately, even though tethered flight tests demonstrated stability in pitch and yaw, the vehicle lacked the responsiveness in roll necessary to maintain stable hovering flight due to the slow response of the linear servos responsible for varying the pitch amplitude. These servos are in the process of being replaced with those providing a three times faster response. Free flight tests will be conducted following the servos integration.

SUMMARY AND CONCLUSIONS

This paper presents the development of a VTOL-capable, UAV-scale cyclocopter, with a gross weight of 17 pounds. The vehicle features a dual-rotor configuration with one

Fig. 18: Gimbal test stand experiments.

Fig. 19: Tethered flight tests.

cyclorotor positioned at the front and the other at the rear of the vehicle. The rotors feature a novel split-blade design. Each rotor is capable independent control of the pitch amplitude and phase of each blade half. The current attitude control strategy uses a combination of differential RPM between the front and rear rotors, differential pitch amplitude between the left and right halves of the rotors, and differential cyclic pitch phase between the left and right halves of the rotor for pitch, roll, and yaw control. To date gimbal test stand and tethered flight experiments were conducted. Specific conclusions from this study are as follows:

1. A systematic experimental parametric study was conducted to investigate the effects of rotor geometry and blade kinematics on aerodynamic performance. Based on the study a set of design parameters were established to create an optimized cyclorotor. The vehicle's rotors feature a 6 in blade chord, 18 in rotor diameter, and 18 in blade span. The 18 in blade span is divided equally between two 9 in blade halves featuring NACA 0020 airfoils with nominal pitch amplitudes of $\pm 35°$.

2. Centrifugal forces result in large transverse bending loads on the blades thus minimizing the blade weight and maximizing blade stiffness is crucial. A lightweight composite foam composite rib assembly with carbon fiber rod stiffeners was utilized to create blades with the lowest ratio of blade weight to blade area achieved to date.

3. A compact and mechanically simple pitch mechanism was created using a four-bar linkage system. The central control link's length and radial position was varied independently by linear and rotational servos providing both pitch amplitude and phase control for each blade half.

4. Independent control of each cyclorotors RPM and each blade halves' pitch amplitude and phase introduced redundancy into the control system. The present attitude control strategy relied on differential

RPM, pitch amplitude and cyclic phase to achieve pitch, roll, and yaw control.

5. Present gimbal and tethered flight tests demonstrated sufficient control authority exists in pitch, roll and yaw. The current electronic systems provided fast enough response rates to maintain stability in pitch and yaw, but the reduced speed in the current linear servos resulted in roll instabilities. Following integration of new servos free flight tests will be conducted.

Author Contacts:

Adam Kellen adamkellen@tamu.edu
Moble Benedict benedict@tamu.edu

ACKNOWLEDGMENTS

This work is supported by Army/Navy/NASA sponsored Vertical Lift Research Center of Excellence (VLRCOE) (2016 – 21) led by the University of Maryland with Dr. Mahendra Bhagawat as Technical Monitor.

REFERENCES

[1]Hein, B., and Chopra, I., "Hover Performance of a Micro Air Vehicle: Rotors at Low Reynolds Number," *Journal of the American Helicopter Society*, Vol. 52, No. 3, April 2007, pp. 254-262.

[2]Lind, A. H., Jarugumilli, T., Benedict, M., Lakshminarayan, V. K., Jones, A. R., and Chopra, I., "Flow field Studies on a Micro-Air-Vehicle-Scale Cycloidal Rotor in Forward Flight," *Experiments in Fluids*, Vol. 55, November 2014, pp. 1 – 17.

[3]Benedict, M., Jarugumilli, T., Lakshminarayan, V. K., and Chopra, I., "Effect of Flow Curvature on the Forward Flight Performance of a MAV-Scale Cycloidal Rotor," *AIAA Journal*, Vol. 52, (6), 2014, pp. 1159 – 1169.

[4]Jarugumilli, T., Benedict, M., and Chopra, I., "Wind Tunnel Studies on a Micro Air Vehicle-Scale Cycloidal Rotor," *Journal of the American Helicopter Society*, Vol. 59, (2), April 2014, pp. 1–10.

[5]"Cyclogyro History," World-Wide Web document, http://www.rotoplan.narod.ru/historye.htm, May 19, 2007.

[6]Sachse, H., "Kirsten-Boeing Propeller," Technical Memorandum 351, National Advisory Committee for Aeronautics, January 1926.

[7]Kirsten, F. K., "Cycloidal Propulsion Applied to Aircraft," Transactions of the American Society of Mechanical Engineers, Vol. 50, (12), 1928, pp 25–47.

[8]J. Boschma and M. McNabb, "Cycloidal Propulsion for UAV VTOL Applications," Naval Air Warfare Center-Aircraft Division, Tech. Rep., 1998

[9]Runco, C., Coleman, D., and Benedict, M., "Development of the World's Smallest Cyclocopter," Proceedings of the 72nd Annual National Forum of the American Helicopter Society, West Palm Beach, FL, May 17–19, 2016.

[10]Elena, S., Hrishikeshavan, V., Benedict, M., Yeo, D., and Chopra, I., "Development of Control Strategies for a Twin-Cyclocopter in Forward Flight," *Journal of the American Helicopter Society*, Vol. 61, No. 4, October 2016, pp. 1 – 9.

[11]Benedict, M., Jarugumilli, T., and Chopra, I., "Effect of Rotor Geometry and Blade Kinematics on Cycloidal Rotor Hover Performance," *Journal of Aircraft*, Vol. 50, (5), September-October 2013, pp. 1340–1352.

[12]Walther, C., and Benedict, M., "Experimental and Computational Studies to Understand Unsteady Aerodynamics of Cycloidal Rotors in Hover at Ultralow Reynolds Numbers," Proceedings of the 7rd Annual National Forum of the American Helicopter Society, Fort Worth, TX, May 9–11, 2017.

[13]Benedict, M., Jarugumilli, T., and Chopra, I., "Experimental Optimization of MAV-Scale Cycloidal Rotor Performance," *Journal of the American Helicopter Society*, Vol. 56, (2), April 2011, pp. 022005-1-022005-11.

[14]Benedict, M., Ramasamy, M., Chopra, I., and Leishman, J. G., "Performance of a Cycloidal Rotor Concept for Micro Air Vehicle Applications," *Journal of the American Helicopter Society*, Vol. 55, (2), April 2010, pp. 022002-1 - 022002-14.

[15]Benedict, M., Ramasamy, M., and Chopra, I., "Improving the Aerodynamic Performance of Micro-Air-Vehicle-Scale Cycloidal Rotor: An Experimental Approach," *Journal of Aircraft*, Vol. 47, (4), July-August 2010, pp. 1117–1125.

[16]Lee, C. H., Min, S. Y., Lee, J. W., and Kim, S. J., "Design and Experiment of Two-Rotored UAV Cyclocopter," Proceedings of the 29th Congress of the international council of the Aeronautical Sciences, St. Petersburg, Russia, September 7-12, 2014.

[17]Hwang, I. S., Min, S. Y., Lee, C. H., Lee, and Kim, S. J., "Development of a Four-Rotor Cyclocopter," *Journal of Aircraft*, Vol. 45, (6), November 2008, pp. 2151-2157.

[18]Hrishikeshavan, V., and Chopra, I., "Refined lightweight inertial navigation system for micro air vehicle applications," *International Journal of Micro Air Vehicles*, Vol. 9, (2), 2017, pp. 124–135.

[19]Kellen, A., and Benedict, M., "Effect pf Rotor Geometry on UAV-Scale Cycloidal Rotor Hover

Performance," Proceedings of the 74th AHS Annual Forum, Phoenix, AZ, May 14–17, 2018.

Configuration Redesign and Prototype Flight Testing of an Unmanned Fixed-Wing eVTOL Aircraft with Under-Fuselage Hover Lift and Pusher Wingtip Propulsion System

Philipp Stahl
M.Sc.
Technical University of Munich
Garching/Munich, Germany

Christian Roessler
Dr.-Ing.
Technical University of Munich
Garching/Munich, Germany

Mirko Hornung
Prof. Dr.-Ing.
Technical University of Munich
Garching/Munich, Germany

ABSTRACT

The 'Quad+Tilt' fixed-wing eVTOL configuration is the combination of a quadrocopter in '+' orientation and tiltable wings or wingtips. Its wingtips used for propulsion and control as well as the mounting of lift rotors on the fuselage are two major design features that are addressed in this paper. Inflight power consumption measurements on a small unmanned implementation of this configuration suggest a 4% reduction in cruise flight power consumption accredited to wingtip propulsion. Three alternative wingtip designs for a Quad+Tilt configuration with different control principles were proposed and flight tested. Two of them implement a pusher propeller configuration. Rotor thrust losses and fuselage cross section depict a sensible trade-off for the Quad+Tilt configuration. Taking helicopter tail rotors as an example, the mounting of a Quad+Tilt's hover rotors below the fuselage is examined as a strategy to overcome this trade-off. Verifying measurements on the thrust loss caused by blocking objects in the in- and outflow of a rotor were performed and used for modelling. Crucial topics to realize the advantages of underbody rotor mounting like foreign object damage, rotor strike and noise emission are addressed and solutions proposed. Despite the required measures implemented, a gain of 3% in cruise endurance seems achievable for a 5 kg Quad+Tilt aircraft with lift rotors mounted to the fuselage bottom instead of the fuselage top.

INTRODUCTION

Fixed-wing eVTOL UAV are electrically driven, unmanned aerial vehicles capable of both hover and fast forward flight. Lift during take-off and landing is generated by rotors or ducted fans. After takeoff, they transition into forward flight mode in which a wing produces the lift. Many existing types of fixed-wing VTOL aircraft (see Ref 1, 2, 3) show the wide range of possibilities to combine vertical flight capabilities with efficient forward flight. The question arose which of these configurations is best suited for a certain mission. Using a dedicated conceptual design and mission simulation tool (see Ref 4), a benchmarking of typical fixed-wing eVTOL configurations was performed. A configuration that combines a quadrocoper in '+' orientation with tiltable wings or wingtips, called 'Quad+Tilt' in this paper (see Figure 1 and Figure 2), showed significant performance benefits compared to its contenders.

Figure 1. Boeing's 'Phantom Swift' as a Representative for a 'Quad+Tilt' Configuration ©Boeing

Figure 2. Autel Robotics's 'Dragonfish' as a Representative for a 'Quad+Tilt' Configuration ©dronerush.com

To keep power consumption in hover low and, by that, enable increased battery energy density, the majority of lift thrust is efficiently generated by hover optimized, fuselage mounted fans or rotors. The small amount of thrust generated by the wingtip propulsion system is vectored to provide yaw control in hover, as well as forward thrust and roll control in cruise. Differential thrust on the wingtip powertrains is used to control roll in hover and yaw in cruise flight. Aircraft control

Presented at the Vertical Flight Society Autonomous VTOL Technical Meeting and Electric VTOL Symposium, Mesa, Arizona, UAS, Jan. 29-31, 2019.

is realized with a minimum number of effectors and thus minimal system mass. The optimization of the wingtip powertrain for the cruise condition, the beneficial interaction between propeller or fan swirl and wingtip vortex (see Ref 5, 6, 12) and a wing clean of nacelles enable efficient cruise flight.

As visible in Figure 1 and Figure 2, the integration of lift powertrains is challenging for Quad+Tilt configurations. Lift fans take valuable space inside the fuselage and have diameter limitations, while rotors mounted outside the fuselage suffer from thrust loss as their stream tube is blocked by the fuselage. Further, the tiltable wingtip carrying powertrains is a sophisticated aircraft system.

This paper first validates the performance benefits of wingtip mounted propulsion with inflight measurement data of a prototype aircraft. To implement a tilt-wingtip with integrated powertrain, three designs are proposed, evaluated and flight tested. The paper further experimentally examines and models the thrust loss of blocked rotors. Especially, an unconventional mounting of rotors under the fuselage is addressed.

The performed measurements and proposed designs primarily have unmanned, fully electric aircraft with take-off masses up to 25 kg in mind.

ASPECTS OF TILT-WINGTIP PROPULSION AND CONTROL

The positioning of cruise powertrains on tiltable wingtip sections combines the principle ideas of

- using the actuators and powertrains for aircraft control and propulsion in both hover and cruise flight to reduce system weight
- using the wing span lever arm to achieve
  - roll and yaw hover control moments with little thrust, respectively power, and thus enable a powertrain optimization for cruise
  - very high hover control moments using the full thrust capability of the wingtip powertrains if superior agility and gust handling is required
- keeping the wing clean of the drag, mass and manufacturing related penalties of nacelles or control surfaces
- using the beneficial interaction of propeller swirl and wingtip vortex

Opposed to that stand challenges like

- robust mechanical implementation of the tilt-wingtip
- power distribution to the remote wingtip location (cable mass, ohmic losses, risk of electro-magnetic compatibility issues)
- significant contribution to the aircraft's roll and yaw inertia (aircraft dynamic behavior)
- handling of 'one engine inoperative' scenario
- trade-off between wing geometry parameters and pitch trim effort due to cruise thrust vector above/below center of gravity (CoG)

Wingtip Propulsion Benefits

While most above points are obvious and well describable in conceptual design tools (like used in Ref 4), the presence and extent of wingtip propulsion benefits was of special interest.

An inflight experimental attempt was made to evaluate the expected positive impact of wingtip vortex and propeller swirl interaction. As the positive effect is ascribed to the reduction of the wingtip vortex (a source of induced drag) by a counter-rotating propeller swirl, the approach is to compare the power consumption for the same aircraft, in similar flight conditions between flights with propellers rotating 'inward blade up' (counter-vortex) and flights with propellers rotating 'inward blade down' (vortex-wise). The 'inward blade up' should exhibit the lower power consumption of both. According to Ref 12, the overall efficiency of an aircraft with only the wingtip motor nacelle mounted (no forward thrust) is right in the middle of counter-vortex and vortex-wise wingtip propulsion. Subtracting the end plate effect of the wingtip motor nacelle yields the overall efficiency of an aircraft without any wingtip device. Measurability of the difference in power consumption is expected best for airspeeds close to stall speed. Here, the vortex and induced drag almost gains maximum intensity relative to other drag components. The recorded power consumption P and airspeed data v was averaged and translated with the aircraft's total weight G into the overall efficiencies $\frac{L}{D}\eta$ by

$$\frac{L}{D}\eta = \frac{G\,v}{P}$$

Table 1. Test Flight Data on Wingtip Propulsion Benefit

Day-Flight		1-1	1-2	2-1	2-2
Prop Rotation w.r.t. Wingtip Vortex		counter	co	co	counter
Test Duration [min]		22.7	25.3	28.0	28.7
Airspeed [m/s]	av.	17	16.5	17.7	16.9
	dev.	0.4	0.4	0.4	0.7
Power [W]	av.	117.7	132.8	127.4	120.3
	dev.	14.9	23.1	12.2	22.6
$\frac{L}{D}\eta$ [-]	abs.	4.30	3.70	4.16	4.21
	rel.	1.162	1	1	1.011

Table 1 summarizes the flight test data and results. The aircraft depicted in Figure 3 (span 2.1 m, mass 3 kg, wing aspect ratio 14) was used to obtain the data. The propellers providing forward thrust are each mounted at the wingtips before the leading edge in a tractor configuration (see Figure 4). In total, four test flights were performed on two days, one after another with just a short break to change the propeller direction and the battery. Early-morning environmental conditions with minimal horizontal and vertical air movement

Figure 3. Aircraft to Evaluate Wingtip Propulsion Benefits and Wingtip Designs in Flight

were chosen. Airspace restrictions only allowed a trajectory of earth-fixed, 15° nominal bank angle circles. Throughout the trajectory, constant altitude and an airspeed of 17 m/s were controlled by an autopilot. These parameters result in a nominal lift coefficient of 0.58. The trajectory shape and a slight daytime dependent increase in wind speed seem to cause an increased deviation in power consumption for the later flight of each day. The higher variations in power consumption may impair the overall efficiency. Looking at the figure of relative $\frac{L}{D}\eta$, an increase of 16.2% and 1.1% (despite the higher variations in power consumption of the later flight) in overall efficiency for the aircraft with counter-vortex rotating propellers can be observed against the aircraft with propellers rotating in the direction of the wingtip vortex. If one conservatively (not subtracting the nacelle end plate effect) assumes the overall efficiency of an aircraft without wingtip propulsion right in-between the efficiencies of wingtip propulsion with different propeller directions, counter-vortex wingtip propulsion would achieve an increase in overall efficiency $\frac{L}{D}\eta$ in the order of 4% over an aircraft without wingtip propulsion. Due to the underlying assumptions, precision of data acquisition and deviations in the data records, the statements made have to be considered with care. They will differ further for different aircraft and flight conditions.

Wingtip Design for Quad+Tilt Configurations

The key enabler for a Quad+Tilt's propulsion and control concept is its wingtip. To understand the origin of the challenging wingtip design, a brief outline of the requirements for actuation, bearing and power supply is given:

- sufficient roll and yaw authority in hover and cruise flight (tilt angle range of typically >140°)
- dimensioned for the following load cases
  - engine thrust
  - oscillating propeller unbalance (fatigue!)
  - wingtip aerodynamic forces
  - movement related inertial and precession forces
  - operator handling
  - ground contact
- low aerodynamic drag
- stable against aero-elastic divergence or flutter
- integration within given aerodynamic surfaces
- good accessibility for integration, maintenance and repair

Three wingtip designs, namely 'tiltable tractor', 'tiltable pusher' and 'fixed pusher', are proposed.

'Tiltable Tractor' Wingtip The 'tiltable tractor' wingtip design (see Figure 4) combines the tilting of an outermost

Figure 4. 'Tiltable Tractor' Wingtip

wing section with the vectoring of the propeller thrust. In hover, the downwash is only minimally blocked by the wingtip section. In a vertical position, the propeller cannot touch the ground during a turnover while the trailing edge is more robust against ground strike. In cruise, both the deflection of the wingtip surface as well as the thrust vectoring generate a roll moment. The propeller swirl is partly straightened by the wingtip surface, however can still interact positively with the lift-induced wingtip vortex (see chapter 'Wingtip Propulsion Benefits'). However especially for the deflected wingtip, major parts of the wing vortex already form at the gap between inner wing and wingtip. As a consequence, the intensity of the wingtip vortex positively interacting with the tractor propeller swirl is reduced, and so are the interaction benefits. As well, control effectiveness of the wingtip section is decreased. A gap sealing e.g. by an endplate is advised. The propeller's swirl increases the angle of attack on the wingtip section. Under certain flight conditions, this might lead to stall and its related drawbacks. The tilt axis at quarter chord is behind the wingtip CoG to avoid flutter issues. The oscillating loads coming from blade unbalance or transverse flow effects depict a potential source of fatigue fracture. Main weak point are often the actuator gears. To reduce the loads to a minimum the distance between propeller plane and tilt axis is minimized.

'Tiltable Pusher' Wingtip The 'tiltable pusher' wingtip design (see Figure 5) implements thrust vector tilting for a pusher configuration. The motivation is that propeller efficiency benefits were obtained for propellers impinged with inflow swirl induced by a wingtip vortex (see Ref 6) and partly laminar flow over the wingtip surface reduces aerodynamic drag. However, simply moving the propeller aft of the trailing edge of the 'tiltable tractor' wingtip would result in several problems:

- The clearance between the ground and the propeller is reduced. A contact between a rotating propeller and the ground is significantly more critical than the trailing edge of the 'tiltable tractor' touching the ground.
- The increased distance between propeller plane and shaft axis promotes unbalance and static moment loads on the actuator.
- The aft CoG of the wingtip section promotes flutter.

Figure 5. 'Tiltable Pusher' Wingtip

Furthermore, the gap between wingtip and inner wing and its related drawbacks remain. To solve or mitigate these issues, the motor is mounted on a shortest possible U-bracket that is directly driven and supported by the actuator. A tiltable aerodynamic surface, the related gap flow and flutter tendency is omitted. Thrust vectoring is solely used for direct roll moment generation in cruise. Thus, roll authority in cruise is dependent on the cruise thrust level and, thus, on the flight state. For the aircraft depicted in Figure 3, roll rates around 37°/s are calculated for horizontal cruise and asymmetric tilt angles of 45°/-20°. If roll control authority is insufficient, conventional ailerons may be added. In hover, the propeller operates in sufficient distance to the wingtip surface such that no noteworthy thrust blocking is expected. The omission of a tilt shaft and its bearing reduce wing structural mass and complexity.

'Fixed Pusher' Wingtip The idea behind the 'fixed pusher' (see Figure 6) wingtip is to build up the thrust vector by two separate powertrains instead of tilting a single one. The tilt actuator is replaced by a 'roll thruster' electric motor which represents a significant gain in system robustness. In hover, yaw is controlled by differential thrust of the horizontal cruise powertrains. For fast control response and to avoid frequent switching of the spin direction, a forward thrust offset has to be set. This must be compensated by a slight nose-up attitude of the aircraft. The implied force fighting means increased power consumption in hover. The roll thruster can principally be designed with small diameter as it just has to provide little roll control thrust. Their installation close to the wingtip surface makes them suffer of thrust blocking. In combination

Figure 6. 'Fixed Pusher' Wingtip (bottom view)

with their small diameter, again, the ratio between thrust and power is unfavorable. The ground clearance of the horizontal thrust propeller is further reduced over the 'tiltable pusher' wingtip. As with the 'tiltable pusher', no aerodynamic control surface for roll control in cruise is available. The roll thrusters must be used for this purpose. As before, additional ailerons are imaginable to provide aerodynamic roll control in cruise. In forward flight, parasitic drag of both the spinning and stopped roll thruster rotors is present. The missing tilt shaft, bearing and a smaller extent of complex geometry reduce mass. The roll thruster powertrain as well is lighter than an actuator.

Table 2. Wingtip Mass Break Down

Wingtip Version	Tiltable Tractor	Tiltable Pusher	Fixed Pusher
Cruise Motor, Propeller, Adapter	100 g	100 g	100 g
Roll Thruster Rotor, Adapter, Motor, ECU	-	-	28 g
Actuator	46 g	46 g	-
Wingtip Structure	69 g	38 g	25 g
Tilt Shaft, Inner Wing Load Introduction	21 g	-	-
Inner Wing Extension	-	10 g	12 g
Wire Harness	58 g	58 g	58 g
Total (one wingtip)	294 g	252 g	223 g
Mass Saving w.r.t. 'Tiltable Tractor' in % of Aircraft Weight (two wingtips)	-	1.7	2.8

Table 2 breaks down the mass of the different wingtip versions as they were manufactured for flight testing on the aircraft depicted in Figure 3. The 'fixed pusher' is the lightest wingtip and saves 2.8% in aircraft overall mass compared to the 'tiltable tractor'. The 'tiltable pusher' can still save 1.7% of the total aircraft mass.

All wingtips need to be tested in flight. Especially for the pusher versions, partly significant compromises have to be made in operability (reduced propeller clearance) and controllability (dismissal of aerodynamic roll control). These must proof to be feasible in first place and furthermore be justified by enhanced effectiveness of favorable wingtip propulsion interactions. The aircraft depicted in Figure 3 was fitted with fuselage mounted rotors to enable VTOL.

Wingtips in Hover Flight Test In hover, very good yaw authority is achieved with the 'tiltable tractor' wingtip. Position control can be achieved without changing the pitch attitude respectively wing's angle of attack, solely by rotating the wingtips over their range of 20° backwards and 120° forward from vertical position. Both provide the aircraft with good wind and gust handling capabilities in hover. The thrust share between wingtip propellers and fuselage rotors can be reduced down to 20% while still achieving sufficient controllability in hover. This contributes to a low absolute power consumption of the non-hover optimized wingtip propellers. The 'tiltable pusher' wingtips behaved likewise from controllability and power consumption point of view. The reduced propeller clearance resulted in occasional ground strikes. As expected, the roll thrusters on the 'fixed pusher' wingtip had to perform wide changes in RPM to stabilize the roll axis in hover. Due to the motor's high RPM, they were well audible out of the homogenous spectrum of the other powertrains. Despite very aggressive control gains and elevated idle RPM on the roll thrusters, it was not possible to realize a control behavior as tight as with the tiltable wingtips. Therefore, the roll thrusters likely lack in available thrust. Yaw control via differential thrust, in contrast, was surprisingly good in the general context of high yaw axis inertia of the aircraft. To avoid force fighting, it was tried to set the RPM of the horizontal motors to zero if no yaw control input is given. The increased response time due to the start-up time of the electric motors leads to unsatisfactory yaw control performance. The required force fighting for yaw control, the wide variation in thrust for roll control in combination with the inefficient thrust generation of the roll thrusters becomes apparent in the aircraft's average power consumption for static hover (see Table 3) as well as in the peak power consumption during hovering manoeuver flight. The average power loading during stationary hover flight is 26% higher for the Quad+Tilt test aircraft equipped with 'fixed pusher' wingtips instead of 'tiltable pusher' wingtips.

Table 3. Hover Efficiency for the Test Aircraft with Different Wingtip Designs

Wingtip Version	Tiltable Pusher	Fixed Pusher
Power Loading [W/N]	13.3	16.8

In high agility maneuver flight, the temperature of the roll thrusters approached the allowable limit. For practical use, the sluggish roll control in combination with the smallest propeller clearance lead to frequent cruise propeller ground strikes and damaged blades on the 'fixed pusher' wingtip. Here, the 'tiltable pusher' proved to be significantly less critical. The higher clearance and roll attitude controllability comparable with the 'tiltable tractor' wingtip helped to reduce the occurrence of propeller strikes. The more parallel attitude of the propeller plane with regard to the ground also reduces the extent of damage. A ground strike of the trailing edge on the 'tiltable tractor' showed to be uncritical.

Wingtips in Cruise Flight Test In cruise, roll control with the 'tiltable tractor' behaved as expected thrust dependent. Maximum feasible roll rates with cruise speed of 17 m/s and corresponding thrust setting ranged between 75 and 130°/s. But even at low speed, windmilling conditions (representing slight reverse thrust), no control reversal is present and sufficient roll rates of 25°/s can be achieved. Excellent yaw authority and damping is solely provided by differential thrust. In case of one engine inoperative and the other still in cruise thrust setting, the aircraft immediately introduces sustained stall spirals. With stopped propellers, controllability and directional stability is sufficient to e.g. perform an emergency landing. For the 'fixed pusher', it was not possible to achieve roll rates beyond 7°/s using full throttle on one of the roll thrusters. A likely reason is the roll thruster's rotor operating in stall conditions. At a tip speed of around 25 m/s and a cruise speed of 17 m/s, the retreating blade barely generates lift. The advancing blade faces an increased angle of attack due to the wingtip vortex and the horizontal inflow speed. Therefore, the occurrence of stall is likely. Hence, a limited directional control of the aircraft was solely possible via differential cruise thrust and yaw-roll coupling. With the 'tiltable pusher' wingtips mounted, sufficient roll rates in the order of 37 to 63°/s could be achieved in cruise flight with asymmetric deflections of –20° and +45°. As expected, the roll authority proved to be strongly thrust dependent. Roll rates of 22 to 28°/s could still be achieved during descent flight with repeated roll doublets and a glide angle of -14°. One must be aware of the increased drag of the performed flight maneuvers (adverse yaw and sideslip). At idle thrust, no roll control reversal occurred. Regarding the noise emission, no significant differences among the wingtip versions could be noticed by the flight test crew. As well, the emitted noise was rated as low enough for typical mission scenarios. Dedicated measurements on the pusher wingtips' power consumption in cruise could not yet be performed.

REDUCTION OF THRUST LOSS DUE TO ROTOR BLOCKING

Most fixed-wing VTOL aircraft have to accept thrust losses as their rotors' in- or outflow field is blocked by structure that

Figure 7. Representative of the Popular Fixed-Wing VTOL Configuration 'QuadXCruise' ©AltiUAS

holds the rotor and its engines. As the cross section of the blocking structure is typically small (see Figure 7), no relevant thrust blocking losses arise. Quad+Tilt configurations use their fuselage to hold its engine and rotor. While weight and drag of nacelles can be saved, thrust losses due to the larger blocking cross section of a fuselage are significantly increased. The consequence is a higher required gross thrust and hence increased power consumption. This, in turn, causes secondary effects like

- increased powertrain mass (rotor, motor, engine control unit, wire harness)
- mass, power consumption and aerodynamic drag of improved powertrain cooling
- reduced battery energy density due to higher required battery power capability

To reduce thrust losses at a given blocking cross section, the tail rotor of most helicopters is mounted in a pusher configuration to the tail fin (see Ref 9). This principle applied to the Quad+Tilt configuration leads to rotors mounted below the fuselage structure.

In the history of rotorcraft aviation, the approach to mount rotors to the bottom of a fuselage is unused. The concerns are obvious:

- close ground operation with the risk of foreign object damage (FOD) and restrictions for the blade movement and the aircraft's attitude
- limitations for operation with spinning rotor and insufficient passenger/operator safety
- typically increased noise emission of rotors with disturbed inflow
- complicated landing gear design

In the context of unmanned electric fixed-wing VTOL aircraft, these issues either vanish or can be solved by reasonable design compromises.

Avoidance of Rotor Strikes Issues arising from close ground operation of the rotors relate to the impact of the rotor blade with the ground or objects that were blown up by the rotor downwash. Due to the multirotor control principle and high stiffness of the hingeless blades with comparatively small diameter, the rotor's coning, tilting and flapping movement is marginal. The required clearance between rotor and fuselage to allow for the vertical amplitudes of the blade tips are well arrangeable in aircraft design. Ref 11 finds that the clearance between rotor and fuselage only plays a minor role for the aerodynamic drag of the stopped rotors in cruise flight. Furthermore, certain pitch and roll angles must be possible for the aircraft on the ground without a rotor ground strike. Besides an unobstructed payload field of view and the limitation of rotor tip vortex-landing gear noise interaction, the provision of this freedom of movement is the main driver for landing gear height and position. Here, the flatness of the VTOL pad (e.g. high grass) is as well relevant. Fortunately, the mass for a VTOL landing gear is reduced compared to conventional take-off and landing. It however must be pointed out that for an underbody rotor vehicle, a failure of the landing gear means a rotor ground strike. The mass and drag penalty of a higher and more robust VTOL landing gear will however be still moderate. Most transition VTOL aircraft's ability to control their position to a certain degree independent from pitch angle can help to reduce ground clearance required for VTOL in wind or from sloped surfaces. The consequences of dust whirl-up are mitigated by the nature of unmanned electric vehicles. An electric engine does not need to ingest the contaminated air for operation, hence can be enclosed easier. Furthermore, brown-out is irrelevant for autopiloted vehicles. In case of a harmful object uptake into the rotor or a rotor ground strike, an electric multicopter's rotor system is highly robust and damage-tolerant. For most missions however, a favorable VTOL site can be found or prepared with no effort.

Pusher Rotor Noise Mitigation Propellers and rotors in pusher configuration are known for their increased noise emission. In hover flight, the induced velocity of the rotor produces vertical flow around the fuselage cross section. The aft flow field that enters the rotor contains uneven velocity distributions out of potential flow, boundary layer or separation effects. This disturbed inflow produces pressure fluctuations on the rotor blades which are a source of pusher noise. The hover optimized rotors of the Quad+Tilt configuration work with low disk loading and therefore induce low vertical flow velocity. This helps to limit the unevenness of the inflow. A further source of pusher noise is the interaction of the rotor blades' pressure distribution with the fuselage geometry. For both mentioned pusher noise sources, the distance of the rotor plane to the object that blocks the inflow is crucial. In terms of the shape of noise propagation, reflections on the fuselage bottom may yield noise maxima in a wedge below the fuselage. Combined with the general in plane noise maxima of a rotor, this unfortunately corresponds with the position of operators and spectators. Human noise perception highly depends on the frequency of the noise (see Ref 8). Below 1 kHz, the perceived loudness for the same sound pressure level progressively decreases with frequency. The relevant frequency derives from the frequency the blade passes the blocking object. The low rotational speed of Quad+Tilt's hover optimized rotors are favorable also in this regard. Not having performed more in-depth noise modelling, these simple considerations and easily tuneable parameters to further reduce the pusher noise component (distance fuselage-rotor plane, rotational speed of rotors) raise hopes that noise might not evolve as a show stopper. Finally however, it must be evaluated during flight testing if a satisfactory noise level can be achieved.

Rotor Thrust Blocking of the Fuselage Rotor thrust blocking is a driver towards narrow and streamlined cross sections of Quad+Tilt fuselages (see Figure 2). A rectangular or elliptical fuselage shape however better matches the components' geometry and thereby reduces unusable space inside the fuselage. Besides a reduced volume and a better volume-to-surface ratio, fuselage drag benefits from rounded cross sections which are less prone to flow separations under sideslip. Also, more freedom can be given to the designer to create an appealing look of the fuselage. One obvious and easy to model source of thrust loss is an object's aerodynamic drag in the rotor induced flow field. Other reasons are the interaction of pressure distributions of rotor and object or the disturbance of a uniform rotor inflow velocity distribution. Using simple actuator disk theory, thrust loss due to parasitic drag on a rotor in hover can be expressed by

$$D = \frac{\varrho}{2}\, v(z)^2\, c_D\, S(z) \qquad \text{eq. I}$$

$$v(z) = k_v(z)\, v_{disk} \qquad \text{eq. II}$$

$$S(z) = 2\, r(z)\, b \qquad \text{eq. III}$$

$$\varrho\, r(z)^2\, \pi\, v(z) = \varrho\, R^2\, \pi\, v_{disk} \qquad \text{eq. IV}$$

$$v_{disk}{}^2 = \frac{T}{2\, \varrho\, R^2\, \pi} \qquad \text{eq. V}$$

$$\frac{D}{T} = \frac{1}{4}\, c_D\, \frac{S_{disk}}{A_{disk}}\, k_v(z)^{\frac{3}{2}} \qquad \text{eq. VI}$$

Air density ϱ is assumed constant, as actuator disk assumes an incompressible fluid. Drag depends on the encountered flow velocity $v(z)$ and the effective cross section $S(z)$ inside the rotor flow field (eq. I, refer to Figure 8). Both are dependent on the separation z to the rotor disk which is normalized by the rotor radius R and is positive for positions aft of the rotor disk. $k_v(z)$ describes the flow velocities with respect to the induced velocity v_{disk} (eq. II). Airflow $v(z)$ is generally accelerated along the z-coordinate. An object extending infinitely across the rotor disk is assumed (eq. III). Using conservation of mass (eq. IV) and momentum theory's the induced velocity for a hovering rotor (eq.V), thrust loss relative to the unblocked thrust is dependent on the drag coefficient, the blocking ratio S/A and the velocity distribution. As this is just a simple representation of one thrust loss component, measurements are used to validate, update and calibrate the model. Therefore, data from helicopter tail rotors (see Ref 9) and own measurements on a 17 inch two blade multicopter UAV rotor (see Ref 10) are used. Measurements on the UAV rotor were taken with objects placed both in the in- and outflow. To represent possible fuselage shapes, a thick wing section with symmetrical airfoil and a flat plate were each placed with different distances in front and aft of the rotor. The objects were rotor-centered and extended far beyond the rotor diameter. Blocking ratio and disk loading of the rotor were kept constant. Figure 9 depicts the blocking ratio-normalized relative thrust loss $\frac{T_{loss}}{T}\frac{A_{disk}}{S_{disk}}$ for all measurements. As indicated by eq. VI, the different drag coefficients of airfoil and flat plate lead to a difference in thrust loss. Neither the absolute values nor the ratio of the thrust loss between flat plate and airfoil fits the measurements when using eq. VI with typical drag coefficients (known e.g. from Ref 13) and $k_v(0) = 1$. Besides the mentioned other effects, one must consider that objects like airfoils generate additional induced drag as the swirl of especially the aft rotor flow causes an angle of attack to the object. The tail rotor was blocked by a fin of identical cross section, but different

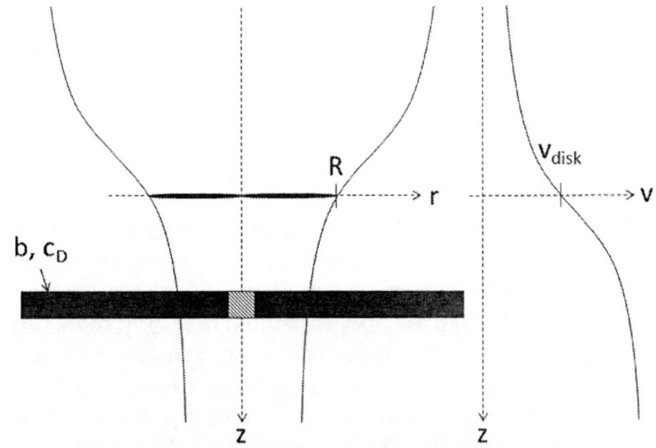

Figure 8. Schematic of Rotor Flow Field

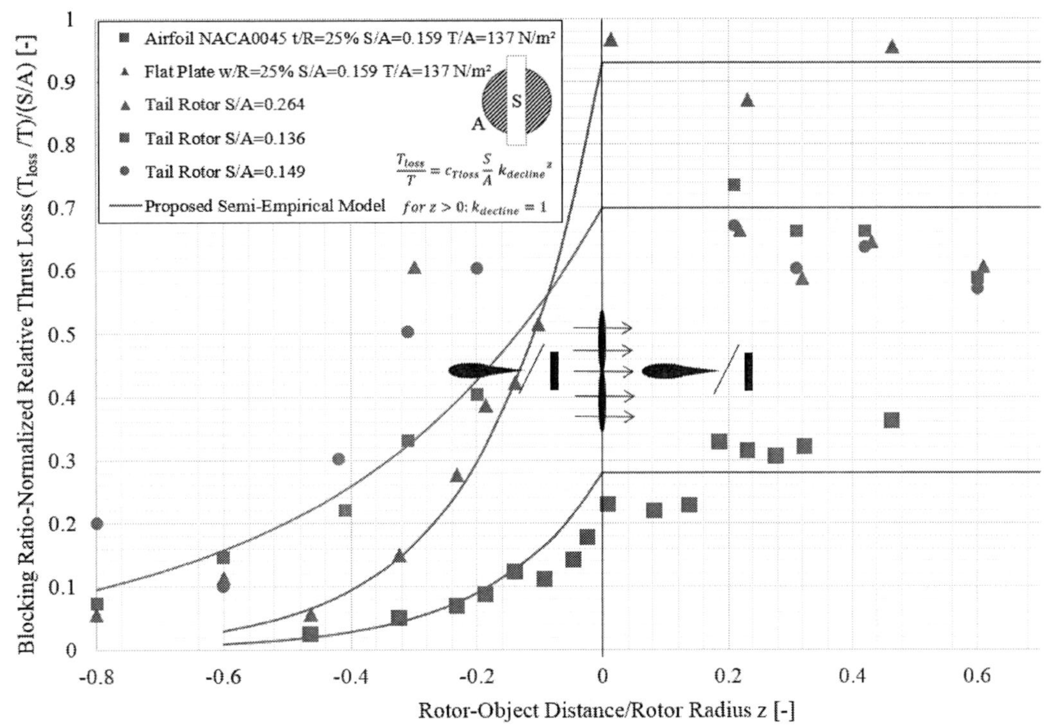

Figure 9. Measurements on In- and Outflow Blocking of a Rotor

blocking ratios. The good alignment of these measurements, at least for the aft rotor side, confirm eq. VI's linear dependency on blocking ratio. One would expect better alignment of flat plate with tail rotor measurements as their blocking objects have comparable cross sections. The placement of the blocking object with respect to the rotor disk is however not comparable. The velocity distribution within the disk of a z-coordinate is not uniform. Especially for aft rotor flow, the velocities have maxima at outer radius positions. Blocking of rotor disk regions with above-average velocities (as the flat plate does compared to the tail rotor fin) shows increased thrust loss. The flow velocity increase $k_v(z > 0)$ after the rotor would indicate an increase in drag. Looking at all aft rotor measurements, no such trend is visible. For simplicity, one can assume a constant drag, or better, thrust loss coefficient c_{Tloss}. This aligns with the approach of Ref 9. Based on given data, thrust loss due to blocking on the inflow side is always smaller than due to outflow blocking. Thrust loss decreases differently fast for tail rotor and UAV rotor measurements. It is likely affected by the present thrust loading. Taking tail rotor and UAV rotor data separately, the measurements show good consistency as identical thrust loadings and velocity trends exist. Eq. VII proposes a simple semi-empirical model for thrust blocking losses which combines information from eq. VI and Figure 9.

$$\frac{T_{loss}}{T} = c_{Tloss} \frac{S_{disk}}{A_{disk}} k_{decline}{}^{z} \qquad \text{eq. VII}$$

while $k_{decline} = 1$ for $z > 0$. $k_{decline}$ and c_{Tloss} must be calibrated from measurements. Further data is required to model the likely dependency between $k_{decline}$ and thrust loading. As a conclusion, the mounting of hover rotors in a pusher configuration leads to a significant reduction in blocking thrust losses for a rectangular fuselage cross section. 68% of the tractor configuration's thrust loss can be saved. The required rotor-fuselage separation of 0.2 R is small enough to be implementable in aircraft design. A further increase of rotor-fuselage separation depicts an effective means to e.g. allow for wider fuselages without additional rotor thrust loss.

Adding to the aspects of rotor tip movement and pusher noise generation, the study on thrust blocking again underlines the importance of rotor-fuselage separation for an underbody mounted rotor and the possibility to implement it with low penalties.

Overall Performance of the Under-Fuselage Pusher Rotor Configuration The modifications required to implement an under-fuselage pusher rotor configuration (like elevated landing gear and longer rotor shafts) entail mass and drag increase that face the benefits like reduced hover power consumption and reduced hover powertrain mass. To see the performance impact of these changes, simple estimation models were used to compare the cruise endurance of the under-fuselage pusher rotor configuration against its tractor rotor pendant. The VTOL version of the test aircraft depicted in Figure 3 equipped with a fuselage with rectangular cross sections serves as geometry baseline. Except for the rotor placement, the landing gear geometry and the rotor shaft lengths, the geometry of the compared aircraft versions is identical. The mass of the compared aircraft is each 5 kg. Non-powertrain components' mass and structural masses as well as cruise powertrain efficiency are identical between the aircraft. Linear, power-dependent powertrain sizing is used and calibrated with the test aircraft data. As well, the test aircraft's aerodynamic performance is used as baseline. Battery energy density is considered independent of power consumption. The thrust loss model is derived from measurements in Figure 9 ($c_{Tloss} = 0.93$, $k_{decline} = 300$). The airfoil shaped landing gear struts are elevated by the maximum height for the fuselage on the underbody pusher rotor aircraft. The rotor shafts are elongated to a separation of 0.2 R and housed with low drag fairings. The additional mass and drag of both modifications is modelled with the respective material densities and shape-dependent drag coefficient found in Ref 13. The modelling of the modifications can be considered conservative. The mission consists of a 2 min hover phase and a horizontal cruise phase with identical airspeed for which the residual energy of the battery is used.

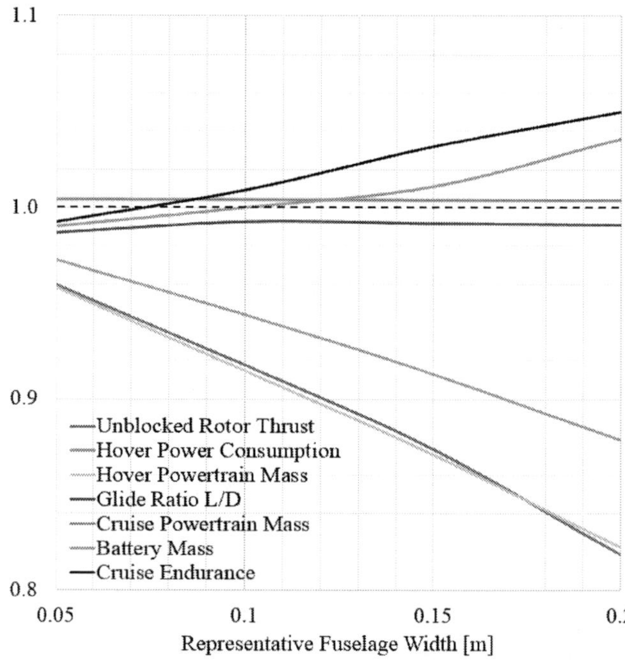

Figure 10. Relative Comparison of Under-Fuselage Hover Rotor Aircraft Over Its Tractor Pendant

Figure 10 depicts the change of selected performance relevant properties for different fuselage widths. The properties of the under-fuselage pusher rotor aircraft are normalized by the properties of the tractor rotor aircraft. Increasing fuselage width leads to increased occurrence of thrust blocking. Hence, lower gross thrust levels are required for the pusher rotor. Hover powertrain mass and hover power consumption decrease. As can be seen on the mass available for battery, additional landing gear and rotor shaft weight are just compensated by reduced hover powertrain mass for fuselage widths beyond 0.1 m. Already at this fuselage width, the small penalty in glide ratio and hence cruise power consumption is overcompensated by a higher energy available for cruise, as less energy is spent during the hover phase. As a result, the under-fuselage hover pusher aircraft achieves, with the exception of fuselage widths below 0.075 m, better cruise endurances. For a fuselage width of 0.15 m, as it is required to fit a typical payload of the examined aircraft, the improvement is around 3%. Conclusively, the trade of elongated landing gear and rotor shaft mass and drag against lower thrust blocking is beneficial for typically required fuselage widths. The trade-off gets increasingly beneficial for wider fuselages, especially if the low hover power consumption can still maintain the usage of high energy density batteries.

CONCLUSIONS

Motivated by a performance comparison of various unmanned fixed-wing eVTOL configurations, major effects and challenges of the superior Quad+Tilt configuration were addressed. Firstly, the benefits and implementation of wingtip propulsion providing roll and yaw control in both hover flight were approached. Secondly, the integration of Quad+Tilt's fuselage mounted lift rotors was investigated as a sensitive design trade-off is present here. The cross section of the fuselage causes a considerable amount of rotor thrust loss which triggers further performance compromising effects.

One aspect, which is difficult to model with wingtip propulsion, is the beneficial interaction between wingtip vortex and the counter-rotating swirl of the wingtip mounted propeller. Therefore, an inflight analysis using different spin directions of the wingtip propellers was attempted. It yielded A 4% reduction in cruise flight power consumption compared to an aircraft without wingtip propulsion was found. Three alternative designs were proposed to implement the control and propulsion tasks of a Quad+Tilt wingtip. The 'tiltable tractor' wingtip tilts an outermost wing section that carries the powertrain. It provides yaw and roll moments by a combination of thrust vectoring and the deflection of aerodynamic surfaces. The 'tiltable pusher' and 'fixed pusher' wingtips try to implement a pusher configuration as further wingtip propulsion benefits seem feasible. The 'tiltable pusher' omits the aerodynamic control component and solely relies on thrust vectoring. The 'fixed pusher' uses a second, perpendicular powertrain to substitute a tilt mechanism. This wingtip as well has no possibility of aerodynamic control moments. Flight testing of the wingtips shows that, despite its lowest mass, the 'fixed pusher' wingtip can overall not be recommended as it lacks control authority, requires the highest power in hover flight and is prone to damages due to propeller strikes. The 'tiltable pusher' wingtip seems to provide sufficient propeller clearance and can be compared to the 'tiltable tractor' in terms of hover controllability. As it lacks aerodynamic control surfaces, its roll authority is dependent on the thrust level. The 'tiltable tractor wingtip' provides good control authority, also in critical flight conditions. It is currently recommended over the other two wingtip versions. Further evaluation of pusher wingtip propulsion benefits may eventually justify the usage of the 'tiltable pusher' wingtip.

Taking helicopter tail rotors as an example, a mounting of Quad+Tilt's hover rotors below the fuselage is examined as a strategy to overcome the sensitive thrust blocking trade-off. Verifying measurements on the thrust loss caused by blocking objects in the in- and outflow of a rotor were performed. They revealed the distance of rotor to fuselage as critical to achieve low thrust blocking on a pusher configuration. The crucial topics to realize the advantages of underbody rotor mounting are foreign object damage, rotor strike and noise emission. Elevating the landing gear is, besides the inherent advantages of using sealed electric motors, fully automated flight control and small diameter, rigid rotor hinging, a measure that solves foreign object damage and rotor strike issues. The risk of unacceptable loudness is as well mitigated with sufficient rotor-fuselage clearance and a low blade pass frequency which is perceived as agreeable for the human ear. To quantify the overall aircraft performance, an aircraft with the under-fuselage mounted pusher rotor and the therefore required changes was compared to an otherwise identical aircraft with rotor mounted on top of the fuselage in a tractor configuration. The trade of higher landing gear and higher rotor-fuselage clearance for reduced thrust blocking results in a 3% gain in cruise endurance for the examined 5 kg pusher rotor aircraft. Up to now, the considerations on mounting rotors below the fuselage of a Quad+Tilt configuration yield no show-stopper. As a next step, such an aircraft must demonstrate in real-life operation if its rotor mounting still proves to be sensible.

Author contact:

Philipp Stahl philipp.stahl@tum.de

Christian Roessler christian.roessler@tum.de,

Mirko Hornung mirko.hornung@tum.de

ACKNOWLEDGMENTS

The author wants to thank the project team members Sebastian Oberschwendtner, Thomas Seren and Moritz Thiele for their support, as well as all students that contributed to fixed-wing VTOL aircraft development at Institute for Aircraft Design/Technical University of Munich.

REFERENCES

[1]Poisson-Quinton, P., "Introduction to V/STOL Aircraft Concepts and Categories", AGARDograph 126, 1968

[2]Hirschberg, M. J., "V/STOL: The First Half-Century", *Vertiflite*, March/April 1997.

[3]Anderson, S., "Historical Overview of V/STOL Aircraft Technology", NASA Ames Research Center: NASA Technical Memorandum 81280, 1981.

[4]Stahl, P., Seren, T., Rößler C., and Hornung, M., "Development and Performance Comparison of Optimized Electric Fixed-Wing VTOL UAV", ICAS 2018 Proceedings, Belo Horizonte, September 2018.

[5]Miranda, R. L., "Aerodynamic Effects of Wingtip-Mounted Propellers and Turbines", A86-31826, AIAA, 1986.

[6]Patterson, J. C., and Bartlett, G. R., "Evaluation of Installed Performance of a Wing-Tip-Mounted Pusher Turboprop on a Semispan Wing", NASA Technical Paper 2739, 1987.

[7]Prouty, R.W., *Helicopter Aerodynamics*, Vol. 1, Eagle Eye Solutions, Lebanon, 2009, pp. 147-150.

[8]Müller, G., Möser, M., *Taschenbuch der Technischen Akustik*, Springer Verlag , Berlin Heidelberg, 2004, pp. 104, 585-638.

[9]Johnson, W., *Rotorcraft Aeromechanics*, Cambridge University Press, Cambridge, 2013, pp. 149-150.

[10]Rieger, C., "Experimentelle Untersuchung der Auswirkungen von Objekten im Zu- und Abstrom von Propellern für Multirotor UAV", Institute of Aircraft Design, Technical University of Munich, Munich, 2017.

[11]Stahl, P., Roessler, C., Hornung, M., "Benefit Analysis and System Design Considerations for Drag Reduction of Inactive Hover Rotors on Electric Fixed-Wing VTOL Vehicles", Paper AIAA-2018-4150, AIAA Aviation and Aeronautics Forum Proceedings, Atlanta, GA, June 25-29, 2018.

[12]Dimchev, M., "Experimental and numerical study on wingtip mounted propellers for low aspect ratio UAV design", Faculty of Aerospace Engineering, Delft University of Technology, Delft, 2012.

[13]Hoerner, S. F., "*Fluid-Dynamic Drag*", Hoerner Fluid Dynamics, Bakersfield, CA, 1965.

Comparative eVTOL UAV Design Optimization

Thomas Seren Research Assistant

Christian Roessler Research Associate

Mirko Hornung Head of Institute

Institute of Aircraft Design – Technical University of Munich – Munich, Germany

ABSTRACT

This paper describes a method to simulate and optimize eVTOL UAV. The simulation includes aerodynamics, flight mechanics, structural mechanics, electrical propulsion and flight performance calculations. The optimization is performed in the form of sensitivity studies, varying configuration parameters such as aspect ratio or the cruise speed. The propulsion system simulation is based on a database approach with commercially available components and a method is presented on how to efficiently incorporate and analyze such a database. Three different configurations are compared: a tiltless multirotor VTOL with a designated cruise motor, a fully tiltable multirotor VTOL and a VTOL with tiltable wingtip propulsion. The first and the last one are shown to have the best performance depending on the required cruise speed.

INTRODUCTION

The market of unmanned aerial vehicles (UAV) is constantly growing. Especially in the segment for miniature and small UAV, this is happening fast, and a lot of different solutions are available on the market, or are currently under development. Aircraft systems in this size have the advantage of reduced complexity and cost, which allows a variety of different types to be built and operated. This applies especially to vertical take-off and landing (VTOL) UAV. These combine the flexibility of taking-off and landing in small spaces with the efficiency of fixed-wing aircraft during the cruise flight. The combination of different layouts for hovering with different layouts for the cruise flight, however, leads to a very high number of possible UAV designs.

There are a lot of different types of VTOL UAV already in operation. It is however difficult to compare them, as they all use different concepts for hover, cruise and transition in between. Also, most are designed for specific missions, which makes it even harder to compare them.

A method is needed to be able to compare different VTOL concepts. The idea is to get a better understanding of the configurations and their advantages and disadvantages. Also a way to better understand the influences of various design parameters onto the aircraft performance is wanted. Reliable data is needed, to decide which configuration is the best for a certain mission.

For solving this, a design tool has been developed. With this tool, it is possible to optimize different aircraft configurations for the same missions. This allows to directly compare different aircraft configurations and choose the best. The wanted outputs are sensitivity study plots, aircraft performance plots and three dimensional aircraft visualizations.

Presented at the Vertical Flight Society Autonomous VTOL Technical Meeting and Electric VTOL Symposium, Mesa, Arizona, UAS, Jan. 29-31, 2019.

OVERVIEW

The overall aircraft design process is implemented in this tool. It calculates the aircraft's endurances for certain missions.

Simulated Mission

The simulated mission is to transport a fixed payload mass, with different stationary flight condition requirements and a defined maximum take-off mass (MTOM) while maximizing the endurance.

A typical flight consists of the following five phases:

- Vertical take-off
- Transition
- Cruise
- Retransition
- Vertical Landing

In this research only vertical take-off and landing and cruise have been analyzed.

During vertical take-off and landing the combined static thrust of all propulsion systems contributing to the hover thrust, has to be equal to the aircraft's weight. For maneuvering, the static thrust has to be equal to a multiple of the aircraft's weight. This is done, to cover vertical acceleration, climb and the additional thrust needed for maneuvering during hover.

Cruise is separated into four different flight conditions, where three are used for the flight performance calculation and one is used for the stall speed calculation. The flight performance conditions are steady horizontal flight, climb and turn with a moderate bank angle. For the stall speed calculation a turn with a more extreme bank angle is analyzed.

During the steady horizontal flight, the aircraft's weight is compensated by the aerodynamic lift and drag is compensated by the thrust.

During climb, the aerodynamic lift is slightly decreased, but the thrust increased, as it needs to additionally compensate the downhill slope force.

During turns the aircraft's weight is increased by the centrifugal force which leads to a higher lift requirement and thus a higher drag which needs to be compensated by increased thrust.

Table 1 shows the mission parameters, which were used as basis for the simulation.

Table 1: Mission Parameter

Parameter	Value
MTOM	5 kg
Payload Mass	885 g
Payload Length	210 mm
Payload Height	100 mm
Payload Width	100 mm
Cruise Altitude	800 m
max. Altitude	1800 m
Climb Speed	3 m/s
Hover Time	1 min
nz_{max}	4
Turn Bank Angle	15°
Stall Bank Angle	45°

Configurations

There are three different types of motors used:

- Cruise
- Hover
- Hybrid

Cruise motors are only used to provide thrust during the cruise phase. They are typically installed in a fixed position i.e. not tiltable.

Hover motors are only used to provide thrust during the hover phase. They are designed to be fixed or to be tiltable. Fixed motors position the propellers in a position, that minimizes the cruise drag. If they are tiltabe, they are combined with folding propellers. When tilted during the cruise phase, the propeller folds and thus reduces the drag.

Hybrid motors are used for both phases, i.e. to provide thrust during hover and cruise. For this they need to be tiltable to switch between operating points.

Three different configurations have been analyzed:

- tiltless multirotor with designated cruise motor (4+1)
- fully tiltable multirotor (2+2)
- tiltable wingtip propulsion (2X2)

The **4+1** configurationis the combination of a quadrocopter and a conventional aircraft. It has four hover motors providing the complete hover thrust, integrated in two underwing nacelles, with integrated tailbooms. Additionally, there is one single motor used to provide the thrust for horizontal flight. It is integrated in the center fuselage and designed to be a pusher configuration. The hover propellers remain in the free airstream during cruise. The separation of the nacelles depends on the propellers' diameters, so that there is no danger of the propellers to touch each other.

This configuration features a T-tail with two vertical tailplanes sitting on the twin-tailbooms.

Comparable configurations are the ICARUS IBIS1, seen in figure 1 or the UAVs by JOUAV, seen in figure 2.

Fig. 1: ICARUS IBIS1[1]

Fig. 2: JOUAV CW-30[2]

The main idea behind this configuration is to have an individual system for takeoff and landing, which is already well known and consequently should be easy to implement. Furthermore the single cruise powertrain can by optimized for the cruise operating point only and should therefor reach a higher overall efficiency.

There are two big drawbacks of this configuration. One is the fact, that the hover propellers are in the free airstream during cruise which increases the parasitic drag. Also, the two big nacelles and the center fuselage tend to be heavier than a single one and also increase the parasitic drag. But as the nacelles are located off center they reduce the bending and torsion load of the wing, which can be built slightly lighter. It is expected, that

[1] http://icarus.drone.services/wp/wp-content/uploads/2017/08/ibis-45.jpg, accessed 15/01/2019

[2] https://www.unmannedsystemstechnology.com/wp-content/uploads/2017/09/TerraHawk-CW30-LiDAR-mapping-drone.jpg, accessed 15/01/2019

the saved weight compensates the drag drawback to a certain extend.

As the 4+1 configuration is already widely in operation, it is used as baseline concept, to which all other configurations have to be compared

The **2X2** configurationis also a combination of a quadrocopter and a conventional aircraft. The difference to the the 4+1 configuration is, that it has only four motors, not five. They are integrated into two nacelles, that are mounted at the wings. The two front motors are used as hybrid motors, i.e. they provide thrust for the hover phase and during horizontal flight. The two rear motors are used as pure hover motors, providing only thrust for the hover phase. All four motors are tilted during cruise: the forward facing to provide the needed thrust and the backward facing to enable the propellers to fold in the airstream and reduce the drag.

This configuration features a V-tail.

Comparable configurations are the Germandrones Songbird, seen in figure 3 or the Quantum Systems Tron, seen in figure 4.

Fig. 3: Germandrones Songbird[3]

Fig. 4: Quantum Systems Tron[4]

The **2+2** configuration has two main hover motors integrated in one single fuselage. Additionally, it has two tiltable motors at the wingtips, integrated into an aerodynamically shaped pylon. The wingtip motors are used in three different situations. They provide some thrust for hover, they are used for maneuvering during the hover phase and they provide the complete thrust for the horizontal flight. The hover propellers remain outside of the fuselage for the cruise phase; i.e. they increase the drag. This is an important factor, which has to be considered.

Comparable configurations are the Boeing Phantom Swift, seen in figure 5 or the AUTEL DragonFish, seen in figure 6.

Fig. 5: Boeing Phantom Swift[5]

Fig. 6: AUTEL DragonFish[6]

The idea behind this configuration is two have the hover motors to carry nearly the complete aircraft's weight, with the two wingtip motors providing additional thrust for maneuvering and leveling during the hover phase. By this the hybrid motors can be optimized to be more efficient during cruise, while sacrificing efficiency during the hover phase.

This configuration features a conventional tail.

More details on this configuration have been described by Stahl et al. (Ref. 1).

IMPLEMENTED DISCIPLINES

The tool covers aerodynamics, flight mechanics, structural mechanics, electrical propulsion and flight performance. The different design disciplines are implemented in different levels of complexity in order to use the minimum needed level of detail and the highest allowable simplifications.

Aerodynamic Model

The aerodynamic model is based on airfoil polar data, which are calculated with XFOIL (Ref. 2).

The polar data are generated in a reynolds number range from 10000 to 100000 in steps of 10000 and from 100000 to

[3] https://germandrones.com/wp-content/uploads/2017/09/sb2.png, accessed 15/01/2019

[4] https://www.munich-startup.de/wp-content/uploads/2017/01/Quantum-1061.jpg, accessed 15/01/2019

[5] https://breakingdefense.com/wp-content/uploads/sites/3/2015/01/Boeing-VTOL-X-Plane-Concept-DARPA.jpg, accessed 15/01/2019

[6] https://www.dronerush.com/wp-content/uploads/2017/01/Autel-Robotics-prototype-full-CES-2017-1024x576.jpg, accessed 15/01/2019

1 300 000 in steps of 100 000. During the analysis any values inbetween the precalculated points are interpolated linearly.

For a conversion between the values coming from the airfoil polar data to the wing data and vice versa simplified approaches are used.

Lift: For the lift, the approach by Lowry and Polhamus (Ref. 3) and Quabeck (Ref. 4) is used. For small wing sweep angles and low Mach numbers it can be simplified, leading to a wing's lift slope $C_{L\alpha}$, only dependent on the wing's aspect ratio Λ and the airfoil lift slope $C_{l\alpha}$.

$$C_{L\alpha} = \frac{C_{l\alpha} \cdot \Lambda}{2 + \sqrt{\Lambda^2 + 4}} \quad (1)$$

This approach can be applied to the absolute lift coefficient (C_L & C_l) and the maximum lift coefficients (C_{Lmax} & C_{lmax}) as well (Ref. 4)

$$C_L = \frac{C_l \cdot \Lambda}{2 + \sqrt{\Lambda^2 + 4}} \quad (2)$$

$$C_{Lmax} = \frac{C_{lmax} \cdot \Lambda}{2 + \sqrt{\Lambda^2 + 4}} \quad (3)$$

For reasons of flight safety, there needs to be a margin between the stall speed and the cruise speed at a certain banked angle. This limit is used to iterate the wing area at which this condition is exactly matched. This approach leads to the smallest possible aircraft configurations, which in return reduce the aircraft's structure's weight. So the stall speed needs to be properly calculated.

From the lift calculation the following equation for calculating the stall speed can be derived:

$$v_{stall} = v_\infty \cdot \sqrt{\frac{C_L}{C_{L,max}}} \quad (4)$$

with the stall speed requirement

$$v_{stall} \overset{!}{\leq} v_\infty - \Delta v_{stall} \quad (5)$$

The Lift Coefficient (C_L) is calculated during the trimming of the aircraft. The stall speed margin Δv_{stall} is a mission requirement of 3 m/s.

Drag: The aircraft's drag consists of several components:

- Wing and Empennage Drag
- Fuselage and Nacelle Drag
- Antenna Drag
- Propeller Drag
- Landing Gear Drag
- Interference Drag

The aircraft's drag D_{total} is calculated by summing up all drag components D_c

$$D_{total} = \sum D_c \quad (6)$$

To not underestimate the complete drag, a drag safety factor of 1.15 is applied to the total Drag.

$$D_{corr} = 1.15 \cdot D_{total} \quad (7)$$

For the wing and the empennages the drag is assumed to consist of two components. The airfoil drag and the induced drag:

$$C_D = C_{D0} + C_{Di} \quad (8)$$

The airfoil Drag C_{D0} is derived from the airfoil polar data. The induced Drag C_{Di} is calculated based on the Lift coefficient and the aspect ratio:

$$C_{Di} = \frac{C_L^2}{\pi \cdot \Lambda \cdot e} \quad (9)$$

where e is the Oswald factor (Ref. 5, Equation 12.48, page 456), which can be calculated with

$$e = 1.78 \cdot \left(1 - 0.045 \cdot \Lambda^{0.068}\right) - 0.64 \quad (10)$$

The drag of fuselages and nacelles is calculated with the method of Raymer (Ref. 5, Chapter 12.5, page 429ff.).

The antenna drag is calculated with empirical equations for cylinders with the method of Hoerner (Ref. 6, page 13-19 Chapter XIII).

The landing gear is simulated as small wings which don't produce any lift, with its drag calculated as the airfoils zero lift drag.

Depending on the exact configuration interferences between different parts have to be considered. These are Interference between:

- wing and fuselage
- wing and nacelle
- empennage and fuselage
- two empennage parts
- landing gear and fuselage

The interference drag is calculated based on the method presented by Hoerner (Ref. 6, Chapter VII.4)

The hover propellers, which are in the free airstream during the cruise phases, produce drag which needs to be taken into account. The drag consists of three components: friction drag, propeller shaft drag and interference drag. For the friction drag, a virtual propeller shape consisting of one ellipsoidal plate per blade is assumed. The shaft is simulated as a cylinder. To take into account the interference drag, the sum of the single drag components is multiplied with a factor of 1.5.

The drag calculation has been verified with the work presented by Stahl (Ref. 7).

Flight Mechanics

The flight mechanics are based on a static stability requirement at a design point. Trimming at this point defines the overall geometry of the aircraft. For all other flight conditions the needed trimming and the related lift and drag coefficients are calculated. In order to calculate all relevant flight conditions, they need to be analyzed from a flight mechanics point of view. During cruise three flight conditions are taken into account.

- steady horizontal flight
- steady turn
- steady climb

For a steady, trimmed flight, a force and moment equilibrium has to exist. The relevant force equilibrium is analyzed in the aircraft's z-axis.

$$\sum F_z \overset{!}{=} 0 = n_z \cdot m \cdot g - L \tag{11}$$

$$L = n_z \cdot m \cdot g \tag{12}$$

The total lift is the sum of the wing's lift and the tail's lift.

$$L = L_w + L_t \tag{13}$$

For the longitudinal stability the moment equilibrium needs to be analyzed around the aircraft's y-axis. The relevant components come from the aircraft's mass, the wing's and tail's lift and their moments.

$$\sum M_y \overset{!}{=} 0 = -x_{CG} \cdot n_z \cdot m \cdot g + M_{0w} + M_{0t} + L_w \cdot x_{Nw} + L_t \cdot x_{Nt} \tag{14}$$

For a static longitudinal stability a certain *stability margin* σ_l is needed. It is defined as the normalized difference of the positions of the center of gravity and the neutral point. For the simulated mission it is set to 15%.

$$\sigma_l \cdot l_\mu = x_{CG} - x_N \tag{15}$$

The neutral point is defined as the point around which the total pitching moment is constant, independently of the angle of attack. This can be interpreted as the point at which any additional lift, following a change of the angle of attack, is applied. With known component neutral points of the wing and the tailplane and their lift's slope, the neutral point can be calculated.

$$x_N = \frac{x_{N,w} \cdot C_{L\alpha,w} + \frac{q_t}{q} \cdot \frac{S_t}{S} \cdot x_{Nt} \cdot C_{L\alpha,t} \cdot \left(1 - \frac{\partial \varepsilon_t}{\partial \alpha_w}\right)}{C_{L\alpha w} + \frac{q_t}{q} \cdot \frac{S_t}{S} \cdot C_{L\alpha t} \cdot \left(1 - \frac{\partial \varepsilon_t}{\partial \alpha_w}\right)} \tag{16}$$

These equations are used to calculate the tailplane's neutral point, the position of the center of gravity and the needed lift coefficients for each flight condition.

The steady, horizontal flight is used to do the basic trimming, i.e. calculating the position of the tailplane and the center of gravity. The tailplane's lift coefficient and relative size are used as variation parameters in the simulation making them known values. The wing's lift coefficient is calculated with

$$C_{Lw} = \frac{n_z \cdot m \cdot g}{q \cdot S} - \frac{q_t}{q} \cdot \frac{S_t}{S} \cdot C_{L,t} \tag{17}$$

To calculate the position of the center of gravity, equations (14) to (16) are combined.

$$x_{N,t} = \left[\sigma_l \cdot l_\mu - x_{N,w} + \frac{q_t \cdot S_t \cdot x_{N,w} \cdot C_{L,t} - (M_{0w} + M_{0t})}{n_z \cdot m \cdot g} + \right.$$
$$\left. + \frac{x_{N,w} \cdot C_{L\alpha,w}}{C_{L\alpha,w} + \frac{q_t}{q} \cdot \frac{S_t}{S} \cdot C_{L\alpha,t} \cdot \left(1 - \frac{\partial \varepsilon_t}{\partial \alpha_w}\right)}\right] /$$
$$/ \left[\frac{q_t \cdot S_t \cdot C_{L,t}}{n_z \cdot m \cdot g} + \frac{\frac{q_t}{q} \cdot \frac{S_t}{S} \cdot C_{L\alpha,t} \cdot \left(1 - \frac{\partial \varepsilon_t}{\partial \alpha_w}\right)}{C_{L\alpha,w} + \frac{q_t}{q} \cdot \frac{S_t}{S} \cdot C_{L\alpha,t} \cdot \left(1 - \frac{\partial \varepsilon_t}{\partial \alpha_w}\right)}\right] \tag{18}$$

For static flight, with $n_z = 1$, equation (18) leads to the tail's neutral point.

Inserting $x_{N,t}$ calculated with equation (18) in equations (15) and (16) leads to the center of gravity, needed to fulfill the static stability requirement.

With the center of gravity and the tail's neutral point, the flight mechanic data for the longitudinal motion is defined. With this information, the flight characteristics for the other relevant flight states, turn and climb, can be calculated.

Turn: Two different turns need to be analyzed: one to calculate the flight performance during a typical turn and one to calculate the stall speed at an extreme bank angle. The forces acting during a steady turn can be seen in figure 7. The force

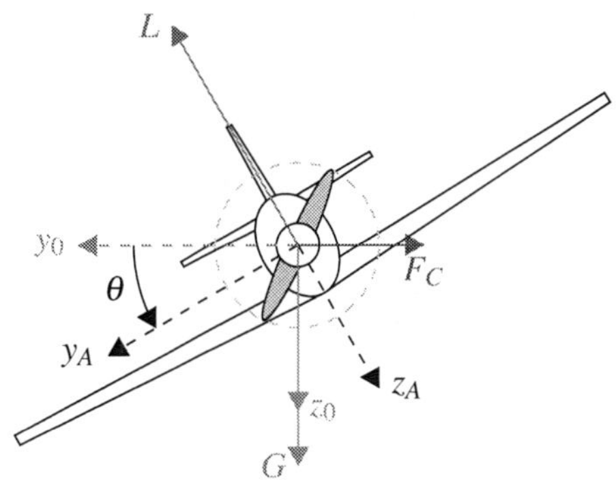

Fig. 7: Forces on a Banked Aircraft

equilibrium in the turn plane leads to the following z acceleration.

$$n_{z,turn} = \frac{1}{\cos\theta} \tag{19}$$

The wing and tail lift coefficients for the turn are calculated with

$$C_{Lt,turn} = \frac{(x_{Nw} - x_{CG}) \cdot \frac{1}{\cos\theta} \cdot m \cdot g + M_{0t} + M_{0w}}{(x_{Nw} - x_{Nt}) \cdot q_t \cdot S_t} \tag{20}$$

$$C_{Lw,turn} = \frac{m \cdot g}{q \cdot S \cdot \cos\theta} \cdot \left(1 - \frac{x_{Nw} - x_{CG}}{x_{Nw} - x_{Nt}}\right) - \frac{M_{0t} + M_{0w}}{q \cdot S \cdot (x_{Nw} - x_{Nt})} \tag{21}$$

Climb: The forces acting during the climb can be seen in figure 8. Here the thrust not only compensates the drag, but

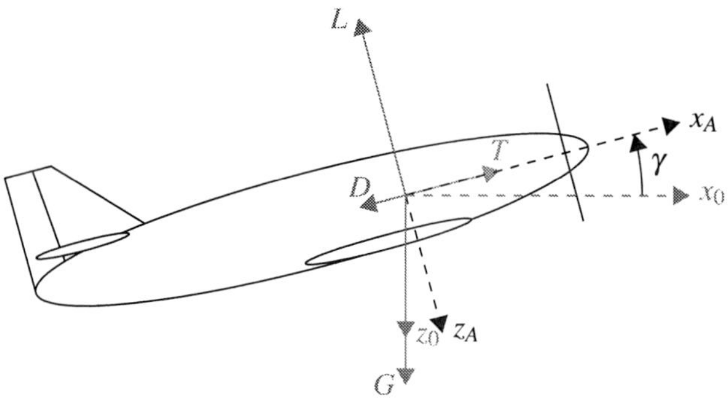

Fig. 8: Forces on a Climbing Aircraft

also the downhill-slope force, which reduces the aerodynamic lift, leading to the following n_z

$$n_{z,climb} = \cos\gamma \tag{22}$$

This leads to the following equations to calculate the lift and wing coefficients during climb.

$$C_{Lt,climb} = \frac{(x_{Nw} - x_{CG}) \cdot \cos\gamma \cdot m \cdot g + M_{0t} + M_{0w}}{(x_{Nw} - x_{Nt}) \cdot q_t \cdot S_t} \tag{23}$$

$$C_{Lw,climb} = \frac{\cos\gamma \cdot m \cdot g}{q \cdot S} \cdot \left(1 - \frac{x_{Nw} - x_{CG}}{x_{Nw} - x_{Nt}}\right) - \frac{M_{0t} + M_{0w}}{q \cdot S \cdot (x_{Nw} - x_{Nt})} \tag{24}$$

Directional Stability: For the directional stability a vertical tailplane is needed which provides weathercock stability to counteract the destabilizing yawing $M_{z\beta}$ moment, when the aircraft is disturbed. A simplified approach using the relative tailplane volume is used and the vertical tailplane's size is calculated with

$$S_{VTP} = \frac{c_{VTP}}{b \cdot S_{ref} \cdot (x_{NP,wb} - x_{NP,HTP})} \tag{25}$$

The tool includes an iteration process, to position the center of gravity in the right place, by changing the battery's position.

Structural Mechanics

The structural model is oriented towards practicability of building a prototype, rather than minimizing the components' masses.

Fuselage and nacelle: The fuselages and nacelles are simulated as simple hard shell constructions, consisting of pure carbon fiber reinforced plastic (CFRP) with a wall thickness of 0.9 mm.

Wing: The wing is built as a sandwich shell. It has an outer, torsion carrying, CFRP layer, a Sandwich core and on the inside a single layer of thin glass. Figure 9 shows a representation of all structural components inside the wing. This sand-

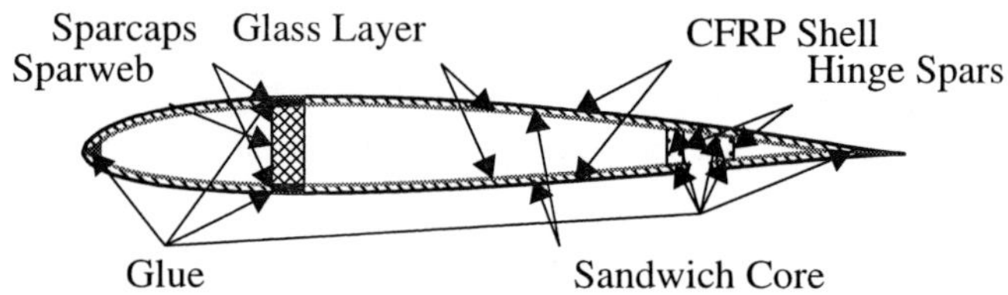

Fig. 9: Wing Structure

wich structure avoids buckling of the shell under the needed flying conditions. It also allows a good manual handling of the parts without any danger of damaging them. Above all, it is a lightweight solution.

The bending is assumed to be completely covered by a single spar, located at 25% of the wing chord.

The wings torsion is calculated at a maximum speed with the airfoils maximum pitching moment coefficient $C_{m,max}$. For swept wings an additional twisting moment induced by the lift, calculated with the maximum lift coefficient $C_{l,max}$ is taken into account. By integration the total torsion moment can be derived.

$$M_T(y) = \underbrace{\int_y^s \frac{1}{2} \cdot \rho \cdot C_{m,max} \cdot v_{max}^2 \cdot l(y)^2 \, dy}_{\text{pitching moment}} - \underbrace{\int_y^s y \cdot tan(\alpha_{25\%}) \cdot \frac{1}{2} \cdot \rho \cdot C_{l,max} \cdot v_{max}^2 \cdot l(y) \, dy}_{\text{lift and sweep induced}} \tag{26}$$

The minimum material thickness is calculated by applying the Bredt equation (27) with a maximum wing twist of $\theta_{max} = 2.5°$.

$$I_T(y) = \frac{4 \cdot A_m^2(y)}{\oint \frac{ds}{t(y,s)}} \overset{t(y,s)=t(y)}{=} \frac{4 \cdot A_m^2(y)}{\frac{\oint ds}{t(y)}} \overset{\oint ds = C(y)}{=} \frac{4 \cdot A_m^2(y) \cdot t(y)}{C(y)} \tag{27}$$

With $A_m(y)$, the included area inside the wing's cross subsection and $C(y)$ the subsections circumference. Both are calculated as seen in figure 10, based on the specific airfoil for each position.

The wing bending is calculated with the maximum demanded vertical load factor of $n_{z,max} = 4$.

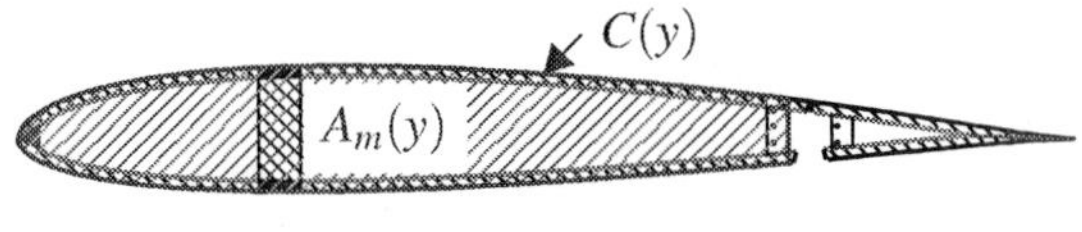

Fig. 10: Wing Torsion

As seen in figure 11 the weight of the wing and the weight of the nacelles with all included components and possible attached tailplanes is used to compensate the wing bending. The beam is assumed to be a cantilever beam. Additionally

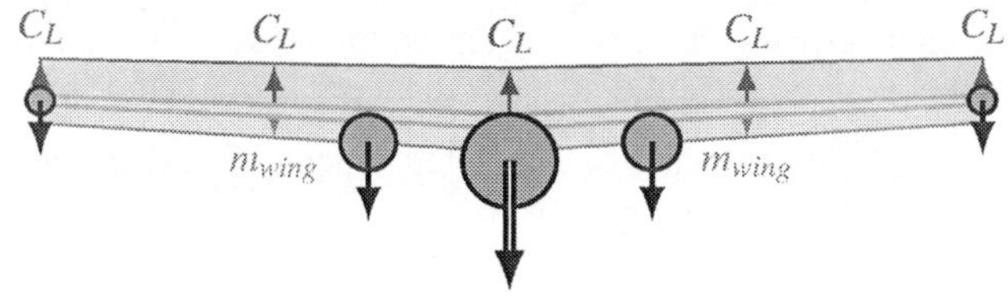

Fig. 11: Wing Loads

an even C_L distribution is assumed. Compared to an elliptical lift distribution, the load center moves slightly outward, increasing the bending moment inside the wing, making the analyzis mathematically more easy and at the same time more conservative.

The spar is assumed to have the full height of the airfoil, at its position of 25% of the wing chord, reduced by the outer shell thickness. The load carrying spar caps have the same thickness as the core material for the outer wing box. So, the width of the spar caps is calculated with the following equations:

$$L(y) = \int_y^s \frac{1}{2} \cdot \rho \cdot C_L \cdot n_{z,max} \cdot v_{cruise}^2 \cdot l(y) \,\mathrm{d}y \tag{28}$$

$$M_B(y) = \int_y^s L(y) \,\mathrm{d}y = \int_y^s \int_y^s \frac{1}{2} \cdot \rho \cdot C_L \cdot n_{z,max} \cdot v_{cruise}^2 \cdot l(y) \,\mathrm{d}^2 y \tag{29}$$

$$\sigma = \frac{|M_B(y)| \cdot z(y)}{I(y)} \overset{!}{\leq} \frac{\sigma_{max}}{SF} \tag{30}$$

With the now known width, the total cross section and weight of the spar can be calculated.

The landing gear has to carry vertical and horizontal loads. It consists of an internal beam structure carrying the loads. It is designed as a truss. The bending is calculated assuming a massive carbon beam.

A maximum horizontal landing impulse of 1 g is assumed. The maximum vertical landing impulse is assumed to be 2 g. The loads can be seen in figure 12.

Electric Propulsion System

The electrical propulsion system covers all needed parts, including lithium ion batteries, electronic speed controllers, connecting cables, brushless DC motors, possibly gearboxes and propellers. An overview of these components and their arrangement can be seen in figure 13. The ESC, the motor and the gearbox combined are called the drivetrain.

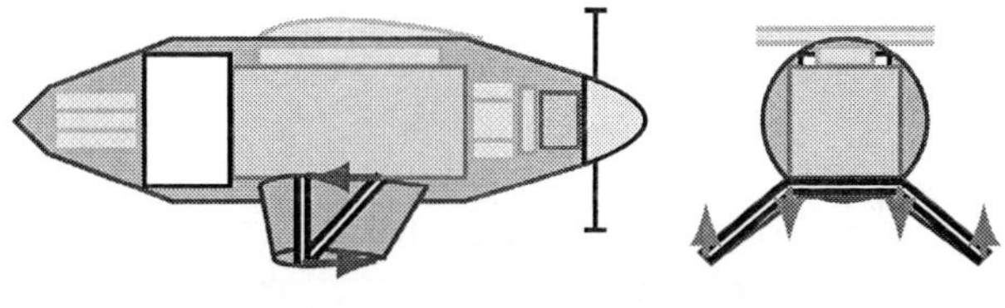

Fig. 12: Landing Gear Loads

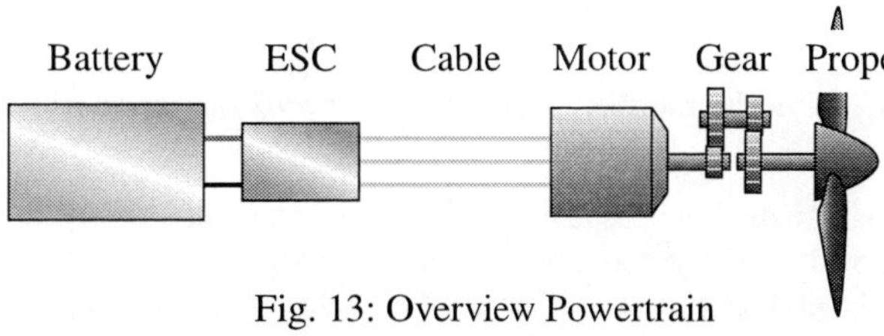

Fig. 13: Overview Powertrain

Battery: The battery is the energy source providing the energy for the propulsion systems, the flight controller electronics and possibly the payload.

A battery model has been implemented, which describes the batteries behavior, i.e. the trend of voltage and capacity versus the current.

It is based on *Peukert's law* (Ref. 8) which describes a general behavior of the effective capacity at different currents. It was originally written for lead batteries but it also can be used for lithium-ion batteries.

$$C_0 = I^{pk} \cdot t \tag{31}$$

with C_0 being the capacity at 1 Ampere and pk the battery dependent Peukert factor. The effective capacity is

$$C(I) = I \cdot t \tag{32}$$

rearranging leads to the description of the effective capacity

$$C(I) = C_0 \cdot I^{1-pk} \tag{33}$$

As the propulsion system does not only need a single current but different currents for different durations, additionally the total effective capacity needs to be calculated. For this a model inspired by the Miner's Rule (Ref. 9), which is used in structure calculation to asses the cumulative damage for structures under changing alternating stresses, is implemented. It is adapted to battery capacity with the following approach:

$$\sum_{i=1}^{k} \frac{I_i \cdot t_i}{C(I_i)} \leq 1 \tag{34}$$

inserting equation (33) it becomes

$$\sum_{i=1}^{k} \frac{I_i \cdot t_i}{C_0 \cdot I_i^{1-pk}} \leq 1 \tag{35}$$

This equation can be used to check if the battery is sufficiently sized.

The batteries are modeled with a constant inner resistance.

$$U(I) = U_{0,cell} - R_{i,cell} \cdot I \tag{36}$$

For a battery consisting of several batteries, the total inner Resistance is calculated with

$$R_{i,bat} = \frac{R_{i,cell} \cdot n_s}{n_p} \tag{37}$$

with n_s being the number of serial cells and n_p the number of parallel cells.

This inner resistance is however the inner resistance which is valid for all energy consumers. This means that a virtual inner resistance has to be calculated for each energy consumer, i.e. for each propulsion system, the flight controller and the payload. This is done with the following approach:

$$U_{eff} = U_{0,batt} - R_{i,batt} \cdot I_{total} \tag{38}$$

$$U_{eff} \stackrel{!}{=} U_{0,batt} - R_{virt,j} \cdot I_j \tag{39}$$

$$R_{virt,j} \cdot I_{comp} = R_{i,batt} \cdot I_{total} \tag{40}$$

$$R_{virt,j} = R_{i,batt} \cdot \frac{I_{total}}{I_j} \tag{41}$$

This results in a virtual inner resistance which is higher than the real inner resistance but leads to the same voltage drop. With this it is possible to calculate all propulsion systems independent of each other.

Electronic speed controller: The electronic speed controller or ESC controls the motor. It transforms the incoming direct current coming from the battery into the rotating three phase power, that the brushless motor needs to rotate.

The ESC Model is based on the work of Rößler (Ref. 10), where the ESC is working as an ohmic resistance at full throttle. At part throttle settings the ESC influences the efficiency of the overall powertrain, and is calculated with

$$\eta_{PT} = \eta_{FT} - \frac{PTF}{U_{ESC,in,PT}} \cdot (1 - \theta) \tag{42}$$

Where PTF is the part throttle factor, which "is an ESC constant that must be measured for different controllers" (Ref. 10). θ is the throttle setting, which is defined by the ratio between the powertrain's shaft power at part throttle and full throttle.

$$\theta = \frac{P_{prop,in,PT}}{P_{prop,in,FT}} \tag{43}$$

The total Powertrain Power (from ESC input to Propeller input) at a part throttle setting can then be calculated by solving the following equation:

$$\eta_{PT} = \frac{P_{prop,in,PT}}{P_{ESC,in,PT}} \tag{44}$$

Cables: The cables are used to transfer the electric energy from the battery to the electronic speed controller. The cables are modeled as simple ohmic resistances. Their total resistance is calculated with

$$R = \rho_{el} \cdot \frac{l_c}{A_c} \qquad [\Omega] \tag{45}$$

With ρ_{el} being the electrical resistivity, l_c the cable length and A_c the cable's cross-section. The cable consist of a copper core with an electrical resistivity of $\rho = 1.68 \times 10^{-8}\,\Omega \cdot \mathrm{m}$.

The cable cross-section is selected on the basis of the maximum continous current that needs to be transferred (Ref. 11).

The length of the cables is calculated depending on the position of the motors on the aircraft.

Motor: The motor transforms the incoming electric energy into mechanical energy, which is then passed to a gearbox or directly to the propeller. Even though the used motors are only brushless motors, the mathematical model is based on an approach for brushed motors. But since the relevant key-parameters are derived empirically, the error that is made is negligible. The model has been presented by Schenk (Ref. 12) and Lundstöm (Ref. 13). The mathematical model describing the motor has only a few parameters to describe the motor's behavior. The motor's rotational speed is calculated with

$$\omega = (U_{motor} - I \cdot R_{i,motor}) \cdot K_v \tag{46}$$

with U_{motor} being the motor input voltage, I the input current, $R_{i,motor}$ the motor's inner resistance and K_v the rotational speed constant.

One way to describe the motor's torque is

$$Q = \frac{I - I_0(U_{motor})}{K_v} \tag{47}$$

Where $I_0(U_{motor})$ is the idling current which is dependent on the Voltage. Typically one idling current (I_0) at a certain voltage (V_0) is given. This is not enough to describe the motor's behavior just yet. Therefore a second description is used

$$Q = Q_i - Q_l \tag{48}$$

This describes the motor's torque as the difference between the inner torque Q_i and the loss torque Q_l. The inner torque is described with a linear dependency on the current, described by the torque constant K_e which is the reciprocal value of the rotational speed constant K_v.

$$Q_i = K_e \cdot I = \frac{I}{K_v} \tag{49}$$

The loss torque is described with a linear dependency on the rotational speed.

$$Q_l = K_l \cdot \omega \tag{50}$$

Combining equations (46) to (50) the loss coefficient K_l can be derived

$$K_l = \frac{I_0}{(U_0 - I_0 \cdot R_{i,motor}) \cdot K_v^2} \tag{51}$$

With these equations the motor behavior can be described by constants that are typically given by manufacturers. To take the supply resistance R_{supply} and the battery voltage U_{batt} into account some additional rearranging is needed. The result is the following equation, describing the torque behavior depending on the rotational speed.

$$Q = K_e \cdot \frac{U_{batt}}{R_{i,t}} - \left(\frac{K_e^2}{R_{i,t}} + K_l\right) \cdot \omega_{motor} \tag{52}$$

with

$$R_{supply} = R_{i,batt} + R_{i,ESC} + R_{i,cable} \tag{53}$$

and

$$R_{i,t} = R_{i,batt} + R_{i,ESC} + R_{i,cable} + R_{i,motor} \tag{54}$$

Gearbox: A gearbox transforms the incoming torque and rotational speed from the motor. This allows having different rotational speeds for motor and propeller, which can possibly lead to a higher efficiency of the overall propulsion system even if the gear itself introduces some mechanical losses.

The gearbox is modeled with the gearbox ratio GR, describing the fraction of input speed to output speed, and the gearbox's efficiency η_{gear}. The gearbox transforms the power and rotational speed as follows

$$P_{gear,out} = \eta_{gear} \cdot P_{gear,in} \tag{55}$$

$$\omega_{gear,out} = \frac{\omega_{gear,in}}{GR} \tag{56}$$

which leads to the torque transformation

$$Q_{gear,out} = \eta_{gear} \cdot GR \cdot Q_{gear,in} \tag{57}$$

Propeller: The propeller transforms the incoming mechanical energy to thrust which is used to propel the aircraft.

A crucial point for the propulsion system calculation is the propeller data. In general, propeller data is available in form of C_T– and C_P/C_Q–Tables. C_T is the Thrust Coefficient, C_P the Power Coefficient, C_Q the Torque Coefficient. They are dimensionless coefficients, describing the propeller's characteristics. These values are typically calculated or generated at different rotational speeds and different advance ratios. The advance ratio J is an equivalent to the ratio of air speed to the propellers tip speed:

$$J = \frac{v_\infty}{n \cdot D} = \frac{\pi \cdot v_\infty}{\omega \cdot r_{prop}} \tag{58}$$

When combining a propeller with a motor the absolute values of thrust and torque are needed.

$$T = \rho \cdot n^2 \cdot D^2 \cdot C_T = \frac{4}{\pi^2} \cdot \rho \cdot \omega^2 \cdot r_{prop}^4 \cdot C_T \tag{59}$$

$$P = \rho \cdot n^3 \cdot D^5 \cdot C_P = \frac{4}{\pi^3} \cdot \rho \cdot \omega^3 \cdot r_{prop}^5 \cdot C_P \tag{60}$$

$$\begin{aligned} Q &= \frac{P}{2 \cdot \pi \cdot n} = \frac{\rho \cdot n^2 \cdot D^5}{2 \cdot \pi} \cdot C_P = \\ &= \frac{P}{\omega} = \frac{4}{\pi^3} \cdot \rho \cdot \omega^2 \cdot r_{prop}^5 \cdot C_P \end{aligned} \tag{61}$$

Since C_T and C_P are depending on the advance ratio, the rotational speed ω needs to be calculated depending on the flying speed and the advance ratio:

$$\omega = \frac{\pi \cdot v_\infty}{r_{prop} \cdot J} \tag{62}$$

The propeller data used for this analysis is derived from the measurements performed by the University of Illinois (Refs. 14, 15).

Propulsion system performance calculation: The output torque of the gearbox is described by combining equations (52) and (57)

$$Q_{gear} = \eta_{gear} \cdot GR \cdot \left[K_e \cdot \frac{U_0}{R_{i,t}} - \left(\frac{K_e^2}{R_{i,t}} + K_l\right) \cdot \omega_{gear} \cdot GR\right] \tag{63}$$

The intersection of the curves described by equations (61) and (63) leads to the full throttle Torque Q_{FT} and rotational speed ω_{FT} for a certain airspeed.

The current is calculated with

$$I = \frac{U_0 - \omega_{gear} \cdot GR \cdot K_e}{R_{i,t}} \tag{64}$$

The powertrain's full throttle output power is calculated with

$$P_{prop,in,FT} = Q_{FT} \cdot \omega_{FT} \tag{65}$$

The ESC's input power is calculated with

$$P_{ESC,in,FT} = \left(U_0 - R_{i,batt} \cdot I_{FT}\right) \cdot I_{FT} \tag{66}$$

This leads to the powertrain's full throttle efficiency

$$\eta_{Powertrain,FT} = \frac{P_{prop,in,FT}}{P_{ESC,in,FT}} \tag{67}$$

For a part throttle calculation, the needed power for a certain thrust has to be calculated. The propellers power requirement for a certain thrust at a defined speed can be derived from the propeller curves (equations (59) and (60)). Using the part throttle model (equation (44)), the powertrain's power requirement can be calculated with

$$I_{PT} = \frac{2 \cdot P_{prop,in,PT}}{U_\theta \overset{+}{(-)} \sqrt{U_\theta^2 - 4 \cdot R_{i,batt} \cdot \eta_{FT} \cdot P_{prop,in,PT}}} \tag{68}$$

with

$$U_\theta = U_0 \cdot \eta_{FT} - PTF \cdot (1 - \theta) \tag{69}$$

Propulsion system preselection The applied overall UAV design process needs an early selection of the propulsion system. This is the reason why commercially available components are used for the propulsion system design, integrated in this tool.

The combination of a parameter variation and a propeller and motor database leads to high number of possible combinations. Analyzing all combinations every time would be very inefficient and time consuming. Therefore a propulsion system preselection is performed.

For the preselection the thrust and the motor run time are varied in a regular grid. For each grid point all motors and propellers are combined and their respective power requirement is calculated with the presented method. Based on the needed energy a virtual, rubber sized battery is generated. For each grid point a reduced selection is kept, based on the mass of the components (including the rubber sized battery) and on the overall efficiency is kept. Figure 14 shows an exemplary pattern, with the "'x"' describing the preselection grid points. In this example the thrust has been varied in 5 N-steps between 10 N and 35 N and the cruise time has been varied in 10 min-steps between 40 min and 120 min.

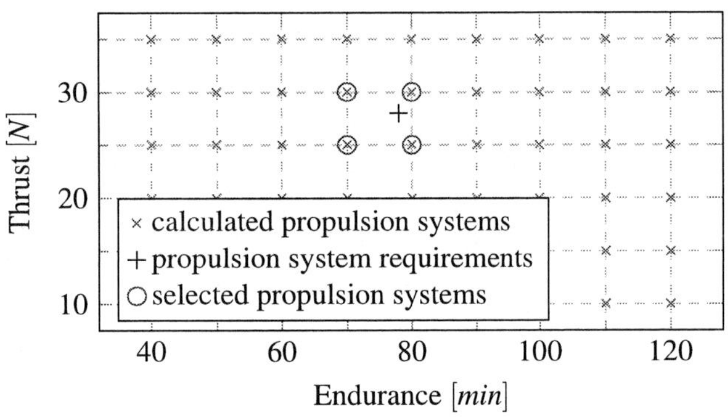

Fig. 14: Propulsion System preselection

In the simulation, the exact thrust requirement is calculated based on the drag. The cruise time is a iteration parameter. With the known values, the nearest grid point combinations are selected, to be analyzed at this point. In the example in figure 14, the thrust requirement is 28 N with a cruise time of 78 min, marked with a cross. The selected grid points are marked with the circles.

All these selected propulsion systems are then recalculated with the effective cruise time and thrust requirement.

This process significantly reduces the simulation time and allows to use discrete propulsion systems rather than generic models only.

SIMULATION TOOL

The overall simulation process uses parameter variations of different geometric and performance parameters. These result in sensitivity studies, which can be used to find an optimum within these parameter ranges. This approach has been developed during an practical course, which is discribed by Rößler (Ref. 16).

A total of 15 Parameter are implemented to be varied:

- Battery Type
- Number of Serial Cells
- Number of Parallel Cells
- Cruise Speed
- Hover Time
- Thrust Ratio
- Aspect Ratio
- Taper Ratio
- Airfoil
- Airfoil Thickness
- Airfoil Camber
- HTP Surface Ratio
- HTP Lift Coefficient
- Motor Cable Cross Section
- Maximum Propeller Diameter

The performance calculation is the core of the simulation. It starts with the wing size and tailplane calculation, where the wing, HTP and VTP size and position are calculated. It continues with the motors' positioning, the respective cable length calculation and the fuselage and nacelles sizes and positions. The aircraft's drag is calculated depending on the wing, HTP, VTP, nacelle and fuselage size, as described in the aerodynamic model section. The aircraft's propulsion systems are calculated depending on the drag and cruise speed, as described in the electric propulsion system section.

The structure calculation for the wing and tailplane is calculated based on their dimensions and the position and masses of the propulsion systems and nacelles. First the HTP is calculated. It is followed by the VTP, which might be influenced by the HTP, when a T-Tail configuration is analyzed. Finally the wing, which is influenced by the tailplanes for a twin tailboom configuration, is calculated. The calculation starts with the torsion, which is carried by the shell skin of the wing/tailplane, and then the bending, which is carried by the spar inside of the shell skin.

Finally the center of gravity is calculated and if needed an adjustment of the battery position is used.

The overall process is implemented as an iterative process, based on the false position method by Press et al. (Ref. 17). It constantly adjusts the endurance, the dimensions of all structural elements and the position of the center of gravity. As soon as all these components converge, and the center of gravity is at the position required by the flight mechanics calculation, the result is kept.

RESULTS

The outputs of the simulation process are sensitivity studies and detailed information on the configurations which lead to the longest endurance. The sensitivity studies are used to understand the influence of different parameters on the result. With the detailed information and outputs, the results become tangible.

Detailed Results

Table 2: Performance Comparison

Parameter	Unit	Configuration 4+1	2X2	2+2
Endurance	[%]	**100.0**	73.1	110.8
Aerodynamic Efficiency	[%]	**100.0**	107.3	110.9
Hover Power	[%]	**100.0**	117.8	89.3
Cruise Power	[%]	**100.0**	93.8	86.7

The results for the three different configurations, that lead to the best endurance, can be seen in table 2. The 2+2 configuration has an endurance that is 10% longer compared to the baseline 4+1 configuration and the 2X2 configuration has an endurance, that is 27% shorter. The 2+2 configuration has the highest aerodynamic efficiency, which has been expected, since it has the lowest wetted surface with only one central fuselage.

Table 3: Variation Parameter Comparison

Variation Parameter	Unit	Configuration 4+1	2X2	2+2
Cruise Speed	$[\frac{m}{s}]$	20	20	20
Aspect Ratio	[−]	14	14	16
Battery Type		HE	HD	HE
Serial Cells		10	10	10
Airfoil Camber	[%]	5	5	3
Max. Prop. Diameter	[cm]	50	30	50

Table 3 shows a few selected variation parameters for these configuration, that lead to each individual best result. It can be seen, that they are all very similar. With the cruise speed being the same for all three, they can be very good compared, as they work in similar Reynolds number regimes.

Table 4: Geometry Comparison

Parameter	Unit	Configuration 4+1	2X2	2+2
Wing Surface	[%]	**100.0**	97.7	108.8
Wingspan	[%]	**100.0**	98.9	111.5
Overall Length	[%]	47.0	43.0	62.6

Their geometric properties can be seen in table 4. Since they are all sized to the same criteria, to fulfill the stall speed margin their wing areas are in the same region.

Figure 15 gives the mass breakdown of the three configurations. Here already one disadvantage of the 2X2 configuration can be seen, that leads to the reduced endurance: it has the highest fixed masses, which leads to the lowest battery weight. This results in a reduction in capacity and thus a reduction in endurance. The combination of the small battery size and the high power requirement of this configuration leads to the ne-

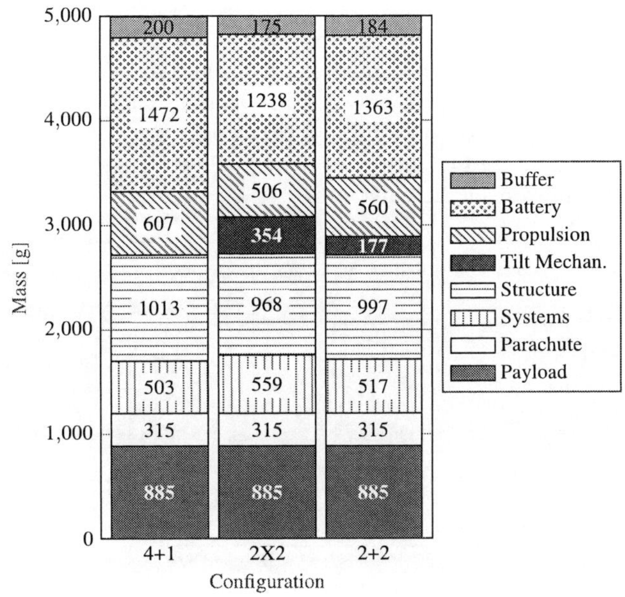

Fig. 15: Mass Breakdown

cessity to use high drain batteries instead of high energy batteries, which further reduces the endurance.

Also the advantage of the 4+1 configuration's fixed mass can be seen, allowing for a higher battery mass. This compensates for the low aerodynamic efficiency and enables the configuration to perform very well.

The combination of the intermediate fixed mass of the 2+2 configuration and it's high aerodynamic efficiency lead to the highest performance of all three configurations.

Sensitivity Studies

The influence of some variation parameters on the endurance is further analyzed. This helps to further understand the different configurations and what potential they have.

Figure 16 shows the influence of the wing's aspect ratio on the endurance. The trend in this figure shows the expected behav-

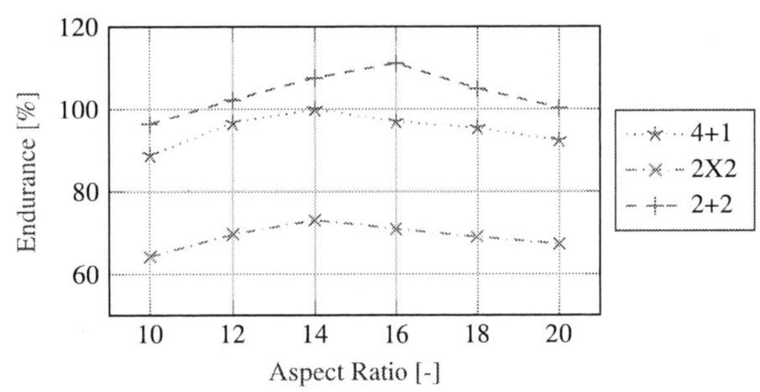

Fig. 16: Aspect Ratio Influence

ior for the aspect ratio influence, as a low aspect ratio leads to a lower aerodynamic efficiency, whereas a high aspect ratio results in higher structural masses, reducing the battery size. These two effects lead to a maximum for all configurations.

For the 2+2 configuration the maximum is at a higher aspect ratio compared to the other two configurations. This is due to the bending moment reduction of the wing, induced by the wingtip propulsion systems and the reduced airfoil camber for this configuration, which also reduces the wing's torsional moment. These two factors combined allow for a higher aspect ratio wing at the same mass.

Figure 17 shows the influence of the cruise speed on the endurance. For the endurance a lower cruise speed is favored, but

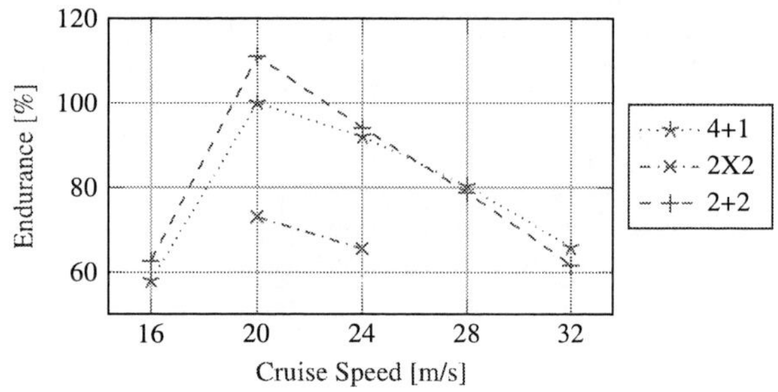

Fig. 17: Cruise Speed Influence

not too low. This trend comes from two effects. The first effect is the increasing power requirement with increasing speed. The second effect is an increasing wing size and thus an increasing structural weight with a decreasing speed. The combination of these two lead to the endurance maximum.

At increasing speeds the 4+1 configuration's endurance gets longer than the 2+2 configuration's. At higher speeds, the hybrid propulsion of the 2+2 configuration decreases in efficiency as its two design points become further apart. As the 4+1 configuration's propulsion is optimized only for cruise it is not affected by this. The same effect can be seen in figure 18. The maximum range of the 4+1 configuration is

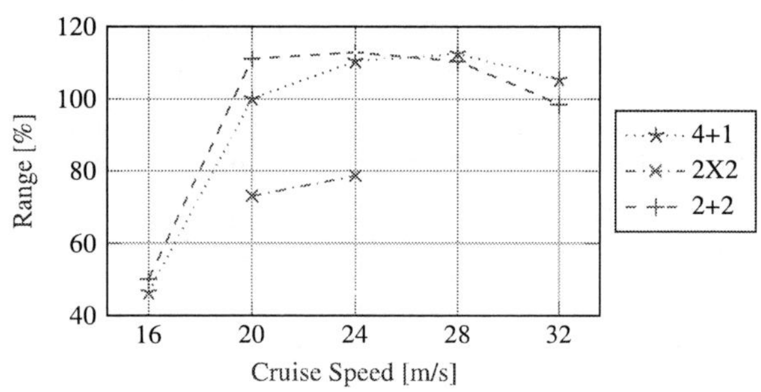

Fig. 18: Cruise Speed Influence

reached at higher speeds than the maximum range for the other two configurations, making it more suitable for missions with a higher speed requirement.

Figure 19 shows the influence of the battery type on the endurance. For the 4+1 and 2+2 configurations the better battery type is the high energy type. These configurations have a sufficiently sized battery in combination with low maximum power requirements, which allow the use of the high energy cells, which have a limited maximum power they can provide.

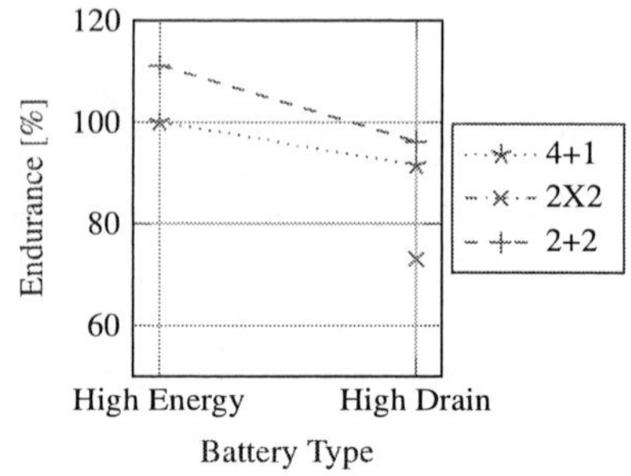

Fig. 19: Battery Type Influence

The 2X2 configuration has, as already described, the lowest battery mass in combination with the highest maximum power requirements. Therefore it only works with high drain batteries, that can provide a high maximum power at the cost of a lower capacity.

Figure 20 shows the influence of the number of serial cells on the endurance. A higher number of serial cells increases the

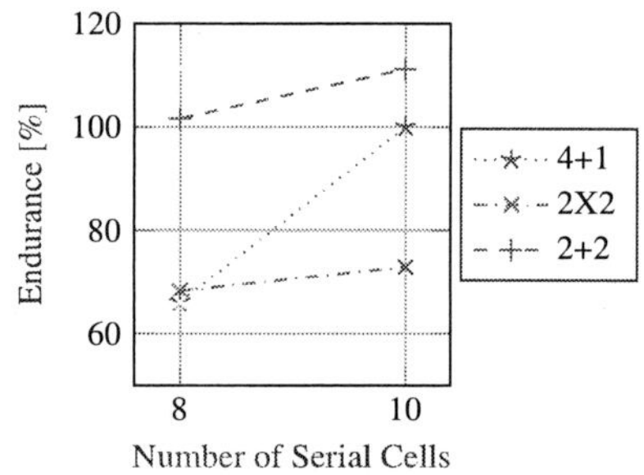

Fig. 20: Number of Serial Cells Influence

system's voltage, which in turn reduces the currents. Reduced current lead to less loss voltages and thus less energy loss due to internal resistances. This leads to a benefit of higher voltages. These are however limited by the used electronic components which need to be able to cope with the high voltage.

CONCLUSIONS AND OUTLOOK

A method to simulate and optimize eVTOL UAV has been developed. The disciplines aerodynamics, flight mechanics, structural mechanics, electrical propulsion and flight performance calculation have been implemented in different levels of complexity. The methods are partly already widely accepted methods or a validated with tests, which are partly still ongoing. Based on the performed tests and the comparison with flying UAV, the results are judged to be sufficient reliable, to be used for a comparison of the analyzed configurations.

With the selected levels of complexity the overall simulation runs efficient and allows to vary a wide range of different design parameters. This leads to the creation of sensitivity studies, which enable a better understanding of the influences on the performance of the different configurations.

Basing the propulsion system simulation on a database approach with commercially available components, allows the selection of propulsion systems already in the early concept phase of the design process. Efficiently incorporating and analyzing such a database has been done by a preselection method. This allows the required propulsion system to be selected very early in the design process and to be analyzed in more detail for a later more detailed design phase in the overall UAV design process. The disadvantage of this method is the lack of availability of propeller data in a big scale. Solving this needs either a big number of wind tunnel tests or a reliable method to simulate propellers in a wide range of advance ratios and also at static thrust, which is not available at the moment.

Three different configurations have been compared: a tiltless multirotor VTOL with a designated cruise motor (4+1), a fully tiltable multirotor VTOL (2X2) and a VTOL with tiltable wingtip propulsion (2+2).

For the presented mission, the 2+2 configuration has the highest performance. However if there is a requirement for higher speeds, the 4+1 configuration would be the preferred one. The 2X2 configuration has a too complex and too heavy tilt system and too inefficient propulsion systems, which make it perform worse by a margin, than the other two configurations.

These results are of course only valid for the presented configuration.

In general it is advantageous to have a separation of the propulsion systems requirement or at least not to have two diametrical different requirements for them, as is the case for the 2X2 configuration. A good design, such as the 2+2 configuration can compensate for the slight drawback of a multi-objective optimized propulsion system.

The next steps will be to adapt this method to incorporate even more different configurations to compare them too. Also verifying the methods for different take-off weights will be paramount, to be able to apply the tool to different mission requirements.

Author contact: Thomas Seren thomas.seren@tum.de,
Christian Roessler christian.roessler@tum.de,
Mirko Hornung mirko.hornung@tum.de

ACKNOWLEDGEMENTS

The author wants to thank the project team members Sebastian Oberschwendtner, Philipp Stahl and Moritz Thiele for their support and contribution to fixed-wing VTOL aircraft development at the Institute for Aircraft Design / Technical University of Munich.

REFERENCES

[1]P. Stahl, T. Seren, C. Rößler, and M. Hornung, "Development and Performance Comparison of Optimized Electric Fixed-Wing VTOL UAV Configurations," in *31st Congress of the International Council of the Aeronautical Sciences*, 2018.

[2]M. Drela and H. Youngren, *XFOIL 6.9 User Guide*. http://web.mit.edu/drela/Public/web/xfoil/xfoil_doc.txt, 2001. [online; accessed 31. January 2017].

[3]J. G. Lowry and E. C. Polhamus, "A method for predicting lift increments due to flap deflection at low angles of attack in incompressible flow," 1957.

[4]H. Quabeck, *Design, Leistung und Dynamik von Segelflugmodellen*. HQ Modellflugliteratur, 1994.

[5]D. P. Raymer, *Aircraft Design: A Conceptual Approach*. American Institute of Aeronatuics and Astronautics, Inc., fifth ed., 2012.

[6]S. F. Hoerner, *Fluid-Dynamic Drag*. Published by the Author, 1965.

[7]P. Stahl, C. Rößler, and M. Hornung, "Benefit Analysis and System Design Considerations for Drag Reduction of Inactive Hover Rotors on Electric Fixed-Wing VTOL Vehicles," in *AIAA Aviation 2018*, 2018.

[8]W. Peukert, *Über die Abhängigkeit der Kapazität von der Entladestromstärke bei Bleiakkumulatoren*. Elektrotechnische Zeitschrift, 1897.

[9]M. A. Miner, *Cumulative Fatigue Damage*. Journal of Applied Mechanics, 1945.

[10]C. Rößler, *Conceptual Design of Unmanned Aircraft with Fuel Cell Propulsion System*. Verlag Dr. Hut, 2012.

[11]Federal Aviation Administration, *Acceptable Methods, Techniques, and Practices - Aircraft Inspection and Repair*, pp. 11–21 – 11–66. AC-43.13-1B, 1998.

[12]H. Schenk, *Motormodellierung/Theorie*. https://www.rc-heli.de/board/attachment.php?s=93f3df5e388a5edb0d418b60f3d0ec4f&attachmentid=153491&d=1357461119, 2005. [online; accessed 2018-01-15].

[13]D. Lundström, K. Amadori, and P. Krus, "Validation of models for small scale electric propulsion systems," 48th AIAA Aerospace Sciences Meeting Including the New Horizons Forum and Aerospace Exposition, 2010.

[14]J. B. Brandt and M. S. Selig, *Propeller Performance Data at Low Reynolds Numbers*. 49th AIAA Aerospace Sciences Meeting, AIAA Paper 2011-1255, 2011.

[15]R. W. Deters, G. K. Ananda, and M. S. Selig, *Reynolds Number Effects on the Performance of Small-Scale Propellers*. AIAA Aviation and Aeronautics Forum and Exposition (Aviation 2014), AIAA Paper 2014-2151, 2014.

[16]C. Rößler and M. Hornung, "Electric VTOL UAV Configuration Studies and Optimization within Aircraft Design Practical Course at Technical University of Munich," in *31st Congress of the International Council of the Aeronautical Sciences*, 2018.

[17]W. H. Press, S. A. Teukolsky, W. T. Vetterling, and B. P. Flannery, *Numerical Recipes*. Cambridge University Press, third ed., 2007.

Analysis Methods for Tilting Wing and Tailsitter e-VTOL Configurations

Todd R. Quackenbush
Senior Associate

Daniel A. Wachspress
Senior Associate

Glen R. Whitehouse
Associate

Michael K. Yu
Staff Associate

Continuum Dynamics, Inc., Ewing, NJ 08618

ABSTRACT

Several current conceptual designs for advanced eVTOL aircraft feature tiltwing or tailsitter configurations that involve strong proprotor/wing interactions. This paper describes an assessment of the capability of contemporary analysis tools to address some of the key aerodynamic modeling issues faced by these vehicles. The particular focus here is on correlation studies with available data on public domain configurations to provide "lessons learned" regarding the appropriateness of various levels of aerodynamic analysis for modeling novel flight vehicles of current interest. The comparisons shown here center chiefly on the application of a comprehensive rotorcraft aerodynamic model to representative configurations, with selected comparisons to alternate CFD-based analysis methods.

NOTATION

A	propeller/rotor disk area, ft^2
c	chord, ft
C_L	lift coefficient, L/qS
$C_{L,T}$	lift coefficient including thrust component, $(L + T \sin \alpha)/qS$
C_D	drag coefficient, D/qS
C_P	power coefficient, $P/\rho\pi R^2(\Omega R)^3$
C_T	thrust coefficient, $T/\rho\pi R^2(\Omega R)^2$
$C_{T,fs}$	free stream thrust coefficient, $T/\rho\pi R^2(V_{fs})^2$
C_X	horizontal force coefficient, $(T \cos \alpha - D)/qS$
d	diameter, ft
D	drag, lbf
J	axial advance ratio, V_{fs}/nd
L	lift, lbf
n	propeller rotation frequency, hz
P	rotor power, ft-lbf/sec
q	free stream dynamic pressure, lbf/ft^2
R	blade radius, ft
T	thrust, lbf
V_{fs}	free stream speed, ft/sec
α	angle of attack, deg
ρ	air density, slugs/ft^3
μ	advance ratio, $V_{fs}/\Omega R$
σ	rotor solidty, $N_b c/\pi R$
Ω	rotor rotation frequency, rad/sec

INTRODUCTION

Several current conceptual and experimental designs for advanced full-scale V/STOL aircraft as well as related eVTOL designs (Figure 1) feature tiltwing or tailsitter configurations, whose aerodynamics intrinsically involve strong proprotor/wing interaction. As a consequence of recent and current work by the authors supporting several similar vehicle designs, assessment of the capability of current-generation analysis tools for addressing some of the key modeling challenges faced by these configurations is ongoing. Predictive challenges for these configurations have been generally well understood for many years [1-3] but have persisted, given both their intrinsic complexity and the slow maturation of relevant analysis tools. The work reflected here builds both on recent activity supporting particular advanced full-scale vehicles as well as ongoing support of several novel eVTOL designs. However, it does not concern itself with specific new designs, but rather with assessing the readiness of current tools for capturing relevant public domain data sets.

Figure 1: Advanced V/STOL and eVTOL concepts under study - clockwise from upper left: Airbus A^3 Vahana; Google 'Project Wing' tailsitter; NASA 'Greased Lightning' UAV tiltwing; Terrafugia TF-2; NASA X-57; and the conceptual tail-sitter aircraft for the DARPA TERN program (Northrop Grumman).

Presented at the Vertical Flight Society Autonomous VTOL Technical Meeting and Electric VTOL Symposium, Mesa, Arizona, UAS, Jan. 29-31, 2019.

As noted by several authors (e.g., [2]), enhancement of rotorcraft aerodynamic modeling tools has been a continuing requirement in recent decades because of the wide range of configurations of interest in meeting important VTOL mission needs. Many prior efforts to develop vertical lift aircraft with high speed capabilities have encountered difficulty in meeting their performance and/or payload targets, owing to excessive power and fuel requirements, high weight-empty fraction, or poor cruise efficiency. Ongoing technology developments (see [4, 5] among many others) are supporting efforts to overcome these problems and develop advanced VTOL aircraft with either high speed or long range capability by using a wide range of lift and propulsion concepts. These efforts require moving beyond conventional helicopter designs to various forms of compound rotorcraft or other VTOL variants. A complementary development is the proliferation of novel eVTOL aircraft for a wide range of sensing, passenger transport or delivery missions, where the ability to both hover and achieve more efficient forward flight is required.

One candidate solution involves tilting open rotor configurations, featuring lifting proprotors for hover, converting to axial flight for forward propulsion. Tiltrotors are clearly the most well-developed example of this class of vehicles (Figure 2), but tiltwing and tailsitter concepts have been considered in past development efforts (Figure 3) and can provide many comparable capabilities; they are being revisited in recent studies and tests (Figure 1), given that the growing role for unmanned VTOL systems removes some prior objections to consideration of this option.

Figure 2: XV-15 tiltrotor demonstrator (left) and the V-22 Osprey (right).

Figure 3: Past tiltwing/tailsitter concepts – clockwise from upper left: LTV XC-142; Vertol VZ-2; Lockheed XFV-1; Convair XFY-1.

As noted above, tiltrotor configurations have been the most successful and widely applied high speed VTOL air vehicle design. After being shown to be technically feasible with the McDonnell XV-3 and the Bell XV-15 [6], the tiltrotor concept has now seen many thousands of hours of successful operation following the introduction of the V-22 into service in 2007. Pending further prototype flight testing, the Leonardo AW609 is projected to extend these capabilities into the civil field in the near future.

While tiltrotor configurations have enjoyed considerable success, tiltwing designs have also been strongly advocated for manned missions [2]. The design challenges for tiltwing vehicles have many similarities to tiltrotors, though they also pose distinct challenges – particularly in terms of strong interactions between the propeller slipstream and the lifting wing and the sensitivity of these effects to forward speed and propeller thrust. Such issues have long been known, and became evident during the development of the XC-142 aircraft during the 1960s [2, 7, 8]. Many of the same design challenges apply to tailsitter aircraft, which have been studied for a variety of missions in recent years (e.g., [9, 10]; see also Figure 3).

Design analysis methods that could capture both the aerodynamics of open rotors typically used on these aircraft as well as the characteristic interactions of the rotor and wing would be highly desirable; they could address the many pressing questions facing groups tasked with conceptual and preliminary design of tailsitter and tiltwing concepts, and weighing their capabilities against the better-known challenges of tiltrotor designs.

ANALYSIS METHODS

The work described here was undertaken as an offshoot of recent activity involving application of computational design analyses of these classes of aircraft. This activity has indicated the desirability of having at a minimum appropriate "mid-fidelity" models applied as early as possible in the conceptual and preliminary design process; it has also made clear some of the trades involved in the fidelity, generality, ease of use, and design analysis cycle time for their employment[11-13].

The analysis of advanced tilting rotor VTOL concepts can be addressed using several different levels of modeling tools, including: potential flow/comprehensive analyses; "mid-fidelity" CFD tools; and advanced URANS CFD methods. The suitability of each depends on the analysis goals, the setup/CPU time available, the range and type of cases to be studied, and the level of departure of the configuration in question from past design practice.

The particular software tool featured here for validation and assessment studies was CHARM – a comprehensive rotorcraft code featuring a lifting surface/free vortex wake model of the lifting rotors and an inviscid panel model of the airframe; as will be noted below, supporting calculations were also executed using CGE – a cut-cell Cartesian grid Euler CFD model with an actuator disk rotor model.

Regarding CHARM, analysis and design work with this widely-used model has verified its accuracy for conventional rotor and propeller configurations in hover and axial flight [13] [14-16], also its the fast panel fuselage and lifting surface wing models in CHARM [17] can be used to account for airframe effects on performance predictions. The major component capabilities of CHARM include: fast panel/fast vortex methods that make first order "full vehicle" calculations (similar to Figure 4) relatively quick and routine; surface meshing requirements that are more straightforward than for viscous CFD applications, though roughly the same as for the CGE Cartesian grid CFD models; and a knowledge base on both open rotor and ducted propeller modeling in hover and axial flight. Limitations of this model include: its assumption of inviscid flow and the consequent restriction to cases with known separation lines; limited applicability in high angle of attack and sidewash conditions; and a reliance on airfoil data and other semi-empirical models.

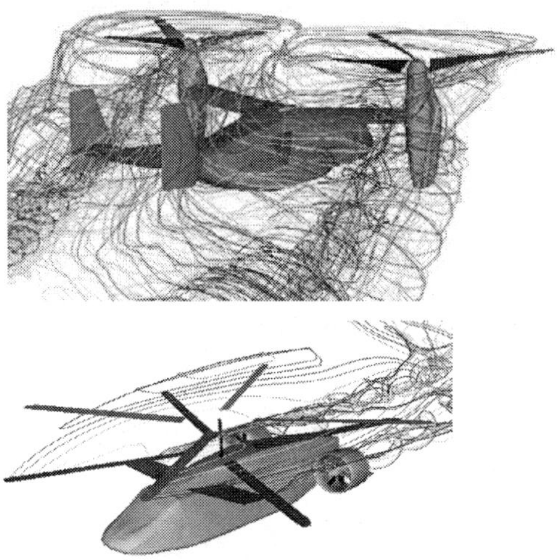

Figure 4: CHARM models of the V-22 (above) and the AVX CCH compound helicopter (below).

The Cartesian Grid Euler (CGE) analysis [18, 19] was developed at CDI and is a steady/unsteady cut-cell Euler solver that features automatic grid generation and the ability to handle non-watertight non-singular geometries. It is similar in general terms to other Cartesian grid models [20-22], but includes features that make it well suited to rotorcraft applications, e.g., an actuator disk model for analysis of lifting rotors. It is applicable to complex configurations (see Figure 5) and allows efficient studies of rotor/propeller effects on airframes and wings [22]. It also allows studies of downwash and download, bluff body drag, and operation of the lifting airframe in ground effect. CGE requires more CPU time than CHARM and a roughly equivalent level of setup time, but much less in each area than viscous URANS models using body-fitted meshes(e.g., [23]), and so qualifies as a "mid-fidelity" CFD tool.

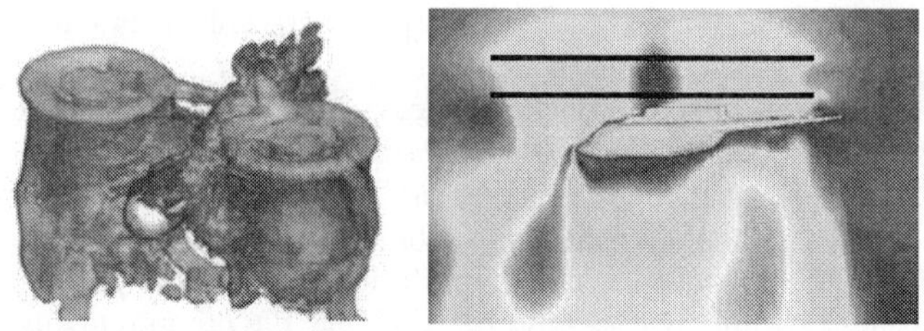

Figure 5: Rotorcraft modeled with CGE and its embedded actuator disk rotor model – right: vorticity field around a generic tiltrotor aircraft; left- downwash field near a notional coaxial helicopter.

VALIDATION STUDIES

Prior validation studies for general fixed and rotary wing applications for these CHARM and CGE are described in [14-19, 22]. Here we focus on particular cases that are relevant to analysis of proprotor/wing interaction problems. Figure 6 illustrates several rotor and rotor/wing systems for which relevant data is available One set of test results for a propeller of this class is the VZ-2 proprotor described by [24]. This design featured a 3 bladed propeller with a diameter of 9.5 ft, an activity factor per blade of 178, and a constant NACA 0009 airfoil. Figure 7 shows the propulsive efficiency in axial flight at zero shaft angle of attack predicted by CHARM compared to measured data for this design as a function of advance ratio for two different blade pitch settings.

Figure 6: Top left – VZ-2 tilting propeller model [24]; top right – pivoting propeller from [25]; bottom left – JVX proprotor bottom right – JVX proprotor [26]; wing/propeller wind tunnel model [27].

Regarding more recent designs specifically tailored for tilting proprotor systems, a key challenge for such configurations is predicting performance of individual rotor systems in the critical hover and axial flight conditions. An important target case is the JVX rotor system, a 2/3 scale V-22 configuration tested at NASA/Ames in the 1980s. Figure

8 shows the measured hover performance at the NASA/Ames Outdoor Aerodynamic Rotor Facility (OARF)[28], as well as the CHARM prediction. Additional testing of the JVX rotor was conducted in axial flight at the NASA/Ames 40x80-ft wind tunnel[26]. Figure 9 shows the computational model of the JVX rotor while Figure 10 shows the generally good accuracy available with CHARM for this data set in axial flight.

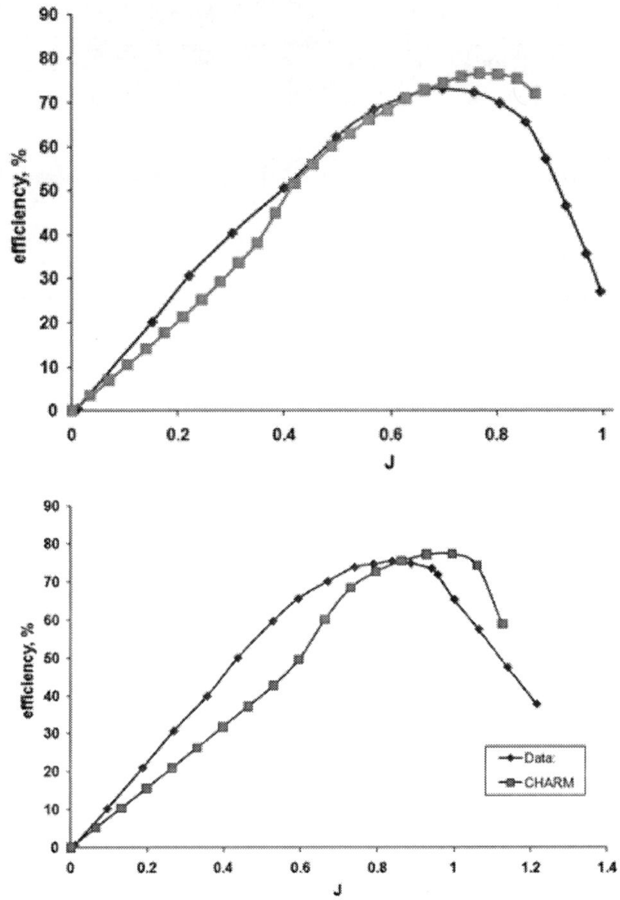

Figure 7: Propulsive efficiency for the VZ-2 proprotor at two blade pitch settings (β=24 above, β =30 deg. below)

One key feature of tilting rotor performance is its behavior in flight conditions other than axial flight. Comparisons of CHARM to a range of data for the XV-15 full scale proprotor at a constant onset speed of 100 kts but at variable shaft angle of attack were also undertaken (compared with data from [29]). Figure 11 shows the CHARM rotor/wake model for these calculations, while Figure 12 shows the measured and predicted rotor performance.

As is evident from these comparisons (and others available in the literature), CHARM provides a generally very capable model in terms of predicting rotor performance for rotor systems typical of those used on tilting rotor systems. Alternate models can be applied to analyze isolated rotor performance, though a recent study [30] has provided a discussion of the application of CHARM compared to alternative full CFD models such as FUN3D, indicating the trades in computer time and accuracy for the respective models.

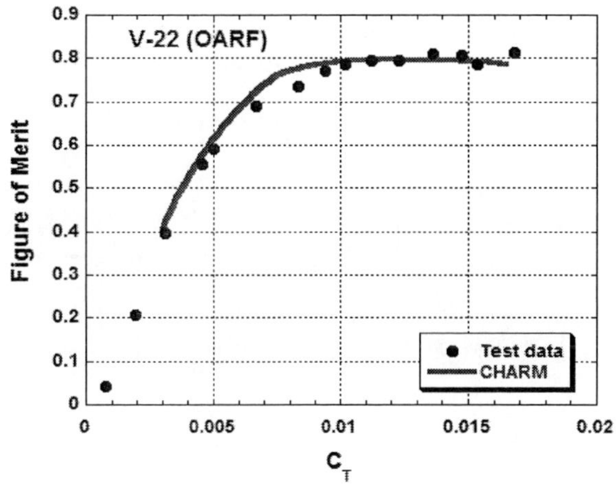

Figure 8.: Measurement and prediction of JVX (a .658 scale V-22) hover performance using CHARM.

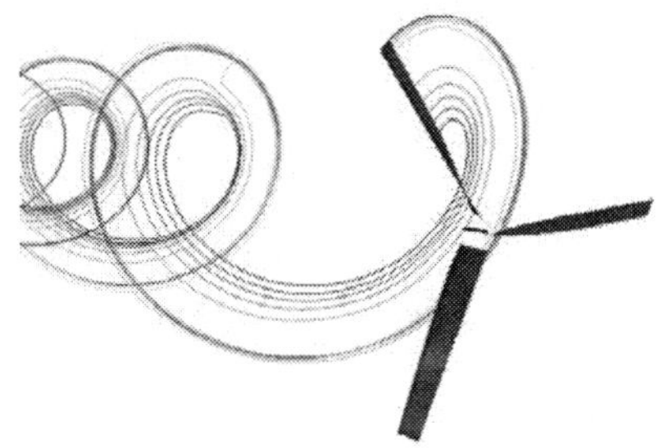

Figure 9: Axial flight model of the JVX rotor.

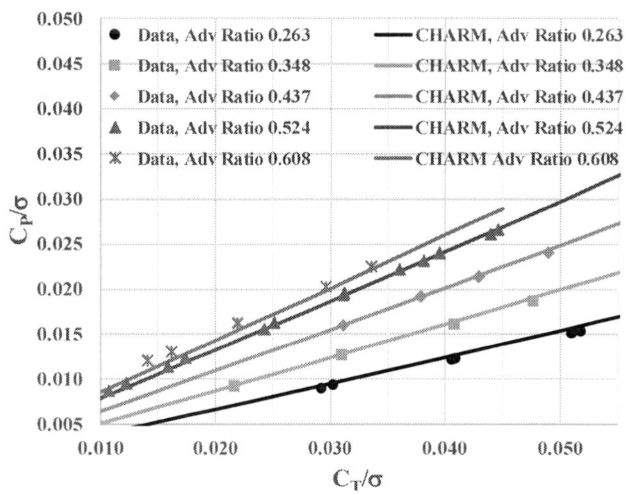

Figure 10: Prediction of the JVX rotor performance in axial flight as predicted by CHARM; power vs. thrust for a range of tip speeds and onset flow speeds shown.

While very important to the assessment of tiltrotor configurations, rotor aerodynamics are only one of the important mechanisms involved in full vehicle aerodynamics. Tiltrotors such as the V-22 (Figure 2) feature strong wake/airframe interactions, which can lead to significant airframe download and other strong interactions. The CHARM model can in principle address a wide range of cases for such configurations, though cases involving large scale flow separation should be addressed with models such as CGE, as indicated in recent studies of hover download [12, 31, 32].

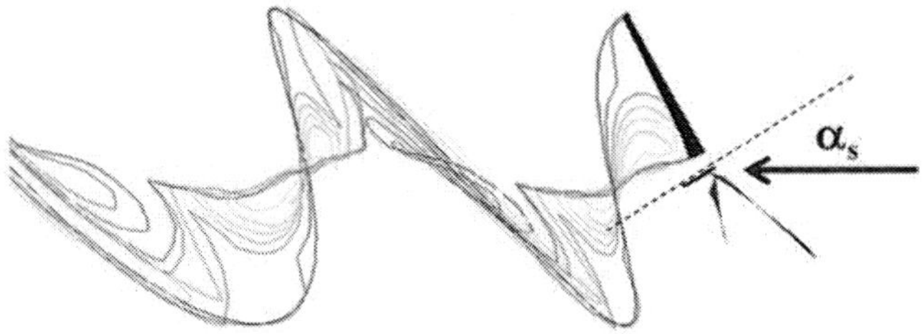

Figure 11: CHARM model of the XV-15 proprotor in forward flight at an angle of attack of α_s.

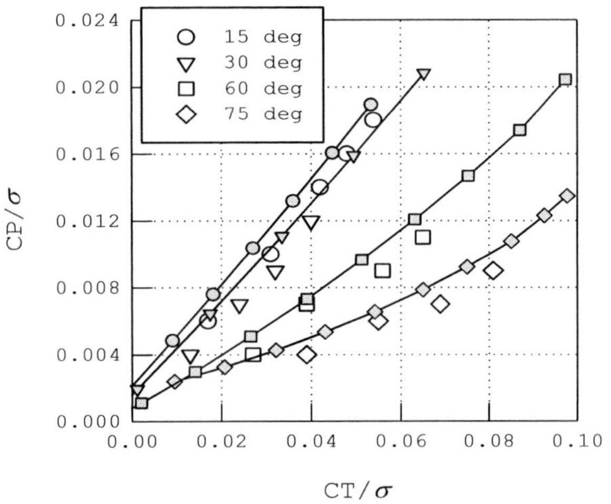

Figure 12: Predicted and measured performance for the XV-15 rotor over a range of shaft angles of attack [29]: open symbols = measurement; filled symbols = CHARM.

The available literature includes several experimental studies that provide insight into the more general problem of propeller wing interaction, including both cases with predominantly attached and separated flow. For present purposes it was of interest to compare the predictions of the CHARM model to one such case, a general research wing/propeller configuration described by Grunwald [33] and Traybar [34] (Figure 13). These sources include a wide range of speed and angle of attack cases for this model for comparison and comparisons of CHARM a representative subset of available data points helps identify the range of applicability of the model.

The two references just cited contain a significant volume of useful data on this generic configuration, which was tested first at the NASA 30x60 ft wind tunnel at Langley Research Center (Figure 14) and then on a moving carriage at the Princeton Dynamic Model Test Track. Not only did results obtained within the latter allow for assessments of repeatbility of the wind tunnel data but also permitted very low speed conditions to be tested with no concerns regarding wall flow separation or recirculation effects at low speed.

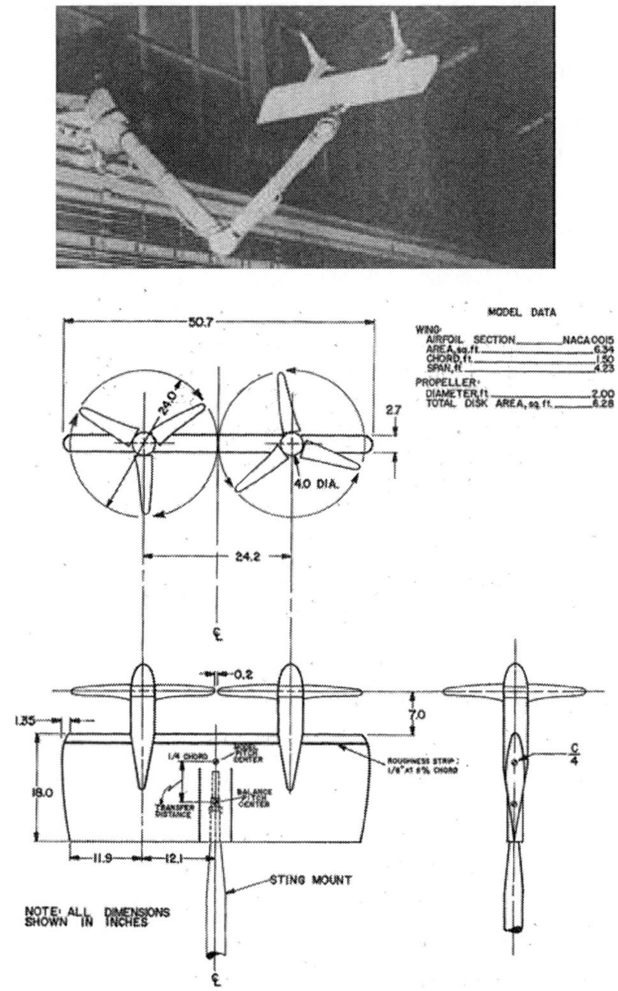

Figure 13: Test track experiment (above) and design details (below) for a wing/rotor interaction model (below) used for the study of tiltwing/tailsitter aerodynamics.

As indicated in Figure 13, the model consisted of an untwisted, constant chord wing 4.2 ft. in span and 1.5 ft in chord that used a NACA 0015 airfoil section. It also included two three-bladed, 2.0-ft diameter propellers, and a wide variety of operating conditions were studied. Predictions using the CHARM model were compared to measured lift coefficients using including thrust component ($C_{L,T}$) and horizontal force coefficient (C_X) (Figures 15 and 16). Results are shown in Figures 17 through 22 for three different values of wing angle of attack; note that these results used the vortex lattice option shown in Figure 16. Here, the blade pitch is held constant with constant RPM while carriage speed is varied to produce a variation in thrust coefficient. It is seen that for angle of attack of 40 and 60 deg., generally good accuracy is achieved. At 70 deg., however, correlation begins to degrade somewhat, though even at 80 deg angle of attack generally good accuracy is observed, a moderately suprising outcome given the inviscid wing model in CHARM.

Figure 14: Wind Tunnel testing in the NASA/Langley 30x60 tunnel [33] .

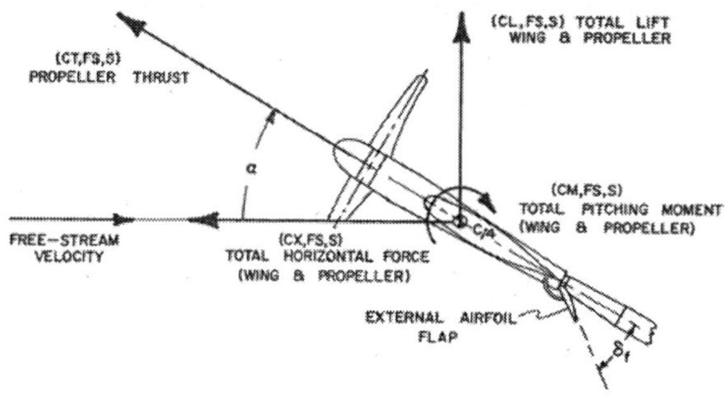

Figure 15: Schematic of force resolution used in the experiments of [34].

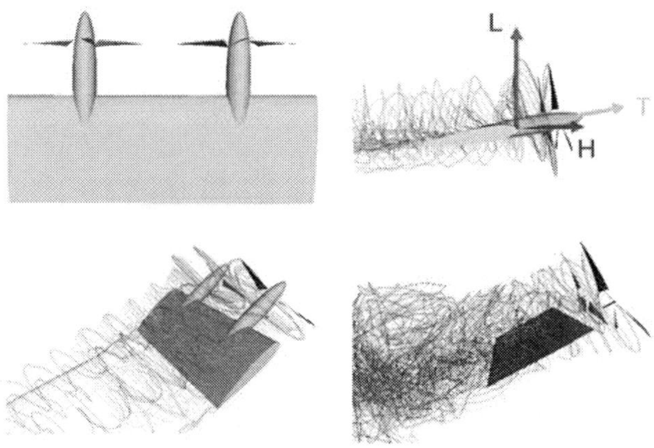

Figure 16: Initial model of this wing/propeller combination in CHARM in use for ongoing valdiation and assessment studies; top left – full panel model, top view; top right – side view showing major force components; bottom left – oblque view showing prop wing interaction; bottom right – thin vortex lattice modeling option, bypassing surface panelization.

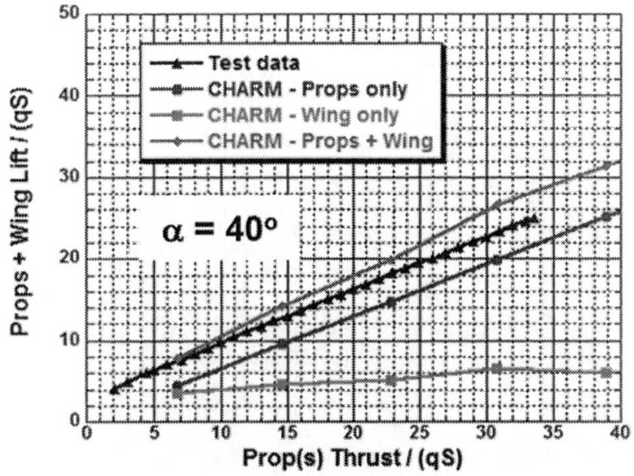

Figure 17: Measured and predicted lift force for the Traybar model: angle of attack 40 deg.

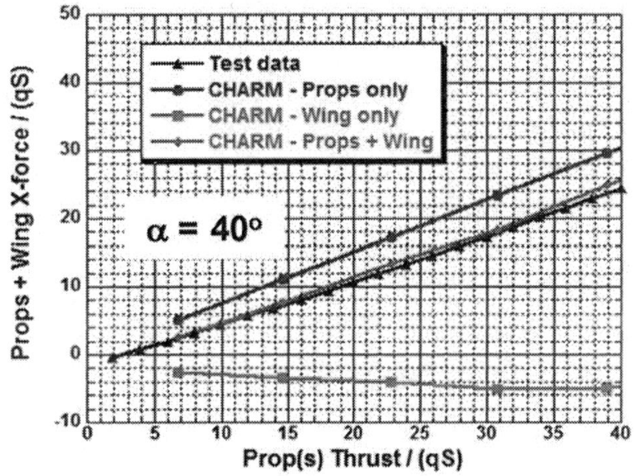

Figure 18: Measured and predicted X-force for the Traybar model: angle of attack 40 deg.

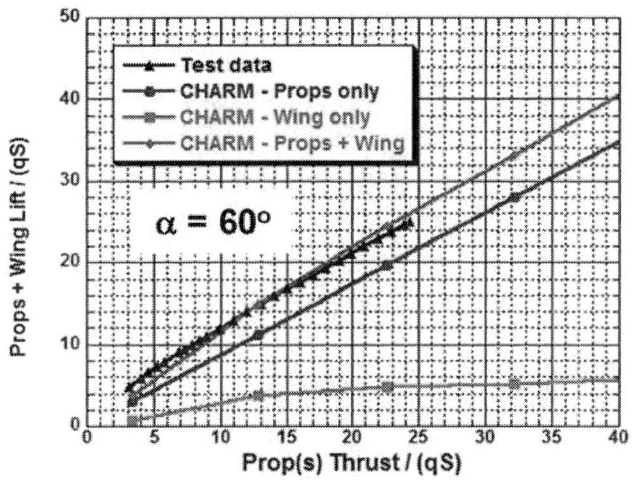

Figure 19: Measured and predicted lift force for the Traybar model: angle of attack 60 deg

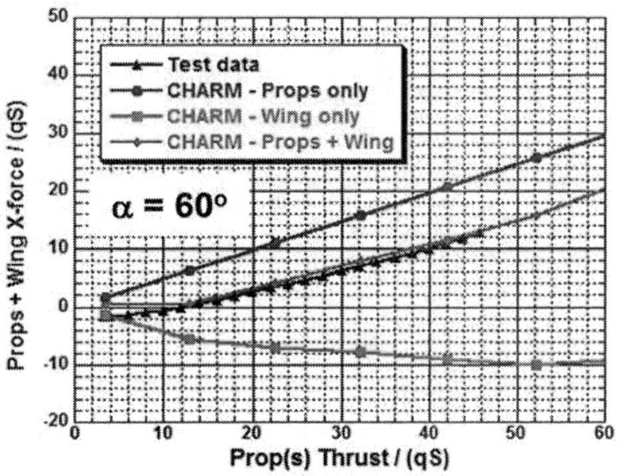

Figure 20: Measured and predicted X-force for the Traybar model: angle of attack 60 deg

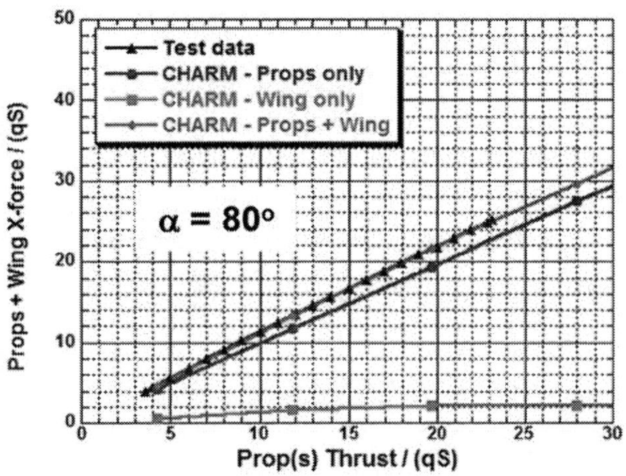

Figure 21: Measured and predicted lift force for the Traybar model: angle of attack 80 deg

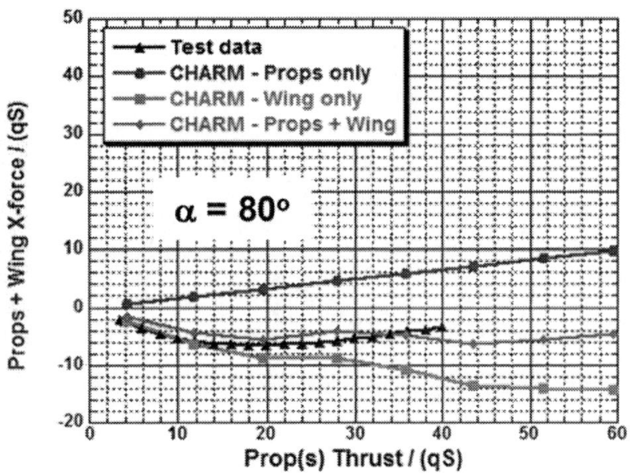

Figure 22: Measured and predicted X-force for the Traybar model: angle of attack 80 deg

At high angles of attack CHARM's fidelity becomes largely dependent on lookup table data for wing drag, though a reasonable additional inference is that propeller downwash helps maintain attached flow at high angles of attack, improving the performance of CHARM. To explore this issue and the overall nature of the flow over the model, the more general CGE analysis was applied to the Traybar experimental configuration to study the effect of propeller wash on attached flow. The surface geometry for CGE (see Figure 23) was developed based on dimensions/drawings provided in [34]. Precise dimensions of the nacelles were not available, and approximations were made for these components and fairing with the wing geometry. Numerical sensitivity studies including mesh resolution were performed as part of the analytical investigation.

For the CGE analysis, this numerical study involved specification of the minimum Cartesian cell size near the surface and rate at which the mesh is coarsened moving away from the surface. The rotors are modeled as actuator disks that were located relative to the surface mesh based on reported dimensions for the experimental configuration. The actuator disk model is implemented by specifying the blade geometry and desired thrust for each rotor. During the CGE solution procedure, the rotor speeds are adjusted to trim the thrust, which is analogous to the experimental procedure used.

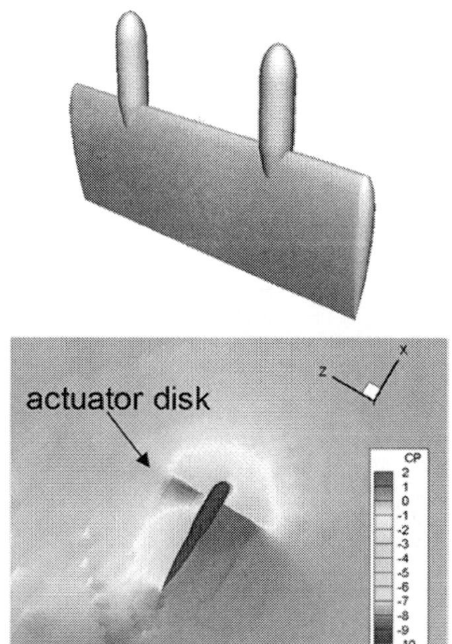

Figure 23: Top - Traybar test configuration surface geometry used for CGE analysis of high angle of attack cases; bottom – pressure coefficient over the combined rotor/fuselage condition, showing the effect of the actuator disk propeller model on surface loading

Computational results were obtained for a range of angle of attack and thrust conditions that span the experimental data in [34], though attention here was focused on the 80 deg angle of attack case. Flow visualization shown in Figure 24 indicates that the addition of propeller wash even at this very high angle of attack can produce attached flow, a circumstance that helps explain the good fidelity observed for the inviscid CHARM model in this case

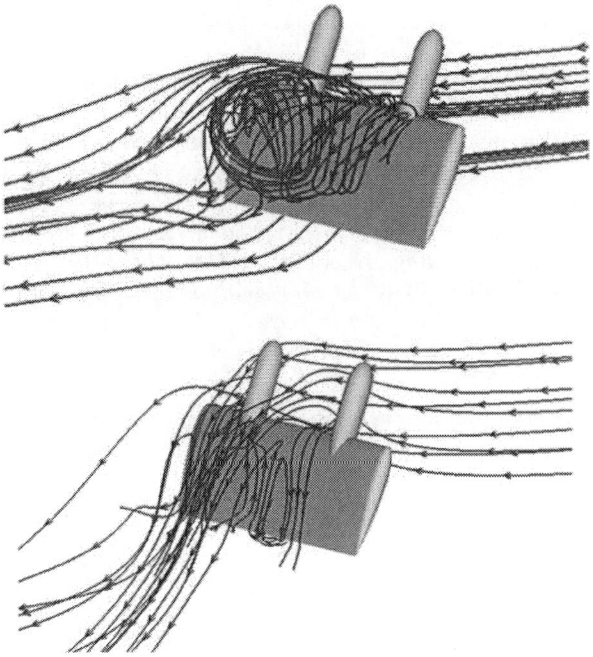

Figure 24: Computed flow streamlines at $\alpha = 80$ deg for the Traybar model using CGE: above – results with no propeller wash; below – results with propeller effects added.

An additional correlation study built on a configuration studied in recent work [35], modeling the wing/ propeller interaction for the NASA X-57 Maxwell (see Figure 1). Figure 25 compares the computed wing/propeller interaction using two modeling options – a kinematically prescribed wake and a dynamically free wake model. (Note: as discussed in [35] CHARM here uses an approximate model definition, in that full details were not available for the propeller planforms or the airfoil characteristics of the prop blades and main wing. Therefore, this illustrative calculation used standard propeller planforms obtained from other work at CDI and simple aerodynamic tables, e.g., the main wing was modeled with NACA 0012 airfoils with an appropriate shift in zero lift angle to account for camber.)

Figure 26 shows CHARM predictions of propeller/wing augmented lift compared with CFD solutions for the X-57 Maxwell from [36]. However, it is seen that even without a precise X-57 prop and wing definition, the CHARM analysis with free wake still matched the fundamental lift augmentation predicted by the CFD codes, both in terms of magnitude and roll-off in slope at high angles of attack.

Of interest here also is that the CHARM analysis significantly over-predicted the lift augmentation when a prescribed wake was used; for the prescribed wake, the high lift prop rotation rate had to be decreased from the 4550 RPM level used in the CFD solutions in order to match the lift augmentation. Also, the prescribed wake did not predict the roll-off in slope at high angles of attack. This result provides useful guidance regarding the need to retain free wake modeling when applying CHARM to recover their lift augmentation behavior of tilting wing aircraft.

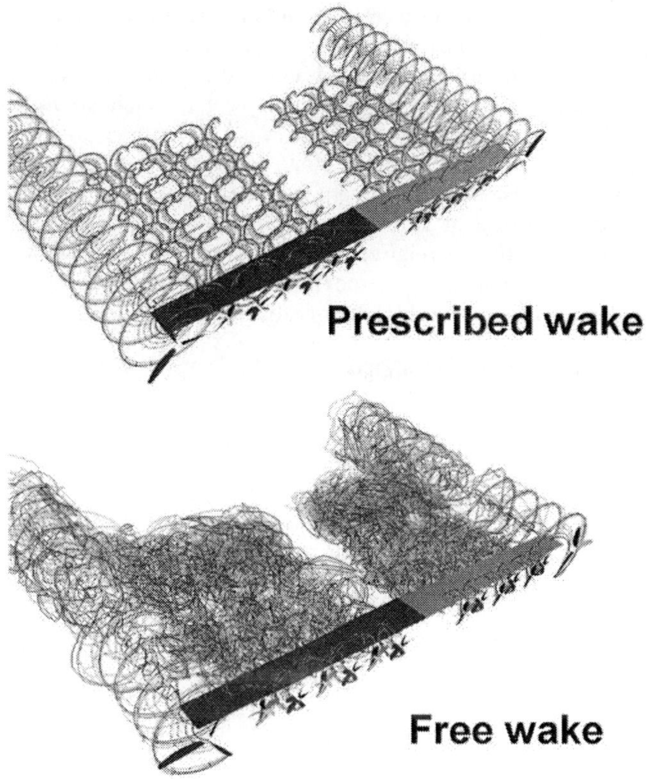

Figure 25: Computed wake geometry for the CHARM model of the X-57 Maxwell propeller+wing combination: above – prescribed wake; below – free wake.

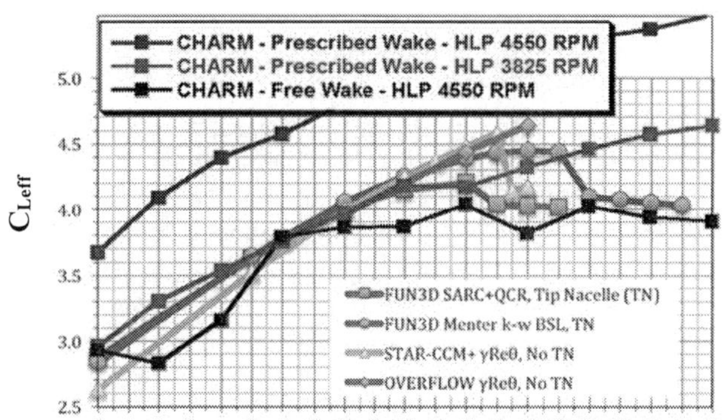

Figure 26: Wing lift coefficient as a function of angle of attack for CHARM with several modeling options, compared to the predictions of several standard high-fidelity CFD analyses (original figure from [36]).

SUMMARY AND PROJECTED WORK

To summarize, the work described here was motivated by the desire to assess the capability of contemporary analysis tools for application to aerodynamic design challenges of interest for tiltwing/tailsitter designs. The particular focus here was on comparison of the CHARM comprehensive model to a representative subset of publicly available data. The principal observations to date are that, in addition to providing good correlation with available isolated rotor data for proprotors typical of such aircraft, CHARM produces encouraging correlation with both lift and horizontal force data on simple but representative wing/rotor configurations, even up to very high wing angles of attack, for operation out of ground effect. These favorable results, combined with the relative ease of setup of the CHARM model make it an attractive point of departure for early stage design studies of this class of aircraft

Targeted CFD studies, however, indicate that the propeller wash produces largely attached flow in the cases studied here, so a more extensive set of comparisons considering cases with less propwash immersion appears to be called for. In addition, several other propeller and propeller/wing interaction data sets are available and are the subject of ongoing correlation studies to more fully define the capabilities of the existing model. Finally, extension of this work to include the potentially complex interactions of propeller and wing flows with the ground plane is highly desirable.

Author contact: Todd Quackenbush, Continuum Dynamics, Inc. (todd@continuum-dynamics.com)

REFERENCES

1. Lindenbaum, B., *"V/STOL Concepts and Developed Aircraft – Volumes I and II"*. 1986, AFWAL TR-86-3071: Universal Energy Systems and Air Force Systems Command.
2. Harris, F.D., *"Introduction to Autogyros, Helicopters, and Other V/STOL Aircraft: Vol III"*. 2012, NASA SP-2012-215959.
3. Borst, H., *"Review of the V/STOL Aircraft with Tilt-Propellers and Tilt-Rotors"*. The Aeronautical Journal, 1968. **72**(693): p. 817-830.
4. Warwick, G., *VTOL Visions*. Aviation Week and Space Technology, 2013. **175**(9).
5. Keen, E.B. and B.S. Tenney, *"A Conceptual Design Study of a Vertical Lift Family of Systems"*, in *American Helicopter Society Aeromechanics Conference*. 2016: San Francisco, CA. p. Jan 2016.
6. Maisel, M.D., D.J. Giulianetti, and D.C. Dugan, *"The History of the XV-15 Tilt Rotor Research Aircraft from Concept to Flight"*. Vol. NASA/SP-2000-4517. 2000.
7. Anderson, S.B., *"Historical Overview of V/STOL Aircraft Technology"*. NASA TM 81280, March 1981.
8. Goodson, K.W., *"Comparison of Wind-Tunnel and Flight Results on a Four Propeller Tilt-Wing Configuration"*. NASA TN-D 3938, May 1967.
9. Moore, M., *NASA Puffin Electric Tailsitter VTOL Concept*, in *10th AIAA Aviation Technology, Integration, and Operations (ATIO) Conference*. 2010, AIAA Reston, VA. p. 13-15.
10. Scott, R.C. *A Design Survey of Unmanned Tactical Rotorcraft for Efficient Hover and Cruise*. in *17th AIAA Aviation Technology, Integration, and Operations Conference*. 2017.
11. Gaffey, T., et al., *"Aeromechanics of the Coaxial Compound Helicopter"*, in *55th AIAA Aerospace Sciences Meeting, AIAA Paper 2015-1411*. 2015.
12. Quackenbush, T.R., et al., *"Analysis of Download and Rotor/Airframe Interactions in Hover"* in *American Helicopter Society 73rd Annual Forum* 2017: Fort Worth, TX.
13. Quackenbush, T.R., et al., *"Analysis Methods for Advanced V/STOL Rotorcraft"*, in *72nd Annual Forum of the American Helicopter Society*. May 2016: West Palm Beach, FL.
14. Wachspress, D.A., Quackenbush, T. R., and Boschitsch, A.H., *"First-Principles, Free-Vortex Wake Model for Helicopters and Tiltrotors"*, in *59th Annual Forum of the American Helicopter Society*. May 2003: Phoenix, AZ.
15. Wachspress, D.A., Quackenbush, T.R., and Boschitsch, A.H., *"Rotorcraft Interactional Aerodynamics with Fast Vortex/Fast Panel Methods"*. Journal of the American Helicopter Society, 2003. **48**(4).
16. Quackenbush, T.R., et al. *"Aeromechanical Analysis Tools for Design and Simulation of VTOL UAV Systems"*. in *60th Annual Forum of the American Helicopter Society* 2004. Baltimore, MD.
17. Boschitsch, A.H., W.J. Usab, and R.J. Epstein, *"Fast Lifting Panel Method"*, in *AIAA 14th Computational Fluid Dynamics Conference*. 1999, AIAA-0999-3376: Norfolk, Virginia.
18. Usab, W.J., and Boschitsch, A. H., *"Next Generation Preliminary Design Technology"*. June 1999, CDI Report 99-06 (Final Report Contract NAS1-99025).
19. Keller, J.D., et al. *"Computational Fluid Dynamics for Flight Simulator Ship Airwake Modeling"*. in *Interservice/Industry Training, Simulation, and Education Conference (I/ITSEC) 2007*. Orlando, FL.
20. Aftosmis, M.J., M.J. Berger, and J.E. Melton, *"Robust and Efficient Cartesian Mesh Generation for Component-Based Geometry"*, in *35th Aerospace Sciences Meeting and Exhbit*. 1997, AIAA-1997-0196: Reno, NV.
21. Aftosmis, M.J., M.J. Berger, and J.J. Alonso, *"Applications of a Cartesian Mesh Boundary-Layer Approach for Complex Configurations"*, in *44th AIAA Aerospace Sciences Meeting*. 2006, AIAA-2006-0652: Reno, NJ.
22. Whitehouse, G.R., T.R. Quackenbush, and A.H. Boschitsch. *"Variable Fidelity Preliminary Design Tools for Advanced Vertical Flight Vehicles"*. in *66th*

Annual Forum of the American Helicopter Society. 2010. Phoenix, AZ.

23. Anon., *FUN3D-Analysis and Design*. 2013, NASA http://fun3d.larc.nasa.gov
24. Yaggy, P. and V. Rogallo, *"A Wind-Tunnel Investigation of Three Propellers Through an Angle of Attack Range from 0 deg to 85 deg"*. 1960, NASA TN-D-318, May 1960. p. 83.
25. McLemore, H.C. and M.D. Cannon, *"Aerodynamic Investigation of a Four Blade Propeller Operating through an Angle of Attack Range from 0 to 180 degrees"*. 1954, NACA TN 3228, May 1954.
26. Acree, C., *"JVX Proprotor Performance Calculations and Comparisons with Hover and Airplane-Mode Test Data"*, in *NASA Technical Memorandum 2009-215380, April 2009*. 2009.
27. Kuhn, R.E. and J.W. Draper, *"Investigation of the Aerodynamic Characteristics of a Model Wing-Propeller Combintation and of the Wing and Propeller Separately at Angles of Attack up to 90 degrees"*. 1956, NACA TN 3304, Jan 1956.
28. Felker, F.F., Signor, D.B., Young, L.A., and Betzina, M.D., *"Performance and Loads Data from a Hover Test of a 0.658-Scale V-22 Rotor and Wing"*. NASA-TM-89419, April 1987.
29. Johnson, W., *"An Assessment of the Capability to Calculation Tilting Prop-Rotor Aircraft Performance, Loads and Stability"*. 1984, NASA-TP-2291,Mar 1984.
30. Wachspress, D.A., Whitehouse, G.R., and Quackenbush, T.R., *"Predicting the Influence of Blade Tip Shape on Hovering Rotor Performance with Comprehensive Analyses"*, in *53rd AIAA Aerospace Sciences Meeting, AIAA Paper 2015-1248*. 2015.
31. Quackenbush, T.R., et al., *"Further Assessment of Analysis Methods for Rotor/Airframe Interaction in Hover"*, in *2018 AIAA Aerospace Sciences Meeting, AIAA Paper 2018-1783*. 2018. p. 1783.
32. Quackenbush, T.R., Whitehouse, G.R., Silbaugh, B.S. and Danilov, P.V., *"Analysis of Rotor Configuration and Ground Proximity Effects on Hover Download"*, in *American Helicopter Society 74th Annual Forum*. 2018.
33. Grunwald, K.J., *"Experimental Study of Wind Tunnel Wall Effects and Wall Corrections for a General-Research V/STOL Tilt Wing Model with Flap"*. 1965, NASA TN-D 2887.
34. Traybar, J.J., *"Analysis and Comparison of Aerodynamic Data Obtained in a Large Wind Tunnel and a Moving-Model Track Facility for a General Research Tilt-Wing VTOL Configuration"*. 1969, USAAVLABS TR 69-74, Sept. 1969.
35. Wachspress, D.A. and J.A. Horn, *"Distributed Electric Propulsion Aircraft Comprehensive Analysis and Design Tool"*. 2018, CDI Report 18-06, Final Report to NASA Contract NNX17CA22P.
36. Deere, K.A., et al. *Comparison of High-Fidelity Computational Tools for Wing Design of a Distributed Electric Propulsion Aircraft*. in *35th AIAA Applied Aerodynamics Conference, AIAA Aviation forum*. 2017. Denver, CO.

NexGen UAS Take-off and Landing Methods and Sizing Studies

Robin Preator
Aerodynamics Engineer
The Boeing Company
Ridley Park, PA, USA

Elia Gianopulos
Configuration Design Engineer
The Boeing Company
Ridley Park, PA, USA

Brian Babie
Aerodynamics Engineer
The Boeing Company
Ridley Park, PA, USA

ABSTRACT

A comprehensive investigation into the costs and benefits of a wide variety of aircraft platforms and take-off and landing methods was performed to address future US Army unmanned aerial vehicle needs as part of the Next Generation Tactical UAS Technology Demonstration Phase 1 Concept Study. Rotary-wing and fixed-wing platforms were evaluated as part of the study using a common set of tools, processes and evaluation criteria. Multiple take-off and landing approaches were also evaluated including short, runway based take-off and landing, assisted launch and recovery and vertical take-off and landing operations. Sensitivities to key mission parameters were explored in terms of their impacts on mission effectiveness and system costs. Reducing the take-off and landing footprint was found to improve operational capability and mission effectiveness but at increased system costs. Vertical take-off and landing platforms were found to have the highest aggregate mission effectiveness and the conventional helicopter was found to offer the greatest value with consideration for system costs for the missions studied and metrics used for the study.

INTRODUCTION

The Boeing Company explored take-off and landing methods and sizing sensitivity studies for the Next Generation Tactical Unmanned Aircraft System (NexGen UAS) Technology Demonstration Phase 1 Concept Study. A key objective of this research was to identify how to maximize persistent system capability of an individual UAS while providing a launch and recovery capability that minimizes manning and logistics footprint to enable expeditionary maneuver. Another key objective of this research was to support requirements definition for the NexGen UAS program, helping to identify "knees in the curve" for relevant performance parameters to help maximize potential capability and value for the fielded system.

Typically, design studies for new unmanned aircraft systems compare potential configurations of a single type or attempt to optimize the capability of a single system concept with a single take-off and landing method. This study compared multiple platform types including both fixed-wing and rotary-wing vehicles with a common set of tools, processes and metrics. Variations in take-off and landing methods were also explored. The resulting designs were compared in terms of their sizing and performance characteristics, operating footprint, transportability, required ground support equipment, mission effectiveness and system costs.

Presented at the Vertical Flight Society Autonomous VTOL Technical Meeting and Electric VTOL Symposium, Mesa, Arizona, UAS, Jan. 29-31, 2019.

The views and conclusions contained herein are those of the authors and should not be interpreted as necessarily representing the official policies or endorsements, either expressed or implied, of the U.S. Government.

To address the question of take-off and landing approaches, tools were developed and analyses were performed to compare multiple UAS take-off and landing approaches for relevant platform types to perform the same mission. The take-off approaches considered included short take-off (STO) on a prepared runway, catapult launch and vertical take-off (VTO). The landing approaches considered included run on landing on a prepared runway, arresting cables, nets and other fixed devices, parachutes and vertical landing. The aircraft platforms considered for this study included the fixed-wing airplane, the conventional helicopter, the tiltrotor and a tandem tailsitter. Baseline aircraft sizing was established for each configuration to common mission requirements to identify the impacts of variations in takeoff and landing approaches. For the second task, sensitivities to variations in key design and performance parameters were explored in terms of their impacts to mission effectiveness and system costs.

For a given set of mission requirements, the fixed-wing airplane configuration will generally provide lighter weight, less complexity, lower cost and better cruise efficiency relative to rotary-wing configurations provided the use of a runway of sufficient length. However, when a short runway is required, the size, cost and complexity of fixed-wing configurations increase. And when runway independence is required, the overall complexity and costs increase for the total aircraft system including any necessary peculiar ground support equipment.

The helicopter offers the smallest weight and power penalty for achieving organic vertical flight capability driven by lower complexity and rotor disk loading relative to other fielded VTOL configurations. These penalties enable full independence from a runway or any off-board equipment to launch or recover the aircraft. However, the rotor borne, or edge-wise flight of the helicopter exhibits far lower cruising efficiency relative to typical fixed-wing conventional takeoff and landing airplanes. The conventional, single main rotor helicopter in this study is considered a surrogate for the potential range of helicopter configurations, including the tandem rotor, coaxial, lateral twin and the synchropter (intermeshing).

Rotorcraft configurations that employ wing borne flight in cruise typically offer significantly improved cruise efficiency relative to rotor-borne cruise, but suffer penalties for wing weight, download, design complexity for multi-mode flight and low speed maneuverability. The tiltrotor and the tailsitter are used in this study to represent all concepts that use rotary wing propulsion for hovering lift and cruise thrust, and a wing (or wings) for cruise lift. The lack of a manned requirement enables the tailsitter configuration which provides reduced weight and complexity relative to the tiltrotor configuration.

This paper provides an overview of the approach taken, including the key tools and processes used, a summary of key findings on take-off and landing methods including assisted launch and recovery approaches and a summary of key findings on sizing sensitivities to mission performance parameters including an investigation of mission effectiveness and system costs.

METHODOLOGY

The goal of this research was to perform a comprehensive study to determine relative differences in mission performance and effectiveness, operational impacts and system costs while evaluating multiple aircraft platform types and take-off and landing methods. A Multidisciplinary Design, Analysis and Optimization (MDAO) environment was established to achieve this goal, linking together aircraft sizing analysis, take-off and landing calculations, operational footprint and transportability models, mission effectiveness assessments and cost modeling.

Aircraft Sizing Models

Existing sizing tools were leveraged for this study including the well vetted, Boeing developed Helicopter Sizing and Performance Computer Program (HESCOMP) and V/STOL Aircraft Sizing and Performance Computer Program (VASCOMP). Taken together, HESCOMP and VASCOMP are capable of evaluating a wide variety of rotary-wing and V/STOL concepts.

The conventional helicopter configuration sizing was performed with HESCOMP, while the tiltrotor and tailsitter configuration sizing was performed with VASCOMP. Fixed wing sizing analysis utilized VASCOMP for aircraft size, weight and mission performance calculations and a take-off and landing time history analysis that was integrated into the MDAO environment using Phoenix Integration MODELCENTER software. A process flow for the fixed-wing platform sizing and assessment models is provided in Figure 1.

The MDAO environment consists of a basic aircraft sizing loop where aircraft size, weight and fuel requirements are established followed by assessments of Measures of Performance (MOP), Measures of Effectiveness (MOE), operational footprint, and system costs. The Fixed-wing MDAO environment includes tools for estimating take-off and landing field length calculations and an assessment of peculiar ground support equipment required for launch and recovery operations, which are not included in the VTOL concept MDAO environment.

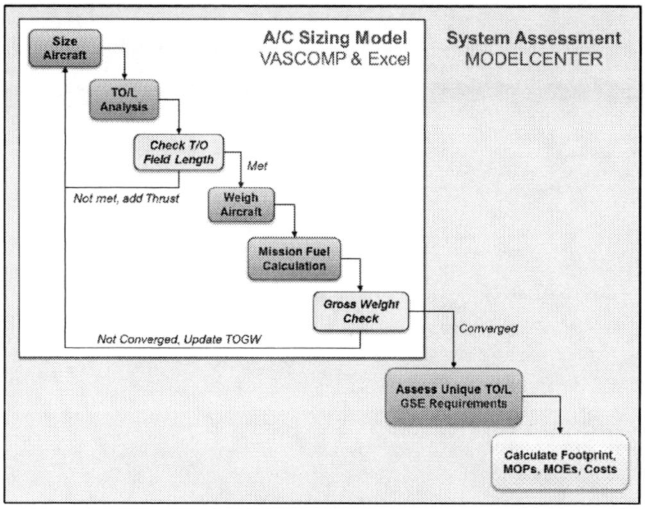

Figure 1. Fixed-Wing MDAO Environment for NexGen UAS Study

All of the aircraft concepts were sized assuming current state of the art technology for parametric weight trends, common fuselage payload volume, drag trends based on historical data for the configuration type and a common rubber engine model with engine weight and fuel burn trends based on historical data ranging from small reciprocating engines to medium turboshaft and turboprop engines. All concepts carried a common structural load factor except for the fixed-wing assisted launch and recovery concept which carried a significantly higher limit load assumption to account for the large loads incurred during take-off and landing events based on research into similar systems.

Take-off & Landing Models

A time history analysis was developed to evaluate traditional fixed-wing take-off and landing and assisted launch and recovery maneuvers depicted in Figure 2. Key inputs for the tool include basic aircraft sizing information and environmental information. Key outputs include required cleared take-off and landing field lengths for a given obstacle height and launch and recovery equipment requirements. The tool includes basic airframe aerodynamic models accounting for high lift devices and a representative propeller map to give a thrust to power relationship. All analysis presented assumed the need to clear an obstacle 50 feet in height.

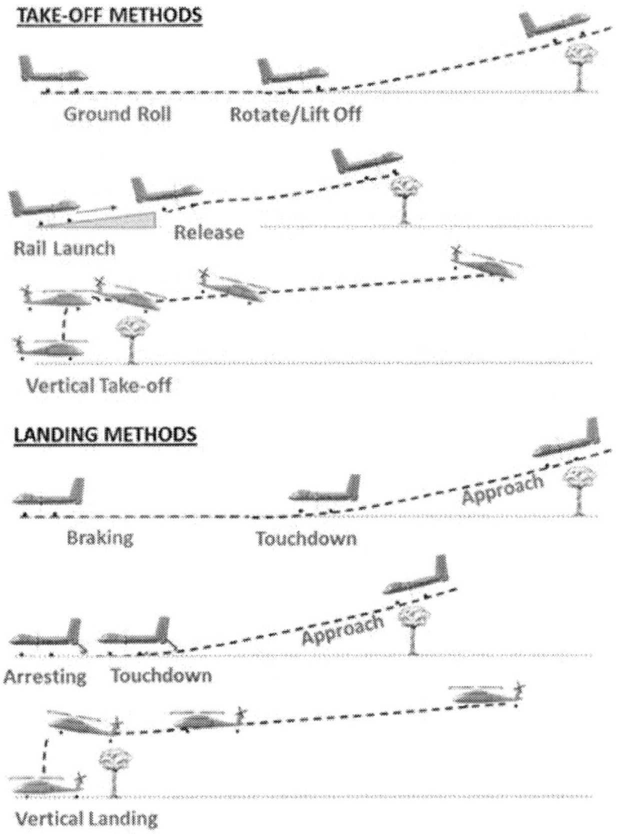

Figure 2. Take-off and landing methods

For assisted launch analysis, the launch event was modeled as a steady acceleration along the length of the catapult launch rail up to the aircraft minimum flight speed, which was assumed to be 1.2 times the stall speed. The length of the catapult was scaled to maintain consistent average acceleration to estimated RQ-7Bv2 Shadow catapult requirements, which were derived based on basic weight and dimensional information available in the public domain. Therefore, an aircraft with a higher stall speed would require a longer catapult. Catapult launch energy is calculated in the model, driven by aircraft mass, minimum flight speed and rail angle. Catapult weight was derived based on calculated launch energy requirements, referencing available catapult weight data. All analysis of maneuvers requiring runways assumed rolling and braking friction consistent with wet concrete. An additional run-off distance of 100 feet was included in fixed-wing landing field lengths. Rejected take-offs and landings including wave-off or engine failure were not modeled. Safety considerations would likely lead to longer cleared field requirements.

VTOL designs assumed a baseline hover out of ground effect (HOGE) capability with power margin and minimal cleared field lengths consistent with a brief vertical climb over an obstacle. An alternate longer field VTOL case was evaluated with sizing criteria based on hover in ground effect (HIGE) capability. The take-off and landing distance for the HIGE case was based on criteria for a US Army and Air Force VFR limited use helipad with same direction take-off and landing based on Reference 1 and tailored to clear a 50 foot obstacle.

Operational Footprint & Transportability Assessment

An operational footprint, ground and air transport model was developed as part of the study to allow a rapid assessment of a given configuration and sizing point, and the model was embedded within the MDAO environment for each platform type. The model (summarized in Figure 3) utilized four primary sets of input parameters: conceptual aircraft dimensions, military transportation vehicle data, the RQ-7 Shadow unit composition, and candidate aircraft minimum take-off and landing distances.

The model yielded the type and quantity of transport trucks required to transport a given sized UAS configuration, the number of C-130s required to move an operational UAS unit, and the total operational footprint required by each candidate UAS configuration.

The model allowed transportability trends for a single platform type as well as comparisons of different platforms sized to equivalent mission requirements. The model also was able to highlight when an aircraft became too large to be transported by conventional military ground transporters or within a C-130 transport aircraft.

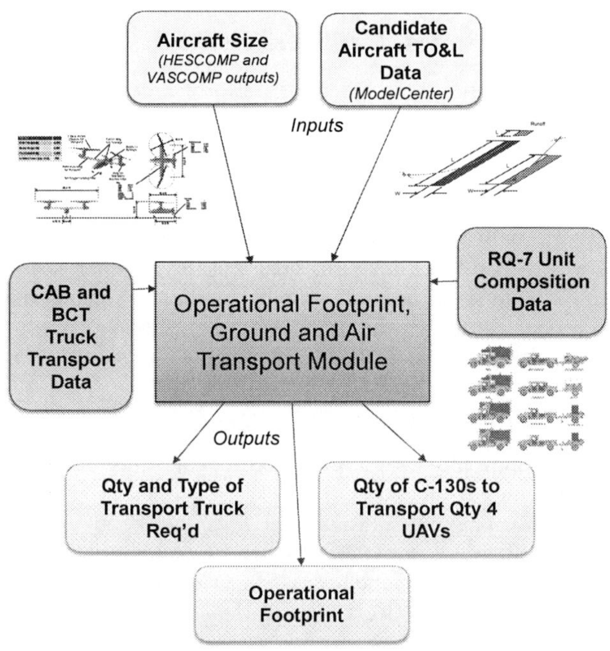

Figure 3. Operational Footprint Ground and Air Transport Model

Mission Effectiveness Modeling

Boeing developed a design effectiveness tool that has been used across multiple programs. The tool contains over 30 key system attributes (KSA) relevant to a NexGen UAS that would either support the US Army Combat Aviation Brigade (CAB) or the Brigade Combat Team (BCT). The tool is the fall-out of an extensive Quality Functional Deployment (QFD) conducted in 2016 using a diverse group of subject matter experts. The resulting tool has multiple attributes across categories including responsiveness, survivability, situational awareness, payloads, system operability, and supportability. For each configuration case, the KSAs were scored based on sizing, performance, take-off analysis and operational footprint tools in the MDAO environment. A total weighted score for the CAB and BCT was then evaluated within the design effectiveness tool to represent an overall measure of mission effectiveness.

To evaluate the responsiveness MOPs a spreadsheet tool representing 24/7 persistence of a NexGen UAS was developed. The model contains a high level examination of how many aircraft would be required to provide a 24 hour coverage period over an area of operation (AO). The parameters which contribute to the model are aircraft mission performance information, the turnaround time between missions and an assumed mission reliability. Given a sized aircraft and a fixed combat radius, the number of aircraft required and total number of sorties are calculated to provide the desired coverage at the AO.

Cost Modeling

A weight and complexity driven parametric approach was used for the cost estimating to support the NexGen UAS study. For each aircraft platform type, a SEER-H cost estimating model was derived based on complexity factors and empty weight information from initial sizing for each subsystem. Additional inputs to the model included labor rates, learning curves, prototype and production quantities, and estimated software development costs.

These estimates were exported to Excel to integrate them into the MDAO study environment. Aircraft and launch and recovery equipment unit price and development costs were then scaled based on the weight outputs from the aircraft sizing and take-off analysis models. Operating costs, in terms of dollars per flight hour were estimated based on relevant historical data using the unit price by system and mission fuel burn rate to derive the total price per flight hour.

TAKE-OFF & LANDING METHODS

Understanding the sensitivity to variations in take-off and landing methods on platform performance capabilities, operational impacts, mission effectiveness and system costs was a key goal of the study. The take-off approaches considered included short take-off (STO) on a prepared runway, catapult launch and vertical take-off (VTO). The landing approaches considered included run on landing on a prepared runway, arresting cables, nets and other fixed devices, parachutes and vertical landing.

Point of Departure Sizing

A Point of Departure (POD) mission (Figure 4) to support the CAB was defined to serve as a baseline capability set for the conceptual design study. The POD mission represents an intelligence, surveilance and reconnaissance (ISR) type mission with significant coverage at the AO. Mission parameter selections were made based on feedback from the NexGen UAS community. The POD payload weight of 600 pounds was determined by examining requirements for a variety of relevant mission roles. Sensitivities were examined above and below each of the key mission parameters.

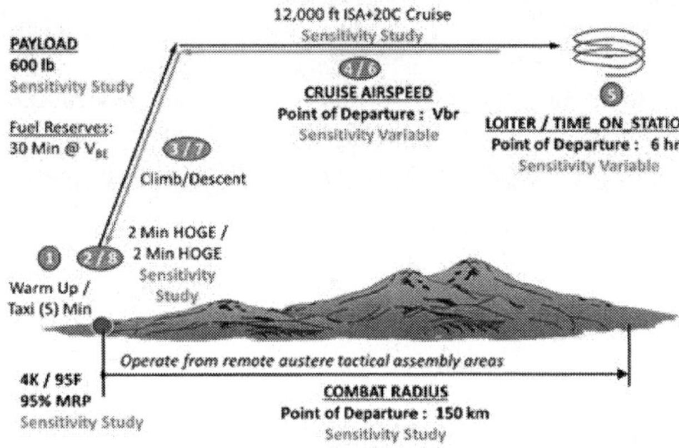

Figure 4. POD mission profile used for NexGen UAS sizing and sensitivity studies

Two fixed-wing and three rotorcraft platforms were evaluated for take-off and landing and sizing sensitivity studies for the NexGen UAS study including a fixed-wing short take-off and landing (STOL) aircraft, a fixed-wing assisted launch and recovery aircraft, a conventional helicopter, a tiltrotor and a tandem tailsitter (Figure 5).

Figure 5. NexGen UAS Aircraft Configurations

The fixed-wing STOL POD design featured a large, high aspect ratio wing to minimize landing distance and two high thrust propellers to minimize take-off distance. The overall cleared field length for the STOL POD design was approximately 1,500 feet driven by landing distance.

The fixed-wing assisted launch and recovery POD design assumed catapult rail launch and landing on a short runway with arresting gear and cables, similar to the RQ-7B Shadow. The assisted launch and recovery aircraft also had a large, high aspect ratio wing to minimize approach and landing distance, but a lower thrust-to-weight ratio as the catapult launcher provides the required acceleration capability for short take-off. The overall cleared field length for the assisted launch and recovery POD design was approximately 1,100 feet, driven by approach and landing distance.

The three rotorcraft POD configurations were generally sized to balance the goals of minimizing gross weight and operational footprint while providing enough installed power for vertical take-off and landing operations at high hot conditions. Powerplant sizing was driven by take-off power requirements with maximum cruise speed and rate of climb as fall-out capabilities. A comparison of the POD sizing results for the five aircraft platforms is provided in Figure 6.

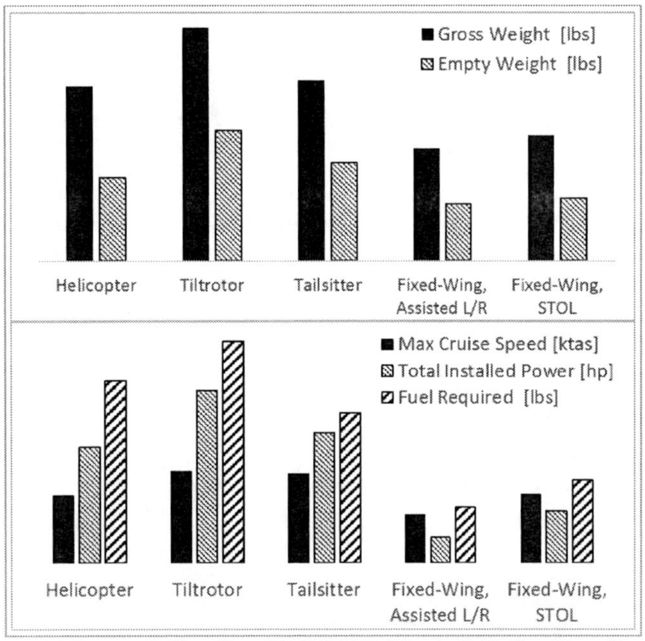

Figure 6. NexGen UAS POD Aircraft Sizing Results

The fixed-wing assisted launch-recovery configuration was found to be the lightest weight aircraft design (not considering ground support equipment) with the least amount of fuel and installed power required to meet the POD mission capability. However, the minimal installed power enabled by the assisted launch resulted in the lowest maximum cruise speed and rate of climb capability. The STOL fixed-wing POD sizing resulted in higher weight, fuel and installed power requirements relative to the assisted launch concept with a noticeable increase in maximum cruise speed and rate of climb capability.

The tiltrotor configuration was found to be the heaviest configuration with the highest fuel and installed power requirements, but exhibited the greatest maximum cruise speed and rate of climb capability as a result of the excess power. The tailsitter configuration, which offers reduced design complexity relative to the tiltrotor with similar cruise efficiency posessed a significantly lighter design than the tiltrotor with similar maximum cruise speed capability. Finally, the conventional helicopter was the lightest of the rotorcraft concepts with maximum cruise capability comparable to the STOL fixed-wing configuration.

Take-off & Landing Comparisons

Fixed-wing STOL sizing trends explored variations in take-off thrust-to-weight ratio and wing loading. Conventional runway take-off and landing techniques were assumed for this analysis. Increases in wing loading (reduction in wing size) resulted in increases in both take-off and landing field length. A wing loading of 15 pounds per square foot was selected for the point of departure configuration to keep required field length below 1,500 feet. Increases to thrust-to-weight ratio improve acceleration capability and decrease take-off field length requirements, but these increases had no benefit on landing field length. A thrust-to-weight ratio of 0.4 was selected for the point of departure configuration to limit take-off field length. The impact to aircraft gross weight to achieve a 1,500 foot runway was a roughly 35% increase relative to that of a minimum gross weight design with a longer runway. This increase was driven by the weight of a larger wing and powerplant and increased fuel burn associated with a heavier, more powerful aircraft. These parameters were held fixed during sizing sensitivity studies. A complex flap system and brakes were required to achieve this short field capability, further adding to the aircraft complexity and cost impacts.

A sample of fixed-wing assisted launch and recovery sizing trends are described in Figure 7. A carpet plot of launch catapult weight with variations in take-off thrust-to-weight ratio and wing loading is provided. Catapult assisted launch and runway recovery with arresting hook and cables was assumed for this analysis. Increases in wing loading (reduction in wing size) resulted in a faster minimum flight speed which increased catapult stored energy requirements, catapult length and catapult weight. Increasing wing loading also resulted in increased approach distances and overall landing field length. A wing loading of 15 pounds per square foot was selected for the POD sizing to limit landing field length and launcher size and weight. Increases to thrust-to-weight ratio were not shown to have a significant impact on field length but did result in increases to aircraft gross weight, therefore a low thrust-to-weight ratio of 0.2 was selected for the point of departure assisted launch and recovery fixed-wing configuration to minimize aircraft weight and cost. These parameters were held fixed during sizing sensitivity studies. In order to limit the approach distances for the assisted launch and recovery aircraft, a complex flap system was also assumed.

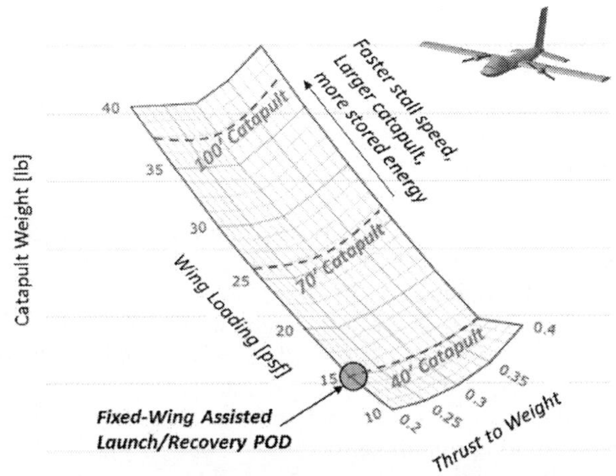

Figure 7. Catapult Sizing Trends based on Aircraft Performance Parameters

Assisted recovery devices including nets, wires and rail type devices were also considered. These types of recovery approaches can eliminate the need for a runway, but they result in additional ground support equipment and can result in significant aircraft loads during recovery that are comparable to crash loads. As aircraft size and gross weight increase, these loads increase and there is likely a practical limit to vehicle size for this type of recovery although it was not determined as part of this study. Another consideration for these types of devices is that the recovery maneuver requires additional precision relative to a conventional runway landing. This provides an additional challenge for an autonomous aircraft and the system would likely have a significant failure rate for the recovery maneuver. For that reason, failure modes need to be considered including rejected approaches. An analysis of cleared field length required for various heights of a fixed recovery device was performed with and without margin for a rejected approach as shown in Figure 8.

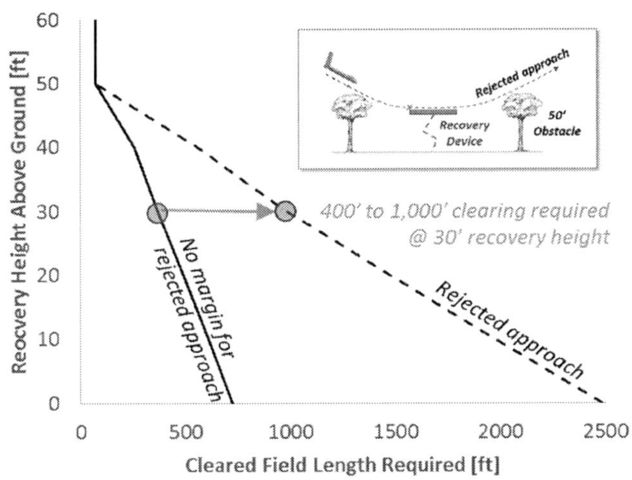

Figure 8. Cleared Field Length vs. Height Above Ground for Recovery Device

It is shown that for a recovery at a 30 foot height, a 400 foot cleared field is required to clear a 50 foot obstacle, driven by approach distance. This increases to a 1,000 foot field length when considering a rejected approach with a climb out starting at the recovery point. Failed recoveries could also result in significant damage to the aircraft and/or the recovery device which would impact operations.

Two types of parachutes were considered as an assisted recovery approach to reduce cleared field length requirements: a recovery parachute and a drogue parachute. A recovery parachute that would deploy in flight was considered not operationally suitable based on the lack of control over landing location and significant weight and structural impacts to the aircraft. A drogue parachute deployed after touchdown to slow the aircraft may be used to reduce landing distances on a runway, but likely do not offer any significant field length benefits when compared with arresting gear approaches.

The impact of cleared field length on system costs is shown in Figure 9. The costs were expressed in terms of platoon unit price. For this analysis an aircraft platoon consists of 4 air vehicles and 2 launchers (if required). It is shown that as field length decreases, the cost of the fixed-wing designs tend to increase until they eventually spike as some practical limit is reached in the 1,000 to 1,500 foot field length range. The VTOL concepts show minimal cleared field lengths and platoon unit prices that are competitive with the shorter field fixed-wing designs with the helicopter providing the lowest cost.

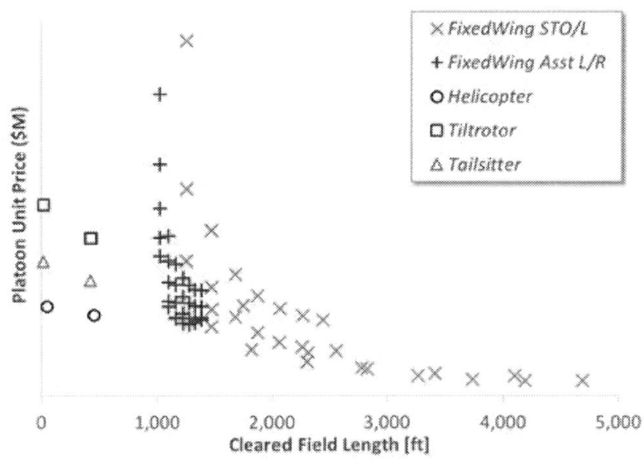

Figure 9. System Costs vs. Cleared Field Length for multiple aircraft platforms

Operational Footprint Comparisons

With the candidate aircraft sizing around the POD mission (Figure 4), the operational footprint model was used to assess the comparative strengths or limitations of the different configurations. The results, shown in Figure 10 and Figure 11 present clear configuration impacts on where and how the aircraft can be employed. The operational footprint for each aircraft platform type included a mission planning and control area, a parking and staging area, a runway of some form, additional runway if assisted launch/recovery, and runoff at the end of the runway for obstacle clearance.

The tailsitter configuration was omitted for clarity but had similar footprint characteristics to the tiltrotor configuration shown.

The area required for each of the five configurations for mission planning and control, i.e. where the maintenance, command and control, and support equipment are placed showed little variation between aircraft platform types. Similarly the aircraft parking and staging areas differed only based on the geometries of the candidate aircraft. The fixed wing STOL aircraft required the most parking space due to its large wing span. The assisted launch aircraft required similar but slightly less space. The VTOL configurations, tiltrotor, tailsitter and helicopter, required the least parking space.

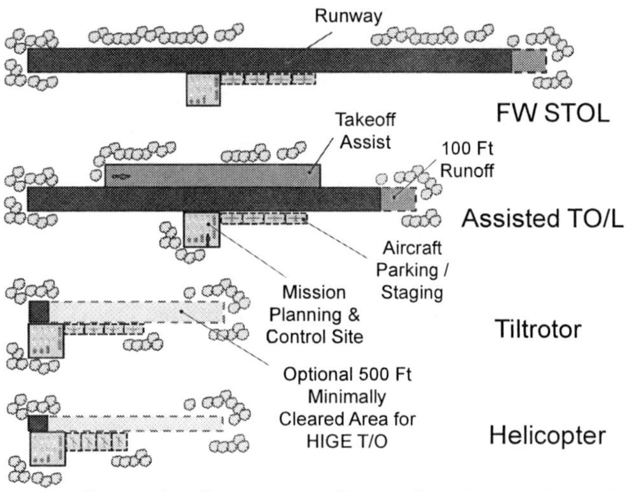

Figure 10. Operational Footprint Comparison of POD Configurations

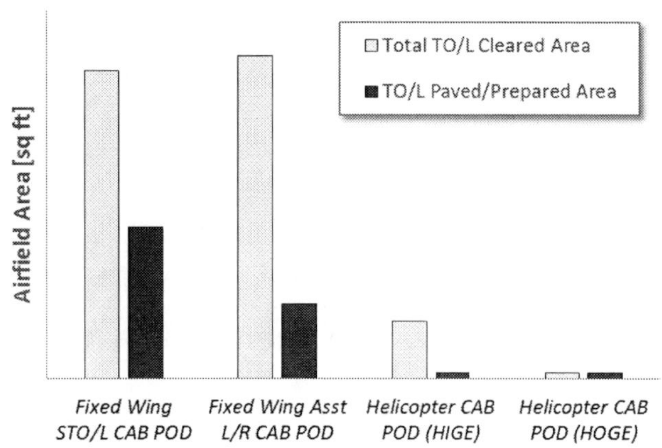

Figure 11. Airfield Area Comparison for Fixed-Wing and VTOL Platforms

The largest difference between the configurations, lay in the runways, and areas required for launch and recovery. Surprisingly, the assisted launch and recovery aircraft required more overall area to operate than the conventional TO/L concepts due to the need for a dedicated launcher site, recovery cable or net mechanisms, and having an ability to launch and recover aircraft in a staggered or interspersed manner. Both of these concepts required significantly more runway space than the VTOL candidates, where the 'runway' was little more than a single parking spot having clearance around the sides and ends of the aircraft and their rotors to allow for VTOL operations.

Figure 11 presents a comparison of the aircraft platform types and the airfield areas required to operate them, illustrating the significant difference in required operational area for the VTOL aircraft as compared to the fixed wing aircraft both in terms of total cleared area and paved or prepared areas. The helicopter data shown is a reasonable surrogate for the tiltrotor and tailsitter platforms, whose operational footprint dimensions are similar, and far less than the fixed wing aircraft.

The key insight from these comparisons is that it is far easier to find an existing area of the size required to operate a VTOL platform than finding an area of sufficient size to set up and operate either an assisted launch and recovery type or a conventional TO/L aircraft. If an operational site needs to be prepared then it will take far less time to prepare an area for the VTOL aircraft than it would for the fixed wing concepts, greatly impacting operational capability and readiness.

SIZING SENSITIVITIES

Sizing sensitivities to variations in key mission performance parameters were evaluated for the helicopter, tiltrotor, tailsitter and fixed-wing configurations described in the previous sections. Parameter variations included payload, weight, mission radius, time on station, cruise speed, cruise altitude and take-off conditions. The POD mission, provided in Figure 4, was used as the basis for these sensitivities. When running sizing sensitivities, basic sizing parameters such as rotor disc loading and wing loading were held fixed throughout the study rather than attempting to optimize the configuration at every point.

Among the key attributes evaluated during the sizing sensitivity studies were mission effectiveness and system costs. An effectiveness versus cost comparison is presented for all sensitivity data to assess the relative value of differing configurations and capabilities.

Sensitivity to Mission Parameters

A sample of the comparison of sizing sensitivities of the five configurations due to variations in mission radius is provided in Figure 12. Parameters presented in Figure 12 include aircraft gross weight and the weighted mission effectiveness score over a range of mission radius from 50 to 450 km. The sensitivity data show that the aircraft weight increases with mission radius for all configurations, which is primarily driven by an increase in fuel weight. Increases in gross weight with increased mission radius for the fixed-wing configurations is shown to have a shallower slope than the gross weight increases of the rotorcraft configurations, driven by superior cruise efficiency characteristics.

The mission effectiveness score is also shown to increase for all configurations as they move into higher scoring bands of mission radius. The fixed wing aircraft exhibited lower relative effectiveness scores than the VTOL solutions, primarily driven by larger operational footprint and site preparation times. The system cost trends for the platforms generally exhibited increases consistent with the weight of the aircraft.

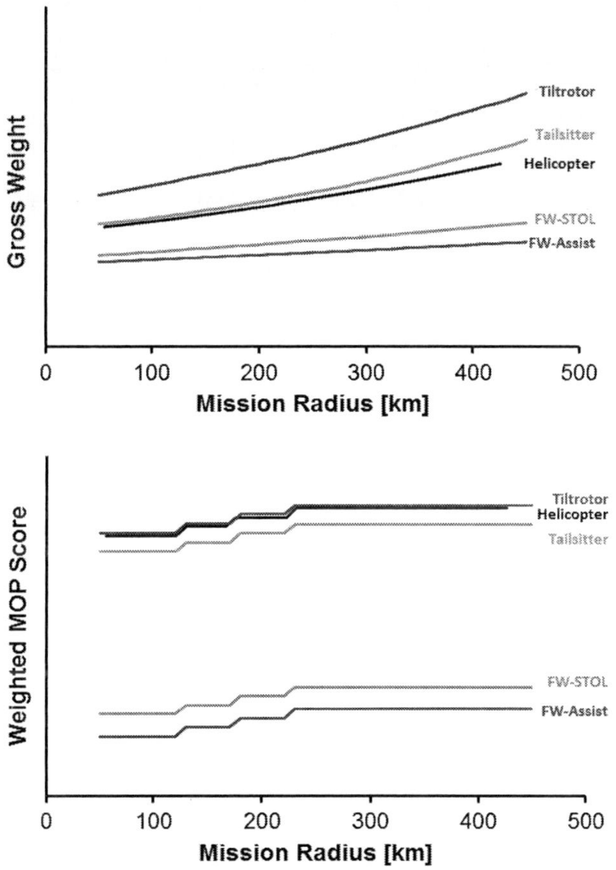

Figure 12. Gross Weight & Mission Effectiveness Sensitivities due to Variation in Mission Radius

Sample sizing sensitivity results due to the variation of maximum cruise speed are presented in Figure 13. Vehicle gross weight and the weighted mission effectiveness score are presented to show the effects of variation in maximum cruise speed. The range of cruise speeds studied varied based on the capability of each configuration. The sensitivity data show that the aircraft weight increases nonlinearly with airspeed for all configurations. Of particular interest is the notable uptick in size for the tailsitter and tiltrotor configurations above 200 knots, where engine power ratings for these configurations become driven by cruise speed rather than vertical take-off and landing requirements. The variation in cost parameters is generally consistent with the gross weight of each vehicle.

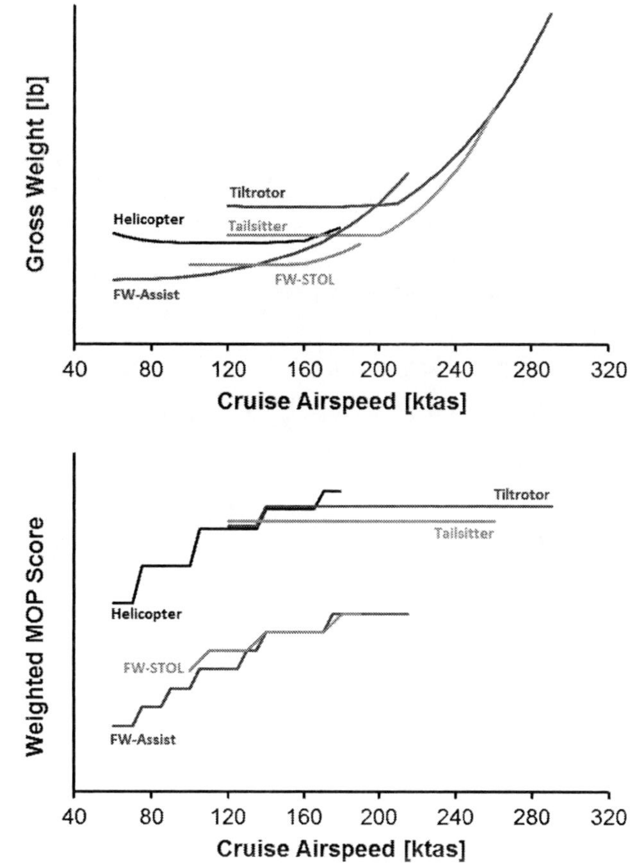

Figure 13. Vehicle Gross Weight & Mission Effectiveness Sensitivities due to Variation in Cruise Airspeed

Results from the sizing sensitivity to payload are shown in Figure 14 for a range of payloads between 50 and 1,050 lbs. Aircraft gross weight and the weighted mission effectiveness have been used to demonstrate the effects of changing payload during the sizing mission. The sensitivity data show that the weight increases nearly linearly with payload for all configurations, with the exception of the tailsitter. It is suspected that the tailsitter configuration would benefit more than other configurations from optimization at the higher gross weights based on the coupled nature of the rotor and wing sizing for transition flight.

At a payload in the neighborhood of 300 to 500 pounds, all of vehicle sizes increase to the point where none are transportable in the trucks considered for this study which results in a decreased mission effectiveness score based on the need to ferry the aircraft to the operational area. The mission effectiveness score is shown to increase for all configurations as they move into higher scoring bands of payload weight and airspeed capability, with decreases when each aircraft is no longer transportable by truck. The variation in cost parameters is generally consistent with the gross weight of each vehicle.

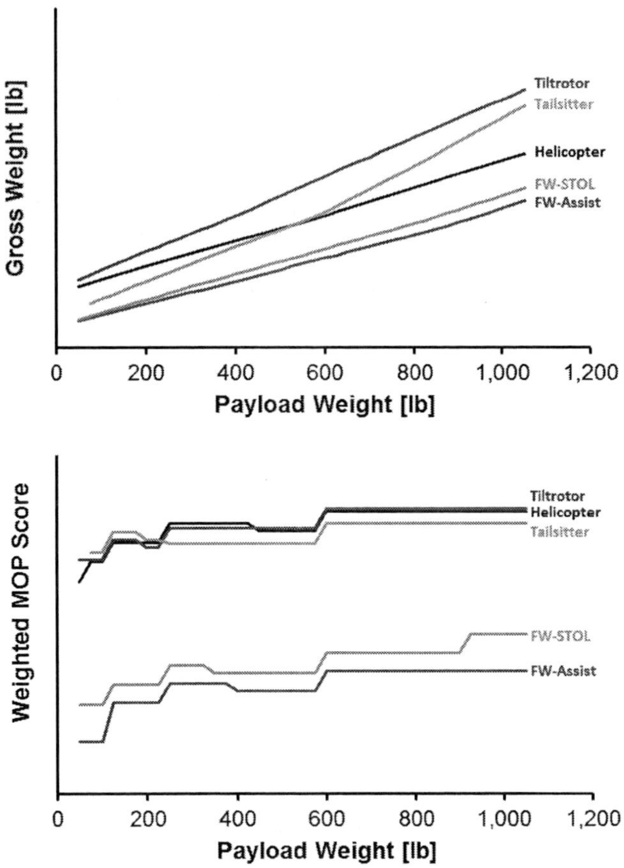

Figure 14. Vehicle Gross Weight & Mission Effectiveness Sensitivities due to Variation in Mission Payload

Effectiveness vs. Cost

A cost-effectiveness comparison for sizing sensitivity trend data is provided in Figure 15, displaying normalized mission effectiveness and normalized platoon cost. The normalized effectiveness is based on the weighted MOP scores of the configurations, normalized by a similar assessment of the RQ-7Bv2 aircraft. Similarly, the normalized platoon cost is based on the platoon unit price (4 air vehicles and 2 launchers if required) normalized by the estimated platoon unit price of the RQ-7Bv2, which was estimated based on size, weight and cost predictions from the sizing and analysis models used for this study.

The scatter plots of all available trend data for the 5 configurations in Figure 15 shows that the helicopter had the most cases to the left of the even value line with the tailsitter configuration showing high value especially for lighter weight, reduced capability cases. The fixed-wing configuration cases showed the lowest normalized mission effectiveness, driven by operational impacts including total operating footprint and responsiveness. Effectiveness is shown to flatten as size increases while costs continue to rise. This effect is a result of a mission effectiveness model that did not place added value on mission capabilities above those thought to be relevant for the NexGen UAS application. These assumptions can be updated in future work depending on the application and value proposition.

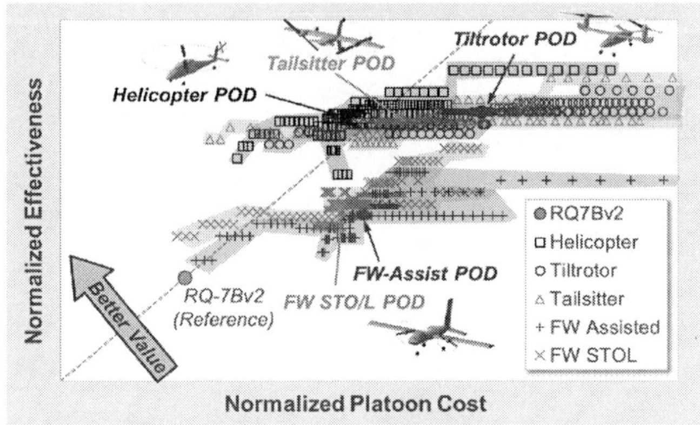

Figure 15. Cost-Effectiveness Comparison for all Sizing Sensitivity Trend Data

CONCLUSIONS

A comprehensive study on take-off and landing methods and sizing sensitivities was performed for a wide array of aircraft platform types for the US Army's NexGen UAS program. The study considered various measures of mission effectiveness and system costs for VTOL, STOL and runway independent fixed-wing platforms. From this research, the following conclusions can be drawn:

1. Rotorcraft configuration sizing resulted in higher gross weight, installed power and fuel requirements than fixed-wing configurations, but exhibited cruise speed and rate of climb capability greater than the assisted take-off fixed-wing configuration while offering greater operational flexibility
2. Analysis of the baseline fixed-wing assisted take-off and landing approach showed that approach and landing with arresting gear drove cleared field length requirements of greater than 1,000 feet
3. Recovery devices such as raised wires, nets or rails can eliminate the need for a runway but greatly increase aircraft structural loads and consideration of failed approach can lead to cleared field length requirements in excess of 1,000 feet
4. System costs increase as cleared take-off and landing field length is decreased for fixed-wing configurations reaching some practical limit where costs are equal to or greater than those of VTOL configurations
5. The VTOL and fixed-wing configurations showed similar sizing sensitivity to mission payload increases, but the VTOL configurations were more sensitive than fixed-wing designs to increases in mission range and endurance driven by lower cruise efficiency
6. The VTOL configurations were shown to have the greatest mission effectiveness, driven by greater responsiveness and operational flexibility with the

smallest cleared field and shortest site setup requirements

7. Cruise speeds beyond those of typical helicopters is costly for all of the aircraft platform concepts studied, including fixed-wing platforms
8. The conventional helicopter configuration offered the best value approach of all of the configurations analyzed when considering mission effectiveness and total system costs

Author contact: Robin Preator robin.m.preator@boeing.com Elia Gianopulos elia.n.gianopulos@boeing.com Brian Babie brian.m.babie@boeing.com

ACKNOWLEDGMENTS

Effort sponsored by the U.S. Government under Other Transaction number W15QKN-16-9-1001 between Vertical Lift Consortium, Inc. and the Government. The US Government is authorized to reproduce and distribute reprints for Governmental purposes notwithstanding any copyright notation thereon.

REFERENCES

[1]U.S. Army Corps of Engineers, Naval Facilities Engineering Command, Air Force Civil Engineer Support Agency, "Unified Facilities Criteria: Airfield and Heliport Planning and Design," UFC 3-260-01, 2008.

Precision Docking for Multi-UAV Cooperative Transportation

Kevin Webb
PhD Student
Mechanical Engineering
Georgia Institute of Technology
Atlanta, Georgia, USA

Jared Elinger
PhD Student
Mechanical Engineering
Georgia Institute of Technology
Atlanta, Georgia, USA

Jonathan Rogers
Associate Professor
Aerospace Engineering
Georgia Institute of Technology
Atlanta, Georgia, USA

ABSTRACT

Collaborative lift missions for unmanned aerial vehicles (UAVs) may support a breadth of practical applications including package delivery, emergency rescue, and military logistics. The development of highly autonomous modular vertical lift aircraft capable of cooperatively transporting large loads may reduce logistics costs and improve safety compared to the use of conventional manned aircraft. This paper outlines the development of a docking mechanism designed to enable autonomous multi-UAV payload pickup and delivery. The mechanism consists of a static ring attached to the payload and a set of onboard actuated legs. The design of the dock leads to unique advantages in terms of reliability, simplicity, and robustness to positioning error. A variety of indoor and outdoor flight tests demonstrate the robustness of the docking mechanism and sets the stage for development of a fully-autonomous multi-vehicle cooperative lift system.

INTRODUCTION

Next generation payload delivery missions will be performed in large part by autonomous aircraft. When considering operation autonomous logistics aircraft, the docking mechanism that allows the aircraft to connect to and release payloads is a critical element. It is important that such a mechanism reduce or eliminate the need for active human oversight and operate successfully and reliably across a wide range of environmental conditions.

A wide variety of research has explored precision autonomous landing of small UAVs, which in an outdoor environment proves to be a difficult task particularly when vehicles are subjected to environmental effects such as wind disturbances and poor visibility. Nevertheless, future missions will require small aircraft to robustly and reliably dock with payloads in the presence of real-world environmental disturbances. As a result, docking devices must possess either active or passive positioning capabilities that mitigate terminal positioning error. Robust active positioning solutions in the form of high-bandwidth controllers have the potential to drastically drive up logistical cost of UAS. For example, several authors have considered use of a robotic arm that locates and connects to the aircraft in flight, placing it in position in the dock (Ref. 1, 2, 3). Such an approach will lead to high costs and complexity. An alternative lies in passive centering approaches, which use geometric constraints to guide the aircraft into the dock through mechanical forces. Such a design alleviates the demand high-bandwidth control and highly precise onboard sensors, while providing sufficient reliability under foreseeable environmental conditions. A variety of work has investigated passive terminal positioning, with some work focusing on more novel geometric passive design (Ref. 4), and other work blending passive and active mechanisms (Ref. 5).

This paper describes a unique and novel docking mechanism concept designed with some specific objectives in mind. First, the docking stations were designed to easily attach to many payload types, expanding their serviceablity across a va-

Presented at the Vertical Flight Society Autonomous VTOL Techincal Meeting and Electric VTOL Symposium, Mesa, Arizona, UAS, Jan. 29-31, 2019.

riety of applications. Furthermore, the mechanical design leverages passive terminal positioning techniques, increasing landing efficacy while relieving demand on the active control system. Finally, the completed system lays the groundwork for collaborative lift, with a central controller on the payload communicating docking commands to each UAV. Following numerous indoor flight tests focused on a single UAV at a time, the system was verified through outdoor multi-UAV flight tests.

DESIGN

Physical Design

Figure 1 shows the passive positioning scheme employed here, wherein an insertion cone (created by actuated legs on the aircraft) and reception ring onboard the payload passively and precisely position the aircraft concentrically with the dock. This design uses similar geometry to that discussed in (Ref. 6), but removes the need for additional grasping mechanisms. In the current work, this design (previously proposed in (Ref. 7)) is expanded upon to include a payload attachment mount, a folding beacon arm, and sufficient cantilever clearance for the aircraft propellers. These elements are considered in the design shown in Figure 2. The 4-bolt standard mounting plate allows for easy attachment to a variety of payloads. Any number of docks may be attached in any arbitrary configuration to a given payload (an autonomous flightworthiness determination algorithm for such configurations was proposed in (Ref. 8)). In this way, the cooperative lift capability is not solely governed by geometric constraints of the payload. The beacon is included below the docking ring to enable terminal tracking into the dock using an image-based visual servoing scheme, to be described later in this paper. The folding beacon arm provides compact storage and transportation while maintaining the necessary offset for image based visual-servoing control. Once the folded conical legs of the UAV are fully inserted into the reception ring, the legs extend outward at 45 degrees from the horizontal plane, locking the vehicle in place. The reception ring is designed to fit just around the leg joints for a firm hold in the docked position.

The current implementation of this actuating leg design was implemented on a standard Parallax ELEV-8 v3 quadcopter (Ref. 9), displayed in Figure 3. This model was chosen for early proof-of-concept demonstrations for its mechanical simplicity as well as its research-grade flight controller, which offers multi-core processing and plug-and-play compatibility with XBee RF modules, simplifying the communication necessary for multiple autonomous landings as well as providing a means of data collection.

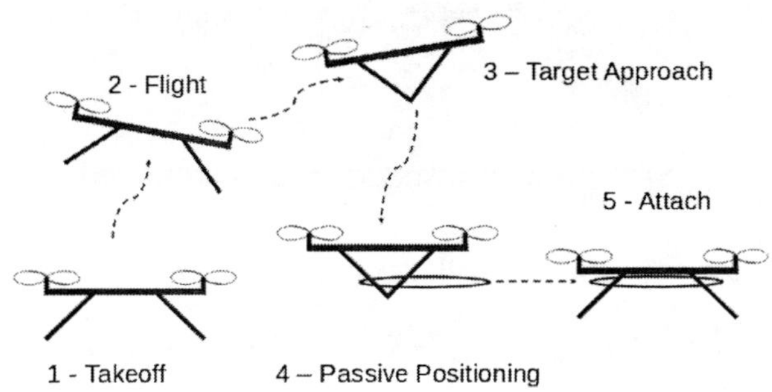

Fig. 1: Passive positioning and attachment mechanism for a single UAV (Ref. 7).

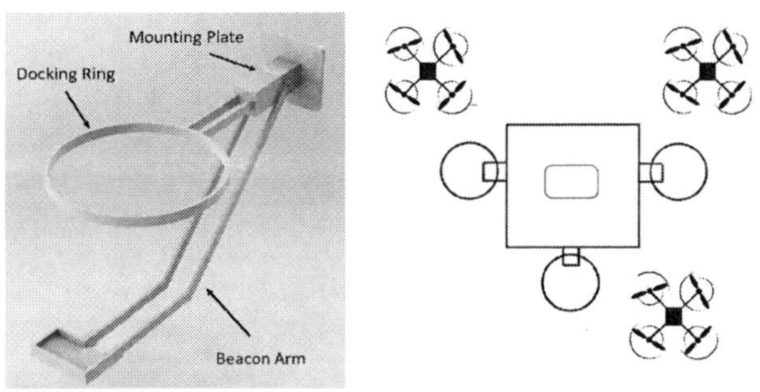

Fig. 2: CAD rendering of modular dock design (left) and top-view illustration of payload setup (right).

Fig. 3: Modified Parallax ELEV-8 UAV (Ref. 7)

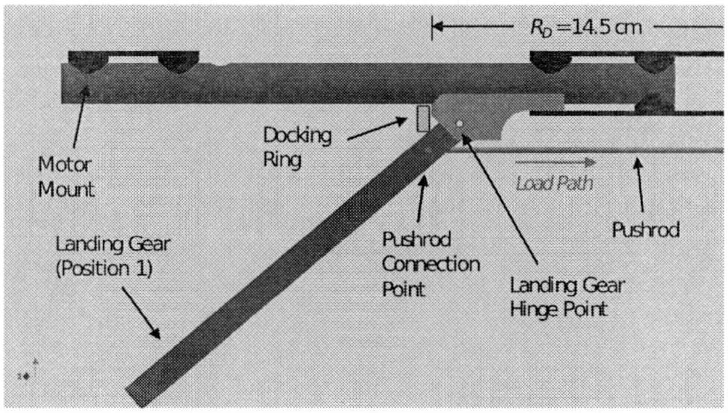

Fig. 4: Side view of a single folding actuated leg.

Among the peripheral devices used onboard the prototype aircraft are a central servo motor for leg actuation and an Arduino-Pixy camera pair mounted on a custom 3D printed gimbal. The camera and gimbal are used to track the beacon during terminal docking. The legs are constructed from aluminum L-bars which fit into 3D printed joints, shown in Figure 4. A custom mount holds the central servo motor, which is connected to the legs through a steel pushrod. In the fully extended position, the loads produced on each of the pushrods run directly through the line of action of the central servo motor. Thus, as vehicles transport the payload, the weight of the load is transferred through the pushrod to the center of servo horn. This eliminates any static rotational load on the servo which would have to be counteracted.

Control Scheme

Terminal control of the UAVs during docking is performed in three hierarchical loops: a PID inner stabilization loop that tracks roll and pitch angle commands, an outer-loop PID controller that computes desired roll and pitch angles from gimbal pointing angles, and an altitude control loop that determines the descent rate based on the error in the pan and tilt of the gimbal used to locate the beacon.

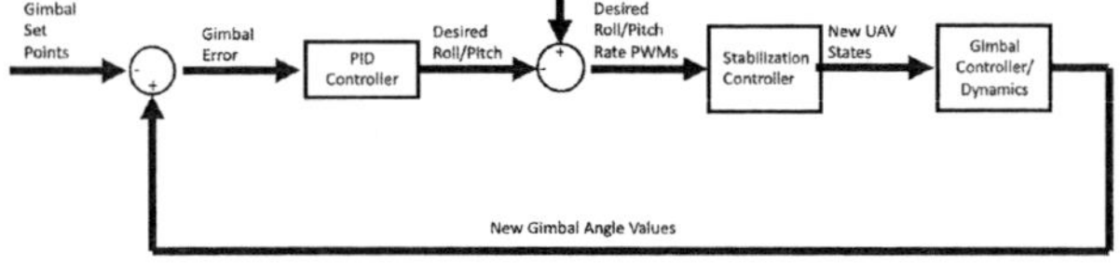

Fig. 5: UAV landing controller block diagram

Horizontal Positioning of the UAV

Roll and pitch commands are tracked by the inner flight stabilization loop. The outer loop control uses an image-based visual servoing scheme based on feedback from a Pixy camera (Ref. 10) and infrared beacon pair. This camera is mounted to the vehicle via a two-axis gimbal. The control scheme proceeds as follows. First, a controller onboard the gimbal mount acts to center the beacon in the camera field of view. The gimbal pan and tilt angles that result are used as input to the outer-loop controller, which, through a PID controller, drives the desired pitch and roll angles to be equal in magnitude and opposite in direction to the gimbal tilt and pan angles, respectively. This concept is shown in Fig. 6. A standard inner loop PID controller (Ref. 11) tracks these desired pitch and roll angles, driving the aircraft toward the beacon. Note that this control formulation assumes that the gimbal dynamics are much faster than the aircraft dynamics (which is indeed the case), and that the gimbal pitch and roll angles are small. Gimbal angles of zero mean the UAV is directly over the beacon, which is desired as the vehicle descends vertically into the dock. Note that in reality, a small calibration is needed as the gimbal is not exactly centered on the UAV and the beacon is not centered below the ring. Thus, a small offset is needed with the error angle of the beacon not desired as exactly zero, but with some small value found through calibration.

The infrared beacon system used here is manufactured by IR-Lock (Ref. 12). The infrared frequency emitted from the IR-Lock beacon corresponds to a dip in the solar spectrum, meaning that its signature can be identified reliably in many outdoor lighting conditions, including bright and sunny days. These beacons have a maximum detection range of 20 meters and that range drops to 14 meters when viewed from an angle of 30 degrees. Note that the Pixy camera is placed on the two-axis gimbal in order to allow the aircraft to track the beacon at higher positional offsets (it was found that a fixed camera lost sight of the beacon regularly during docking trials).

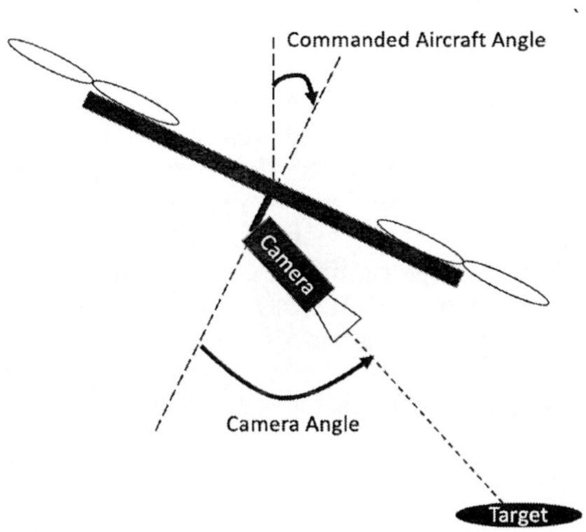

Fig. 6: Camera angle diagram for roll/pitch control (Ref. 7)

Vertical Decent Control

While the horizontal positioning scheme keeps the UAV centered over the beacon, the camera angle relative to the body frame is used to determine the desired descent rate. The decent rate is given in Eq. (1) where b is the maximum descent rate, b/a is the largest acceptable angle error which defines the edge slope of the accuracy cone, E is the maximum of the gimbal pan and tilt angles as defined in Eq. (2) and dt is the time step between control updates. In Eq. (2), ϕ_g and ψ_g are the roll and pitch angle of the gimbal respectively. Note that the b and a parameters in Eq. (1) are tuning parameters.

$$dz = (-aE + b) * dt \tag{1}$$

$$E = max(|\phi_g|, |\psi_g|) \tag{2}$$

In general, as this error grows, the commanded descent rate decreases. If the maximum error exceeds a threshold value, the vehicle is commanded to maintain altitude until the horizontal positioning error is decreased. Figure 7 shows a visualization of the descent rates using this control scheme in the vertical space above the dock, where the color denotes the descent rate if the vehicle is positioned in that location (red for slow, green for fast). For most flight tests performed here, the center of this cone (zero error) was programmed to produce a descent rate of 30 cm/s. This descent rate controller is an adaptation of the vision-based landing scheme discussed in (Ref. 13), (Ref. 14).

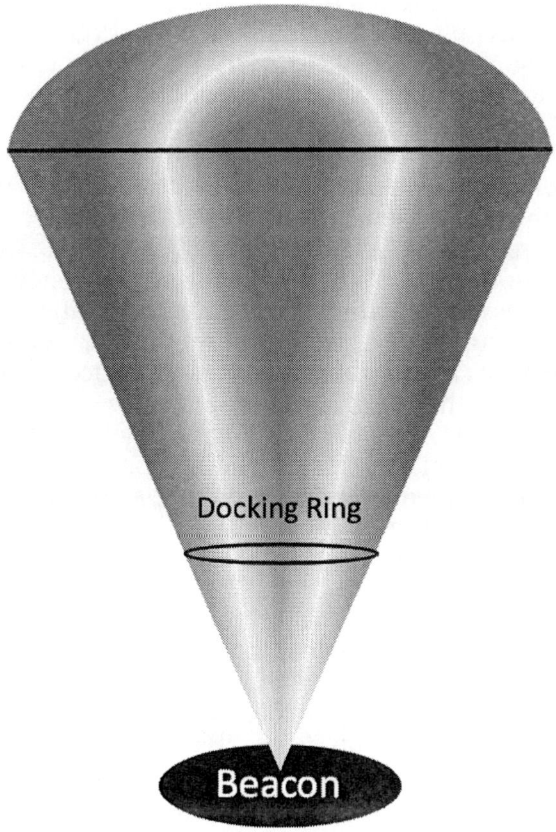

Fig. 7: Descent rate based on accuracy cone (Ref. 7)

Note that the resolution and noise characteristics of the onboard altitude sensor play a significant role in determining whether the vehicle tracks the desired descent rate, and whether successful docking is achieved. In the flights described in this work, a barometric altimeter sensor was used providing somewhat low-resolution and noisy feedback. This enforced a limitation on performance and caused occasional descent outside of the designated descent cone.

MULTI-AGENT BEHAVIOR

The previously discussed methods are sufficient for the control of a single docking UAV. However, when multiple vehicles are commanded to dock in quick succession to a shared payload, additional factors must be accounted for. For example, sufficient space is required between UAVs in the flight space above the payload in order to avoid collisions. Additionally, downwash from UAVs in queue to dock can affect the performance of the vehicle currently docking. In order to mitigate problems inherent in multi-vehicle operation, in the flights reported here each vehicle is held at a unique angular offset over its beacon prior to docking. This provides ample space over the payload while each UAV is waiting to dock. When the docking sequence is initiated, the vehicles are commanded to land one at a time. The selected UAV reduces its angular offset to zero while beginning descent. When the onboard limit switch is depressed signalling a successful dock, the corresponding beacon is turned off by the central controller, and the next UAV is signalled. Additionally, this limit switch commences the outward actuation of the legs, immediately locking the UAV in place. These landing commands are generated by a master device on the payload which wirelessly transmits a command to each UAV to begin docking.

For the purposes of this demonstration, the master device on the payload consists of two major components. First, an additional Parallax ELEV-8 flight controller is used to easily communicate with the UAVs via an XBee-PRO 900HP module. Second, a simple MOSFET transistor circuit controls the sequencing of each IR-Lock beacon.

RESULTS

Autonomous landing of three UAVs was successfully performed outdoors with moderate wind conditions and clear visibility. The primary objective of these tests was to demonstrate consecutive docking of three UAVs beginning from queued hover. In order to illustrate the system's versatility, the three docks were mounted to a standard cardboard box lined with pegboard panels. These demonstrations were recorded on video, and flight data was captured for numerous indoor landings, validating operation of the control algorithm. Still frames from an outdoor landing sequence of three UAVs on the box payload are displayed in Figure 8. The first frame shows the descent of the first UAV with the Pixy fixed on its corresponding infrared beacon. During this phase, all three infrared beacons are powered simultaneously. The next two frames show the same for the second and third UAVs, respectively, with each beacon being powered off following the docking of its corresponding UAV. The final frame shows all three UAVs after docking with full leg extension.

Figures 9-12 show time histories of the vehicle and controller states for a single example indoor docking maneuver. The commanded descent rate, shown in Figure 9, is inversely proportional to the absolute camera angle error within an acceptable margin, demonstrating desired performance of the aforementioned vertical descent rate control scheme where higher descent rates coincide with small camera angles. As shown in Figure 10, the roll angle of the Pixy camera and the roll angle of the vehicle are roughly inversely correlated. This indicates proper operation of the visual servoing control scheme which seeks to drive to zero the gimbal pan and tilt angles when the camera is pointed toward the beacon.

Commanded vehicle roll and pitch are shown in Figures 11 and 12, respectively, and exhibit similar inverse characteristics with respect to the gimbal angles. The finite differencing scheme used to compute the derivative error leads to noise in the commanded angles, as seen in Figures 11-12. However, inclusion of a derivative term in the control law is necessary to improve damping in the feedback system. A low-order low pass filter may be used to mitigate the effects of this noise, although no filter was used in the trials conducted here and adequate docking performance was still achieved.

Fig. 8: Outdoor docking demonstration.

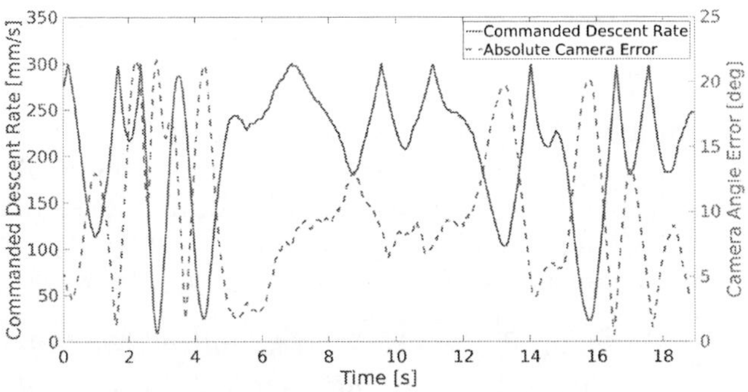

Fig. 9: Commanded descent rate and maximum absolute camera angle error.

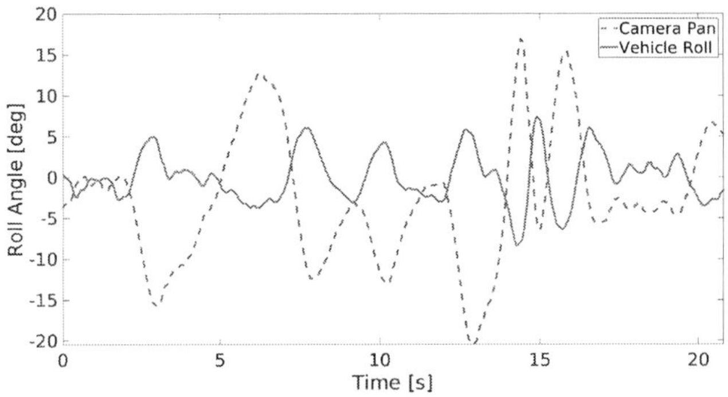

Fig. 10: Camera pan and vehicle roll during landing sequence.

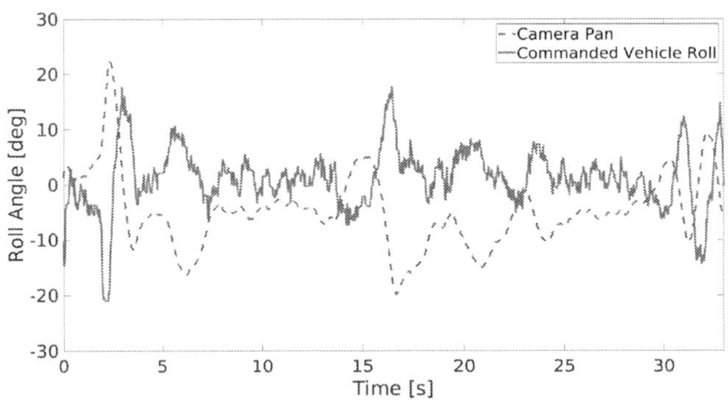

Fig. 11: Commanded roll corresponding to camera pan error.

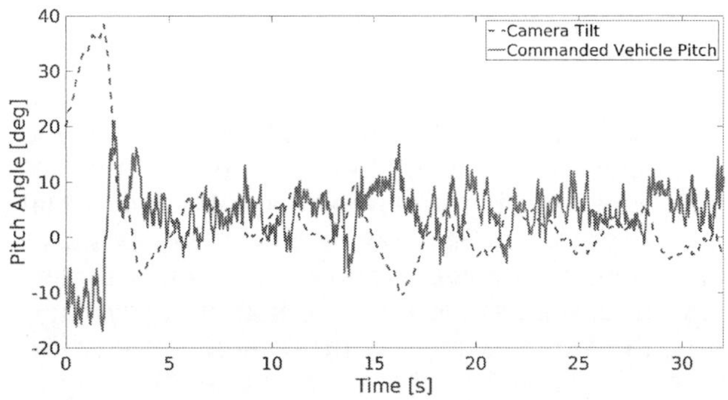

Fig. 12: Commanded pitch corresponding to camera tilt error.

FUTURE WORK

Since the demonstration outlined in this paper, substantial modifications have been made to the UAV's control system. These modifications primarily consist of sensor improvements meant to improve tracking performance. For example, LIDAR has replaced the altitude readings from the onboard barometer, producing more reliable descent rates and altitude hold during hover. Previously, noise in the barometer readings had caused difficulties in holding a precise altitude in hover and caused a failure mode wherein the UAV dipped below the docking ring outside of the designated descent cone. In addition, differential GPS has been installed, providing absolute positioning which will allow each vehicle to locate payloads autonomously and follow waypoints with high precision. With this capability, autonomy can be extended to the point of takeoff such that each vehicle finds the payload and enters its corresponding queueing position without a human in the loop.

Concurrent efforts are directed toward demonstrating collaborative lift of a known payload. For this stage, one central flight controller attached directly to the payload will send throttle and yaw commands to each individual UAV. Since each vehicle will be rigidly docked to the package, it will be sufficient to treat each vehicle as a single thruster device. In this formulation, yaw control is simplified because in such a scheme, it is possible for each thruster device (i.e., vehicle) to generate any thrust within its minimum and maximum limits with zero associated yaw moment. While each of the prototype aircraft discussed here have a lift capacity of roughly 5 pounds depending on the amount of installed hardware, future missions will employ the same control techniques with larger aircraft capable of about 60 pounds of lift.

The final vision for this modular UAS is full autonomy from takeoff to payload delivery. While the focus of this work is on demonstration of the visual servoing control scheme for individual aircraft docking, development of a collaborative multi-vehicle control algorithm is currently underway. In collaborative flight with an attached payload, the vehicles must jointly stabilize the complete vehicle with likely uncertainty in the weight, mass center, and moment of inertia characteristics of the payload. Robust control laws are being developed for this cooperative flight control problem. Both indoor and outdoor flight experiments are planned to evaluate performance in real-world winds and in the presence of varying levels of uncertainty in payload inertial parameters.

ACKNOWLEDGMENTS

This material is based upon work supported by the National Science Foundation Graduate Research Fellowship Program under Grant No. DGE-1650044. Any opinions, findings, and conclusions or recommendations expressed in this material are those of the authors and do not necessarily reflect the views of the National Science Foundation.

REFERENCES

1 Arlton, P., and Arlton, D., "UAV Launch and Recovery System." U.S. Patent 2009/0314883, Dec. 24, 2009.

2 Lovell, G., Hui, E. C.K., and Umbreit, M., "Stabilized UAV Recovery System." U.S. Patent 2009/0294584A1, May 8, 2012.

3 Baarsma, J. H., *Docking a UAV using a Robotic Arm and Computer Vision* M.S. Thesis, Dept. of Robotics and Mechatronics, Univ. of Twente, Enschede, The Netherlands, 2015.

4 Byun, Y., Song, J., Song, W., and Kang, B., "Conceptual Study of a Smart Docking System for VTOL-UAV", *ASCEJ. Aerosp. Eng.*, Vol. 29, (2), Sep. 2015, pp. 1-9.

5 Mullens, K., Burmeister, A., Wills, M., Neson, T., and Deneweiler, T., "Development of a UGV-Mounted Automated Refueling System for VTOL UAVS", Space and Naval Warfare Systems Center, San Diego, CA, Jan 2006.

6 Stefansson, T., *A UAV Mechanism for Autonomous Landing and Transportation of Cargo*. PhD Thesis, Royal Institute of Technology, 2014.

7 Papayanopoulos, J., *Autonomous UAV Precision Item Pickup.* M.S. Thesis, Georgia Institute of Technology, 2017.

8 Warner, J., and Rogers, J., "Autonomous Flightworthiness Determination for Modular Vertical Lift Vehicles" *Journal of Aircraft*, Vol. 55 (5), Mar. 2018, pp. 1-15.

9 *Parallax Incorporated*, ELEV-8 v3 Quadcopter, Rocklin, CA, 2018. [Online] Available: https://www.parallax.com/product/elev-8.

10 Pixy-Cam Documentation [Online] Available: http://cmucam.org/projects/cmucam5/documents.

11 Beard, R., and McLain, T., *Small Unmanned Aircraft*, Princeton, NJ, USA: Princeton University Press, 2012, Chapter 6.

12 *IR Lock LLC*, MarkOne Beacon, Atlanta, GA, 2018. [Online] Available: https://irlock.com/collections/markone.

13 Jeffrey, J., Yanev, A., Biaz, S., and Murray, C., "UAV Quadcopter Landing on a Displaced Target", Tech. Rep., Auburn University, Auburn, AL, Technical Report # CSSE14-05.

14 Al-Sharman, M. K., *Autonomous Takeoff and Precision Landing using Integrated GPS/INS/Optical Flow Solution.* M.S. Tesis, Sharjah, United Arab Emirates: American University of Sharjah, 2015.

New Architectures for VTOL All-Electric Crafts Using Enhanced Specific Power Electric Motor

Oved S.F. Zucker
President
Polarix Corporation
San Diego, CA, U.S.A.

Christopher Liu
University of California, San Diego
San Diego, CA, U.S.A.

ABSTRACT

The characteristically higher thrust required by VTOL crafts over winged crafts is usually addressed by increasing the fan's diameter. For applications where the VTOL duration is short relative to cruising duration, a smaller diameter fan can be used, albeit at much higher power and much lower efficiency to provide the VTOL thrust if it comes without weight penalty. For this a higher specific power source is needed.

Polarix has conceived and developed a key enabler for the design of a new class of VTOL manned and unmanned UAV class crafts. It is an electric motor with specific power exceeding 50kW/kg. It uses turn-less electro-mechanical elements each with their own dedicated inverter, which results in an extremely high specific power (kW/kg). Active centering (dynamic bearings) is inherent to the design. The highly redundant motor architecture provides enhanced reliability, which makes it almost foolproof to winding or inverter failure. The combination of such a motor with a high specific power Li-ion batteries with specific power exceeding 15kw/kg allows us to provide the VTOL thrust with a small weight penalty.

INTRODUCTION

The thrust required for vertical takeoff is 5 to 10 times the thrust requires for cruising because of the absence of aerodynamic lift during this part of the flight. That larger thrust is normally delivered quite efficiently by using an outsized propeller or fan such as in the helicopter. In turn, the much larger diameter distorts the aerodynamic shape of the craft resulting in poor aerodynamics so necessary for efficient cruising.

This larger VTOL thrust can be delivered with a small propeller or fan but with much poorer efficiency due to the much higher power requirement. The poorer efficiency is irrelevant for the takeoff time duration, which is quite small relative to the cruising time duration. However, the much larger power is problematic since it entails a much bigger power plant. The power plant size depends on both the total power and its specific power. With a major increase in the power plant specific power, the larger power can be delivered with a smaller weight penalty to a smaller propeller or fan which will allow us to maintain the aerodynamic sleekness with its attendant higher L/D ratio geometry and thus cruise efficiency and still provide the VTOL function.

DUCTED FAN

In this paper we describe a ducted fan propulsion system driven by a rim motor in a 'tail sitter' configuration of a high L/D craft (Figure 1).

Figure 1. The VTOL craft in its take-off position. The full take-off thrust is produced with a modest fan size which does not conflict with the aerodynamics needed for efficient cruising.

Presented at the Vertical Flight Society Autonomous VTOL Technical Meeting and Electric VTOL Symposium, Mesa, Arizona, UAS, Jan. 29-31, 2019.

The layout of the ducted fan and rim motor is shown in Figure 2. The fan blades and rim motor are part of the same assembly. In the zoomed view on the right side, the conductors and magnets of the motor are shown laid around the circumference of the fan. This will be described in greater details later in the paper.

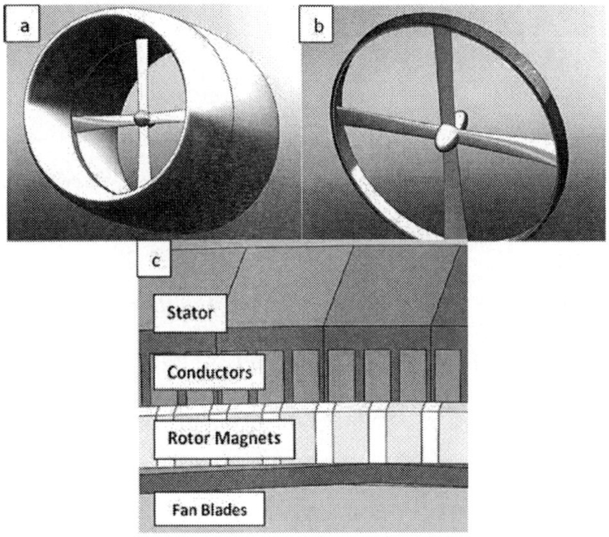

Figure 2: (a) The Ducted Fan Rim motor with the duct assembly on the left, (b) the fan's rim with its embedded permanent magnets on the right and (c) the enlarged rotor and stator with the individual motor elements distributed around the periphery on the bottom.

The key craft requirements are given in Table 1 describing a nominal 600kg all-electric craft capable of VTOL function, loitering at 100m with velocity of 30m/s and L/D of 6.

Table 1. VTOL Key Parameters.

MTO W (kg)	Craft (dry) Mass (kg)	L/D	Loiter Height (m)	Velocity (m/s)	VTOL Accel (g)	T/O Thrust (kN)
600	222	6	100	30	0.1	6.47

The fan and duct calculations follow Piolenc, Ducted Fan Analysis (2001) [1] and uses the standard atmosphere model for pressure, density and Mach velocities at varying heights. Figure 3 shows the required fan area as a function of power to provide the takeoff thrust (Table 1). Our fan analysis procedure is as follows: with the VTOL thrust already determined as 6.5kN, we choose a takeoff power and fan area from. Higher powers will result in a smaller fan area and are well within the capabilities of our motor. With the fan area thus established, we solve for the cruise power. With the cruise power, takeoff power, craft parameters, and present and near future battery technologies, we estimate the total mission duration. The following section will walk through each of the calculations, values, and analysis of the UAV. The equations used are:

$$T_{TO} = \text{MTOW} * g * (1 + a_{VTOL})$$

$$A = \frac{T_{TO}^3}{4\rho P_{TO}^2}$$

Figure 3 shows the relation between fan area and takeoff power. As seen by the curve, the larger the power, the smaller the fan. With our motor technology, we have the potential to push the limits for power, which yields us small fan area(s). Here the resulting total fan area is 0.85m^2 (about 1m diameter). Also the tip velocity of the fan is kept below Mach 0.5, which corresponds to 2700 RPM.

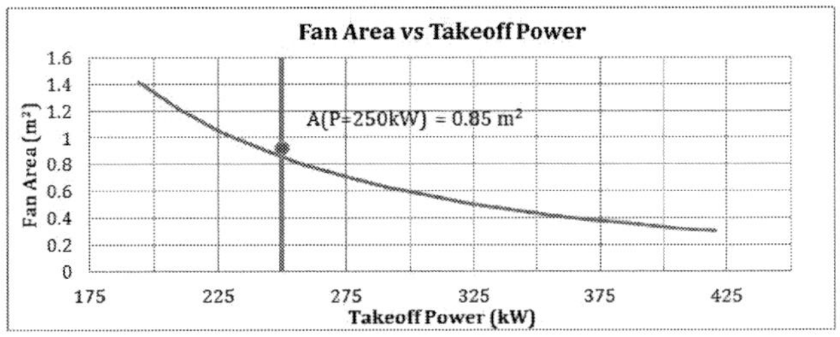

Figure 3: Fan Area vs Takeoff Power.

With the fan area established we solve for the cruise power using the following equations:

$$T_{cruise} = MTOW * g * \frac{1}{L/D_{ratio}}$$

$$P_0 = T_{cruise} * v_{cruise}$$

$$\eta_{cruise} = \frac{1}{\frac{T_{cruise}}{2A\rho v_{cruise}^2} + 1}$$

$$P_{cruise} = \frac{P_0}{\eta_{cruise}}$$

$$K_{swirl} = \frac{\tau}{\dot{m}}$$

$$P_{swirl} = \dot{m} * K_{swirl}^2$$

The resulting loitering parameters at 100m elevation and velocity of 30m/s yield a loitering/cruising power of 44kW not including swirl. The swirl calculations, see equations above, used the vector tip speed to be less than the Critical

Mach number of the airfoil to be 0.61, which corresponds to a tip velocity of 200m/s or 3600 RPM. The calculated swirl power added an additional 1% to the cruise/loiter power.

PROPELLERS

Polarix's motor concept allows for multiple propellers in the same duct but whose calculations are not included here.

MISSION REQUIREMENT AND RESULTING POWER PLANT POINT DESIGN

Here we have optimized the fan design to maximize mission duration time; however, a similar analysis can easily be made to maximize mission range or velocity or some other combination of the two. Using a 20% hub area, this analysis has shown that for a ducted area of $0.85m^2$, we need 250kW for takeoff power and 44kW for cruising/loitering power. This provides us with the motor requirement discussed in the following section.

POLARIX'S MOTOR SYSTEM

The specific power physics limits

Figure 4 shows an idealized fundamental electromechanical device applicable to all motors. In it, a horizontal gap separates the stationary and moving parts above which are two permanent magnet poles and below are the nested three 3-phase windings embedded in soft iron carrying 3-phase shifted periodic currents. The currents distort the magnets field in the gap to create a shear force that causes relative motion of velocity v. The magnetic field path from stator to rotor makes it intuitively obvious that the vertical dimension which accounts for all the motor components (iron copper and magnets) is comparable to the horizontal dimension b - the width of a single pole pair in the direction of motion. This holds true for all high power density configurations.

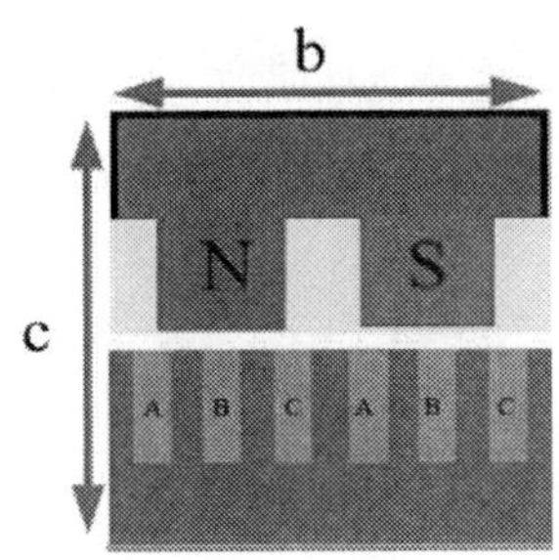

Figure 4. The fundamental motor configuration to calculate power density.

The expression for the power density, shows a linear dependence with B, H, v, (the flux density, magnetic field intensity, and the gap relative velocity), and inversely proportional to the two poles size b.

$$\frac{P}{volume} \cong \frac{BHAv}{Ac} = BHv/b$$

b is in the denominator since the vertical dimension c, which encompass the entire bulk of the motor is always proportional and similar in size to the two pole dimension b.

It is clear that in order to increase the power density of motors, B, H, and the velocity v needs to increase. But B is fixed by material properties, H depends on the conductor currents which has to be chosen with dissipation in mind, and a too high a velocity will require a gear box whose weight and inertia will defeat the purpose. Reducing the pole size b, will greatly improve the power density but it too is limited by dissipation considerations.

Figure 5a shows a typical multi-turn winding motor where the winding path outside the gap region limits the pole size reduction and adds additional volume, inductance (reactive power) and magnetic and electromagnetic shielding. This is a characteristic of all multi turn motors to variable degrees [2].

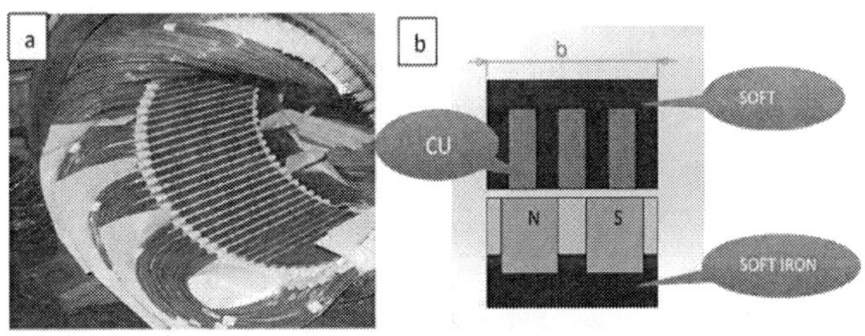

Figure 5: (a) A typical motor, (b) showing the effects of turns on pole size and total volume, and the new motor concept described here.

In this concept as shown in Figure 5b, we change the motor topology that allows the smallest possible pole size by configuring each pole pair as three single conductors shorted at their end which eliminates turns, where each is driven by their own inverter as seen in Figures 6 and 7. This eliminates all turns and reduces by a factor of two, the pole width for the same current density and resulting magnetic field intensity H. Pole sizes in the few mm range are now possible with appropriate magnetic field strength H for high power density and appropriate efficiency and cooling.

As an example using the following values, 1T, 20kA/m, 150m/s, 6mm respectively, for B, H, v, and b, yields a power density of 500 MW/m3, or ~62.5kW/kg.

Figure 6 shows the basic topology in both cylindrical and for clarity in planar geometry. In the cylindrical geometry we see the arrangement N-S magnet pairs on the rotor where each two poles correspond to three conductor elements. Figure 7 depicts the integration of inverter with the electro-mechanics. Figure 7a shows the transition from three nested windings of conventional motors to two Turn-Less Structure circuits (TLS) each with their own inverter [3]. Note that there are no wire crossings in this topology. Figure 7b shows triple conductors of the TLS energized by three 3-phase currents. The TLS is but three conductors shorted at one end. Its integration with its dedicated inverter of the same width is termed CTS for controlled turn less structure (Figure 7c). Figure 7d shows the details of how the TLS and inverter are connected. Note that the width of the inverter has to match the width of the TLS. As Figure 7d shows, the H-bridge folded over a micro channel cooler to eliminate wire crossing. The three upper switches are shown on the top with the other three switches below not seen. Figure 7e shows an ensemble of CTS elements constituting a cylindrical motor and Figure 7f shows the present design for a 100A/phase inverter of 6mm width (upper part only). With today's high current low voltage MOSFETs (100Amp 15V, 0.1cm^2 area) the inverter's current per unit width (the value of the magnetic field intensity H) matches the TLS motor requirement. [4][5] At present the TLS and inverter width is 6mm and length d of 2cm.

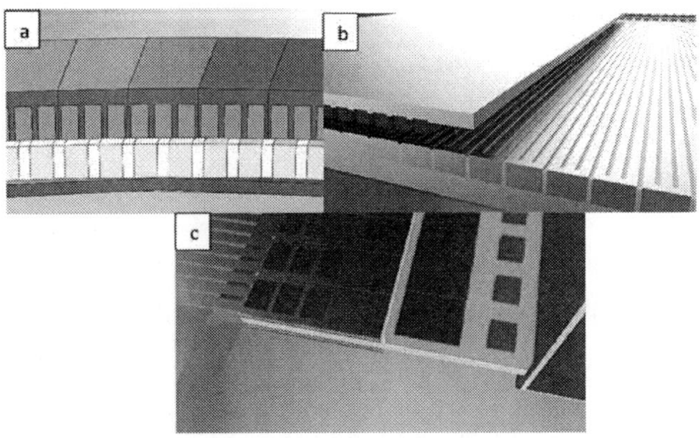

Figure 6. The Polarix motor, (a) with details (b), and (c) shown in planar configuration. Each three conductors are shorted at one end, with a dedicated 3-phase inverter at the other. The rotor has a north-south magnet pair per three conductor elemental motor pair.

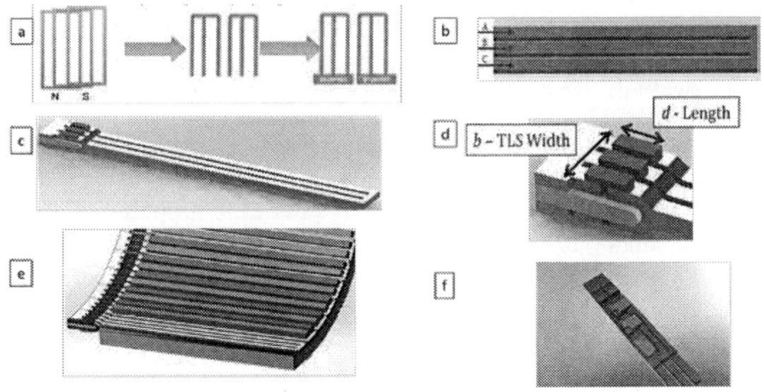

Figure 7. Details of the motor concept, a)shows how a single conventional 3 phase nested winding is configured a 2 Turn Less Structures (TLS); b) shows a single TLS with three, 3 phase current driving it; c) shows a coupled inverter to the TLS, here termed CTS for Controlled Turn-less Structure; d) shows details of this coupling where the inverter is folded over a 2 sided micro channel cooler with the upper three MOSFETs showing; e) shows the arrangement of multiple CTS elements around a motor stator periphery constitute a cylindrical motor, also seen are the alternating polarity magnets on the rotor periphery; f) shows the present physical design of the inverter as seen from above.

Figure 8a shows the upper and lower sides of our present inverters. The three MOSFETs on upper side and their current sensing resistors are visible. Figure 8b shows the view from above of eight inverters mounted on the micro-channel cooler in a planar (also disk rotor) configuration. Figure 8c shows the corresponding alternate polarity permanent magnets on the rotor section.

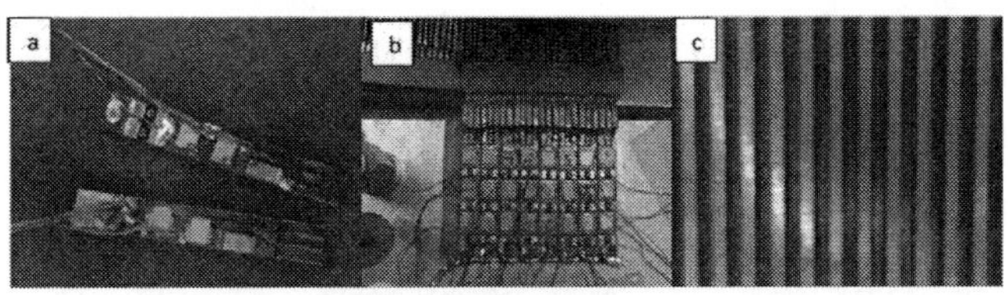

Figures 8a-c: Two sides of the of the 6.5mm wide H-bridge (8a), 8 assembled H-bridges with sandwiched microchannel cooler (8b), and permanent magnets arrayed on the (rotor) (8c).

Analysis of the motor has been performed for a variety of diameters and surface velocities. It is understood that in a rim driven fan motor both the diameters and the peripheral velocities are approximately the same.

An analysis of the specific power as a function of gap velocity for a nominal 50cm radius motor with TLS width of 1cm, length of 10cm, (Figure 7e) is given in Figure 10. The rotor/stator gap is fixed at 1mm at operating speed.

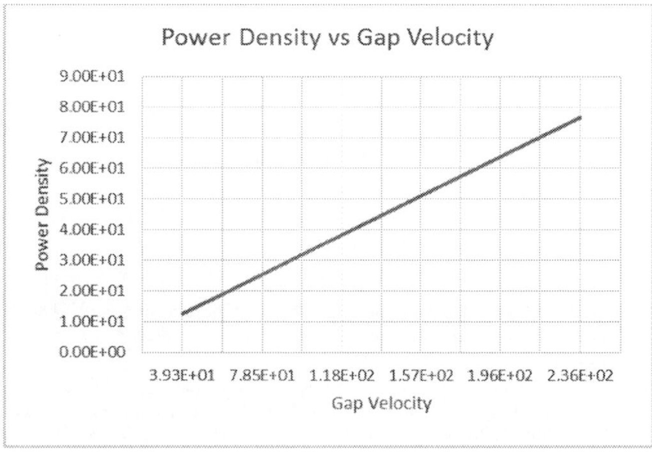

Figure 9: Power density vs. gap velocity.

The graph in Figure 9 shows the gap velocity ranging from 40m/s to 240m/s with the resulting specific power varying from10kW/kg to 80kW/kg. The corresponding power is given in Figure 10.

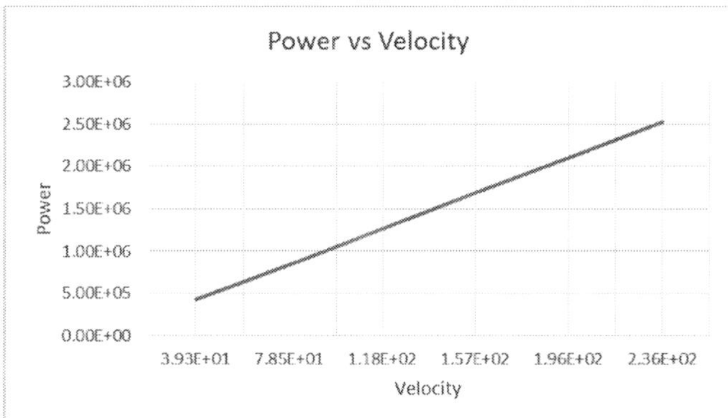

Figure 10: Power vs. velocity.

BATTERIES

As discussed above the concept reduces the fan diameter to comply with an aerodynamic craft by increasing the fan power by a factor of 5 to 10 higher than the cruising power. We have shown the motor technology that can generate the power at high specific power. But the source of the electric power requires similar specific power as the motor and with enough energy to execute the VTOL function.

Figure 11 shows a battery Ragone plot [7] of where present and future battery technology lie. From it we extracted the values in Table 2 which shows the specific energy and power for batteries with enhanced energy density and power density respectively for both present and future three years from now. The values are from With the Saft battery at 14kW/kg it takes but 18kg to supply the 250kW takeoff power.

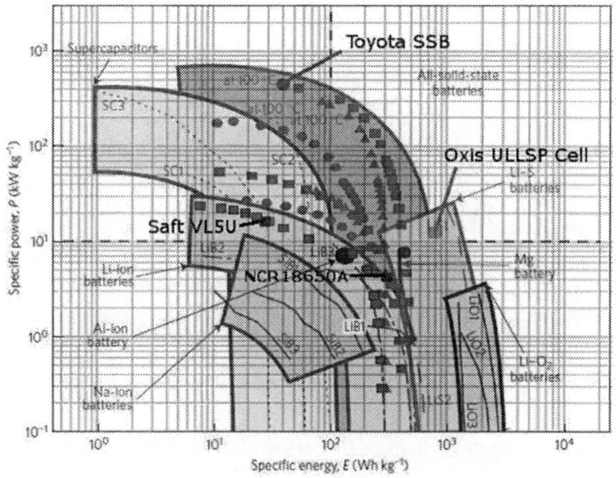

Figure 11: Ragone Plot.

Ragone plot for the major battery technologies showing our present choice for Saft high power and Panasonic 18650 high energy as representative of present technology the black dots represent technologies and f=values for both high specific power and high specific energy projected to be available in the next three years and are used in the table and the pie charts.

Table 2. Specific Energy and Power of Present and Advanced Batteries.

	Company	Model	Specific Energy (J/kg)	Specific Power (W/kg)
Basic	Saft	VL5U	9.72E+04	1.40E+04
	Panasonic	18650	8.10E+05	5.87E+02
Advanced	Toyota (SSB)	Kato	1.37E+05	4.50E+04
	OXIS (LiS)	ULLSP	1.80E+06	2.00E+03

For the flight characteristics given in Table 3, the duration and range for a 600kg craft using both present and advanced battery technology we have the following weight distributions as given in the pie charts in Figure 12.

Table 2. Flight Performance for Present and Advanced Batteries.

	Height (m)	Speed (m/s)	T/O Accel (g)	T/O Power (kW)	Loiter Power (kW)	T/O Thrust (kN)	Loiter Thrust (kN)	Duration (hours)	Range (km)
Basic	100	30	0.1	250	44.4	6.5	0.98	1.22	132
Advanced								3.29	355

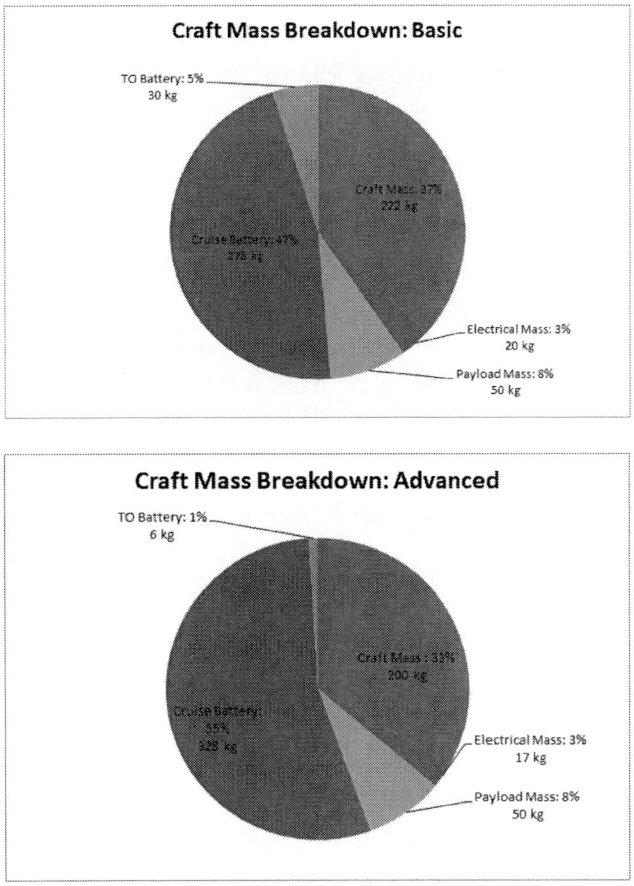

Figure 12. Craft Mass distribution for the basic and advanced battery technology.

CONCLUSIONS

We have shown how the development of a new motor technology with enhanced specific power, in conjunction with batteries with enhanced specific power, it is possible to provide the VTOL function with a small diameter propulsion fan and thereby retain the high efficiency cruising and attendant range available in a high L/D craft.

Author contact:

Oved Zucker oved.zucker@polarixcorp.com

Christopher Liu liuchris10@gmail.com

REFERENCES

[1] F. Marc de Piolenc, George E. Wright, Jr., "Ducted Fan Design". 2001-06-01.

[2] Fitzgerald and Kingsley, Electric Machinery, 7th Edition, McGraw Hill.

[3] Zucker, Oved. Electric motor/generator with multiple individually controlled turn-less structures US Patent # 9,755,468, September 5, 2017.

[4] Zucker, Oved, "New Topologies in Electro-mechanical Systems – A New Paradigm," presented at IEEE Pulsed Power Conference, Texas, USA, June 1, 2015.

[5] International Rectifier, "DirectFET N-Channel Power MOSFET," IRL6283MTRPbF datasheet, September 24, 2014.

[6] Kato, Yuki, et al. "High-Power all-Solid-State batteries using sulfide superionic conductors." Nature Energy, vol. 1, no. 4, 2016, p. 16030., doi:10.1038/nenergy.2016.30.

[7] Kato, Yuki, et al. "High-Power all-Solid-State batteries using sulfide superionic conductors." Nature Energy, vol. 1, no. 4, 2016, p. 16030., doi:10.1038/nenergy.2016.30.

[8] Zucker, Oved, et al. "Advanced Inverter Arrays for Controlled Turn-less Motor (CTM), Semiconductor Limits and MEMS Manufacturing" IEEE Power Modulator and High Voltage Conference, in press.

Algorithmic Icing Detection for On-Demand Electric Aircraft

Robert McKillip, Jr.
Senior Associate
Continuum Dynamics, Inc.
Ewing, NJ 08618

Andrew Kaufman
Senior Associate
Continuum Dynamics, Inc.
Ewing, NJ 08618

Todd Quackenbush
Senior Associate
Continuum Dynamics, Inc.
Ewing, NJ 08618

ABSTRACT

On-demand electric aircraft, if they are to realize their full potential, will be required to provide mobility in almost all weather conditions. Even if these vehicles are not certificated for operations into know icing conditions, they will need sufficient regime recognition to depart icing conditions if they are inadvertently encountered in their operations. The work described here represents the application of a patented icing detection algorithm to electric aircraft that exploits the rich information content associated with the use of electric motors as primary propulsion systems. This algorithmic approach for the assessment of aircraft icing incorporates model-based detection filters that discriminate icing events from other propulsion system anomalies, and can serve as a trigger for a mitigation strategy that could include triggering deicing systems or maneuvers that depart the icing environment. Current progress in the development of this system is described, along with results from some simulated icing testing.

INTRODUCTION

New classes and categories of aircraft are under development, which are poised to revolutionize short-duration air travel. The impetus for this work comes from advances in electronics and controls, and increases in electric motor power densities (Moore, Ref. 1). Concepts for reexamining VTOL operations from city centers to outlying suburbs are being proposed by both ground-vehicle-based ride scheduling companies and commercial aircraft manufacturers (Refs. 2,3). Some single and dual passenger aircraft are undergoing flight evaluations in foreign countries, and testing activities are underway on configurations here in the U.S.

As these aircraft are integrated into the transportation options available for use, they will encounter weather conditions that may challenge the design performance of these vehicles. On-demand aircraft will need to operate in inclement weather, and may encounter icing conditions during flight, even if inadvertently. Being able to detect the onset and growth of icing, and understand its effects, will allow designers to include options on these vehicles for mitigating these effects, whether they be operation of anti-icing or de-icing systems, or executing evasive maneuvers, including precautionary landings.

Many proposed concepts for short-distance travel are designed to use propeller/rotor systems for direct lift, to avoid the requirement for conventional runways and allow for operation from heliports/vertiports near commercial centers. While some concepts include provisions for rotating these propeller/rotor systems away from the vertical to transition to wing-borne lift during cruise, many still maintain a near-vertical rotor orientation and operate similarly to a helicopter, with the rotating wing providing both lift and forward propulsion through disk/airframe tilting. Thus, these systems will potentially encounter a wide range of inflow conditions, and thus may accrete ice in markedly different ways.

Also, since these platforms are often characterized by incorporating distributed propulsion/lift systems, the physical size of the propeller/rotor systems is often significantly smaller than conventional helicopters of the same weight class, and thus, physical space for on-blade icing detection, and anti-ice thermal protection, is limited. The effort described here plans to map a patented algorithmic icing detection scheme, previously developed at CDI for use on the Navy's V-22 Osprey tiltrotor, to the propeller/rotor systems for this category of air vehicle, with extensions to the method to use measurements on the electric motor itself to infer both icing level and icing growth during flight operations. Since these vehicles, due to their dependence on VTOL operations, are dominated by lifting propeller/rotors, the present study is limited to the use of the coupled motor and controller for determination of the level of icing on this important component for these aircraft.

ALGORITHMIC ICING DETECTION

Past CDI efforts in the development of an icing detection system for the V-22 tiltrotor aircraft for the Navy resulted in the creation, and ultimately patenting, of an algorithmic process for processing multiple sensor data using a fault detection filter methodology (McKillip, Refs. 4,5). Shown in Figure 1, the icing detection filter system structures the model-based estimation of V-22 performance such that estimation errors generate specific trends, or "modes" of response, in the combined directivity of the resulting estimation state values. The methodology borrows from research in the controls literature where algorithmic

Presented at the Vertical Flight Society Autonomous VTOL Technical Meeting and Electric VTOL Symposium, Mesa, Arizona, UAS, Jan. 29-31, 2019.

consistency checking of a process model against observed sensor data can be used to extract information concerning possible failures in actuators or sensors within a control loop. This technique is designed to provide maximum separation of dynamic "modes" of the tracking filter's response to unambiguously indicate the likelihood of a failed component in the controller. For the rotorcraft application described here (a V-22 tiltrotor), the system "faults" were postulated within the filter design as discrete icing events on various components, such as the rotor system, the tail planes, and the main wing. Each type of icing accretion generates its own performance (trim) and handling qualities (stability and control) "signature", that is related to how the aerodynamic efficiency of that component degrades with increasing ice accumulation.

Testing of this system using the Navy's V-22 full-mission simulator at the Patuxent River Naval Air Station showed that the algorithm was robust to tracking perturbations from maneuvering flights (by not generating any "false positives") and yet was able to pick up degradations in performance and flying characteristics due to simulated accreted icing events. This same detection filter structure was able to track icing shedding events that arose from the rotor system as the ice debonded from the blade leading edge and left the airframe (shown in Figure 2). Since this detection filter methodology is suitably generalized to address virtually any pre-postulated system "fault," it can be adapted to address the propeller/rotor condition monitoring function desired here for detection of icing levels and accretion rate for on-demand aircraft platforms.

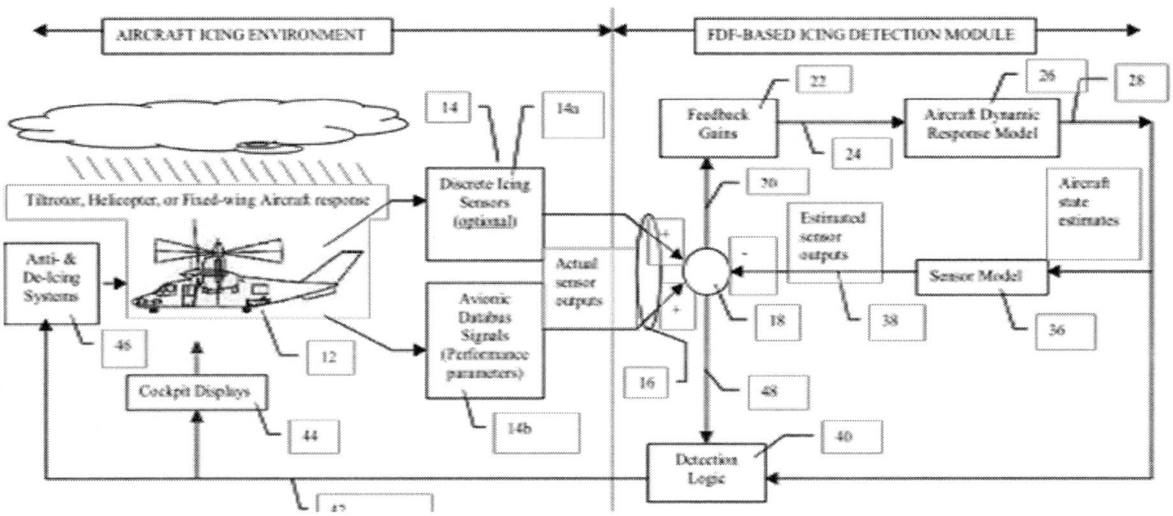

Figure 1. Schematic diagram of signals and processing steps in the icing detection system.

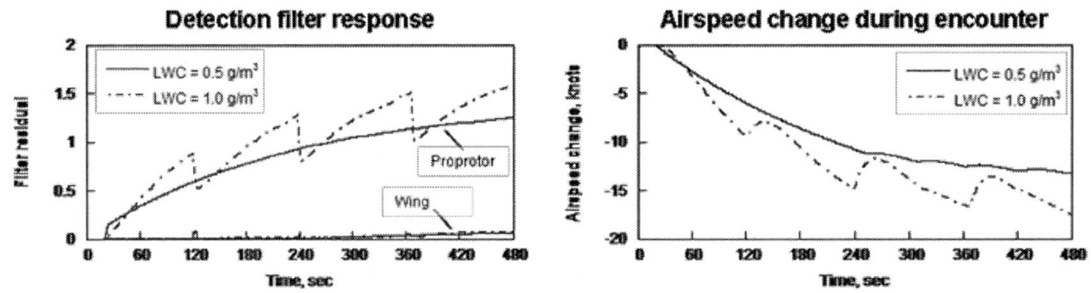

Figure 2. Tracking performance of icing "fault" detection algorithm for V-22 simulation in two different liquid water content (LWC) conditions.

Since this detection filter approach is a model-based algorithm (because it uses a process model to predict sensor outputs), it can be readily integrated onto on-demand aircraft platforms that have known performance response changes to icing events. Unlike the V-22 application just cited, however, these on-demand aircraft may not have an internal databus (such as MIL-STD-1553) that has connectivity to a wide array of onboard sensors and systems. Instead, an additional monitor of aerodynamic performance is required, and that is to be the electric motor controller itself, as this unit will be able to infer attached rotor/propeller performance via the imposed torque load on the motor as it attempts to respond to flight vehicle command inputs. Shown in Figure 3 is a similar schematic of how the proposed Icing Detection Filter would couple with the onboard motor controllers of an on-demand aircraft platform for detection of accreted ice.

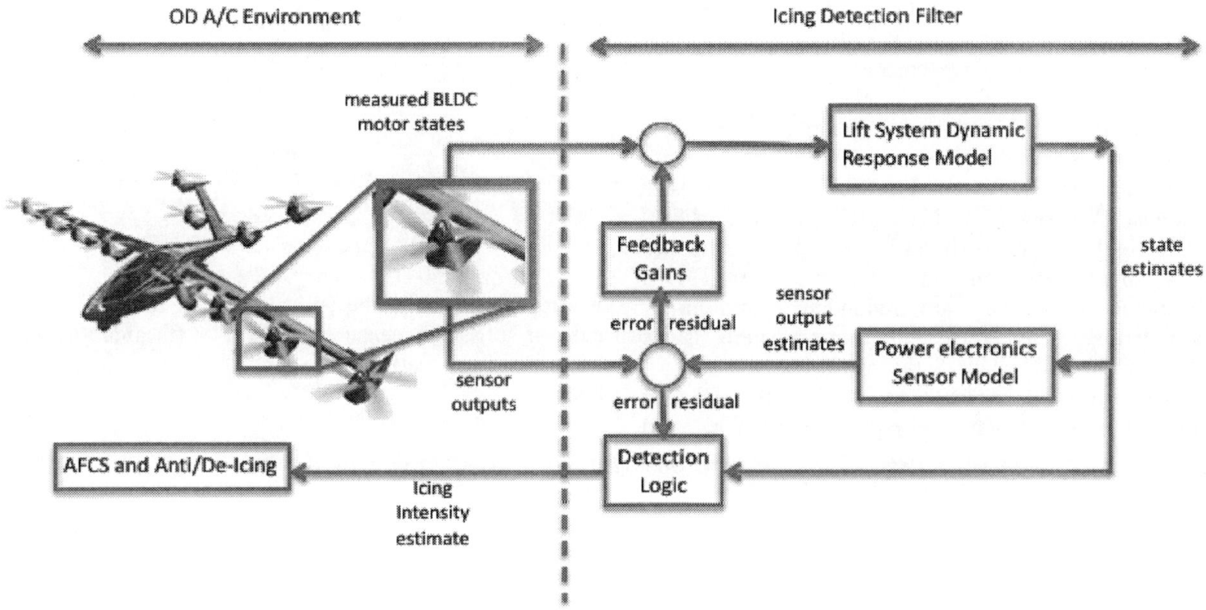

Figure 3. Schematic of Icing Detection Filter processing local information on each electric lift system motor component.

Proper representation of the performance degradation of the lift/propulsion system is clearly needed to provide the predictive model for use with the Icing Detection Filter. If such a model does not exist, it may be generated using a combination of aerodynamic analysis tools, coupled with ice accretion models. Fortunately, best-of-breed physics-based aerodynamic analyses are currently in hand at CDI that can address this, building on extensive prior experience with a wide range of complete-airframe models involving both fixed-wing and rotating-wing components. A central pillar of this activity is the CDI CHARM (Comprehensive Hierarchical Aeromechanics Rotorcraft Model) (Wachspress and Quackenbush, Ref. 6; Quackenbush et al., Ref. 7, 8), a code originally developed for helicopters but one that has also been applied to multiple fixed wing and compound configurations. It should be noted that CHARM has been the key tool in directly designing the rotor blades of the Shiebel S100, Carson S-61, and VH-3D.

CHARM captures the aeromechanics of complete airframes using a well-developed unstructured unsteady fast panel analysis and free rotor wake model. The software can also integrate multiple lifting and propulsive elements into a single model for evaluation of full-configuration aerodynamics. CHARM has been operated not only as a stand-alone design analysis but also as a modular component of 6-dof modeling tools, such as the FST (Flight Simulation Toolbox) environment previously developed at CDI. Integrated CHARM/FST models have been used to analyze full-vehicle systems (e.g., the Northrop Grumman FireScout and the NASA/General Atomics Predator B/Altair) quantifying the nonlinear, time-varying effects of aerodynamics and structural mechanics on handling qualities and control system design. Moreover, other modular variants of CHARM are available (e.g., the CHARM/SI Shipboard Interaction) model that can capture complete effects of rotorcraft/ship interaction. (Keller and Wachspress, Refs. 9, 10). The CHARM aeromechanics model can be used to generate the piecewise-linear models for design of the filter structure, as well as serving as a full nonlinear simulation tool for evaluating the detection filter's performance. This fact means that the condition monitoring function could be designed without initial reference to actual baseline aircraft operational data, a very powerful feature that allows this methodology to be extended to a variety of on-demand aircraft configurations and sizes.

Detection Filter Background

Detection filter (DF) systems are designed to identify when an anomalous event occurs in a dynamic system and to indicate what component or subsystem has generated that event. The typical tradeoff facing the DF system designer is to properly balance the DF logic to respond rapidly to events while minimizing the number of "false alarms" that may be generated by noise present within or external to the system being monitored. In general, the use of DF systems is a means to extend the useful life of a system without the (possibly great) increases in cost associated with use of additional/redundant or more reliable sensors, actuators, or system components. Designers of DF systems have accepted that the component lifetime for a system is finite and wish to incorporate knowledge of how various sub-systems may malfunction, into a total scheme for performance enhancement (or extension).

Interest in DF usage as a means of extracting more utility from vehicles and systems has steadily increased with attendant increases in the use of feedback control on these same systems. Willsky (Ref. 11) and Isermann (Ref. 12) provide excellent surveys on the state of design approaches for DFs and fault detection and isolation systems (FDI). Methods range from hardware redundancy and inter-component voting to analytical redundancy, where use is made of the interrelationships between the various sensors and the dynamics of the "plant", to "intelligent" FDI that incorporates learning through the use of neural networks and expert systems. More recently, DF approaches for determining anomalous responses for aircraft have been formulated for closed-loop system response (Szaszi, Ref. 13) and UAV platforms that include separation of control effects from wind disturbances (Freeman and Balas, Ref. 14). Of paramount importance in the design of practical FDI systems, however, is that the proper attention has been paid to the tradeoff of complexity of the FDF vs. the increased performance it provides to the overall system.

The DF system for implementation here is an extension of that first proposed by Beard (Ref. 15), and later extended by Messerole (Ref. 16) to nonlinear systems. The main feature of the method consists of constructing a state estimator for the dynamic system that selects a feedback gain on the measurement residual to make the overall estimator sensitive to pre-specified types of system failures/anomalies. Through careful selection of the gain matrix, one may force the measurement residual to remain fixed in a particular direction or a specific plane for a given failure mode. A simple example for a linear time-invariant system from Beard (Ref. 15) serves to illustrate the approach. For a system described in state variable format, with x representing a vector of state variables, u a vector of control inputs to the system, and A and B representing the perturbations in the state time derivatives (marked by the dot superscript) to state and control changes, one has:

$$\dot{x}(t) = A\,x(t) + B\,u(t)$$

with a vector of measurements (z) linearly related to the state variable through a matrix C according to:

$$z(t) = C\,x(t)$$

a detection filter is designed of the form:

$$d\hat{x}/dt = A\,\hat{x} + D\,r(t) + B\,u(t)$$

where the "hatted" quantities represent the current estimates of the state variables, and $r(t) = [\,z(t) - C\,\hat{x}]$ the measurement residual, and with D chosen to accentuate any failures that may occur in the system. For example, if a bias of magnitude v suddenly appears on the i-th actuator, the system becomes:

$$\dot{x}(t) = A\,x(t) + B\,[\,u(t) + v\,ei\,]\text{, or}$$

$$\dot{x}(t) = A\,x(t) + B\,u(t) + v\,bi,$$

where bi represents the i-th column of matrix B. For the case of full state measurement, $C = I$ (the identity matrix), and thus the equation governing the measurement residual becomes:

$$\dot{r}(t) = [\,A - D\,]\,r(t) + v\,bi,$$

so if D is chosen = $[\,\sigma I + A\,]$, then the residual dynamics become:

$$\dot{r}(t) = -\sigma\,r(t) + v\,bi\text{, or,}$$

$$r(t) = \exp\{-\sigma\,(t-t_o)\}\,r(t_o) + v\,[\,1 - \exp\{-\sigma\,t\}]\,bi/\sigma.$$

Thus, after the initial transient dies out, the steady residual maintains a fixed direction aligned with vector bi. Note that its magnitude decreases as the bandwidth of the detection filter increases, thus trading off speed of response with sensitivity to noise in the system.

The above technique has been extended in the literature to handle a wide variety of failure types that may be present in the actuation systems, system dynamics, or sensors, for systems that do not possess full state feedback. For systems that have observation matrices C of rank less than the number of system states, some limitations may exist on the capability to isolate all potential system failures. This limitation is somewhat akin to the observability of the system, and typically poses no problem, as the sensors have been installed to provide sufficient information for achieving good automatic control by design. The primary limitation of the above technique, then, is that the selection of feedback matrix D is deterministic, and that the system representation is both linear and time-invariant.

Meserole (Ref. 16) applied the above technique to detect faults in the sensors and control system for an F100 aircraft gas turbine engine and included nonlinear effects in the engine performance map through re-linearization of the system about a new operating point. He found that the success of the detection filter did not strongly depend upon the accuracy of the linearized representation of the rather nonlinear character of the engine response. The method used to select feedback gains for the residuals included a combination of transformation techniques and least-squares methods for eigenvalue/eigenvector assignment. Eigenstructure control over the dynamics representing the measurement residuals assures that each failure mode exhibits a unique and identifiable deviation in the measurement residual whose magnitude may change with time but whose shape (direction) does not.

Eigenstructure assignment techniques are well established in the context of feedback control design and may be performed directly using least-squares methods (Andry, Ref. 17) or through asymptotic properties in specification of the cost function to be optimized (Park, Ref. 18). The least-squares methods (sometimes called the "direct approach"), do not result in guaranteed stability margins for the resulting feedback controllers. Along with providing a feedback structure with stabilizing gain values, the asymptotic approach lends itself to other methods of gain adjustment in order to enhance robustness of the resulting feedback system. In particular, use of frequency-weighted cost functions can significantly improve the robustness of both control systems and recursive filters to unmodeled (or poorly modeled) high-frequency dynamic components.

For systems that are time-invariant, the filter gains would form constants in the "D" matrix above, but if the technique is to be applied to a nonlinear, time-varying system, then the filter structure would need to track these variations. Use of an Extended Kalman Filter formulation allows these variations to be represented in the filter, by using the full nonlinear and time-varying form of the system dynamics in the state propagation and measurement equations, but a linearized representation of the dynamics and observation matrix in the state estimation error covariance equations. The linearized system and measurement matrix are formed by using the current state estimate (and time) to compute the matrix elements of "A" and "C" above. This may be done through numerical linearization of the system, where small perturbations are introduced to the system model and the resulting outputs used to compute the parametric sensitivity of each matrix component.

Motor Performance Modeling

Brushless motors are used almost exclusively for providing lift system power on a variety of UAVs and now for personal/on-demand aircraft, due to their improved power to weight ratio over brushed motors and their higher reliability. However, their use in flight-critical applications has spawned a large body of development in the production of fault-tolerant control, which includes mitigating both electrical faults (such as winding short circuits or breaks; decreased insulation on wiring; poor contacts and connectors) and mechanical issues (bearing wear; bearing roller and cage failures; rotor shaft deformation; heat rejection problems), if only for a limited time prior to complete failure. Methods for tracking these anomalies include monitoring covariance changes in online estimators (Aghili, Ref. 19), fuzzy logic monitoring of system parameters (Abramov, Ref. 20), particle filter estimators (Brown, Ref. 21), and even model-based fault detection using measured motor voltages, currents, and rotation speed (Moseler, Ref. 22).

Icing accretion on the propeller/rotor that is coupled to these motor drives would appear as a torque anomaly that does not align the torque/thrust/shaft speed curve along anticipated "surfaces" in multi-dimensional state space. This offset, once determined from icing analysis, would provide an additional "failure mode" for tracking in a fault detection filter scheme that would isolate this change as a separate fault, one due to ice accretion. Since several researchers on BLDC motor control and sensing have suggested that the motor operating state may be inferred from observations of armature currents and applied voltages with speed changes, this approach would be used to assess the anomalous behavior propeller/rotor blade icing would present to the motor controller as an unknown torque load.

Icing Performance Degradation

Icing on rotors or propellers is much more complicated than ice formation on the leading edges of fixed-wing aircraft. For rotating blades, the incident velocity increases along the span from root to tip and therefore the water captured and the local aerodynamic heating changes along the blade. The thinner blade sections are also more efficient at capturing water droplets than fixed-wing airfoil sections. Additionally, for glaze ice, the blade rotation and local airflow can affect the motion of the liquid water film on the blade. Rotors, generally, are more complicated than propellers because the local flow, angle of attack and velocity not only change with span but also with azimuthal location (except in hover).

All of these factors can have a profound impact on icing, especially when coupled with environmental conditions. Testing of an OH-58 tail rotor to simulate a main rotor documented that icing changed along the span of a rotor blade depending on temperature (Miller and Bond, Ref. 23). At cold temperatures (-26C) rime ice was formed along the entire blade span, at -15C the ice type gradually changed from rime inboard to glaze outboard, and at -6.7C the ice was predominately glaze, but no ice formed on the outer 15 to 20 percent of the blade. Furthermore, changing the liquid water content (LWC) or median volume diameter (MVD) moved the rime to glaze transition, and increasing rotor rpm increased the glaze type of icing.

On demand mobility vehicles are expected to fly at lower altitudes and will be more vulnerable to icing. Icing occurs most frequently at lower altitudes and has greatest impact on small aircraft that fly slowly. The probability of encountering super-cooled water droplets that can lead to icing is most frequently encountered at lower altitudes and is terrain and weather dependent (Green, Ref. 24). Review of icing related accidents have identified hundreds of accidents over a period from 1978 through 2002, with a decline of accidents to a rate of 15-20 per year. Since most of the accidents involve general aviation aircraft, it can be expected that the rate of icing accidents may increase with the advent of on-demand vehicles unless something is done to identify icing conditions.

A recent test of a mini UAV, "Project SOPHIA," (Meteomatics, Ref. 25) illustrated how quickly small UAVs can ice up and lose the ability to remain aloft. Testing was conducted in outdoor, indoor and icing tunnel conditions. For comparison purposes, ice amounts less than 4mm in thickness were considered "much ice." The results of the tunnel tests indicated that noticeable icing started within 100 seconds and that the UAV was unable to fly after about 300 seconds of exposure. Icing reduced the thrust of the propellers and increased the weight of the UAV. It was noted that the start of icing could be detected by an increase in power during flight, which supports the proposed fault detection filter development.

A significant amount of work has been done to model ice accretion on rotors. The earliest work was done in 1985 based on experimental work (Flemming and Lednicer, Ref. 26) and developed empirical correlations. A numerical model was developed (Briton, Ref. 27) to predict ice shapes and the corresponding degradation of rotor performance. LEWICE was used for icing prediction on 2D blade sections. A collaborative effort between CIRA and Boeing looked at ice accretion with LEWICE and a Reynolds-average Navier Stokes solver and the MULTI-ICE code developed by CIRA. At Penn State University, ice accretion on rotors have been studied both experimentally and numerically (Brouwers, et.al, Ref. 28, Palacios et. al., Ref. 29). Their model is based on a LEWICE type accretion model and a low fidelity performance model that was focused on performance of de-icing solutions. Several publications have documented collaborative work between Georgia Tech, Sikorsky and NASA since 2009. These efforts involve an integrated approach for ice accretion and performance for rotors in hover and forward flight (Bain et. al., Refs. 30, 31; Nucci et.al. Ref. 32; Rajmoahn, et.al., Ref. 33). An iterative coupling is performed between a structural dynamics code and CFD code and LEWICE 3D is called to compute the ice accretion in 3D. A joint effort between NASA and Boeing has modeled rotorcraft icing using OVERFLOW-RANS and LEWICE3D for rotors in hover and forward flight (Narducci et.al., Ref. 34; Kreeger, et.al., Ref. 35). These approaches typically compute ice accretion in 2D strip-theory fashion with varying degrees of sophistication for the flow field calculation and rotor performance calculations. A similar approach has been undertaken for modeling the icing accretion process to form the "anomaly model" to be used by the fault detection filter in the work described here.

Concept Summary

This investigation is undertaking the transition of an algorithmic approach for icing detection, developed for tiltrotor aircraft, to these emerging new classes of on-demand personal air vehicles. The algorithm focuses on the monitoring of electric motor operational parameters to infer both icing levels and growth rates on the propeller/rotor systems that provide much of these aircraft's mobility. Technical objectives being addressed in the effort are: (1) determination of icing scope for these class of vehicles; (2) cataloging performance impacts of the various types of icing on the electrical lift/propulsion systems; (3) design of the detection filters to track performance anomalies due to icing events; (4) validating these designs using experimental tests on simulated icing of components; (5) and assessment of implementation details to accommodate this capability on these electrically powered aircraft.

ICING PERFORMANCE IMPACT

Icing performance degradation modeling for this effort is currently following two paths, each with different levels of detail and required modeling inputs. In the first, a full representation of the components necessary to track supercooled water droplets is being developed, to allow for particle tracking in the complex flowfield environment surrounding this class of vehicles. Similar work was done in the past to model airborne chemical and biological agent interaction with aircraft flowfields (Quackenbush, Ref. 36), but in this implementation, detailed physical modeling of droplet behavior is needed to represent rollback and freezing effects, depending upon environmental conditions and thermal characteristics of the impacted surfaces. Much of this detailed physical modeling is captured within the LEWICE 2D and 3D series of software tools, while the flow field analysis and particle trajectories incorporate both CDI's CHARM analysis and internal codes for Lagrangian particle tracking. This analytical approach represents a time-intensive approach, as it must include sufficient numbers of particles and time intervals to generate averaged effects within this highly unsteady flow environment.

The second modeling approach represents more of a "surrogate" style of accretion modeling, and incorporates the correlation work of Fleming (Ref. 26) for the determination of icing growth and the resultant local loss of aerodynamic performance. In this model, average flow field assessments are taken near the lifting and nonlifting surfaces, and then the correlation equations from Ref. 26 are applied over time following first icing exposure to estimate local drag increases and lift and moment changes as icing forms on the lifting surfaces. This alternate modeling paradigm has the advantage of only needing the generation of the local flow environment in order to apply the appropriate icing collection correlation equations based upon environmental conditions being modeled. While this latter approach can provide estimates of performance degradation more quickly, the use of correlation relationships on conventional helicopter configurations for multi-rotor, articulating lift/propulsion systems may be questionable. Thus, use of the more detailed modeling approach is anticipated to support the development of these alternative expedited prediction relationships.

Representative configurations of eVTOL aircraft and other UAVs are being examined in this comparison work, and an example of the level of wake detail needed for some of this analysis can be seen in Figure 4, where the wake trailing from the multiple rotors on a Joby S-2-like vehicle can be seen. The propulsors differ from conventional propellers and rotor systems in that they have a lower aspect ratio, smaller diameter, and a large number of blades on each to keep the effective solidity high.

In addition to these two analysis developments, additional accretion modeling is being performed on configurations on which some additional testing has been performed. These include propeller configurations that were not used in the formulation of the correlation equations described above, Such modeling work provides validation of the development of a toolchain in the prediction of performance degradation of rotating components following an icing encounter.

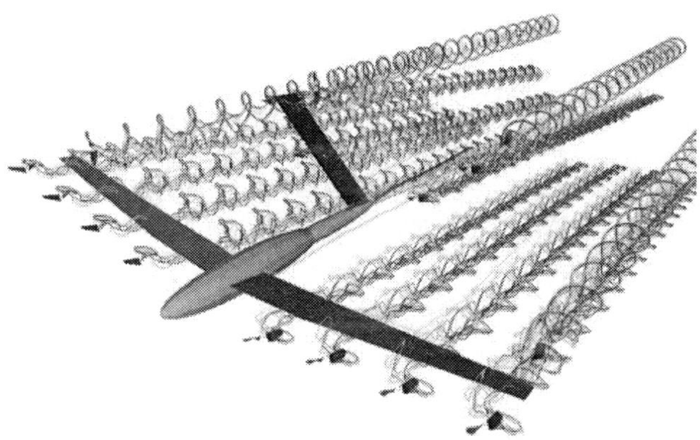

Figure 3. OpenVSP model of Joby-like configuration with propeller trailing wakes.

DETECTION FILTER DESIGN

Icing detection filters being evaluated for this effort can span different levels of vehicle systems, from isolated electric motor performance tracking, through the consideration of the complete vehicle performance. As would be expected, as the level of modeling detail grows, so does the dynamic model and thus the order of the filter needed to provide anomaly detection for icing-related events. For isolated motor performance, the effects of icing appear as torque increase anomalies that are distinct from other postulated fault conditions in the literature, particularly as they tend to be gradual changes in local drag and lift on the blades as icing levels build (with the exception of shedding events). For full vehicle icing response tracking, consideration of vehicle trim changes as aerodynamic efficiency decreases, which may include weight growth with added ice on the airframe.

Design of the filter uses eigenvalue assignment techniques, as was done previously for the V-22 aircraft, as these are relatively easy to use and provide sufficient designer control of anticipated filter performance and tracking behavior. Initial trials, however, have resulted in a change to the assignment of filter (observer) feedback gains that help create the desired eigenstructure without the same level of operator adjustment. The description of that algorithm follows.

Given a linearized dynamic system described by the state equations:

$$\dot{x} = Ax + Bu + \sum_i f_i d$$

where x is the perturbations in the state vector, u is a perturbation control input vector, and d is a scalar time-varying disturbance value that perturbs the system response. The sensors for the system can be described by a linear output relation with the state vector as:

$$y = Cx$$

with y as the vector of measurement outputs from the system. As was described earlier, the detection filter uses a state observer to feed back a weighted difference between the measured outputs and the estimated outputs (with carets over them), with those weights (gains) adjusted to give the observer desired response modes that respond to (or, track) pre-postulated disturbances or errors. Calling this matrix collection of gains L, we thus have the detection filter equations as:

$$\dot{\hat{x}} = (A - LC)\hat{x} + Bu + Ly$$

$$\hat{y} = C\hat{x}$$

If one then subtracts the above expression for the observer from the linear dynamic system equations, what remains is the equation for the estimation error dynamics, where that error is the actual state subtracting the estimated state values:

$$\dot{e} = (A - LC)e + \sum_i f_i d$$

Now if the feedback gain matrix L is selected to have one of the eigenvectors of the closed-loop observer dynamics match the forcing vector f_i, a nonzero value of d will result in a nonzero vector combination of the estimation error dynamics vector e. If λ is the eigenvalue associated with the forcing vector turned eigenvalue f_i, then the two will satisfy the eigenequation:

$$(A - LC)f_i = \lambda f_i$$

Rewritten this becomes:

$$Af_i - \lambda f_i = LCf_i$$

And if we apply a pseudo-inverse to the right of both matrix expressions, we can isolate an expression for the gain matrix L:

$$L = (Af_i - \lambda f_i)(Cf_i)^+ + E(I - (Cf_i)(Cf_i)^+)$$

Where the (+) superscript indicates a matrix pseudo-inverse, and the E matrix is arbitrary. E multiplies the null space of the pseudo-inverse matrix and can be ignored here. That term represents the unobservable space associated with the dynamics equation upon use of the feedback gain matrix L. This algorithm, detailed by Park (Ref. 18), can be repeated for a second cycle to place another eigenvector/eigenvalue, using the modified A and C matrices:

$$L_1 = (Af_1 - \lambda f_1)(Cf_1)^+$$

$$A_1 = A - L_1 C$$

Initial testing of the filter on the motor simulation is underway, using the dynamic simulation model shown above. Early observations show that a full-state observer may be very fast indeed, as the dominant electrical time constants of the motor system are very small, relative to the aircraft dynamics (and icing accretion rate), and thus may potentially be treated as algebraic in nature. Simplification of the motor equations to support the lower-bandwith process of icing accretion and performance degradation, as it impacts motor response, is still underway and will be reported later.

EXPERIMENTAL VALIDATION

Development of the icing detection filter needs to include the features associated with the sources of available data, most notably the motor controllers for these brushless motors that typify the drive units for the rotors and propellers on these eVTOL aircraft. To demonstrate such system integration, a simplified test environment was constructed whereby a propeller was directly coupled to a brushless motor and motor controller, with additional mechanical attachments made to isolate and measure total thrust and torque on this propeller. As this was not an actual icing test, simulated "ice" was added to the leading edge of the propeller to represent a somewhat typical "dual horn" configuration of icing growth that could occur in rime ice conditions (Figure 4). This approach for simulating icing accretion has been used in prior NASA studies as well.

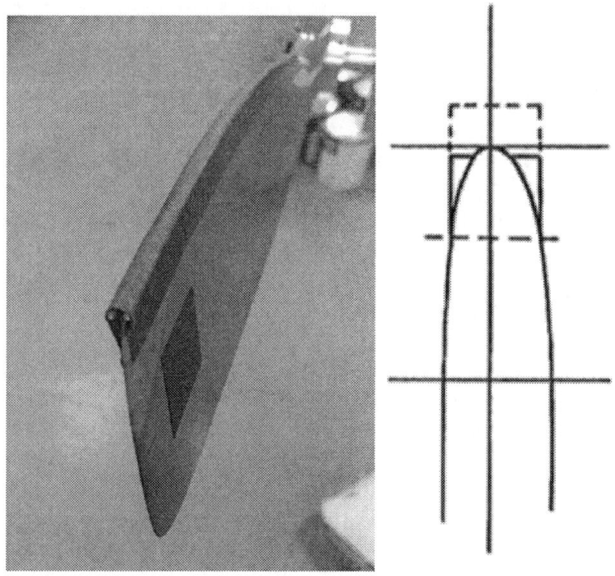

Figure 4. Simulated "dual horn" icing growth using conformal tape-attached small gauge wires on upper and lower leading edge (left); simulated icing shape for accretion correlation studies (Ref 26).

Tests were run for full-span simulated icing on the leading edge, 2/3 span inboard length, 1/3 span inboard, and a clean configuration. These scenarios loosely simulate icing effects following a shedding event that would remove outboard accreted ice. Data collected on this propeller with simulated icing were thrust and torque measurements for rpm sweeps, with motor controller data extracted via a serial interface to an embedded microcontroller. Strain gauge balance data was amplified, offset and filtered, and then collected via an on-board A/D converter in a separate microcontroller, providing a second serial data stream, with both combined via software running on a Raspberry Pi 3B+. This computational platform was chosen to demonstrate the minimal computational requirements needed for hosting the icing detection filter algorithm.

Figures 5 and 6 show the variation of torque and thrust with rpm for various levels of simulated icing. As might be expected, the full-span simulated icing requires markedly larger torque values for the same rpm operation, with the torque increase falling off rapidly as the outer portions of the simulated icing is removed. Thrust changes with icing are much less pronounced, as the leading edge ice has a much more significant impact on sectional drag then on local lift changes (except near stall conditions). Figure 7 shows the variation of thrust and torque for this range of rpm, and the shift in nominal slope of the data for the different simulated icing conditions is apparent. While such a trend would be readily applicable for tracking with a detection filter, both measurements are not intrinsic to the motor controller data source, and thus the goal of this testing program was to identify an optimal means of reconstructing this trend from the available serial data.

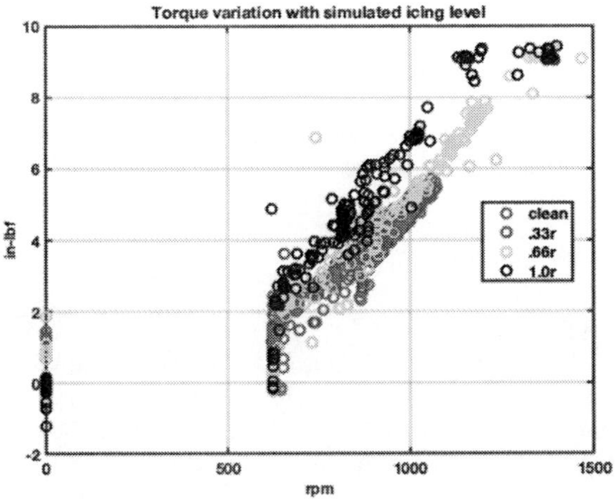

Figure 5. Torque variation with rpm, different levels of simulated leading edge ice.

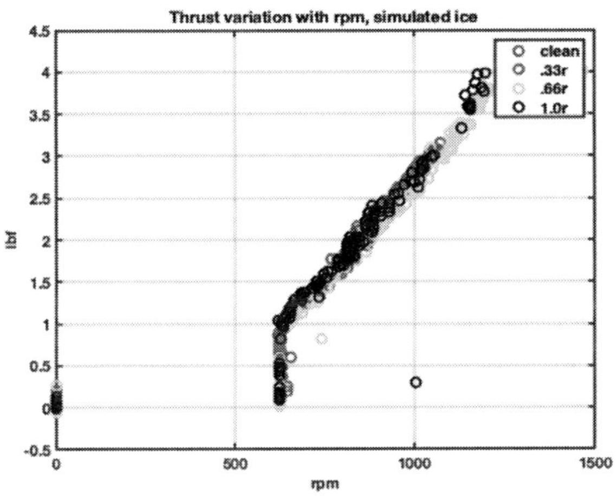

Figure 6. Thrust variation with rpm, different levels of simulated leading edge ice.

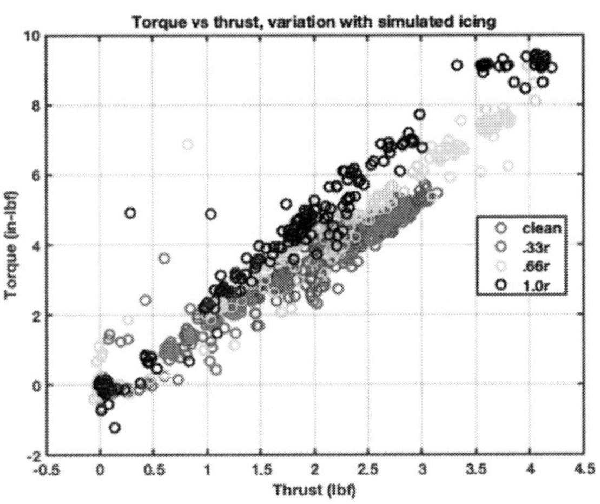

Figure 7. Torque variation with thrust for different levels of simulated icing.

The particular motor controller used in this test provides a wealth of data, but is considered typical of electronics of this power handling capability. The serial outputs include measurements of supply voltage and ripple, averaged current, fractional total power, input command, and electrical rpm (based on number of poles in the motor). Figure 8 plots the output power percentage for different icing levels versus shaft rpm, and when compared to Figure 5, is demonstrates an equivalent means of extracting the torque being required to overcome the drag from the simulated icing impact. Also, since the motor controller is operating (for this case) as a power-command system, the relationship between the input command percentage and the motor controller output power is linear, and invariant with external torques from the simulated icing. As a result, either the integrated power measurement, or the command signal applied, could be used as a surrogate for torque measurement for input to the detection filter.

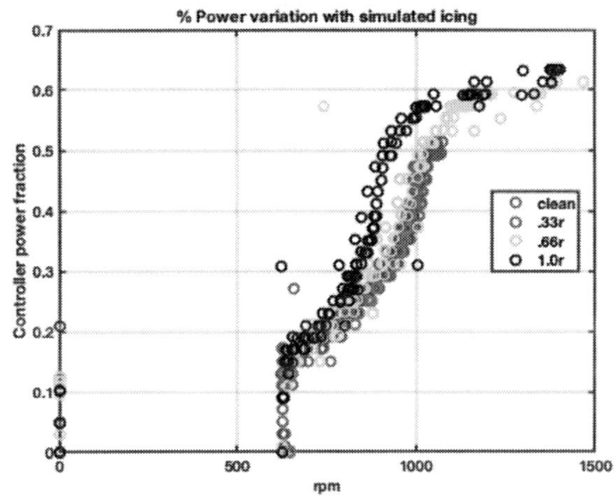

Figure 8. Measured motor controller power with rpm, for different levels of simulated leading edge ice.

An icing detection filter was formulated to track this behavior with leading edge icing, but since the testing was performed on static simulated icing shapes, no time variation of the filter performance is shown here. Additional work on filter formulation for isolated propellers and full eVTOL vehicles is underway, and will be reported in a future paper.

CONCLUSIONS

The patented icing detection algorithm from earlier work on icing detection for the V-22 Osprey (and other tiltrotor aircraft) is being extended to provide similar icing accretion indication (severity and location) for eVTOL and UAS vehicles. The detection approach leverages already available data sources within the electric motor and controller prime movers, without requiring additional external aspirated sensors with their associated weight and placement issues. Continuing work will advance this work to ultimately validate its approach in both icing tunnel and flight trials in natural icing conditions.

Author contact: Robert McKillip, Jr., bob@continuum-dynamics.com.

ACKNOWLEDGMENTS

The authors acknowledge the additional support provided by Dr. Jeffrey Keller and Mr. Raymond Tiberge, both of CDI, in this effort. This work is funded by a NASA SBIR grant, with Mr. Michael King of NASA Glenn Research Center, as technical monitor.

REFERENCES

1. Moore, M., "Aviation Frontiers: On-Demand Aircraft," 10th AIAA Aviation Technology, Integration, and Operations Conf., Fort Worth, TX, Sept. 2010.

2. Uber, "Uber Elevate: Fast-Forwarding to a Future of On-Demand Urban Air Transportation," Oct. 27, 2016, https://www.uber.com/info/elevate.

3. Joby Aviation, www.jobyaviation.com, accessed February 2018.

4. McKillip, R., Jr., et al., "Aircraft Icing Detection System," U.S. Patent No. 6,304,194, Oct. 16, 2001.

5. McKillip, R., Jr., Keller, J., and Kaufman, A., "Algorithmic Icing Detection for the V-22 Osprey", Proc. AHS Flight Controls and Crew Systems Specialists' Meeting, Philadelphia, PA, October 2002.

6. Wachspress, D., and Quackenbush, T., "First-Principles, Free-Vortex Wake Model for Helicopters and Tiltrotors," Proceedings of the 59th Annual American Helicopter Society Forum, Phoenix, AZ, May 2003.

7. Quackenbush, T. R., et al. "A Comprehensive Hierarchical Aeromechanics Rotorcraft Model (CHARM) for General Rotor/Surface Interaction," Final Report to NASA under contract NAS2-14342 (CDI Report No. 99-03), Continuum Dynamics, Inc., Jan. 1999.

8. Quackenbush, T., Keller, J. and Whitehouse, G., "Analysis Methods for Advanced V/STOL Configurations," Proceedings of the 72nd Annual American Helicopter Society Forum, West Palm Beach, FL, May 2016.

9. Keller, J. and Wachspress, D., "High Fidelity Real-Time Rotor Wake Module with Shipboard Interactions," CDI Report No. 12-11, Continuum Dynamics, Inc., May 2012.

10. Keller, J., Wachspress, D., and Hoffler, J., "Real Time Free Wake and Ship Airwake Model for Rotorcraft Flight Training Application," Proceedings of the 71st Annual American Helicopter Society Forum, Virginia Beach, VA, May 2015.

11. Willsky, A. S., "A Survey of Design Methods for Failure Detection in Dynamic Systems," Automatica, Vol. 12, 1976, pp. 601-611.

12. Isermann, R., "Process Fault Detection Based on Modeling and Estimation Methods - A Survey," Automatica, Vol. 20, 1984, pp. 387-404.

13. Szaszi, I., Marcos, A., Balas, G., and Bokor, J., "Linear Parameter-Varying Detection Filter Design for a Boeing 747-100/200 Aircraft," J. Guidance, Control and Dynamics, Vol. 28, No. 3, May-June 2005.

14. Freeman, P., and Balas, G., "Analytical Fault Detection for a Small UAV," Proc. Guidance, Navigation and Control Conf., AIAA Infotech, 2013.

15. Beard, R., "Failure Accommodation in Linear Systems Through Self-reorganization," PhD Thesis, Dept. of Aeronautics and Astronautics, M.I.T., Cambridge, MA, 1971.

16. Meserole, J. S., Jr., "Detection Filters for Fault-Tolerant Control of Turbofan Engines," Ph. D. Thesis, Dept. of Aeronautics and Astronautics, M.I.T., Cambridge, Mass., June 1981.

17. Andry, A. N., Shapiro, E. Y. and Chung, J. C., "Eigenstructure Assignment for Linear Systems," IEEE Transactions on Aerospace Electronic Systems, September 1983, Vol. AES-19, pp. 711-729.

18. Park, J. and Rizzoni, G., "An Eigenstructure Assignment Algorithm for the Design of Fault Detection Filters," IEEE Transactions on Automatic Control, Vol 39, no. 7, July 1994, pp. 1521-1524.

19. Aghili, F., "Fault-Tolerant Torque Control of BLDC Motors," IEEE Trans. on Power Electronics, Vol. 26, No. 2, Feb. 2011.

20. Abramov, I., Nikitin, Y;, Abramov, A., Sosnovich, E., and Bozek, P., "Control and Diagnostic Model of Brushless DC Motor," J. of Electrical Engineering, Vol. 65, No. 5, 2014.

21. Brown, D., et al., "Particle Filter Based Anomaly Detection for Aircraft Actuation Systems," Ch. 4, in K. Valavanis, "Applications of Intelligent Control to Engineering Systems," Springer Science, 2009.

22. Moseler, O., Isermann, R., "Application of Model-Based Fault Detection to a Brushless DC Motor," IEEE Trans. on Industrial Electronics, Vol. 47, No. 5, Oct. 2000.

23. Miller, T., and Bond, T., "Icing Research Tunnel Test of a Model Helicopter Rotor," NASA TM-101978, 1989.

24. Green, S., "A Study of US Inflight Icing Accidents and Incidents, 1978 to 2002," AIAA Aerospace Sciences Meeting, Reno, NV, Jan. 9-12, 2006.

25. Meteomatics, "Project SOPHIA: Study of Propeller Icing Hazard in Mini-UAV Aviation," Final Report Presentation, May 1, 2017, https://www.meteomatics.com, accessed Feb. 2018.

26. Flemming, R. J., and Lednicer, D. A. "High Speed Ice Accretion on Rotorcraft Airfoils," NASA CR-3910, 1985.

27. Briton, R., "Development of an Analytical Method to Predict Helicopter Main Rotor Performance in Icing Conditions," NASA CR-1890110, Jan. 1992.

28. Brouwers, E. W., Palacios, J. L., and Smith, E. C. "The Experimental Investigation of a Rotor Hover Icing Model with Shedding," *AHS 66th Annual Forum.* American Helicopter Society, Phoenix, 2010.

29. Palacios, J. L., Brouwers, E. W., Han, Y., and Smith, E. C. "Adverse Environment Rotor Test Stand Calibration Procedures and Ice Shape Correlation," *AHS 66th Annual Forum.* American Helicopter Society, Phoenix, 2010.

30. Bain, J., Deresz, R., Sankar, L., Egolf, T. A., Flemming, R., and Kreeger, E. "Effects of Icing on Rotary

Wing Loads and Surface Heat Transfer," *49th AIAA Aerospace Sciences Meeting*. Orlando, Florida, 2011.

31. Bain, J., Lakshmi, S. N., Egolf, T. A., Flemming, R. J., and Kreeger, R. E. "Progress Towards Modeling the Effects of Ice Accretion on Rotorcraft Performance in Hover and Forward Flight," *35th European Rotorcraft Forum*. Hamburg, Germany, 2009.

32. Nucci, M., Bain, J., Sankar, L., Kreeger, E., Egolf, T. A., and Flemming, R. "Assessment of the Effects of Computational Parameters on Physics-based Models of Ice Accretion," *48th AIAA Aerospace Sciences Meeting*. AIAA, Orlando, 2010.

33. Rajmohan, N., Bain, J., Michael, N., Lakshmi, S., Flemming, R., Egolf, T. A., and Kreeger, R. "Icing Studies for the UH-60A Rotor in Forward Flight." AHS Aeromechanical Specialists' Conference Proceedings, 2010.

34. Narducci, R., Orr, S., and Kreeger, R. E. "Application of a High-Fidelity Icing Analysis Method to a Model-Scale Rotor in Forward Flight." Proc. AHS Annual Forum, Virginia Beach, VA, May 2011.

35. Kreeger, R. E., Sankar, L., Narducci, R., and Kunz, R. "Progress in Rotorcraft Icing Computational Tool Development." SAE Technical Paper 2015-01-2088, 2015.

36. Quackenbush, T.R., et al. (1997) "LDTRAN/CB: A Model for the Assessment of the Impact of the Chemical/Biological Environment on Military Rotorcraft Operations", Presented at the AHS 53rd Annual Forum, Virginia Beach, VA., April 29-May 1, 1997.

Fleetwide Integrated Prognostics Health Management for Emerging eVTOL Operations

Andrew Vechart
Principal Aerospace Engineer
Sentient Science
Buffalo, NY, USA

Jacek Kawecki
Elevate Vehicle Components Lead
Uber Technologies
San Francisco, CA, USA

ABSTRACT

Anticipation of a revolution in personal travel fueled by electric vertical take-off and landing aircraft is leading to the development highly innovative platforms. Health management systems and technologies, such as digital twins, physics-based models, and machine learning-based models, will play key roles as OEMs and operators manage safety and reliability of these brand-new platforms. Key to enabling health management is capture and storage of appropriate data streams. For distributed electric propulsion aircraft, it is anticipated that algorithms based on electrical signals, especially electrical current signatures, provides a pathway towards diagnosing a myriad of mechanical, electrical, and magnetic faults within key aircraft subsystems. Maturing the prognostics capability of health management algorithms will further enable the adoption and growth of urban air mobility, as have knowledge into future component health state enables robust decision making for operations, sustainment, and supply chain management support of these vehicles.

INTRODUCTION

Anticipation of a revolution in personal travel fueled by eVTOLs (electric Vertical Take-Off and Landing aircraft) is leading to the development highly innovative platforms. Companies developing clean-sheet designs for eVTOLs have a unique opportunity to strategically address safety and reliability of these vehicles, leveraging state-of-the-art technologies and lessons learned from the adjacent rotorcraft and commercial airline industries. To achieve scalability of the eVTOL/aerial on-demand mobility industry, from the outset OEMs must manage safety and reliability as operations tend towards greater autonomy. This will partially be achieved by incorporation of a robust automated methodology for managing the health of the vehicles and their critical components. This paper presents a study of the life cycle of data supporting a Fleetwide Integrated Prognostics Health Management architecture using Uber Elevate conceptual aircraft and operations.

Most publicly shown eVTOL concepts use distributed electric propulsion (DEP) powering schemes. One advantage of a DEP approach is that it inherently allows for redundant design. However, DEP aircraft require flight control systems (FCS), sometimes referred to as Fly-By-Wire systems, to maintain stable flight. FCS' require sensors that can be leveraged by Health Management. This effort will not include analysis of sensors required by FCS' but it is the overall intent to maximize the utility of onboard sensors.

Presented at the Vertical Flight Society Autonomous VTOL Technical Meeting and Electric VTOL Symposium, Mesa, Arizona, UAS, Jan. 29-31, 2019.

KEY COMPONENT HEALTH MANAGEMENT DATA NEEDS

Since the 1990's, Health and Usage Monitoring Systems (HUMS) have had a significant positive impact on safety and availability for rotorcraft operations (Ref. 1). One report on helicopter safety states "the introduction of HUMS is unanimously judged to be the most significant isolated safety improvement measure during the last decade" (Ref. 2). Another study conducted by the UK Civil Aviation Authority (CAA) estimates that 65% of rotor/drive train failure related accidents could be prevented by HUMS (Ref. 3). HUMS has become so ubiquitous in rotorcraft that standards have been written specifically for its airworthiness approval. FAA Advisory Circular (AC) 29-2C "Certification of Transport Category Rotorcraft" and AC 27-1B "Certification of Normal Category Rotorcraft" both contain miscellaneous guidance sections for airworthiness approval for rotorcraft HUMS. These references specifically address HUMS installation, credit validation, and Instructions for Continued Airworthiness (ICA). The US Army, the largest rotorcraft operator in the world, also provides guidance for HUMS (as well as other condition-based maintenance technologies) in ADS-79E-HDBK "Aeronautical Design Standards Handbook Condition Based Maintenance System for US Army Aircraft".

The benefits offered by HUMS have been achieved using a combination of damage signature detection and usage tracking to anticipate or diagnose failures in flight-critical parts of the aircraft. Similar efforts are applied in commercial fixed-wing aircraft as well; NASA has put significant effort into maturing system, subsystem, and component health management technologies that contribute to improved air vehicle safety (Ref. 4) (Ref. 5). It is expected that this wealth of knowledge developed over the last few decades will feature prominently in aerial on-demand mobility vehicles with the primary goal of ensuring the safety of people both in the air and on the ground.

These new eVTOLs will also be able to take advantage of emerging technologies for safety improvement. Digital clones, or digital twins, are becoming increasingly capable as connectivity, processing, and storage continue to improve for air vehicles. A joint paper from NASA and the US Air Force indicates, "This paradigm shift, the Digital Twin, integrates ultra-high fidelity simulation with the vehicle's on-board integrated vehicle health management system, maintenance history and all available historical and fleet data to mirror the life of its flying twin and enable unprecedented levels of safety and reliability" (Ref. 6). Digital clones encapsulate detailed design knowledge and experience of the OEMs and component experts into virtual representations of physical assets and maximize the knowledge obtainable from installed sensors to understand the current and future health state of components. Given the extreme sensitivity of eVTOLs to added sensor weight, maximizing the benefit from installed sensors is highly desirable. Digital clones are also beneficial as urban air mobility expands to new markets – estimated component performance may be assessed virtually and subsequently tracked as new missions are undertaken. Uber sees value in evaluating impact of HUMS, digital clones, and other health management technologies on aircraft fleet-level reliability, safety, and operating cost.

Another broadly popular area of research in the area of health management that will undoubtedly play a role in urban air mobility is use of machine learning and artificial intelligence to develop improved diagnostics and new prognostics capabilities. The US Army, for example, has been focusing on developing new machine-learning based diagnostics for rotorcraft sustainment using the large database of historical HUMS data that has been collected over several years (Ref. 7).

Common to the emerging and traditional health management approaches for aircraft components is the need for data, specifically the right data in the right place at the right time. From a health management perspective, the right data is the data that is sensitive to the physical phenomena of interest. That physical phenomena may be a key operational state characteristic of the aircraft (such as vertical acceleration) or a damage signature of a faulted component (such as increasing vibration due to a spalled bearing race). Existing applications indicate that many data sources may be required to enable a comprehensive health management approach, including, for example:

- operational state data
- performance data
- environmental data
- commanded data
- sensor data
- diagnostics data
- built-in test results

eVTOL subsystems will employ basic fault detection technologies as a part of their normal function. They will also collect data that could be used for more sophisticated health management of the vehicle components. However, the data that is collected for normal operation of the subsystem may not be accessible for use for health management. Addition of health management functionality to a subsystem controller may place an undue burden on the certification of the controller, for example. Nevertheless, use of health management technologies can improve safety and operational efficiency. This section explores whether additional sensors should be added to the vehicle strictly for the purposes of health management focusing on a few key subsystems.

Motors, Drives, and Controllers

Basic Fault Detection

The motors for this application are likely to be permanent magnet synchronous motors (PMSM) using (e.g. SiC MOSFET) inverter drives and electronics comprising a controller. To achieve noise and efficiency requirements advanced control schemes will be employed that require measurements of each of the phase currents along with the rotor position. Many controllers in these applications also monitor the temperature of the drive and motor. The primary purpose of these functions is to protect the controllers, drives, and motors from damage to ensure operability. Certain faults are likely to be detected and annunciated by the motor controller as part of its normal function. The faults that fall into this category are listed in Table 1.

Table 1 – Basic Motor/Drive/Controller Fault Protections

Faults
Motor phase overcurrent
Motor overtemperature
Drive overtemperature
Rotor position sensor failure
Input undervoltage
Input overvoltage
Input parameters out of range
Internal circuit error
Communication bus error
Software error
Overload
Motor stall

Addition of Current Sensors for Health Management

Though motor controllers will measure phase currents to implement a control strategy, it is possible that these current measurements will not be accessible for analysis of the health of the motor and drive. Integrating a comprehensive health monitoring strategy into a controller may create an undue certification burden on both the controller function and the health monitoring function. However, having measurements of the individual phase currents of the motor as well as the input current to the drive can enable diagnosis and prediction of many types of motor faults. After initiation these faults may not be severe enough to trigger the basic fault protections described in the previous section. While it may be possible for the motor to function as required to drive propulsors in a satisfactory fashion despite some of these faults, progress of these faults can lead to substantially more damage. These faults generally lead to dynamic forces that manifest as fatigue-inducing stresses.

The types of faults that can be detected by monitoring the phase currents include electrical, magnetic, and mechanical faults. Electrical and magnetic faults, such as inter-turn short circuits, line-to-line short circuits, demagnetizations, and open circuits are detectable by current measurements sampled at twice the product of motor rotation frequency and number of pole pairs in the motor, roughly on the order of hundreds of Hz. Left uncorrected, short circuit type faults lead to localized heating and periodic current surges that lead to additional thermal, electrical, and mechanical stresses on the drive, windings, rotor, and power supply system. These types of faults manifest in disparate current spectra (differing frequencies and/or amplitudes) within the three phases powering the motor. Faults detectable using current samples collected at moderate rates are shown in Table 2 (Ref. 8) (Ref. 9).

Table 2 - Motor/Drive/Controller Electrical Faults Detectable Using Moderate Current Sample Rates

Faults
Inter-turn short circuit
Demagnetization
Loose or cracked magnets
Line-to-line short circuit
Open circuits

Increasing the current measurement sample rate to the order of the drive switching frequency, roughly in the 10-20 kHz range, enables additional fault detection. This higher sampling rate gives insight into the switching dynamics of the drive components (e.g. MOSFETs) that can be leveraged to diagnose degradation of these drive components. Similarly, current measurements at these higher rates are able to capture intermittent type faults, such as arc faults. Series arcs manifest as short sporadic durations of reduced current consumption due to temporary disconnections within the electric circuit due to, for example, broken wires, loose connectors, and pushed or bent connector pins. Faults detectable by higher current sampling rates are given in Table 3 (Ref. 10) (Ref. 11).

Table 3 - Motor/Drive/Controller Electrical Faults Detectable Using High Current Sample Rates

Faults
Switching device degradation
Loose connectors
Broken wires
Pushed/bent connector pins

Current analysis can also be used to diagnose mechanical faults within a motor/propulsor system. Mechanical faults that cause fluctuations or differences in air gap flux values will manifest as fluctuations in stator currents and be reflected in the input phase current measurements. The types of mechanical faults that can be diagnosed by having this current data are given in Table 4 (Ref. 12) (Ref. 13).

Table 4 - Motor/Drive/Controller Mechanical Faults Detectable Using Current Samples

Faults
Rotor eccentricity
Mass imbalance (rotor, propulsor)
Bent shaft
Shaft misalignment
Bearing fault (spalling, lubrication problems)
Rotor rub

There are some common current sensing technologies that may be used to facilitate the above functions, including Hall effect-based and fluxgate-based sensors. Shunt resistor based current sensing may be excessively intrusive to add to the system and introduces additional losses in efficiency. Despite sensor availability, there may be issues with accessibility to the current carrying elements between the motor drive and motor itself, especially in the cases where tightly integrated motor/drive packages are utilized. In this case some of the above fault protections may be realized using a current sensor on the DC input or return line to/from the drive though with much lower sensitivity and resolution and greater computational complexity. Alternatively, the path towards realizing the above benefits lies in including health management functionality within the motor controller itself, which has access to phase current measurements.

Addition of Vibration Sensors for Health Management

Many different types of motor faults will introduce dynamic forces during rotation of the rotor that manifest as vibrations that can be measured by, say, accelerometers. This includes electrical and magnetic faults as well as mechanical faults. Specifically, the types of faults that have been shown to affect the vibration spectrum of motors in previous applications are given in Table 5 (Ref. 14).

Table 5 - Motor/Drive/Controller Faults Detectable Using Vibration Data

Faults
Inter-turn short circuit
Demagnetization
Rotor eccentricity
Mass imbalance (rotor, propulsor)
Bent shaft
Stator lamination shorts
Line-to-line short circuit
Open circuits
Shaft misalignment
Bearing fault (spalling, lubrication problems)
Rotor rub

The sensitivity of vibration data to these types of faults generally depends on the integrity of the load path between the sensor and the vibrating components. Often accelerometers are placed as close to a bearing as possible on the case of a machine preferably with a direct load path between the bearing outer race and the accelerometer. However, for the concept eVTOL vehicles that have been shown, as well as for some of the motor candidates, such a convenient load path may not be available. Especially for outrunner-style motors where the shaft of the motor is stationary and the case rotates, alternative placements of the accelerometers are necessary, such as on a frame holding the motor in place. Flexibility in the load path may prevent collection of high integrity vibration measurements and reduce the ability of the health management function to achieve the protections listed above.

For applications where it is practical to have both current and vibration data related to a motor, using both data streams improves the ability to disambiguate between different types of faults. This proves useful because different fault types may require different levels of urgency with respect to required maintenance actions. In other words, perhaps a mass imbalance can wait longer to correct than a line-to-line short circuit; having the ability to disambiguate between these two faults types is useful from an operations and sustainment standpoint.

Other Sensors

Other data streams may be useful for health management of a motor/drive/controller system. Adding a rotor position sensor supports certain algorithms that transform three phase current signals into different stationary frames before conducting fault analysis. A tachometer (or differentiation of a rotor position signal) is useful to support vibration spectrum analysis, especially for transient events. Additionally, tachometer or position sensors can be used to support rotor balance calculations.

Temperature sensors may also help with fault prediction and diagnosis. These may be useful on the switching devices (increasing "on-resistance" causes Joule heating), in various positions within the motor (shorts or rotor rub causing excessive heating), and near the bearings (incorrect lubrication leads to excessive heat generation).

Torque and load (e.g. strain gauge) sensors are useful for usage-based health estimation of the motor and its subcomponents as well as for overload monitoring.

Gearboxes

Basic Fault Detection

The gearboxes discussed in this section are those that may be used between motors and propulsors on eVTOLs. Given the unique designs of these vehicles it is likely that any gearboxes that are employed will take on architectures and perhaps even be constructed using materials that are not common in applications in aerospace or other industries. However, it is not expected that these gearboxes will be multi- or continuously variable-speed; instead, thrust will be modulated by varying motors speed, varying propulsor pitch, or both. Such gearboxes are completely mechanical devices, not requiring any sensors to operate properly. As such, a gearbox would not have any inherent fault detection capabilities by themselves.

Addition of Oil Debris Monitoring (ODM) Sensors for Health Management

In aerospace applications, gearboxes are often equipped with ODM sensors. In most cases, these are chip detectors that provide a cockpit indication when enough metallic particles build-up on a sensor to close a circuit. More sophisticated sensors are available that also differentiate between ferrous and conducting non-ferrous materials and estimate particle sizes. The premise of these sensors is that as contact surfaces at gear meshes or within bearings wear, pit, or spall, they produce particles that flow through the lubrications system. Correspondingly, these sensors are only appropriate for a certain type of gearbox lubrication system. A gearbox chip detection in rotorcraft is a serious event. Often, if a chip detection is indicated to a pilot, the aircraft is to be landed as soon as possible. Faults detectable with ODM are given in Table 6 (Ref. 15) (Ref. 16).

Table 6 – Gearbox Faults Detectable Using ODM Sensors

Faults
Gear and bearing wear
Gear and bearing pitting
Gear and bearing spalling wires

Addition of Vibration Sensors for Health Management

One of the primary functions of HUMS in rotorcraft is to implement vibration monitoring of dynamic drivetrain components, mainly using accelerometers. Accelerometers reflect any impacts at damage sites and differences in contact stresses in the various parts of the gearbox. As a result, vibration monitoring has proven quite effective for gearbox condition monitoring, not only in aerospace but in other industries as well. Typically, a single-axis accelerometer is installed at specific bearing locations on the gearbox housing. In the types of gearboxes envisioned for eVTOL applications as described above, it is likely that one or two accelerometers would be used per gearbox, at one or both of the input or output shaft bearings. In addition, many vibration-based diagnostics leverage time synchronous averaging, where time-series signals are averaged to increase the signal-to-noise ratio of the signals of interest. This approach requires a once-per-revolution signal to accompany the vibration data from something like a magnetic or optical tachometer signal. Faults detectable using vibration data are given in Table 7 (Ref. 17) (Ref. 18) .

Table 7 - Gearbox Faults Detectable Using Vibration Data

Faults
Gear and bearing wear
Gear and bearing pitting
Incorrect lubrication
Gear tooth breakage or loss
Eccentricity
Housing cracking
Misalignment
Gear and bearing spalling
Gear and bearing scuffing
Contaminated lubricant
Imbalance
Shaft cracking
Bent shaft

Other Sensors

A few other sensing technologies have been applied to gearbox diagnostics. For example, temperature sensors may be used to detect increased friction due to damaged components or contaminated or incorrect (over or under) lubrication. In other relatively recent research, acoustic emission (AE) sensors have been shown to be sensitive to gearbox damage (Ref. 19). However, publicly available information suggests that AE for diagnostics remains in the realm of research and is awaiting practical application.

Another practical health monitoring strategy for gearboxes involves the application of motor current signature analysis (MCSA) to permanent magnet synchronous motors (PMSM). While MCSA is typically used for identifying faults in squirrel cage-based induction motors, researchers have shown that torque modulations due to faulty driven machinery will modulate the load of the PMSMs accordingly, showing promise for monitoring the health state of gears and bearings (Ref. 20). In other words, if the motor current sensing strategy discussed in the "Motors, Drives, and Controllers" section is implemented, a path may exist for implementing gearbox diagnostics as well.

Actuators

Basic Fault Detection

It is anticipated that electromechanical actuators (EMA) will be utilized in DEP aircraft. One of the most critical potential functions of EMAs is movement of control surfaces in flight. In this section, it is assumed ball-screw EMAs are used in a control surface movement application. Ball-screw EMAs include the screw, the nut with ball return channel, a motor, and a motor controller. Some further include gearing between the motor and ball-screw. Substantial motions may not be commanded to the EMA in flight. As a result, pre-flight flight controls checks may help elucidate some of these faults. With a motor controller involved that has access to speed, position, current, voltage, and possibly temperature information, a basic level of fault protection is likely to be available for EMAs. Furthermore, given the application of moving control surfaces, more severe faults (e.g. jamming) of the actuator may be determined by discrepancies between commands and measured surface positions. The types of faults that may be detectable (e.g. with controller functionality or using a reasoner) using sensors installed for proper operation of an EMA and for use in flight controls is given in Table 8.

Table 8 - Basic Actuator Fault Protections

Faults
Motor phase overcurrent
Motor overtemperature
Drive overtemperature
Motor position sensor failure
Low input voltage
Input overvoltage
Jamming
Input parameters out of range
Internal circuit error
Communication bus error
Software error
Overload
Motor stall
Breakage leading to floating actuator

Addition of Current Sensors for Health Management

The EMA controller will provide detection of some of the more obvious faults. As with the motors discussed earlier, having motor current measurement data available for diagnostic algorithms may provide detection capability for additional faults, especially ones that may cause larger problems if left uncorrected. While the list of potential faults that could be detected using motor current information is long (Table 9), disambiguation of the faults may be difficult in a health management setting (Ref. 21) (Ref. 22).

Table 9 - EMA Additional Fault Detectable Using Current Data

Faults
Inter-turn short circuit
Demagnetization
Loose or cracked magnets
Switching device degradation
Loose connectors
Rotor eccentricity
Mass imbalance (rotor)
Bent shaft
Spalling/pitting/scuffing/wear of the motor, gear, and screw support bearings
Jamming
Line-to-line short circuit
Open circuits
Broken wires
Pushed/bent connector pins
Shaft misalignment
Bearing fault (spalling, lubrication problems)
Rotor rub
Spalling/pitting/scuffing/wear of the screw, return channel, or balls
Spalling/pitting/scuffing/wear of the gears
Breakage leading to floating actuator

Other Sensors

It is assumed actuator position data is available for health management purposes, along with corresponding position commands. If not available, adding a position sensor for health management purposes would be useful for diagnosing faults such as jams, wear leading to increasing backlash, broken linkages, and sluggish dynamic response due to motor faults or increased friction from spalling or lubrication issues (Ref. 23).

Several studies published on EMA fault detection incorporate accelerometers on the actuator. However, while vibration-based gear and bearing condition indicators are well-defined for continuously rotating systems, EMAs in a flight control capacity generally experience small movements and few or fractional rotations of the screw. Examining the general vibration energy, normalizing by the load, can provide some insight into bearing faults, spalling/pitting/wear of the screw, balls, or return channel, and jams of the return channel. However, it seems inconclusive that adding accelerometers to monitor vibration of the actuator significantly improves fault detection capability beyond helping narrow a fault ambiguity set (Ref. 22).

Published studies on health management for EMAs also incorporated temperature sensors on the motor housing and/or nut. Specifically, incorporation of temperature sensors on the motor helped detect motor current increases and nut temperature increases due to the increased friction from an actuator jam (Ref. 24).

INTEGRATION FOR EFFICIENCY

All-electric urban air mobility vehicles will be hypersensitive to the typical aerospace evaluation metrics of size, weight, and power (SWaP) given the limitations on battery capacity and charging times. Consequently, PHM implementations for these vehicles need to avoid installation of duplicate sensors and data management systems which would increase the weight of the vehicle and increase the power requirements on the batteries. In addition, from an operational standpoint it is advantageous to reduce the amount of duplicate data that needs to be transferred off a vehicle for various purposes, including PHM.

Common Data Needed for PHM

HUMS and/or FDM systems aboard rotorcraft today record parametric information required to estimate flight regimes encountered during flights as well as the amount of time spent in each flight regime. This is useful for usage-based health-state estimation of components on-board the aircraft as supported by load surveys conducted by the aircraft in each of the identified flight regimes. It is anticipated that a similar capability will be required for eVTOL. Furthermore, models for component health state prediction may rely on some of the same data. For example, vehicle linear and angular accelerations can be leveraged to estimate the loads borne by bearings supporting propulsors, informing physics-based life prediction models for those bearings. A vehicle design enabling health management should make this type of data available for in-flight and/or post-flight analysis. Types of information that may be used across PHM functions for various aircraft components are given in Table 10.

Table 10 – Common Data Needed for PHM

Data
Air speed
Altitude
Attitude (roll/pitch/yaw)
Relative wind angles (angle-of-attack, sideslip angle)
Linear and angular accelerations
Outside air temperature and pressure
Effector commands
Effector measured outputs
Center of gravity / vehicle weight
Aircraft configuration data
Battery SoC / SoH

Data Sharing

Sharing this type of data has proven non-trivial in existing aerospace applications for several reasons. Recent advances in machine learning and artificial intelligence have caused the perceived value of data to increase tremendously. To maximize capture of this value data owners are disinclined to make that data available outside of their own organizations. Data ownership is divided amongst original equipment manufacturers (OEM), operators, maintenance, repair, and overhaul (MRO) organizations, and regulators.

Sometimes data ownership is not clearly defined at all (Ref. 25).

It is not just the potential value of data that prevents owners from making it freely available. Data could be used to determine design, performance, or reliability information considered proprietary and highly sensitive to OEMs and suppliers. Releasing operational or passenger information may be a privacy concern. Data preparation, management, and delivery may impose an unwanted burden on an organization.

Despite these difficulties, there are and have been organized efforts to reduce the barriers to sharing information. The Open Group Future Airborne Capability Environment (FACE) Consortium is working to produce open standards aimed at making military avionics products more interoperable, interchangeable, and secure (Ref. 26). An industry group leveraged U.S. Navy funding to develop the Open System Architecture for Condition-Based Maintenance (OSA-CBM) aimed at providing a standard architecture for condition-based maintenance systems (Ref. 27). It is expected that data sharing issues will continue to be a challenge going forward, but not one that is insurmountable.

LEVERAGING PROGNOSTICS TECHNOLOGIES

Benefits of Prognostic Capabilities

Having a component health prognostics capability can provide significant benefits to eVTOL operators. Prognostics in the context considered here can be understood to mean either prediction of a future component failure or evolution/degradation of a component given a known or estimated current health state. Prognostics is inherently uncertain, as producing a prediction of the future health state of an in-service component requires assumptions about the future operating conditions the component will experience. However, having a usage-based probabilistic remaining useful life estimate of in-service components is a big advantage over simple mean-time-to-failure data.

Having a quantitative estimate of the time until a component will need to be replaced provides significant decision support in multiple areas. First, an operator is able to better plan for required vehicle maintenance. If the vehicle needs to be repaired at a maintenance depot, having more time to plan for locating the vehicle at the depot and having the correct personnel and parts on-hand is helpful. Correspondingly, operators are able to better plan for repairs that may take the vehicle out of operation for significant periods of time.

Taking a view of component health across a fleet, further benefits are realized. Supply chain managers are better able to plan purchase timing and quantities of replacement components. Similarly, prediction of emerging fleet-level reliability issues provides supply chain managers and OEMs time to evaluate and consider alternate suppliers that may alleviate similar issues in the future. Relatedly, operators and maintenance service providers are able to better establish budgets with increasing visibility into future repair needs.

Other Considerations Related to PHM Implementation

While it is generally acknowledged that health management technologies offer an opportunity to improve aircraft safety (Ref. 2), this comes at a cost. Addition of sensors, wiring, and signal conditioning electronics reduces vehicle operating capacity and increases operating costs (in addition to the initial cost of the health management system). Burden is placed on on-board computing capacity, storage, and data buses to handle data and run algorithms. Some sort of ground infrastructure is also generally required to facilitate data retrieval off of the aircraft, to manage the data, and to make it available to decision-makers.

There are also support costs associated with implementing PHM. To effectively realize benefits of leveraging health management, data and algorithm outputs must be regularly downloaded and examined. Technologies can help automate and expedite this process as much as possible. Ultimately, however, human decision makers must complete the feedback loop, determining if any actions are required as a result of health management indications. In addition, as health management technologies mature or aircraft configurations change, updating on-board health management software must be managed.

Health management is an ever-evolving field of study. The technologies are not perfect and generally have to balance sensitivity to failures with the potential for false alarms. This may especially be an issue in the early stages of eVTOL employment when highly accurate usage profiles are still being developed. False alarms can lead to extraneous inspections, reduced aircraft availability, or premature replacement of components.

This investment in health management must be considered in the context of the safety and economic benefits offered by the various technologies to determine which technologies to pursue.

Digital Clones for Prognostics

As alluded to earlier on, digital clones, or digital twins, provide an opportunity for significantly improved management of in-service components for new air vehicles. Digital clones that directly estimate damage initiation and evolution for in-service components provide live health state estimation for components without needing to sense damage directly. This is significant because likely component damage can be predicted before any damage signature is picked up by health management sensors.

As an illustration, consider the motors driving eVTOL propulsors. The manufacturer-specified life of the motor is often limited by the life of the bearings supporting the rotor. The bearing life, in turn, is computed using industry standards that require assumptions and simplifications about the application as well as limited reliability testing that only partially represents in-service conditions experienced by the motor. A high-fidelity digital clone of the motor can be "flown" along with the in-service motor using in-fight measurements of the conditions experienced by the physical motor. This digital clone could then signal any impending faults to be expected in the fleet based on real usage data before motor current signature analysis or vibration data would provide any indication. Part of this process is to apply "future" loading conditions to the digital clone to observe the estimated damage evolution. This capability also allows an operator to explore alternative future mission profiles to observe the effect on the rate of damage evolution, enabling many of the benefits discussed above.

Successful application of digital clones requires that the necessary data streams are available to subject the digital clone to the same operating conditions experienced by the fielded component. Direct measurement of relevant quantities provides the best opportunity for obtaining high-accuracy digital clone outputs. With reference to motors and gearboxes, digital clones that include damage mechanics models for bearings require information on bearing temperature, operating speed, torques, axial loads, and bending moments imparted on the bearing. While preferable, it may not be practical to measure all of these quantities in flight. However, they may be estimated from other measured and available quantities. Regime recognition capabilities informed by vehicle velocity, attitude, altitude, linear and angular accelerations, and effector commands, all of which are easier to measure, can be used to estimate the required quantities to execute the digital clone.

CONCLUSIONS

Prognostic health management technologies are expected to feature prominently in eVTOL operations to achieve safety, reliability, and availability goals.. The work herein identifies relevant concepts eVTOL manufacturers and operators should consider to enable efficient PHM implementation.

1. The concept of HUMS for eVTOLs differs in important ways from current fixed and rotary wing aircraft. For the critical components considered in this paper, electrical current signature analysis has been identified as a method to detect a myriad of faults. This contrasts with HUMS for rotorcraft, where vibration monitoring features prominently. This is a reflection of the focus on DEP architectures.
2. Constraints on battery capacity and charging rates causes hypersensitivity of eVTOL urban air mobility vehicles to traditional aerospace SWaP-C metrics. As a result, integrating health management technologies into the vehicle design as much as possible is desired. Correspondingly, maximizing data sharing to minimize data duplication (e.g. avoiding extra sensors, unneeded data bus traffic, and extra storage) is paramount.
3. Prognostic technologies provide several potential benefits to support adoption and growth of urban air mobility. Digital clones especially offer a method to use actual operational data to estimate the health state of components, predict damage progression and required maintenance activities, and to conduct exploratory analyses of impacts of various future mission profiles on vehicle reliability. Uber believes there is considerable potential in HUMS and digital clones in reducing operating cost and increasing aircraft safety. The use of these technologies requires further investigation to quantify their benefits including discussions with regulators for certification credit.

Author contact: Andrew Vechart
avechart@sentientscience.com

REFERENCES

[1]Clark, E., Edwards, C., Perry, P., Campbell, G. and Stevens, M., "Helicopter Safety in the Oil and Gas Business," in *IADC/SPE Drilling Conference*, Miami, 2006.

[2]Hokstad, P., Jersin, E., Hansen, K. G., Sneltvedt, J. and Sten, T., "Helicopter Safety Study 2, Volume I: Main Report," SINTEF Industrial Management Safety and Reliability, Trondheim, 1999.

[3]Clark, E., Edwards, C., Perry, P., Stevens, M., Sheffield, E. R. and Campbell, W. A., "Helicopter Safety in the Oil and Gas Business," Society of Petroleum Engineers, Miami, 2006.

[4]Reveley, S. M., Briggs, L. J., Evans, K. J., Jones, M. S., Kurtoglu, T., Leone, M. K., Sadifer, E. C. and Thomas, A. M., "Commercial Aircraft Integrated Vehicle Health Management Study," NASA, Cleveland, 2010.

[5]Hunter, W. G., Ross, W. R., Berger, E. D., J. D. L., Mah, W. R., Perey, F. D., Schuet, R. S., Simon, L. D. and Smith, W. S., "A Concept of Operations For An Integrated Vehicle Health Assurance System," NASA, Cleveland, 2013.

[6]Glaessgen, E. and Stargel, D., "The Digital Twin Paradigm for Future NASA and U.S. Air Force Vehicles," in *53rd Structures, Structural Dynamics, and Materials Conference*, Honolulu, 2012.

[7]Wade, D., Wilson, A., Reddy, A. and Bharadwaj, R., "Validating Machine-learned Diagnostics Classifiers in Safety Critical Applications with Imbalanced Populations,"

in *Annual Conference of the Prognostics and Health Management Society*, Philadelphia, 2018.

[8]Vinson, G., Combacau, M., Prado, T. and Ribo, P., "Permanent Magnets Synchronous Machines Fault Detection and Identification," in *IECON 2012 - 38th Annual Conference on IEEE Industrial Electronics Society*, Montreal, 2012.

[9]Rohan, A., Rabah, M. and Ho Kim, S., "An Integrated Fault Detection and Identification System for Permanent Magnet Synchronous Motor in Electric Vehicles," *International Journal of Fuzzy Logic and Intelligent Systems,* vol. 18, no. 1, 25 March 2018. pp. 20-28,

[10]Raj, N., Mathew, J., G, J. and George, S., "Open-Transistor Fault Detection and Diagnosis Based on Current Trajectory in a Two-level Voltage Source Inverter," in *Global Colloquium in Recent Advancement and Effectual Researches in Engineering, Science, and Technology (RAEREST 2016)*, Kottayam, 2016.

[11]Valko, A., Petrella, J., Murray, D., Frederick, D. and Alves, K., "Solid-State Secondary Power Distribution," Federal Aviation Administration, Renton, 2014.

[12]Varal, P., Patole, V., Shinde, S., Deshmukh, P. and Srivastava, D., "Bearing Fault Analysis of Permanent Magnet Synchronous Motor (PMSM) Using Current Signature Analysis," *International Research Journal of Engineering and Technology,* vol. 04, no. 04, April 2017. pp. 410-415,

[13]Rosero, J., Romeral, L., Rosero, E. and Urresty, J., "Fault Detection in Dynamic Conditions By Means of Discrete Wavelet Decomposition for PMSM Running Under Bearing Damage," in *2009 Twenty-Fourth Annual IEEE Applied Power Electronics Conference and Exposition*, Washington D.C., 2009.

[14]Singh, G. and Al Kazzaz, A. S. S., "Induction Machine Drive Condition Monitoring and Diagnostic Research - A Survey," *Electric Power Systems Research,* vol. 64, no. 2, February 2003. pp. 145-158,

[15]Dempsey, J. P., "A Comparison of Vibration and Oil Debris Gear Damage Detection Methods Applied to Pitting Damage," NASA, Cleveland, 2000.

[16]Howe, B. and Muir, D., "In-Line Oil Debris Monitor (ODM) for Helicopter Gearbox Condition Assessment," Defense Technical Information Center, Fort Belvoir, 1998.

[17]Edwards, T., Patterson, A., Ballentine, E. and Bayoumi, A., "Dependence of Vibration Characteristics on Grease Service Levels in AH-64D Intermediate Gearbox," in *AHS Forum 70*, Montreal, 2014.

[18]Lebold, M., McClintic, K., Campbell, R., Byington, C. and Maynard, K., "Review of Vibration Analysis Methods for Gearbox Diagnostics and Prognostics," in *Proceedings of the 54th Meeting of the Society for Machinery Failure Prevention Technology*, Virginia Beach, 2000.

[19]Elasha, F., Greaves, M. and Mba, D., "Planetary Bearing Defect Detection In A Commercial Helicopter Main Gearbox With Vibration And Acoustic Emission," *Structural Health Monitoring,* vol. 17, no. 5, 11 January 2018. pp. 1192-1212,

[20]Adrakani, D. H., Liu, Z., Lee, J., Bravo-Imaz, I. and Arnaiz, A., "Motor Current Signature Analysis for Gearbox Fault Diagnosis In Transient Speed Regimes," in *IEEE Conference on Prognostics and Health Management*, Austin, 2015.

[21]Babu, S. G., Lingamurthy, A. and Sekhar, A., "Condition Monitoring of Brushless DC Motor-Based Electromechanical Linear Actuators Using Motor Current Signature Analysis," *The International Journal of Condition Monitoring,* vol. 1, no. 1, June 2011.

[22]Balaban, E., Saxena, A., Narasimhan, S., Roychoudhury, I., Koopmans, M., Ott, C. and Goebel, K., "Prognostics Health-Management System Development for Electromechanical Actuators," *Journal of Aerospace Information Systems,* vol. 12, no. 3, 2015. pp. 329-344,

[23]Isturiz, A., Vinals, J., Abete, M. J. and Iturrospe, A., "Health Monitoring Strategy for Electromechanical Actuator Systems and Components. Screw Backlash and Fatigue Estimation," in *Recent Advances in Aerospace Actuation Systems and Components*, Toulouse, 2012.

[24]Balaban, E., Bansal, P., Stoelting, P., Saxena, A., Goebel, F. K. and Curran, S., "A Diagnostic Approach For Electro-Mechanical Actuators In Aerospace Systems," in *2009 IEEE Aerospace Conference*, Big Sky, 2009.

[25]Canaday, H., "Aircraft Data Ownership Debate is Ongoing," 3 March 2017. [Online]. Available: https://www.mro-network.com/technology/aircraft-data-ownership-debate-ongoing. [Accessed 19 December 2018].

[26]Open Group, "Future Airborne Capability Environment," [Online]. Available: www.opengroup.org/face. [Accessed 19 December 2018].

[27]MIMOSA, "Open Standards for Physical Asset Management," [Online]. Available: www.mimosa.org/mimosa-osa-cbm/. [Accessed 19 December 2018].

Multi Objective Optimization of Wiring Harnesses for Overall System Performance of Electric Fixed Wing VTOL UAVs

Sebastian Oberschwendtner
M.Sc.
Technical University of Munich
85748 Garching, Germany

Christian Roessler
Dr.-Ing.
Technical University of Munich
85748 Garching, Germany

Mirko Hornung
Univ.-Prof. Dr.-Ing.
Technical University of Munich
85748 Garching, Germany

Abstract

This paper deals with the impact of the wiring harness of an electric fixed wing VTOL UAV to the flight performance of the aircraft. The main influence parameters are the power loss within the harness and the harness weight. Associated with the power loss, the temperature of the harness wires has to be considered in the design of the harness. This paper shows the importance of a detailed model of the behavior of a wire under load and derives such a model using the manufacturer's data of different wire types. The manufacturer's data is used for the geometry model. A pragmatic thermal model is adapted from design guidelines provided by the FAA. The model is validated and used in an optimization framework to improve the overall hover endurance of a VTOL aircraft. The influence of different configurations is examined for different objectives. As a result, the harness mass tends towards higher masses to minimize the power loss within the harness.

Introduction

Within an UAV design process, the wiring harness as an important subsystem has to be defined. Especially in electric VTOL aircraft, the wiring harness has a significant influence on the overall performance, since for example the power loss within the wiring harness of the hover power train can be up to 10% of the hover power, as is shown later on. Hence, this paper will focus on the propulsions system and wiring harness of a battery powered electric VTOL.

A commonly used design method for the harness wire sizes is to determine the maximum occurring current of the power train and select the minimum wire gauge which can withstand this current, see Reference 1. The wires are selected using different ampacity tables of various standards. These tables give acceptable wire currents for a certain allowable temperature rise of the wire. This method allows for quick sizing of the wiring harness but does not take the voltage loss within the wire into account. The resulting wires will be able to handle the flowing current without physical damage, but the impact on the system performance is not apprehended. The voltage loss of the wires results in an increased power consumption of the whole system.

The next common sizing method takes the voltage drop of the wires into account as described in Reference 2. By defining a maximum allowable voltage drop within the wire, you can determine the maximum resistance of the wire. With this resistance the wire gauge can be selected. With the known voltage drop the power loss within the harness is estimated. The difficulty is to define a reasonable limit for the voltage drop. A small limit will decrease the power loss but increase the harness weight and a higher limit vice versa. In some standards the limit is defined as a fixed percentage of the supply voltage level. In Reference 3 the mass of a wiring harness is minimized with a given fixed voltage drop. As described later, this optimization objective does not provide the best performance for the aircraft, but in this reference the focus is on the routing topology of the harness and not the overall system performance.

The allowable voltage drop is a well suited starting point for a rough sizing of a harness, but to design a harness which gives the optimal system

Presented at the Vertical Flight Society Autonomous VTOL Technical Meeting and Electric VTOL Symposium, Mesa, Arizona, UAS, Jan. 29-31, 2019.

performance, you have to adjust the allowable voltage drop to a value, which compromises between a low power loss and a low harness weight. A low power loss is achieved by selecting thicker wires with lower resistance leading to an increased efficiency of the propulsion system. But with the increased weight of the wiring harness more power is needed to perform the UAV's mission, thus, giving a reduced endurance of the vehicle. Thinner wires result in a lighter harness and decreased hover power, but the efficiency of the propulsive system is reduced due to the higher voltage loss in the wire, which can affect the endurance negatively, but not necessarily. It depends on which of the two mentioned effects is dominant in a certain scenario.

To identify the dominant part, the harness has to be modelled as a subsystem and examined in more detail as it is in the previous mentioned methods. This paper approaches this task by implementing a generic wire model, which is detailed enough to capture the main influences to the harness performance, but does not introduce a level of detail which needs a large amount of computational power. The model gives the mass and dimensions for different wire gauges and isolation classes as well as an approximation of the temperature behavior at different currents. For realistic results, the parameters of the model are calibrated using a wire database, which consists of the data from the catalogues from different wire manufacturers.

With the calibrated model, the influence of the wiring harness on the hover performance of different VTOL configurations is examined and optimized. A basic simulation model for the vehicles is derived and the overall system performance is calculated with varying parameters for the wiring harness. The results are used to get an understanding of the mechanism of the influence of the wiring harness on the system performance and ultimately to choose the optimal harness for the vehicle.

Wire Database

The wire database contains the needed parameters to calibrate the wire model in such a way that it produces realistic results, especially for the mass of the wire. Associated with the mass calibration, the geometric dimensions are calibrated as well.

For the wire database the nominal values in the catalogues of different wire manufacturers are used. The current database has over 200 entries with different wire types and gauges and will be extended in future work. While the database is intended for several different wire types, it currently focuses on wires defined by the military standard MIL-DTL-16878G (Reference 6), since this standard offers a well suited basis for deriving a wire model. The later on mentioned isolation classes are taken from this standard.

Geometry

The scope of the geometry calibration is to determine the isolation thickness $t_{Isolation}$, the wire diameter d_{Wire}, and the diameter of one strand of the conductor d_{Strand} for different gauges and isolation types. The notation of the wire geometry can be seen in Figure 1.

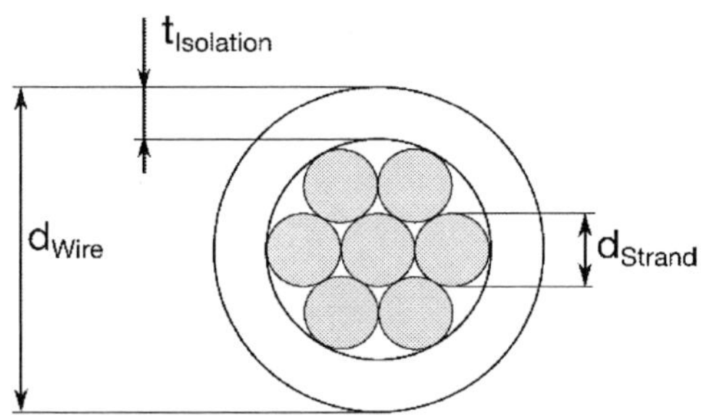

Figure 1: Notation of Wire Geometry

To determine the parameters, following information, which is usually provided by the wire manufacturers, is used:

- Wire Gauge
- Stranding (Number of Strands x AWG of Strands)
- Minimum Outer Diameter $d_{Out,min}$
- Maximum Outer Diameter $d_{Out,max}$
- Nominal Diameter of Conductor d_{inner}
- Maximum Resistance per Length
- Maximum Mass per Length
- Isolation Material and Type

The mean of the maximum and minimum outer diameter results in the nominal wire diameter. With the wire diameter and the nominal diameter of the conductor, the mean isolation thickness for one entry in the database can be derived as follows:

$$t_{Isolation} = \frac{d_{Out,min} + d_{Out,max}}{4} - \frac{d_{inner}}{2} \qquad (1)$$

By calculating the thickness for each entry and computing the mean value for the thickness of the different isolation classes the values as shown in Table 1 are derived.

Table 1: Mean Isolation Thickness of Different Isolation Classes

Isolation Class	Material	Thickness	Standard Deviation
ET	PTFE	0.149 mm	6.2%
ET+[a]	PTFE	0.185 mm	0.0%
E	PTFE	0.276 mm	13.7%
EE	PTFE	0.429 mm	16.4%
NeG[a]	Silicone	0.647 mm	53.6%
EG[a]	Silicone	0.779 mm	23.9%

[a] Not a standard isolation class from MIL-DTL-16878G

By evaluating the standard deviation of the mean isolation thickness for each isolation class, one can observe that the approach of assuming a constant thickness for every gauge is not valid. For the classes 'ET' and 'ET+' the isolation thickness can be presumed constant. The other isolation classes show a thicker isolation with increasing wire gauge. For the first version of the wire model, the thickness is fixed at the constant thickness shown in Table 1 for each class. This results in an error for larger wires, but as the comparison between the database and the wire model will show, this error is tolerable.

The actual conducting area of a nominal wire gauge is calculated by taking the stranding of the wire into account. A wire with one solid strand has a conducting area which equals the area within the isolation, but a wire with several strands shows a reduced conducting area, because of the spacing between the individual strands. In Figure 1 this effect becomes evident. The area within the isolation is not completely filled with the conductor strands. With the shown hexagonal packing, the maximum possible ratio of conductor area to the inner area of the wire is approx. 78%. Assuming that the conductor strands are in direct contact with each other and the inner diameter of the wire is therefore three times the diameter of the strands, the filling ratio μ can be calculated as follows:

$$\mu = \frac{7 * \left(\frac{d_{Strand}}{2}\right)^2 * \pi}{\left(\frac{3 * d_{Strand}}{2}\right)^2 * \pi} = \frac{7}{9} \approx 0.777 \qquad (2)$$

Every other packing method in the 2D plane has a lower filling ratio than the hexagonal one. Only a solid strand can achieve a filling ratio of 100%.

To determine the filling ratio for the wire model, which is achieved by the wire manufacturers for different stranding, the filling ratio for each entry in the database is calculated using the nominal inner diameter, the gauge of the strands, and the number of strands in the wire. The area of the strands is compared against the inner area with the nominal diameter and the mean value for different stranding is computed. The diameter of one strand in dependency of its gauge is derived using the following formula as defined in Reference 4:

$$d_{Strand} = 0.127mm * 92^{\frac{36-AWG}{39}} \qquad (3)$$

With the resulting diameter the area of one conductor strand is calculated and ultimately the filling ratio. In Table 2 the mean filling ratio for the wire database is given. Since the military standard, on which the database focuses, defines preferred stranding for each wire gauge, the filling ratio is split into wires with preferred stranding and wires with non-preferred stranding. The last entry includes all wires in the database.

Table 2: Mean Filling Ratio with Different Stranding

Type of Stranding	Filling Ratio	Standard Deviation
Preferred Stranding	0.769	7.7%
Non-Preferred Stranding	0.748	6.0%
All Wires	0.759	7.1%

The wires with preferred stranding show the highest filling ratio, which almost reaches the maximum filling ratio described in (2). The difference compared to the theoretical value occurs because the stranding is made of more strands than the seven strands of the hexagonal packing, which decreases the filling ratio. In addition, the strands do not align in a geometrically perfectly matter, but there can be small gaps between the strands or the strands are not perfectly round. These imperfections are applicable to the other examined wire types as well. As a conclusion, the mean filling ratio of the wires in the database plausibly fits the theory of the filling ratio with some limitations and can be used in the derived wire model.

Resistance

The specific resistance of the wires is derived from the manufacturer data as well. The military standard MIL-DTL-16878G specifies a maximum DC resistance, but for a more practical model the manufacturer values are taken into account. By knowing the stranding and the resistance per length $\frac{R}{L}$ of the wire, you can compute the specific resistance of the conductor with (4).

$$\varphi_{Conductor} = n_{Strands} * \frac{d_{Strand}{}^{2}}{4} * \pi * \frac{R}{L} \qquad (4)$$

The diameter of the strand follows (3). The mean specific resistance for all copper wires in the database is 0.0186 $\frac{\Omega * mm^2}{m}$ with a standard deviation of 6.3%. Compared to the literature value of Reference 5 for copper of 0.0172 $\frac{\Omega * mm^2}{m}$, the specific resistance is increased by 8.1% due to the purity of the copper alloy and the used alloy itself.

Mass

As mentioned before the wire model should be able to predict correct wire masses. The previous defined parameters all contribute to this property. The last parameter needed for the mass calibration is information about the density of the wire material.

A single conductor wire usually consists of two materials, the conductor and the isolation. The data of the manufacturers only specifies the mass per meter of the whole wire. This one detail is not sufficient to compute the two densities of the wire materials separately. Since the conductor material is a known alloy with no major variation in the density, the density of the conductor is assumed as its nominal value. In this paper only wires with a copper conductor are examined. The density of copper is set to 8940 $\frac{kg}{m^3}$ throughout the further calculations according to Reference 5.

With the fixed density of the conductor, the conductor area, the area of cross section of the isolation, and the mass per meter of wire $\frac{M}{L}$, the density of the isolation material using formula (5) can be calculated.

$$\rho_{Isolation} = \frac{\frac{M}{L} - \rho_{Conductor} * A_{Conductor}}{A_{Isolation}} \qquad (5)$$

The results for the mean density for the two isolation materials currently present in the database are shown in Table 3.

Table 3: Mean Density for Different Isolation Materials

Isolation Material	Density	Standard Deviation
PTFE	2274.2 $\frac{kg}{m^3}$	21.4%
Silicone Rubber	1347.8 $\frac{kg}{m^3}$	16.4%

The mean density of the PTFE material settles around the nominal density of 2200 $\frac{kg}{m^3}$ (Reference 5), but with a significant standard deviation. For a more conclusive result, the database will be extended to get more data for the calculation. Because of the low confidence in the current mean value, the standard density for PTFE is used in the further calculations.

For the silicone material the densities do not vary as much. As the later investigation will focus on the military standard PTFE wires, the research for the silicone material is not complete as it is presented in this paper. Therefore no nominal value for the density of the silicone rubber is considered. This requires additional effort, as many variations of the silicone rubber material exist, which are manufacturer specific and not directly provided by the catalogues.

Wire Model

With the collected parameters, the generic wire model can be implemented and calibrated to produce realistic results. The model is mainly used to predict the resulting mass in dependency of the isolation class and gauge. The other main task is to comprehend the thermal behavior of a wire under different loads, since the wire heating is one of the main sizing restrictions for every method mentioned in the introduction.

The model is implemented using object oriented programming. The basic class is a material class with all material specific properties. With the material, a class for the conductor and a class for the isolation material are created. A wire then consists of one conductor and one isolation class. On an aircraft level, wires are combined in a separate class to harness links.

This chapter describes the implementation of the wire model and how the data captured in the database is used for calibrating the model.

Geometry

The basis for all following wire properties is the geometry of the wire. The modeling follows the structure shown in Figure 1. The dimensions are split into the size of the isolation and the conductor.

At creation of a wire object the isolation class needs to be specified. With the isolation class, the parameters for isolation thickness are selected as presented in Table 1. The thickness is constant for every wire gauge for first quick simulation results.

To be more accurate the isolation thickness will be modelled in more detail in future work.

By specifying the nominal wire gauge for the wire object, the conductor area and diameter are calculated using the stranding and filling ratio. The stranding consists of the preferred stranding for each gauge as defined in Reference 6. There is an option for solid strands, which results in a filling ratio of 100%. Hence, the conductor diameter equals the wire gauge in this case.

For stranded wires the filling ratio is assumed to be 76.9%, as the wire database suggests for the preferred stranding. Using the filling ratio, you can calculate the diameter of the inner area of the wire by solving (2) for the diameter of the inner area of the wire. This results in (6), which is implemented in the model.

$$d_{inner} = \sqrt{\frac{n_{Strands}}{\mu}} * d_{Strand} \qquad (6)$$

For the calculation of the concrete wire mass and resistance, the model contains the length of the wire as an additional input parameter.

Resistance

The resistance of a wire is one of the most significant parameters, which determines the current handling capability of the wire. Hence, an accurate knowledge of the resistance is crucial for a wire model, which should predict the thermal behavior under a load current.

The wire model computes the resistance using the actual conductor area of the gauge and the specific resistance of the conductor material. The conductor area is the area of one strand times the number of strands in the wire. The specific resistance is the value derived from the wire database as described in the previous chapter. With these parameters the resistance of a wire can be calculated straightforward with (7).

$$R_{Wire} = A_{Conductor} * L_{Wire} * \varphi_{Conductor} \qquad (7)$$

This is the DC resistance at room temperature T_0, which is defined in this case as 20 °C. For higher wire temperatures the resistance will increase. The temperature behavior of the specific resistance is apprehended with (8).

$$\varphi_{Conductor}(T) = \varphi_0(T_0) * [1 + \alpha(T - T_0)] \qquad (8)$$

The specific resistance of the conductor is multiplied with a factor, which is dependent on the scaled difference of the conductor temperature T to the reference temperature T_0. The scaling factor α is called the temperature coefficient of resistance and is a material specific constant. For copper it is approx. $3.9*10^{-3}$ according to Reference 5.

Mass

The wire mass is the combined mass of the conductor and the isolator. Based on the wire gauge and stranding, the area and diameter of the conductor is calculated. With the area, the length of the wire, and the density of the conductor material, the conductor mass can be derived.

With the filling ratio and stranding, the diameter of the conductor is obtained by (6). This diameter is the inner diameter of the isolation. By selecting the isolation class of the wire, the isolation thickness is known and the isolation area of the cross section can be calculated. With the density of the isolation material, which results from the evaluation of the wire database, the mass of the isolation in dependency of the wire length is computed. The isolation mass and the conductor mass sum to the total mass of the wire.

Thermal Model

In this draft of the wire model, the thermal behavior is completely modeled according to the design procedure described in Reference 2. This procedure takes the wire temperature into account for the sizing of a wire. Charts are provided to determine the temperature rise of a wire at a certain current.

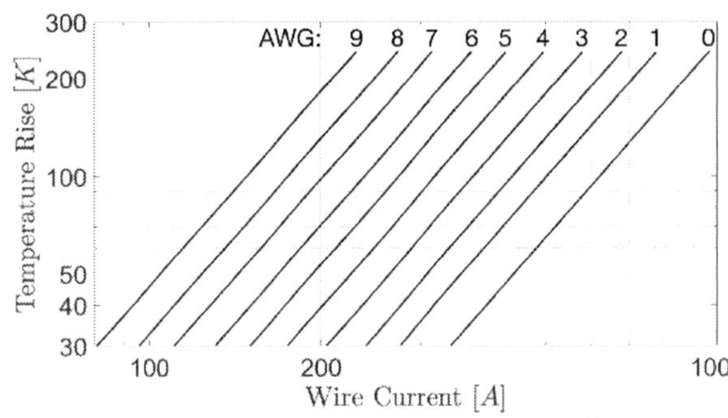

Figure 2: Wire Temperature Rise for Wires from AWG9 to AWG0

Figure 2 shows an examplary digitized chart for wire gauges from AWG9 to AWG0. The wire current and temperature rise are plotted using a logarithmic scale. The reference data is only available as a scanned chart. For the wire model, the charts are digitized using following initial function for the correlation of current and temperature:

$$T(I) = a * I^4 + b * I^3 + c * I^2 \quad a, b, c \in [0, \infty[\qquad (9)$$

The initial function is chosen in such a way, that it gives a zero temperature rise with zero wire current and it is monotonously rising outside the positive interpolation interval. The second property is important for correct optimization results. If the wire temperature decreases with increasing current, the optimization would result in a nonrealistic optimum.

The parameters a, b, and c are obtained by a least squares curve fitting with three points, which are measured in the original charts. The charts only provide data for gauges with an even gauge number. For completeness, the temperature curves for uneven gauge sizes are computed additionally as an interpolation of the data of two even wire gauge numbers. The results for each wire gauge from AWG26 to AWG0000 are saved in a separate function, which uses (9) to calculate the wire temperature with a given wire current and gauge. This function is already used to compute the curves shown in Figure 2.

By combining the temperature information with the knowledge of the temperature dependent resistance, the wire resistance for different currents can be derived. In Figure 3, the resistance for an AWG18 wire is plotted against the wire current from zero current to the maximum rated current for this gauge. In this case, the maximum rated temperature of the wire was assumed to be 150°C. This results in a maximum current of approx. 28A at an ambient temperature of 20°C. This current is determined by calculating the wire temperature with (9) for each current until the maximum rated temperature is breached.

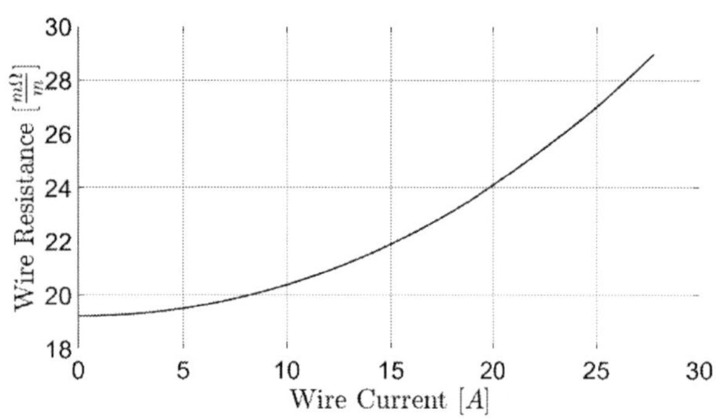

Figure 3: Wire Resistance in Dependency of the Wire Current

With the rising current in the wire, the power loss within the wire increases. This results in a temperature rise of the wire, which increases the resistance. In the calculation of Figure 3 it is assumed that a current source provides the load, which means that the flowing current is not dependent on the wire resistance.

At no current, the wire resistance equals the nominal value at room temperature of approx. $19\frac{m\Omega}{m}$. At the maximum rated current, the wire resistance is increased to approx. $29\frac{m\Omega}{m}$. This is a growth of approx. 50% compared to the initial value. This shows that the temperature dependency of the resistance cannot be neglected in the sizing of wires.

Comparison to Database

For validation purposes, the results of the wire model are compared against values of the database. Since the model is calibrated using the database, the results should be similar. Nevertheless, this validation is useful to find systematic errors in the modeling of the wire.

Since the mass prediction is one crucial parameter for the optimization and mostly relies on calibrated values, the mass per length is used to compare the model prediction to actual values in the database. In Figure 4 the specific mass of wires with the isolation class 'EE' is shown. The point data of the database is represented as circles. The prediction of the wire model is plotted as a dashed line. A logarithmic scale is used for the specific mass to see all the relevant areas.

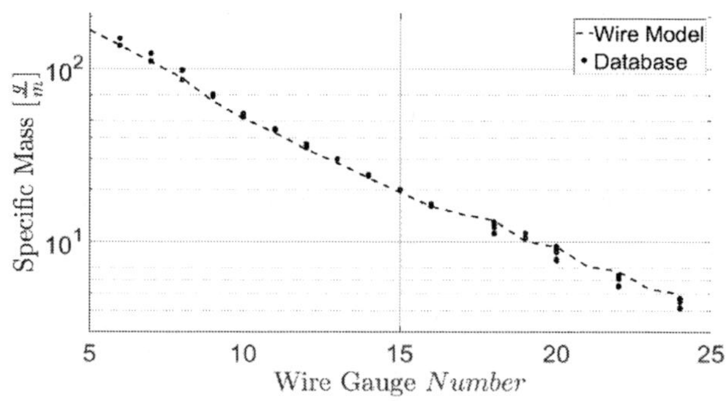

Figure 4: Comparison of Specific Mass, Wire Model vs. Database

As you can see, the wire model calculates the wire mass with a good fit to the database values for each wire gauge. The wire model is therefore sufficient to predict the wire mass for this isolation class. The comparison with other isolation classes yields similar results, but is not presented here since the later optimization will focus on the 'EE' isolation class.

Harness Optimization

The wire model is used to build up the links of a wiring harness for an eVTOL. The performance of the eVTOL is simulated taking the harness loss into account. By varying the harness parameters, the influence of the parameters to the flight performance can be derived.

Assumptions

The simulation and optimization use the assumptions, which are described in this chapter, to simplify the problem for a faster implementation of the simulation framework to get first results.

Mass Distribution

The simulation framework does not include a sophisticated aircraft model in this stage, which can be used to estimate the mass components of the aircraft. Therefore, some straightforward assumptions are made.

The take-off mass (TOM) of the aircraft is split into the payload mass (PLM), the harness mass, the battery mass, and the remaining mass, which is called operational empty mass (OEM). The last mass includes the propulsion mass, the avionics mass, and the structure mass. The distribution can be seen in Figure 5.

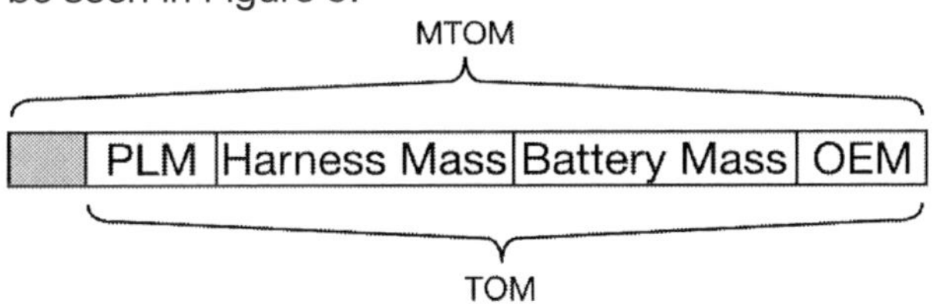

Figure 5: Mass Distribution for Simulation Model

The PLM and OEM are assumed to be constant. The values for these fixed masses are derived from a 5kg eVTOL, which is designed at the institute. This "Quad Tilt" configuration is described in a later chapter.

The harness mass results from the variation of the harness parameters. Hence there is a need for the mass prediction of the wire model.

The battery mass is split into the discrete masses of the cells it consists of. Taking the additional mass for assembling the battery pack into account, the mass of one cell is assumed to be 50g. The battery consists of serial and parallel cells. The serial cells define the voltage level of the pack and the parallel cells the resulting capacity. The number of parallel cells times the number of serial cells times the mass per cell is the resulting battery mass, see (10).

$$M_{Battery} = n_P * n_S * M_{Cell} \quad (10)$$

The battery voltage is a fixed input parameter for the simulation, which defines the number of serial cells needed in the battery. The number of parallel cells is the step size for the battery mass with the fixed voltage. The simulation sizes the battery according to the available mass. The aircraft mass cannot exceed the maximum take-off mass (MTOM). The available battery mass is therefore the MTOM without the OEM, the PLM, and the harness mass. The parallel cells are increased until the maximum number of parallel cells, which fit into the available mass, is reached. This process ultimately defines the available battery capacity for the aircraft.

The resulting TOM depends on the varying battery and harness mass and is less or equal to the MTOM. In the first calculations, the TOM was fixed to the MTOM to isolate the influence of the wiring harness to the system performance. While this is not a practical approach as such, the argument is, that the configurations are designed for the use with the MTOM and the difference between the TOM and MTOM is compensated with a higher PLM. In a later step, the actual TOM is used for the performance calculation. Both approaches affect the influence of the wiring harness differently and the optimization results for both are presented.

Performance Parameters

The simulation framework is intended to be capable of representing different configurations and capturing their performance parameters. In this paper the focus is on the influence of the wiring harness to the flight performance. Thus, the configurations are considered to be equal in efficiency, only the wiring harness and its efficiency are different for each configuration. The investigated configurations are described in the next chapter.

In Table 4 the assumed values for the fixed parameters are listed. As mentioned before, the fixed masses are derived from the "Quad Tilt" configuration, which is designed for a MTOM of 5kg. The wingspan of 2.0m is also taken from this design, see Reference 7.

Table 4: Fixed Parameters for Performance Simulation

Parameter	Variable	Assumed Value
Aerodynamic Efficiency	E	10
Wingspan	B	2.0 m
Maximum Take-Off Mass	MTOM	5.0 kg
Payload Mass	PLM	0.8 kg
Operational Empty Mass	OEM	2.8 kg
Hover Propulsion Efficiency	μ_{Hover}	$20.0\,\frac{W}{N}$
Cruise Propulsion Efficiency	$\mu_{Propulsion}$	60 %
Cruise Speed	v_{cruise}	$20.0\,\frac{m}{s}$

The power consumption during a mission is split into the hover power and the cruise power. To calculate the cruise power, the aerodynamic efficiency and the given cruise speed are used, see (11).

$$P_{Cruise} = \frac{TOM * g * v_{cruise}}{E} \qquad (11)$$

$$P_{Propulsion} = \mu_{Propulsion} * P_{Cruise} \qquad (12)$$

For the consumed electric power of the cruise propulsion the efficiency of the drive train has to be taken into account with (12). The efficiency includes the efficiencies of the speed controller, of the motor, and of the propeller. The assumed efficiency of 60% is chosen conservatively based on experience.

The hover power is calculated using an efficiency value, which directly gives the needed electrical power for a given TOM, see (13). This efficiency also includes losses within the drive train and is again with 20.0 $\frac{W}{N}$ conservatively chosen based on experience.

$$P_{Hover} = \mu_{Hover} * TOM * g \qquad (13)$$

The calculated powers are constant for all the configurations and do not include the power loss within the wiring harness.

Configurations

The purpose of the different configurations is to examine the influence of different wire lengths to the system performance. Since the discussed power calculations do not take scaling effects into account, it is not sufficient to scale the wingspan to vary the wire length.

Quad Tilt

The 'Quad Tilt' is a configuration development at the Institute of Aircraft Design at the Technical University of Munich (Reference 7). It consists of two main hover power trains, which produce approx. 85% of the lift during hover. The remaining 15% of the needed trust is provided by two tiltable propulsion systems in the wingtips, which are also used to produce all the needed trust in cruise flight. The hover propulsion is not tiltable. In Figure 6 the flying demonstrator is shown.

Figure 6: 'Quad Tilt' Demonstrator

The important dimensions of this configuration are shown in the top view in Figure 7.

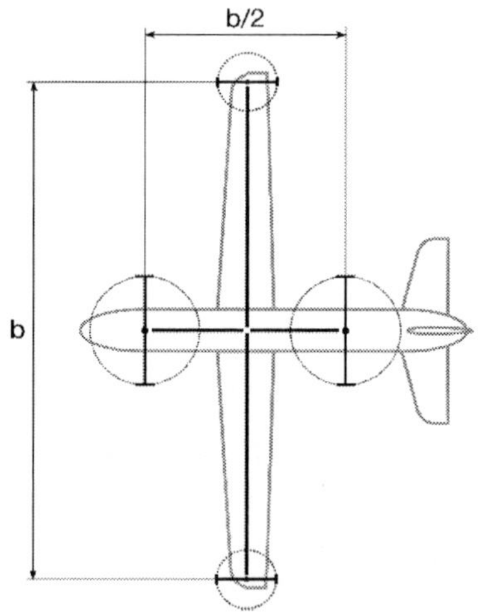

Figure 7: Top View of the 'Quad Tilt'

The distance of the wingtip propulsion is approx. the wingspan. The hover power trains are spaced with half of the wingspan. The placement of the motors is symmetric to the center of gravity.

The routing topology of the wiring harness is indicated by the solid black lines. This topology is the basis of the calculation of the wire length during the optimization.

Quad H

The 'Quad H' configuration is based on the common concept using a quadrocopter basis, where the needed lift during hover is provided by all motors equally. During cruise two or more motors are tiltable and used as the cruise propulsion. In this case the four motors are placed within pods in the wing, see Figure 8.

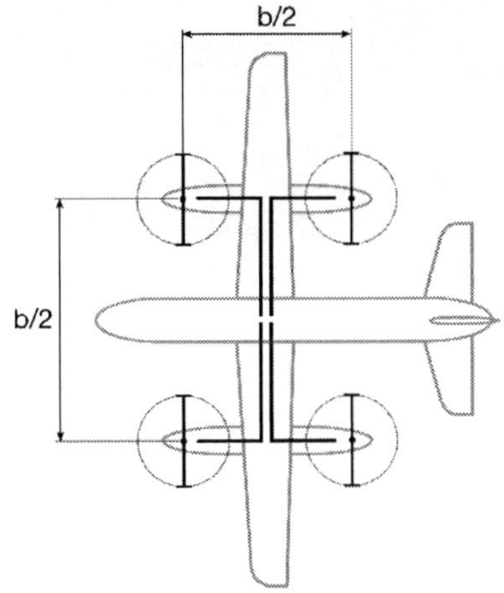

Figure 8: Top View of the 'Quad H' Configuration

The distance of the pods is equal to the distance of the motors in the flight direction and is defined as half of the wingspan. The wiring harness is first routed along the pods and then along the wing to the battery.

Quad +

The 'Quad +' configuration is another derivative of the quadcopter basis. The motors are placed in the wing and the fuselage, see Figure 9.

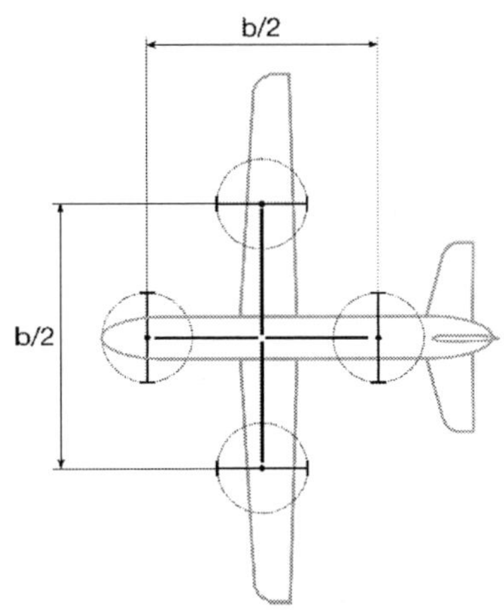

Figure 9: Top View of the 'Quad +' Configuration

The distance of the motors is half the wingspan and symmetric to the center of gravity. The wiring is routed directly from the propulsion system to the battery.

Optimization Procedure

Using the derived wire model and the simplifications described in the previous chapter, the harness optimization can be performed. The optimization is implemented in such a way, that multiple objectives can be addressed. This is achieved by using an objective function, which can be set to different optimization targets.

In this version, the optimization consists of a parameter variation, which stores only the optimal parameter set. The algorithm calculates the aircraft performance for every parameter set, which range is defined before the optimization. The parameter set which yields the optimal performance, is then chosen as the optimal solution. The parameter set consists of the wire gauges for different harness links. A routing optimization is not performed in this optimization.

While this method needs a lot of computational power, the simulation model is fast enough to make this approach feasible. The small search space defined by the finite amount of wire gauges and the low count of harness links contribute to the low computation time as well. When the search space is increased a different optimization method has to be used, but for the scope of this paper, the parameter study is sufficient.

In the following, an overview of the main loop of the optimization will be given and important subroutines will be described.

Main Loop

The main loop of the harness parameter optimization can be seen as a flowchart diagram in Figure 10. Before the main loop is started, the range of parameters, which are varied, are defined along with the optimization targets. At the beginning of the main loop, the varied parameters are set to the corresponding values according to the iteration loop. With the updated properties of the harness, the new available battery mass is computed and the battery is sized accordingly.

Knowing the new mass distribution and battery capacity, the current aircraft performance is computed taking the harness power loss into account. A detailed description of this routine is given in the next chapter.

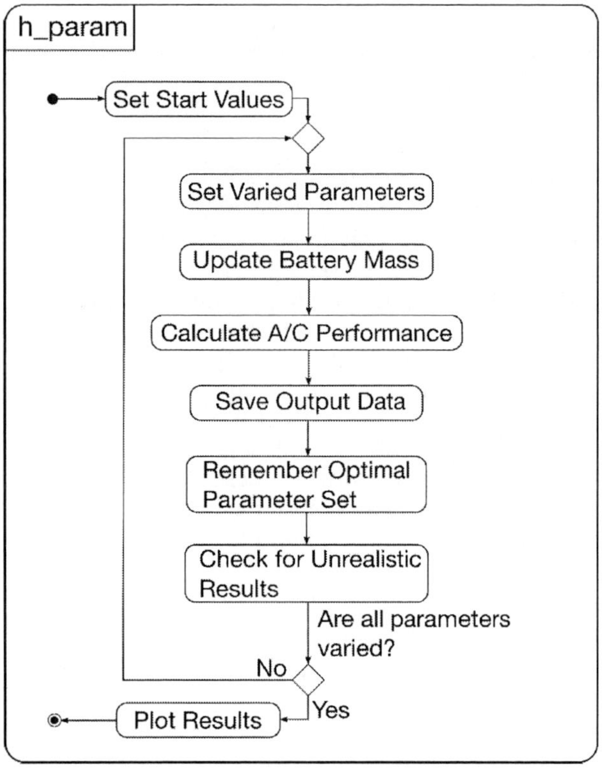

Figure 10: Flowchart for the Harness Parameter Optimization

After getting the new aircraft performance, the properties, which should be plotted after the optimization, are saved. This avoids memory overhead and makes sure, that only information of interest is saved.

The resulting aircraft or harness performance is rated with the objective function and checked for unphysical results. After all parameters are varied, the saved optimum parameter set is displayed along with the saved output data.

Aircraft Performance Calculation

The aircraft performance is calculated using the power consumption of the propulsion system. The aircraft power consists of two parts. Firstly, the electrical power which is needed by the propulsion system itself. Secondly, the power loss within the harness. The sum of the two powers is the total power the battery has to provide.

The propulsion power is computed as described in the chapter on the assumptions for the performance parameters. In the flowchart in Figure 11, the first routine is to set the efficiencies for the propulsion systems. These are fixed to the values as described before. Nevertheless, the function is prepared for supporting varying efficiencies for different operation points.

After calculating the harness power loss based on the electric power of the propulsion system, the aircraft performance for different mission types is computed.

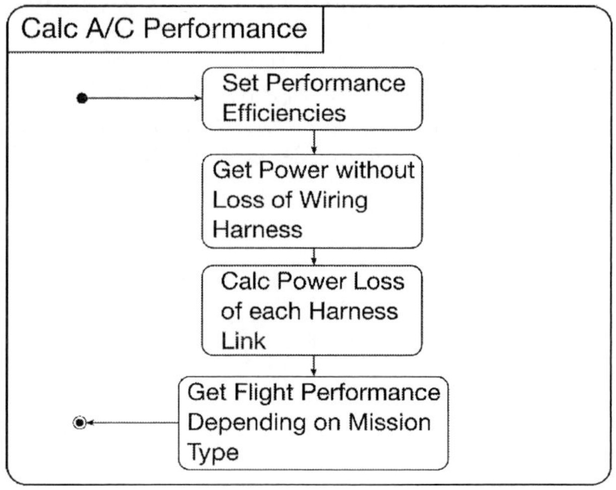

Figure 11: Flowchart for Aircraft Performance Calculation

The performance is based on a linear approach using the battery capacity. The total extracted power of the battery is transformed to a current by assuming that the battery voltage is fixed to the nominal value. With the resulting current, which is assumed to be constant, the time until the capacity of the battery is depleted can be calculated. This time is the basis for the endurance and range calculation of the aircraft.

Harness Power Loss Calculation

The calculation for the harness power loss assumes, that the electrical power for the propulsion is constant regardless of the power loss within the harness.

The power loss is dependent on the wire current and resistance. The wire current and resistance are dependent on each other. With an increased wire current, the wire temperature rises and therefore the wire resistance. The increased resistance results in a higher voltage loss at the given current. This voltage loss increases the power loss of the harness, which results in a higher wire current. As you can see, the power loss calculation is an implicit function and needs an iteration loop to solve for the wire current, as it is shown in Figure 12.

This iteration loop converges because the power loss and the resistance are not dependent on the current in the same order. The power has a linear dependency on the current. The resistance is linear dependent on the wire temperature, but the wire temperature follows the initial function of (9), which has an order of 4.

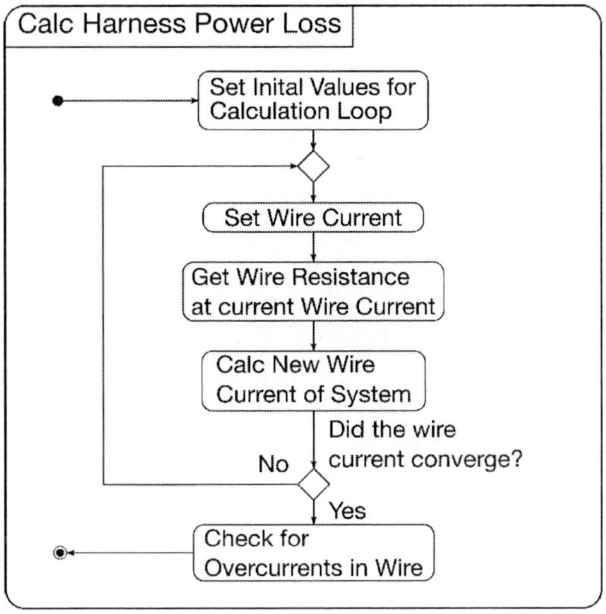

Figure 12: Flowchart for the Harness Power Loss Calculation

After the wire current reached a convergent point, it is checked whether the current is above the maximum rated current for the used wire gauge. Overcurrent situations are saved and given to the objective function to mark the current parameter set as unphysical.

Objective Function

The value of objective function is the actual parameter, which is optimized and in this case minimized. By selecting different values, which affect the function, the optimization can be adjusted flexible to optimize the harness for different target performances. The general notation of the objective function is shown in (14).

$$f_{Objective} = \sum_{k=1}^{N} w_k * P_k \qquad (14)$$

The value of the objective function is the sum of all the target parameters P_k, where N is total number of objectives. Before summing, every parameter is multiplied with a specific target weight w_k.

The weight sets the priority of the parameter. By increasing the magnitude of the weight, the target parameter has more influence to the value of the objective function and the optimization will result in a solution, which favors this parameter. The sign of the weight determines whether the parameter is minimized or maximized. Since the objective function is always minimized, a positive sign results in a minimization of the parameter and a negative sign accordingly to a maximization.

Once the value of the objective function for a harness configuration is determined, only configurations, which do not have a physical error like overcurrent situations, are considered in the selection of the optimal harness.

Optimization Results

The harness optimization is applied to the three described configurations. For this optimization, only a hover flight mission is considered due to time restrictions. In addition, only the harness links used during hover flight are varied and optimized. The links are assumed to consist of wires of the isolation class 'EE' with a rated maximum temperature of 150°C and every link has the same gauge. The ambient temperature is set to 20°C. The gauge of the links is the only parameter, which is varied for the optimization. The gauges start from AWG0 and end at AWG24. The routing is fixed as described for each configuration.

Two separate objectives are used for the objective function. One is the minimization of the harness mass and the other is to maximize the hover endurance. The two resulting harnesses for each configuration are compared to each other in the following discussion of the results.

In addition the optimization is performed separately for each assumption for the TOM. One harness is derived for the TOM fixed at the MTOM and one harness for the actual resulting TOM.

Results of the Harness Optimization

The evaluation of the optimization starts with the presentation of the output of one optimization run for the 'Quad Tilt' configuration and discusses the different influence parameters of the harness to the system performance.

The plots always contain a triangle as a marker for the optimal performance and a dashed line, which shows the boundary from physical to nonphysical results. Which objective and TOM assumption is used, is indicated in the captions of the diagrams.

Temperature Influence

In Figure 13 you can see the resulting static wire temperature of the harness during the hover flight with different wire gauges.

The temperature of the wires rises significantly with smaller wires and approaches the maximum rated temperature. The last gauge, which can be operated below the rated temperature, is marked with the dashed line.

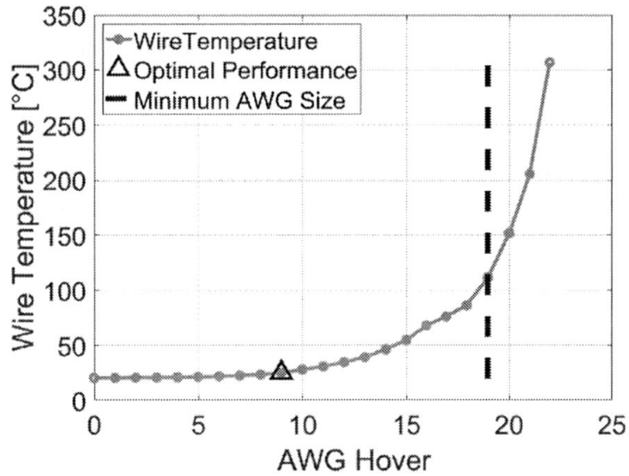

Figure 13: Resulting Wire Temperature, TOM = MTOM, Objective = Endurance

The influence of the temperature rise to the hover endurance of the aircraft can be seen in Figure 14.

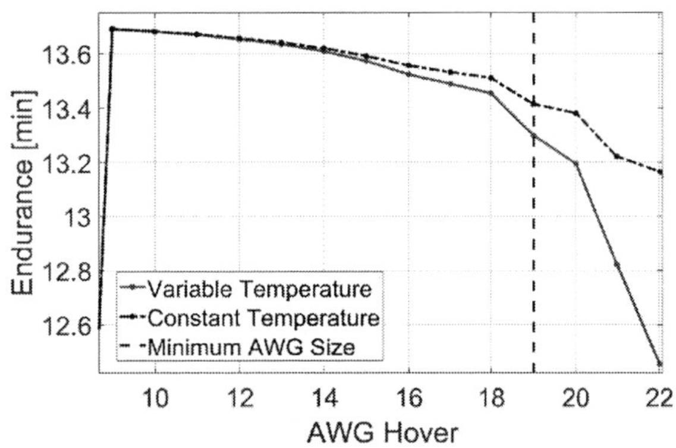

Figure 14: Influence of the Wire Temperature Rise to the Endurance, TOM = MTOM

The endurance shown with the dashed line is derived assuming that the temperature of the wires stays at the ambient temperature. The endurance displayed with the solid line takes the temperature behavior of Figure 13 into account.

When the wire temperature does not differ from the ambient temperature at higher wire diameters, the difference in endurance is negligible. As the wire temperature increases with smaller diameters, the temperature aspect gains more influence due to the increased wire resistance and the associated increase in the power loss within the harness, see Figure 15.

This results in a difference in endurance of approx. 0.6% at the smallest physical possible wire gauge.

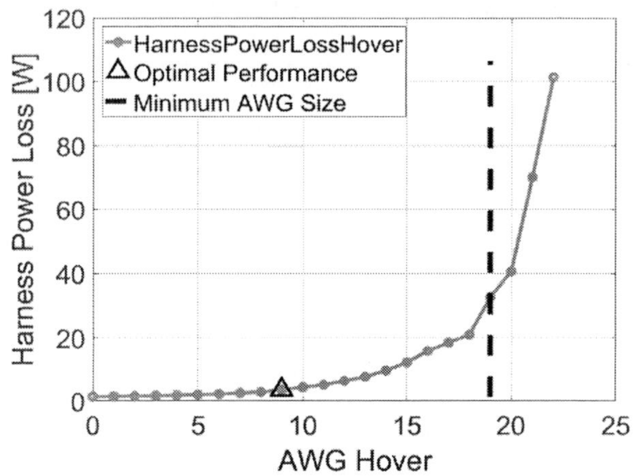

Figure 15: Harness Power Loss, TOM = MTOM, Objective = Endurance

While this difference could be neglected, it is considered in the following optimizations for consistency.

Masses

The mass distribution of the subsystems is one of the main influences to the aircraft's performance. In Figure 16 you can see the resulting total harness mass for different wire gauges.

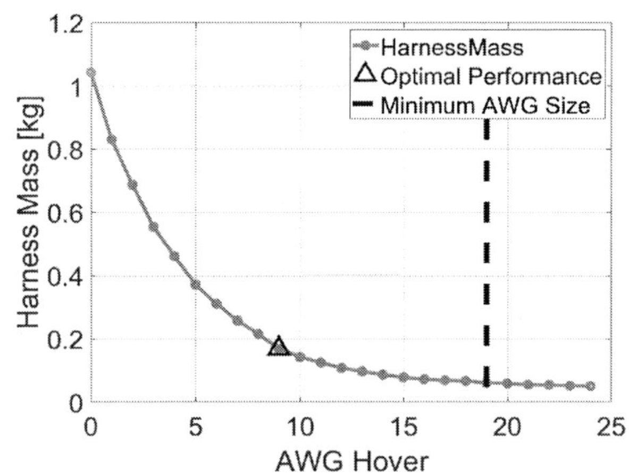

Figure 16: Resulting Harness Mass, TOM = MTOM, Objective = Endurance

The mass shows a nonlinear behavior with decreasing gauge of the wires. The harness mass with a wire gauge of AWG9 gives an optimum for the endurance. Before addressing the reasons for this optimum, the behavior of the battery mass has to be examined.

The battery mass is shown in Figure 17. The discrete steps for the battery mass are visible. Each step represents the increase of the mass by the mass of one parallel pack with the specified serial cells. The additional pack is only fitted, if the available mass allows it. At the beginning of the variation interval of the wire gauge, the harness mass is approx. 1kg, leaving an available battery mass of approx. 0.3kg. As the harness mass decreases with thinner wires, the available battery mass increases. Once the available mass is high enough to fit another parallel pack of the battery, the battery mass increases by this discrete step. The maximum battery mass, which can be fitted in this configuration is 1.2kg.

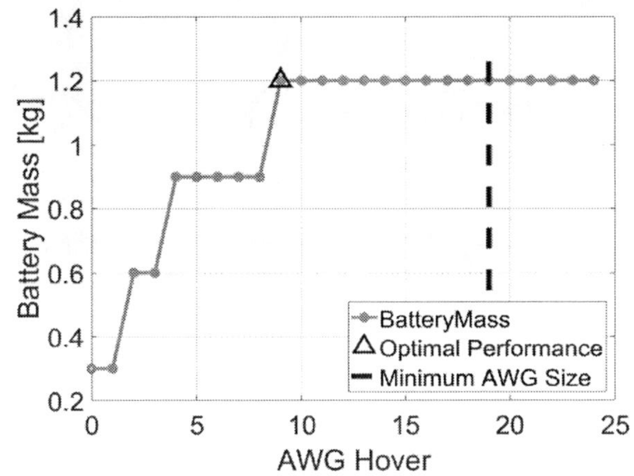

Figure 17: Battery Mass, TOM = MTOM, Objective = Endurance

Endurance

In this evaluation of one optimization run, the focus is on the maximization of the endurance of the 'Quad Tilt' configuration. The resulting endurance for a TOM, which equals the MTOM, is displayed in Figure 18.

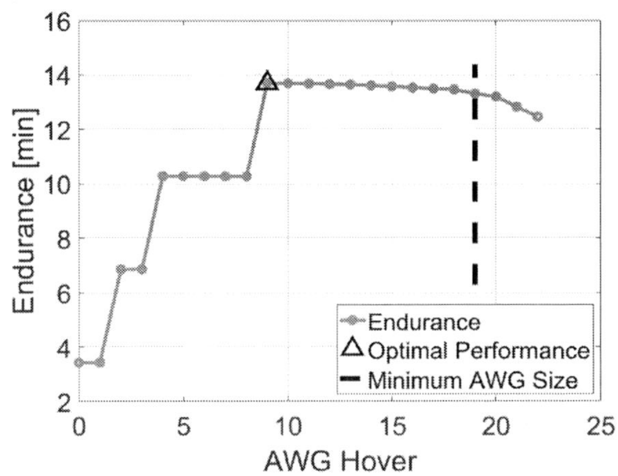

Figure 18: Resulting Endurance for the 'Quad Tilt' Configuration, TOM = MTOM, Objective = Endurance

Starting with a higher wire gauge, the endurance is dominated by the steps in the battery mass. The battery mass directly defines the battery capacity and the capacity directly the hover endurance. With the low power loss of the harness relative to the hover power, the impact of the battery capacity is the critical factor for this region of wire gauges.

As the wire gauges get thinner, the battery mass reaches a maximum and the influence of the harness power loss increases as shown in Figure 15. The resulting effect on the endurance is shown in Figure 19, in which only the region starting at AWG9, where the battery mass has reached its maximum, is plotted.

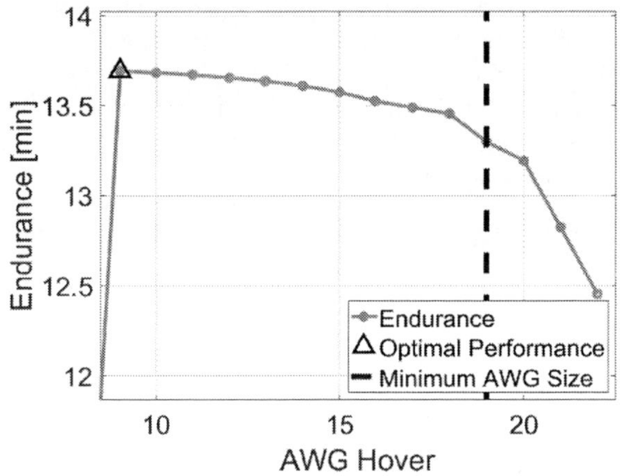

Figure 19: Endurance with Maximum Battery Mass, TOM = MTOM, Objective = Endurance

In this region the endurance starts to decrease due to the increased power consumption caused by the increasing power loss in the harness. At a wire gauge of AWG19, the maximum current handling capacity of the wires is reached.

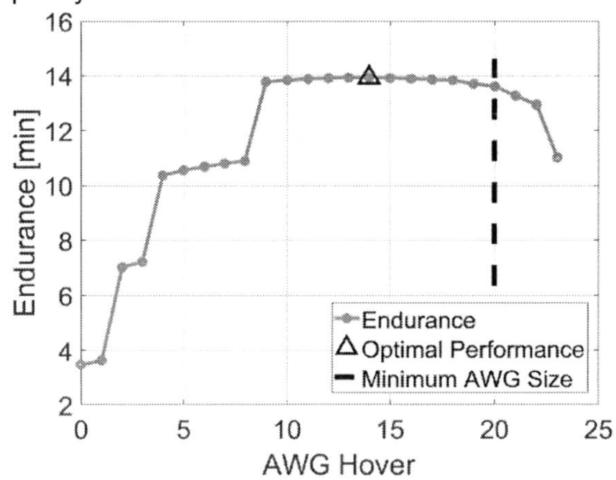

Figure 20: Resulting Endurance, TOM = Actual TOM, Objective = Endurance

The optimum endurance is obtained right after the battery mass has reached its maximum at AWG9. At this point the maximum battery capacity is reached with a minimal loss in the wiring harness. These are the two main opposing objectives in this optimization. The battery capacity has to be high to allow for an increased endurance. For a high battery capacity, the available battery weight has to be maximized, resulting in a minimization of the harness mass. In addition, the harness power loss has to be minimized using thicker wire gauges, which leads to a maximization of the harness mass. This shows the multi objective character of the optimization.

By calculating the aircraft performance taking the actual TOM, which varies for each gauge size, into account, the progression of the endurance changes, see Figure 20. At thicker gauges, the influence of the available battery mass is still dominating, but since the TOM is depending on the combination of battery mass and harness mass, the harness mass starts to affect the endurance over the whole variation interval. With this effect the importance of the harness mass minimization grows compared to the minimization of the harness power loss. Therefore, the optimum wire gauge is shifted to a smaller diameter, in this case AWG14.

It is to be noted, that the optimum gauge is not the thinnest possible gauge. At the end of the variation interval, the influence of the harness power loss exceeds the influence of the harness mass.

In the later discussion, the performance of a harness optimized for endurance is compared against the performance of a harness, which is sized taking only the maximum current handling capability of the wires into account. This results in a harness, which consists of the smallest possible wire gauge. In the optimization, this behavior is achieved by changing the optimization objective to the minimization of the harness mass. The results are not shown in separate plots, since the dependencies of the performance parameters shown in the previous figures do not change. Only the operating point with the optimal performance shifts to the smallest possible wire gauge, which is represented by the dashed line in those figures.

Influence of the Configuration

This chapter discusses the results of the harness optimization for the different configurations. The influence of the two discussed objectives – endurance vs. harness mass – is shown as well as the difference of the two assumptions for the TOM.

Harness Mass

To compare the optimal harness masses of the different configurations, the harness mass is analyzed relatively to the TOM, which is used for the performance calculation. There are two ways the TOM is derived. The first is to fix it at the MTOM, called "Fixed TOM" in the following. The other way is to use the actual TOM calculated with the system masses. This is called "Fixed Payload Mass" hereinafter.

First, the harness mass for the fixed TOM is evaluated. The graph of the relative harness masses of all configurations optimized with the two mentioned objectives is shown in Figure 21.

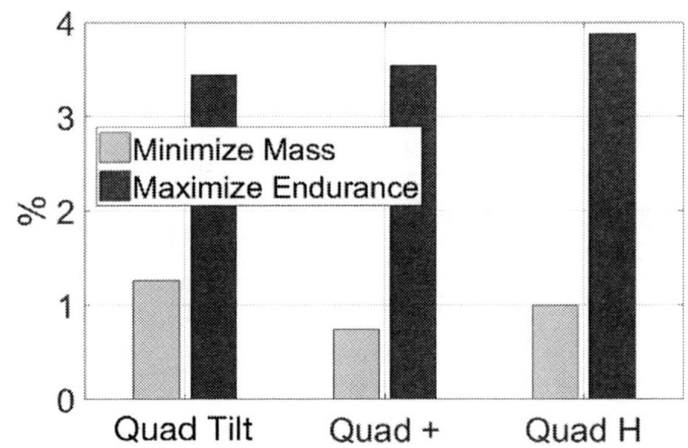

Figure 21: Relative Harness Masses, TOM = MTOM

As expected, the harness masses for the minimized harness masses are lower than the harness masses optimized for endurance. For the mass optimization, the harness mass varies from 0.75% to 1.25% for the different configurations.

The 'Quad Tilt' has the heaviest harness of 1.25% of the TOM with the mass optimization, because of the lift distribution among the hover motors. The main lift is produced by the two motors in the fuselage. To satisfy the boundary for the maximum wire current, the wires for this motors need to be bigger compared to the other configurations. Since all the wires in the harness are assumed to be of the same size, the low power wires to the wingtips are oversized in this case, which results in the higher mass.

The wires for the "Quad +" configurations can be sized smaller due to the smaller single power of one motor, because the hover lift is evenly distributed among the four hover motors. The difference between the two quadrocopter configurations is explained by different total harness lengths of each configuration. The harness length of the "Quad H" configuration is longer than the "Quad +" harness, resulting in a higher relative mass of this harness.

For the harnesses optimized for endurance, the harness mass is dominated by the length of the harness links which carry the majority of the electrical power. In this case the 'Quad Tilt' has an advantage as the main power is provided only by two motors or respectively two harness links. The other configurations use all four harness links for the distribution of the main power. The "Quad H" configuration has the longest harness, resulting in the highest harness mass, followed by the harness mass of the "Quad +" configuration and the 'Quad Tilt' configuration.

In Figure 22, the results of the performance calculation using the actual TOM with the fixed PLM can be seen.

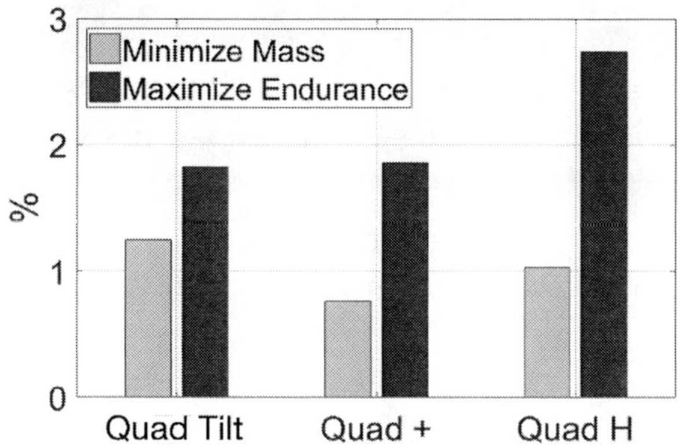

Figure 22: Relative Harness Masses, TOM = Actual TOM

In this optimization run, the results for the minimization of the harness mass are the same as in the previous one. The harness mass for the optimal endurance is still higher, but since the influence of the harness mass to the endurance is increased, the optimum harnesses show a decreased weight of approx. 2%-3% of the TOM.

The 'Quad Tilt' and "Quad +" configuration share a similar value for the mass. The "Quad H" configuration has the highest mass due to the relatively long length of the harness links.

Harness Power Loss

The evaluation of the harness power loss is done in relation to the needed hover power of the propulsion systems. The results for the case of the fixed TOM are plotted in Figure 23.

The same mechanisms as in the evaluation of the harness mass apply here as well. By the minimization of the harness mass, the resistances of the harness links are increased and inherent the total power loss. As the length of the links increases with the different configurations, the power loss increases accordingly. Thus, the 'Quad Tilt' shows the lowest power loss of approx. 3.5%, followed by the "Quad +" with approx. 10% and the "Quad H" with approx. 13.5%.

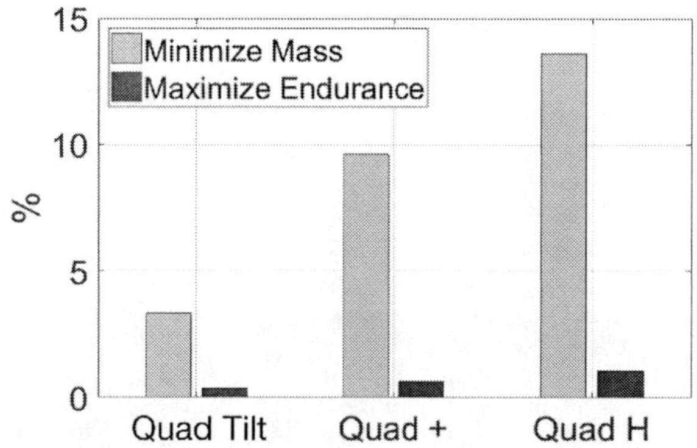

Figure 23: Relative Harness Power Loss, TOM = MTOM

With the optimization for endurance, the power losses are significantly decreased and do not vary as much over the configurations. The power loss is always below 1% of the hover power. The dependency of the harness length is still visible.

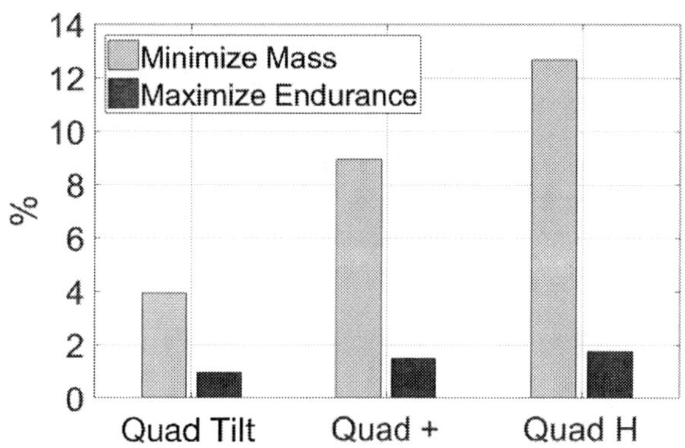

Figure 24: Relative Harness Power Loss, TOM = Actual TOM

When the TOM calculation is switched to the fixed PLM, the relative power loss for the mass minimization decreases slightly, see Figure 24. The behavior of the harness optimized for endurance, shows an increase in the power loss, but stays below 2%. The harness length affects the power loss in both cases.

Hover Endurance

Since the hover endurance is the objective of one of the optimizations, it is not feasible to compare results of the optimization runs with the different objectives. Therefore, only the influence of the modeling of the TOM is discussed.

In Figure 25 the hover endurance for each configuration and TOM model is shown. The influence of the harness power loss is mirrored in the endurances. The harness, which shows the lowest

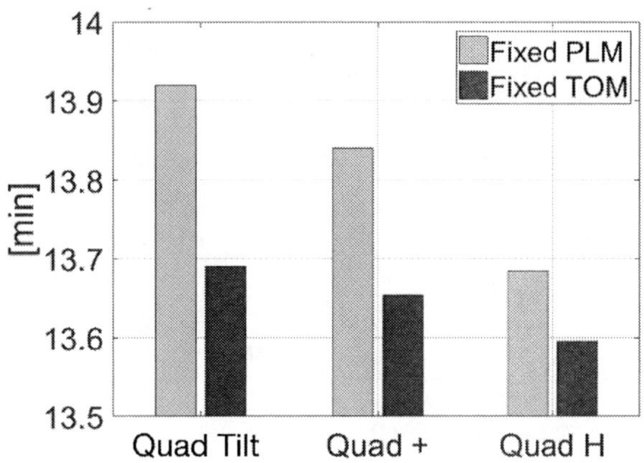

Figure 25: Maximum Endurance, Objective = Endurance

power loss, achieves the highest endurance. The endurance for the fixed PLM is higher than the endurance for the fixed MTOM, because the first assumption results in a lower mass than the MTOM and therefore the total hover power is decreased, resulting in a higher endurance with the same configuration.

The improvement of the endurance optimization relative to the endurance of the mass optimized harness is shown in Figure 26.

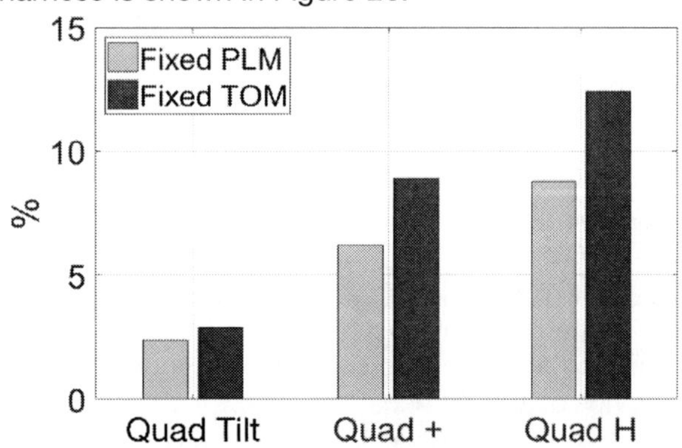

Figure 26: Improvement of Endurance

The endurance is improved for both assumptions for the TOM. For the fixed TOM, the harness power loss can be reduced further than for the fixed payload, therefore the endurance improvement is higher for these cases.

The dependency on the harness length is here visible as well. By plotting the endurance improvement relative to the harness length you get Figure 27. As a result, the endurance improvements of the configurations are in the same order of magnitude for each model of the TOM. For the fixed TOM the improvement is approx. 1.5% per meter of wire and the improvement for a fixed PLM is approx. 1% per meter of wire.

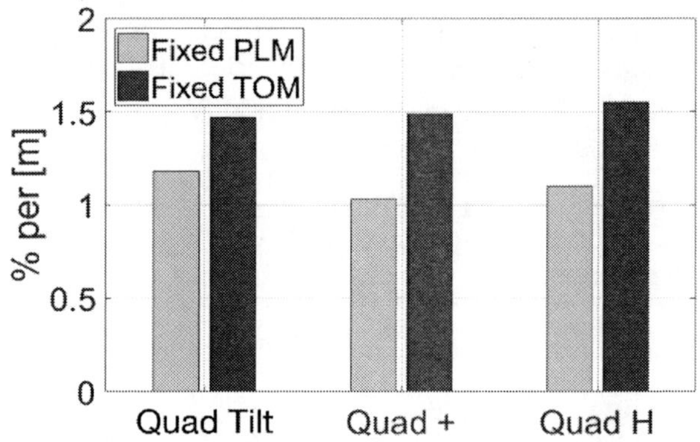

Figure 27: Endurance Improvement Relative to Harness Length

Conclusion

After the evaluation of the optimization results the main conclusions of this paper are:

1. Using the wire database, the filling ratio of the strands is compared against the theoretical possible filling ratio. The filling ratio is the ratio of the area of the conductor strands relative to the total inner area of a wire. It describes how much of the available cross section is used by the strands of the wire. The theoretical maximum of the used area is approx. 77%. The database shows that the manufactured wires reach that value very closely.
2. The mean resistance of the wire in the database does not reach the nominal value of copper, but settles at approx. 0.0186 $\frac{\Omega * mm^2}{m}$.
3. The wire model assumes a constant isolation thickness for different wire gauges of an isolation class. The evaluation of the wire database shows, that this assumption is not valid except for the classes 'ET' and 'ET+'. A comparison to the database values shows, that the model is sufficient for a first prediction of the wire mass of other isolation classes.
4. The implementation of the pragmatic thermal model for wire shows the influence of the wire current on the wire resistance. When the current increases the power loss within the wire increases. The power loss is dissipated as heat resulting in a temperature rise of the wire. With the increased temperature, the resistance of the conductor material rises. This leads to an increase of the wire resistance of up to 50% at the maximum rated temperature compared to the nominal resistance at ambient temperature.

5. The optimization of wiring harnesses for maximum hover endurance results in harnesses, which tend to higher harness masses to decrease the power loss within the harness. The available battery mass limits the harness mass. Once the battery mass has to be reduced to fit the increased harness weight, the endurance of the aircraft is significantly reduced.
6. The evaluation of the optimization results shows an increase of the hover endurance for all three examined configurations. The increase is dependent on the total harness length of the configuration. For a fixed TOM at the MTOM, the increase of the endurance is approx. 1.5% per meter of wire and for the variable TOM the endurance increases by approx. 1.0% per meter of wire.

Authors contact:
Sebastian Oberschwendtner
sebastian.oberschwendtner@tum.de,
Christian Roessler christian.roessler@tum.de,
Mirko Hornung mirko.hornung@tum.de

References

[1]European Standard, Current ratings for electrical cables with conductor EN 2083, EN 2853, Brussels, 2005

[2]Federal Aviation Administration, ACCEPTABLE METHODS, TECHNIQUES, AND PRACTICES – AIRCRAFT INSPECTION AND REPAIR, AC-43.13-1B, Chap. 11, 1998, pp. 11-1 – 11-272

[3]Combettes, S., Sontheimer, T., Rougemaille, S., and Glize, P., Multi-Agent Cooperation for Optimizing Weight of Electrical Aircraft Harnesses, EUMAS 2014, LNAI 8953, pp. 301-315, 2015, doi: 10.1007/978-3-319-17130-2_20.

[4]ASTM International, Standard Nominal Diameters and Cross-Sectional Areas of AWG Sizes of Solid Round Wires Used as Electrical Conductors, B 258-02, United States

[5]O'Neil, M.J., The Merck Index: An Encyclopedia of Chemicals, Drugs, and Biologicals, 14th ed. New Jersey: Merck, 2006.

[6]Department of Defense of the United States of America, DETAIL SPECIFICATION WIRE, ELECTRICAL, INSULATED, GENERAL SPECIFICATION FOR, MIL-DTL-16878G, Columbus, 2000

[7]Stahl, P., Seren, T., Roessler, C., Hornung, M., DEVELOPMENT AND PERFORMANCE COMPARISON OF OPTIMIZED ELECTRIC FIXED-WING VTOL UAV CONFIGURATIONS, ICAS, Belo Horizonte, 2018

Resolving the conflict between ship design and UAV Launch and Recovery deck limits; the development of enhanced dynamic interface study tests

Dr. Bernard Ferrier
bferrier@hoffmanengineering.com
SETA (NAVAIR PMA 266)
Hoffman Engineering Corp
Arlington, VA

Robert Ernst
robert.ernst@navy.mil
Chief Engineer
PMA 266
Patuxent River, MD

Ajay Sehgal
Ajay.sehgal@wyle.com
Chief Engineer
KBRwyle
Lexington Park, MD

ABSTRACT

With more than 10 years of MQ-8 Fire Scout Unmanned Air Vehicle (UAV) at-sea deployment on various ship classes, a wealth of ship integration and operations data has been harvested. In general, these various platforms are the results of the Naval Architect's efforts to design a vessel that performs effectively, safely, and within budget for all communities on-board. However, rarely is there a feedback loop in the design or early acquisition phase offering confident environmental parametric aircraft/ship interface data to the ship deck community users. For the most part, aircraft-deck interface parameters, such as, deck motion and air wake field characteristics, are the Navy's to define. In the case of the UAVs, this defined data is essential for safe and efficient air vehicle launch and recovery events. With mono-hull legacy vessels, certain common seakeeping properties have open ocean performance similarities permitting the test and evaluation engineers to practice standard test procedures with a minimum of modification. For a variety of reasons, this model does not hold with the latest ship acquisitions requiring revisions in testing procedures. Unlike manned flight operations, UAVs require more precise environmental information to make a successful recovery. Enhanced dynamic interface (DI) testing is designed to better understand the effects the new ship classes (i.e. Littoral Combat Ship (LCS)) have on the deck environment. The enhanced study aspects, beyond the standard Naval Air Training and Operating Procedures Standardization (NATOPS) envelope expansion evaluation, are composed of a number of analyses measuring and documenting the responses of the ship to the maritime climate and of the air vehicle in operation, but particularly in recovery mode. Naval Air Systems Command (NAVAIR's) Program Management Office (PMA 266), Naval Sea Systems Command (NAVSEA's) Seakeeping and Full Scale Departments have combined with NAVAIR's Dynamic Interface (AIR 5.1.6.11) and Air Wake (AIR 4.3.2.1) engineering agencies to detail these enhanced at-sea and simulation tests. The article summarizes selected testing and computational methodologies covering ship motions, air wake and their impacts on the air vehicle. The study objectives and performance criteria, providing deck environment definitions directed to the UAV operational algorithms, are discussed and illustrated. Data results are presented as a function of helicopter on-deck stability to encountered deck forces in progressively higher sea conditions.

LIST OF ACRONYMS

AS	Airborne Subsystem
AV	Air Vehicle
CFD	Computational Fluid Dynamics
CONOPS	Concept of Operations
DI	Dynamic Interface Study
DIPES	Deck Interface Pilot Effort Scale
EI	Energy Index

FFG	Guided Missile Frigate
LCS	Littoral Combat Ship
LPD	Landing Period Designator
LSO	Landing Signal Officer
L&R	Launch and Recovery
MCS	Mission Control Station
NATO	North Atlantic Treaty Organization
NOAA	National Oceanic and Atmospheric Administration
NATOPS	Naval Air Training and Operating Procedures Standardization
NAVAIR	Naval Air Systems Command
NAVSEA	Naval Sea Systems Command
PMA	Program Manager AIR
PEO	Program Executive Officer
SHOL	Ship-helicopter operational limit
UCSD	University of California, San Diego
UAS	Unmanned Aircraft System
UAV	Unmanned Aerial Vehicle
WOD	Wind over Deck

INTRODUCTION

The current US Navy Concept of Operations (CONOPS) for the deployment of the MQ-8 Fire Scout Unmanned Air System (UAS) is to have the system based aboard, amongst other vessels, the Littoral Combat Ships (LCS). Both platforms (UAS and LCS) were developed independently of each other. Since the UAS is primarily operated as a ship-based system, each aircraft system is required to undergo a ship suitability assessment in the form of Dynamic Interface (DI) testing, also known as helicopter-ship qualification analysis, or envelope expansion trials. DI testing evaluates all aspects of shipboard helicopter compatibility including the adequacy, effectiveness, and safety of shipboard aviation support facilities and/or procedures for all ship-based helicopter types. These empirical at-sea interface analyses have been conducted using established protocols for nearly forty (40) years. The ship platforms are designed independently of the air vehicle. Testing in the legacy Navy has occasionally presented the test team with some unique challenges to a legacy platform. With the recent acquisition of new classes of ship platforms, like the LCS, the DI test teams have been confronted by deck responses that are departures from the legacy ships. This article describes some of the trial procedures and solutions developed to improve MQ-8 operations on the Navy's new platforms.

BACKGROUND

The Navy and Marine Corps Multi-Mission Tactical UAS Program Management Office (PMA-266) of Naval Air Systems Command (NAVAIR) is headquartered at Patuxent River, Maryland, with military and civilian personnel stationed at locations across the United States and around the world.

NAVAIR is organized into eight "competencies" or communities including: program management, contracts, research and engineering, test and evaluation, logistics and industrial operations, corporate operations, comptroller and counsel. It provides support and resources (people, processes, tools, training, mission facilities, and core technologies) to Naval Aviation Program Executive Officers (PEOs) and their assigned program managers, who are responsible for meeting the cost, schedule, and performance requirements of their programs. NAVAIR's affiliated PEOs are:

- PEO for Tactical Aircraft Programs, PEO (T)
- PEO for Air ASW, Assault and Special Mission Programs, PEO (A)
- PEO for Unmanned Aviation and Strike Weapons, PEO (U&W)
- PEO for Joint Strike Fighter, PEO (JSF)

PMA-266 is part of PEO (U&W). The process of converting a research and development project into an acquisition program of record and then shepherding the program into service typically requires many years. The PMA is responsible not only to track the acquisition program's performance requirements but to adapt over time the acquisition program with current mission applications and conditions. This certainly includes system growth and technology evolution. One of the primary responsibilities in PMA-266 portfolio includes the MQ-8 Fire Scout program.

DYNAMIC INTERFACE

Dynamic Interface (DI) is defined as the study of the relationships between an air vehicle and a moving platform. It is performed to reduce risks and maximize operational flexibility **[1].** Globally, DI is concerned with the effects that one (1) free body has in respect to another free body. Historically, this means the effects that a ship may have on a recovering or launching of air vehicles. However, more recent studies have concluded that the same principles apply to other motion related activities, such as, the boarding of Landing Crafts and/or the Landing Craft Air Cushion (LCAC) vessels into the wells of Amphibious Warfare Ships, the docking of submarines or the launching of unsophisticated missiles.

DI is divided into two broad categories: analytical and at-sea experimental analysis. The analytical provides for a mathematical analysis and solution to flight operations. The methods are not mutually exclusive. Neither method alone can produce a comprehensive and timely solution of the deck interface problem **[2].**

The traditional approach for DI is the experimental approach; which investigates operational launch and recovery of vehicles, engaging and disengaging of rotors, vertical replenishment and helicopter in-flight refueling envelopes. However, the experimental approach primary focus is on launch and recovery envelope development and expansion. "Shipboard suitability testing" assesses the adequacy, effectiveness, and safety of shipboard aviation. Testing methodologies and procedures have been standardized by such departments at NAVAIR (Patuxent River).

Dynamic Interface UAV Envelope Expansion

DI testing of unmanned vehicles is not straight forward. The methodology of replacing piloted evaluations with operator estimates and the corresponding test criteria, is established prior to actual testing. However, it is often found that assumptions applied early in testing may need real-time adjustments to adapt to circumstances found during the test.

Naval air training and operating procedures standardization (NATOPS) or ship-helicopter operational limit (SHOL) limits are determined based on wind and ship motion data as assembled by conducting various launches, approaches, and recovery events. Central in the SHOL determination is the pilot evaluation. The unmanned configuration during testing normally is instrumented to provide, amongst other parameters, oleo compression and deflection, torque monitoring along with indications of precise weight on wheels or skids. The number of landings per condition becomes almost academic since there is no piloted variation or technique since the piloting is the same as originally programmed. The settled approach is to assess system performance using multiple launch and recovery cycles but only one recovery is required to justify an envelope expansion.

Table 1 displays the primary components affecting the DI rating scales. While the other DI functions may be directly or indirectly measured, visual cues assessments are entirely empirical and can be subjective. For UAS evaluations, visual cues are replaced by instrumental analysis **[3].**

Table 1 – Dynamic Interface Measures

o	**Relative Wind Speed and Direction**
o	**Ship Air wake and Turbulence**
o	**Ship Motion**
o	**Visual Cues or Instrumented (UAS)**

The behavior of an air vehicle to the deck environment of a given ship platform, is the primary objective of the DI envelope expansion tests. In the piloted case, handling qualities, pilot workload, flight control positions and aircraft limitations are used to define the progress of shipboard launch and recovery envelope development (control margins, torque, rotor speed, power margins, excessive pilot workload, inadequate visual cues, inadequate clearance between aircraft and ship structure). Pilots assign ratings to both the approach and landing, and the takeoff and departure sequences for a specific wind over deck (WOD) condition, in accordance with guidelines set forth in the Deck Interface Pilot Effort Scale (DIPES). Data is evaluated after each flight and plotted on a polar plot. Data points are interpolated to fill in the missing data. The missing data is primarily due to weather conditions which prevents the ship from obtaining the various wind speeds.

If comments and ratings are favorable (DIPES-1, 2, or 3) for both the approach and landing sequence and the takeoff and departure sequence, under a specific WOD speed and direction, the ship will be maneuvered to produce new relative WOD conditions for the next sequence. A sample wind envelope is displayed in figure 1 **[4].**

SH-60B Wind Limits for DDG-51 Class Ships

Figure 1 – Typical Wind Envelope

A significant portion of the envelope expansion testing procedures centers on aircraft responses to near ship air wake. These are based on assumptions developed over many trials concerning the ship's responses to the maritime environment. These responses create the deck conditions in which the air vehicle is tested. When these assumptions are shown to be invalid or inaccurate, it becomes necessary to revise the basic testing assumptions. This is exceptionally important when testing and operating unmanned vehicles. In the course of conducting the unmanned sea tests it became evident that greater emphasis is needed on the simulation side to test for contingencies normally diagnosed by a pilot to deal with an unplanned event. In addition, a greater understanding of the control logic is needed representing a pilot decision making process. Several metrics are used to develop the relationship between the simulated and real world.

REFINING THE DEFINITIONS OF THE UAV x SHIP'S BEHAVIOUR

As explained earlier, the current CONOPS assigns the MQ-8 to any suitably equipped air capable ship, but designates LCS vessels, in particular. LCS has two variant classes of ships, the Freedom (mono-hull) and Independence (trimaran) class. As part of the Navy deployment process, a ship suitability assessment is conducted in the form of DI testing. DI testing evaluates all aspects of shipboard helicopter suitability and compatibility. A significant component parameter centers on the ship's motion characteristics, the ship's control and environmental conditions which control those motions.

In the course of MQ-8B / C and LCS envelope expansion, DI trials were held on both classes of LCS ships, which unexpected ship motions have been measured and recorded. These unexpected motions were almost exclusively observed to be large roll angles. These roll motions may have been induced by a seaway, ship control system interactions or may have been induced by non-linear interaction of the seaway with the ship. Since seaway measurements were not made during these trials, there was no way of knowing with certainty the cause of these motions **[5].**

On further study, it was shown that ship motion responses, in most conditions, were small, generally in phase with each other and motion periods consistent with the encountered seaway. This implies that the motions were generally linear, uncoupled responses to the encountered seaway. Roll motions recorded during some of these events were not consistent with the other motions and were large enough to limit MQ-8B operations. It has been hypothesized that these large roll motions may be due to: undetected oblique swell exciting roll motions, interactions of the ship control system with the encountered waves causing roll in response to steering nozzle activity or some other roll phenomena due to the LCS designs. Both variants demonstrate a lack of roll damping. These two designs, initially intended to operate at high speed, lack roll damping appendages normally installed for low speed seakeeping performance. This can lead to unexpectedly large roll motions under the right combination of wave field, ship speed and possibly autopilot control.

In addition to ship motion, the unsteady aerodynamic environment in the vicinity of the landing deck can have a profound impact on the aircraft's flight characteristics. The on deck measurements of the air wake velocities at key locations over the flight deck during operating conditions match those of the aircraft recovery, and can provide valuable information to evaluate aircraft expected responses during recovery.

The implications for Fire Scout operations encountering a resonance response for roll motion, particularly on recovery, is a concern for the PMA. The need for a complete understanding of the ship's motion characteristics are being driven by ship operational considerations: limited manpower availability, a requirement to reduce the time needed to execute aircraft launch and recovery (L&R) and/or to maintain deck tempo. To facilitate the upcoming ship-suitability assessment, a parallel effort was initiated to determine the feasibility of applying simulations and plug-and-play mission module options in support of unmanned DI at-sea testing and experimentation. Initial analysis indicates traditional linear ship motion models do not predict the presence of roll. Real time measurements appear to offer one of the few choices available to assign a cause to observed, adverse roll motions.

Legacy Launch and Recovery

On legacy ships like Guided Missile Frigate (FFG-8) (Oliver Hazard Perry) aircraft stability at touchdown on or near the deck in real-time is calculated using ship motion as a function of the aircraft model **[6]**. The aircraft model is considered an extension of the ship. The aircraft experiences ship transferred forces and moments, which create rectilinear and angular accelerations on the air vehicle. The accelerations can be numerically integrated to determine the position and attitude of the helicopter relative to the ship as a function of time for various ship motions. This is the inspiration to use the Energy Index (EI) today, to measure and predict deck motion to complete launch and recovery events **[7]**. Figure 2 displays EI based measures from a test of the MQ-8B Fire Scout aboard USS MCINERNEY (FFG 8).

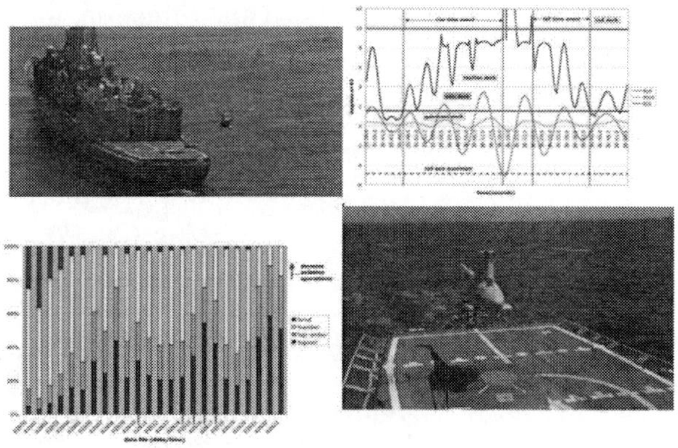

Figure 2 – Motion Characterization

The Quad chart shown in Figure 2 contains a time history trace containing rise and fall time events along with the corresponding ship pitch and roll traces. For an eventual auto-land system to function, an auto-land command would be sent to the air vehicle during hover in an appropriate designated position over the deck. Assuming a descent rate similar to other maritime helicopters, the aircraft touches down well within the rise time of the ship. Still referring to Figure 2, the lower left corner of the Quad chart displays a typical 24-hour period of ship motion recordings showing the distribution of deck energies per hour recording, and the hours in which flight operations occurred.

The time required to raise the deck from minimal motion to unacceptable motion is called the rise-time. The rise-time is a thumb print characteristic of the ship's response and rarely changes. In terms of the energy index scale, it is defined as the period of time that is measured from the end of a green signal to the positive side of the red line. The rise-time is mirrored by a drop time, which is the time period measured from the negative side of the red line to the negative side of the green line. A typical trace as recorded, is displayed in Figure 3.

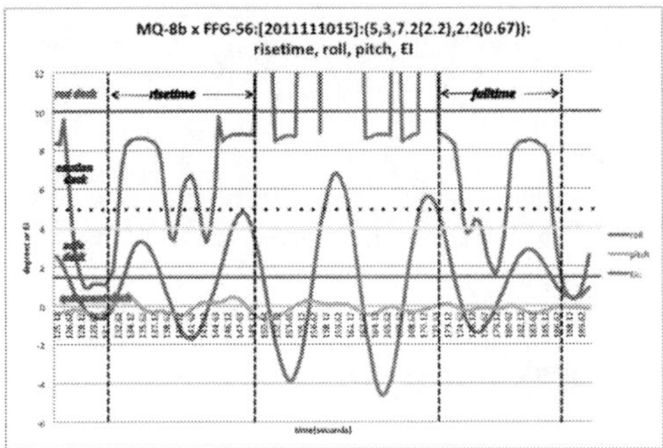

Figure 3 – Rise-time events in FFG file

The example shows a roll axis divergence eventually exceeding the MQ-8B deck roll limit of 5.0 degrees. In this case the time it took for the deck to rise from quiescence was 18 seconds. There appears to be a 1° roll list to port whilst pitch appears to be essentially in trim. The impulse converges in a fall time measuring 15 seconds. Interestingly, the index exceeded 10.0 prior to the roll axis exceeding the max deck motion limit. Figure 4 displays the deck motion rates for the same time period.

The Y' axis diverged like roll, but is 1/4 cycle out of phase with roll. There was sufficient energy on the second limit crossing to cause a roll strong enough to exceed deck limits. This is another example of the axial coupling between Y' and roll. During this test, the ship managed to record at various points 18.9° roll, 4.7° pitch, 2.3 m/s Z' vertical rate, and 2.7 m/s lateral rate maximum deck displacement.

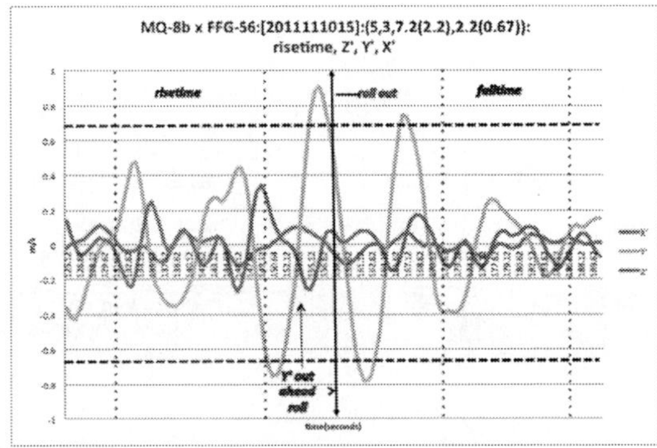

Figure 4 – Corresponding rise-time rates

Launch and Recovery on New Ship Designs

Recorded shipboard data are often compared with related information from other sources such as the National Data Buoy Center run by National Oceanic and Atmospheric Administration (NOAA). The sea surface and atmospheric data frequently explains the recorded ship data.

As typical in envelope expansion trials, the first class of evolutions are constrained both in environmental and ship motion limits. Over time envelope expansion trials, limits were relaxed as the UAV System proved to operate safely in increasingly difficult conditions. As envelope expansion continued to develop, unexpected ship motion recordings and unusual air wake observations began to accumulate. Figure 5 displays a typical hour file recorded in which excessive motions were recorded. Of particular note (green) were the extent of roll cycles in what appeared to be low sea conditions. Figure 6 displays a recent typical newer Fire Scout (MQ-8C) recovery **[8]**.

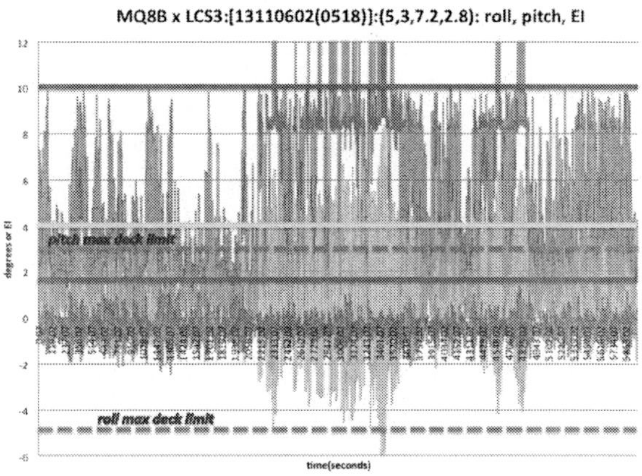

Figure 5 – Typical Time History Trace

Figure 6 –MQ-8C on Final

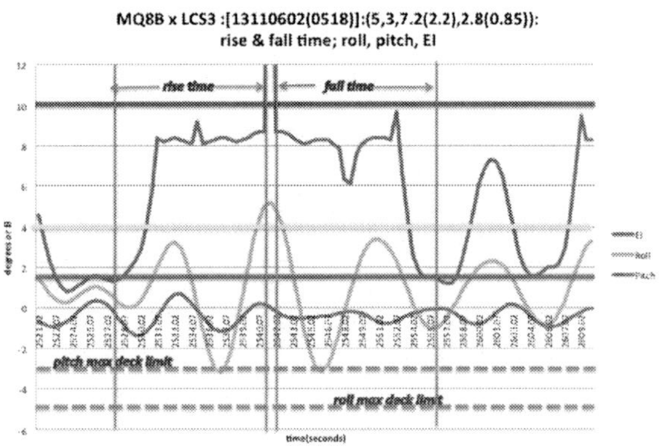

Figure 7 – Rise / Fall time events recorded

Examining the related traces (figures 5, 7, 8), and lining up the graphs (figure 9), it would appear a rare event occurred in which all the degrees of freedom appear to be in phase, but only roll contains sufficient energy to move the ship beyond the MQ-8 landing limits.

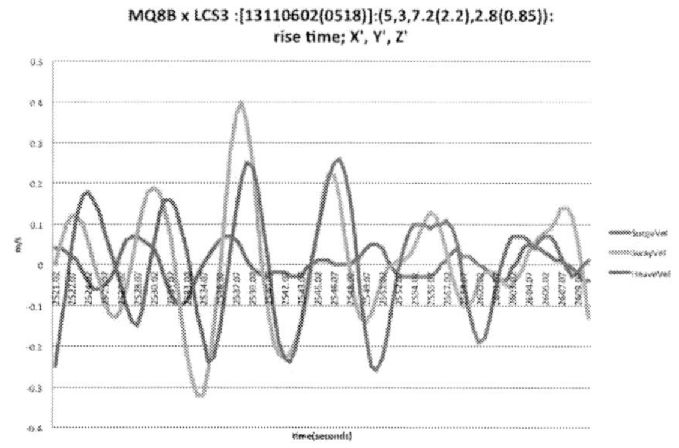

Figure 8 – Corresponding Rise time rates

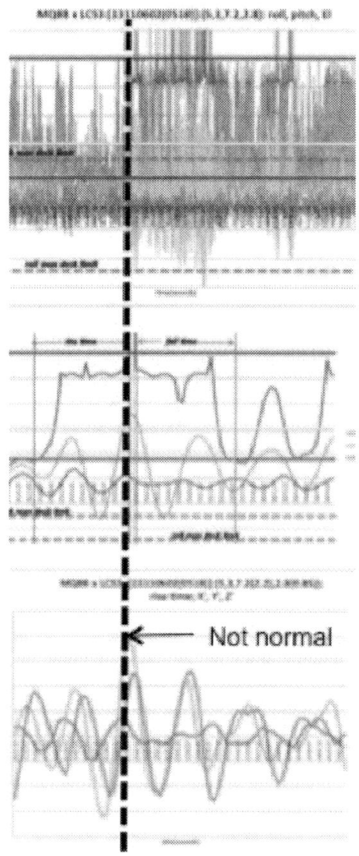

Figure 9 – All DOF rise time axes in-phase

On closer inspection, it was determined this event was not an isolated case. Figure 10 displays other examples of rise time events composed of in-phase axial motions even those that are diametrically opposed.

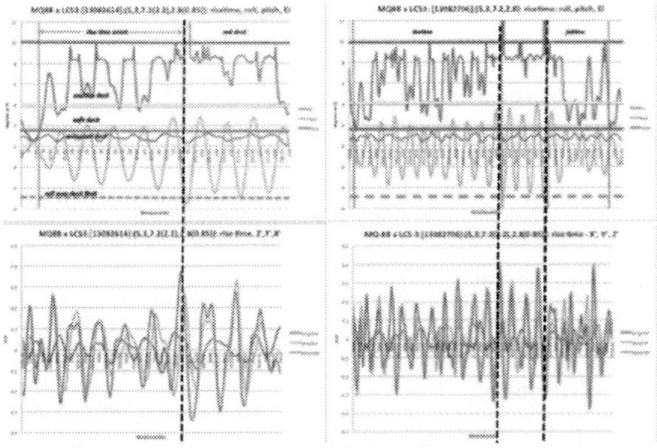

Figure 10 – Similar excessive roll events

Table 2 lists the motion statistics for one of the LCS DI trials.

Table 2 – DI Trial Motion Statistics

- **Total hours recorded at sea: 240**
  - ***Sea State: 0 – 3*** Ship handling: 0 – 5 Ship Speeds: 2 – 15 knots
  - Max motions: roll – 9.9° pitch – 2.5° Z' – 0.7 m/s Z'' – 1.0 m/s²
    Y' – 0.6 m/s Y'' – 0.6 m/s² X'- .28 m/s X'' - .29 m/s²
  - MQ-8 deck limits: roll – 5°, pitch – 3°, Z'– 2.2 m/s(7.2'/s), Y'– 0.86m/s(2.8'/s)
- **The number of rise time events: 252 (non slam)**
  - Average minimum rise time recorded: 10 seconds
  - Computational minimum rise time: 6.3 seconds
  - No of Slam events: 0 (non aviation);
  - ***Excessive ship motion axis causing rise time events***:
    - ***Roll :*** ***100.0%***
    - Y, Z' : 0.0%
    - Pitch 0.0%

This went unnoticed until mugs fell over; raised eyebrows These events did not interfere generally with the DI

Red flags

- **The number of launch and recovery events recorded = 177 (86 launch; 86 recovery; 5 wave-off): 64 piloted serials; 22 autonomous serials;**
  - Launch: 55 green, 20 gr-amber, 25 amber, 0 red events
  - Recovery: 60 green, 15 gr-amber, 25 amber, 0 red events
- ***Other: ship's power failure events (some occurring during flight ops). System robustness was a requirement for the success of this test.***

According to the NOAA wave charts based on local wave buoy data, the test box off of California was experiencing sea states 0 – low sea state 3. Yet the LCS platform recorded a 10° roll. There was motion along the other degrees of freedom but nothing to support a 10-degree roll. Lateral and vertical rates never exceeded the respective deck limits of the MQ-8. Based on the statistics table, only roll exceeded deck limits of the MQ-8B.

These classes of ship appear to be very stable and capable of maintaining a strict adherence to a heading and speed. This is a basic requirement for aircraft at-sea launch and recovery. However, in order to maintain a lock on heading, jet-pod corrections are made, sometimes in rapid succession. These corrections may be linked to the ship's autopilot or ride control or both. The impact on UAV flight operations which follow the deck, as seen in figure 11, may cause corrections in the air vehicle glide slope or flight path. These rapid corrections are informally termed ship induced oscillation (SIO).

Figure 11 – Fire Scout respond in roll cycle

The new classes of ship may not resemble several of the air wake and ship motion characteristics of the legacy Navy. Study of these characteristics are necessary in order to better program the UAV system to accommodate these behaviors.

DEFINING SHIP CHARACTERISTICS

With the definition of the LCS in digital form, it should be possible to adjust the various programmed approaches and procedures to accommodate the ship's motion and over-deck vortex structures leading to improved operational efficiency; specifically, less time the ship has to remain on course for flight ops and maintain personnel at flight quarters.

With a better digital model of the ship's behavior in various sea conditions, it should be possible to integrate deck motion into aircraft simulations that will allow more preliminary DI testing in a lab environment and reduce the number of test points required to be completed in flight.

With the validation of the digital ship model, aircraft systems software may be updated to accommodate the ship's behavior on and over the deck leading to safer launch and recovery events through higher sea conditions.

With the platform fully defined, anticipation of the expected deck conditions may lead to better launch and recovery process planning improving unnecessary heavy deck landing or over-torque of air vehicle engines. This improves the quality and system services impacting maintainability and reliability of the system. Improved ship platform definitions will lead to a shorter learning curve for operators.

There are other non-legacy type vessels being introduced into the fleet to which MQ-8 Fire Scout is expected to be deployed. The trials which have been planned, developed, and executed in the LCS case may prove to be a blue print to more efficiently characterize the ship's environmental responses to improve successful system deployment.

The enhanced DI tests are designed to reduce the number of unknown related functional variables that have not been verified by the standard NATOPS trials. Careful analysis of the seaway to include currents, swells, directions, and speeds along with significant wave heights and their periods, are required to gain the complete picture or definition. To this point, much of the analysis has been using motions in the ship coordinate system without attentive documentation of the ship's heading and speed relative to any existing seaway.

Envelope expansion DI trials are expected to be conducted with an MQ-8C on board the LCS class vessels later this year. The test instrumentation supporting this task include ship motion recording devices. The recordings from these trials will join the library of ship motions data already in existence. Coupled with these recordings, careful attention will be paid to maintaining an eye on the ships speeds and headings. This is combined with external seaway recordings both from "nearby" anchored wave buoys managed by NOAA and close up wave buoy information recorded by University of California San Diego's (UCSD) Scripps Institution of Oceanography.

In addition, airflow over the deck and the disturbances created by the ship's motion behavior will be documented using traditional anemometer towers placed strategically on the flight deck. It is hoped the vortex structures developed by certain relative wind angles and speeds passing over the ship's solid structures, might be identified. The recorded data will be used to further study this air wake using Computational Fluid Dynamics (CFD) analysis.

SUMMARY AND CONCLUSIONS

Parts of the MQ-8 future improvements involve the enhancement of the associated system components, such as that being done for the L&R segment. The primary goal for conducting DI analysis is to expand existing operating envelopes and increase AV's availability thereby improving overall naval effectiveness. The objective of DI is to determine the maximum safe air vehicle/ship platform operational limitations. Given an air/ship system and inherent operational limitations, DI strives to increase tactical flexibility for any set of environmental conditions. Modeling and Simulation complements DI testing to delineate system limitations. The calculated system limitations provide experimental DI with the necessary data to more effectively set testing strategy to probe expected limiting conditions.

This paper provides a discussion of the improvements associated with the MQ-8 system focusing on the L&R process and equipment. These improvements are driven by a need for L&R systems requiring few or no people to operate; systems requiring reduced time to execute high tempo L&R on the deck; systems that provides improved reliability whilst being simple to maintain. One potential problem that could seriously affect MQ-8's L&R on both of the LCS class ships is the large magnitude of roll experienced in seemingly benign sea conditions. This phenomenon should be further investigated to determine its cause, whether by a swell or parametric conditions, so that it possibly can be avoided.

There has been a great deal of interest in this project from its inception. The team looks forward to providing a status update at the conclusion of the development and start of deployments.

ACKNOWLEDGEMENTS

This paper is dedicated to the loving memory of Dr. Peter J. F. O'Reilly and Dean Carico, founding fathers of the DI discipline. The authors gratefully acknowledge the support afforded by Captain E. Soderberg, USN, PMA266, SEA 851 Senior Test Engineer Andy Silver and the contributions made by CDR David Eccles, USN (ret), Assistant Program Manager for Engineering PMA266, Wyle Program Manager CDR Steven "Rocky" Rauch, USN (ret), and Wyle Technical Editor, Monica Rauch.

REFERENCES

1) *Ferrier, Dr. B et al (2016). "Helicopter-Ship Motion Interface Prediction using Simulation Technology Developments and Trial Results from a Dedicated Royal Navy Trial". Proceedings of the Warships Conference. **RINA**. Bath.*

2) *Ferrier, Dr. B., Ernst, R., Sehgal, A., Nelson, J. (2014). "MQ-8C (Fire Scout) Launch and Recovery Program Improvements". Proceedings of the American Society of Naval Engineers Launch & Recovery Symposium. Alexandria*

3) *Ferrier, Dr. B., Ernst, R. (2017). „Fire Scout Launch and Recovery Consideration in Unexpected Ship Roll Motion Conditions". Naval Engineers Journal. **Vol. 129**, No 4. Arlington.*

4) *Ferrier, Dr. B. et al (2006). « Improved VTOL UAV Shipboard Suitability by Autonomous Deck Recovery Operation". Proceedings of the American Helicopter Society. Phoenix.*

5) *Ferrier, Dr. B (2014). MQ-8B/C x LCS-4 Deck Characterization using NAVAIR LPD. QR140530. Arlington.*

6) *Ferrier, Dr. B. et al (1998). « Simulation and Testing of the Landing Period Designator (LPD) Helicopter Recovery Aid". Naval Engineering Journal. **Vol. 110**, No 2. Alexandria.*

7) *Nelson, J. & Danielson, J. (2013). MQ-8B Fire Scout Initial Ship Suitability Tests Aboard USS McInerney (FFG 8). NAWCADPAX/RTR-2009/118. Lexington Park.*

8) *France, W. N., et al. (2001). "An Investigation of Head-Sea Parametric Rolling and its Influence on Container Lashing Systems". Proceedings of the SNAME. New York.*

IGBT Robustness Capabilities and Setting Appropriate Protection Thresholds in High Power Propulsion Inverters

Satyavrat R. Laud
Staff Product Marketing Manager
Renesas Electronics America
Milpitas, CA, U.S.A.

Phillip E. Thibodeau
President
HSV Consulting LLC
Huntsville, AL, U.S.A.

ABSTRACT

High-voltage power semiconductor devices such as IGBTs (Insulated Gate Bipolar Transistors) and FRDs (Fast Recovery Diodes) are at the core of most high-power inverter modules used in propulsion and traction motor drive applications. SiC (Silicon Carbide) MOSFETs are presently making their way into propulsion inverter designs and offer some distinct advantages such as faster switching speeds and a higher operating temperature range. However, flight heritage, cost, and device availability still strongly favor silicon IGBTs. While there is a wealth of parametric static and dynamic performance-related information available in most IGBT product datasheets, which is often the focus of most application engineers and users, the robustness aspects of the device are often not as well-understood or analyzed from a user or application perspective.

It is the attempt of this paper to shine a light on aspects such as short circuit withstand, high current turn-off withstand capability, consideration of transient high dv/dt events, and significance of a high thermal capacitance for the device in the context of short circuit events. In this paper we also explain how these withstand capabilities are tested for by IGBT device manufacturers. For a circuit designer, understanding these device capabilities are critical for withstanding severe and abnormal transient conditions as can be expected in aviation and other demanding propulsion system applications.

NOTATION

IGBT: Insulated Gate Bipolar Transistor

FRD: Fast Recovery Diode

SC WT: Short Circuit Withstand Time (us)

Esc: Energy dissipated during Short Circuit (J)

RBSOA: Reverse Bias Safe Operation Area

Cres: Reverse Transfer Capacitance (pF)

dv/dt: rate of change of voltage across the device (V/ns)

di/dt: rate of change of current through the device (A/ns)

Vce: Collector to Emitter voltage (V)

Ic: Collector current (A)

Zthjc: Transient Thermal Impedance, die to heat sink

T_j: Instantaneous Die (IGBT) Temperature

T_hs: Maximum Heat Sink Temperature

INTRODUCTION

A survey of the literature and application notes and product datasheets published by device manufacturers discuss certain key robustness aspects of IGBTs. The capability of the device to withstand extreme application conditions such as a short circuit event in an inverter (Fig. 1) is one such aspect.

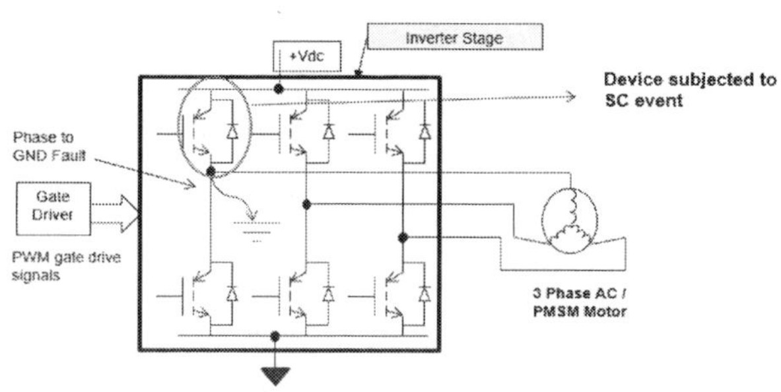

Fig. 1: Simplified representation of an SC event in an Inverter

Another severe application requirement for the device to be able to withstand is a high current turn-off event, where the peak of the current well exceeds the guaranteed continuous current rating of the device. In both these cases, the electrical and thermal stress on the silicon die can lead to failure, unless

Presented at the Vertical Flight Society Autonomous VTOL Technical Meeting and Electric VTOL Symposium, Mesa, Arizona, UAS, Jan. 29-31, 2019.

the device is safely turned off within a specified time duration. Reference 1 is an excellent resource in this regard. In particular, with respect to a short circuit event, while the duration aspect of the event is often the focus and a key consideration of the circuit designer when selecting devices or setting protection thresholds, the energy dissipation in the device (Esc) under such a condition, is of equal if not even more significance. This is a relatively less highlighted and studied aspect. Similarly, in case of a high current turn-off event, the device must be able to withstand an overlap period of simultaneous high voltage and high current application. This leads to the well-known RBSOA definition (Fig. 2), tested and guaranteed by device manufacturers.

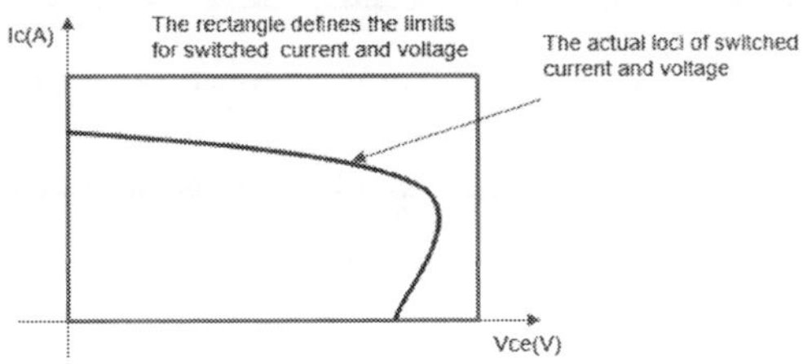

Fig. 2: A simplified representation of an RBSOA for an IGBT

If the loci of the voltage and current waveforms stay within the defined RBSOA, the device manufacturer can guarantee its safe turn-off capability. However, a less well-known and understood aspect of such events is the slew rate of the current and the associated impact on the dv/dt across the device. An excessively high di/dt can result in a high dv/dt which merits consideration from the circuit designer, and which unfortunately tends to be overlooked. In this paper, while the relatively well-known robustness aspects of IGBTs are discussed with example data, the relatively lesser known aspects are also highlighted and discussed. Finally, some areas for future study are noted in the conclusion.

SHORT CIRCUIT ROBUSTNESS

As an example of a short circuit rating from a device perspective, an example IGBT SC WT specification is shown in table 1.

Table 1. Short Circuit Rating, RJP6511JWT

tsc (us), min	Test Conditions	Comments
5	VDC=400V, Rg=5Ohm, Tj=150C, Vge=15V	Guaranteed by design and characterization

The key ratings and specifications of this IGBT are shown in table 2 for reference.

Table 2. RJP6511JWT, Key Ratings, Ta = 25C

Rating/Specification	Value	Units	Remarks
BVces	650	V	Ic<=100uA

Table 2. (Contd.) RJP6511JWT, Key Ratings

Vce(sat), max	1.6	V	Ic=200A
Ic,DC	200	A	Limited by Tj (Tj,max<=175C)
Ic,pulsed	400	A	Limited by Tj (Tj,max<=175C)

Short Circuit Detection

Based on the SC WT rating, a circuit designer must ensure that the short circuit event is detected and the device is turned off within the withstand time specified. Typically, such detection and protection schemes need to be implemented at the gate driver circuit level itself to ensure a sufficiently fast response time. A key requirement of these protection schemes is to set an appropriate short circuit event detection threshold. There are two main types of design approaches used for this purpose; a Vce based approach, or an Ic based approach. The Vce sensing based approach is also known as the 'desaturation' detection method since it is based on the IGBT coming out of saturation mode in the event of a short circuit, with the device being driven by a nominal +15V gate to emitter voltage signal. During the short circuit event, the DC bus voltage is effectively applied across the collector to emitter of the device, and as the Vce starts to increase, a suitable threshold (typically in the 6 to 10V range) can be set to detect the event and initiate turn-off. The Ic based approach is simply based on sensing the level of the instantaneous collector current and setting a threshold based on circuit considerations, while keeping in mind the device's current ratings. As soon as the current exceeds the set threshold (such as 3 times rated Ic), the turn-off can be initiated. The Vce sensing method has the benefit of lower cost and relative simplicity of implementation at the gate driver level, while the Ic sensing approach has the advantage of enabling a more precise detection threshold at the cost of additional components such as current shunts. In either case, the device's output characteristics (Vce vs. Ic) must be made available to the circuit designer, particularly under elevated temperature conditions, to determine a suitable voltage or current threshold at which to initiate turn-off of the device when a short circuit event occurs. A typical output characteristic curve for the IGBT RJP6511JWT packaged in a prototype assembly is as shown in Fig. 3.

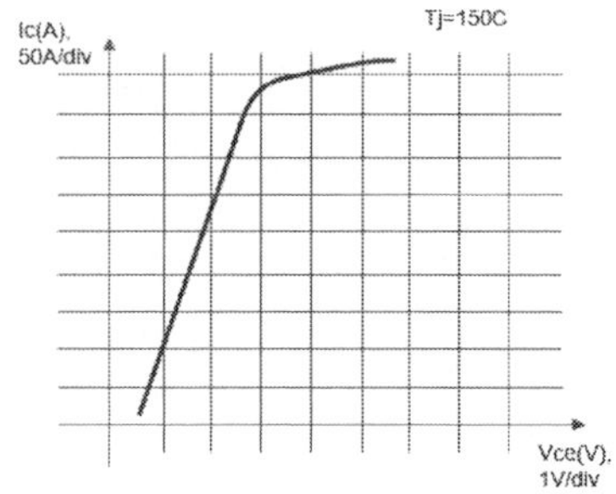

Fig. 3: Typical Vce vs. Ic for IGBT RJP6511JWT

It should also be noted that a blanking time is generally required from a noise filtering perspective and to avoid nuisance tripping of the inverter, which tends to be in the 1-2 usec range, from the instant the voltage or current threshold is exceeded until the device turn-off is initiated. This blanking time, along with the additional time required for the turn-off of the device, requires the device to be able to withstand the short circuit event for a total duration of 3-4 usecs.

Short Circuit Testing

Since circuit designers rely on the SC WT rating specified by the IGBT manufacturers, it is critical that the device be tested and characterized for withstanding such events. An example test circuit used by Renesas for short circuit testing of IGBTs is shown in Fig. 4.

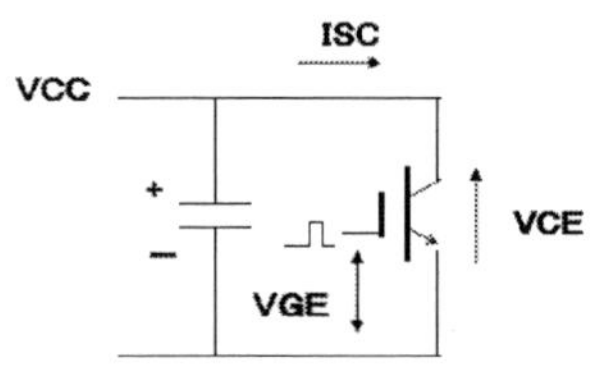

Fig. 4: Simplified SC WT Test Circuit

Further, the test conditions should reflect worst case operating conditions for the device, such as an elevated junction temperature close to maximum rated temperature, a nominal gate drive signal, and typical values of turn-on and turn-off gate resistances. It is essential that designers pay close attention to these test conditions and further verify with the device manufacturer their application's operating conditions vis-à-vis those specified for the SC WT on the datasheet. Another key aspect of a short circuit event is the energy dissipation in the device during the event. While the SC WT is a key parameter that helps inform the circuit designer as to the allowable time from a short circuit event detection till the time the device is turned off, it is also almost equally important to understand the energy dissipated in the device during the event. As can be seen from a sample test waveform shown in Fig. 5, the energy level during a short circuit event can be on the order of several Joules, and if the circuit components and conductors in the device's current path are not sized appropriately, it could cause a given component to fail catastrophically.

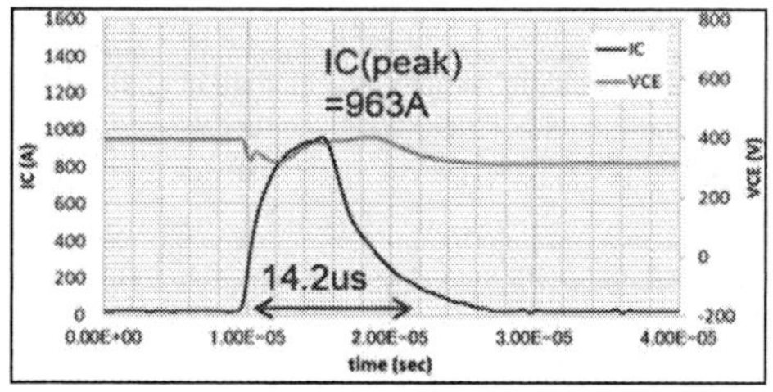

Fig. 5: An example SC waveform for RJP6511JWT

Significance of a low Transient Thermal Resistance

Another important consequence of the energy dissipation in the device during short circuit is the instantaneous rise in junction temperature. Since there is a finite and limited thermal capacitance of the die itself, it must be ensured by suitable die attach and substrate design that there is a sufficiently low transient thermal resistance of the device from silicon to heat sink. Based on Fig. 5, it is seen that the Esc for the RJP6511JWT is on the order of 3J or so. Considering variation across production lots, a 20% margin is added to this value, and the required transient thermal impedance for the device for a 5us duration is estimated as shown in equation 2.

$$Zthjc = (T\_j - T\_hs) / (Esc / SC\ WT),\ \text{deg C/W} \quad (1)$$

$$= (175C - 75C) / (3.6J / 5us),\ \text{deg C/W} \quad (2)$$

$$= 0.000139\ \text{deg C/W} \quad (3)$$

The estimated transient thermal impedance can be referenced by the thermal engineer as part of the heat sink design. It is mentioned here that the assumptions of heat sink temperature used in the equations 1-3 are based on typical liquid cooled heat sinks. In the case of air-cooled heat sinks, the assumption would likely be higher, but the underlying principles of the physics and the calculation methodology remain the same.

REVERSE BIAS SAFE OPERATION AREA

While a device's short circuit ratings and its considerations are critical for a circuit designer when selecting a device and designing the inverter, another almost equally important robustness attribute for a device is its high current turn-off capability. In most traction motor inverter applications such as a propulsion system of an aircraft or vehicle, there arise conditions where the device is expected to safely turn-off a current that is twice the rated continuous current, or even higher still. The circuit designer thus must know the capability and limits of the device with respect to such extreme conditions. The RBSOA is also helpful to the circuit designer in terms of understanding the margin available with respect to the maximum amplitude of the PWM current pulses in the case of sinusoidal modulated inverters. To help the circuit designer in this regard, IGBT device manufacturers test and characterize and publish a turn-off safe operation area, more commonly termed a Reverse Bias Safe Operation Area (RBSOA).

RBSOA Testing

From a device robustness characterization perspective, the standard method typically used by most IGBT manufacturers is the well-known Double Pulse Test. A simplified test circuit as used by Renesas is as shown in Fig. 6.

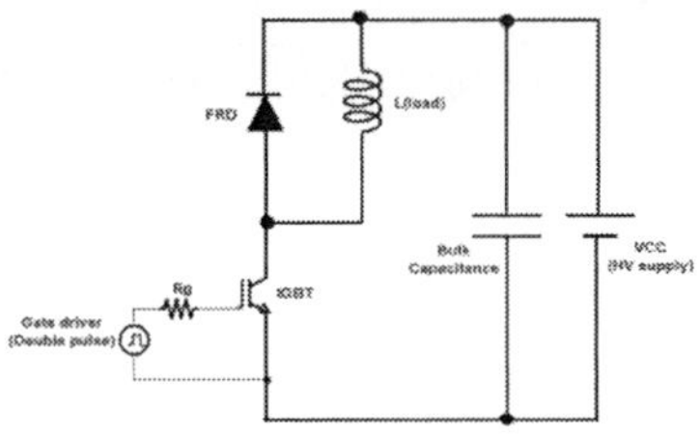

Fig. 6: Double Pulse Test Circuit

The main intent of this test is to set up conditions as would be expected to occur in the case of inverters driving inductive loads such as traction or propulsion motors. Further, the peak current through the IGBT at turn-off can be programmed to be a desired level such as two times or three times the DC rating. For example, for the RJP6511JWT, under specified test conditions as shown in Table 3, the characterized RBSOA is as shown in Fig. 7.

Table 3. RBSOA Test Conditions, RJP6511JWT

Test Conditions	Ic=400A, VDC=500V, Tj=150C, Rg=5Ohm, L=50uH

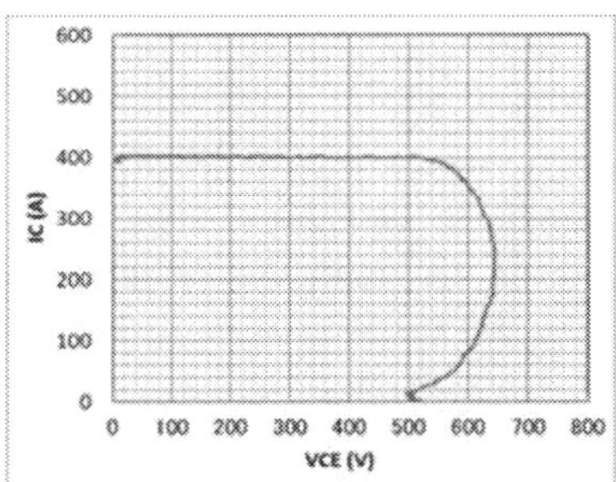

Fig. 7: RBSOA for RJP6511JWT

The associated turn-off waveform under the RBSOA conditions is as shown in Fig. 8.

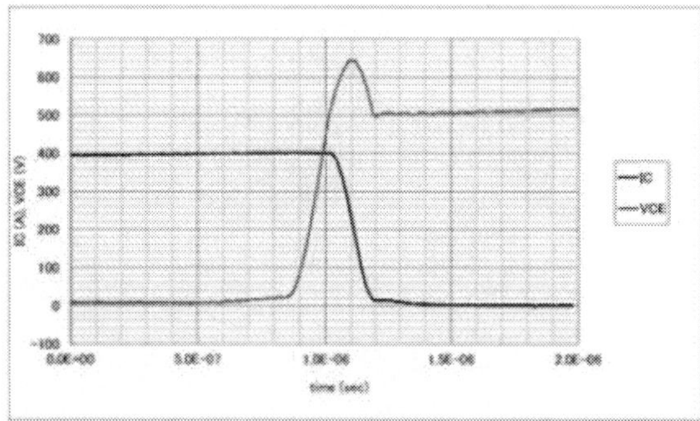

Fig. 8: RBSOA: Switching Waveforms of Ic and Vce

RBSOA: Transient dv/dt and di/dt

The example waveform while specific to the test conditions and set up by the device manufacturer, also provides some useful information and guidance to the circuit designer. Among these are the information regarding the rate of rise of the collector voltage (dv/dt) and the rate of fall of the collector current (di/dt). Such information can help guide the choice of gate resistance values and the design of the circuit layout with the goal of minimizing the parasitic inductances in the path of the current. For example, the higher the di/dt, the higher will be the peak of the voltage overshoot at the collector. It must be ensured that this overshoot does not exceed the minimum breakdown voltage rating specified for the device. There are also implications of excessive di/dt resulting in high frequency ringing effects on the waveforms, which may then result in adverse effects from an Electro-Magnetic Interference (EMI) perspective. Many excellent references are available in the literature on this subject such as Reference 2. The implication of an excessive dv/dt also has implications from an EMI perspective, particularly radiated EMI; however, a parasitic phenomenon due to high dv/dt needs to be well-understood by the circuit designer. In particular, for a multi-phase inverter application with complementary PWM (Pulse Width Modulation) switching of the two IGBTs, one high side, and the other low side, in each phase, if the dv/dt of the high side device when turning off is excessive, it can couple a charge flow into the gate of the low side device, and if the voltage rise at the gate of the low side exceeds the gate to emitter threshold voltage, it can result in a parasitic turn-on of the low side device. This will then result in a low impedance condition across the high voltage DC bus, and in some cases, can cause devices to fail. This phenomenon can be better understood by means of Fig. 9 which shows a simplified equivalent circuit model of the complementary IGBT (the low side), at the time when the high side device is being turned off, and the low side device is being held off by the gate driver circuit.

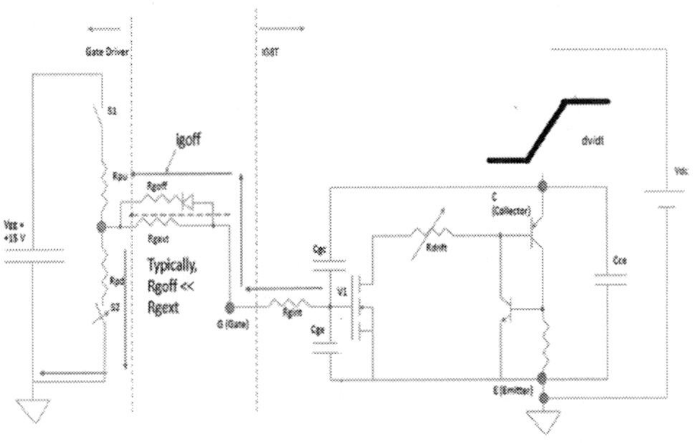

Fig. 9: Simplified Equivalent Circuit Model of low side IGBT with the high side device (not shown) being turned off

The parasitic gate to collector capacitance is represented as Cgc. If the dv/dt of the high side device is excessive, it can result in a relatively large instantaneous current flow through the internal gate resistance of the IGBT and through the low side gate driver impedances. This can then result in a spike at the low side IGBT gate and under extreme conditions can result in turn-on of the device. Using basic circuit theory, the allowable turn-off dv/dt for a given IGBT and under specific gate drive circuit conditions can be estimated as shown in equations 4-6.

$$Igoff = (Cgc * dV/dt) \quad \ldots\ldots\ldots\ldots (4)$$

But from the equivalent circuit model,

$$Igoff \sim (V1\text{-}0.7V) / (Rpd + Rgoff + Rgint) \quad \ldots\ldots\ldots\ldots (5)$$

Or,

$$dV/dt = (1 / Cgc) * ((V1\text{-}0.7V) / (Rpd + Rgoff + Rgint)) \;..(6)$$

Based on current IGBT technology levels, a suitable value of dv/dt in the 3-5 V/ns range is suggested. This value can then be further tuned as necessary based on switching loss, and EMI considerations, and with equation (6) as a reference.

CONCLUDING REMARKS

The robustness aspects of an IGBT have been highlighted in this paper. Specifically, the two most critical aspects of short circuit withstand and high current turn-off have been discussed with reference to a state of the art device, and keeping in view aviation and automotive vehicle propulsion inverter applications. As an area for future study, a deeper analysis of the short circuit withstand capability from an energy and junction temperature perspective is suggested. Such a study should include thermal modeling of the IGBT and consider aspects such as non-uniform current densities over the die active area. Another area of further research and study is the development of specific and advanced packaging methods including double side cooling and sintered attachment techniques. Packaging is a critical aspect of a product's long-term reliability, and has particular significance to demanding applications such as aviation and automotive propulsion inverters. Reference 3 has published relevant results with regards to the benefits of sintered IGBT die from a thermal impedance and reliability perspective.

Author contact information:

Satyavrat Laud: Satyavrat.laud@renesas.com

Phillip Thibodeau: phil@hsv-consulting.com

ACKNOWLEDGMENTS

The authors would like to sincerely thank the Renesas IGBT Design and Applications Team for support on providing bare die samples and on the characterization of the packaged prototype devices. Without such support, this paper would not have been possible.

REFERENCES

[1]M. Trivedi, K. Shenai, "Failure mechanisms of IGBTs under short-circuit and clamped inductive switching stress", IEEE Transactions on Power Electronics, Volume: 14, Issue: 1, Jan 1999 (pages 108-116).

[2]Pierre Lefranc, Dominique Bergogne, "State of the art of dv/dt and di/dt control of Insulated Gate Power Switches", Proceedings of the Conference Captech IAP1, Power Supply and Energy Management for Defence Applications, Jun 2007, Bruxelles, Belgium. pp.1-8, 2007.

[3]Francois Le Henaff, Gustavo Greca, Et al., "Reliability of Double Side Silver Sintered Devices with various Substrate Metallization" PCIM Europe 2016, Publisher VDE, Print ISBN: 978-3-8007-4186-1.

Challenges for Vehicle and Occupant Safety in Autonomous Electric Vertical Take-off and Landing (eVTOL) Vehicle Crashworthiness

Justin Littell

Structural Dynamics Branch

NASA Langley Research Center

ABSTRACT

The burgeoning electric Vertical Take-off and Landing (eVTOL) vehicle industry has generated a significant level of enthusiasm amongst aviation designers, manufacturers, researchers, and consumers. This industry is determined on changing the urban transportation paradigm from traditional ground-based vehicles (cars, taxis, buses) to air-based eVTOL vehicles which can be summoned, much like how conventional taxi services work currently. These new eVTOL vehicles are designed to be small, lightweight and operate autonomously without user intervention. One of the major players in this new industry is the Uber peer-to-peer ridesharing company, which is leading the development through its Elevate network.

There are many unknowns as to how the industry will mature. The logistics of creating a completely new category of vehicle along with a completely new set of rules for which to operate the vehicles is complex, and there are many known and unknown barriers to overcome. Some (of many) known barriers include airspace management, ground logistics, physical space, and, of course, the vehicle itself, and there are many eVTOL vehicle manufacturers (OEMs) and organizations working these problems presently.

This report will focus on one major barrier as related to the level of safety specifically within the framework of the eVTOL vehicles themselves. A high level of safety is necessary for the vehicles to gain acceptance by the public. An overview of current levels of transportation safety and some extrapolation into how eVTOL vehicles might compare is first presented. Next, a discussion categorizing the major differences between *Crash Prevention* and *Crash Mitigation* as it pertains to eVTOL vehicle safety is included with identification of current deficiencies in the ongoing discussions within various trade organizations. The report then expands a framework for specific ideas that could be used for *Crash Mitigation* to improve vehicle safety through a crashworthy systems level approach which includes supporting rationale.

Finally, a discussion into the regulatory approach and potential guidelines as they pertain to new eVTOL vehicles is presented. Accordingly, when discussions are presented, much of the supplemental data will be taken from sources pertaining to either General Aviation (GA) aircraft, rotorcraft, or transport category aircraft, due to the lack of data from eVTOL vehicles themselves, with references cited. It is expected that the eventual eVTOL aircraft performance, expected operations and anticipated future regulations will consist of some combination thereof.

The Public Perception of Safety in Current Transportation Systems

During 2017, there were zero fatalities in commercial air travel in the United States [1]. This fact is a result of the regulations put in place by the Federal Aviation Administration (FAA) to ensure that flying on a commercial airline is safe. The regulations for 14 CFR Part 121 – "Air Carrier Certification", which cover the operations for all U.S. commercial carriers, and 14 CFR Part 25 – "Transport Category Airplanes" which cover the airplanes themselves, have created an inherently safe environment for the general public to fly. These regulations have been continuously updated since the beginning of commercial air travel, and improvements are continuously being implemented through the study of previous accidents, aircraft/airline research and/or public input. Updated regulations lead the aircraft manufacturers (OEMs) to create new designs or modify existing fleets to ensure that improved safety is included. Additionally, flying on commercial airlines/airplanes for the past 80+ years has ingrained a certain amount of acceptance into the public about what expectations will be when flying in commercial air travel. Even when incidents and accidents do occur, the regulators and industry address the problem through new regulations or procedural modification, and the level of safety is maintained or improved.

In a similar fashion, the automotive industry has been implementing safety improvements for the last 50+ years. The addition of the first seat belt laws in the United States [2] in 1968 provided some of the first safety features into automobiles. Over the years, additional notable improvements including the first requirement for the inclusion of airbags [2] in 1998, along with fundamental changes in automotive design (i.e. the "crumple zone"), now provide an unprecedented level of protection never before achieved in automotive transportation. With all of these improvements in safety, the public now generally accepts the automobile as safe and expects that if a crash occurs, these safety features will provide protection from injury. Additionally, various metrics for quantifying the level of safety of automobiles exist [3-4], and OEMs will often promote a positive rating in advertising. The addition of these safety factors, along with campaigns to crack down on intoxicated and now distracted driving, has led to record lows in fatality and injury rates. These rates are down to 1.14 fatalities per 100 million vehicle miles in 2010 versus almost 3.5 per 100 million vehicle miles in 1975 [5], which represents a reduction by a factor of approximately 3.

Problems in Defining eVTOL Safety Standards

There are no current data or metrics for eVTOL aircraft safety because these types of vehicles are not in operation currently. The closest data that can be used for examination is data from self-driving automobiles, which represent the closest analogous autonomy component, and from GA aircraft or small rotorcraft, which represent the closest analogous vehicles.

Self-driving automobiles are being developed by a variety of companies, and use much of the same technology and sensors (GPS, RADAR, LIDAR, cameras, etc.) that the proposed eVTOL aircraft will use. These technologies allow the automobiles to maintain speed, stay in their lane and avoid obstacles - all of which have parallels for use in eVTOL vehicles. However, a recent survey conducted by the American Automobile Association (AAA) found that 63% of drivers still fear driving in a self-driving car [6], due to the highly visible accidents that have recently occurred. Furthermore, there is operator confusion regarding the terms "self-driving", "autonomous", "autopilot" and other terms currently used to identify the various semi- and fully-autonomous modes of cars. Many accidents that occur today are due to the driver's unrealistic expectations for what the vehicle can reasonably accomplish when operating in these modes.

A notable example includes a Tesla Model X crash into a concrete highway barrier while on Autopilot mode, killing the driver [7]. The article notes that Tesla has stated the Autopilot mode is a "driver assist tool", not a replacement and the operator should always be aware of their surroundings. Furthermore, Tesla also claims that the Autopilot modes have known weaknesses, and may not be able to identify stationary objects, stay in lanes if dividing lines end, and other potential obstacles or scenarios for which the computers cannot recognize.

In a second example, on March 18, 2018, a modified self-driving Volvo XC90 struck and killed a pedestrian in Tempe AZ [8]. The vehicle struck a pedestrian crossing the road outside of a designated crosswalk while operating with the computer controlled self-driving system in control. The vehicle first registered observations of an object (the pedestrian) approximately 6 seconds before impact. At 1.3 seconds before impact, the system determined that an "emergency braking maneuver was needed to mitigate the collision," however, the emergency braking maneuvers were not enabled while the vehicle was under the computer control, and instead needed operator intervention to occur. The operator intervened less than one second before the impact, and began braking less than a second *after* impact. As a result of this accident, Uber has decided to remove all of its self-driving cars from testing in Arizona, and has eliminated 300 jobs as a result [9]. NuTonomy, a startup company operating in the Boston area, has also suspended its pilot program [10] due to this accident. The effects of these two tragic accidents are rippling around the autonomous vehicle world, and will negatively impact the perception of the level of safety in these vehicles in the short term.

General Aviation (GA) aircraft and the current conventional fixed and rotary air taxis operating under 14 CFR Part 135 are the two categories most closely aligned to how the anticipated eVTOL vehicles will look and operate. GA is the closest analog to the vehicle designs themselves, while Part 135 operations would be the closest analog to the expected eVTOL operation logistics. A summary of accident data from 2010 shows that accidents from these two categories encompass 274 out of 275 total fatal accidents in the United States [11], with the vast majority being GA (268). Of those 268 GA aircraft, loss of control of the aircraft was the highest contributor at 38%. The cause for the loss of control is not stated, but loss of control can occur due to a number of factors resulting from either pilot error, aircraft failure, weather or a combination thereof. Of the accidents that occurred on Part 135 operating aircraft, the additional pilot experience (average flight hours 5,459 for fixed wing and 11,679 for rotorcraft) needed for operation and the actual amount of hours recorded by this type of operation (approximately 300,000 hours total as compared to over 8 million for personal flying in GA) reduced the number of fatal accidents significantly. Out of the 6 fatal accidents, there were various factors which caused the accident ranging from loss of control on ground or in flight, collision, or a powerplant failure. The results from the Part 135 accident data point to the need for both pilot and/or operator training, along with the recognition that internal factors such as aircraft performance, and external factors such as weather and airspace management can still lead to accidents, even under ideal circumstances.

If problems with autonomy exist for automobiles operating on roads, which are two-dimensional (forward-backward, and left-right) design spaces, the problems will be infinitely more complex in a three-dimensional (now adding ascent-descent) space, for which eVTOL aircraft will exist. Additionally, the problems with autonomy will be compounded with all of the problems associated with the actual process of flying the aircraft, all of which will be under the scrutiny of the public eye. If the two autonomous automobile accidents described above can cause a significant setback for the autonomous vehicle industry, there is a chance that one or two accidents in an eVTOL vehicle can dismantle the entire eVTOL industry.

One major difference between autonomous vehicles in current operation and autonomous conceptual eVTOL aircraft is that the developers of the autonomous vehicles are retrofitting current fleets of vehicles with the autonomous technology. The vehicles themselves have already been manufactured to existing safety requirements, and are not being modified structurally for any additional occupant protection. eVTOL aircraft are starting at either a disadvantage or advantage, depending on the point of view, in developing a new vehicle design essentially from scratch. The major challenge (or opportunity) will be to include both the autonomous technology and occupant protection in something that the public wants to fly.

Deficiencies in Current eVTOL Vehicle Safety Research

The largest challenge right now for eVTOL vehicles is to gain acceptance from the public as being as safe as commercial air travel and automotive transportation currently. To accomplish this, a white paper written by Uber Elevate [12] in 2016 suggests setting a safety goal of twice that of driving a car based on fatalities per passenger mile. Using this metric with an assumed certification using 14 CFR Part 135, Uber states that the current level of safety would need improvement of four times over the current 14 CFR Part 135, which if using the 2010 data included above, means that there will be less than 1.5 fatalities per 300,000 flight hours. Note that Uber does not specify the level of safety as that meeting 14 CFR Part 121 requirements, which are much more stringent.

To achieve that level of safety, Uber lists Distributed Electric Propulsion (DEP) as a major design factor [12]. DEP establishes a redundancy in flight by implementing multiple motors, blades or propellers and batteries onto each vehicle. That way if one were to fail, the vehicle would rely on the remaining functional components to safely land a vehicle. Much research is devoted to battery design, range extension capabilities, hybrid engines, and engine power distribution networks, all of which are encompassed under the umbrella of DEP, and all necessary for continued eVTOL vehicle development.

There are also many other challenge areas that are involved in achieving the required level of safety. Some of these include airspace factors such as queueing and separation [13], ground infrastructure which includes items such as landing pads, security and charging stations, and communications which includes items such as an encrypted communications network to ensure against unauthorized access into the vehicle and ground communications systems.

OEMs and other researchers from various areas of the eVTOL world are aggressively examining all of the above listed items. Taken together, these items can be grouped into an overarching category called "*Crash Prevention*." In crash prevention, the underlying philosophy is that all of the developed technology works in harmony to prevent a crash from ever occurring in the first place. While these are noble and necessary goals, the reality is that at some point, a crash will occur, and when that happens, additional separate systems need to be in place to mitigate its effect to provide an outcome where the occupants are able to survive. It is in this area of safety research where much attention is needed. This area of safety can be categorized in an overarching category called *"Crash Mitigation."*

At both the Uber Elevate Summit, now in its 2nd year, and the Electric VTOL Symposium organized by The Vertical Flight Society, now in its 6th year, much time and energy is devoted to safety focusing primarily on the area of crash prevention. The sole focus on the topic of prevention may be intentional since mitigation is primarily associated with vehicle specific performance, and of primary concern to an OEM. However, it could also be an oversight, since crash mitigation performance should be discussed between groups representing all sides of the eVTOL industry (government, regulators, OEMs, etc.). This way, a dialogue can be opened to ensure that requirements for crash mitigation capabilities are discussed and mutually agreed upon between all parties and expectations can be set.

By its nature, the research focused in crash mitigation will deal primarily with the vehicle design itself. This paper has presented a very brief overview for topics already covered under crash prevention, and will turn its focus instead to crash mitigation. One major barrier that already exists from a mitigation standpoint is evaluations of performance must be compared to baselines. Because all of the eVTOL vehicles that are being developed are new designs, there is no heritage in which to establish a baseline for which to compare a crash mitigation design. In that regard, some examples and rationale for crash mitigation concepts developed for existing aircraft – whether it be an airplane or a rotorcraft - and how these concepts could be implemented into an eVTOL vehicle are discussed. This report will not provide an exhaustive list, but rather a few examples that have worked for similar aircraft under similar circumstances.

Current Crash Mitigation Capabilities

The only major safety feature that has been identified for use by multiple eVTOL OEMs for crash mitigation is a Ballistic Recovery System (BRS), which is essentially a parachute that either manually or automatically deploys in the event of an imminent loss of control of the aircraft. Many OEMs have identified the use of the BRS in their design for safety. While this feature is a necessary item in the aircraft's equipage, there are limitations that prevent it from being the *only* safety product. These limitations are written into the literature for the BRS product itself. A statement from Cirrus Aircraft regarding its parachute system (called Cirrus Airframe Parachute System – CAPS) notes: "If CAPS is deployed too close to the ground, the chance of a successful deployment greatly decreases [14]."

When examining the Uber preliminary missions and requirements for eVTOL aircraft [15], eight out of the eleven segments for the nominal mission requirements occur at altitudes below 300 ft. above ground level (AGL), representing a large portion of each flight. While no specifications exist for the lower deployment limit of a BRS, it is conceivable that the capability of a BRS would be greatly diminished or completely negated if deployed at these low altitudes. The protection at a lower altitude level is critical because, according to the NTSB, the majority of thc accidents occur in GA aircraft during the landing phase of the flight [11], which occurs at a low altitude and low aircraft speed.

Additionally, vertical drop testing conducted at NASA Langley Research Center (LaRC) in 2018 [16] has shown that a drop from a height of only 14 ft., which generates an impact velocity of ~29 ft/s, has the potential to cause occupant loading limits that exceed those specified for both GA [17] and Rotorcraft [18]. Tests were conducted on 50th percentile Anthropomorphic Test Devices (ATDs, a.k.a. crash test dummies) seated in certified Transport Category Aircraft seats [19] and subjected to deceleration pulses identified for either Rotorcraft [18] or Transport Category Aircraft [19]. The data shows that the occupant safety limits are approached for occupants when subjected to Transport Category Aircraft [19] accelerations and exceeded for occupants subject to Normal Category Rotorcraft accelerations [18]. While not directly comparable due to the unknowns in eVTOL performance and Regulations (discussed in more detail in *Regulatory Requirements* section of this report), there are two notable items. First, the ultimate impact conditions and/or limits used for evaluating eVTOL safety will likely be similar to what has already been tested, and thus the results from the drop tests indicate the general loading environment. Second, it demonstrates that impacts from even low levels have the capability of causing injury. Figure 1 shows the test setup of the ATD vertical drop testing conducted a NASA LaRC at the 14-foot drop height.

Figure 1 - ATD Drop Testing at LaRC. ATDs at 14-ft drop height

The figure gives an idea of what could be an injurious impact from a very short vertical distance. It is very conceivable that this figure could represent an actual eVTOL vehicle landing in an anomalous condition. In order to protect against these types of scenarios, a serious consideration into crash mitigation other than the BRS is required.

Potential Crash Mitigation Design Capabilities

The crashworthy design of civilian [20] and military [21] aircraft has been investigated extensively. Crashworthiness is typically discussed as taking a "systems level" approach, where the entire structure of the vehicle acts in harmony to maximize the attenuation into the occupant from an impact event and reduce the chance of injury. Figure 2 shows a simple illustration that has been widely used to convey the systems level approach of crashworthiness in the aircraft vehicle design.

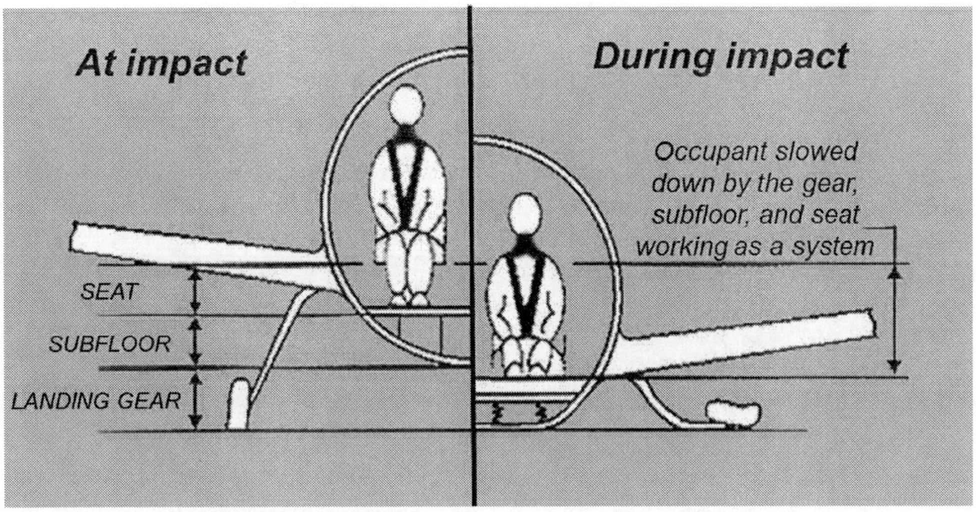

Figure 2 - Systems level crashworthiness approach

In Figure 2, the systems approach first starts with the deformation of the landing gear, then evolves to crushing of underfloor structure and ends with stroking of the seat. All three of these features working in conjunction will provide the maximum amount of energy attenuation, thus reducing the impact loads transmitted into the occupant from the impact event. Some aircraft have incorporated the systems level approach into their design [22-23], while others have incorporated, to as much extent as possible, crashworthy features either in design or in retrofit.

Three potential areas for the inclusion of crashworthiness into vehicles as it pertains to the components identified in Figure 2 will be discussed in this section. Where possible, some specific examples are presented and data, if available, is included. There are numerous other designs and concepts such as external airbags, mass shedding, and occupant ejection (among many others) that have been developed for the use as crashworthiness features on aircraft and aircraft components, however these items are not the focus of this study, and can be researched independently.

Concept 1 - Crushable Landing Gear Tubes

Carbon composite material systems are identified as the leading materials for use in fabrication of eVTOL vehicles due to their significant weight reduction over traditional metallic components and high strength. However, what carbon materials gain in strength, they lose in ductility, which is important when examining the mechanisms that cause failure in these types of materials, especially in a crash event, where it is probable that these structures will fail. Carbon fibers are known to exhibit brittle failure in that fibers will splay or pulverize upon reaching their ultimate stress limit [24]. Some of these deficiencies can be mitigated by using a combination of carbon and a more ductile fiber such as aramid in the composite material system. These hybrid material systems exist currently and are formed by orienting the carbon fibers in the warp direction, and the aramid fibers in the fill direction. Prototype tubular specimens using these hybrid material systems have been manufactured and tested at NASA LaRC, and characteristics such as strength, stiffness, weight and energy absorbing capabilities were examined and compared to traditional carbon fiber material systems [25]. Figure 3 shows an example of two tubes, one fabricated out of traditional carbon weave (left) and the other fabricated out of carbon/aramid hybrid weave (right).

Figure 3 - Composite tube specimens. Traditional carbon (left) and hybrid carbon/aramid (right) weaves

The tubes were hand laid up in a ±45° configuration, having four layers of material with a nominal outer diameter of 3 in. and a nominal height of 6.8 in. Specimen properties, along with some of the material response characteristics are presented in Table 1.

Table 1 - Tube specimen properties

	Carbon tube	Carbon/aramid tube
Height (in)	6.89	6.82
Wall thickness (in)	0.051	0.050
Weight (lb)	0.129	0.131
Static Stiffness (kip/in)	48.8	36.3
Crush avg. load – static (lb)	1,031	1,216
Crush avg. acceleration – dynamic (g)*	N/A	17.0 (between 0.001 and 0.03 s)
**See plots in Figure 4*		

When optimizing the energy absorbing characteristics of the structure as it relates to occupant safety, the main idea is to limit the loads going into the occupants to survivable levels during a dynamic impact event. This concept is accomplished by obtaining a uniform acceleration at sub-injurious levels over as large a time span as physically feasible to maximize impact energy absorption. Using these requirements as design factors, a good design would include large crush distances, stable crush (to ensure uniformity in the acceleration limits) and crush acceleration levels tolerable to human occupants. All three can be optimized through the design of materials, geometry, or both. Using the two described crush tubes as a first step in this optimization process, Figure 4 shows the acceleration levels between the carbon woven tube and the carbon/aramid hybrid woven tubes when undergoing a dynamic crush test.

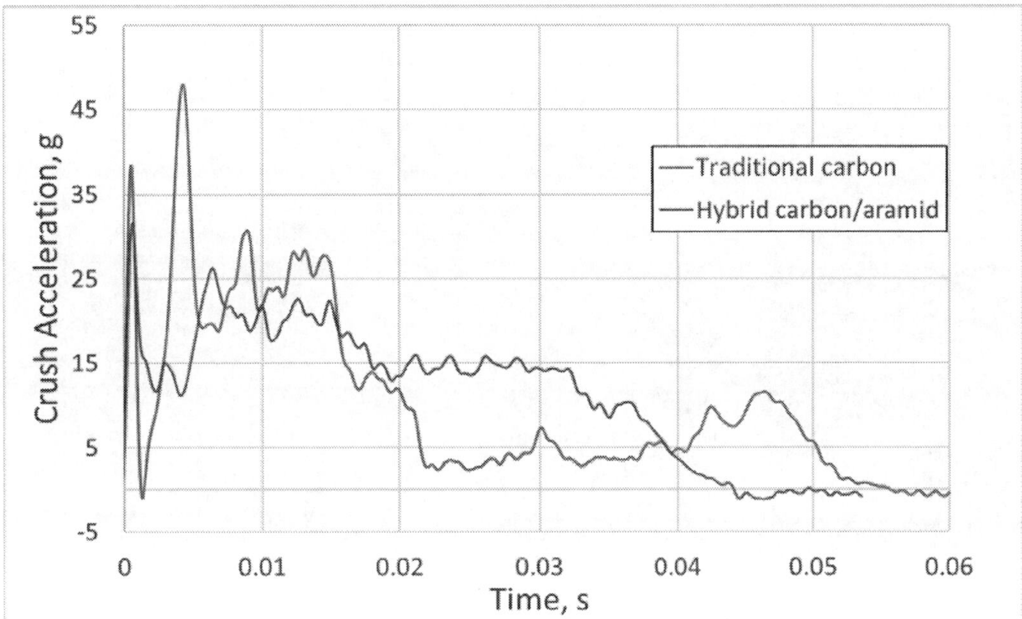

Figure 4 - Crush Tube response

The difference in the response of the materials is illustrated in Figure 4. Both materials exhibit an initial spike which is typically called the initiation or peak load and can be mitigated through various triggering mechanisms (which were not included in the study, but could be optimized for actual components), then settle into a post-peak crush response. The traditional carbon specimen drops to zero g, then exhibits a second spike at approximately 48 g after which it settles into a sustained crush response of 25 g until 0.017 s after impact. After this time, the response goes unstable caused by the specimen collapse, which is depicted by the sustained acceleration response below 5 g between 0.02 s and 0.04 s after impact. The slight bump at the end of the curve is the equivalent of a specimen "compaction" resulting from the failed tube walls resisting further motion of the impactor, causing the acceleration to rise slightly between 0.04 s and 0.05 s after impact. This result is contrasted by the hybrid carbon/aramid material response. After the initial spike, the material settles into an approximate 15 g sustained crush response with a brief rise to an average of 21 g between 0.005 s and 0.015 s after impact. An average of 17 g is sustained through the specimen crush for approximately 0.04 s, which is approximately twice as long as the response for the traditional carbon. Figure 5 shows the post-impact illustration of the crushed specimens.

Figure 5 - Post-impact-test crushed specimens. Traditional carbon (left) and hybrid carbon/aramid (right)

The main differences between the all carbon tube and the carbon aramid tube is the ductility of the aramid provides a mechanism for the structure to fold during the crushing, as opposed to the all carbon tube, which suffers a catastrophic failure and signs of buckling near its base. The folding in the hybrid system allows for a controlled and predictable failure leading to a sustained crush acceleration, as opposed to the traditional carbon, which exhibits instability and a crush response that is unpredictable.

These are not final designs of these specimens, but rather are initial developmental concepts. Further optimization of the carbon/aramid specimens must be conducted to lower the initial peak stress, achieve longer crush strokes, and exhibit a more uniform crush response. Instead, the main takeaway from this study should be that by changing something as simple as the material used in fabrication from a traditional carbon material system to a hybrid carbon/aramid material, the crashworthiness capability of the structure is significantly increased. The second concept will examine the aircraft structure itself, with the focus specifically in the subfloor or underfloor area of the vehicle.

Concept 2 - Energy Absorbing Subfloor

The inclusion of an energy absorbing underfloor or subfloor structures to replace the traditional beam and frame construction is addressed next. If the design already includes a traditional beam and frame structure, these components can be replaced with energy absorbing frames and/or keel beams. (Note: If the vehicle does not have a traditional underfloor structure, then items like foams [26] can be placed between the bottom skin and seat to provide some energy absorbing capabilities that would not be present otherwise.) Examples of traditional carbon woven subfloor have been demonstrated to be crashworthy by optimizing the geometry [27] or materials [28-29]. The designs that optimize the materials are described in detail in this section. Note that both of these designs were developed as stand-alone retrofit in existing fleets of aircraft. The first example is shown in Figure 6, where a Kevlar ® wrapped polymethacrylimide (PMI) foam keel beam component [28] was developed and retrofit into a GA aircraft fuselage section.

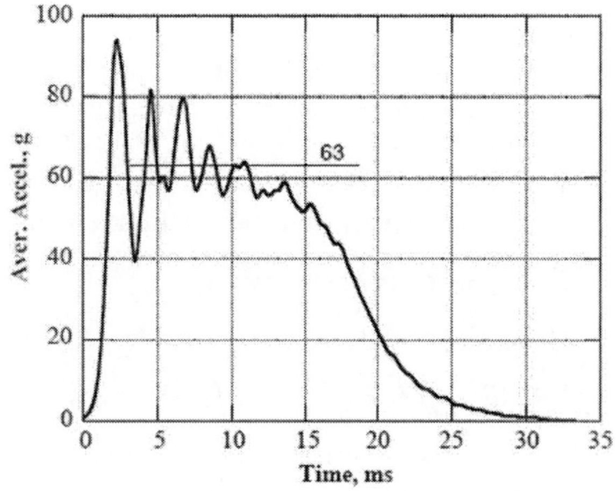

Figure 6 – Energy Absorbing Keel Beam. Installed in mock-up aircraft (left) and component level crush test acceleration (right). *Reprinted from [28]*

The energy absorbing keel beam identified in Figure 6 was developed for GA aircraft operating under a different set of limitations, and thus the 63 g average crush stroke would likely be considered too high of a limit for eVTOL vehicles. However, the actual number is insignificant, and the focus should be on the technology identified for use. A second subfloor retrofit is shown in Figure 7, which was manufactured from the same hybrid carbon/aramid material described above, in a geometrical configuration colloquially known as the "conusoid." In this design, the composite was fabricated into a conical beam section, which was then retrofit as an underfloor frame section into a drop test of a CH-46 fuselage section [29].

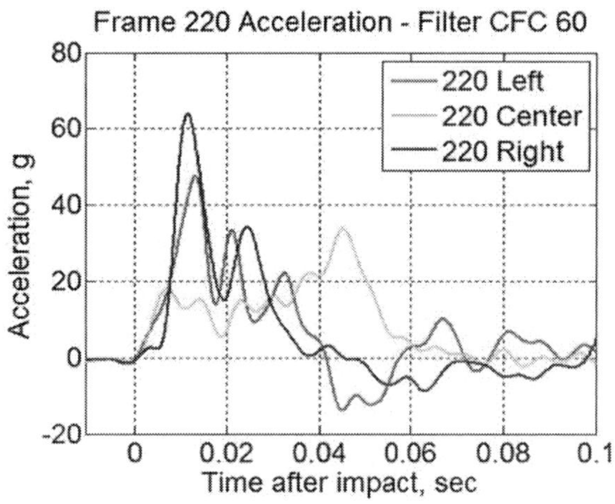

Figure 7 - Conusoid retrofit into barrel section drop test (left), and resultant drop acceleration data (right)

The data plotted in Figure 7 shows the effect of adding the conical EA beam member. During the first 0.020 s of impact, the accelerations recorded in the left (red curve) and right (blue curve) portions of the fuselage showed large spikes in acceleration of over 60 g immediately after impact because they track the deceleration of the fuselage itself, while the center section of the conusoid beam member (green curve) recorded a generally sustained, mostly constant acceleration of under 20 g. The inclusion of a conusoid subfloor attenuated the impact acceleration levels by a factor greater than 3 for the first 0.040 s. The spike at the end of the green curve at approximately 0.050 s after impact was a resultant of the conusoid fully compacting similar to the behavior exhibited in the carbon tube member from concept 1. As with the crush tube concept, the performance of the conusoid can be noted as an impact attenuator proof of concept; however, the specific performance as it relates to eVTOL vehicle impact attenuation would need to be optimized in order to maximize its energy absorbing capabilities. The third and final concept that will be addressed in this report will focus on the seat and restraint systems the occupant will use.

Concept 3 – Seat and Restraints

A third concept that can be incorporated in the crashworthy design of an eVTOL aircraft is an adequately designed seat/restraint system. The seat design serves two critical functions. The first is to absorb some of the kinetic energy at impact through the use of seat stroking mechanisms. A good background into the various seat stroking mechanisms used in a seat attenuation is provided in [30]. These mechanisms are used in seats for aircraft ranging from agricultural [31] to military [32]. The second function is to keep the occupant restrained during the impact and prevent the occupant from experiencing secondary impacts (i.e. striking an internal aircraft component such as the yoke). To mitigate secondary impacts, concepts like the design of the seat with bolsters, multi-point restraint systems, cockpit airbag systems [33] and automatic pretensioners [34] can be used. Research has determined that a conforming seat, along with the right type of restraint systems will significantly improve occupant projection during a crash [35]. Additionally, the NTSB found that when airbags were integrated into general aviation restraint systems, "aviation airbags can mitigate occupant injuries in severe but survivable crashing in which the principal direction of the force is longitudinal [36]." If the seat is molded directly as a part of the aircraft itself, the chances for potential stroking mechanisms to be included may be diminished; however, the correct shape and placement of the restraint system will still play a major role preventing secondary impact, so the seat should not be overlooked in the design process.

With these three concepts taken together, an effective systems-level crashworthy vehicle response can be achieved. In a notional vehicle, if the landing gears are able to absorb 20% of the impact energy while limiting the loads to 20 g then only 80% of the initial energy is going into the floor structure of the vehicle. If the floor was able to attenuate another 20% of energy and limit the loads going into the seat to 15 g, then the seats only have to take 60% of the energy. If the seats are able to absorb 20% of the energy, then the occupant may only be experiencing 40% of the initial impact energy, which will play a major role in the determination of occupant survivability. Finally, if crashworthiness is considered during the design process of the aircraft like those identified in [22-23], then the design, weight, cost and other factors can be optimized to ensure maximum utilization of crashworthy features and maximize the chances for occupant survival while minimizing the initial cost. However, if the designs are not considered in vehicle development, but rather are included instead as a retrofit item (either voluntarily or as dictated through a regulatory change such as a FAA Advisory Circular), there could be a significant cost and/or weight increase for only marginal crashworthiness improvements. Thus, it is imperative for crashworthiness to be considered during the design process of the vehicle development.

Proposed Concepts for Regulatory Requirements

There are no current regulations that exist specifically for eVTOL aircraft. Prototype aircraft that are currently flying (either at a sub- or full-scale level) have been granted certification as an experimental aircraft for testing purposes [37], on a case-to-case basis due to the lack of overarching policy.

Uber suggests using a performance-based standard, similar to what has been issued for the 14 CFR Part 23 Amendment 64 standards update [38] as the way of certifying eVTOL vehicles. Under the new performance based standards for Part 23 aircraft, for a new aircraft design, the applicant can certify establishing that the aircraft can meet a defined level of safety through its performance, rather than by specific design criteria dictated by the regulation. Stated more generally, a performance-based standard promulgates the expected performance of the design but does not stipulate the "how". This change in paradigm allows for OEMs to meet the expected performance in a variety of ways of their choosing, and will also hopefully allow for the certification process to move much quicker, allowing for a large number of uniquely designed aircraft to become certified in less time.

However, there must be ways to quantify the "levels of safety" and "expected performance." Because no baseline standard for eVTOL aircraft exist now, some example metrics can then be found using some of the regulations previously developed for other aircraft types. One way would be to use the existing crashworthy requirements from 14 CFR Part 2x.562 (where the "x" represents either Part 23, Part 25, or Part 27) "Emergency Landing Dynamic Conditions." A second example would be to parallel what was recently completed for Part 25 Transport Category Aircraft. The FAA formed an Aviation Rulemaking Advisory Committee (ARAC) in 2016 specifically for the development of airframe level crashworthiness regulations [39] to address the FAA's concern that the level of crashworthiness in transport category aircraft be maintained for new aircraft made with primarily composite materials and/or unique design features. The rationale behind the FAA's logic for new regulations for aircraft using composite materials and/or novel designs can be directly applied since eVTOL OEMs are advertising the exact same two design features – composite materials and novel designs. In this case, new eVTOL regulations would be comparable to transport category aircraft rules, should the FAA adopt them.

However, current transport category aircraft using composite and/or novel designs are currently certified under Special Conditions (SC) - which also can be used as guides for eVTOL certification. One example SC for the Boeing 787-8 [40] stipulates that in order to receive certification, four criteria must be passed: 1 – Retention of items of mass, 2 – Maintenance of acceptable acceleration and loads experienced by the occupants, 3 – Maintenance of survivable volume, and 4 – Maintenance of egress paths. This SC stipulates that the structure must be evaluated at ranges of vertical descent speeds up to 30 ft/s, and would presumably be evaluated either by a drop test or a combination of test and analyses. Other SCs [41-43] stipulate that crashworthiness must be met on aircraft sections either through "analysis using validated tools or by direct-test evidence" in conditions of up to 30 ft/s, indicating that a full-scale drop test and/or testing with the help of computer analyses are necessary to gain certification.

Taken all together, if a performance-based regulation for eVTOL vehicle certification is established, then the OEMs are free to develop and utilize whichever features they chose as a part of a systems level crashworthiness vehicle development program. In order to pass the certification, a set of criteria defining occupant loads, egress paths or others would need to be established in which to evaluate the vehicle design, and the method of completing certification would fall into the use of testing and/or analyses at some level. While this scenario is still hypothetical at present, it does provide a notional framework for the certification of eVTOL vehicles.

Finally, as of October 15, 2018, the European Aviation Safety Agency (EASA) has issued a SC for "Small-category VTOL aircraft [42]," which is the first regulatory issuance of its kind. The SC is applicable for aircraft with passenger seating of 5 or less and a maximum take-off mass of 2,000 kg or less. These constraints fall in line with Part 23 GA aircraft or small rotorcraft currently. The SC describes consensus standards as acceptable means of compliance based on a performance based specification, similar to what has recently been achieved through the Amendment 64 update for Part 23 aircraft. Specific to crashworthiness, the SC stipulates that aircraft must protect each occupant against injury under "likely flight, ground and emergency landing conditions…by not exceeding established human-injury criteria for human tolerance." Additionally, there is text describing occupant protection from secondary impacts, protection from baggage contents and the maintenance of egress paths. In this way, the SC contains ideas that are similar to the SCs for the larger composite aircraft described previously, however the detail is presumably left to the consensus standards.

Conclusion

This report describes the many ongoing issues categorized under the general topic of "safety for eVTOL vehicles." The major barriers that need to be overcome in establishing a viable marketplace for eVTOL vehicles is the public perception of autonomous vehicles, especially with the more-recent high profile accidents, infrastructure and vehicle safety features, and lack of clear regulations and standards. Because the industry is still in its infancy, there are many unknowns as to how it will eventually mature.

The current perception of "driverless cars" which is the closest parallel to how autonomous eVTOL vehicles will eventually operate is still unfavorable. 63% of people fear a self-driving car, and the only way to change that number is to continue testing and to perform improvements to the both the vehicles and infrastructure, with a major emphasis on vehicle safety.

Generally, when OEMs and researchers discuss eVTOL safety, their discussions largely fall under the category of crash prevention. Crash prevention involves subjects such as redundant power systems, airspace management, secure communication and others primarily focused on the prevention of a crash, which are all necessary for the continued maturation of the eVTOL industry. They miss, however, the second important safety concept of crash mitigation. The unfortunate fact is that at some point an accident is going to occur, and when that happens, the mitigation of the severity of that crash by the vehicle itself cannot not be overlooked. It is important that both crash prevention research along with crash mitigation research must be working in conjunction with each other if the intent is for the perception of eVTOL vehicles to be twice as safe as current automobile standards.

Under *Crash Mitigation*, some example conceptual designs are presented which detail concepts such as crushable tubes, energy absorbing subfloors and seat/restraint systems. The crush tube example shows that by simply changing a material from a carbon based system to a hybrid carbon/aramid based system, the sustained crush duration doubles, and the crush levels are maintained thru a stable crushing mechanism. The inclusion of a subfloor structure achieves similar results by limiting the impact levels to defined standards, and the addition of the correct seat and restraint system will further lower the acceleration limits into the occupants, all while keeping the occupant for experiencing a secondary object impact. All of these (and many other examples which are not mentioned) concepts working together will develop into a "systems-level" crashworthiness approach.

Furthermore, safety testing must be completed on these vehicles similar to that currently conducted on automobiles in order to achieve levels of safety above the current automobile standards. This type of testing should be embraced by the OEMs since it will provide critical data on vehicle design, along with informing the public of the work being completed in these areas. The public understands the necessity of "crash testing" through various public service announcements and manufacturer advertisements, and generally responds positively toward a vehicle with a high safety rating. Additionally, full-scale testing is the only way to fully evaluate a full systems-level approach to vehicle safety. If the market strategy is to manufacture eVTOL vehicles at production numbers that rival automotive manufacturers, then there should be a significant number of vehicles available that can be specifically designated for use in testing. This strategy is unlike the current aviation industry market model, where, for example, only 136 B787 airplanes were manufactured by Boeing during 2017 [43], and more like the automotive model where almost 900,000

Ford F-150 trucks, for example, were manufactured [44]. Uber predicts that in the long term, a production volume of 5,000 units per year is feasible [12].

Safety as a requirement will eventually fall into some type of new regulation(s) dictated by the FAA. These regulations will hopefully take on a performance-based criteria to allow for rapid market development, however the specifications to design the vehicles is still unknown at present, though there are examples from which to look. A first attempt has been addressed by EASA in their issuance in October 2018 of the first SC for these types of vehicles.

Finally, there is additional research that should be conducted in the area of bird and other object strike. Birds have the capability of causing major accidents on transport aircraft [45], and regulations have been in place for conventional airplane [46] and rotorcraft [47] for certification. Bird/object strike was not presented in this report, however a detailed discussion on bird strike as it affects eVTOL aircraft especially in the areas of crash mitigation will be the subject of a future publication.

It is hoped that this report is the beginning of discussions in the area of eVTOL safety, not the end. The eVTOL industry is ambitious in its journey toward certification and market acceptance and that only happens when all of the challenges have been addressed, and the industry is working in harmony toward this exciting future.

References

1. Shepardson, D. "2017 Safest Year on Record for Commercial Passenger Air Travel: Groups." Reuters. https://www.reuters.com/article/us-aviation-safety/2017-safest-year-on-record-for-commercial-passenger-air-travel-groups-idUSKBN1EQ17L Accessed May 29, 2018.
2. Department of Transportation. "Standard No 208; Occupant Crash Protection." 49 CFR § 571.208. Amended October 1, 2011.
3. NHSTA. Ratings. https://www.nhtsa.gov/ratings. Accessed May 29, 2018.
4. IIHS. Vehicle Safety Ratings. http://www.iihs.org/iihs/ratings. Accessed May 29, 2018.
5. Glassbrenner, D. "An Analysis of Recent Improvements to Vehicle Safety." DOT HS 811 572. June 2012.
6. Stepp, E. "More Americans Willing to Ride in Full Self-Driving Cars." https://newsroom.aaa.com/2018/01/americans-willing-ride-fully-self-driving-cars/. Accessed May 23, 2018.
7. Steward, J. "Tesla's Autopilot was Involved in another Deadly Crash." March 30, 2018. https://www.wired.com/story/tesla-autopilot-self-driving-crash-california/. Accessed May 23, 2018.
8. NTSB. Preliminary Report Highway HWY18MH010. https://www.ntsb.gov/investigations/AccidentReports/Reports/HWY18MH010-prelim.pdf Accessed May 29, 2018.
9. Associated Press. "Uber self-drives Robo-cars out of Arizona after fatal crash." https://www.cbsnews.com/news/uber-ends-self-driving-car-arizona-after-tempe-fatal-crash-elaine-herzberg/ Accessed May 29, 2018.
10. Edelstein, S. "NuTonomy Suspends Boston Self-Driving Car Pilot After Fatal Uber Crash." http://www.thedrive.com/tech/19493/nutonomy-suspends-boston-self-driving-car-pilot-after-fatal-uber-crash Accessed May 29, 2018.
11. NTSB. Review of U.S. Civil Aviation Accidents 2010. NSTB/ARA-12/01. October 2012.
12. Uber Elevate. "Fast-Forwarding to a Future of On-Demand Urban Air Transportation." October 27, 2016. https://www.uber.com/elevate.pdf. Accessed May 23, 2018.

13. Prevot, T. "Uber Elevate Panel on Challenge of Urban Air Mobility Airspace." AHS 5th Annual Transformative VTOL Workshop. Jan 16-19, 2018. San Francisco CA.
14. Cirrus Aircraft. "Guide to the Cirrus Airframe Parachute System (CAPS)." https://cirrusaircraft.com/wp-content/uploads/2014/12/CAPS_Guide.pdf. Accessed May 23, 2018.
15. Uber. "Uber Elevation Mission and Vehicle Requirements." https://s3.amazonaws.com/uber-static/elevate/Summary+Mission+and+Requirements.pdf. Accessed May 23, 2018.
16. Littell, J.D. and Annett, M.S. "The Evaluation of Anthropomorphic Test Device Response Under Vertical Loading." AHS International 74th Annual Forum & Technology Display. May 14-18, 2018. Phoenix, AZ.
17. Federal Aviation Administration. "Emergency Landing Dynamic Conditions." 14 CFR § 23.562. Amended December 2, 2011.
18. Federal Aviation Administration. "Emergency Landing Dynamic Conditions." 14 CFR § 27.562. Amended November 13, 1989.
19. Federal Aviation Administration. "Emergency Landing Dynamic Conditions." 14 CFR § 25.562. Amended May 17, 1988.
20. Hurley, T.L and Vandenburg, J.M. "Small Airplane Crashworthiness Design Guide." AGATE-WP3.4-034043-036. April 12, 2002.
21. Simula Inc. "Aircraft Crash Survival Design Guide Vols 1-5." USAAVSCOM TR 89-D-22A. December 1989.
22. Boitnott, R. L.; Jackson, K. E.; Fasanella, E. L.; Kellas, S.: "Full-Scale Crash Test of the Sikorsky Advanced Composite Airframe Program Helicopter," Proceedings of the American Helicopter Society Forum 56, Virginia Beach, VA, May 2-4, 2000.
23. Terry, J.E., Hooper, S.J., and Nicholson, M. "Design and Test of an Improved Crashworthiness Small Composite Airframe." NASA CR-2002-211774. August 2002.
24. Hamada, H. et al. "Comparison of Energy Absorption of Carbon/Epoxy and Carbon/PEEK Composite Tubes." Composites Vol. 23, No. 4. Pp 245-252. July 1992.
25. Littell, J.D. and Puthnam, J.B. "The Evaluation of Composite Energy Absorbers for use in UAM VTOL Vehicle Impact Attenuation." VFS 75th Annual Forum and Technology Display. May 13-16, 2019. Philadelphia, PA.
26. Fasanella, E.L. et al. "Mult-Terrain Impact Testing and Simulation of a Composite Energy Absorbing Fuselage Section." AHS International 60th Annual Forum & Technology Display. June 8-10, 2004. Baltimore, MD.
27. Farley, G.L., and Jones, R.M. "Energy-Absorption Capability of Composite Tubes and Beams." NASA TM 101634. September 1989.
28. Kellas, S. and Knight, N.E. "Design, Fabrication, and Testing of Composite Energy-Absorbing Keel Beams for General Aviation Type Aircraft." NASA CR-2002-212133. December 2002.
29. Littell, J.D., Jackson, K.E., Annett, M.S., Seal, M.D. "The Development of Two Composite Energy Absorbers for Use in a Transport Rotorcraft Airframe Crash Testbed (TRACT 2) Full-Scale Crash Test." AHS International 71st Annual Forum & Technology Display. May 5-7, 2015. Virginia Beach, VA.
30. Desjardins, S.P. "The Evolution of Energy Absorption Systems for Crashworthy Helicopter Seats." AHS International 59th Annual Forum & technology Display. May 6-8, 2003. Phoenix AZ.
31. Kellas, S. "Energy Absorbing Seat System for an Agricultural Aircraft." NASA CR-2002-212132. December 2002.

32. BAE Systems. Aircraft Seating and Survivability Systems. https://www.baesystems.com/en-us/product/aircraft-seating-systems. Accessed May 25, 2018.
33. BAE Systems. Cockpit Air Bag System. https://www.baesystems.com/en/download-en/20170217225403/1434577483404.pdf. Accessed May 31, 2018.
34. Harris, A., and Bark, L. "Performance of CH-46 Pilot Seats in NASA Full-Scale Crash Test." AHS International 70th Annual Forum & Technology Display. May 20-22, 2014. Montreal, Quebec, Canada.
35. Kaul, A. et al. "A Revolution in Preventing Fatal Craniovertebral Junction Injuries: Lessons Learned from the Head and Neck Support Device in Professional Auto Racing." J. Neurosurgery Spine. 25:756-761. 2016.
36. NTSB. "Airbag Performance In General Aviation Restraint Systems." NTSB/SS-11/01. 2011.
37. Swartz, K.I. "Charging Forward New eVTOL Concepts Advance." Vertiflite. July/August 2017. Pp 24-29.
38. Federal Register. "Revision of Airworthiness Standard for Normal, Utility, Acrobatic and Commuter Category Airplanes." Vol. 81, No. 251. December 30, 2016.
39. Federal Register. Federal Aviation Administration, Aviation Rulemaking Advisory Committee, Transport Airplane and Engine Issues, Vol 80. 2015.
40. Federal Register. "Special Conditions: Boeing 787-8 Airplane; Crashworthiness." Vol 72.No. 186. Wednesday, September 26, 2007.
41. Federal Register. "Special Conditions: Airbus A350-900 Airplane; Crashworthiness, Emergency Landing Conditions." Vol 79. No. 43. Friday July 25, 2014.
42. EASA. "Vertical Take-off and Landing Aircraft: Proposed Special Condition for small-category VTOL aircraft." SC-VTOL-01. October 15, 2018.
43. Boeing Company. The Boeing Company 2017 Annual Report. https://s2.q4cdn.com/661678649/files/doc_financials/annual/2017/2017-Annual-Report.pdf. Accessed May 30, 2018.
44. Ford Authority. "Ford F-Series Sales Numbers." http://fordauthority.com/fmc/ford-motor-company-sales-numbers/ford-sales-numbers/ford-f-series-sales-numbers/. Accessed May 30, 2018.
45. NTSB. "Loss of Thrust in Both Engines after Encountering a Flock of Birds and Subsequent Ditching on the Hudson River US Airways Flight 1549." Accident Report NTSB/AAR-10/03. May 4, 2010.
46. Federal Aviation Administration. "Bird Strike Damage." 14 CFR § 25.631. Amended April 8, 1970.
47. Federal Aviation Administration. "Bird Strike." 14 CFR § 29.631. Amended July 1, 1996.

Aerodynamic Modeling of Coaxial Counter-Rotating UAV Propellers

Moritz Thiele M.Sc. — **Martin Obster** M.Sc. — **Mirko Hornung** Prof. Dr.-Ing.

Institute of Aircraft Design – Technical University of Munich – Munich, Germany

ABSTRACT

While coaxial rotor systems experience complex aerodynamic effects, an application to eVTOL UAV can entail positive consequences for the overall system. Aerodynamic calculation of coaxial counter-rotating rotors is carried out and the knowledge gained is used to analyze a wingtip pusher propeller configuration. The aerodynamical model based on the BEMT is adapted to the use cases presented and includes azimuthal inflow and induced velocity components as well as the Prandtl tip loss factor. Wake contraction is calculated using an empirical model for the tip vortex. Moments acting on the rotor hub and oblique inflow can be considered by the discretization of the rotors at arbitrary azimuthal positions. Subsequently, the methods were validated against measurement data from literature and were found in good accordance. Based on aerodynamic parameters, the implemented methods and configurations can be used for rotor and propeller design and modifications.

NOTATION

Abbreviations

AoA .. Angle of Attack .... [−]
BEMT Blade Element Momentum Theory .... [−]
ESC ... Electronic Speed Controller .... [−]
eVTOL electrical Vertical Take-Off and Landing .... [−]
FVA ... Free Vortex Analysis .... [−]
MTOW Maximum Take-Off Weight .... [−]
SARF . Synthesis and Analysis Rotor Framework .... [−]
UAV .. Unmanned Aerial Vehicle .... [−]

Letters

A ..... Rotor swept area .... $[m^2]$
b ...... Wingspan .... $[m]$
c ...... Sectional chord length .... $[m]$
c_d ..... Sectional drag coefficient .... [−]
c_l ..... Sectional lift coefficient .... [−]
C_P .... Rotor Power coefficient .... [−]
c_P ..... Sectional Power coefficient .... [−]
C_Q .... Rotor Torque coefficient .... [−]
C_T .... Rotor Thrust coefficient .... [−]
c_T .... Sectional Thrust coefficient .... [−]
D ..... Drag .... $[N]$
F ..... Prandtl tip loss factor .... [−]
FM ... Figure of Merit .... [−]
L ..... Lift .... $[N]$
l ...... Sectional lift .... $[N]$
$l1$ ..... Lower rotor inner area .... [−]
$l2$ ..... Lower rotor outer area .... [−]
N_b .... Number of Blades .... [−]
P ..... Power .... $[W]$
R ..... Rotor tip radius .... $[m]$
r ...... Radial coordinate .... $[m]$
rpm ... Revolutions per minute .... $[\frac{1}{min}]$
rps .... Revolutions per second .... $[\frac{1}{s}]$
T ..... Thrust .... $[N]$
t ...... Time .... $[s]$
U ..... Velocity Component .... $[\frac{m}{s}]$
V ..... Total Velocity .... $[\frac{m}{s}]$
w ..... Induced velocity .... $[\frac{m}{s}]$
x_{wing} .. Distance wing leading edge to rotor disk .... $[m]$

Greek Symbols

α ..... Angle of attack .... $[rad]$
β ..... Blade or sectional pitch .... $[rad]$
η ..... Propulsive efficiency .... [−]
Γ ..... Circulation .... $[\frac{m^2}{s}]$
λ_∞ .... Tip speed ratio .... [−]
Ω ..... Rotational speed .... $[\frac{rad}{s}]$
ϕ ..... Inflow angle .... $[rad]$
Ψ_w .... Wake angle .... $[rad]$
ρ ..... Density of air .... $[\frac{kg}{m^3}]$
σ ..... Rotor solidity .... [−]
θ ..... Blade azimuth angle .... $[rad]$
θ_b ..... Angle between two adjacent blades .... $[rad]$
Θ_{lin} ... Linear Blade Twist .... $[rad]$

Indices

θ ...... Azimuthal direction .... [−]
c ...... Contracted Wake .... [−]
Hub .... Rotor Hub .... [−]
$Inflow$ .. Inflow on Rotor Disk .... [−]
r ...... Radial direction .... [−]
s ...... Wake Propagation .... [−]
Tip .... Rotor Tip .... [−]
u/l .... Upper / Lower rotor .... [−]
wt ..... Wingtip .... [−]
z ...... Axial direction .... [−]

Presented at the 8th Biennial Autonomous VTOL Technical Meeting & 6th Annual Electric VTOL Symposium 2019, Mesa, Arizona, USA, Jan 29–31, 2019,

INTRODUCTION

The increasing use of civil and military eVTOL UAVs requires highly safe, reliable and efficient UAV configurations. Key components concerning UAV reliability are the motors used for hover propulsion. While inherently redundant electrical motors, possibly wired with 6 redundant phases, are still rare and heavy, a coaxial rotor system can enable higher redundancy while also maintaining a small areal footprint.

By providing sufficient lift in a small installation space this rotor arrangement may be used to reduce failure probabilities and meet safety requirements. Still, complex aerodynamic effects occur between the two rotors which consist of flow interaction and turbulent vortex structures.

Calculation of coaxial rotors has been the topic of several publications with most approaching the subject using a solver of the underlying Navier-Stokes equations as written by Ho (Ref. 1) and Barbely (Ref. 2). The theory used here is based on the model by Leishman and Ananthan (Ref. 3) for coaxial rotors in hover and axial flight. However, losses due to swirl and the effect on blade design as well as the influence of the lower rotor on the upper rotor are not incorporated in these calculations. This theory was used and improved to include the circumferential induced velocities proposed by Modarres and Peters (Ref. 4) as well as Giovanetti (Ref. 5). Those conclusions are adapted to calculate the local performance values at arbitrary azimuthal positions separately using the propeller calculation environment SARF, first described in (Ref. 6) and refined in (Ref. 7) to also consider oblique or arbitrary quasi stationary inflow. The goal was to use an underlying Blade Element Momentum Theory (BEMT) solution that differs from that proposed by Adkins and Liebeck (Ref. 8) which is already included in the SARF environment in also being valid while in hovering flight.

Validation for coaxial rotors relies mostly on the work of Harrington (Ref. 9) and Dingeldein (Ref. 10). Both use a full scale coaxial rotor in a wind-tunnel. Further validation of the method used here should also be carried out using the work of McAlister (Ref. 11) who used smaller rotors which are well suited for the intended purpose of designing VTOL UAV with this publication's calculations.

A major point of interest besides the calculation of classical coaxial rotor systems is the assessment of certain other configuration layouts which experience similar flow effects as the coaxial counter-rotating rotors.

For a eVTOL UAV to reach high endurance it is necessary to transition into a cruise flight state where the lift produced is supplied by wings as opposed to hover rotors on a multicopter. With electrical propulsion it is easy to mount a motor at the wing tip gaining an undisturbed flow over a very large part of the wing which increases efficiency. An exemplary concept for this kind of propulsion is the EVIATION "Orca" UAV. Additionally the wingtip propulsion can be used for yaw authority and thus eliminate the need for a separate rudder.

A wingtip mounted propeller can be used in a pusher or a tractor configuration. The pusher configuration promises to deliver some aerodynamic advantages where the propeller is operating in the wingtip vortex which positively influences the local inflow angle on the propeller blades. Additionally, by working against the wingtip vortex the propeller is actively reducing the wing's induced drag.

This alignment of the propeller and vortex present at the wingtip can be described similarly to a coaxial rotor and will be discussed below.

TECHNICAL APPROACH

The coaxial rotor system is described as two separate rotors operating in close proximity. The aerodynamic model is defined in a modular manner to enable a transfer of the model to configurations of equivalent modes of action like wingtip mounted pusher propellers.

Aerodynamic Modeling

Each rotor is modeled individually using a modified Blade Element Momentum Theory (BEMT) that includes additional azimuthal swirl components primarily following the approach developed by Giovanetti (Ref. 5). Subsequently the models of the two rotors are coupled while also considering the mutual interference effects and arbitrary inflow phenomena.

Blade Element Momentum Theory Following the classical BEMT approach each blade is discretized in a certain number of blade sections along the radius. These sections are described using the parameters shown in the upper half of figure 1. Additional information used for computation include the airfoil type and polars, chordlength and local blade sweep.

A rotor with rotational speed Ω is operating on a UAV in an arbitrary flight state. The inflow V_{Inflow} observed is examined in a local coordinate system turning with a rotor blade and can be written as the sum of three perpendicular components in axial (z), azimuthal (θ) and radial (r) directions. The magnitude and direction of V_{Inflow} is calculated using the functionality provided in SARF which was developed by Ost (Ref. 12).

$$V_{Inflow} = \sqrt{U_z^2 + U_\theta^2 + U_r^2} \quad (1)$$

The component in radial direction U_r is not considered for the BEMT but will be used for the calculation of oblique inflow phenomena.

To calculate the local sectional forces $\frac{dL}{dr}$ and $\frac{dD}{dr}$ the local relative inflow velocity V_{rel} has to be determined for each radial and azimuthal position according to the velocity vectors shown in the lower half of figure 1 by

$$V_{rel} = \sqrt{U_z^2 + (\Omega r + U_\theta)^2 - w^2} \quad (2)$$

This enables the calculation of the relative local flow angle ϕ:

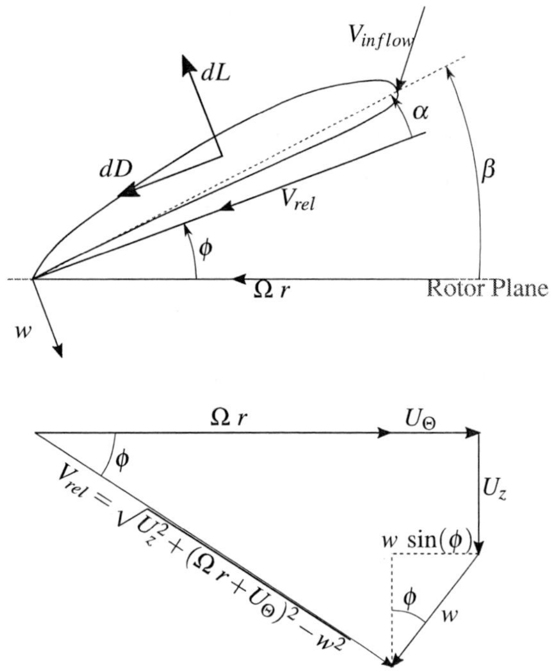

Fig. 1: Geometry of the flow at a blade section, including axial and swirl components of velocity

$$\sin\phi = \frac{U_z \cdot V_{rel} + w \cdot (\Omega r + U_\theta)}{U_z^2 + (\Omega r + U_\theta)^2} \tag{3}$$

$$\cos\phi = \frac{(\Omega r + U_\theta) \cdot V_{rel} - w \cdot U_z}{U_z^2 + (\Omega r + U_\theta)^2} \tag{4}$$

According to (Ref. 4) the sectional lift $\frac{dL}{dr}$ and profile drag $\frac{dD_{profile}}{dr}$ which are acting perpendicular and parallel to V_{rel} are obtained with a momentum balance.

$$\frac{dL}{dr} = 4\pi\rho F\, [U_z + w\cos\phi]\; w\, r \tag{5}$$

$$\frac{dD_{profile}}{dr} = \frac{1}{2}\rho N_b\, V_{rel}\; c\; c_d \tag{6}$$

The Prandtl Tip loss factor F is incorporated to consider tip losses.

By resolving this force in axial and azimuthal components and normalizing all velocities and lengths denoted with an overbar by $\Omega\, r_{Tip}$ and r_{Tip} respectively according to (Ref. 5), the equations of the non-dimensional local coefficients c_T and c_P are obtained. These have to be integrated in radial and azimuthal directions and lead to the thrust and power coefficients of the rotor C_T and C_P.

$$C_T = \int_0^{2\pi}\int_{\bar{r}_{Hub}}^{1} 4F\; (\bar{U}_z + \bar{w}\cos\phi)\; \bar{w}\cos\phi\; \bar{r} \\ - \sin\phi\; \frac{N_b \bar{c} c_d}{2\pi}\; \left(\bar{U}_z^2 + (\bar{r} + \bar{U}_\theta)^2 - \bar{w}^2\right)\; d\bar{r} \tag{7}$$

$$C_P = \int_0^{2\pi}\int_{\bar{r}_{Hub}}^{1} 4F\; (\bar{U}_z + \bar{w}\cos\phi)\; \bar{w}\sin\phi\; \bar{r}\; (\bar{r} + \bar{U}_\theta) \\ + \cos\phi\; \frac{N_b \bar{c} c_d}{2\pi}\; \left(\bar{U}_z^2 + (\bar{r} + \bar{U}_\theta)^2 - \bar{w}^2\right)\; (\bar{r} + \bar{U}_\theta)\; d\bar{r}\,. \tag{8}$$

In order to assess the rotor performance it is necessary to measure the efficiency. While this model is valid for the hovering as well as an axial or forward flight state, it is important to distinguish between those flight states when assessing the propulsive efficiency.

When operating in hover, the Figure of Merit (FM) is an adequate efficiency metric, however it is not valid in axial or forward flight ($U_\infty \neq 0$) and can only be used for the comparison of rotors with a similar disk loading (DL). It is defined, while hovering, as the ideal power required divided by the actual power required and can be calculated with:

$$FM = \frac{C_T^{3/2}}{2C_P} \tag{9}$$

For axial and forward flight the propulsive efficiency is given by

$$\eta = \frac{TU_\infty}{TU_\infty + P_i + P_0}\,. \tag{10}$$

The thrust T and power $P = P_i + P_0$ as the sum of profile power and induced power are defined based on the non dimensional coefficients:

$$T = C_T \rho A \Omega^2 r_{Tip}^2 \tag{11}$$

$$P = C_P \rho A \Omega^3 r_{Tip}^3 \tag{12}$$

Calculation of the Induced Velocity A remaining variable for the calculation is the induced velocity w. The induced velocity is produced by the lift of the rotor blades and is aligned in the opposite direction of the lift vector.

The base for the model used here is the BEMT where one key characteristic is the assumption that all radial annuli are independent and do not influence each other. From this the induced velocity at each blade section can be calculated by relating it to the local lift coefficient produced by the rotor. Depending on the relation between the rotational speed, the axial and the oblique inflow, it is possible for certain sections to produce a negative lift at certain azimuthal positions which has to be considered when calculating the total thrust and torque produced by the rotor.

The Blade Element Theory yields an expression for the normalized lift produced at each blade section:

$$\frac{d\bar{L}}{d\bar{r}} = \left(\frac{N_b \rho \bar{c} c_l}{2}\right)\; \left(\bar{r}^2 + \bar{U}_z^2 - \bar{w}^2\right)\,. \tag{13}$$

This can be extended to also consider azimuthal velocities U_θ and connected with the normalized equation (5) to:

$$4\pi\rho F\ [\bar{U}_z + \bar{w}\cos\phi]\ \bar{w}\ r = \\ = \left(\frac{N_b \rho \bar{c} c_l}{2}\right)\left((\bar{r}+\bar{U}_\theta)^2 + \bar{U}_z^2 - \bar{w}^2\right) \tag{14}$$

This quadratic equation of the induced velocity can be solved using the standard approach:

$$\bar{w}_{1,(2)} = \frac{-b \pm \sqrt{b^2 - 4ac}}{2a} \tag{15}$$

with the parameters a,b and c being:

$$\begin{aligned} a &= 8F\pi\bar{r}\cos\phi + \bar{c}N_b c_l \\ b &= 8F\pi\bar{r}\bar{U}_z \\ c &= -\bar{c}N_b c_l\left((\bar{r}+\bar{U}_\theta)^2 + \bar{U}_z^2\right) \end{aligned} \tag{16}$$

The positive solution of equation (15) results in the induced velocity of a rotor for a given inflow angle ϕ and local lift coefficient c_l. The induced velocity will be negative and thus pointed upward where negative lift is produced on the blade. As the local inflow angle ϕ, the resulting AoA α and the local lift coefficient c_l are themselves dependent of the induced velocity an iterative approach is used to solve the system for each rotor with a start value of the induced velocity that is required.

This start value is calculated based on the model proposed by Betz (Ref. 13) which describes the optimum inflow distribution for a maximum efficiency propeller. From this inflow distribution on an ideal rotor the postulated normalized induced velocity distribution over the blade can be extracted to be:

$$\bar{w}(r) = \frac{w(r)}{\Omega r_{Tip}} = \bar{w}_0\ cos(\phi) \tag{17}$$

with $\bar{w}_0 = \frac{w_0}{\Omega\ r_{Tip}}$ being a normalized nominal induced inflow velocity corresponding to an infinite rotor radius.

Relating the induced velocity to the inflow of the optimal rotor yields:

$$\bar{w}(r) = \frac{\bar{w}_0 r}{\sqrt{(\bar{V}_{rel} + \bar{w}_0)^2 + r^2}}\ . \tag{18}$$

To get an appropriate sectional start value for $\bar{w}(r)$ a constant value $\bar{w}_0 = 0.07$ for an optimal rotor in static conditions (hover) is taken from literature (Ref. 4). As this value is merely used to calculate the start value of $\bar{w}(r)$, it is also used for a non hovering operational state.

An iterative process for the calculation of rotor performance values can be set up for the given geometry of a rotor and its external inflow and is shown in the flowchart depicted in figure 2.

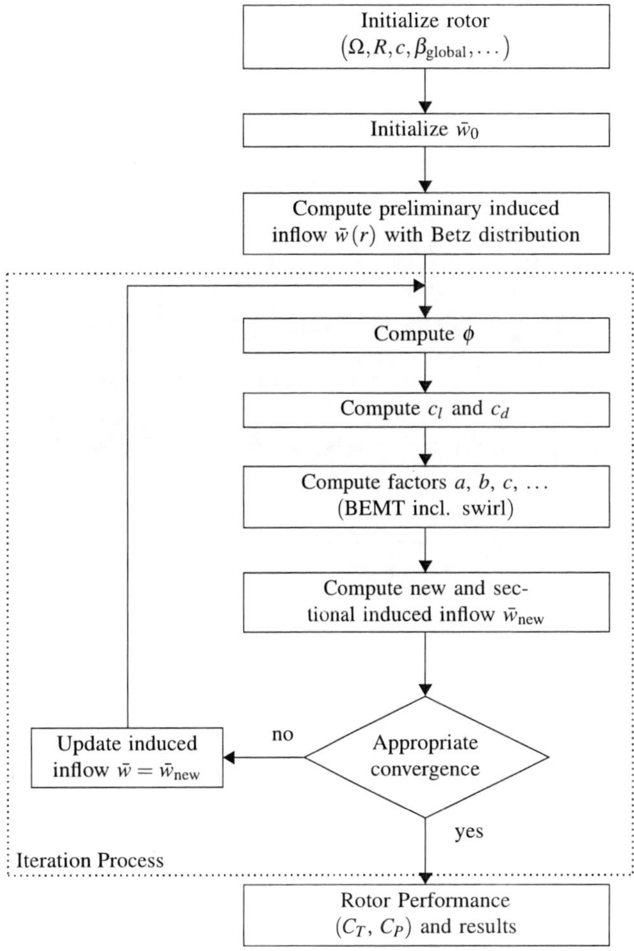

Fig. 2: Iterative process for the calculation of a single rotor

After the initialization of the rotor operating on a UAV an initial $\bar{w}_0$ is chosen to calculate $\bar{w}(r)$ using the Betz distribution. In the following iterative process the airfoil coefficients c_l and c_d are calculated for the local AoA α which depends on the inflow angle ϕ which is calculated in equation (3) and (4) considering all inflow velocities including the induced velocity. The resulting induced velocity $\bar{w}(r)$ can be determined with these coefficients and can be compared to the previous value until appropriate convergence is acquired. During this iteration process the Prandtl tip loss factor F is neglected to prevent an overshoot of the induced velocity of the outermost sections. It is again included in the calculation when the rotor performance values C_T and C_P are being determined using equations (7) and (8).

This model for a single rotor in an arbitrary flight state is then adapted to a coaxial rotor system.

Application to the Coaxial Rotor System A coaxial rotor system as shown in the schematic figure 3 is characterized by two distinct rotors separated by the distance z_{ul} in axial direction. When modeling this system the mutual influence of the two rotors has to be considered. The lower rotor (denoted by the subscript l) is operating in part in the slipstream of the upper rotor (subscript u).

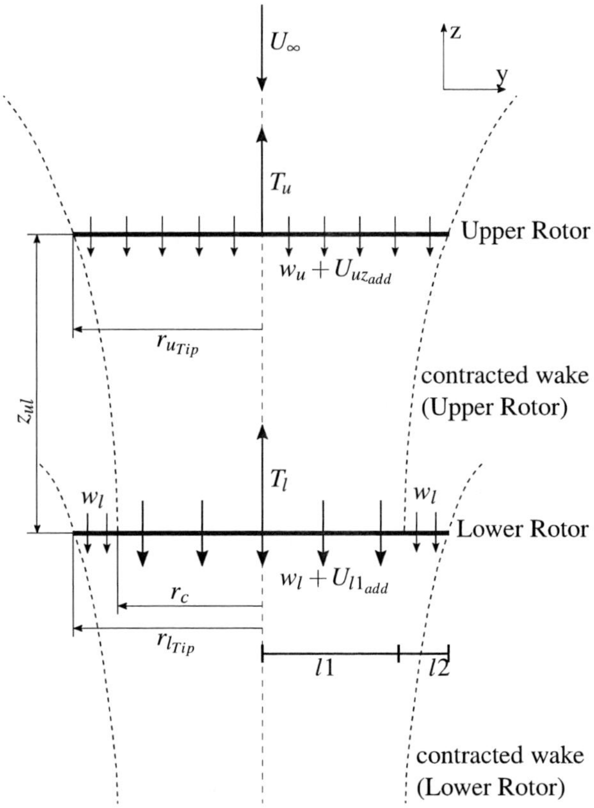

Fig. 3: Schematic of a coaxial rotor system with acting velocities

The upper rotor is experiencing an additional induced flow velocity $\bar{U}_{uz_{add}}$ due to its location in the induced inflow velocity field of the lower rotor. This velocity is only considered in axial direction following a model by McAllister (Ref. 11) which is based on the Biot-Savart Law. It is generated by a helical vortex ring originating at the tip of the lower rotor blades. The axial additional velocity acting on the upper rotor at the radial position $r = 0$ at a distance $\bar{z} = \bar{z}_{ul}$ from the lower rotor plane is calculated with this model by

$$\frac{\bar{U}_{uz_{add}}}{\bar{U}_{lz|\bar{z}=0}} = 1 + \left(\frac{|\bar{z}|}{\sqrt{1+\bar{z}^2}}\right)^k sign(\bar{z}) \tag{19}$$

With $\bar{U}_{lz|\bar{z}=0}$ being the average induced axial velocity of the lower rotor.

$$\bar{U}_{lz|\bar{z}=0} = mean(\bar{w}_{lz}) \tag{20}$$

Following (Ref. 5) the value of the empirical parameter k is set to 0.5 as the upper influenced rotor resides above the lower rotor. The velocity $\bar{U}_{uz_{add}}$ is considered as an uniform additional axial inflow velocity influencing the entire upper rotor plane. The reason behind this approach is the expanded area of inflow drafted by the dashed lines in figure 3 originating from the momentum balance of the lower rotor.

Meanwhile the wake of the upper rotor will contract while traveling in the direction of the lower rotor. When the lower

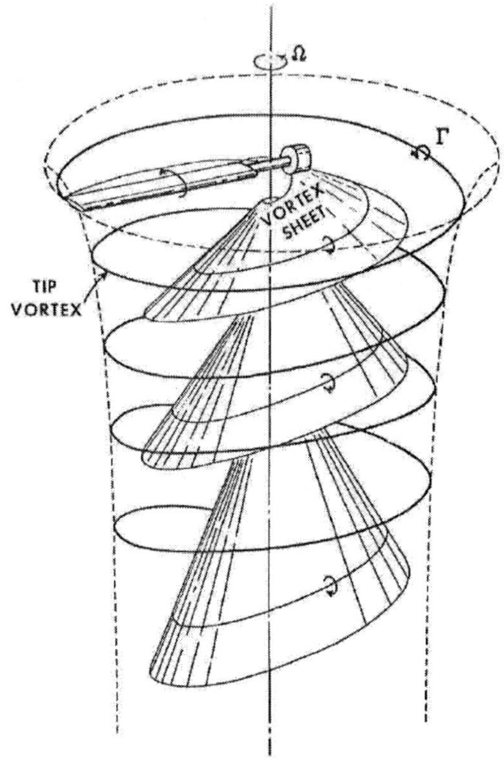

Fig. 4: Schematic rotor wake structure (Ref. 14)

rotor has a radius $r_{l_{Tip}}$ greater than r_c as shown in figure 3 the inner portion $l1$ of the lower rotor with the radius r_c spanning the area $A_c = r_c^2\pi$ will be affected by an additional inflow velocity $\bar{U}_{l1_{add}}$ from the contracted wake while the outer portion $l2$ will experience no influence by the upper rotor.

The additional inflow originates from the velocity induced by the upper rotor $\bar{w}_u$ and is calculated in axial and azimuthal direction with a model by Giovanetti (Ref. 5) where $\bar{w}_{uz}$ and $\bar{w}_{u\theta}$ are the velocities induced by the upper rotor at its rotor plane.

$$\bar{U}_{l1z_{add}} = \bar{w}_{uz}\left(\frac{A}{A_c}\right) \tag{21}$$

$$\bar{U}_{l1\theta_{add}} = \bar{w}_{u\theta}\left(\frac{A}{A_c}\right)^{3/2} \tag{22}$$

The calculation of radius r_c is based on a simple generalized empirical model developed by Landgrebe (Ref. 14) for rotors in hover producing static thrust.

The model by Landgrebe divides the wake of a rotor in two distinct components as depicted in figure 4:

1. A tip vortex originating from the roll up of the vortex sheet at the tip of the rotor

2. A vortex sheet being shed from the trailing edge of the rotor blade over the whole radius.

Through experiments Landgrebe developed equations describing the axial and radial coordinates of those two wake structures based on the wake angle Ψ_w which is defined as the angle the rotor blade traveled since shedding the particular wake vortex element.

To calculate the radius of the contracted wake of the upper rotor when impacting the lower rotor only the tip vortex model of (Ref. 14) is used and modified to include the propagation of the wake vortex tube.

As the model was developed for a rotor in hover, the wake is assumed to experience a stretching effect caused by the inflow velocity U_z. This modification only considers the axial propagation distance $\bar{z}_S$ of the vortex tube as a lateral propagation is not assumed to change the wake contraction behavior but rather move the contracted wake in a lateral direction which is considered together with the oblique inflow considerations.

According to Landgrebe, a vortex element shed by the blade will travel downstream and inward, but not in azimuthal direction with the velocity induced by the rotor while the blade will continue rotating. As soon as the following blade has passed over the position of the element, the downwash produced by this blade will increase its axial velocity.

Thus, the normalized axial coordinate $\bar{z}_{Tip}$, defined with the positive direction pointing upward towards the inflow, is calculated differently when the following blade has traveled the angular distance between two blades $\Psi_w = \theta_b = \frac{2\pi}{N_b}$. This leads to the equations:

$$\bar{z}_T = \begin{cases} k_1 \Psi_w - \bar{z}_S \ \text{ for } 0 \leq \Psi_w \leq \theta_b \\ (\bar{z}_T)_{|\Psi_w=\theta_b} + k_2 \ (\Psi_w - \theta_b) - \bar{z}_S \ \text{ for } \Psi_w \geq \theta_b \end{cases} \tag{23}$$

with constants $k1$ and $k2$ defined in (Ref. 14) as:

$$k1 = -0.25 \left(\frac{C_T}{\sigma} + 0.001 \Theta_{ulin} \right) \tag{24}$$

$$k2 = -(1.41 + 0.0141 \Theta_{ulin}) \sqrt{\frac{C_T}{2}} \tag{25}$$

where Θ_{ulin} is the linear blade twist calculated by

$$\Theta_{ulin} = \frac{\beta_{Tip} - \beta_{Hub}}{r_{Tip}} \tag{26}$$

For each case, the additional distance $\bar{z}_S$ will expand the wake depending on the inflow velocity. It is calculated using the age of the wake which is is dependent on the angular velocity of the rotor and the wake angle: $t = \Psi_w / \Omega$.

$$\bar{z}_S = \bar{U}_z \, \bar{t} = \bar{U}_z \, t \, \Omega = \bar{U}_z \Psi_w \tag{27}$$

As the axial distance $\bar{z}_T = \bar{z}_{ul}$ between the rotors is known and constant, equation (23) can be solved for the wake angle Ψ_w. Depending on the angle traveled by the blades until the vortex element reached $\bar{z}_{ul}$ this leads to:

$$\Psi_w = \begin{cases} \dfrac{\bar{z}_T}{k_1 - \bar{U}_z} & \text{for } 0 \leq \Psi_w \leq \theta_b \\[2ex] \dfrac{\bar{z}_T - (\bar{z}_T)_{\Psi_w=\theta_b} + k_2 \cdot \theta_b}{k_2 - \bar{U}_z} & \text{for } \Psi_w \geq \theta_b \end{cases} \tag{28}$$

This angle can be used in the formula of the radial tip coordinate given in (Ref. 14)

$$\bar{r}_c = r_{cmax} + (1 - r_{cmax}) \ e^{-\lambda \Psi_w} \tag{29}$$

With the parameter r_{cmax} being the maximum contraction radius found to be $r_{cmax} = 0.707$ by (Ref. 15) and (Ref. 3) and the parameter λ relating $\bar{r}$ and C_T is:

$$\lambda = 0.145 + 27 C_T \tag{30}$$

Subsequently the wake contraction radius $\bar{r}_c$ of the upper rotor at the position of the lower rotor can be calculated and used for the calculation of the additional inflow velocities.

This enables the application of the equations (7) and (8) to coaxial rotors.

The inflow velocities U_z and U_θ have to be modified to incorporate the additional velocities.

For the upper rotor $\bar{U}_{uz_{add}}$ is calculated using equation (19) and added to the axial inflow. The azimuthal inflow $\bar{U}_{u\theta}$ is not modified. The integration in radial direction is carried out for the whole blade $[r_{inner}, r_{outer}] = [r_{Hub}, r_{Tip}]$.

The calculation of the lower rotor has to be split at radius r_c. For the inner part $l1$, the additional velocities calculated in equations (21) and (22) have to be added to the axial and azimuthal inflow velocities. This part is integrated in radial direction from the hub to the contraction radius: $[r_{inner}, r_{outer}] = [r_{Hub}, r_c]$. The outer part $l2$ is calculated without any additional inflow velocities and is integrated from r_c to the blade tip: $[r_{inner}, r_{outer}] = [r_c, r_{Tip}]$.

This leads to the velocities $\bar{U}^{u/l}_{z_{total}}$ and $\bar{U}^{u/l}_{\theta_{total}}$ respectively depending on the rotor part calculated.

$$\bar{U}^{u/l}_{z_{total}} = \bar{U}^{u/l}_{z} + \bar{U}^{u/l}_{z_{add}} \tag{31}$$

$$\bar{U}^{u/l}_{\theta_{total}} = \bar{U}^{u/l}_{\theta} + \bar{U}^{u/l}_{\theta_{add}} \tag{32}$$

which can be written into the final equations for C_T and C_P by inserting the respective inflow velocities for the upper (u), lower inner ($l1$) or lower outer ($l2$) rotor.

$$C_T = \int_0^{2\pi} \int_{\bar{r}_{inner}}^{\bar{r}_{outer}} 4F \ (\bar{U}_{z_{total}} + \bar{w} \cos\phi) \ \bar{w} \cos\phi \ \bar{r} - \sin\phi \ \frac{N_b \bar{c} c_d}{2\pi} \ (\bar{U}^2_{z_{total}} + (\bar{r} + \bar{U}_{\theta_{total}})^2 - \bar{w}^2) \ d\bar{r} \tag{33}$$

$$C_P = \int_0^{2\pi} \int_{\bar{r}_{inner}}^{\bar{r}_{outer}} 4F\ (\bar{U}_{z_{total}} + \bar{w}\cos\phi)\ \bar{w}\sin\phi\ \bar{r}\ (\bar{r} + \bar{U}_{\theta_{total}})$$

$$+\cos\phi\ \frac{N_b \bar{c} c_d}{2\pi}\ (\bar{U}^2_{z_{total}} + (\bar{r} + \bar{U}_{\theta_{total}})^2 - \bar{w}^2)\ (\bar{r} + \bar{U}_{\theta_{total}})\ d\bar{r} \quad (34)$$

Similar to the single rotor above, the efficiency of the coaxial system has to be considered. As both rotors can take a different share of the total thrust e.g. when both rotors are tuned to balance the total torque $Q_{coax} = 0$ this has to be considered in the combined efficiency metric.

Leishman and Ananthan (Ref. 16) define this metric combining hovering condition and axial flight and thus also being valid for $U_\infty = 0$:

$$\eta_c = \frac{C_T \lambda_\infty + C_{Pu_{ideal}} + C_{Pl_{ideal}}}{C_T \lambda_\infty + C_{Pu} + C_{Pl}} \ . \quad (35)$$

With the tip speed ratio $\lambda_\infty = U_\infty / \Omega r_{Tip}$, the ideal power coefficients for both rotors are given by:

$$C_{Pu_{ideal}} = C_{Tu} \left(\frac{1}{2} \sqrt{\lambda_\infty^2 + 2C_{Tu}} - \frac{\lambda_\infty}{2} \right) \quad (36)$$

$$C_{Pl_{ideal}} = C_{Tl} \left(\frac{1}{2} \sqrt{\lambda_\infty^2 + 2C_{Tl}} - \frac{\lambda_\infty}{2} \right) \quad (37)$$

Using these equations and the iterative process described above enables the assessment of a coaxial rotor system. The numerical approach will be described below.

Numerical Modeling

The mathematical model described above has to be adapted to fit the numerical scheme present in the SARF environment (Ref. 7).

In this environment a rotor is defined in an object-oriented hierarchical class structure shown in figure 5 where all information on geometry, location in the UAV and environmental conditions is stored.

Multiple rotors on a UAV such as a coaxial rotor can be described separately to incorporate the application of the developed methods to the calculation of a wingtip pusher configuration.

Discretization and Integration Schemes As described above an arbitrary flight state will lead to an inflow that will vary in radial and azimuthal direction.

In azimuthal direction the rotor parameters are calculated at a variable number of azimuthal positions evenly distributed over one rotor revolution. The local varying flow conditions are considered at each position in a quasi static manner where

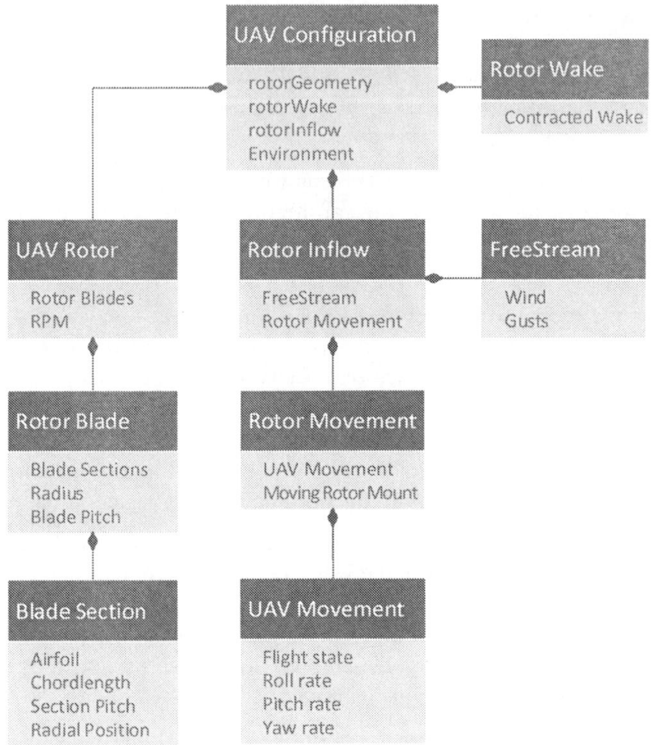

Fig. 5: Hierarchical class structure to represent the UAV

each state is considered static and no relaxation effects between different azimuthal position occur. The resulting forces can be resolved locally to calculate moments acting on the rotor hubs or can be integrated over the whole annulus.

Thrust and torque produced by the whole rotor are integrated first in azimuthal direction from $\theta = [0, 2\pi]$. These averaged values are then integrated in radial direction for the global performance values.

Inherent to the BEMT the discretization of the rotor in radial direction is done by placing multiple radial blade sections along the radius. The structure depicted in figure 5 permits the placement of an arbitrary amount of sections at arbitrary radial positions.

This flexibility is needed for additional sections that have to be placed in the lower rotor at the position where the wake of the upper rotor impinges on the lower rotor.

As shown in figure 6 the thrust- and other coefficients are calculated by trapezoidal integration between each section placed in radial direction. Due to the rapid change in axial and azimuthal inflow velocities at the radius r_c, the local thrust coefficient c_T will experience a step at this radius. By placing an additional section at $r = r_c$ the accuracy of the integrals of formulas (33) and (34) will be greatly increased as an integration boundary can only be considered at an existing section.

When including this additional section in the discretization all geometrical parameters will be interpolated between the two adjacent original sections. However, in case two different airfoil shapes with different airfoil coefficients c_l and c_d are present, a simple interpolation will not suffice. Instead the

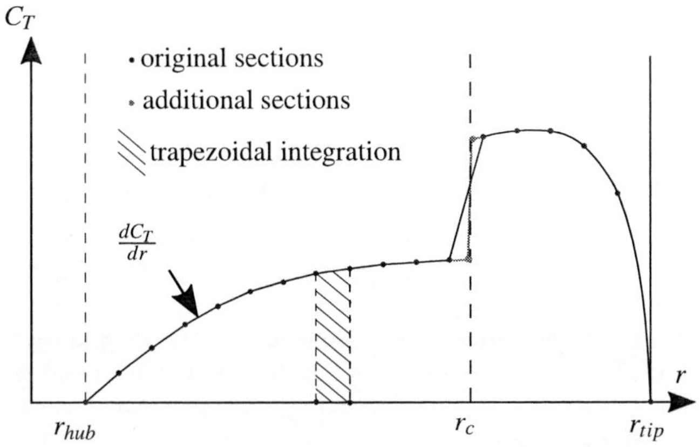

Fig. 6: Trend of dC_T/dr on the lower rotor of a coaxial rotor system, with inserted additional sections

airfoil coefficients of the new interpolated airfoil will be determined using an algorithm developed by Frank in (Ref. 17).

Iteration Scheme of the Coaxial System The iterative scheme shown in figure 2 is used for the individual rotors of the coaxial system. Because of the mutual influences another outer iteration loop has to be carried out.

After initializing both rotors the initial flow state of the upper rotor will be calculated first, according to figure 2 while not considering the additional inflow velocity induced by the lower rotor. With the results the wake contraction radius r_c of the upper rotor will be calculated at the position of the lower rotor $\bar{z}_{ul}$ using formula (29).

Using this radius, the additional inflow velocities $\bar{U}_{lz_{add}}$ and $\bar{U}_{l\theta_{add}}$ will be calculated and added to the nominal inflow acting on the lower rotor using equations (21) and (22). Here, the scheme depicted in figure 2 is executed for the inner and outer part separately. Once convergence on both parts is achieved the additional velocity $\bar{U}_{uz_{add}}$ is determined by formula (19) and the iteration for the upper rotor is started anew.

This process is repeated until all additional induced velocities converge at every section and azimuthal position. Then the performance values of the individual rotors and the coaxial system can be determined.

Application to Wingtip Pusher Propellers

The knowledge gained in the previous sections is now applied to another rotor configuration shown in figure 7 that experiences equivalent flows: When mounting a pusher propeller at the wingtip of a fixed-wing UAV the propeller operates in the slipstream of the separating vortex.

This tip vortex depicted in figure 8 starts at the leading edge of the wing with the span b and expands with further distance to the wing. It will consist of a vortex core with radius r_{core}, where the induced axial velocity will resemble rigid body rotation and an outer part where the induced velocity is dependent on the vortex strength Γ according to the Biot-Savart-Law.

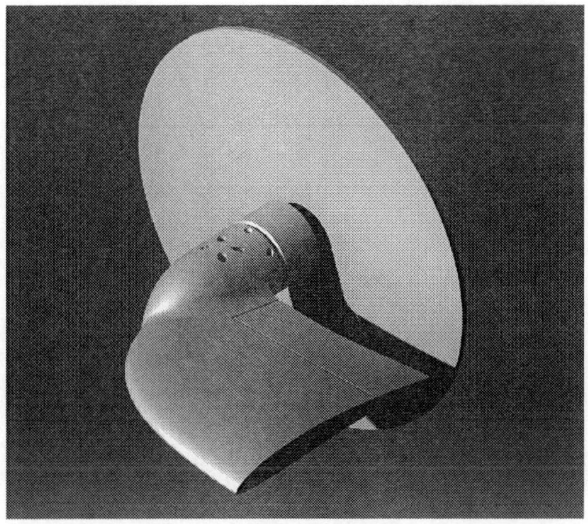

Fig. 7: Visualization of a wingtip mounted pusher propeller at a fixed-wing UAV (TUM-LLS, 2018)

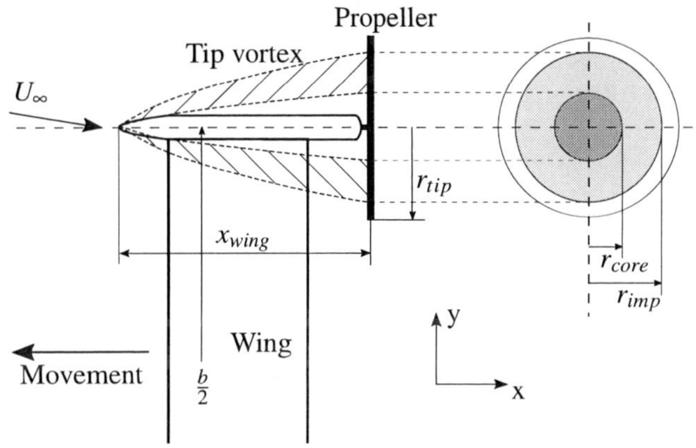

Fig. 8: Draft of a wingtip-mounted pusher propeller, with an inflow U_∞

As the distance between the propeller and the leading edge x_{wing} will be rather small, the vortex expansion will generally be less than the total propeller radius. When using a simple Biot-Savart model for the induced azimuthal velocity the influenced area expands infinitely. However it is also possible to define a limit of the influenced area which leads to an outer boundary r_{imp} where no influence is exerted for radii greater than r_{imp}. Figure 8 also shows an inclined flow U_∞ impacting the wing which can be experienced when the UAV is in yawed flight. This phenomenon impacts the wingtip vortex in a similar manner than an inclined inflow to a coaxial rotor system.

The system will induce an additional axial and azimuthal velocity on the part of the propeller disk which resides in its area of influence. Together these phenomena can be applied to the coaxial rotor model described above where the upper rotor is replaced by the wingtip vortex.

Using either the model described below or any other suitable method to determine the additional inflow velocities on the wingtip propeller $\bar{U}_{wtz_{add}}$ and $\bar{U}_{wt\theta_{add}}$, the total velocities $\bar{U}_{wtz_{total}}$ and $\bar{U}_{wt\theta_{total}}$ can be calculated. These are used in the scheme shown in figure 2 for the lower rotors inner and outer area separated by r_{imp} to get the performance values with the equations (33) and (34).

The influence of the propeller on the wing is not considered here but can be assessed similarly to the influence exerted on the upper by the lower rotor of the coaxial system when using a more sophisticated model to calculate the vorticity on the wing.

Velocities Induced on Pusher Propeller To determine the velocities induced on the propeller, Prandtl's simple lifting line theory is used. Following the approach of Beguin (Ref. 18) and analogous to the discretization used for the rotor blades, the total lift L produced by the wing of a UAV is calculated by integrating the sectional lift $l(y)$:

$$L = \int_{-\frac{b}{2}}^{\frac{b}{2}} l(y)\, dy \tag{38}$$

The Kutta-Joukowski Theorem states that lift can only be produced by a bound vortex $\Gamma(y)$ on the wing. It is related to the lift distribution by:

$$\Gamma(y) = \frac{l(y)}{\rho U_\infty} \tag{39}$$

The local lift can be expressed by the lift coefficient c_l

$$l(y) = \frac{1}{2} c_l\, c(y) \rho U_\infty^2 \tag{40}$$

and this term substituted in equation (39) to relate the bound vorticity to the local lift coefficient:

$$\Gamma(y) = \frac{1}{2} c_l\, c(y) U_\infty \tag{41}$$

This bound vorticity can be used to calculate the influence on the propeller.

Wing Vortex Influence The additional velocities acting on the propeller $\bar{U}_{wtz_{add}}$ and $\bar{U}_{wt\theta_{add}}$ are calculated based on a model by Garmann and Visbal (Ref. 19).

The azimuthal component originating from the rotational vorticity is induced according to the vortex core model by Lamb-Oseen used by Bhagwat and Leishman (Ref. 20) inside the core radius r_{core} and decreases exponentially following equation 42 outside the core.

$$\bar{U}_{wt\theta_{add}}(r) = \frac{\Gamma}{2\pi r}\left[1 - \exp\left(-\frac{r^2}{r_{core}^2}\right)\right] \tag{42}$$

The core radius r_{core} of the vortex according to the core model is expanding while traveling downstream of the wing and is given at the location of the propeller x_{wing}:

$$r_{core} = \sqrt{r_{core0}^2 + 4\alpha\nu \frac{x_{wing}}{U_x}} \tag{43}$$

Here, α is the Oseen Parameter of the vortex core model and is fixed to $\alpha = 1.25643$. The initial vortex core radius is dependent on the geometry where the vortex originates and is set to the trailing edge thickness of the wing.

Knowing the azimuthal component it is possible to determine the axial component of the wingtip vortex.

$$\bar{U}_{wtz_{add}} = U_\infty + \Delta u \exp\left(-\frac{r^2}{r_{core}^2}\right) \tag{44}$$

The axial velocity deficit Δu is defined by Garmann and Visbal (Ref. 19) being proportional to the maximum azimuthal velocity $\max(\bar{U}_{wt\theta_{add}})$ using a swirl parameter q which is set to be greater than $\sqrt{2}$:

$$q \approx 1.567 \frac{\max(\bar{U}_{wt\theta_{add}})}{\Delta u} \tag{45}$$

With this model being based on the Biot-Savart-Law there is no outer boundary r_{imp} limiting the influence of the wingtip vortex. However when a model is used where an outer influential boundary is present it can be easily incorporated as described above.

Computational Results and Validation

The results of the model are validated using the HR1 rotor which was examined by Harrington (Ref. 9). Although the HR1 is a full scale coaxial helicopter rotor it is used for validation as detailed measurement data as well as a detailed description of the geometry are publically available. It is also used by many other publications concerning coaxial rotors and thus is well suited to compare different methods. After determining the correctness of the proposed models the application to wingtip pusher propellers is shown by comparing the pusher performance to that of a propeller operating in undisturbed flow.

Coaxial Validation Leishman and Ananthan (Ref. 3) used the HR1 measurements to validate their BEMT model which does not consider azimuthal components of the induced velocity. This was done using the deviation of a free vortex analysis (FVA) as the original data collected by Harrington does not include sectional thrust and torque coefficient measurements. Leishman determined the correct calculation of the FVA by comparing the global thrust and torque measurements with the measured data by Harrington. As the data was in good agreement with the measurements, he compared the sectional values obtained by the FVA with his implementation of the BEMT. While the prediction of the BEMT for the thrust coefficient was only slightly under predicted, the torque coefficients were slightly more underestimated. As it is desired to validate the sectional values of the BEMT used here these results will be used for comparison rather than the original measured data by Harrington.

Harrington used two geometrically equal two bladed rotors for the measurements as shown in figure 9. They consisted of untwisted blades with linear taper and 4-digit NACA airfoils.

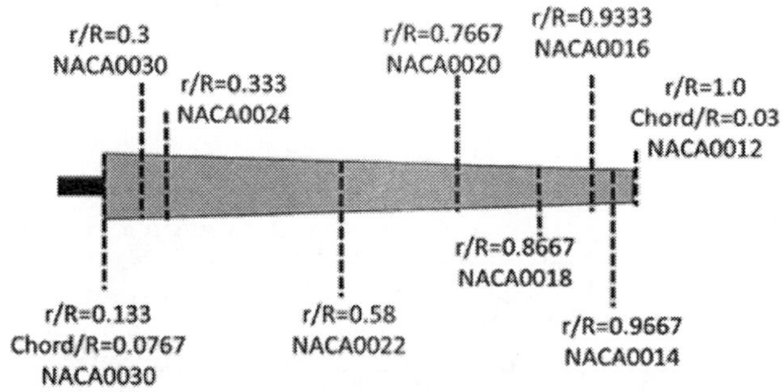

Fig. 9: Blade planform of Harrington Rotor 1 (Ref. 2)

Table 1: Geometric parameters of the Harrington Rotor 1 in hover

Parameter	Value
R_{Tip}/R_{Hub}	3.81m/0.507m
N_b	2
c	linear taper [0.0767;0.03]
z_{ul}	0.71m
V_{Tip}	152.4m/s
σ per rotor	0.027
Twist	none
Airfoils	4 digit NACA
Direction upper rotor	Clockwise
Direction lower rotor	Counter-clockwise

The detailed geometry is listed in table 1. The blades are mounted on a variable pitch hub and able to adjust their global pitch angle β_{global}

The operating conditions used by Leishman are characterized by both rotors operating at torque balance where the total torque coefficient $C_Q = C_{Qu} + C_{Ql}$ is zero. This is achieved with the upper rotor producing more thrust ($C_{Tu}/C_{Tl} = 1.25$) and both rotors operating at the tip speed indicated in table 1.

When replicating these operating conditions with the model described above the data can be compared. Figure 10 shows the thrust coefficient of the upper rotor over the radius.

The sectional positions of the calculation are marked with a cross starting at the rotor hub. At a radial position of $\bar{r} \approx 0.8$ one can clearly see two sections being closer together. This is due to the wake contraction radius r_c where two additional sections are added directly left and right of r_c as discussed in figure 6. Due to the implementation, the sections are also added to the upper rotor. Furthermore the data generated by Leishman is cut off at a radial position of $\bar{r} = 0.2$ whereas this model is calculated up to the rotor hub.

The trend is in very good agreement with the reference data. There are small deviations for the outermost sections which can be explained by a different tip loss behavior.

Subsequently figure 11 shows the thrust coefficient of the lower rotor. One can clearly see the jump where the wake of the upper rotor impinges on the lower rotor and the two sections placed at this position.

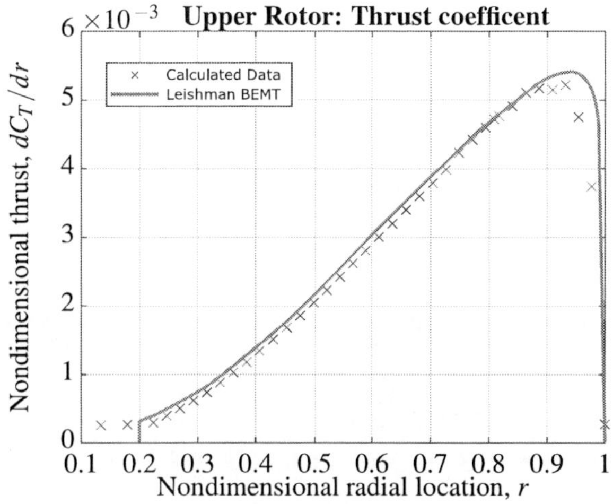

Fig. 10: Thrust dC_T/dr over r/R of the Upper HR1 in hover, comparison with data from Leishman

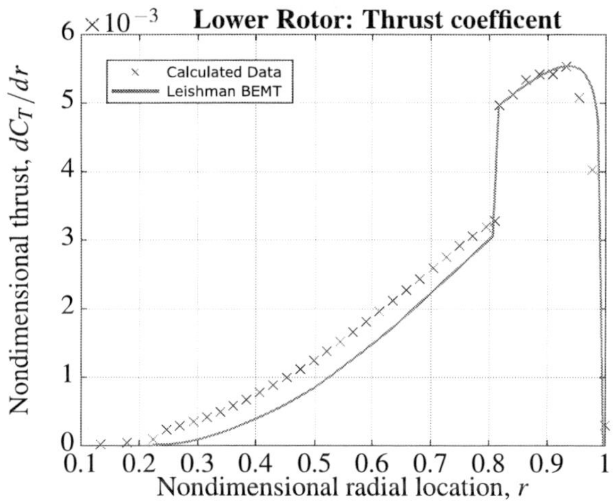

Fig. 11: Thrust dC_T/dr over r/R of the lower HR1 in hover, comparison with data from Leishman

The magnitude of the contraction radius r_c is determined and the adjacent sections are adjusted in every iteration. The progression of the radius, which can be seen in figure 12, is very stable and settles rather quickly. The converged result of $\bar{r}_c = 0.809$ has a difference of less than 1% to the value of $\bar{r}_c = 0.811$ determined by Leishman.

The overall trend of the lower thrust coefficient is also in good agreement with the data. The thrust coefficient is slightly overpredicted at the inner sections which can be explained by the additional consideration of azimuthal induced velocities

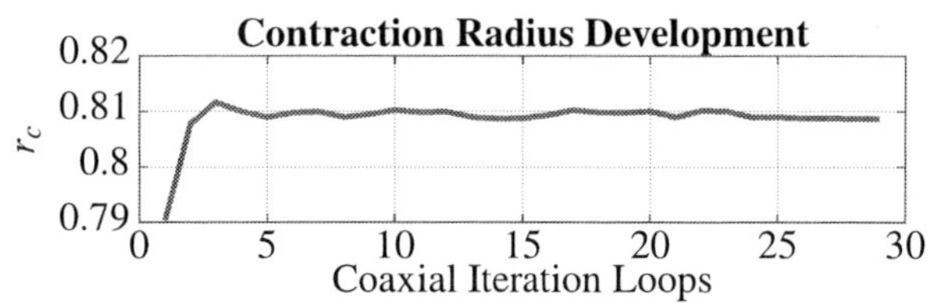

Fig. 12: Development of the Contraction Radius r_c with each iteration.

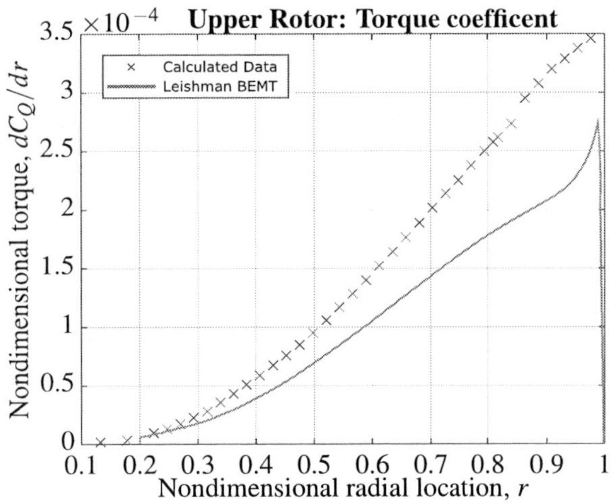

Fig. 13: Torque dC_Q/dr over r/R of the upper HR1 in hover, comparison with data from Leishman

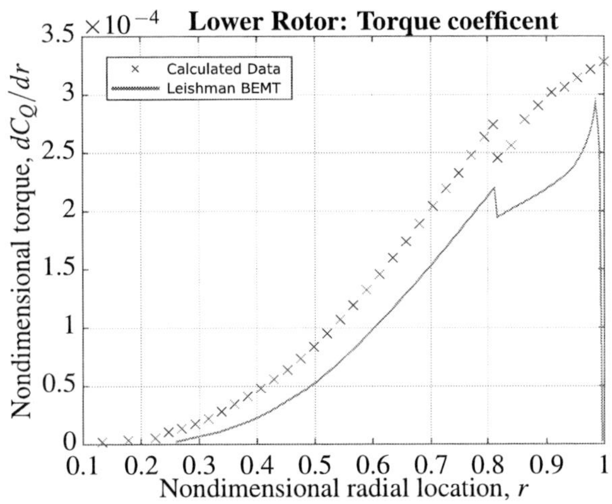

Fig. 14: Torque dC_Q/dr over r/R of the lower HR1 in hover, comparison with data from Leishman

which lowers the local inflow angle ϕ and thus increases the local AoA α. This will produce a higher thrust coefficient which can be seen here.

As the outer sections are not affected by the additional induced velocity there is no overprediction. The model rather experiences similar behavior to the upper rotor where a different consideration of the tip losses leads to small deviations.

The torque coefficient corresponding to the power consumption of the rotors is also considered. Figure 13 shows the torque coefficient of the upper rotor and figure 14 of the lower rotor. Due to the additional azimuthal velocity components the local inflow is tilted compared to the BEMT by Leishman. This increases the torque required to turn the rotor greatly.

As described in (Ref. 3) the reference data used here predicts less thrust than the FVA that was tuned to match the experimental results. The overestimation of the torque coefficient calculated here is compensating the underestimation residing in the underlying data.

Overall the data show a good agreement with the measurements and also show the effect of the additional azimuthal velocities.

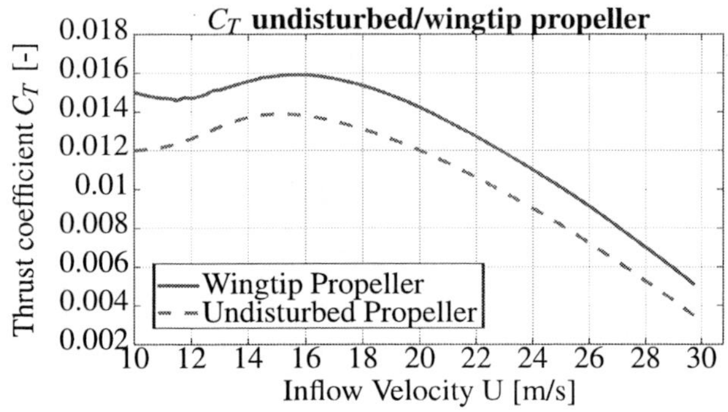

Fig. 15: Rotor thrust coefficient C_T over wingtip inflow velocity

Wingtip Propeller Performance As discussed above, when analyzing a pusher propeller operating behind the wingtip of a fixed wing the azimuthal velocities that are additionally affecting the rotor will reduce the inflow angle ϕ and increase the AoA α. This leads to a higher thrust generation compared to a propeller with an undisturbed inflow. However, as seen in figure 14 it will also greatly increase the required torque and thus the power consumption.

The usefulness of this effect depends entirely on the rest of the configuration and the propeller that is used and cannot be predicted here as the efficiency of a powertrain (ESC, motor and propeller) is not varying in a linear manner. Simply reducing the throttle setting will usually not result in an overall reduction of required energy. However, the wingtip effect can have other consequences.

In general, when designing a small eVTOL UAV it is often desired to operate in a certain regulatory environment which depends on the MTOW. This leads to a design goal where the MTOW is a fixed value that cannot be exceeded. Additionally, for any task except the transport of variable payloads the payload mass is also fixed which results in any residual mass possibly being invested in battery mass.

The pusher configuration will reduce the induced drag of the wing by acting against the vortex. This leads to a smaller thrust required. Additionally the higher thrust produced can lead to a smaller powertrain being sufficient for the design mission. Investing this mass reduction in more battery mass can deliver more energy than the increased energy consumption of the propeller compared to the undisturbed flow will use. In this case a wingtip pusher propeller is beneficial to the UAV design.

These considerations can also be seen in figure 15. For this calculation a 8x6 propeller with a diameter of 8 inch and a blade pitch of 6 inch is used while each blade is turned by an additional pitch of 10 degrees. As described above the model used for the wingtip vortex does not have an outer vortex boundary and the core radius is rather small, thus the whole annulus of the propeller is affected by the vortex with a radially decreasing induced velocity.

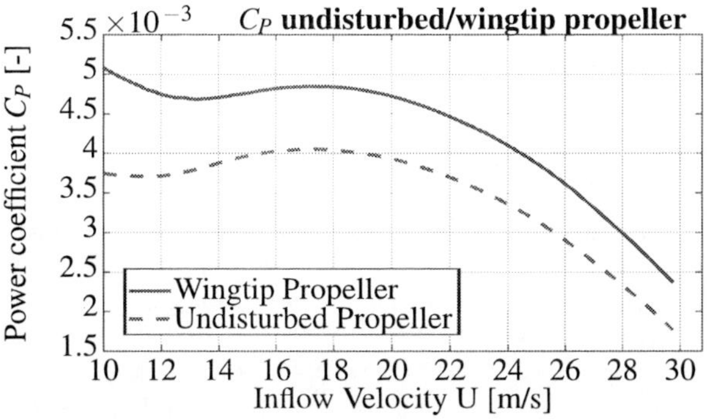

Fig. 16: Rotor Power coefficient C_P over wingtip inflow velocity

Figure 15 shows the thrust coefficient of an undisturbed propeller operating with free inflow as a dashed line. This data was calculated using the process described in figure 2 for a single rotor. The other line depicts the same propeller behind a wingtip with an impinging vortex.

The wingtip pusher propulsion system is operated at a constant $rpm = 7200$ for different cruise velocities. According to equation (39) the vortex strength is a function of the lift produced by the wing and decreases with an increasing incoming velocity. This can be seen as the thrust coefficient difference between the wingtip and the undisturbed propeller is decreasing with a faster inflow velocity.

The propeller exhibits a reduced C_T in the slow flying region as the pitch is quite high which results in a high AoA and subsequently stalled sections near the hub producing less thrust. The thrust coefficient increases with the inflow velocity while more sections are operated at optimal AoA. It can clearly be seen that the azimuthal velocity components caused by the wingtip vortex increase the local AoA and thus increase the thrust coefficient.

However the required torque is also higher than for the undisturbed inflow which is shown in figure 16. An increased AoA is leading to greater profile drag which increases the power coefficient C_P. However as stated above when the required thrust is fixed, the rotor can be operated at a lower rotational speed which redeems additional power.

Whether this process results in less power consumed by the propeller or the energy consumption staying constant depends on the propeller and the airfoils that are used. Figure 17 shows the relative efficiency difference $\Delta\eta$ in percent between the two operating states over the nominal operating rpm range at an exemplary inflow velocity of $U_x = 14\frac{m}{s}$:

$$\Delta\eta = \frac{\eta_{wt} - \eta_{free}}{\eta_{wt}} \quad [\%] \tag{46}$$

The propeller efficiencies are almost equal over a wide rpm range and also over a wide velocity range. When the rotational speed is reduced, there are no gains in required energy for this specific propeller. However the propeller operating behind the wingtip is still reducing the induced drag of the wing and thus reducing the overall thrust required by the UAV configuration

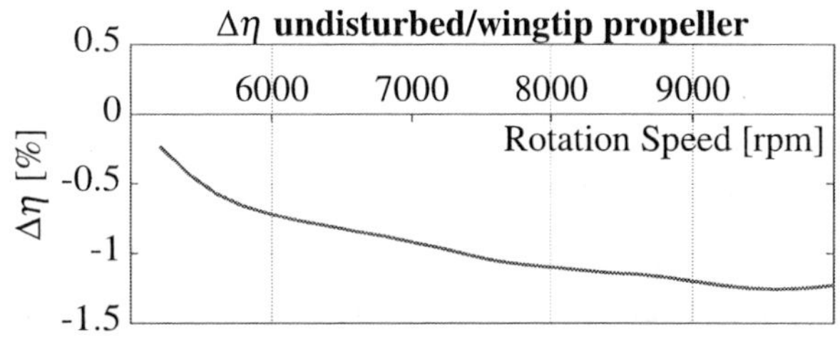

Fig. 17: Rotor efficiency difference $\Delta\eta$ over different rpm at $U_x = 14\frac{m}{s}$

which is beneficial for the endurance. A different propeller with different airfoils that are optimized for this flight state might also gain efficiency when operated behind a wingtip and thus further increase the endurance of the UAV.

CONCLUSIONS

The models developed show very promising results for the calculation of various flow states related to coaxial propulsion. The scheme devised for the single rotor (figure 2) can be the baseline calculation scheme for an individual rotor operating in hover, axial climb and cruise flight with prevalent oblique rotor inflow.

Because the impact of a vortex structure existing in the inflow can be calculated, various other rotor configurations can be assessed by applying this scheme to coaxial counter-rotating rotors or wingtip propulsion.

However, it is important to heed the constraints of the models which can also be addressed in future work on the topic. The interference between a rotor and its environment is only considered using the induced velocities. There is no calculation of a true wake vortex structure. Thus, there is also no consideration of wake turbulence.

As a result, the models proposed show only a limited applicability to coaxial co-rotating rotors like the hover rotors used for the *UBER Common Reference Model 001*. An extension of the calculations for this scenario would enable a detailed assessment of the advantages and disadvantages of the rotational direction.

With the calculations based on the BEMT the model uses a well established discretization basis for the geometry and show good agreement with the reference data.

This enables the fast calculation of a configuration for a given flight state without the need for manual geometry meshing activity. Thus this model can easily be incorporated in an optimization strategy with the aim to develop optimal rotor geometries for a given UAV and its mission.

Author Contact: Moritz Thiele moritz.thiele@tum.de – Martin Obster martinobster@gmx.de – Mirko Hornung mirko.hornung@tum.de

REFERENCES

[1]Jimmy C. Ho, Hyeonsoo Yeo, and Mahendra Bhagwat. Validation of Rotorcraft Comprehensive Analysis Perfor-

mance Predictions for Coaxial Rotors in Hover. *Journal of the American Helicopter Society*, 62(2):1–13, 2017.

[2]Natasha Barbely, Luke Novak, and Narayanan Komerath. A Study of Co-Axial Rotor Performance and Flow Field Characteristics. *AHS Technical Meeting on Aeromechanics*, 2016.

[3]J. Gordon Leishman and Shreyas Ananthan. An Optimum Coaxial Rotor System for Axial Flight. *Journal of the American Helicopter Society*, 53(4):366, 2008.

[4]Ramin Modarres and David A. Peters. Optimum Actuator-Disk Performance in Hover and Axial Flight by a Compact Momentum Theory with Swirl. *Journal of the American Helicopter Society*, 60(1):1–10, 2015.

[5]Eli B. Giovanetti and KennethC. Hall. Minimum Loss Load, Twist, and Chord Distributions for Coaxial Helicopters in Hover. *Journal of the American Helicopter Society*, 62(1):1–9, 2017.

[6]M. Thiele. *Entwicklung eines Programms zur Auslegung und Berechnung von instationär nicht planar angeströmten Rotoren*. Semesterthesis, Technical University Munich-TUM, München, 2015.

[7]M. Thiele. *Erweiterung und Validierung eines Rotortools mit Vorbereitung einer Konfigurationsstudie*. Masterthesis, Technical University Munich- TUM, München, 2016.

[8]Charles N. Adkins and Robert H. Liebeck. Design of optimum propellers. *Journal of Propulsion and Power*, 10(5):676–682, 1994.

[9]R. D. Harrington. Full-scale-tunnel investigation of the static-thrust performance of a coaxial helicopter rotor. (NACA-TN-2318), 1951.

[10]Richard C. Dingeldein. Wind Tunnel Studies of the performance of Multirotor Configurations. (NACA-TN-3236), 1954.

[11]K. W. McAlister, C. Tung, O. Rand, V. Khromov, and J. S. Wilson. Experimental and Numerical Study of a Model Coaxial Rotor. *Journal of the American Helicopter Society*, 62, 2006.

[12]J. Ost. *Parameterbasierte Modellierung von Rotor-Anstroemungszustaenden*. Bachelorsthesis, Technical University of Munich, Garching, 2017.

[13]A. Betz. Schraubenpropeller mit geringstem Energieverlust. Mit einem Zusatz von l. Prandtl. *Nachrichten von der Gesellschaft der Wissenschaften zu Göttingen, Mathematisch-Physikalische Klasse*, 1919:193–217, 1919.

[14]Anton J. Landgrebe. The Wake Geometry of a Hovering Helicopter Rotor and Its Influence on Rotor Performance. *Journal of the American Helicopter Society*, 17(4):3–15, 1972.

[15]Eli B. Giovanetti. *Optimal Aerodynamic Design of Conventional and Coaxial Helicopter Rotors in Hover and Forward Flight*. ProQuest Dissertations And Theses, Duke University Press, dissertation abstracts international, volume: 77-05(e), section: b.; 193 p. edition, 2015.

[16]J. Gordon Leishman and Shreyas Ananthan. Aerodynamic Optimization of a Coaxial Proprotor. *62th Annual Forum and Technology Display of the American Helicopter Society International*, page 64, 2006.

[17]M. Frank. *Geometriebasierte Abschätzung von unbekannten Profilbeiwerten*. Bachelors Thesis, Technical University of Munich, Garching, 2018.

[18]Benoit Beguin. *Development and analysis of an elastoflexible morphing wing*. Dissertation, Technical University of Munich, München, 2014.

[19]D. J. Garmann and M. R. Visbal. Interactions of a streamwise-oriented vortex with a finite wing. *Journal of Fluid Mechanics*, 767:782–810, 2015.

[20]M. J. Bhagwat and J. G. Leishman. Generalized Viscous Vortex Model for Application to Free-Vortex Wake and Aeroacoustic Calculations. In Vertical Flight Society, editor, *58th Annual Forum and Technology Display of the American Helicopter Society International*, 2002.

Scaling Investigation of Electric Quasi-Quadrotor Hover Endurance

Budhyant Venepalli
Graduate Student

D. Stefan Dancila
Associate Professor

Department of Mechanical and Aerospace Engineering
The University of Texas at Arlington
Arlington, Texas 76019

ABSTRACT

Motivated by current developments in distributed propulsion electric vehicles, an effort is made in the present study to analytically investigate the upper bound of hover endurance achievable by electric quasi-quadrotors at various mass scales, in the range of 10^{-1} kg to 10^4 kg total vehicle mass. A feasible rotor propulsion pod design space is determined considering physics-based conceptual design constraints defined by Mach number and the Reynolds number in order to ensure effective and energy efficient generation of lift. A conceptual level scaling optimization for two-bladed rotor quasi-quadrotors was carried out at constant blade tip speed. A momentum theory model and a blade element theory model, respectively, were used to calculate the hover endurance for an electric propulsor pod, defined as a vertically stacked configuration of fixed pitch carbon fiber rotor, driven by a permanent magnet motor, and powered by a Li-ion battery. Results show that on the one hand, individual propulsor pods which are too small cannot be optimal in hover, as the low Reynolds number has adverse effects on aerodynamics. On the other hand, pods that are too large cannot be optimal in hover as the blade mass penalty increases. It is consequently found that an optimal propulsor pod size exists, for which the hover endurance is optimal. Hence, results show that modular endurance optimal design is achievable for larger vehicles using multiple optimal pods while preserving the optimal endurance.

NOTATION

Roman Symbols

b number of blades

c chord of the airfoil, m

c_{d0} mean drag coefficient of airfoil

g acceleration due to gravity, m/s^2

m multiplicity factor of the quasi-quadrotor

t thickness of airfoil as percentage of chord

A area of the rotor, m^2

C_T coefficient of thrust

K_A airfoil area correction factor

M_T total mass, kg

M Mach number

P hover power, W

R radius, m

Re Reynolds number

T thrust, N

Greek Symbols

β specific power for motor W/kg

ϵ specific energy for battery Wh/kg

κ induced power correction factor

μ dynamic viscosity of air, kg/ms

ρ density of air, kg/m^3

σ solidity of the rotor

τ hover endurance, hours

Λ aspect ratio of blade

Ω rotational speed of rotor, rad/s

Subscripts

B battery

EM electric motor

R rotor

Presented at the Vertical Flight Society Autonomous VTOL Technical Meeting and Electric VTOL Symposium, Mesa, Arizona, UAS, Jan. 29-31, 2019.

INTRODUCTION

Vertical Take-Off and Landing (VTOL) aircraft represent a category of flight vehicles which have unique capabilities such as runway independent operations and hover, enabling missions which are not possible with fixed wing aircraft.

Recent progress in battery and electric motor technologies have motivated investigations and the development of numerous electric flight vehicles, including VTOL aircraft such as E-Volo, Joby S2, and others. Leveraging the capability of electric motors to operate efficiently over a wide range of RPM, and the simplicity and convenience of Electronic Speed Control (ESC) made possible by modern, high power electronic solid-state devices, and by the capabilities of modern Single Input Multiple Output (SIMO) control systems, distributed propulsion electric flight vehicles rely upon RPM control of fixed pitch propellers, resulting in system level reductions in vehicle complexity, improved system reliability, and increased vehicle maneuverability.

As an example, quadrotors, deriving their lift from two pairs of two counter-rotating propellers, successfully demonstrated and implemented initially at reduced scale, eliminated the requirement for gearbox, transmission, and swashplate mechanism typical of conventional single main rotor/tail rotor helicopter configurations. An extension of the quadrotor concept, referred to in this work as the Quasi-Quadrotor (QQ), consists in a configuration characterized by the substitution of each of the four rotors by a set of identical co-rotating rotors.

Internal Combustion (IC) engines incur weight penalties when multiple engines are used instead of one large engine. Hence, traditionally the use of one larger engine is preferred to multiple engines having the same total combined power. Exceptions are helicopters and aircraft where two or more engines are used to increase redundancy, paying a penalty with increased weight. Unlike IC engines, electric motors do not incur significant weight penalty in substituting one large motor by smaller motors, thereby effectively enabling distributed propulsion.

Ongoing research interest is devoted to investigating the feasibility and effects upon VTOL performance of scaling up electric QQ VTOL. One such effort, Project LIFT! (Ref. 1) is a scalable, modular, all electric multirotor platform, developed and tested by the Boeing company to fill in the need for VTOL vehicles which can lift high payloads and operate at a short range. Hover endurance of project LIFT! was tested at various gross weights (up to 500 lbs.), carrying different payload weights.

Reference 1 emphasizes the fact that there is no required power penalty when dividing one large rotor into multiple smaller rotors of equivalent total area at a constant disk loading, thereby allowing modular design. However, it was reported that the small rotor diameter of individual disks resulted in reduction in aerodynamic efficiency caused by low Reynolds number aerodynamic effects.

Reference 2 summarizes the literature on the effects of Reynolds number on airfoil maximum sectional lift-to-drag ratio and provides results of a computational study, not only on thick conventional airfoils but also on thin cambered shells, between Reynolds numbers of 10^4 and 10^5. Above the Reynolds number of 10^5, conventional airfoils are known to offer high maximum sectional lift-to-drag ratios compared to thin cambered shells. It was shown that below a Reynolds number of 10^5, maximum sectional lift-to-drag ratio of conventional airfoils decreased drastically when compared to cambered shells. Hence, in this study we choose to constrain operating conditions to blade tip Reynolds numbers higher than 10^5.

A scalability study for an unconventional Quadrotor Bi-plane Tailsitter (QBiT) using a generalized design and analysis tool, HYbrid Design and Rotorcraft Analysis (HYDRA) was performed by Govindarajan, Sridharan and Chopra (Ref. 3). Unlike conventional statistical weight estimation methods, physics-based iterative sizing/weight estimation models (Ref. 4) for rotor and support structure were used.

In this study, an effort is made to analytically investigate the maximum hover endurance achievable by an electric QQ VTOL configuration as a function of Maximum Take-Off Mass (MTOM) across several orders of magnitude, from 10^{-1} kg to 10^4 kg, considering present day electric motor technology with a constant specific power of 6 kW/kg and Li-Ion battery with theoretical specific energy of 390 Wh/kg (Ref. 5). Physics-based conceptual design constraints based on Reynolds and Mach numbers were established to ensure the aerodynamic efficiency for electric QQ VTOL scaling.

APPROACH AND METHODOLOGY

The total number of rotors in a QQ VTOL design is defined by the multiplicity factor, m. For m=1, the QQ VTOL is a quadrotor, for m=2, it is an octarotor, and so on. As all the rotors in QQ are assumed to be identical in size and non-interacting, this endurance study reduces to the investigation of the endurance of a single propulsor.

Considering that each fixed pitch rotor is driven by a motor and powered by a Li-ion battery, a vertically stacked configuration allows for cooling, and reduction in support structure and wiring mass. Such a configuration is referred to in this study as a unit electric propulsor pod.

The upper bound of endurance is achieved when the vehicle is analyzed by disregarding payload and the support structure mass. It is further considered that the total vehicle mass is equally distributed between the identical propulsors consequently, the MTOM of a QQ becomes equal to the sum of battery mass M_B, motor mass M_{EM} and rotor mass M_R.

$$M_T = M_B + M_{EM} + M_R \quad (1)$$

Mass Estimation

Battery Mass: Lithium is favored as the anode material for rechargeable batteries due to its small atomic size, light weight and high electro-positivity which is typical of alkali metals. Out of all the Lithium-based rechargeable battery chemistries available, Li-ion chemistry is considered for the present study due to its technological maturity. Therefore, for a given power P, endurance τ and theoretical specific energy ϵ=390 Wh/kg (Ref. 5), the mass of the battery can be determined by

$$M_B = \frac{P\,\tau}{\epsilon} \tag{2}$$

Electric Motor Mass: High power electric motors based on rare earth permanent magnets are developed for specific high-efficiency applications such as electric vehicles. However, the design of motors intended for land-based vehicles does not consider constraints such as specific power, power density, and cooling needs, which are typical for aerospace applications. Siemens built an electric motor with a specific power of 5.2 kW/kg for the Extra 330LE electric plane, at a comparable power the motors available for industrial applications and electric vehicles only offer approximately 1 kW/kg and 2 kW/kg, respectively. (Ref. 6).

Considering proportional variation between the power and mass of the motor, and a constant specific power β of 6 kW/kg as suggested by Walter (Ref. 7), the mass of electric motor including ESC can be determined by

$$M_{EM} = \frac{P}{\beta} \tag{3}$$

Rotor Mass: Results of prior investigations of rotor blade weight scaling trends published in the literature for specific combinations of rotor blade materials and structural concept show a power law dependence on the scaling parameter, R – the rotor blade radius, with an exponent in the range between 2.3 (Ref. 8) to 2.89 (Ref. 9). It is important to note that with change in scale, not only the cross section but also the material demands of the rotor blade change.

Given the conceptual and analytical nature of this study we use a structural concept based on solid carbon fiber blades as a first order approximation. The mass of the rotor is directly proportional to the number of blades b, volume of each blade, and the density of blade material (ρ_b =1600 kg/m^3) used. Considering rectangular, un-tapered solid carbon fiber blades, with a factor K_A accounting for airfoil cross-sectional area $A_{Airfoil}$, the mass of single blade is written as

$$M_b = \rho_b\, A_{Airfoil}\, R \tag{4}$$

$$M_b \cong \rho_b\, (K_A t\, c^2) R \tag{5}$$

Expressing the chord c of the blade in terms of the aspect ratio Λ, and rewriting Eqn. 5, we obtain the mass for blade and the rotor as

$$M_b \cong \frac{\rho_b\, K_A t}{\Lambda^2}\, R^3 \tag{6}$$

$$M_R \cong b\, M_b \cong \frac{\rho_b\, K_A t}{\Lambda^2}\, b\, R^3 \tag{7}$$

Here, K_A is considered to be 0.60 as suggested by Drela and the thickness percentage of the airfoil t, is taken as 12 percent.

Endurance Calculation

Unlike conventional aircraft powered by fuel, the weight of the electric VTOL vehicles does not reduce during flight, given that no part of the vehicle is jettisoned, thereby placing electric VTOL vehicles at a disadvantage. However, this makes the estimation of endurance simpler, using Eqns. 1, 2, 3 and 7, we have

$$\tau_{QQ} = \frac{\epsilon}{P_{QQ}}[M_T - 4mM_R] - \frac{\epsilon}{\beta} \tag{8}$$

Power Calculation

The present study considers two models for the calculation of power required in hover, which are based on

1. Momentum theory, with induced power correction factor, κ, and
2. Blade element theory, with mean airfoil drag coefficient, c_{d0}

Model 1: To analytically investigate the upper bound of hover endurance, we consider momentum theory for the prediction of the power required to hover, with zero rotor mass. Therefore, for an ideal rotor with rotor disk area A, providing a given thrust T, we can determine the power required to hover P_{QQ}, by

$$P_{QQ} = 4\,m\,\frac{\kappa\, T^{3/2}}{\sqrt{2\,\rho_\infty\, A}} \tag{9}$$

Here, we consider the induced power correction factor κ, to be 1.15 and the density of air ρ_∞, at standard atmospheric conditions 1.225 kg/m^3. We can write Eqn. 9 as

$$P_{QQ} = 4\,m\,\frac{\kappa\left(\frac{M_T\, g}{4\,m}\right)^{3/2}}{\sqrt{2\,\rho_\infty\, A}} \tag{10}$$

Model 2: The power required by a real rotor to hover, is equal to sum of the induced power and the profile power. Equation 11 accounts for the induced power provided by model 1 and the profile power based on mean sectional drag coefficient of the airfoil. The total power required by QQ is obtained by

$$P_{QQ} = 4m\,(P_i + P_o) \tag{11}$$

$$P_{QQ} = 4\,m\left(\frac{\kappa\left(\frac{M_T\,g}{4\,m}\right)^{3/2}}{\sqrt{2\,\rho_\infty A}} + \rho_\infty\,A\,(\Omega R)^3\,\frac{\sigma\,c_{d0}}{8}\right) \quad (12)$$

Assuming that the tip speed ΩR remains constant though-out, we express tip speed in terms of tip Mach number M_{tip}, and the speed of sound a_∞. For rectangular blades, solidity σ can be expressed in terms of number of blades and the aspect ratio of the blades. Sectional mean drag coefficient for twelve percent thick airfoils is estimated to be 0.01. Therefore, using Eqn. 12 the power required by QQ is given by

$$P_{QQ} = 4\,m\left(\frac{\kappa\left(\frac{M_T\,g}{4\,m}\right)^{3/2}}{\sqrt{2\,\rho_\infty\,A}} + \rho_\infty\left(M_{tip}\,a_\infty\right)^3\frac{b\,c_{d0}\,R^2}{8\,\Lambda}\right) \quad (13)$$

Conceptual Design Constraints

The feasible design space is established by considering two important aerodynamic parameters, Reynolds number (Re) and Mach number M, in an effort to ensure the aerodynamic efficiency of the rotor. A blade tip Mach number M_{tip}, not greater than 0.3 is chosen to avoid compressibility effects and a blade tip Reynolds number (Re_{tip}) not lower than 10^5 is chosen.

The blade tip Mach number can be expressed in terms of the tip speed and the speed of sound as

$$M_{tip} = \frac{\Omega\,R}{a_\infty} \quad (14)$$

Therefore, Eqn. 14 can be written as

$$\Omega^{max} \leq \frac{M_{tip}^{max}\,a_\infty}{R} \quad (15)$$

Similarly, the blade tip Reynolds number can be expressed in terms of the dynamic viscosity of air μ=1.789 x 10^{-5} kg/(ms), the density of air, the tip speed and the tip chord as

$$Re_{tip} = \frac{\rho_\infty\,(\Omega R)\,c_{tip}}{\mu} \quad (16)$$

Therefore, Eqn. 16 can be written as

$$\Omega^{min} \geq \left(\frac{Re_{tip}^{min}\,\mu\,\Lambda}{\rho_\infty}\right)\frac{1}{R^2} \quad (17)$$

Disk loading can be expressed in terms of the thrust coefficient C_T, the blade tip speed and the area of the disk. Therefore, for a given disk loading, we have:

$$D_L = \frac{T}{A} = \frac{\rho_\infty\,\pi\,R^2(\Omega\,R)^2 C_T}{\pi\,R^2} = \rho_\infty\,(\Omega\,R)^2 C_T \quad (18)$$

Rewriting Eqn. 18 in terms of the blade aspect ratio, blade loading and the number of blades, we have

$$\Omega = \frac{1}{R}\sqrt{\frac{D_L}{\rho_\infty\,{}^{C_T}/_{\sigma}\,\frac{b}{\pi\,\Lambda}}} = \frac{1}{R}\sqrt{\frac{\pi\,D_L\,\Lambda}{\rho_\infty\,b\,{}^{C_T}/_{\sigma}}} \quad (19)$$

RESULTS AND DISCUSSION

Feasible Design Space

Making use of the conceptual physics-based design constraints defined by the Eqns. 15, 17 and 19, the feasible design space is established for the design and scaling of electric QQ. In order to stay in the defined space, low disk loading values have to be chosen.

Figure 1 shows that for the disk loading chosen, there exists a minimum value of radius below which the Reynolds number constraint is violated. Although a range of Ω operable as highlighted in the Fig. 1 exits, as tip speed is considered to be constant, the blade tip Mach number, defined by Eqn. 14 is equal to 0.3.

Absolute Upper Bound of Hover Endurance

An overly optimistic estimation is achieved by neglecting the rotor mass and considering model 1 for the hover power estimation. Even though such considerations result in unrealistic values, they serve as an upper bound of the endurance achievable by a QQ for the given technology. Figures 2 and 3 provide the absolute upper bound of QQ with M_T of 10^{-1} and 10^4kg respectively. Extremely high values of hover endurance are obtained in the case of vehicles with low MTOM. We notice in Fig., 3, that for very small rotor radii, the mass of the motor required becomes equal to or exceeds MTOM due to high power demands and therefore no room for battery is left, resulting in zero endurance.

The power calculated by model 1 is inversely varying with the individual pod rotor radius, i.e., the hover power required for a given MTOM decreases with the increase of the radius. In addition to that, the endurance defined by Eqn. 8 is inversely varying with power. Therefore, for a given total weight and m, a direct relationship exists between the propulsor radius and the absolute endurance. Additionally, for a given value of rotor radius, increasing the number of rotors results in higher values of the endurance.

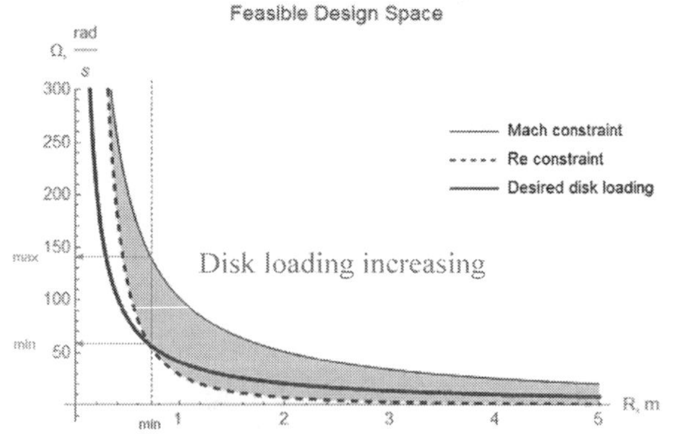

Figure 1. Example of feasible design space describing design bounds

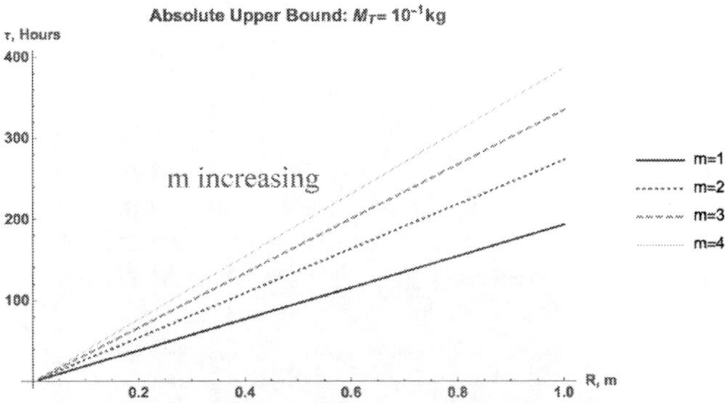

Figure 2. Upper bound of hover endurance achievable by QQ for $M_T = 10^{-1}$kg

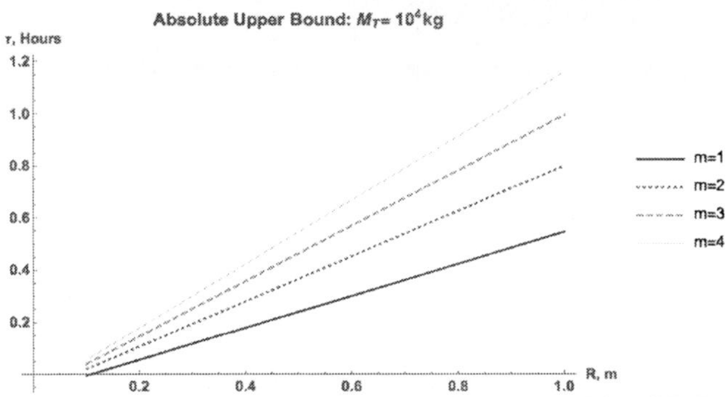

Figure 3. Upper bound of hover endurance achievable by QQ for $M_T = 10^4$kg

Constrained Optimization

The hover endurance of an electric QQ defined by Eqn. 8 and the power estimation model 2, depends on five parameters:

1. Total vehicle mass, M_T
2. Multiplicity factor of quasi-quadrotor, m
3. Number of blades of rotor, b
4. Aspect ratio of the blades, Λ and
5. Radius of individual rotor, R

The Mach number constraint is enforced by considering that the tip Mach number is equal to 0.3 and the Reynolds number constraint is implemented by requiring the tip Reynolds number to be greater than 10^5. Consequently, for the given blade aspect ratio there exists a rotor radius below which Eqn. 17 is violated.

In the work presented in this paper we limit the scaling investigation to two bladed rotors with the blade aspect ratio varying from 5 to 20. The optimization is carried out across several orders of magnitude of M_T by varying individual pod radius and multiplicity factor.

Quasi-quadrotor with $M_T = 10^{-1}$kg: For a quadrotor, results for unconstrained and constrained study across aspect ratios of 5, 10, 15 and 20 are shown in Figs. 4 and 5, respectively.

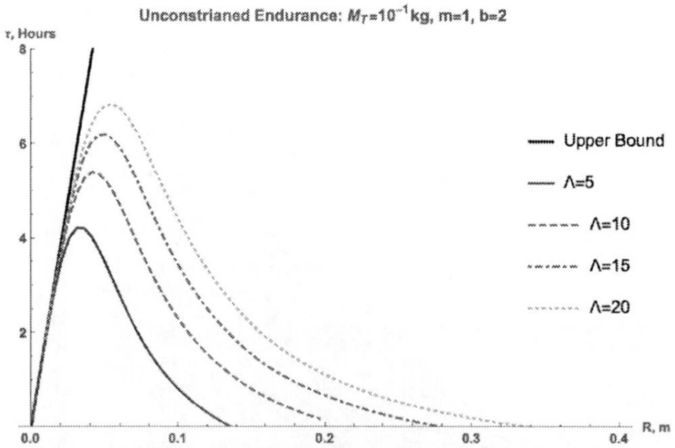

Figure 4. Unconstrained hover endurance: $M_T = 10^{-1}$kg, m=1, b=2

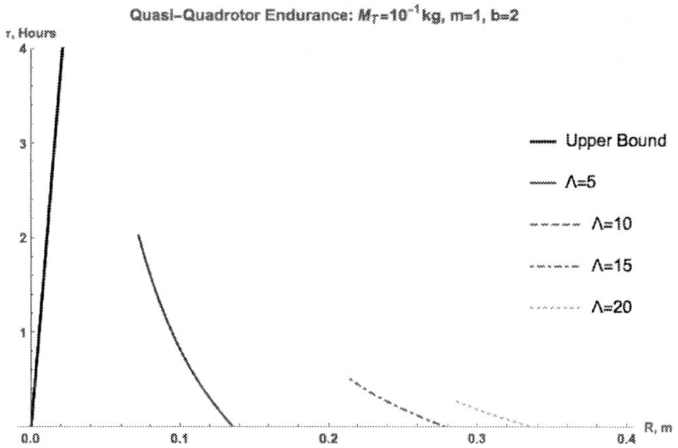

Figure 5. Constrained hover endurance: $M_T = 10^{-1}$kg, m=1, b=2

Although unconstrained results show increase in the endurance with increase in the blade aspect ratio, from constrained optimization it is known due to the low tip Reynolds number, the endurance is decreased with increase in the blade aspect ratio.

Dependence of endurance on aspect ratio is summarized in Fig. 6. The maximum value of endurance is obtained at the lower limit of the blade aspect ratio. Additionally, further increase in m values had adverse effects on the endurance, suggesting that a quadrotor at the lower limit of the considered aspect ratio resulted in the maximum endurance of 2.03 hours.

Quasi-quadrotor with $M_T = 1$kg: Repeating the constrained study at one-kilogram mass resulted in a contour plot as illustrated in Figs. 7. A quadrotor design offered highest endurance of 5.4 hours, at the aspect ratio of 12.5 and radius of 0.178 m.

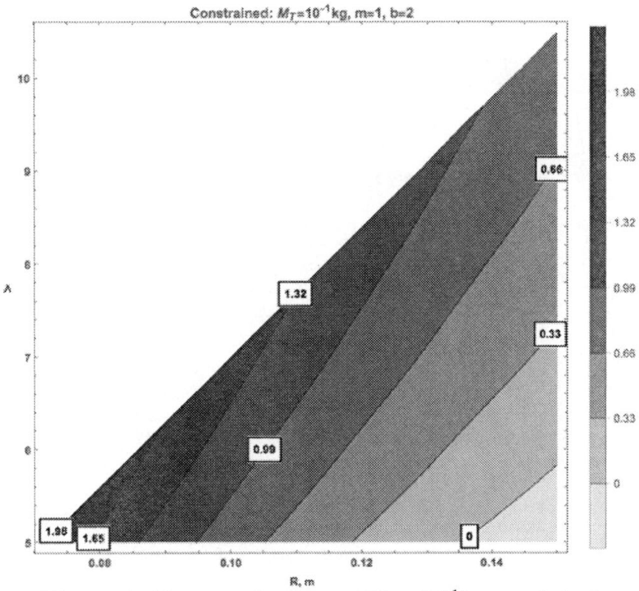

Figure 6. Hover endurance: $M_T = 10^{-1}$kg, m=1, b=2

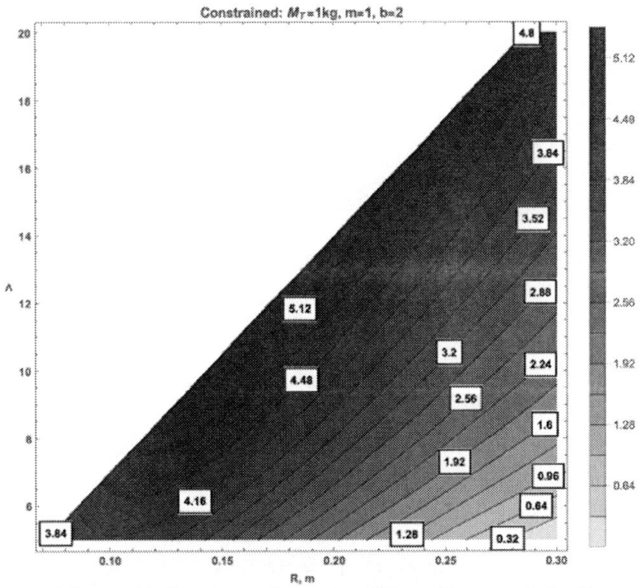

Figure 7. Hover endurance: $M_T = 1$kg, m=1, b=2

Quasi-quadrotor with $M_T = 10$kg: Figure 8 shows the results of the optimization performed for a 10 kg QQ at various aspect ratios. It is observed that with increase in m value, endurance increases slightly until m=3 and then decreases. The endurance was found to be maximum at the upper limit of the blade aspect and the radius of 0.301 m. When compared to the previous case, the endurance value was increase by 1.33 hours.

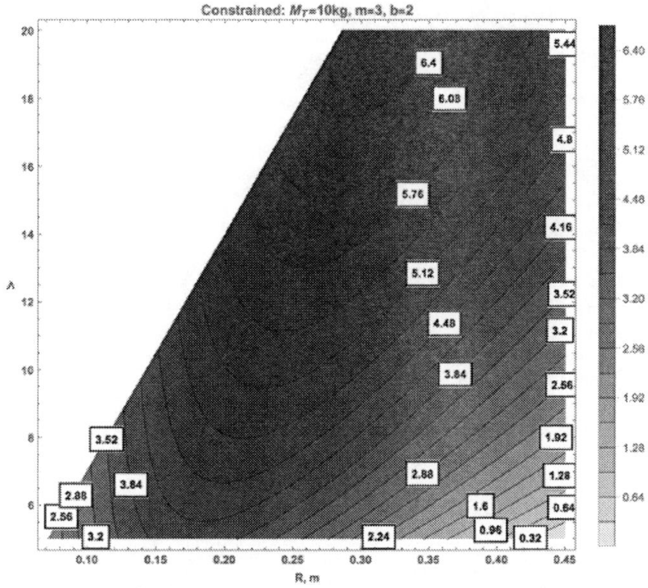

Figure 8. Hover endurance: $M_T = 10$kg, m=3, b=2

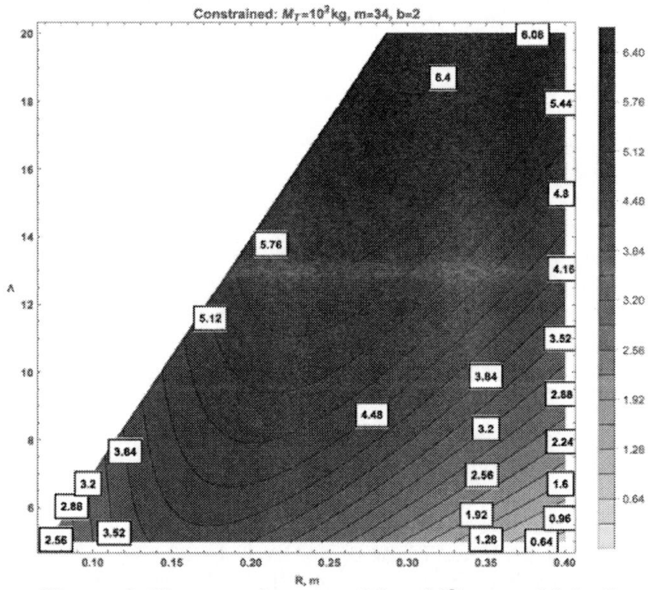

Figure 9. Hover endurance: $M_T = 10^2$kg, m=34, b=2

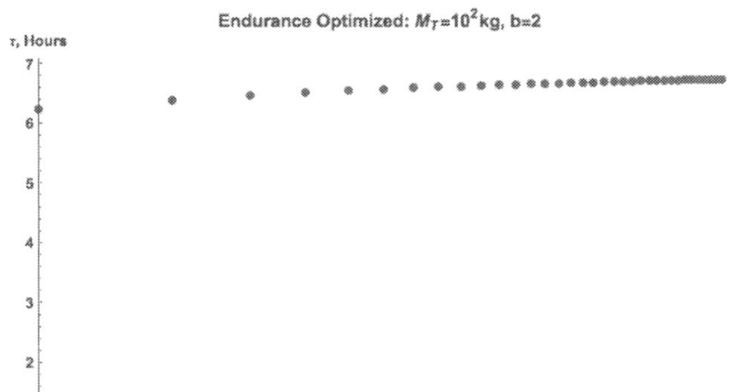

Figure 10. Endurance variation with m: $M_T = 10^2$kg, b=2

Quasi-quadrotor with $M_T = 10^2$, 10^3 and 10^4 kg: Figures 9-14 show the results obtained from the optimization study conducted considering 10^2, 10^3 and 10^4 kg MTOM. While the contour plots shown as Figs. 9, 11 and 13, respectively, show the dependence of endurance on the radius and the aspect ratio of the optimized propulsion pod, Figs. 10, 12 and 14 shows the variation of the endurance with the multiplicity factor for 10^2, 10^3 and 10^4 kg QQ, respectively. It is the 10^4 kg QQ which benefited the most in distributing the propulsion.

It is observed that the radius of rotor, the blade aspect ratio, and the endurance for all the three above mentioned QQ VTOL vehicles is equal. However, the optimized multiplicity factor varies.

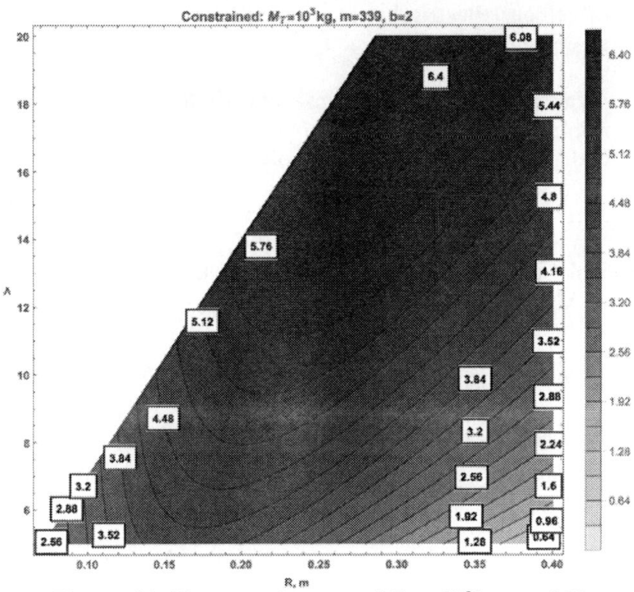

Figure 11. Hover endurance: $M_T = 10^3$kg, m=339

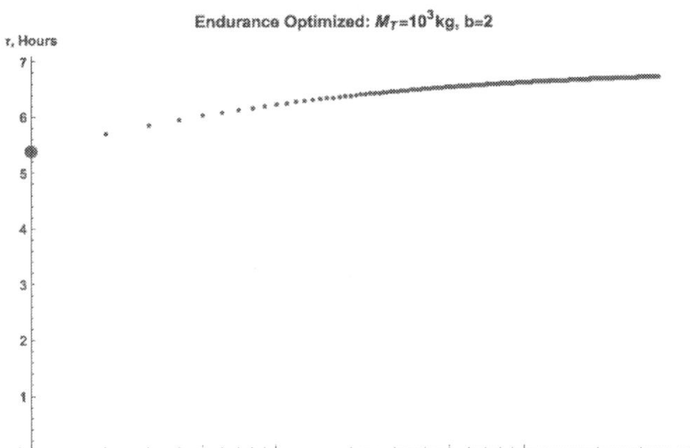

Figure 12. Endurance variation with m: $M_T = 10^3$kg, b=2

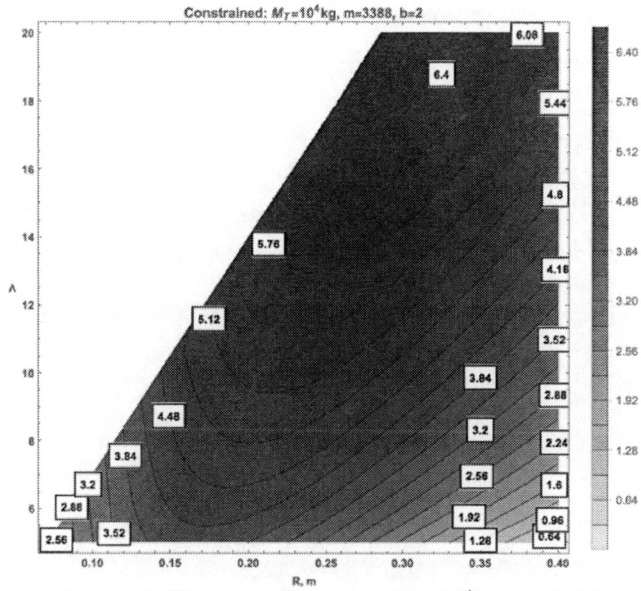

Figure 13. Hover endurance: $M_T = 10^4$kg, m=3388

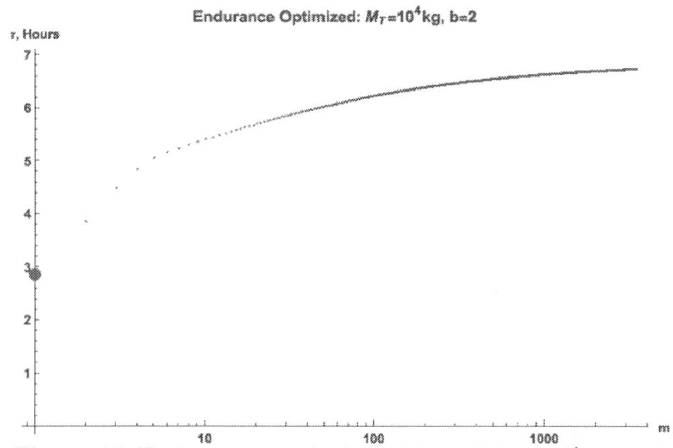

Figure 14. Endurance variation with m: $M_T = 10^4$kg, b=2

Table 1 summarizes the optimized endurance values for two bladed quasi-quadrotors. At low total mass 10^{-1}kg, the blades with an aspect ratio of 5 provide highest endurance and in contrast, at total mass greater than 10kg, the rotor blades with an aspect ratio of 20 provide the maximum endurance. However, the maximized endurance achievable at smaller MTOM is very low when compared to higher MTOM. Moreover, we observe that an endurance optimized pod exists (highlighted in table), allowing any larger vehicle to be conceived by adding optimized pods in multiples of four, until the total weight desired, with no penalty in endurance.

Table 1: Endurance optimized results summary

M_T, kg	m	Λ	R, m	τ, Hours
10^{-1}	1	5	0.07153	2.02389
10^{0}	1	12.4765	0.17848	5.40025
10^{1}	3	20	0.30088	6.72681
10^{2}	34	20	0.28611	6.73393
10^{3}	339	20	0.28611	6.73396
10^{4}	3388	20	0.28611	6.73396

SUMMARY

The present study, using analytical models for hover power required and assuming present day battery and motor technology performance, has established the feasible design space using physics-based constraints, has introduced a methodology to estimate the hover endurance for electric QQ VTOL vehicles, and has identified the absolute upper bound of the hover endurance achievable by a QQ VTOL configuration across several orders of magnitude, M_T varying from 10^{-1} kg to 10^{4} kg. Additionally, by performing a constrained optimization for two bladed electric QQ VTOL across several orders of magnitude of MTOM and considering the tip speed as constant, it is shown in this study that on the one hand, individual propulsor pods which are too small cannot be optimal in hover, as low Reynolds number has adverse effects on aerodynamics, while on the other hand, pods that are too large cannot be optimal in hover as the blade mass is too high. Consequently, it is found that an optimal propulsor pod exists, for which the hover endurance is optimized. Hence, modular design for QQ VTOL configurations can be achieved for larger vehicles leveraging multiple such optimal pods.

Author contact:

Budhyant Venepalli budhyant.venepalli@mavs.uta.edu

REFERENCES

[1]Duffy, M., and Samaritano, T., "The LIFT! Project – Modular, Electric Vertical Lift System," Proceedings of the 71st American Helicopter Society Annual Forum, Virginia Beach, VA, May 5–7, 2015.

[2]Winslow, J., Otsuka, H., Govindarajan, G., and Chopra, I., "Basic Understanding of Airfoil Characteristics at Low Reynolds Numbers (10^4–10^5)," *Journal of Aircraft*, Vol. 55, No. 3 (2018), pp. 1050-1061.

[3]Govindarajan, M., Sridharan, A., and Chopra, I., "A Scalability Study of the Multirotor Biplane Tailsitter using Conceptual Sizing," Proceedings of the 74th American Helicopter Society Annual Forum, Phoenix, AZ, May 14–17, 2018.

[4]Govindarajan, M., Sridharan, A., and Avera, M., "Integration of Physics Based Weight Models into Rotorcraft Design Sizing," Proceedings of the 43rd European Rotorcraft Forum, Milan, Italy, Sep 12–15, 2017.

[5]Hepperle, M., "Electric Flight – Potential and Limitations," NATO Report STO-MP-AVT-209.

[6]"Siemens Electric Flight, accessed January 21st, 2019," www.siemens.com/press/electric-aircraft .

[7]Walter, C, S., "Design of High-Torque-Density Synchronous Drives for Propulsion of Rotary-Wing Aircraft," Ph.D. Thesis, TU Delft, 2016.

[8]Crabtree, J, A., "Weight Estimation for Helicopter Design Analysis," Society of Allies Weight Engineers, 17th Annual Conference, Belmont Plaza Hotel, New York, New York, May 1958.

[9]Prouty, R, W., *Helicopter Performance Stability and Control*, Krieger, Malabar, FL, 1995, p. 663.

[10]Leishman, J. G., *Principles of Helicopter Aerodynamics*, Cambridge University Press, New York, NY, 2000.

Evaluation of Reconfigurable Rotorcraft Concepts As Viable Combined High Efficient Hover and Forward Flight Solutions for Unmanned Aerial Systems (UAS)

Mr. David Alman
Team Leader, Georgia Tech
AHS 35th SDC Graduate Team
School of AE, 270 Ferst Drive
Atlanta, GA 30332-0150

Dr. Daniel P. Schrage*
Professor and Director
VLRCOE, School of AE,
Ferst Drive, Atlanta,
GA 30332-0150
*Principal Author

Mr. Carter Crouch
Team Member, Georgia Tech
AHS 35th SDC Graduate Team
School of Ae, 270 Ferst Drive
Atlanta, GA 30332-0150

ABSTRACT

It is well known that using a rotor is the most efficient hovering device for Vertical Takeoff and Landing (VTOL) aircraft. It is also well known that a wing is the most efficient forward flight device for VTOL aircraft. Configurations using the wing, also as a rotor in hover, have been a concept explored over the past 60 years. However, no reconfigurable stopped rotor/wing concept has successfully been demonstrated in flight test. The goal of the AHS International 35th Annual Student Design Competition (SDC) *Reconfigurable VTOL Aircraft* competition, sponsored by the Army Research Laboratory (ARL) was to design a Group 3 size unmanned VTOL aircraft that achieves high-speed forward flight (relative to current VTOL aircraft) and efficient hover through the use of novel reconfigurable propulsive and lifting devices. The aircraft should have superior performance over a comparably-sized aircraft that does not have reconfigurable systems. In response to the AHS International 35th SDC Request for Proposal (RFP), the Georgia Tech graduate team responded with a truly reconfigurable rotorcraft concept. The RFP stated that the vehicle must be able to reconfigure on its own. It also stated that the proposed aircraft design must have its main lifting device be a reconfigurable system. It also stated that traditional tiltrotors and tail sitters, including multi-rotors, do not meet the "new and novel" requirement. The Georgia Tech graduate and undergraduate teams met this reconfigurable requirements and constraints and finished second in the 35th AHS graduate SDC .However, the two first place winners were both tail sitter configurations which violated the single rotor lifting device reconfigurable requirement. Their coaxial rotor main lifting devices in hover did not reconfigure in forward flight, instead a separate lifting device, a wing, was deployed for forward flight lift. The Georgia Tech Graduate Team proposed a stopped rotor reaction driven Group 3 size UAS, *The Knight Flyer*, with substantial greater forward flight minimum dash speed capability than other concepts. It also addressed the failure of the Boeing X-50A Canard Rotor Wing (CRW) which showed that pitch control at low speeds and during transition was a major challenge. This inclusion met the 35th AHS RFP Design Limitation that designs should be *new and novel* and that prior VTOL designs should only be considered if performance can be improved upon through the use of new technology that was unavailable to the original design. To overcome the X-50 CRW deficiency the Georgia Tech Graduate Team developed an Aerial Coanda High Efficiency Orienting Jet Nozzle (ACHEON). This technology solution can also be introduced using hybrid electric propulsion as a new technology. The Army is now looking for potential solutions for its VTOL next generation tactical UAS (NGTUAS), Groups 3, 4&5, an expressed goal of 460 kts forward flight speed. Achieving this speed requires reconfigurable rotorcraft that combine and utilize a common reconfigurable rotor/wing for hover and forward flight. It also must address the need for enhancing survivability, e.g. by eliminating the rotating signature for forward flight modes of flight. This paper will review the history of reconfigurable rotorcraft using drive, notional VTOL NGTUAS goals and will illustrate how The *Knight Flyer* can serve as a baseline Group 3 UAS and for evaluating stopped rotor as the only viable solution for the NGTUAS notional high speed goal.

INTRODUCTION

In the 35th AHS SDC RFP (Ref.1) it states that the Sikorsky S-72/NASA Rotor Systems Research Aircraft (RSRA) research testbed was envisioned to test rotor systems independently from the fixed wing airframe and propulsion system they were attached to. This allowed the rotor speed to be varied or stopped in flight while maintaining control of the aircraft, creating a "flying wind tunnel" that could test advanced rotor systems. (See www.vtol.org/RSRA for the AHS Vertipedia entries on the S-72.)

It goes on to state that The DARPA / Lockheed X-Wing was a 4-bladed "circulation control rotor" intended to be tested on the RSRA. The X-Wing blades were designed with elliptical airfoil sections and air slots on the leading and trailing edges.

Presented at the Vertical Flight Society's Autonomous VTOL Technical Meeting, Mesa, AZ, USA, January 28-31, 2019.

When the rotor would be stopped in flight, the blown air could be switched from the trailing edge to the leading edge, effectively reversing the airfoil shape so that all four blades were "facing" forward as illustrated in Figure 1. (Ref.1)

Figure 1. Sikorsky Rotor Systems Research Aircraft fitted with the Lockheed X-Wing Circulation Control Rotor (IGOR Sikorsky Historical Archives).

In October 1976, the RSRA achieved its first flight without a rotor, using turbofans for propulsion and fixed wings for lift. Unfortunately, the combined RSRA/X-Wing vehicle never flew as the project was cancelled in 1987 when the government terminated the funding. The X-Wing stopped rotor concept—in its intended final form without fixed wings or turbofans—was a representation of aircraft designers' imagination of what reconfigurable aircraft might look like using 1980's technology. The RSRA is the inspiration for this year's Student Design Competition. Students are asked to design a radical reconfigurable VTOL aircraft technology demonstrator incorporating today's technological advancements. What would it look like? Can it be useful? What technology would enable such a vehicle? (Ref. 1)

The principal author on this paper was involved with overseeing the RSRA/X-Wing research project and was a member of the Technical Review Team that recommended its cancellation. The complexity of this experimental tested was unbelievable. One member of the Technical Review Team said that making this system work would be "as complicated as sending man to the moon". While the ARL authors of the AHS 35th SDC RFP used the RSRA/X-Wing for describing the type of "reconfigurable rotorcraft" they were looking for they missed the boat. The key for this type of "reconfigurable rotorcraft" requires two key technologies: 1) a slowed, stopped or stored rotor and 2) a reaction drive propulsion system.

Early research on sustained autorotative flight type vehicles by the Georgia Tech Montgomery Knight School of Aerospace Engineering in the late 1930s and early 1940s led to the development and testing of a reaction drive rotor commonly referred to as a "tip-jet." Exhaust or bypass air from a turbine engine is vectored out through the rotor. Illustrated in Figure 2 is a picture of Montgomery Knight, School of Aeronautics, with a test stand version of this reaction drive "tip jet" rotor. (Ref.2)

Figure 2. Reaction Drive "Tip-Jet" Research at Georgia Tech in 1940 (Ref.2)

A wind tunnel model of the rotor blade is currently displayed in the entrance lobby of the School of Aerospace Engineering, as illustrated in Figure 3. Pictured in front of the wind tunnel model are key members of the DARPA sponsored Heliplane Gyrodyne Demonstrator Aircraft (HGDA) Program which will be discuss later.

Figure 3. HGDA Team in Front of GT Tip Jet Rotor

A reaction drive rotorcraft (tip-jets used for certain helicopter designs) and more successfully in gyrodynes – a "gyrodyne is a gyroplane with a tip-jet powered rotor for hover and vertical takeoff and landing is a class of VTOL aircraft that uses thrust generated at the blade tip to provide all rotor power required without imparting torque into the airframe. Compressed air is ducted from a compressor, powered by a piston or turbine engine, to the tip of the rotor blade, turned 90° and expanded through a nozzle to produce thrust. (Ref. 3)

There are three primary configurations that provide compressed air to the rotor blade tip. The first is a cold cycle, where the air is compressed by an additional compressor to relatively high pressure ratios and expanded through the reaction drive rotor nozzle. When there are constraints on the allowable pressure and temperature within the rotor blade duct, fuel may be added and combusted at the blade tip to add the necessary energy to the flow. The second configuration is a warm cycle, where air is extracted from mixed engine and fan exhaust. Because there is less energy per unit mass of air, high flow rates are needed to produce adequate reaction drive nozzle thrust. This adversely affects the size of the rotor blade and its associated performance. The third configuration is a hot cycle, where the gas efflux of a turbine engine is ducted through the blade and expanded through the reaction drive nozzle. The primary drawback to this arrangement is the high gas temperature which limited the selection of practical blade materials. Figure 4 are notional rotor tip jet concepts.(Ref.3)

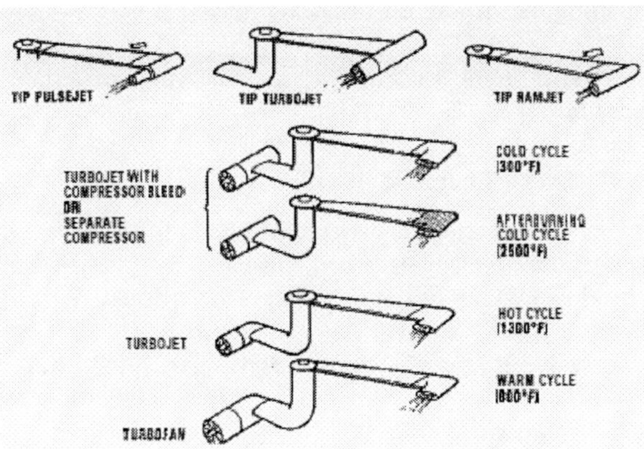

Figure 4. Notional Tip Jet Rotor Types

The reaction drive rotor system has several unique advantages when compared with conventional shaft driven rotors. First, there is no transmission required. There is also no torque imparted into the airframe from the rotor and thus, no anti-torque device is required. This results in a significant reduction of the mechanical complexity and weight of the drive system and thus the aircraft itself. Further advantages can be achieved when the reaction drive rotor is incorporated on a compound rotorcraft such as a "gyrodyne."

Examples of gyrodyne compound reaction driven rotorcraft are the McDonnell XV-1, shown in Figure 5, and the Fairey Rotodyne, shown in Figure 6. Dual use of the propulsion system to power the rotor in hover and produce the thrust in cruise (when desired) is another advantage to this VTOL configuration. Additionally, in forward flight the rotor can autorotate at a slower RPM than typical helicopters, one of the contributors to permit higher cruise speeds than helicopters are capable of. This was the case for both the McDonnell XV-1 and the Fairey Rotodyne which served as baselines for the HGDA Program. (Ref 3)

Figure 5. McDonnell XV-1 Convertiplane

Figure 6. Fairey Rotodyne

Typically the reaction drive rotor system requires a larger hub than a conventional shaft driven rotor with increased design complexity (though not parts count complexity) to accommodate the gas flow. This may result in drag penalties at the higher airspeeds. The solidity of the rotor will also have a direct impact on the operation of the reaction drive. In order to minimize pressure losses through the ducting, the velocity of the gas must be low. The consequence of low gas velocity is a larger duct area for the given mass flow rate. Since the gas is ducted through the blade, the chord has to accommodate both the blade structure and the duct; thereby driving rotor solidity. The hovering Figure of Merit, which is a measure of the ratio of the ideal induced horsepower to the ideal induced horsepower plus the profile power, will begin to decrease as the solidity is increased due to the blade profile drag. Therefore, when the rotor is sized on a similar scale to a conventional shaft driven rotor, there is little volume available for the ducting in the rotor blades. The McDonnell XV-1 and the Fairey Rotodyne included a combustion chamber at the tip of each blade in an effort to maintain reasonable rotor solidity and therefore avoiding penalties. Thermal energy is added to the airflow by burning fuel, thus reducing the mass flow requirements through the rotor blade.

Reaction drive rotors do not require anti-torque devices for powered rotor flight conditions. However, additional means are required to achieve yaw control in hover and low speed flight when a vertical stabilizer is ineffective. When two engines are present, differential thrust or thrust vectoring may provide the necessary yaw control. If only one engine is available, additional yaw thrusters are required; thereby increasing somewhat the complexity of the vehicle system. The McDonnell XV-1 (Figure 5) had two yaw thrusters located at the trailing end of the tail booms. The Fairey Rotodyne had two turboprop engines, allowing each propeller to be feathered separately to produce differential thrust. Since no anti-torque is necessary, the power and complexity of yaw control in a hover is very small by comparison to what is needed for a helicopter. These yaw control aids are only needed in a hover, once forward flight is established. Thus, even though gyroplanes can takeoff and land vertically, yaw control is performed by rudders in the airflow created by engine thrust.

The thrust produced at the reaction drive nozzle is a function of the gas mass flow rate and the pressure ratio across the nozzle. Constraining the mass flow rate leads to an increase in the nozzle pressure ratio or the inclusion of a tip combustion chamber to generate the thrust required by the reaction drive nozzle. The result is high velocity gas exhausting from the reaction drive jet-nozzle ("V_J"). Since the sound power per unit volume is approximately proportional to the eighth power, high gas velocities are undesirable from an acoustic standpoint (Ref.7). The modification of the reaction drive nozzle to include an ejector nozzle has the potential to reduce the acoustic signature of the reaction drive rotor system by entraining free stream air with high speed exhaust gas, resulting in an increase in mass flow rate and a decrease in exhaust gas velocity for a given nozzle thrust (Ref. 7)

There are two more important issues natural to gyrodynes' use of tip-jets in mitigating noise. Tip-jet noise is high pitched which has a very short range – it significantly diminishes as distance from its source increases. And, transition from liftoff to forward flight takes only a few seconds. Once in forward flight, even at very slow speed, power needed for the rotor is dramatically reduced (as is so with all helicopters). Power reduction means lower V_J and since noise is a factor of V_J to the eight power, the V_J only has to be reduced a little in order for noise to reduce a lot.

II. REACTION DRIVE ROTORS – LESSONS LEARNED BOB HEAD

Hughes Helicopters Inc did a lot of innovative development and flight testing of reaction drive rotors in the 1960s. Mr. Bob Head, MDC Research and Engineering Fellow, MDHC, gave an excellent paper on Lessons Learned (Ref.4) which will summarized here

In regard to technology readiness, Mr. Head concludes that the technology is in hand for developing such a reaction drive vehicle and nothing new needs to be invented. However, the need for technology integration and demonstration should not be overlooked:

- **Engines**

- There are small engines ranging from rotary to small turboshafts that can be connected to axial or centrifugal compressors for generating the compressed air needed to drive the cold cycle rotor.

- There are a number of low-bypass-ratio turbofan engines that were designed for fighters, but could have their afterburners removed to power big warm-cycle tip-jet helicopters.

- **Materials**

- High-temperature composites and adhesives are available from which to create hub and blade structures that must withstand the thermal effects of the propulsion gases.

- High-efficiency insulation materials are available for protecting the structure from heat effects.

- High-temperature seal and bearing materials are available.

- **Aero/Thermodynamics**

- The technical aspects of creating jet thrust at the blade tips are well understood and have been verified in the programs in which tip-jet helicopters have flown.

Disadvantages of Reaction Drive Tip Jet Rotors

- **Noise**

Mr. Head's assessment of this disadvantage is somewhat optimistic:

"One might question the noise that a tip-jet rotor generates, but it was shown with the XV-9A Hot Cycle helicopter that this rotor with its three widely spaced sonic jets spouting from the blade tips was quieter than several shaft-drive helicopters that had about the same gross weight."

However, noise was a significant problem on the Fairey Rotodyne and was eventually reduced through the use of various types of ejectors on the tips. Impacts on efficiency wasn't adequately documented. For the DARPA HGDA Program Georgia Tech lead a Phase IB effort to complete a second PDR with a rotor that met the noise constraints.

- **Efficiency**

According to Mr. Head the reason that the sky is not full of tip jet helicopters has primarily to do with the aero/thermodynamic efficiency of the jet drive cycle.

Figure 7 shows that the absolutely best efficiency is 50 percent. From this starting point, the losses for moving the gas through the ducts the tip nozzles must be subtracted. In addition, an allowance for directional control must be included. This is not quite as bad as might be inferred at first glance because a comparable shaft-drive rotor also encounters significant losses, as depicted in Figure 8. In this figure it is seen that losses in the engine's power turbine, the transmission system, the anti-torque requirement, and the directional control result in an overall efficiency of about 70 percent for the shaft-drive machine as opposed to around 44 percent for the jet-drive helicopter. Hence, the best jet helicopter is only about 37 percent less efficient than the conventional machine, not 50 percent.

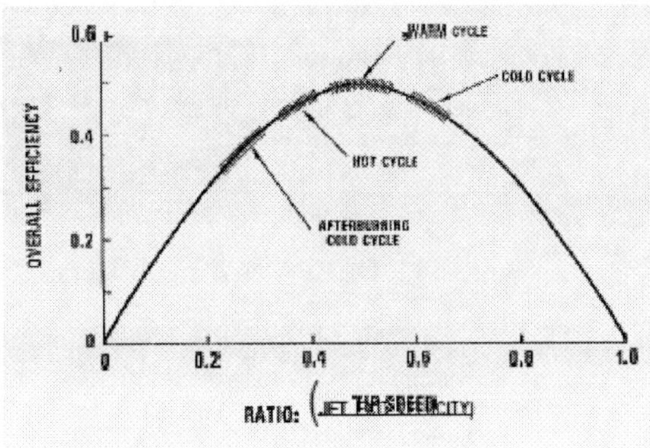

Figure 7. Tip Jet Rotor Efficiency

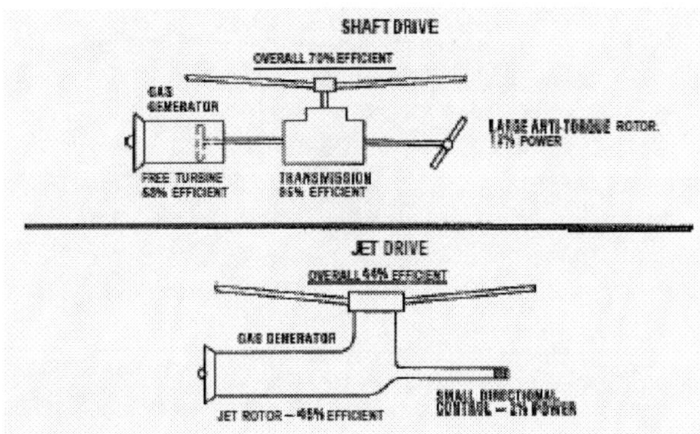

Figure 8. Tip Jet/Shaft-Drive Efficiency

Lessons Learned (Ref.4)

So what have **we** learned from jet helicopter experience that goes back more than 45 years? Mr. Head lists these as follows:

1. "Practical experience with flying prototypes has demonstrated that tip jet helicopters really can fly."
2. "There are a number of rotor drive cycles from which the designer may choose for powering the helicopter, but **only two are really viable - the cold cycle for small helicopters and the warm cycle for large helicopters.**"
3. "Suitable engines are available to power tip jet helicopters in any size range."
4. "There is a wide spectrum of modern structural and insulation materials from which the designer may make a choice."
5. "**Small jet helicopters can be designed with greatly simplified mechanical components in the rotor/drive system**, and the conventional tail rotor can be replaced by a simple exhaust deflector, thus improving maintainability and operational safety. Flying qualities improvements will be realized from the high-inertia main rotor and from eliminating the cross coupling effects of the tail rotor" (which is largely responsible for instability in helicopters – not so in tip-jet helicopters as they are stable enough to be hovered 'hands off').
6. "The **incorporation of a tip jet rotor is the only viable solution, technologically and economically, once the decision has been made to develop a large helicopter**. In the past, the problem has always been that the need for giant cranes for any foreseeable mission description was quite limited, and this would run their unit costs up very high."

So the scenario is: "Let's buy lots of smaller helicopters copters and make do with them, but don't spend money on big helicopters that would be likely to spend most of their life waiting around for suitable cargos." This has happened three times: in the mid-1950s to the XH-28 follow-on to the Hughes Tool XH-17; in the early 1960s to the XHCH-1, which McDonnell Aircraft was developing; and in the late 1970s to Boeing's XCH-62, which was sent to a museum. The only large helicopters that made it into production are the Russian Mil Mi-12 and Mi-26.

As part of NASA's Heavy Lift Rotorcraft System (HLRS) program (Ref 5), Georgia Tech and the University of Maryland were both given studies in 2004-2005 to investigate different rotorcraft concepts and to assess shaft drive and tip drive for slowed rotor compounds as viable alternatives for more attention. There were two questions given by NASA were: that needed to be answered (to the extent possible with the available resources):

- Consider single main rotor, tandem main rotor, and coaxial main rotor configurations for the **slowed-rotor compound** – are there differences significant enough to affect the choice of configurations?
- Consider **shaft drive and tip drive** (single, tandem, and coaxial) for the slowed-rotor compound – **should more attention be paid to tip drive**, i.e. are the problems such as the aerodynamic efficiency and noise solvable with advanced technology.

Both the Georgia Tech and UMD studies showed that the slowed rotor compound tip driven configurations ended up with the lightest weight solutions. The Georgia Tech comparison is illustrated in Figure 9. (Ref.5). It is felt the Slowed Rotor Compound Reaction Drive would be less complex, lighter weight, and generate less cost risk than the ~150,000lbs tilt rotor that the government kept pursuing..

	Tandem	Reaction	Powered SMR	Coax
GW	116,475	88,326	120,408	119,708
Total Power Required (at cruise condition)	10,895	8,693	9,659	9,909
L/D _effective	11.48	10.91	13.39	12.97
EW	79,828	53,273	84,182	83,428
EW/GW	0.69	0.60	0.70	0.70

- Assumption Review
  - DL = 10
  - AR = 10
  - WL = 100
  - $V_{Tip_}$cruise = 400 (ft/s) (μ=1.48)
  - Advance hub drag reduction (-50%)
  - V _cruise = 350 kts
  - Army/NASA kμ approximation

Figure 9. Results of Georgia Tech NASA HLRS Study

III. DARPA RESEARCH ON SLOWED/STOPPED ROTOR RESEARCH W/AND W/O REACTION DRIVE PROGRAMS

Slowed Rotor Compound Concepts offer substantial improvements over conventional helicopters with mostly constant rotor speeds in hover and at cruise speeds. Several research programs sponsored by the DARPA Tactical Technology Office (TTO) have initiated both slowed rotor and combined slowed and stopped rotors concepts, as illustrated in Figure 10. (Ref 6) A summary of each is provided.

Figure 10. DARPA Rotorcraft Research Programs.

The previously mentioned X-Wing Program was initially funded by DARPA and is described in Figure 11 The Canard Rotor Wing (Wing) was an unmanned stopped rotor reaction drive project intended to demonstrate the first stopped rotor in a flight demonstration. The objectives and description of the CRW Program is illustrated in Figures 12 and 13.

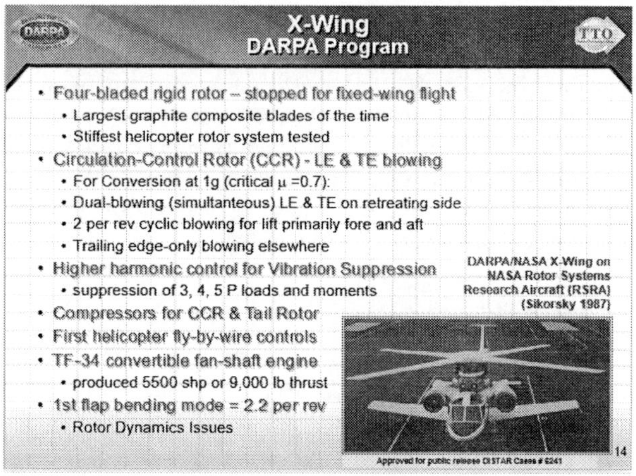

Figure 11. DARPA X-Wing Program

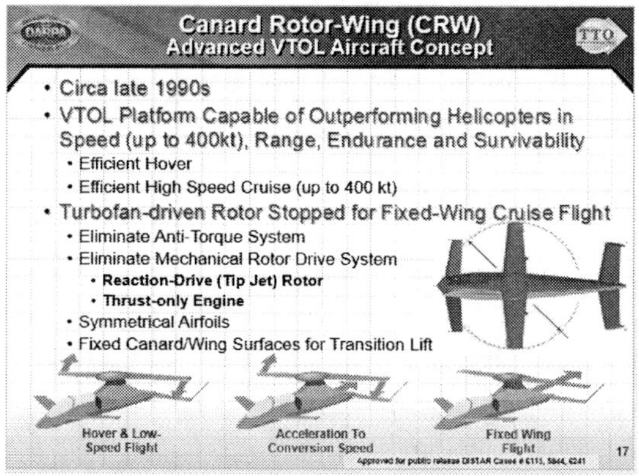

Figure 12. DARPA CRW Program Concept

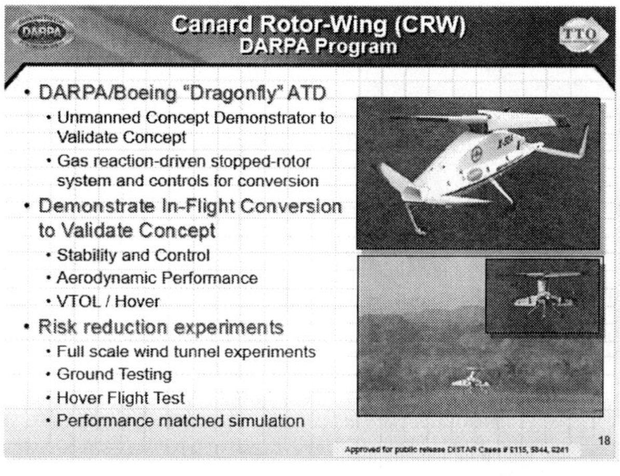

Figure 13. CRW DARPA Program

The DARPA Unmanned Combat Armed Rotorcraft (UCAR) was autonomy oriented and never led to flight testing. It is summarized in Figures 14 and 15.

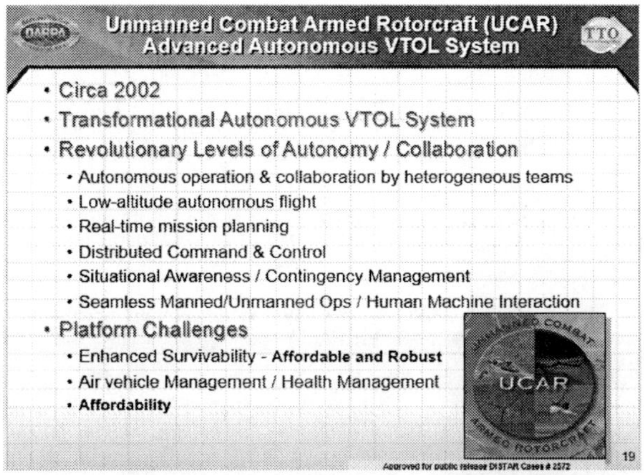

Figure 13. UCAR Program Objectives

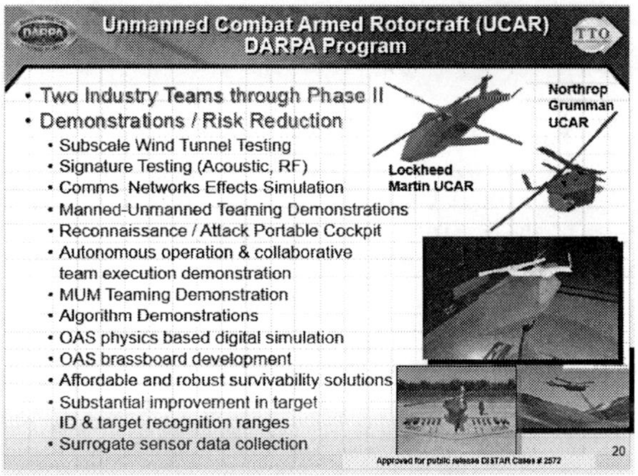

Figure 14. UCAR Program Elements

The A-160 Hummingbird Program, illustrated in Figures 15 and 16, was aimed at long endurance unmanned helicopters and resulted in successful flight testing; although no production.

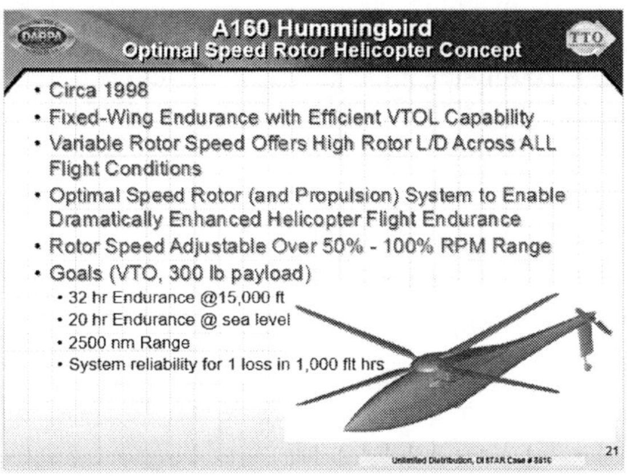

Figure 15. A160 Hummingbird Goals

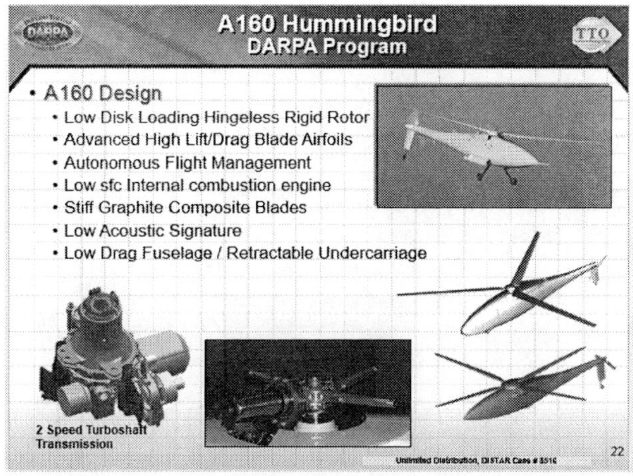

Figure 16. A160 Hummingbird Program Elements

The Heliplane Program was an aggressive four phase program intended to demonstrate in flight a slowed rotor reaction drive to meet the requirements illustrated in Figure 17. It was terminated after Phase IB by Congress, along with the Disc Rotor. Groen Bros Aviation was prime for Phase I. Georgia Tech was the prime for Phase IB to illustrate through preliminary design a noise solution. (Ref. 7)

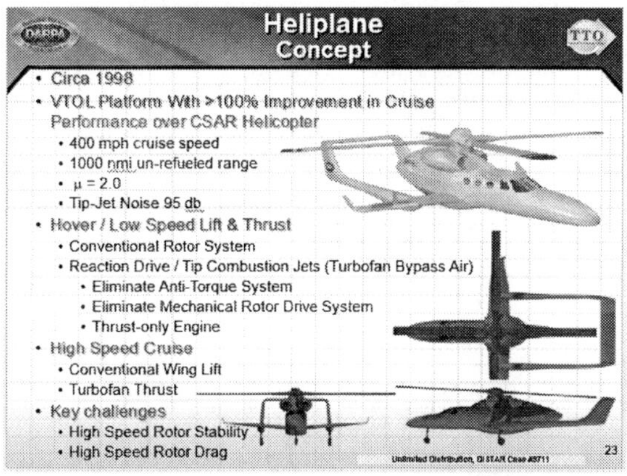

Figure 17. Heliplane Program Goals

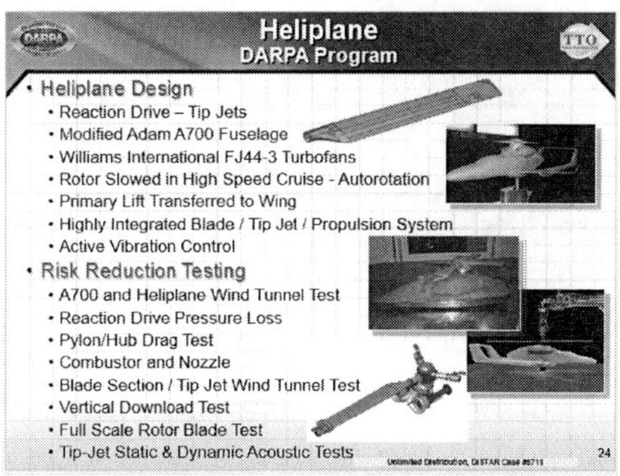

Figure 18. Heliplane Program Elements

OTHER STOPPED/STORED RECONFIGURABLE ROTOR CONCEPTS

The **NASA M-85 Stopped Rotor Wing,** illustrated in Figure 19, used reaction drive and circulation control for hover and low speed flight. It was a NASA internal design for their High Speed Rotorcraft Concept (HSRC) Program. (Ref 8)

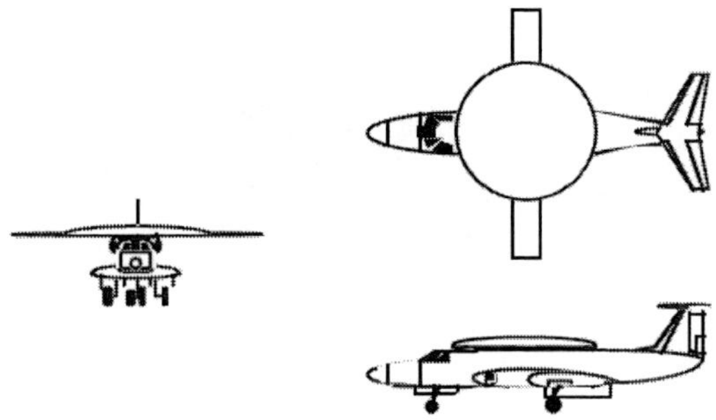

Figure 19. NASA/GT M-85 HSRC Concept

Design of a hybrid reaction drive rotorcraft requires a strongly coupled propulsion-rotor sizing and performance formulation. Since NASA did not have the internal sizing and performance tools for designing and analyzing reaction drive-circulation control Georgia Tech was given a contract to develop these tools which resulted in the PhD Thesis of Dr. Jimmy Tai. Dr. Tai, who was also an important member of the Georgia Tech team during the DARPA Heliplane Program. (Ref. 8)

A follow-on DARPA/Boeing Program somewhat similar to the NASA M-85 Stopped Rotor Wing concept was the DARPA Boeing Rotor Disc Program, illustrated in Figure 20. (Ref.9)

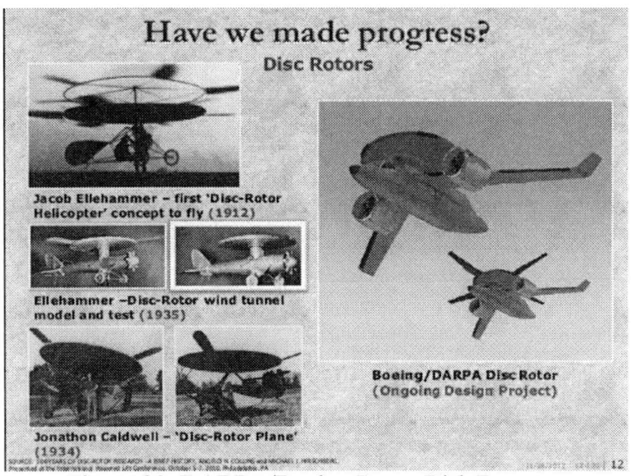

Figure 20. DARPA/Boeing Rotor Disk Program

The Disc-Rotor was a collaboration between DARPA and Boeing, hoping to marry the best features of a helicopter and an airplane, the Disc-Rotor program aimed to develop a new type of aircraft, capable of seamlessly transitioning from hovering like a helicopter to flying like a plane. (Ref 9)

.IV. DOD CLASSIFICATION OF UAS AND NOTIONAL ARMY REQUIREMENTS

The Department of Defense (DOD) Classification of Unmanned Aerial Systems (UAS) is provided in Table 1. (Ref 10)

Table 1: UAVs Classification according to the US Department of Defense (DoD)

Category	Size	Maximum Gross Takeoff Weight (MGTW) (lbs)	Normal Operating Altitude (ft)	Airspeed (knots)
Group 1	Small	0-20	<1,200 AGL*	<100
Group 2	Medium	21-55	<3,500	<250
Group 3	Large	<1320	<18,000 MSL**	<250
Group 4	Larger	>1320	<18,000 MSL	Any airspeed
Group 5	Largest	>1320	>18,000	Any airspeed

*AGL = Above Ground Level
**MSL = Mean Sea Level
Note: If the UAS has even one characteristic of the next level, it is classified in that level.
Source: "Eyes of the Army" U.S. Army Roadmap for UAS 2010-2035

As can be seen Categories range from Group 1 to 5, Gross Weights from 0-20 lbs MGTW to > 1320 lbs to Normal Operating Altitudes from <1,200 AGL to >18,000ft to Airspeeds from <100 to <250kts for Groups 1 to 3 and Any Airspeed for Groups 4 and 5. Recently, the Army Research Laboratory (ARL) hosted an Unmanned VTOL Propulsion Workshop at the Aberdeen Proving Ground, MD on 12 December 2018. (Ref.11). It was focused on ARL propulsion system research studies examining propulsion system configurations for notional unmanned VTOL vehicles in the 100 to 1,000lbs payload class. The research focused on propulsion configurations including:

- Batteries
- Reciprocating Engines
- Turboshaft Engines
- Electric Motors
- Electric Generators
- Hybrid Configurations
- Scalability of propulsion configuration technologies across this vehicle size class

A figure illustrating the spectrum of Useful Load versus Useful Load Weight Fraction and a variety of UAS systems is included in Figure 21.

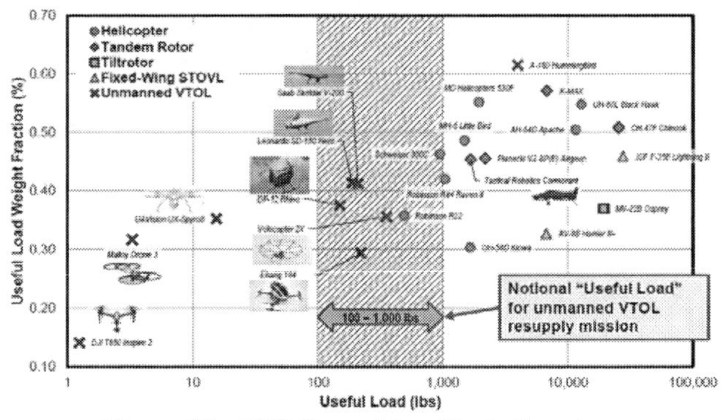

Figure 21. ARL Propulsion Study Spectrum

The challenges in the ARL Propulsion Study were identified as follow:

- Unmanned VTOL in the 100-1,000 lbs. payload class is an enabler for new military mission capabilities.
- Consideration of vehicle platform architecture prompts study of the merits and technical challenges associated with
  - Scaling-down manned rotorcraft propulsion configurations
  - Scaling-up unmanned quadrotor propulsion configurations

Objective of optimizing the propulsion system and vehicle system in the context of multiple parameters

- Endurance, speed, range, maneuverability
- Operation in complex, cluttered environments, ...

These pose technology challenges and opportunities for innovation in propulsion system arrangements and components.

These technical challenges were converted into the specific notional requirements for the ARL Propulsion Study as described below.

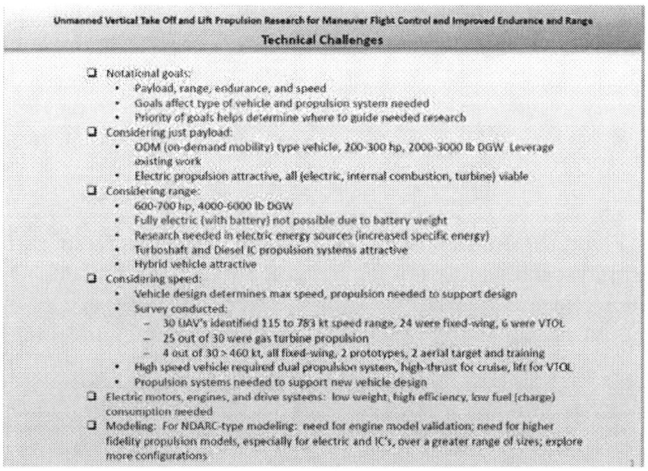
Unmanned Vertical Take Off and Lift Propulsion Research for Maneuver Flight Control and Improved Endurance and Range

Technical Challenges

- ❑ Notational goals:
  - Payload, range, endurance, and speed
  - Goals affect type of vehicle and propulsion system needed
  - Priority of goals helps determine where to guide needed research
- ❑ Considering just payload:
  - ODM (on-demand mobility) type vehicle, 200-300 hp, 2000-3000 lb DGW Leverage existing work
  - Electric propulsion attractive, all (electric, internal combustion, turbine) viable
- ❑ Considering range:
  - 600-700 hp, 4000-6000 lb DGW
  - Fully electric (with battery) not possible due to battery weight
  - Research needed in electric energy sources (increased specific energy)
  - Turboshaft and Diesel IC propulsion systems attractive
  - Hybrid vehicle attractive
- ❑ Considering speed:
  - Vehicle design determines max speed, propulsion needed to support design
  - Survey conducted:
    - 30 UAV's identified 115 to 783 kt speed range, 24 were fixed-wing, 6 were VTOL
    - 25 out of 30 were gas turbine propulsion
    - 4 out of 30 > 460 kt, all fixed-wing, 2 prototypes, 2 aerial target and training
  - High speed vehicle required dual propulsion system, high-thrust for cruise, lift for VTOL
  - Propulsion systems needed to support new vehicle design
- ❑ Electric motors, engines, and drive systems: low weight, high efficiency, low fuel (charge) consumption needed
- ❑ Modeling: For NDARC-type modeling: need for engine model validation; need for higher fidelity propulsion models, especially for electric and IC's, over a greater range of sizes; explore more configurations

These Technical Challenge goals were transmitted for evaluation in a generic UAS cargo resupply mission, as illustrated in Figure 22. The ARL Vehicle Technology Directorate (VTD plan was to evaluate different, VTOL concepts and propulsion systems using the notional target goals in Figure 23. However, a mapping of the speed goal to current UAS in the inventory, as illustrated in Figure 24, shows few if any current UAS can meet the 460kts, so high speed >250kts is not being evaluated in the ARL Study.

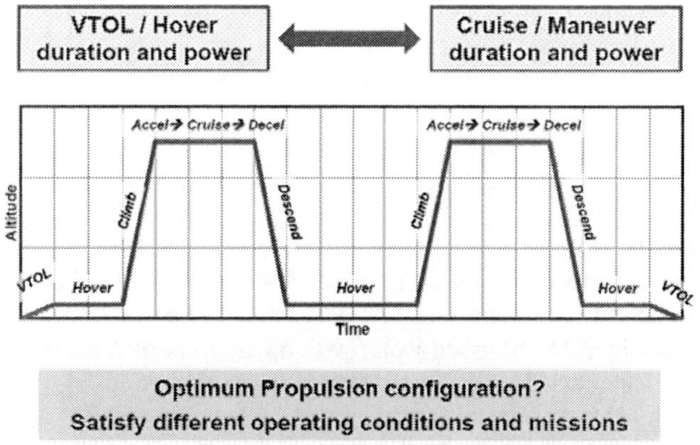

Figure 22. Generic UAS Cargo Resupply Mission

Payload (lb)	100 - 1,000
Speed (kt)	460
Range (mi)	620
Endurance (hr)	4

Figure 23. VTD UAV VTOL Resupply Project Notational Goals

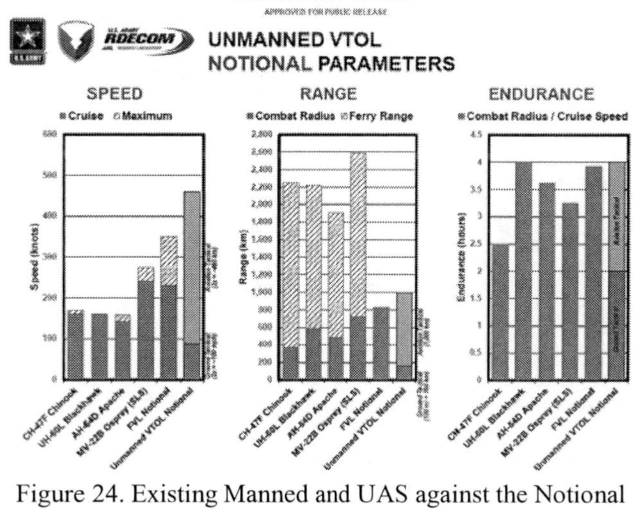

Figure 24. Existing Manned and UAS against the Notional Parameters

In considering speed:

• Vehicle design determines max speed, propulsion needed to support design

• Survey conducted:

☐ 30 UAV's identified 115 to 783 kt speed range, 24 were fixed-wing, 6 were VTOL

☐ 25 out of 30 were gas turbine propulsion

☐ 4 out of 30 > 460 kt, all fixed-wing, 2 prototypes, 2 aerial target and training

• High speed vehicle required dual propulsion system, high-thrust for cruise, lift for VTOL
• Propulsion systems needed to support new vehicle design

In summary it can be seen that considerable more vehicle VTOL concepts need to be identified to address a speed requirement of 460kts for Category 4 and 5 in Table 1. These VTOL vehicle concepts will undoubtedly require stopped or stored rotors. The next section will review the results of the Georgia Tech AHS Graduate Team SDC Reconfigurable Stopped Rotor with a reaction drive system for a Category 3 UAS. It is believed it can be a good baseline starting point for the UAS Categories 4 and 5 conceptual and preliminary design study efforts to include the 460kts speed notional goal if it can be achieved.

V. THE GEORGIA TECH *KNIGHT FLYER*: THE GRADUATE STUDENT ENTRY IN THE 35TH AHS SDC

The Georgia Tech graduate program in Aerospace Systems Design began in the 1980s with a focus on rotorcraft, expanded in the 1990s to fixed wing aircraft and spacecraft. It has been based on an Integrated Product and Process Development (IPPD) Methodology developed by the principal author in the early 1990s. The Practice-Oriented MS degree with IPPD Methods/Techniques is illustrated in Figure 25. (Ref 13)

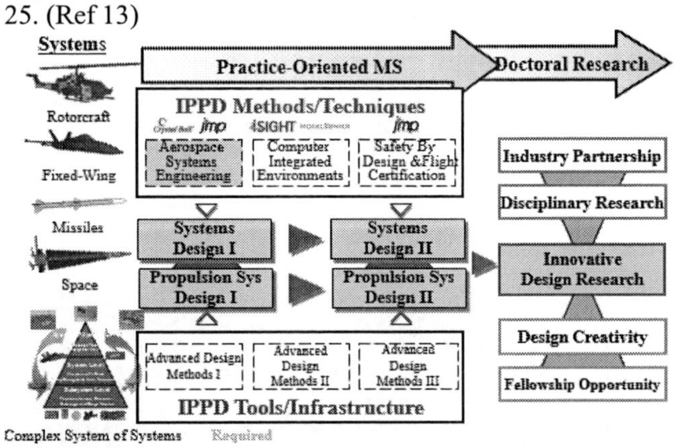

Figure 25. Courses in GT Practice-Oriented MS Program

An illustration of IPPD for rotorcraft preliminary design and the graduate courses involved in implementation are illustrated in Figure 26. Six courses are identified, although few students take all six, especially in recent years where courses for PhD Qualification Exams take a higher priority than participation in student design competitions for many students.
An illustration of how several of these courses supported the Georgia Tech Entry in the AHS 35th SDC will be presented. This led to the conceptual selection of a reconfigurable stopped rotor reaction drive propulsion aircraft, The Knight Flyer, which was then continued for a preliminary design proposal.

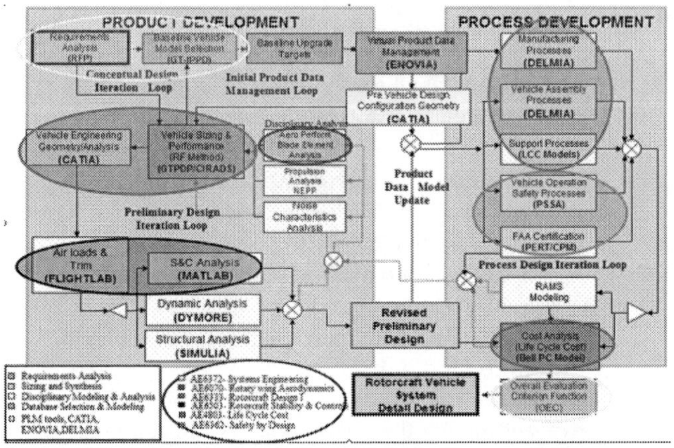

Figure 26. IPPD Framework for Rotorcraft Preliminary Design and Six Supporting Courses

AE 6333 Rotorcraft Design I: Conceptual Design
As can be seen in Figure 26 Vehicle Sizing & Performance analysis is conducted using the Fuel and Power Balance, R_F Method, which illustrated in Figure 27.

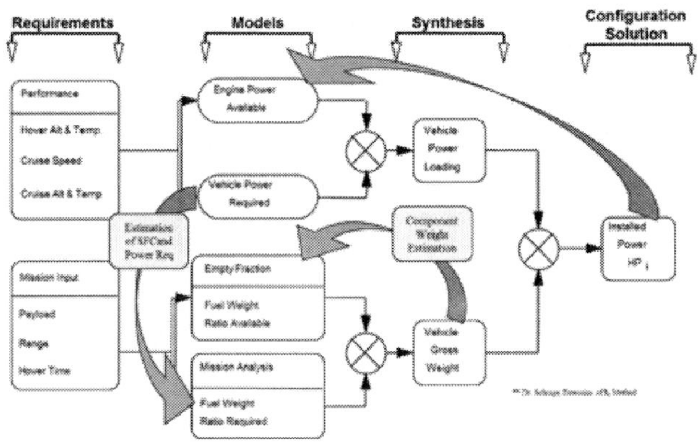

Figure 27. R_F Method for Rotorcraft Conceptual Design
A representative mission for sizing feasible reconfigurable rotor concepts was derived from the AE6372 Aerospace Systems Engineering Course and is illustrated in Figure 28.

Phase	0	1	2	3	4	5
Type	Run-up	Take-off	Accel	Climb	Dash	Desc
Speed (kts)	0	0	$0 \rightarrow V_{tr}$	150	180	—
Alt (ft)	0	$0 \rightarrow 300$	300	$300 \rightarrow 10k$	10k	$10k \rightarrow 5k$
Phase	6	7	8	9	10	11
Type	Loiter	Climb	Dash	Desc	Land	Idle
Speed (kts)	V_{be}	150	180	—	0	0
Alt (ft)	5k	$5k \rightarrow 10k$	10k	$10k \rightarrow 300$	$300 \rightarrow 0$	0

Figure 28. Representative Sizing Mission

Three viable reconfigurable rotorcraft concepts meeting the AHS 35th SDC Rules are illustrated in Figure 26.

Figure 29. Three viable Reconfigurable Concepts

The sizing and power required for the first concept, the Tip-Jet Stopped Rotor Wing (TJSRW) is illustrated in Figure 30.

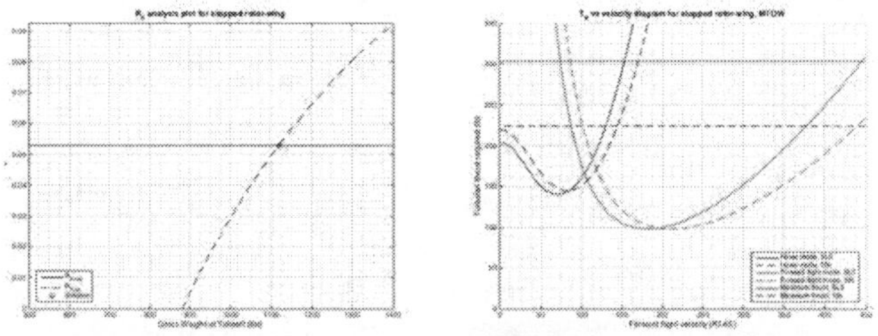

Figure 30. Sizing & Power Required for the TJSRW

A Performance Summary for the TJSRW are in Figure 31.

Performance		
Spec	Units	Value
Hover endurance @ SLS, half fuel	hr	0.48
Hover endurance @ 10k ft, half fuel	hr	0.45
Best range speed @ SLS	kts	239
Best range speed @ 10k ft	kts	278
Cruise range @ SLS, half fuel	nmi	208.04
Cruise range @ 10k ft, half fuel	nmi	242.07
Dash speed @ SLS	kts	446
Dash speed @ 10k ft	kts	440

Figure 31. Performance Summary for the TJSRW

The sizing and power required for the second concept, the Distributed Electric Propulsion Stored Rotor (DEPSR) is illustrated in Figure 32.

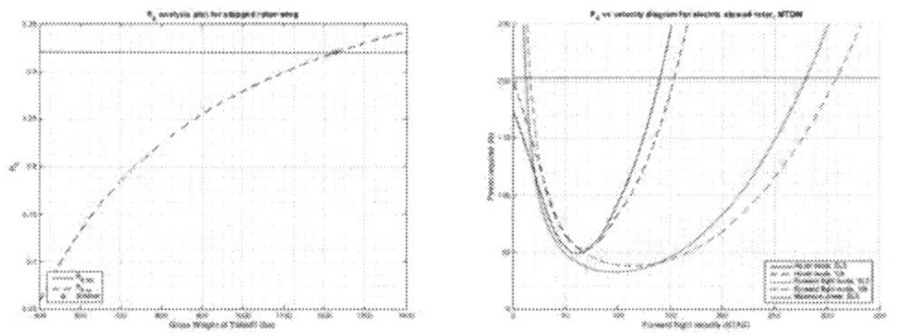

Figure 32. Sizing & Power Required for the DEPSR

A Performance Summary for the DEPSR is in Figure 33.

Performance		
Spec	Units	Value
Hover endurance @ SLS, half fuel	hr	0.28
Hover endurance @ 10k ft, half fuel	hr	0.24
Best range speed @ SLS	kts	128
Best range speed @ 10k ft	kts	149
Cruise range @ SLS, half fuel	nmi	164.24
Cruise range @ 10k ft, half fuel	nmi	164.24
Dash speed @ SLS	kts	281
Dash speed @ 10k ft	kts	309

Figure 33. A Performance Summary for the DEPSR

The sizing and power required for the third concept, the Disc Rotor (DR) is illustrated in Figure 34.

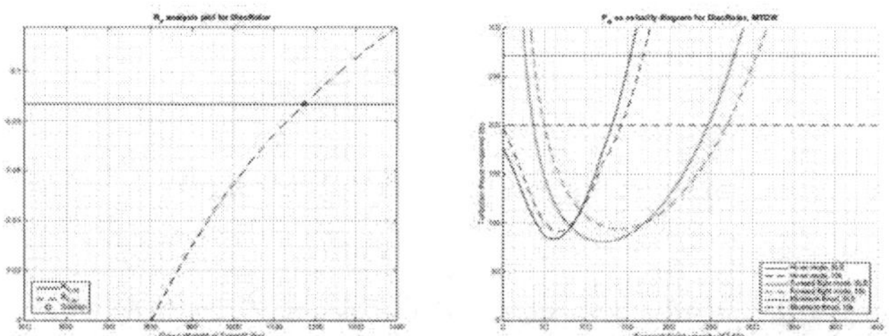

Figure 34. Sizing & Power Required for the DR

A Performance Summary for the DR are in Figure 35.

Performance		
Spec	Units	Value
Hover endurance @ SLS, half fuel	hr	0.48
Hover endurance @ 10k ft, half fuel	hr	0.43
Best range speed @ SLS	kts	159
Best range speed @ 10k ft	kts	185
Cruise range @ SLS, half fuel	nmi	147.09
Cruise range @ 10k ft, half fuel	nmi	147.09
Dash speed @ SLS	kts	279
Dash speed @ 10k ft	kts	270

Figure 35. Performance Summary for the DR

A Multi-Attribute Decision Making (MADM) tool, TOPSIS (Ref 14), was used to do a quantitative assessment of the three concepts. The results are shown in Figure 36 with the TJSRW concept selected.

Results	
Concept	Score
TJ SRW	0.8065
DEP SR	0.3839
DiscRotor	0.2630

- **Tip-jet stopped rotor-wing** concept beats out all others in every category except complexity
  - Consistently ranked first under different weighting scenarios
- Innovative concept with well-documented baseline that presents many opportunities for improvement
  - More robust engine, improved control system, reversible airfoil, etc.
- Rotor and wing are the same physical object
  - Reduces weight and satisfies AHS requirements
- Same propulsion system for hover and forward flight
- No bulky gearbox

Figure 36. TJSRW Chosen as the Best Concept

Typically in the past a 3 View Drawing was sufficient; however today requires a 3D CAD Model as illustrated in Figure 37. Also, External and Internal Views, Figures 38 and 39 are also required for determining weight and balance. For innovative new UAS concepts weight and inertia buildup needs to determine the degrees of freedom and control devices, as illustrated in Figure 40, to calculate the Air Loads and Trim to initiate the Preliminary Design Iteration as illustrated in Figure 26.

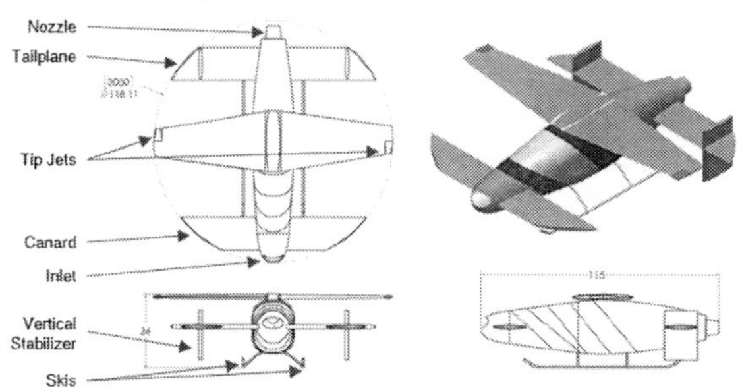

Figure 37. TJSRW 3D View and CAD Model

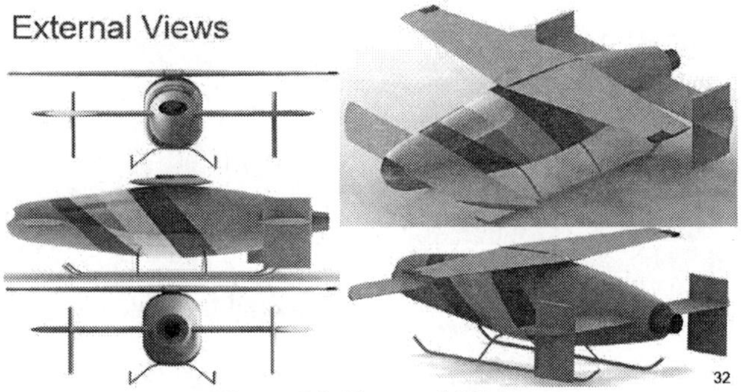

Figure 38. External Views

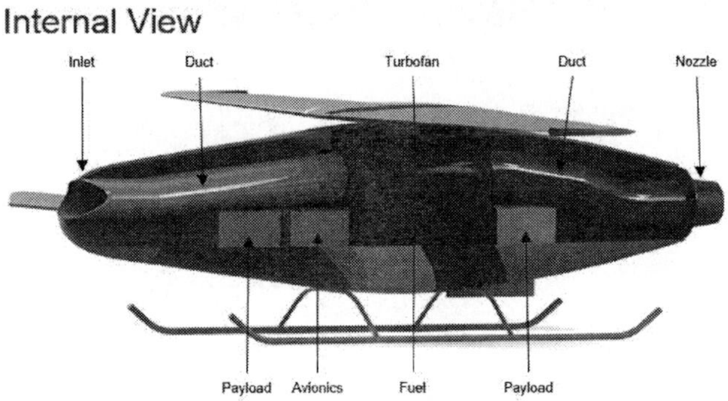

Figure 39. Internal View

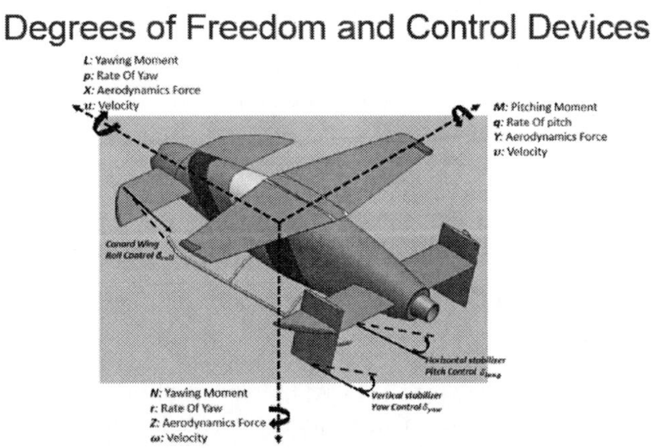

Figure 40. Weights & Inertias Required

The next Product Development iteration is the Preliminary Design Iteration Loop illustrated in Figure 26 which will be summarized.

AE 6334 Rotorcraft Design II: Preliminary Design

To conduct the Preliminary Design Iteration required an integrated design environment with the necessary analysis tools. Illustrated in Figure 41 is a description of the critical analysis tools used for the preliminary design of the Knight Flyer. (Ref 14).

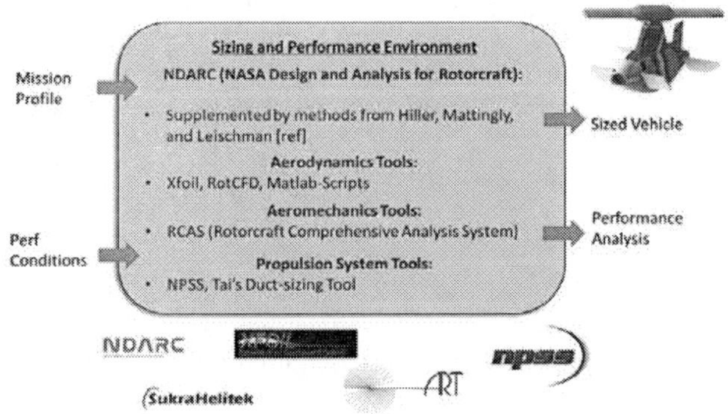

Figure 41. *Knight Flyer* Integrated Design Environment

The conceptual design was updated with a higher fidelity Sizing and Performance Environment including trying to use the NASA Design and Analysis for Rotorcraft (NDARC), supplemented by methods from Hiller, Mattingly, and Leischman. Additional Aerodynamics Tools, i.e. Xfoil, RotCFD and Matlab Scripts, were also used along with Aeromechanics Tools, such as the Rotorcraft Comprehensive Analysis System (RCAS). However, the most critical was the Propulsion System Tools, such as the NASA Propulsion System Simulation (NPSS) coupled with Dr. Jimmy Tai's Duct-sizing Tool. Propulsion Selection and Analysis is critical for reaction drive CRW concepts. Two challenges particular to development of the CRW configuration are coupling between the rotor and power plant and controllability during transition. Each challenge can be addressed through propulsion system design in both hardware innovation and design methodology.

The design approach utilized sized the vehicle using two modules for turbine engine cycle analysis and duct loss calculations. An overview of the Reaction Drive Analysis is presented in Figure 42 below. NASA Design and Analysis of Rotorcraft Theoretical Basis and Architecture (NDARC) and constraints specified in the RFP inform a basic sizing of the vehicle. This provides an initial estimate of the rotor torque required and constrains the geometry of the ducting.

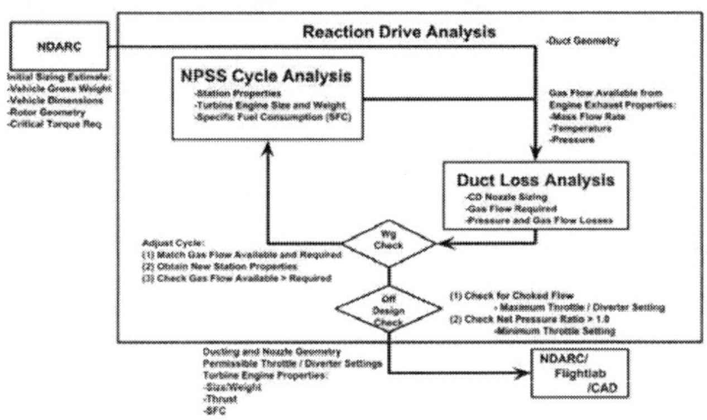

Figure 42. *Knight Rider* Reaction Drive Analysis Modules

The first module of the analysis is the Numerical Propulsion System Simulation (NPSS) which provides turbine engine station properties: mass flow rate, temperature, pressure, Mach number and fuel air ratio (far). These properties and the vehicle sizing estimate feed into the Duct Loss Analysis module. The Duct Loss analysis module calculates flow loses in the system and the resulting gas flow required to fly the vehicle. This gas flow can be compared directly to the NPSS station properties. The NPSS cycle analysis is adjusted manually to meet the gas flow required while minimizing size, weight, and specific fuel consumption. The new station properties are run through the Duct Loss module to confirm that the gas flow is still sufficient after losses. Finally, the chosen NPSS cycle provides thrust and SFC to sizing and performance codes while confirming the size/weight estimate for the power plant. The Duct Loss analysis also assists in sizing the ducting geometry by providing a cross sectional area for a given input Mach number. The results of this design approach are illustrated in Figure 43 as an internal CAD view of the Knight Flyer. Key features are illustrated in Figure 44 More details are in the 35th AHS SDC Georgia Tech Knight Flyer Proposal. (Ref. 12)

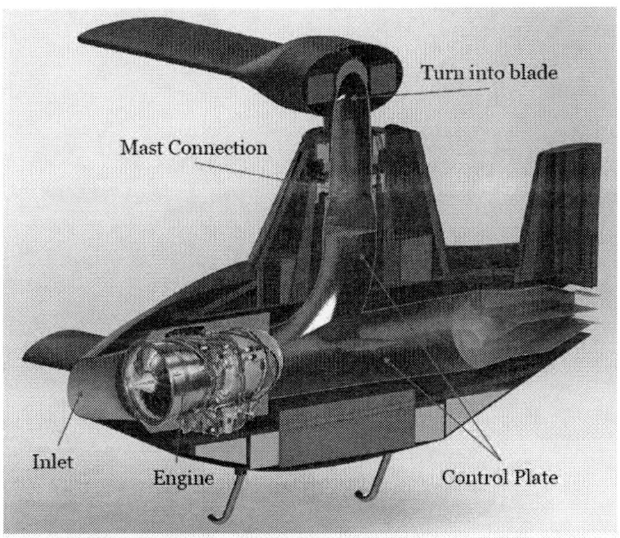

Figure 43. Internal CAD View of Knight Flyer

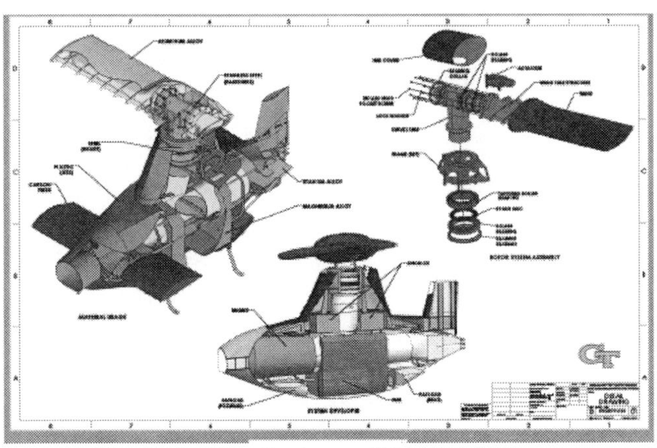

Figure 44. Key Features of the Knight Flyer

The key feature of the Knight Flyer is a unique reaction drive propulsion system that, through the use of control plates, can toggle between forward and vertical flight modes. This allows the vehicle to both hover and achieve high forward flight speeds, the key advantage of this vehicle. It included the design of the nozzle based on the 3D design of an Aerial Coanda high-efficiency orienting jet nozzle (ACHEON) (Ref 15) which can overcome some of the transition control problems encountered with the Boeing XV-50A CRW. The Knight Flyer Tip-Jet Stopped Rotor Wing (TJSRW) concept was motivated by the Boeing XV-50A CRW Dragonfly, illustrated in Figure 45, which came close to becoming the first stopped rotor demonstrated in flight. (Ref.16).

Figure 45. DARPA-Boeing XV-50 Dragonfly CRW

CONCLUSIONS

The most important results of the paper are summarized as follows:

1. Reconfigurable rotorcraft with a reaction drive system can provide a viable solution for combining high efficiency hover with high speed forward flight
2. Reconfigurable rotorcraft with reaction drives may be the best vehicle configurations for the Army NGTUAS and to satisfy the Department of Defense (DoD) UAS Category 3, 4 and 5 VTOL UAS requirements, e.g. max gross weight, operating altitude and speed.
3. Designing and analyzing reconfigurable rotorcraft with reaction drive is a formidable, multidisciplinary design, analysis and optimization (MDAO) task, as evident by the Georgia Tech Graduate Team 35th AHS SDC Proposal.

Author contact: Daniel P. Schrage, daniel.schrage@ae.gatech.edu.

ACKNOWLEDGMENTS

The authors of this paper would like to acknowledge and thank all the other \student team members and contributors to the Georgia Tech 35th AHS SDC Graduate Team

REFERENCES

1. 2017-2018 American Helicopter Society, 35th Annual Student Design Competition, "A Reconfigurable Rotorcraft", sponsored by the Army Research Laboratory (ARL).
2. Scott, R. "Knight, Dutton, Castles, and the Early Rotorcraft Researchers at Georgia Tech", Vertical Flight Society, Forum 68, May 2012.
3. Schrage, D. P., "*What Goes Around, Comes Around-State of the Art Assessment of Reaction Drive, Slowed Rotor, Sustained Autorotative Flight Compound Aircraft*", White Paper, Georgia Institute of Technology, 2015.
4. Head, Bob, "*Reaction Drive Rotors – Lessons Learned*", AIAA 92-4270. Presented at the AIAA Aircraft Design Conference, Hilton Head, S.C. August 22, 1992.
5. Schrage, D. "Reaction Drive Concept Evaluation for HLRS", Georgia Tech, NASA Heavy Lift Rotorcraft Systems Investigation 2004-2005.
6. Newman, D., "DARPA Rotorcraft Round Table – Advanced Rotor Technology Workshop", 12-13 February 2008, Arlington, VA.
7. Moodie, A.M., Schrage, D.P., and Sankar, L.N., "Sizing and Performance Prediction Method for Rotary Wing Ejector Nozzles", Journal of Aircraft, December 2008.
8. Tai, J., "A Multidisciplinary Design Approach to Size Stopped Rotor/Wing Configurations Using Reaction Drive and Circulation Control", Ph.D Thesis, Georgia Institute of Technology, August 1998.
9. Noe, N, "DARPA/Boeing's Disc Rotor – A Copter with retractable rotors", October 26, 2010, Core77 Design Directory
10. Department of Defense (DOD) Classifications of UAS, Wikipedia, Updated 1 November 2018.
11. Army Research Laboratory Unmanned VTOL Propulsion Workshop, 12 December 2018, Aberdeen Proving Ground, MD
12. Georgia Tech Graduate Student Proposal for AHS 35th SDC Competition for a Reconfigurable Rotorcraft, May 2018.
13. Schrage, D. P. "Technology for Rotorcraft Affordability Through Integrated Product/Process Development (IPPD)", AHS Nikolsky Lecture, May 26, 1999, Montreal, Canada.
14. Alman, D, Robinson, J., Cai, T, Crouch, C., Guo, D. Hayhajnh, M. and Khamvilai, T., "Concept Exploration and Selection for a Reconfigurable VTOL Aircraft", Team#5, AE6333: Rotorcraft Design I, Fall 2017.
15. Das, S. S., Páscoa, J. C., Trancossi, M., & Dumas, A, "A. Computational Fluid Dynamic Study on a Novel Propulsive Concept". *Journal of Aerospace Engineering*. 2015.
16. Boeing X-50 Dragonfly, Wikipedia, last edit November 2018.

CFD-CSD Coupled Aeroelastic Analysis of a Highly Flexible Flapping Wing in Hover

Xuan Yang
Graduate Research Assistant
Department of Aerospace Engineering
Texas A&M University

Moble Benedict
Assistant Professor
Department of Aerospace Engineering
Texas A&M University

ABSTRACT

Flapping wings are typically very flexible structures undergoing large deflections, leading to complex nonlinear interactions between aerodynamics and structure. This strong fluid-structure coupling and the complicated flow physics make the performance of such systems extremely difficult to predict through numerical simulations. The challenges arise from the highly unsteady, viscous and vortex-dominated nature of the flowfield leading to strong nonlinearities in aerodynamic loads as well as the accurate representation of the structural dynamics of the flapping-wing. This paper presents a computational framework, where a computational fluid dynamics (CFD) solver is coupled with a nonlinear structural solver to simulate highly flexible flapping wings with significantly improved fidelity. The CFD solver utilized is a compressible Reynolds-averaged Navier–Stokes solver to resolve the complex, highly vortical, three-dimensional flow. The CSD solver accounts for the geometric nonlinearities due to large rotations, in-plane strain, initial curvature and transverse shear deformations and encapsulates the exact deformed shape. A multivariate interpolation scheme, using radial basis functions, is presented, which results in a unified formulation for the fluidstructure interpolation and mesh motion problems. The objective of the study is to understand the unsteady aerodynamic mechanisms that affect the force production on a flexible wing undergoing a hummingbird-like flapping kinematics.

NOTATION

b	Wing span
c	Mean chord length
AR	Aspect ratio of the wing
L,D	Lift, Drag
T	Stroke period
V	Wing tip mean velocity
Ma	Mach number of wing tip
Re	Reynolds number of wing tip
B	Jaumann strain tensor
J	Jaumann stress tensor
D	Displacement vector
K	Stiffness matrix
K_1^0	Initial curvature matrix
K_1	Curvature matrix
N	Shape function Matrix
$\{q\}$	Element displacement vector
T	Transformation matrix
γ_6	In-plane shear rotation angle
γ_5	Out-plane shear rotation angle

2019 Autonomous VTOL Technical Meeting and Electric VTOL Symposium, & 8th Biennial Autonomous VTOL Technical Meeting 6th Annual Electric VTOL Symposium, Mesa, Arizona, USA, Jan. 29-31, 2019

INTRODUCTION

There has been increased interest in understanding biological flight because of the advent of a new class of small, highly maneuverable aircraft called Micro Air Vehicles (MAVs). Over the last two decades, remarkable theoretical and experimental developments have been achieved in this field. However, there is still a disconnect between understanding biological flight and its application to MAVs. The primary reason for this is the fact that it is not trivial to transition many of the key insights we have obtained on natural flight in a meaningful way to improve the flight performance of MAVs. The first step towards this is to formulate design rules and performance analysis tools based on a principled understanding of flapping flight.

Some initial studies on 3D flapping wings at MAV-scale were conducted by Dickinson et al. (Ref. 1) as well as Birch and Dickinson (Ref. 2). Both studies focused on studying a dynamically-scaled Drosophila wing flapping in mineral oil. Aerodynamic force and flowfield measurements helped to improve the understanding of some of the underlying unsteady mechanisms on flapping wings. The leading edge vortex generated during dynamic stall was shown to augment the lift force generated. More recently, Benedict et al. (Ref. 3) performed instantaneous force measurements and flowfield studies on a rigid wing with a bio-inspired planform undergoing prescribed insect-like flapping kinematics with passive pitch rotations. While only limited flowfield experiments were performed, aerodynamic force measurements were obtained for a set of pitch angle variations at a constant flap frequency of 5 Hz. The main objective of the study was to validate the applicability of an inertial force subtraction method in air on a wing with passive wing pitching similar to what would be seen on a flexible flapping wing.

CFD simulation models can help provide a good understanding of the unsteady flow mechanisms and insight into the critical design parameters. The computational approach is to calculate the flow variables by solving the governing equations and associated boundary conditions with a flapping wing numerically. In the last decade, researchers have taken advantage of the advancements in computing facilities and developments in computational techniques to develop solvers that can address the insect flights with varying degrees of success. Sun and Tang (Ref. 4) and Ramamurti and Sandberg (Ref. 5) performed 3D CFD simulations, which were validated against Dickinsons experiments. Bush (Ref. 6) performed immersed boundary framework for Dickinson's Robofly model. Additionally, Lankford et al. (Ref. 7) performed a high-fidelity computational fluid dynamics (CFD) analysis using an in-house unsteady Reynolds averaged Navier-Stokes solver and validated the simulations with the flapping wing experiments described in (Ref. 3). Badrya et al. (Ref. 8) performed computational analysis of a fruitfly inspired MAV with two wings and body in trimmed forward flight. Overall the computational analysis was able to resolve many of the salient flow features captured in the experimental study. Additionally, there was good agreement between the predicted and measured instantaneous force generation for both lift and drag. This provides added confidence in the feasibility of using an inertial force subtraction technique as well as CFD to aid in analyzing the aerodynamic forces produced by a flapping wing at MAV-scale.

Studies by Mayo et al. (Refs. 9, 10) sought to develop a high-fidelity coupled aeroelastic solver and validate its predictability through experimental tests using rigid and flexible flapping wings in a wind tunnel. Particle image velocimetry (PIV) measurements as well as time-averaged force measurements were performed on highly flexible rectangular planform wings undergoing avian-like flap kinematics in forward flight. The measured and predicted flowfields showed good correlation for the range of flap frequencies tested. One of the key conclusions from the studies was that chordwise wing flexibility increased aerodynamic performance and, unlike a rigid wing, allowed for simultaneous production of net positive lift force and propulsive thrust for the wing configurations tested. Work by Lankford and Chopra (Ref. 11) later expanded the study to the hover flight condition performing PIV as well as VICON motion tracking experiments to measure the flowfield and passive wing deformations during flapping. The aeroelastic analysis was able to accurately capture many of the salient flowfield features including the LEV. Additionally, the predicted passive wing deflections were shown to match well with the experimental results, especially at lower flap frequencies. The

Fig. 1. Flapping Wing MAV developed in-house.

coupled aeroelastic analysis was used to explore the effect of varied wing flexibility on wing performance.

Currently, as shown in Figure 1, a hummingbird-inspired MAV has been developed and flight-tested in stable hover at Advanced Vertical Flight Laboratory at Texas A&M University (Ref. 12). This MAV has a 12-inch total wingspan with a gross weight of 62 grams and flaps at about 20 Hz during hover (video Ref). The in-house experimental studies conducted on the flapping wing demonstrate that at high frequencies (20 Hz) required for hovering, the wing undergoes large deflections. Particle image velocimetry (PIV) measurements as well as digital image correlation (DIC) measurements and time-averaged force measurements were performed on these highly flexible wings in hovering flight (Ref. 13).

The wings on the hummingbird MAV are very intricate, anisotropic, flexible structures that undergoes large deflections during flapping flight. It has been observed that hovering hummingbirds and insects exploit the flexibility of their wings to passively achieve highly favorable deflected wing shapes (optimal twist and camber) to generate large magnitudes of lift. These benefits of flexible flapping wings have enamored the researchers and the possibility of utilizing this system in MAVs has recently been addressed. However, because of the small scale, high structural flexibility, non-uniformity in the distribution of mass and stiffness properties, highly nonlinear and unsteady aerodynamics and the tight coupling between aerodynamics and structural dynamics, the analysis of a realistic flexible flapping wing is extremely challenging. Therefore, the computational framework will require a high-fidelity fluid dynamic solver and a geometrically-nonlinear structural model are necessary to accurately resolve the flowfield and capture the wing deflection.

The main objective of this work is to perform high fidelity aeroelastic analysis of extremely flexible flapping wings using a fully coupled CFD-CSD (CSD: Computational Structural Dynamics) analysis. In this work, the CFD model is coupled with an in-house geometrically exact wing structural model (Ref. 20) and the simulation results are validated with experiments. The remainder of the paper is organized as follows. First section will briefly describe the computational solvers, CFD solver and structural solver respectively. The second section describes aeroelastic coupling scheme, including data interpolation and mesh deformation. In the next section, the results from the flapping wing simulations will be presented and compared with experimental data. Finally, conclusions will be offered in the final section.

NUMERICAL FRAMEWORK FOR CFD-CSD SIMULATIONS

Unsteady CFD Solver

A compressible unsteady Reynolds Averaged NavierStokes solver with low-Mach number preconditioning is applied to study the complex, highly vortical, three-dimensional flow of low aspect ratio flapping wings at MAV-scale Reynolds numbers. The wing flaps at a constant 20Hz flap frequency.

OVERTURNS(Overset Transonic Rotor Unsteady Navier-Stokes) is a structured, overset, compressible solver, that solves three dimensional Navier-Stokes equations (Ref. 15). The OVERTURNS solver is implicit in time with dual-time stepping. OVERTURNS was initially used to perform high-fidelity fluid dynamics simulations on full-scale helicopter rotors which operate at much higher Reynolds numbers and Mach numbers than the flapping wing case presented in this study. In order to improve simulation convergence and accuracy, a low-Mach pre-conditioner described by Buelow et al. (Ref. 16) and Pandya et al. (Ref. 17) is implemented in conjunction with a time accurate, dual-time stepping scheme. For spatial reconstruction, the third order MUSCL scheme utilizing Korens limiter was used, with an option to use higher order Compact-Reconstruction Weighted Essentially Non-Oscillatory (CRWENO) schemes (Ref. 18). The inviscid terms were computed using the Roe-Turkel flux difference-splitting scheme and the viscous terms were computed using a second order central scheme. Because Reynolds numbers are low, special treatment is required for turbulence. For the investigated cases all the simulations are purely laminar. A single structured mesh can not preserve the complex flow field features such as tip vortices, trailing edge vortices and far field wake capture. In this case, an overset grid system were used. An implicit hole-cutting technique developed originally by Lee and modified by Lakshminarayan (Ref. 15) is applied in OVERTURNS to transfer information between overset grids.

The code has previously been validated against a number of published results for Re = 10-10,000 and various geometries, with generally good agreement. (Ref. 6) (Ref. 8)

To generate the body-fitted grids for the wings, a 3-D algebraic mesh generator for structured, curvilinear, body-fitted CFD grids was used (Ref. 19). Structured surface definitions can be used as input geometries, mathematical formulations for different body shapes are also available. The volume mesh is generated using a multi-surface algebraic mesh generation procedure with one intermediate surface between the wall and the outer boundary. Initially, a wall normal is calculated at each point on the wall. Secondly, smoothing is applied to this normal with neighbors in all four directions. Thirdly, the smoothened wall normal is projected out at a fixed distance to form the intermediate mesh. Finally, the three points (wall, intermediate and outer) define a quadratic line upon which a given stretching function is used to determine all volume points. A high density of points is used at the wall to capture the viscous effects. This code has been used for various bodies including simplified helicopter fuselages, rotor blades and insect wings (Ref. 19). The resulting body-fitted meshes have an O-O topology. The span-wise section along the wing are shown in Figure 2.a, the clustering near the wall to capture the flow near the boundary layer (Figure 2.b) and the whole O-O volume mesh.

A single structured mesh cannot preserve the complexities in the flowfield such as tip vortices, trailing-edge (TE) vortices, and far-field wake features. In this case, the wing is overset onto a coarser nested background mesh. The nested background meshes are stationary and extended to allow for the wing's motion, and the wing grids translate and rotate according to user-prescribed kinematics, whereas the body mesh is fixed. The overset grid system has 3 grids, as shown in Figure 2.

A O-O type mesh, displayed in Fig. 2.c, was utilized with $195 \times 91 \times 82$ nodes in the wrap-around, spanwise and normal directions respectively. The wing platform geometry for the CFD mesh was created to match the experimental test wing using the dimensions described in Fig. 3. In the simulation, the wing is offset from the flapping axis by 0.3c (c is mean chord length), similar to that of the experimental set-up. The wing kinematics described in Figure 11 are the only grid motions prescribed to the wing mesh. A higher concentration of grid points were placed near the leading edge and trailing edge as well as the wing root and wing tip to better resolve the highly vortical flow expected near those regions.

To capture the vortical flow in the wake, a Cartesian background mesh was implemented with $212 \times 160 \times 148$ nodes in the x, y and z directions respectively. Node points are highly clustered within the initial 3.0c of the wing surface in all directions to improve spatial resolution of the shed wake. The far-field boundary conditions are applied 10c away from the wing surface. All simulations were run for at least 3 consecutive flap cycles with a time step that equates to 0.125 degree of flap angle rotation per iteration (2880 iterations per flap cycle) with Newton-type sub-iterations to minimize factorization errors and achieve adequate residual convergence.

Between the meshes (wings, and 2 background meshes), there are overlap regions, and connectivity/information transfer across the overlapping areas is handled by implicit hole cutting.

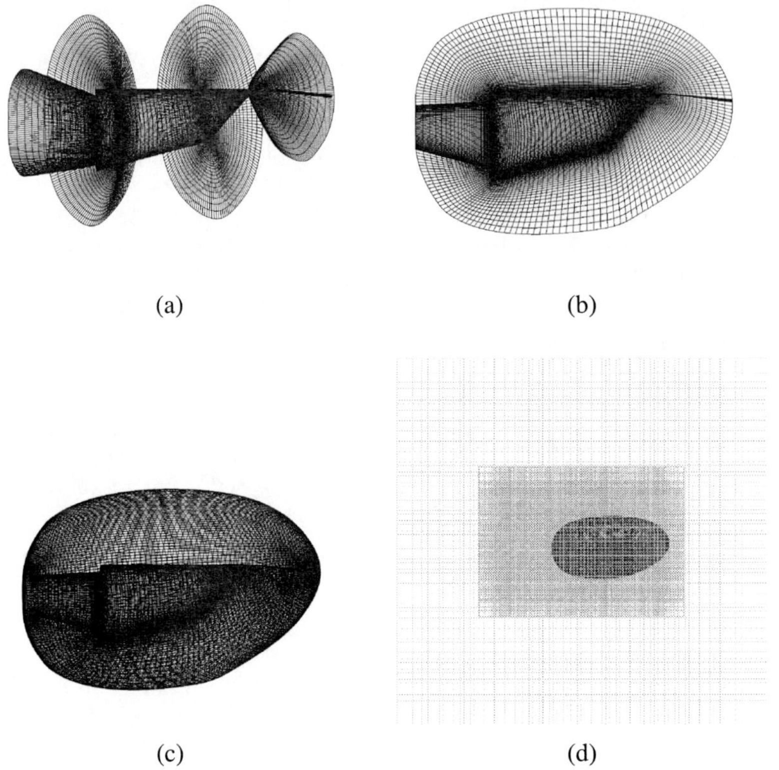

Fig. 2. Wing Mesh: (a) span-wise section mesh, (b) mesh in the normal direction near the leading edge and the trailing edge, (c) volume near-body O-O mesh, (d) overset grids.

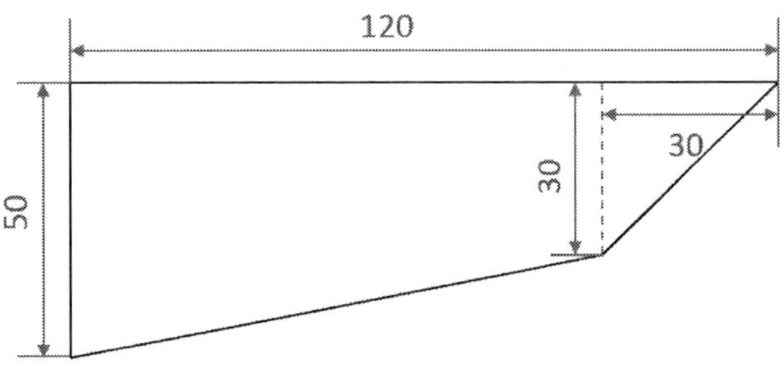

Fig. 3. Wing geometric parameters (unit: mm).

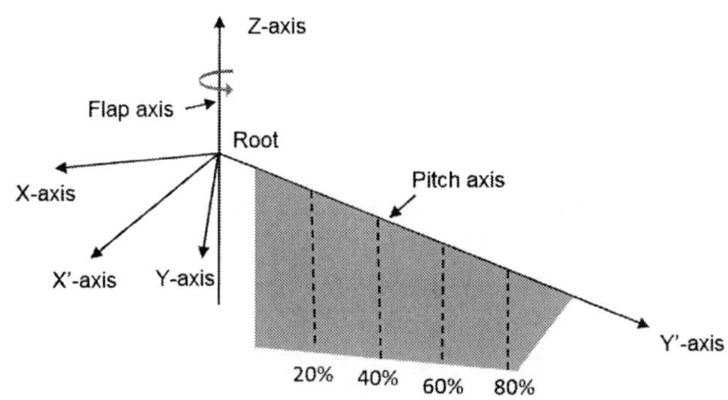

Fig. 4. Coordinate system used

Nonlinear CSD Modeling

HFStruct(Highly Flexible Structure) is a geometrically exact nonlinear finite element solver developed in-house, aims at numerically solving static and dynamic problems of highly flexible structures. The structural analysis is based on the direct integration in time of an initial value problem that describes the dynamics of a set of arbitrarily interconnected bodies. The integral form of extended Hamilton's Principle is used to determine the governing differential equation of deformed shell and beam structure. This principle states that

$$\int_{t1}^{t2} (\delta T - \delta \Pi + \delta W_{nc})dt = 0 \qquad (1)$$

where δT is the variation in kinetic energy, $\delta \Pi$ is the variation of elastic energy, and δW_{nc} is the non conservative work done by external forces.

Geometrically Exact Modeling For a naturally curved shell, three coordinate systems are needed for describing its deformation, as shown in Fig. 5. The *abc* is a fixed global rectangular coordinate system used for reference, the *xyz* is a fixed local orthogonal curvilinear coordinate system used to describe the undeformed shell geometry, and the $\xi \eta \zeta$ is a moving local orthogonal curvilinear coordinate system used to describe the deformed shell geometry (Ref. 21). Moreover,

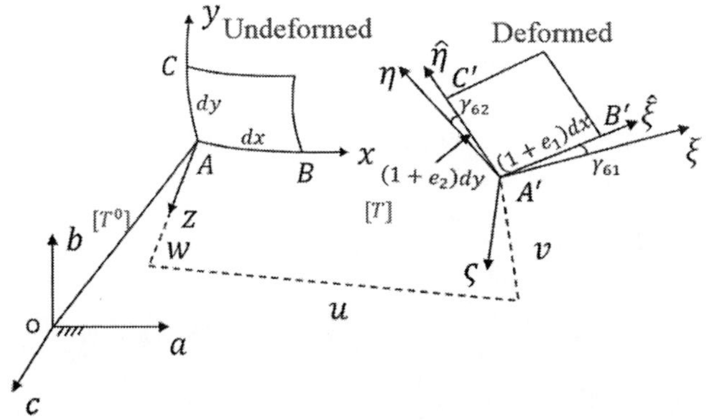

Fig. 5. Undeformed and deformed geometries of differential reference surface with the corresponding coordinate systems (Ref. 22).

i_a, i_b and i_c are unit vectors along the axes a, b and c, respectively; j_1, j_2 and j_3 are unit vectors along the axes x, y and z, respectively; i_1, i_2 and i_3 are unit vectors along the axes ξ, η and ζ, respectively. It can be shown that

$$\{\mathbf{j}_{123}\} = [T^0]\{\mathbf{i}_{abc}\}, \quad \frac{d}{ds}\{\mathbf{j}_{123}\} = [K^0]\{\mathbf{i}_{abc}\} \tag{2}$$

where $[T^0]$ is a known transformation matrix relating the coordinate systems abc and xyz, $[K^0]$ is initial curvature matrices. k_1 is initial twisting curvature, and k_2 and k_3 are the initial bending curvatures.

Moreover, the deformed coordinate system $\xi\eta\zeta$ and the undeformed coordinate system xyz are related by the transformation matrix $[T]$ as

$$\{\mathbf{i}_{123}\} = [T]\{\mathbf{j}_{123}\} \tag{3}$$

differentiating Eq. (3) with respect to s and using Eq. (2) and $[T]^{-1} = [T]^T$ yields

$$\frac{d}{ds}\{\mathbf{i}_{123}\} = [K]\{\mathbf{i}_{123}\} \tag{4}$$

where ρ_1 is the deformed twisting curvature and ρ_2 and ρ_3 are the deformed bending curvatures. Note that ρ_i are not real curvatures because the differentiation is with respect to the undeformed differential length ds, instead of the deformed length $(1+e)ds$. Using Eqs. (3), and (2), one can show that Similar to the plate model, the beam model also comprises of inertial coordinate system and deformed coordinate system. The inter-connectivity between these coordinates is similar to that of the plate model. The Eqs. (2)-(4) hold true for the beam model as well.

Finite Element Formulation For a geometrically exact beam theory, fully nonlinear stress-strain and strain-displacement relations are given by

$$\{J\} = [Q]\{B\} \tag{5}$$

where J is Jaumann stress, and B is Jaumann strain.

$$\{J\} = \{J_{11}, J_{12}, J_{13}\},$$

$$\{Q\} = \begin{bmatrix} \mathrm{E} & 0 & 0 \\ 0 & \mathrm{G} & 0 \\ 0 & 0 & \mathrm{G} \end{bmatrix},$$

$$\{B\} = \{B_{11}, 2B_{12}, 2B_{13}\}$$

where E is Young's modulus, G is shear modulus

For a geometrically exact shell theory, fully nonlinear stress-strain and strain-displacement relations are given by

$$\{J\} = [Q]\{B\} \tag{6}$$

where

$$\{J\} = \{J_{11}, J_{22}, J_{12}, J_{23}, J_{13}\}^T,$$

$$[Q] = \begin{bmatrix} Q_{11} & Q_{12} & Q_{13} & 0 & 0 \\ Q_{12} & Q_{22} & Q_{26} & 0 & 0 \\ Q_{16} & Q_{26} & Q_{66} & 0 & 0 \end{bmatrix},$$

$$\{B\} = \{B_{11}, B_{22}, B_{12}, B_{23}, B_{13}\}^T$$

The way that components of U are approximated defines a specific finite element. Using the finite-element discretization schemes, one can discretize the displacement as

Beam:

$$\{u, v, w, \phi, \gamma_5, \gamma_6\}^T = [N_1]\{q_1^{(j)}\} \tag{7}$$

$$\{q_1^{(j)}\} = \{u_i, v_i, w_i, \phi_i, w_i', v_i', u_i', \gamma_{5i}, \gamma_{6i}, u_k, v_k, w_k, \phi_k, w_k', v_k', u_k', \gamma_{5k}, \gamma_{6k}\}^T$$

Shell:

$$\{u, v, w, \gamma_4, \gamma_5\}^T = [N_2]\{q_2^{(j)}\} \tag{8}$$

$$\{q_2^{(j)}\} = \{\{q^{(l)}\}^T, \{q^{(l+1)}\}^T, \{q^{(l+2)}\}^T, \{q^{(l+3)}\}^T\}^T \tag{9}$$

with each node degree of freedom

$$\{q^{(l)}\} = \{u^{(l)}, u_x^{(l)}, u_y^{(l)}, u_{xy}^{(l)}, v^{(l)}, v_x^{(l)}, v_y^{(l)}, v_{xy}^{(l)}, w^{(l)}, w_x^{(l)}, w_y^{(l)}, w_{xy}^{(l)}, \gamma_4^{(l)}, \gamma_5^{(l)}\}^T$$

where $q_1^{(j)}$ is the displacement vector of the jth beam element and $[N_1]$ is a 6×18 matrix of 1D shape functions, and $q_2^{(j)}$ is the displacement vector of the jth shell element and $[N_2]$ is a 5×56 matrix of 2D shape functions.

Beam:

$$\{U\} = \{u, u', u'', v, v', v'', w, w', w'', \phi, \phi', \gamma_5, \gamma_5', \gamma_6, \gamma_6'\}^T \tag{10}$$

Shell:

$$\{U\} = \{u, u_x, u_y, u_{xx}, u_{xy}, u_{yy}, v, v_x, v_y, v_{xx}, v_{xy}, v_{yy}, w, w_x, w_y, w_{xx}, w_{xy}, w_{yy}, \gamma_4, \gamma_{4x}, \gamma_{4y}, \gamma_5, \gamma_{5x}, \gamma_{5y}\}^T \tag{11}$$

Substituting () and Eq.(9) into Eqs.(10) and (11) yields

$$\text{beam}: \quad U_1 = [D_1]\{q_1^{(j)}\}, \quad [D_1] = [\partial_1][N_1] \tag{12}$$

$$\text{shell}: \quad U_2 = [D_2]\{q_2^{(j)}\}, \quad [D_2] = [\partial_2][N_2] \tag{13}$$

where $[\partial_1]$, $[\partial_2]$ consisting of differential operators.

For beam, the variation of elastic energy is given by

$$\begin{aligned}\delta\Pi &= \int_v \{\delta B\}^T\{J\}dV \\ &= \sum_{j=1}^{N_e}\int_{L^{(j)}}\{\delta q_1^{(j)}\}[D_1]^T[\Psi]^T[\Phi_1]\{\psi\}ds \\ &= \sum_{j=1}^{N_e}\{\delta q_1^{(j)}\}^T[K_1^{(j)}]\{q_1^{(j)}\}\end{aligned} \tag{14}$$

where

$$[K_1^{(j)}]\{q_1^{(j)}\} = \int_{L^{(j)}}[D_1]^T[\Psi]^T[\Phi_1]\{\psi\}ds \tag{15}$$

$$[\Phi_1] = \int_A [S]^T[Q][S]dA \tag{16}$$

where V is the volume, N_e is the total number of elements, $L^{(j)}$ is the length of the jth element, and $K_1^{(j)}$ is the stiffness matrix of the jth beam element.

For shell, the variation of elastic energy is given by

$$\begin{aligned}\delta\Pi &= \int_v \{\delta B\}^T\{J\}dV \\ &= \sum_{j=1}^{N_e}\int_{A^{(j)}}\{\delta q^{(j)}\}^T[D_2]^T[\Psi]^T[\Phi_2]\{\psi\}dA \\ &= \sum_{j=1}^{N_e}\{\delta q_2^{(j)}\}^T[K_2^{(j)}]\{q_2^{(j)}\}\end{aligned} \tag{17}$$

where

$$[K_2^{(j)}]\{q_2^{(j)}\} = \int_{A^{(j)}}[D]^T[\Psi]^T[\Phi]\{\psi\}ds \tag{18}$$

where N_e is the total number of elements, $A^{(j)}$ is the area of the jth element, and $K_2^{(j)}$ is the stiffness matrix of the jth shell element.

So for beam and shell coupling, the variation of elastic energy is given by:

$$\delta\Pi = \sum_{j=1}^{N_e}\{\delta q_1^{(j)}\}^T[K_1^{(j)}]\{q_1^{(j)}\} + \sum_{k=1}^{M_e}\{\delta q_2^{(k)}\}^T[K_2^{(k)}]\{q_2^{(k)}\} \tag{19}$$

with a model including both beam and shell elements, the relation of $q_1^{(j)}$ and $q_2^{(k)}$ with global q can be determined, then $[K_1^{(j)}]$ and $[K_2^{(k)}]$ corresponding elements would be used to form global stiffness matrix $[K]$.

The variation of kinetic energy is given by

$$\delta T = -\int_v \rho(\ddot{u}\delta u + \ddot{v}\delta v + \ddot{w}\delta w)dV \tag{20}$$

for beam and shell coupling

$$\begin{aligned}\delta T &= -\int \{\delta\hat{U}\}^T[\hat{\Psi}]^T[m][\hat{\Psi}]\{\hat{U}\}dA \\ &= -\sum_{j=1}^{N_e}(\delta q^{(j)})^T[M^{(j)}]\{\ddot{q}^{(j)}\} - \sum_{k=1}^{M_e}(\delta q^{(k)})^T[M^{(k)}]\{\ddot{q}^{(k)}\}\end{aligned} \tag{21}$$

similarly, elements from $[M^{(j)}]$ and $[M^{(k)}]$ corresponding to the same global q need to be obtained to form each element in the global $[M]$ matrix.

The variation of non-conservative work due to external distributed loads is

$$\delta W_{nc} = \int_0^L \{\delta U\}^T\{R\}ds = \{\delta q\}^T\{R\} \tag{22}$$

where $\{R\}$ is the global nodal loading vector.

Substituting Eqs. (19), (21), (22) into Hamilton's equation would yield the equations of motion and is given by the form of

$$[M]\{\ddot{q}\} + [C]\{q\} + [K]\{q\} = \{g\} \tag{23}$$

Similar to shell, the static problems in beam theory can be solved using iterative method based on the modified Riks method. For dynamic problems, Eq. (23) can be solved by direct numerical integration using the Newmark-β method.

Newmark-β method is an implicit numerical integration technique used for nonlinear systems with [M] and [K] being displacement dependent and $\{R\}$ being displacement independent. We expand the displacement, velocity and acceleration vectors at $t+\Delta t$ as

$$\{q\}^{t+\Delta t} = \{q\}^t + \{\Delta q\}, \{\dot{q}\}^{t+\Delta t} = \{\dot{q}\}^t + \{\Delta\dot{q}\},$$
$$\{\ddot{q}\}^{t+\Delta t} = \{\ddot{q}\}^t + \{\Delta\ddot{q}\}$$

substituting these equations into Eq. (23) yields

$$[\hat{M}]^t\{\ddot{q}\} + [\hat{C}]^t\{q\} + [\hat{K}]^t\{q\} = \{g\}^t \tag{24}$$

where $[\hat{M}]$, $[\hat{C}]$, and $[\hat{K}]$ are the tangent mass, damping and stiffness matrices at time t, and

$$\{g\}^t = (\{R\}^{t+\Delta t} - [M]\{\ddot{q}\} - [C]\{\dot{q}\} - [K]\{q\}) \tag{25}$$

Assume the velocity vector $\{\dot{q}\}^{t+\Delta t}$ and displacement vector $\{q\}^{t+\Delta t}$ to be

$$\{\dot{q}\}^{t+\Delta t} = \{\dot{q}\}^t + [(1-\alpha)\{\ddot{q}\}^t + \alpha\{\ddot{q}\}^{t+\Delta t})]\Delta t \tag{26}$$

$$\{q\}^{t+\Delta t} = \{q\}^t + \{\dot{q}\}^t\Delta t + [(\frac{1}{2}-\beta)\{\ddot{q}\}^t + \beta\{\ddot{q}\}^{t+\Delta t})]\Delta t^2 \tag{27}$$

solving from Eq. (27) for $\{\ddot{q}\}^{t+\Delta t}$ and substitute into Eq. (26) gives expression of $\{\Delta\ddot{q}\}$ and $\{\Delta\dot{q}\}$, substituting the expression into Eq. (24) yields

$$[\tilde{K}]^t\{\Delta q\} = \{\tilde{R}\}^t \tag{28}$$

where

$$[\tilde{K}]^t = [\hat{K}]^t + \frac{1}{\beta\Delta t^2}[\hat{M}]^t \tag{29}$$

$$\tilde{R}^t = \left(\{\hat{R}\}^{t+\Delta t} - [\hat{M}]\{\ddot{q}\} - [\hat{K}]\{q\}\right)_{\{q\}=\{q\}^t} + [\hat{M}]^t\left(\frac{1}{2\beta}\{\ddot{q}\}^t + \frac{1}{\beta\delta t}\{\dot{q}\}^t\right)$$

One can solve Eq. (28) for Δq and substitute into other equations to obtain $\{\Delta q\}^{t+\Delta t}$, $\{\Delta\dot{q}\}^{t+\Delta t}$ and $\{\Delta\ddot{q}\}^{t+\Delta t}$.

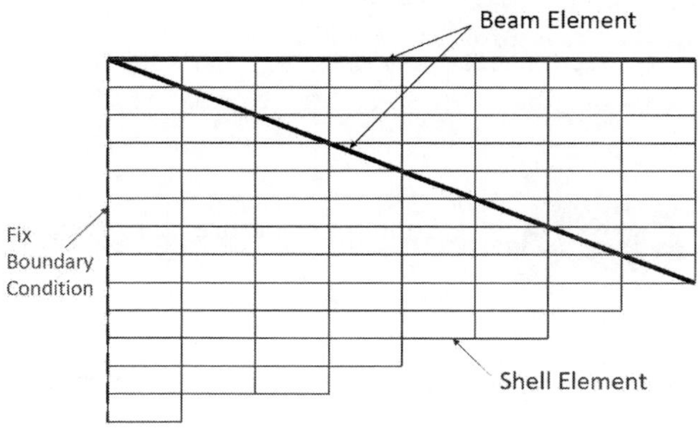

Fig. 6. Experiment setup.

Beam-Shell Coupled Simulation

As shown in Fig. 1, the bio-inspired flapping wing design has the wing structure, which comprises of carbon fiber spars (modeled as beams) and foam membrane (modeled as shell). So it is necessary to develop a structural model that is capable of beam-shell coupled analysis.

Figure 6 shows a cantilevered structure modeled with both beam and shell elements. Both static and dynamic load conditions are simulated, and the result of tip deflection is compared with Abaqus simulation result, as shown in Fig. 7. It shows good agreement between Abaqus and present model predictions.

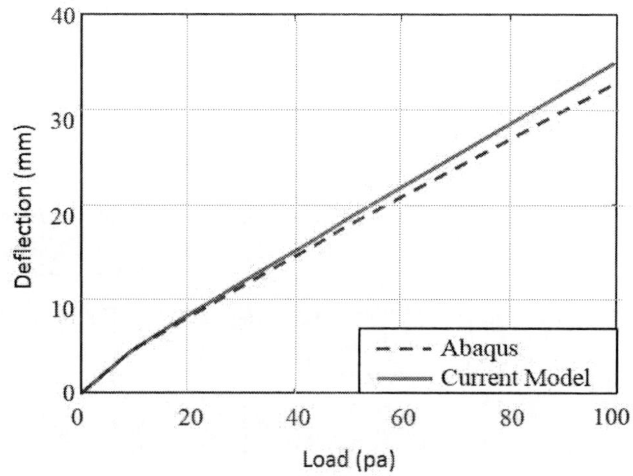

(a) Static load

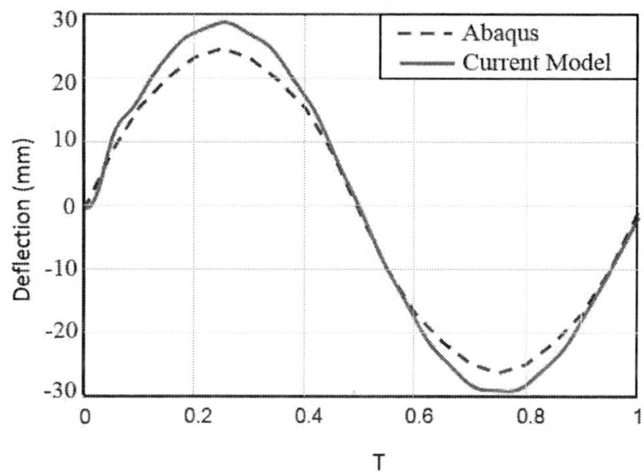

(b) Dynamic load

Fig. 7. Tip deflection comparison between geometrically exact model with Abaqus

AEROELASTIC ANALYSIS

For highly flexible flapping wings, such as the one investigated in the present study, structural deformation and aerodynamic forces are strongly coupled. Strong fluid-structure interaction and the complicated flow regime makes predicting the performance of such a system extremely difficult. In this work, the geometrically exact nonlinear structural model (HFS solver) is coupled with CFD model (OVERTURNS) to develop an aeroelastic analysis.

The aeroelastic coupled solution is based on a time-domain partitioned solution process in which structural solver and CFD solver are solved independently. The coupling algorithm allows for the exchange of structural deformations (from the CSD to the CFD solver) and aerodynamic loads (from the CFD to the CSD solver) at the fluidstructure interface (i.e., the wing surface). A schematic of such a frame work is shown in Fig. 8. In the interface part, both structural solver and aerodynamic solver are called one after the other according to the coupling method adopted for the problem. The solution is time-marched and updated.

As shown in Fig. 9, deflection interpolation and mesh deformation need to be performed at interface. Radial basis function (RBF) based method is used for both data interpolation and volume mesh deformation. Details of this method is given below.

Fluid-Structure interpolation and Mesh Motion based on Radial Basis Function

Examples of the application of RBFs to the information transfer problem are given by Wendland (Ref. 23) where the method is applied to find static aircraft deflections in transonic flow. The general theory of RBFs is presented by Buhmann (Ref. 25) and Wendland (Ref. 26). The solution of an interpolation problem using RBFs begins with the form of the

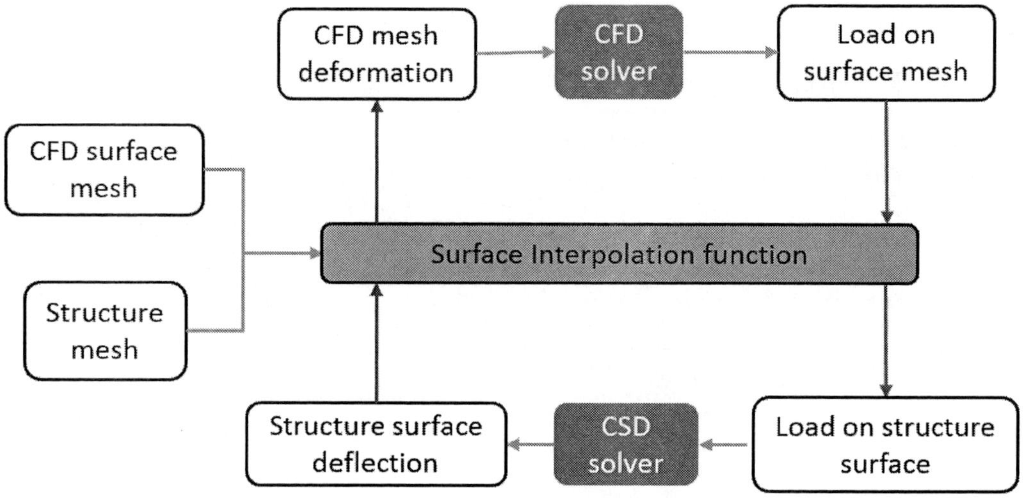

Fig. 8. CFD-CSD solvers information exchange scheme.

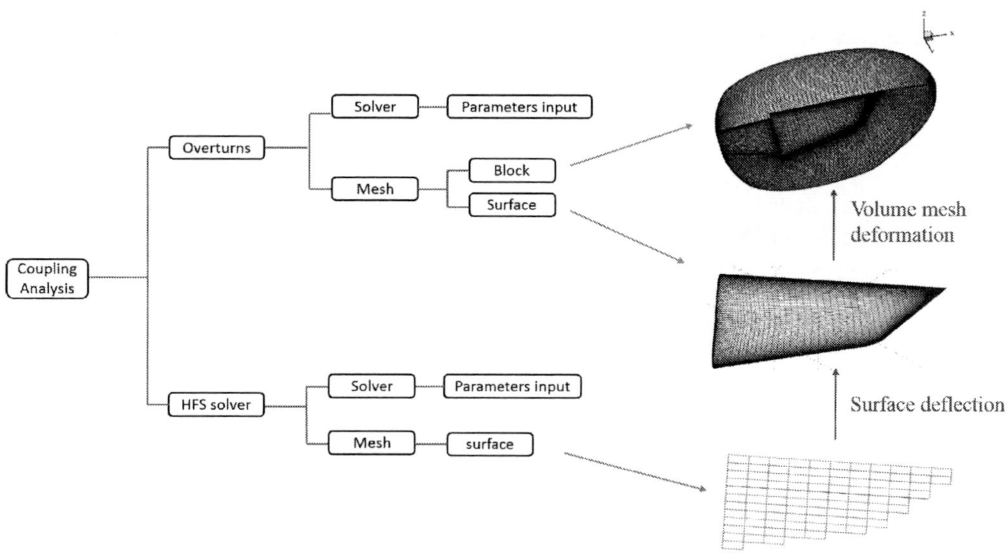

Fig. 9. CFD CSD coupling: interpolation and mesh deformation

required interpolation:

$$s(r) = \sum_{i=1}^{i=N} \alpha_i \phi(\| r - r_i \|) \tag{30}$$

Here, $s(r)$ is the function to be evaluated at location r, ϕ is the form of function adopted, the index i identifies the centres for the RBFs, while x_i is the location of that centre. In this case the centres correspond to mesh points located on the moving surfaces. The coefficients a_i are found by requiring exact recovery of the original function at these points r_i.

$$\phi(\xi) = (1-\xi)^4(4\xi+1) \tag{31}$$

Wendland's C2 function (Eq. (31)) is selected as the basis function since this provides a satisfactory compromise between the quality of the mesh motion, owing to its improved smoothness over the C0 function. Using subscript s to denote a structure surface mesh point, and using the change in position of a mesh point as the variable to be interpolated, the problem is expressed in the following fashion.

$$\begin{aligned} \Delta x_s &= C_{ss} a_x \\ \Delta y_s &= C_{ss} a_y \\ \Delta z_s &= C_{ss} a_z \end{aligned} \tag{32}$$

where

$$\Delta x_s = \{\Delta x_{s_1}, \ldots, \Delta x_{s_N}\}^T \tag{33}$$

$$a_x = \{\alpha_{s_1}, \ldots, \alpha_{s_N}\}^T \tag{34}$$

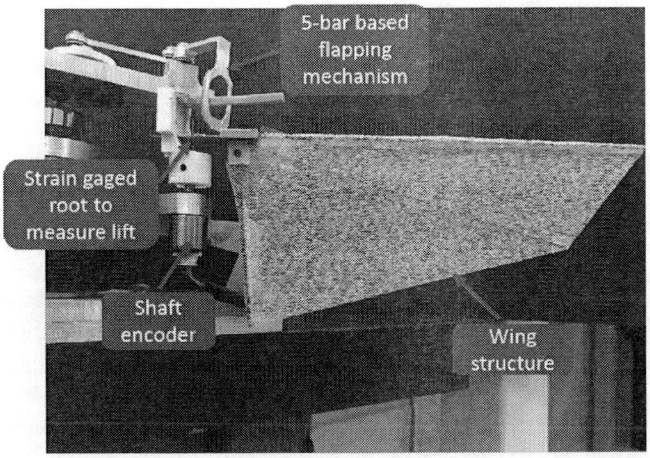

Fig. 10. Experiment setup.

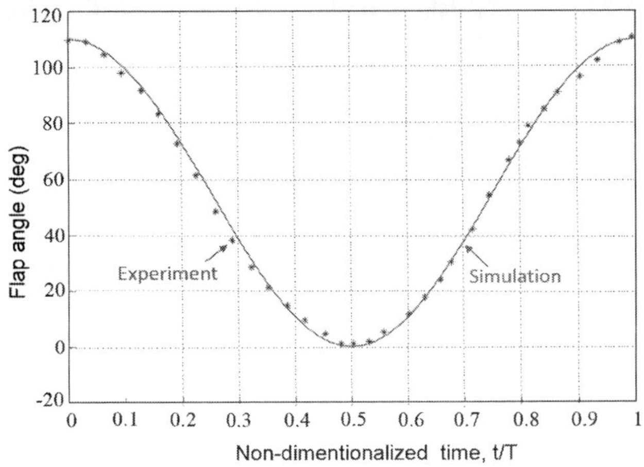

Fig. 11. Flap angle variation during a cycle.

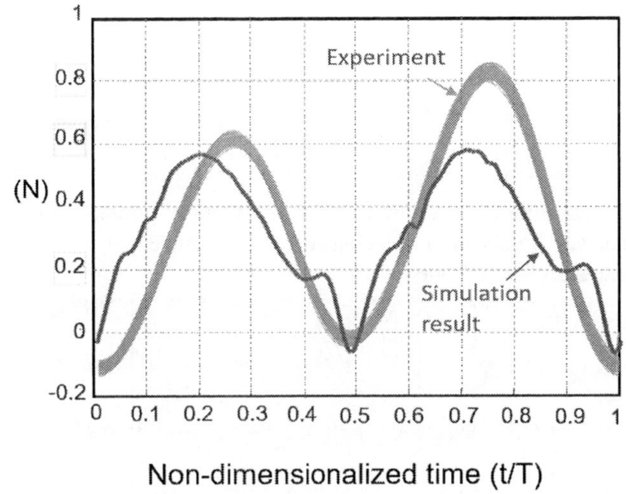

(a) Lift force comparison

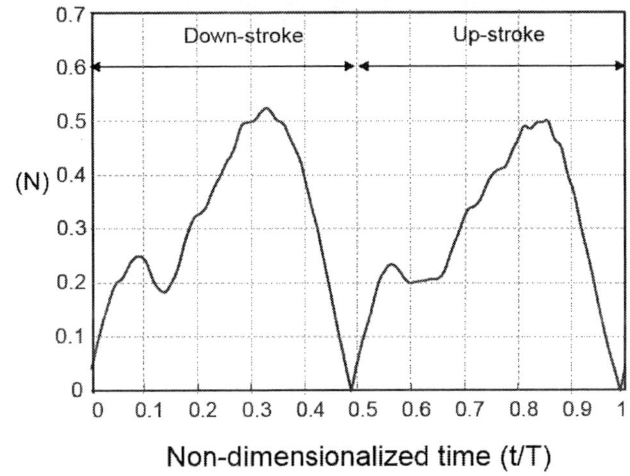

(b) Drag force

Fig. 12. Lift and Drag variation during a flap cycle

$$C_{ss} = \begin{bmatrix} \phi_{s_1s_1} & \phi_{s_1s_2} & \cdots & \phi_{s_1s_N} \\ \phi_{s_2s_1} & \phi_{s_2s_2} & \cdots & \phi_{s_2s_N} \\ \vdots & \ddots & & \vdots \\ \phi_{s_Ns_1} & \phi_{s_Ns_1} & \cdots & \phi_{s_Ns_N} \end{bmatrix} \tag{35}$$

with

$$\phi(s_1s_2) = \phi(\| x_{s_1} - x_{s_2} \|) \tag{36}$$

Exact recovery of the centres gives indicating the basis function evaluated on the distance between points s_1 and s_2 with a support radius of R. If a point lies outside the support radius then it will not be moved by the motion scheme as the basis function is zero beyond this point. To locate the aerodynamic surface points, an analogous matrix A_{as} must be formed, where a indicates an aerodynamic node, exists on the aerodynamic surface.

$$A_{as} = \begin{bmatrix} \phi_{a_1s_1} & \phi_{a_1s_2} & \cdots & \phi_{a_1s_N} \\ \phi_{a_2s_1} & \phi_{a_2s_2} & \cdots & \phi_{a_2s_N} \\ \vdots & \ddots & & \vdots \\ \phi_{a_Ns_1} & \phi_{a_Ns_1} & \cdots & \phi_{a_Ns_N} \end{bmatrix} \tag{37}$$

The positions of the aerodynamic surface points given by the vectors x_a, y_a, and z_a are

$$\begin{aligned} \Delta x_a &= A_{as}a_x = A_{as}C_{ss}^{-1}\Delta x_s \\ \Delta y_a &= A_{as}a_y = A_{as}C_{ss}^{-1}\Delta y_s \\ \Delta z_a &= A_{as}a_z = A_{as}C_{ss}^{-1}\Delta z_s \end{aligned} \tag{38}$$

Using RBFs for mesh deformation is achieved by interpolating displacements through the volume mesh away from the deflected surface. Using subscript a to denote a CFD surface mesh points, and using the change in position of a mesh point as the variable to be interpolated, the problem is written in the following fashion:

$$\begin{aligned} \Delta x_a &= Ma_x \\ \Delta y_a &= Ma_y \\ \Delta z_a &= Ma_z \end{aligned} \tag{39}$$

where

$$\Delta x_a = \{\Delta x_{a_1}, ..., \Delta x_{a_N}\}^T \tag{40}$$

$$a_x = \{\alpha_{s_1}, ..., \alpha_{s_N}\}^T \tag{41}$$

$$M = \begin{bmatrix} \phi_{a_1a_1} & \phi_{a_1a_2} & \cdots & \phi_{a_1a_N} \\ \phi_{a_2a_1} & \phi_{a_2a_2} & \cdots & \phi_{a_2a_N} \\ \vdots & \ddots & & \vdots \\ \phi_{a_Na_1} & \phi_{a_Na_1} & \cdots & \phi_{a_Na_N} \end{bmatrix} \tag{42}$$

To determine the dependence of M volume points on the N surface points the following matrix is formed, where v indicates a volume node:

$$A_{va} = \begin{bmatrix} \phi_{v_1a_1} & \phi_{v_1a_2} & \cdots & \phi_{v_1a_N} \\ \phi_{v_2a_1} & \phi_{v_2a_2} & \cdots & \phi_{v_2a_N} \\ \vdots & \ddots & & \vdots \\ \phi_{v_Na_1} & \phi_{v_Na_1} & \cdots & \phi_{v_Na_N} \end{bmatrix} \tag{43}$$

so that the positions of the volume points given by the vectors δx_v, δy_v, and δz_v are

$$\begin{aligned} \Delta x_v &= A_{va}a_x = A_{va}M^{-1}\Delta x_a \\ \Delta y_v &= A_{va}a_y = A_{va}M^{-1}\Delta y_a \\ \Delta z_v &= A_{va}a_z = A_{va}M^{-1}\Delta z_a \end{aligned} \tag{44}$$

It is clearly that there are two stages for accomplishing the mesh motion. First, interpolation structure surface deflection to CFD surface mesh. This is followed by 'interpolate' deflection from CFD surface mesh to all volume mesh points.

SIMULATION RESULTS

Before discussing the results, it is important to present the reference frame in which the data will be presented. Figure 4 provides a schematic of the coordinate reference frames to be used throughout this paper. The fixed, right-handed global reference frame is defined by the X, Y and Z-axes with the flap axis coinciding with the Z-axis. A body reference frame, denoted by the X' and Y'-axes, rotates with the flapping wing. Note that the Z'-axis coincides with the Z-axis and is not labeled for clarity. The body reference frame moves with the wing such that the Y'-axis remains along the wing span of the undeformed wing as depicted in Figure 4. The down-stroke is defined as the portion of the flap cycle during which negative rotation about the Z-axis (i.e. flap axis) occurs, while the upstroke is when positive rotation about the Z-axis occurs. Pitch axis is along leading edge of the wing.

A picture of experiment setup of the flexible wing tested is shown in Figure 10. Digital image correlation (DIC) experiment were conducted to this wing to measure wing deflection at different instants during the flap cycle (Ref. 13). Figure 11 shows the experimental flap kinematics along with the simplified harmonic variations used for the simulations in this study. The flap frequency used is 20 Hz. Other parameters from kinematics are computed and input into OverTURNS solver. Tip mean velocity, V = 10.7 m/s. Reynolds number, Re = 31,000. Mach number of tip, Ma = 0.0312.

Lift Force Variation Comparison

This section focuses on comparing the measured and predicted aerodynamic lift force produced by the wing. The measured time history of lift forces for the flexible wing is compared with CFD-CSD simulation result in Fig. 12. Note that positive lift force will act along the positive Z-axis. The experimental lift force time history plot is an average of 20 separate trials. Each trial is an average of the lift force variation over 60 flap cycles.

From the CFD-CSD simulation, it can be seen that the lift peak happens a little before mid-stroke, while experimental data shows highest value at mid-stroke. An observation of vortex structure may explain this well, given in 2D Vortex Structure section. Secondly, comparing to experimental results, the CFD-CSD analysis under-predicts the lift force generated by the wing. Additionally, as expected, the predicted results exhibit a much more symmetrical variation in lift force between the up and down strokes with lift being mostly positive throughout the flap cycle. The peak lift force from experiment is 0.6N and 0.82N for down-stroke and up-stroke, respectively. The predicted peak lift force is about 0.55N. The reason for the discrepancy between the simulations and the experiments can be complex. From the DIC measurements, it is seen that the flexible wing takes a complex shape with large dynamic twist and camber variations from root to tip. Similar shape is also shown in simulation result, however, it is uncertain that how sensitive the vortex structure is with respect to surface shape change. Additionally, with symmetric kinematics, it requires further investigation to explain the asymmetry in lift forces between upstroke and downstroke as seen from the experiments.

Wing Deflection Comparison

Digital image correlation (DIC) is an optical measurement technique which calculates the displacement of structural surface features by performing cross-correlations on two images taken at different loading conditions. It is a well established technique that has previously been applied to flapping wings to generate deflections at various locations on the wing. (Ref. 13)

Fig. 13 shows the result of deformed wing comparison between DIC data and CFD-CSD simulation result. Fig. 13(a) shows top view of deformed wing comparison at 5 flap positions. In Fig. 13(b)-(f). The wing is view from tip to root and the deflection is scaled 2.5 times larger in order to make it easier to see the difference. It is that difference between experiment data and simulation result is noticeably large at start and end of half stroke, but at mid-stroke, the deflection agrees quite well.

2D Vortex Structure: Chordwise View

This section gives an overview and comparison of the flow features produced by the flapping wing using PIV and coupled CFDCSD analysis. Of particular interest in the present

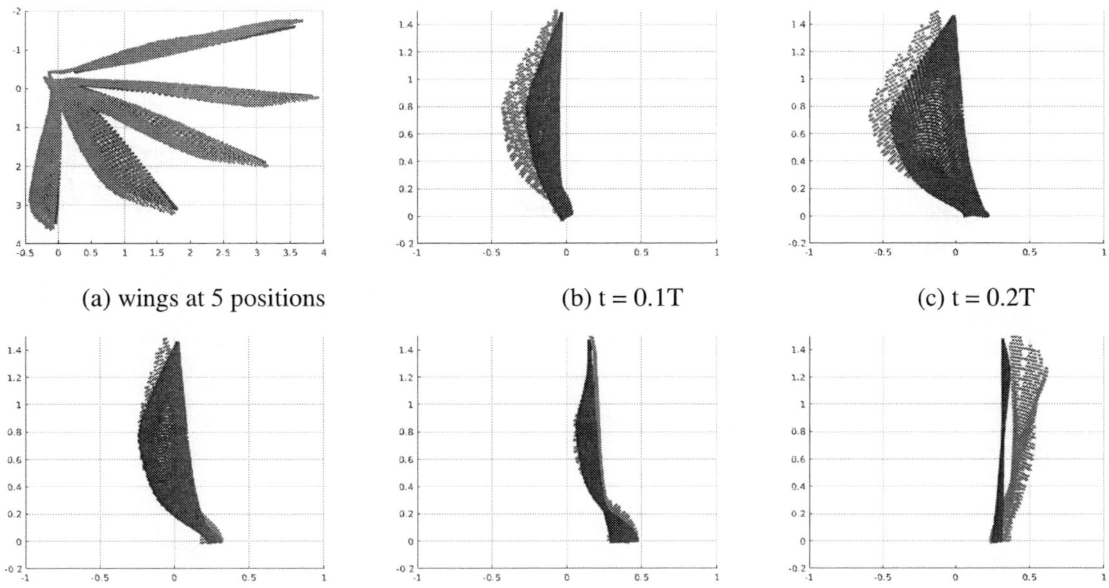

(a) wings at 5 positions (b) t = 0.1T (c) t = 0.2T

(d) t = 0.3T (e) t = 0.4T (e) t = 0.5T

Fig. 13. Wing deflection comparison between DIC data and CFD-CSD simulation result (red: DIC data, blue: CFD-CSD simulation)

work was to gain confidence in the numerical simulation. The measured velocity fields were dominated by separated flow and both large- and small-scale vortices. The details of the experiments are provided in (Ref. 13). Figure 14 shows the experimental setup to measure chordwise and spanwise flow. Figure 15 shows the 2D flowfield around the wing at 70% spanwise location during down-stroke. Each figure is at a different time instant during the downstroke, and the PIV results (chordwise view) and CFD-CSD simulation are shown in upper and lower parts, respectively.

During the early stage of downstroke (t/T = 0-0.15), flow structure is quite simple. Leading edge vortex and trailing edge wake can be seen shed from leading edge and trailing edge of the wing.

During acceleration phase of the wing, the pitch angle rapidly decreases to its midstroke value. Aerodynamically, one would expect separated or detached flow for such steep pitch angles due to deep stall effects on a cambered airfoil in steady flow. However, due to the highly unsteady nature of the flow under investigation, the flow remains attached all the way to the midstroke, leading edge vortex (LEV) continues to grow larger, and the tail of the starting vortex grows longer and extends downward as the vortex moves in that direction.

Note that, in Fig. 12, it shows that the peak of lift occurs a little before t = 0.25T, when wing speed is almost maximum, pitch angle is minimum and LEV is still attached. At mid-stroke, as shown in Fig. 15(e), original LEV shed from leading edge and a new one appears but is relatively small. This explains why lift peak is not at exact mid-stroke. Even though the wing starts slowing down after midstroke, drag still increases until reaching its peak at t = 0.35T, because the wing starts pitching up (increases angle of attack) increasing the drag coefficient. After this point, the reduction in speed reduces drag and this effect quickly dominates the increase in drag coefficient because of the increased angle of attack.

Towards the end of the half stroke, wing speed drops to very low values causing the LEV and trailing edge wake to be unstable and the counter-rotating vortices begin to convect downwards and move away from the stroke plane. At the same time, the lift decreases such that at the end of the stroke, at a geometric angle of attack of 90 deg (t/T = 0.5) the computed and experimental lift values are almost zero. Comparing simulation result and experiment output, we can see vortex structure is quite similar. However, LEV structure is less stable in simulation result. Before mid-stroke, LEV in simulation result grow much larger than the one observed in experiment. At about mid-stroke, original LEV shed from leading edge and a new one appears and grows bigger until the end of the stroke. This LEV get similar size of that observed in experiment. As mentioned before, it seems vortex structure seems quite sensitive to wing shape change, and this influence pressure and wing deflection in turn.

2D Vortex Structure: Spanwise View

Figure 16 shows the flow structures in spanwise direction at different instants during the flap cycle from both PIV experiment and CFD simulation. Note that, for the spanwise experiments, the laser-sheet was kept fixed in the inertial frame and the flowfield was measured at different wake ages in a phase-locked fashion. Similar to the previous case, the upper

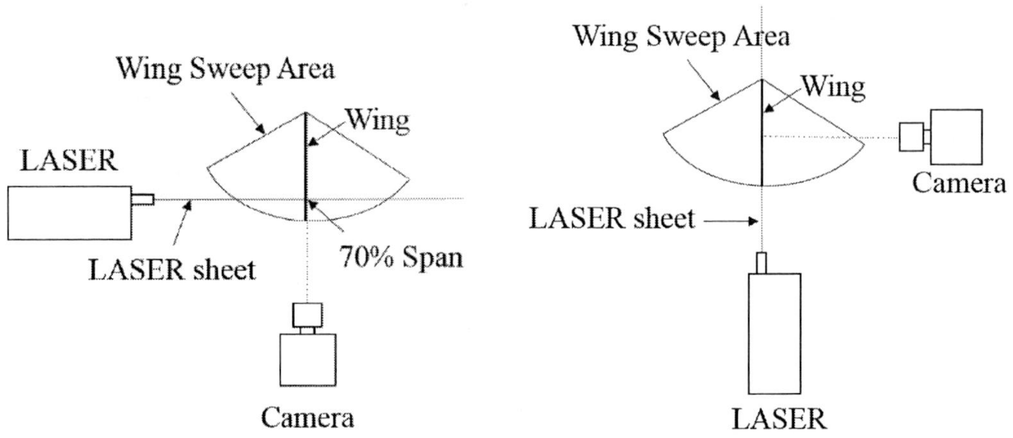

(a) Chordwise flowfield experiment (b) Spanwise flowfield experiment

Fig. 14. PIV experiment setup.

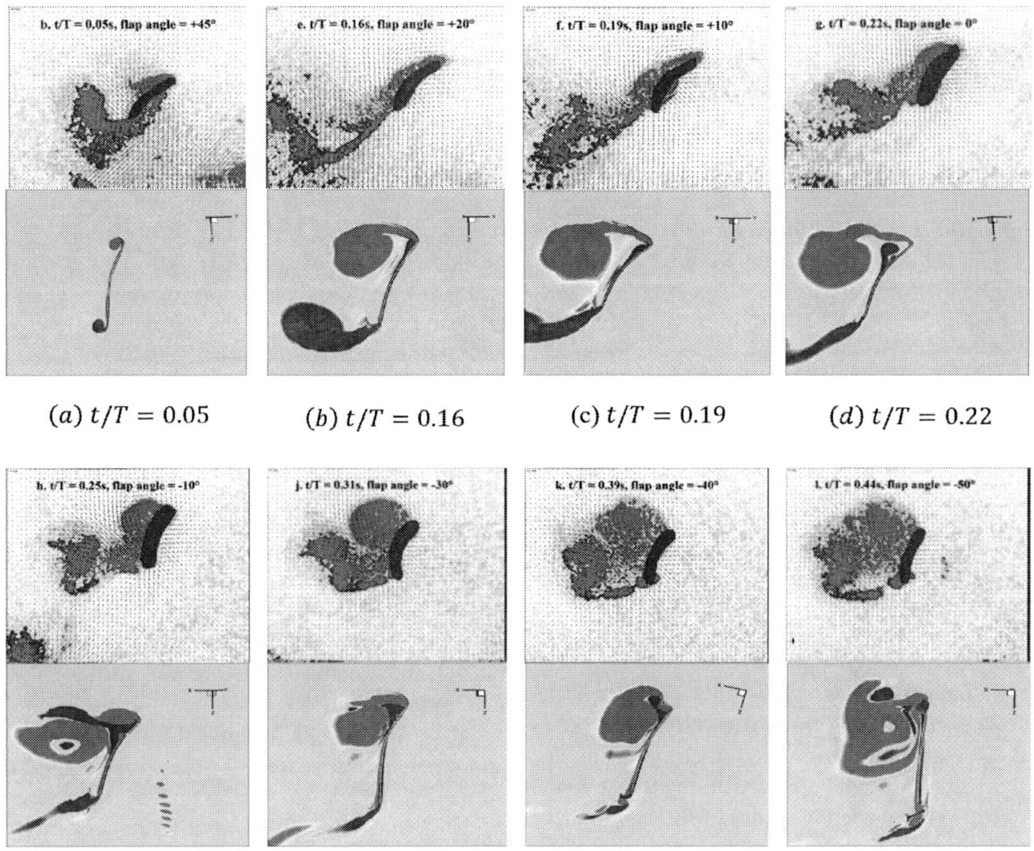

(a) $t/T = 0.05$ (b) $t/T = 0.16$ (c) $t/T = 0.19$ (d) $t/T = 0.22$

(e) $t/T = 0.25$ (f) $t/T = 0.31$ (g) $t/T = 0.39$ (h) $t/T = 0.44$

Fig. 15. Chord-wise flow structure (For each figure: PIV result on top, CFD result on bottom). Red background denotes right-hand vorticity, magenta/blue denotes left-hand vorticity.

part of the figure shows the PIV results and lower part shows the results from the CFD simulations. From PIV results, shed vortices from the previous upstroke motion are visible passing through the plane of interrogation, but relay downward with the inflow far and fast enough such that no interaction between the wing and any previously shed wakes is observed. As the

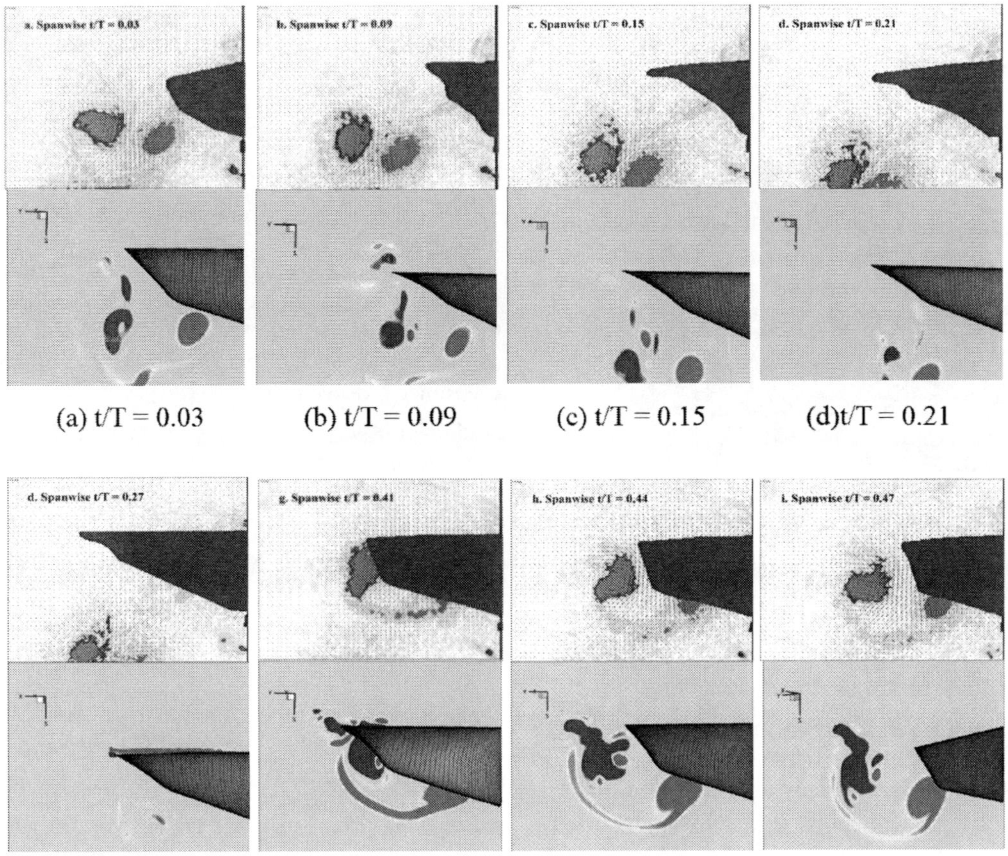

(a) t/T = 0.03 (b) t/T = 0.09 (c) t/T = 0.15 (d)t/T = 0.21

(f) t/T = 0.27 (h) t/T = 0.41 (i) t/T = 0.44 (j) t/T = 0.47

Fig. 16. Span-wise flow structure (For each figure: PIV result on top, CFD result on bottom). Red color denotes right-hand vorticity, magenta/blue denotes left-hand vorticity.

wing passes through the laser plane, a prominent tip vortex is developed and promptly shed. At the same time, root vortex forms and moves quickly to tip direction. Flow field from CFD shows similarity when compared to PIV data. At the start of the stroke, shed vortices from the previous upstroke motion are visible passing through the plane of interrogation, at the same time, strong root vortex forms and shed. After the wing crosses the mid-stroke plane, vortices form at tip area. However, different from the one dominant tip vortex shown in experiment, in CFD simulation result, vortices at tip area are a little more complex. This suggests that tip vortex formed in numerical simulation is not as strong as the one shown in PIV data.

3D Vortex Structure

Figures 17 18 show iso-surfaces around vortex core of the second invariant of the velocity gradient tensor, Q, obtained from the CFD simulation. For a typical flow structure as shown in 17, we can clearly see the main vortex surrounding the wing, leaing-edge vortex, tip vortex, root vortex and wake from previous stroke.

Similar to what was described previously, at the early stage of each stroke, main vortex structure such as leading edge vortex and tip vortex have formed. At t = 0.2T, as shown in Figure 18(b), the vortex sheet that form the root and tip vortices is very distinct.

From Figure 18(a),(b), it can be seen that the strong root and tip vortices shed from the previous half-stroke still is in close proximity of the wing.

The dominant structures during most time of stoke are root vortex, tip vortex and leading edge vortex. These structures are visible in Figure 18(b) to (d). It is notable that leading edge vortex is not attached all the time. Close to tip area, under the influence of tip vortex, leading edge vortex breaks into a lot of small vortex structure. This observation is in consistent with what is shown in chord-wise 2D vortex structure, and could partially explain the reason that lift force of simulation is smaller than experimental result.

As seen from a previous work (Ref. 6), at these low Reynolds numbers, in steady rotation, a concentration of positive (outboard) spanwise velocity is developed just downstream of the trailing edge. The strong radial flow, (at its peak, approximately 50% of the tip velocity) provides a mechanism for the

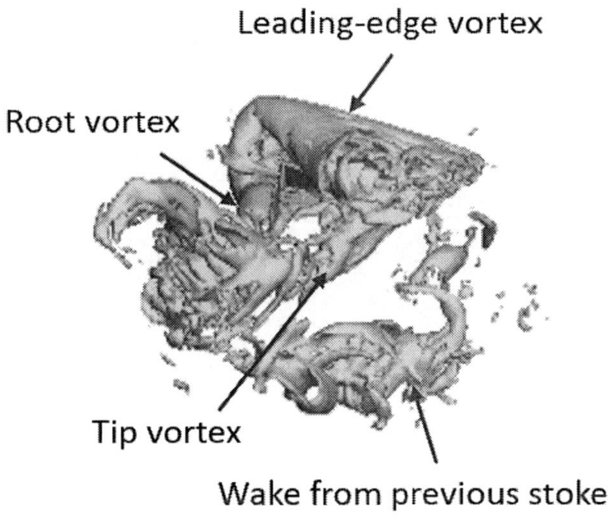

Fig. 17. Vortex Structure.

rapid convection of the root vortex towards the tip.

The strongest structures remaining in the wake at supination are the separated root and tip vortices, and a u-shaped open vortex ring consisting of portions of the secondary tip vortex, the separated LEV, and the separated trailing edge vortex formed during the wing rotation (Figure 18). As the wing begins the upstroke, the leading edge interacts with this shed vortex ring causing dissipation and outboard convection of the portions of the shed vortex ring associated with the wing tip.

Some weaker inboard portions of the shed vortex ring rapidly dissipate, most likely as a result of their interaction with the new attached vortex ring of opposite vorticity. Interestingly, the inboard section of the shed down-stroke LEV gets pushed down along the lower surface of the wing and merges with the trailed root vortex, which itself is being impacted by the trailing edge and pulled upwards along the lower surface of the wing.

SUMMARY & CONCLUSIONS

The objective of the present study was to assess the ability of coupling a 3D unsteady RANS solver and nonlinear structural solver to predict the lift force and flow structure of a flexible wing undergoing hummingbird-like flapping kinematics at MAV-scale Reynolds numbers in hover. The design of the experimental test wing was modeled after the wing of a prototype flapping micro air vehicle (MAV) and experiments that were conducted at flap frequencies of 20 Hz. The computational model was constructed to emulate the flap kinematics of the wing based on the measured deflected shape from the experiments. The force and flowfield results from the CFD-CSD coupled simulation were compared with the experiments data. The simulation result was post-processed to investigate the vortical flowfield around the wing. The key conclusions from the study are enumerated below.

1. The CFD analysis was able to predict the variation of lift force over the course of a flap cycle. However, the solver under-predicted the magnitude of lift produced when compared to the measured values. Additionally, an asymmetry in the magnitude of peak lift force was seen in the experimental data, which is not be captured by the analysis.

2. The CFD model was able to resolve the salient flow features present on the flapping wing. The CFD-CSD coupled model was able to both spatially and temporally predict many of the general flowfield characteristics. What makes difference is that leading-edge vortex in simulation result is not as stable as experiment data. This should be the reason that lead to difference in force result.

Author contact: Xuan Yang, xuan2017@tamu.edu, Moble Benedict, benedict@tamu.edu

REFERENCES

[1]Dickinson, M., Lehmann, F.-O., and Sane, S., "Wing rotation and the aerodynamic basis of insect flight", Science, vol. 284, no. 5422, pp. 19541960, Jun. 1999.

[2]Birch J. M., Dickinson, M. H., "The Influence of Wing Wake Interactions on the Production of Aerodynamic Forces in Flapping Flight", *Journal of Experimental Biology*, Vol. 206, No. 13, July 2003

[3]Benedict, M., Coleman, D., Mayo, D., and Chopra, I., "Experiments on rigid wing undergoing hover-capable flapping kinematics at micro-air-vehicle-scale Reynolds numbers", AIAA Journal, vol. 54, no. 4, pp. 1145-1157, Apr. 2016.

[4]Sun,M.,and Tang, J., "Unsteady aerodynamic force generation by a model fruit fly wing in flapping motion", *Journal of Experimental Biology*, Vol. 205(1),(2002),pp- 55-70.

[5]Ramamurti, R., Sandberg, W. C. ,"A three-dimensional computational study of the aerodynamic mechanisms of insect flight", *Journal of Experimental Biology*,Vol. 205(10),2002 May 15,pp- 1507-18.

[6]Bush, B.L. and Baeder, J.D., "Computational Investigation of Flapping-Wing Flight", Presented at the 37th AIAA Fluid Dynamics Conference and Exhibit, 25-28 June 2007, Miami, FL.

[7]Lankford, J., Mayo, D., and Chopra, I., "Computational investigation of insect-based flapping wings for micro air vehicle applications", *International Journal of Micro Air Vehicles*, vol. 8, no. 2, pp. 64-78, Mar. 2016.

[8]Badrya, C., Sridharan, A., Baeder, J., and Christopher M. Kroninger, "Multi-Fidelity Coupled Trim Analysis of a Flapping-Wing Micro Air Vehicle Flight", *Journal of Aircraft*, 30 January 2017.

[9]Mayo, D., Lankford, J., Benedict, M., and Chopra, I., "Experimental and computational analysis of rigid flapping wings for micro air vehicles", Journal of Aircraft, vol. 52, no. 4, pp. 1161-1178, Jul. 2015.

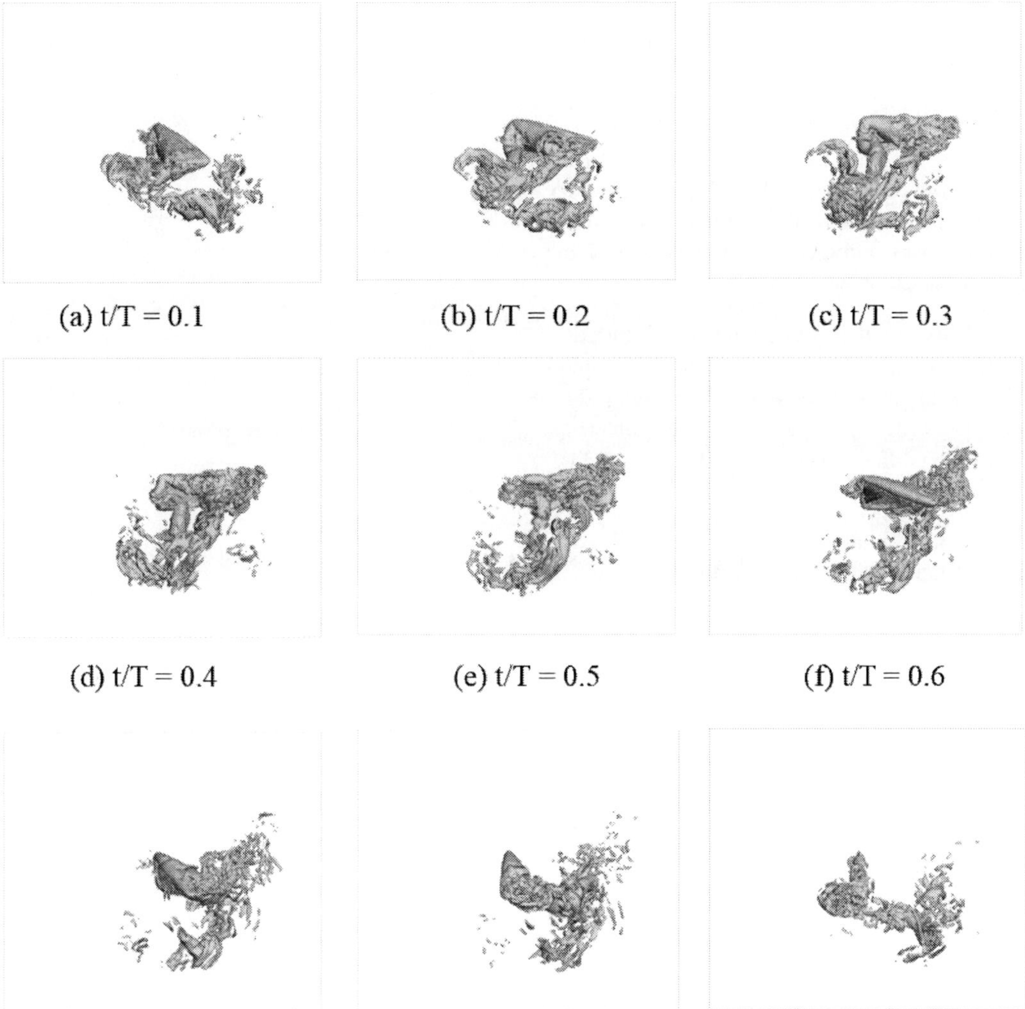

(a) t/T = 0.1 (b) t/T = 0.2 (c) t/T = 0.3

(d) t/T = 0.4 (e) t/T = 0.5 (f) t/T = 0.6

(g) t/T = 0.7 (h) t/T = 0.8 (i) t/T = 0.9

Fig. 18. Vortex structure obtained from the CFD simulation.

[10]Mayo, D., Lankford, J., Benedict, M., and Chopra, I., "Aeroelastic analysis of avian-based flexible flapping wings for micro air vehicles", Journal of the American Helicopter Society, vol. 60, no. 3, pp. 1-18, Jul. 2015.

[11]Lankford, J. and Chopra, I., "A computational and experimental study of insect-based flexible flapping wing aerodynamics and structural deformation", presented at the American Helicopter Society 72nd Annual Forum, West Palm Beach, FL, 2016.

[12]Coleman, D. and Benedict, M., "Design, Development and Flight-Testing of Robotic Hummingbird", American Helicopter Society 71st Annual Forum, Virginia Beach, Virginia, May 5-7 2015.

[13]Coleman, D., Gakhar, K., Benedict, M., Tran, J., "Experimental Studies towards Understanding the Aeromechanics of a Flexible Robotic Hummingbird Wing in Hover", AHS International 73rd Annual Forum, Fort Worth, Texas, USA, May 9-11, 2017

[14]James L. Lankford, and Inderjit Chopra, "Coupled CFD/CSD Study of a Flexible MAV-Scale Flapping Wing in the Hover Flight Condition", AHS 73rd Annual Forum, Fort Worth, Texas, 2017.

[15]Lakshminarayan, V. K., "Computational Investigation of Micro-Scale Coaxial Rotor Aerodynamics in Hover", Doctor of philosophy, Department of Aerospace Engineering, University of Maryland, College Park, MD, 2009.

[16]Buelow, P., Schwer, D., Feng, J., and Merkle, C., "A preconditioned dual-time, diagonalized ADI scheme for Unsteady Computations", presented at the 13th AIAA Computational Fluid Dynamics Conference, Snowmass Village CO, 1997.

[17]Pandya, S., Venkateswaran, S., and Pulliam, T., "Implementation of preconditioned dual-time procedures in OVERFLOW", presented at the 41st AIAA Aerospace Sciences Meeting and Exhibit, Reno, NV, 2003.

[18]Ghosh, D., Medida, S. and Baeder, J. D., "Compact-Reconstruction Weighted Essentially Non-Oscillatory Schemes for Unsteady Navier-Stokes Equations", 42nd AIAA Fluid Dynamics Conference and Exhibit, New Orleans, LA, June 25-28, 2012.

[19]Amiraux, M., "Numerical Simulation and Validation of Helicopter Blade-Vortex Interaction Using Coupled CFD/CSD and The Levels of Aerodynamic Modeling", Doctor of philosophy, Department of Aerospace Engineering, University of Maryland, College Park, MD, 2014.

[20]Yang, X., Sudhir, A., Halder, A., Benedict, M., "Aeroelastic analysis for highly flexible flapping wing in hover", AHS International 73rd Annual Forum, Fort Worth, Texas, USA, May 9-11, 2017

[21]Pai, P. F., "Three Kinematic Representations for Modeling of Highly Flexible Beams and Their Applications", *International Journal of Solids and Structures*, Vol. 48(19), (2011), pp: 2764-2777.

[22]Pai, P. F, Robert D. C., and Zaichun F., "Geometrically Exact Displacement-based Shell Theory", *Thin-Walled Structures*, Vol. 70 (2013): pp 1-18.

[23]Beckert, A., Wendland, H., "Multivariate interpolation for fluidstructure-interaction problems using radial basis functions", Aerospace Science and Technology 5 (MayJune) (2001) 125-134.

[24]Ahrem., R., Beckert, A., Wendland, H., "A meshless spatial coupling scheme for large-scale fluidstructure interaction problems", Computer Modeling in Engineering and Sciences 12 (2006) 121-136.

[25]Buhmann, M., "Radial basis functions", first ed, Cambridge University Press, 2005

[26]Wendland, H., "Scattered data approximation", first ed, Cambridge University Press, 2005

Computational Aeroacoustics of Different Propeller Configurations for eVTOL Applications

Bernadine Passe* James Baeder†

Department of Aerospace Engineering, University of Maryland, MD

ABSTRACT

In the present work, an in-house CFD, GPU-Accelerated Rotorcraft Flow-Field (GARFIELD) solver, was coupled to the Acoustic Code University of Maryland (ACUM), and is presented for aeroacoustic computations of low aspect ratio propellers in hover such as those found on typical eVTOL aircraft. A total of sixteen cases are presented, varying design parameters including twist, planform, number of blades, RPM, and sweep to evaluate their effect on aerodynamic performance as well as the overall sound pressure level (OASPL) experienced by an observer. A plausible nearby boom geometry was implemented to investigate the resulting unsteady airloads and their effect on aeroacoustic performance. Aerodynamic loads were also calculated by a Blade Element Momentum Theory (BEMT) analysis for comparison of the aeroacoustic performance predicted by CFD in GARFIELD. The different propeller configurations are presented to analyze and identify the design parameters that maximize the aerodynamic performance and minimize the noise to achieve a quiet, yet still efficient eVTOL aircraft.

INTRODUCTION

Urban air taxi transport is currently of high interest in the vertical flight community. To have the ability in large urban cities to no longer use a car to get from one place to another would change how we perceive transportation. Hours of productivity are lost by sitting in traffic, not to mention how a single interruption can cause serious delays. An on-demand electric vertical take-off and landing (eVTOL) vehicle, such as that shown in Figure 1 by Aurora Flight Sciences, would greatly improve the convenience of travel. Other companies such as Bell Helicopter and Uber Elevate have teamed up to create their own on-demand urban air taxi concept. Uber Elevate promises to transform a 2 hour drive from the San Francisco Marina to San Jose into just 15 minutes. Incorporating on-demand air travel in cities would increase mobility since fixed routes do not need to be followed and alleviate transportation congestion on the ground suggesting safer commutes. Uber Elevate aims to begin commercial operations in 2023, however, designing such a vehicle to operate in an urban environment raises some challenges (Ref. 1).

Figure 1. An example of an eVTOL aircraft to provide on-demand transportation from Aurora Flight Sciences (Ref. 1)

In order for an urban air taxi to operate in a city, the aircraft needs to be widely accepted by the public. A main concern is the noise produced by the aircraft and how that can effect every day life. Large scale helicopters are perceived as very loud, but often don't operate consistently over large cities. Therefore, the main goal is to design an aircraft that is as quiet as possible, but still efficient, to achieve widespread public acceptance of this future technology.

The goal of the present study is to model various low aspect ratio propellers in hover, such as that found on the Aurora Flight Sciences eVTOL aircraft, to evaluate which design parameters effect the noise. Aeroacoustic analysis for eVTOL vehicles has just begun to be investigated in the vertical flight community. The sources of noise had been rigorously studied in the 1990's by NASA Langley, but it is not yet widely known how exactly to optimize aeroacoustic performance (Ref. 2). The data produced from the cases presented will help advance the effort towards designing an eVTOL vehicle that can be used in the near future in urban environments. Understanding how to design a quieter, efficient aircraft will help gain wide spread acceptance from the public on this new on-demand urban air taxi concept.

*Graduate Student

†Professor

Presented at the Vertical Flight Society Autonomous VTOL Technical Meeting and Electric VTOL Symposium, Mesa, Arizona, UAS, Jan 29-31, 2019.

METHODOLOGY

The methodology section is divided into two parts: I. Aerodynamic Analysis and II. Acoustic Analysis.

I. Aerodynamic Analysis

An in-house Computational Fluid Dynamics (CFD) GPU-Accelerated Rotorcraft Flow-Field (GARFIELD) solver, was used to simulate the low aspect ratio propellers in hover. GARFIELD is a structured, time-accurate, overset solver, based on the Reynolds-Averaged Navier-Stokes (RANS) equations and has been widely tested (Ref. 3). GARFIELD has no limitation of the chosen propeller geometry so variations in chord, taper, twist, and sweep could be considered. The overset meshing feature of GARFIELD allowed for nearby geometries to be included with the propeller to capture the true aeroacoustics. Of the sixteen total cases presented, Cases 8-11 consider a propeller with a nearby boom geometry such as that seen in the Aurora Flight Sciences eVTOL aircraft.

The isolated propeller cases considered were run fully turbulent using the SA-DES turbulence model. A 5th order accurate reconstruction scheme, Weighted Essentially Non-Oscillatory (WENO) was used for the spatial discretization scheme. A finer mesh was placed around the blades to capture nearby transients, and the coarser background mesh was placed far enough away for the wake to blow down and to reach farfield conditions. The propeller cases were run out to at least 12 revolutions to ensure the root vortex had completely blown down to obtain an accurate power prediction. All propellers were trimmed to the same thrust of 1,000N and the blade loading, C_T/σ, was kept constant from case to case for fair comparison. The Figure of Merit, FM, was calculated to assess hovering efficiency to quantify the aerodynamic performance.

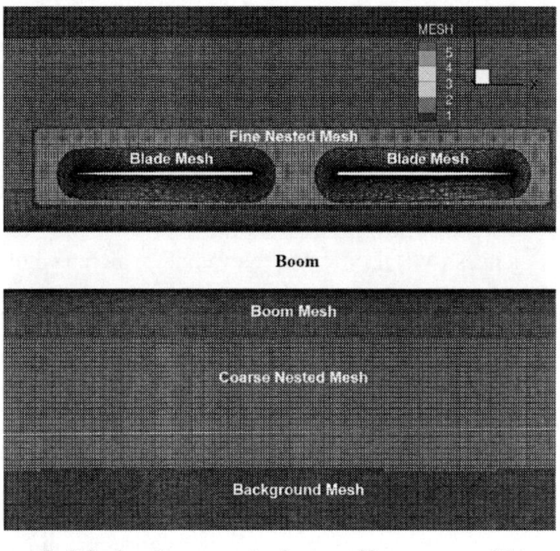

Figure 2. Mesh placement of propeller cases with nearby boom geometry

In order to accurately predict the interactional aerodynamics when incorporating the nearby boom geometry, additional meshes were added while being cognizant of mesh placement and overlap. A second, even finer nested mesh was placed around the blades to capture the resulting unsteady airloads. The relative mesh placement for the propeller cases including the boom geometry is shown in Figure 2.

A Blade Element Momentum Theory (BEMT) analysis was added to quickly assess the aerodynamic performance of propeller geometries of interest. The BEMT analysis was created in MATLAB and was based off of all fundamental equations given in *Principles of Helicopter Aerodynamics* (Ref. 4). No small angle assumptions were made and table lookup of lift and drag as a function of angle of attack and Mach number was used. The Figure of Merit values were much higher than those predicted by CFD, but BEMT properly captures the expected trends taking into account the limitations of the model. Both root and tip losses were included in the BEMT model to include as many approximations for complex effects as possible.

Propeller geometries of interest were first assessed in BEMT to obtain a quick prediction of the aerodynamic performance to evaluate if the case should be further investigated using CFD. Since one propeller revolution takes about 6-8 hours using GARFIELD, it was important to eliminate propeller cases in BEMT and choose only those cases worthwhile testing.

The aerodynamic normal and chordwise forces, C_nM^2 and C_cM^2, were then coupled to the acoustics solver. The normal and chordwise forces from the aerodynamic analysis are needed to convert the aerodynamic forces into pressure fluctuations and eventually into decibel (dB) levels to evaluate the resulting noise.

II. Acoustic Analysis

The in-house Acoustic Code University of Maryland (ACUM) accepts the airloads calculated from CFD (or BEMT) and solves the permeable surface form of the Ffwocs Williams and Hawkings (FWH) equation to obtain the resulting pressure fluctuations (Ref. 5). The FWH equation is well known in the aeroacoustic community and relies on near-field information over a surface that encloses the sources of noise as much as possible (Ref. 6). The main sources of noise considered in this study were thickness and loading noise. Thickness noise is the noise that results from the displacement of air by the volume of the blade element. Thickness noise can be optimized by adjusting the blade's geometry such as thin blade sections or planform sweep. Thickness noise is represented by a monopole source distribution and is important at higher speeds. Loading noise is the noise that results from thrust and torque forces acting on the blade from pressure differences across the surface. Loading noise is represented by dipole sources and can be optimized by decreasing the aerodynamic loads experienced by a blade (Ref. 2). Loading noise is dominant at low to moderate speeds, which is what is focused on in this study. Noise predictions are only as accurate as the aerodynamic interactions modeled, so great care must be taken with the aerodynamics to produce reliable and accurate acoustic results.

The pressure fluctuations found by the FWH equation were then converted into an overall sound pressure level (OASPL) in Decibels (dB) as a function of elevation angle. For an isolated propeller in hover, the noise is only a function of elevation and not the azimuth position of the blade. However, when introducing a nearby body, the unsteady airloads cause the noise to also vary in azimuth.

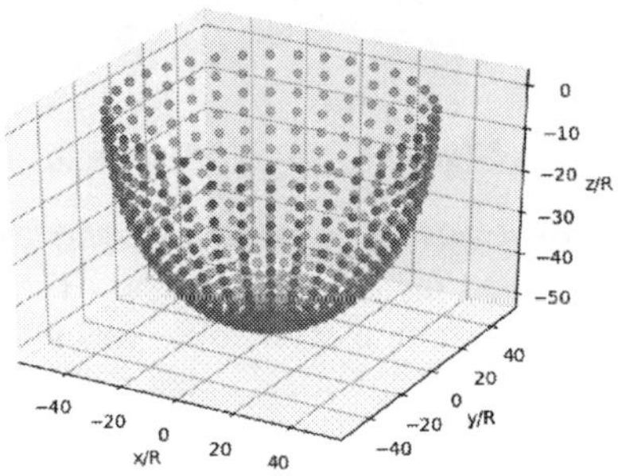

Figure 3. Observer locations placed in a 50 meter radius hemisphere

Polar directivity plots were often generated to see the sound pressure level (SPL) in dB above and below the rotor at observer locations 50 meters away. The observer locations were chosen at various elevation angles above and below the propellers so the observer locations formed a 50 meter radius sphere. In reality, most observers will be below the propellers since the eVTOL aircraft will be flying above most observers in an urban environment. Therefore, these full "sphere" polar directivity plots often only considered the bottom hemisphere, as that contained the observer locations of interest, and the highest levels of noise. The observer locations for the hemisphere are shown in Figure 3 and were discretized by 10° in azimuth and 5° in elevation.

For the propeller cases with the boom geometry, the maximum SPL was first found at a particular elevation, and then at that elevation the azimuth position where the maximum SPL occurred was chosen.

The A-weighted OASPL was also computed and found using a Fast Fourier Transformation (FFT), breaking down the acoustic time history into dB levels at different frequencies. To obtain the OASPL in dB and A-weighted decibels (dBA), the following procedure was used:

$$rms\ pressure = \sqrt{(p)^2_{average}} \tag{1}$$

$$SPL(dB) = 20log_{10}\left(\frac{rms\ pressure}{reference\ pressure}\right) \tag{2}$$

$$R_A(f) = \frac{(12194^2 f^4)(f^2+12194^2)^{-1}}{(f^2+20.6^2)\sqrt{(f^2+107.7^2)(f^2+737.9^2)}} \tag{3}$$

$$AWeight = 20log_{10}(R_A(f)) + 2.00 \tag{4}$$

$$dBA = dB + AWeight \tag{5}$$

Once the dB and dBA values were collected at each frequency they were summed up and put into an overall sound pressure level:

$$OASPL(dB) = 10log_{10}(sum(10^{dB/10}) \tag{6}$$

$$OASPL(dBA) = 10log_{10}(sum(10^{dBA/10}) \tag{7}$$

The OASPL in dB and dBA was compared across all sixteen propeller cases and were used as the metric for acoustic performance.

SUMMARY OF PROPELLER CASES

After preliminary analysis using BEMT, the baseline propeller geometry was chosen as an elliptic planform with linear twist. The elliptic planform was chosen due to the tapered geometry minimizing profile power, and increasing aerodynamic performance. The linear twist was chosen as a baseline since the twisted blade minimized induced power and thus increased aerodynamic performance. The baseline propeller geometry is shown in Figure 4.

Figure 4. Elliptic Planform - Baseline Propeller Geometry

The operating conditions for the baseline propeller case were chosen based off of approximations of what would be expected by a single propeller on the Aurora Flight Sciences eVTOL aircraft. The operating conditions are listed in Table 1.

These operating conditions were kept constant from case to case, except when the RPM and number of blades were changed. The blade loading was kept constant across all cases to ensure they could all be fairly compared to each other.

Once a baseline propeller configuration was decided upon, slight deviations from the baseline were considered to see their relative effects on aeroacoustic performance. An ellipse planform with two different modifications at 90%R to the linear twist distribution, and a case with zero twist, were

Table 1. Operating Conditions for Propeller Cases

Radius	0.925 m
Design Thrust	1,000N
RPM	1800
Number of Blades	2
Speed of Sound	340 m/s
Disk Loading	372.02 N/m
M_{tip}	0.5128
Airfoil	SC1095
C_t/σ	0.12

included to observe any effects of twist in the reduction of noise, which consisted of Cases 1-4 (Case 2 being the baseline case). A summary of these cases are shown in Table 2, with N_b as the number of blades.

Table 2. Overview of propeller test cases with twist variations

Case	Planform	Twist	RPM	N_b
1	Ellipse	N/A	1800	2
2	Ellipse	Linear	1800	2
3	Ellipse	Linear Increased at 90%R	1800	2
4	Ellipse	Linear Zero at 90%R	1800	2

Cases 5-7 were created to investigate the impact on aeroacoustic performance when increasing the number of blades and reducing the RPM. The same elliptic planform was used, with a linear twist distribution. The RPM was reduced from 1800 to 1470, so the same blade loading would be reached if the RPM was kept at 1800, but the number of blades was increased to three. The chord distribution was varied to achieve the same equivalent solidity between cases 5-7 to ensure the blade loading was kept constant. These propeller cases are outlined in Table 3.

Table 3. Overview of propeller test cases with number of blades and RPM variations

Case	Planform	Twist	RPM	N_b
5	Ellipse	Linear	1470	2
6	Ellipse	Linear	1470	3
7	Ellipse	Linear	1800	3

Next, Cases 8-11 implemented a nearby boom geometry to explore the resulting unsteady airloads and their effect on noise. The boom was placed 0.25 rotor radii and 0.50 rotor radii below the propeller to also investigate the effect of the placement of the boom on the noise. The baseline propeller with the boom 0.25R below is shown in Figure 5.

A semi-infinite boom was used for Case 11 where essentially only half the boom was considered. All cases except Case 10, utilized the baseline propeller geometry. Case 10 introduced aft sweep to the elliptic propeller. The aft sweep was accomplished by starting at the 50% spanwise position and sweeping the quarter chord such that with the elliptic planform the trailing edge became straight. The boom geometry being considered was a total of six rotor radii in length. The boom geometry cases are outlined in Table 4.

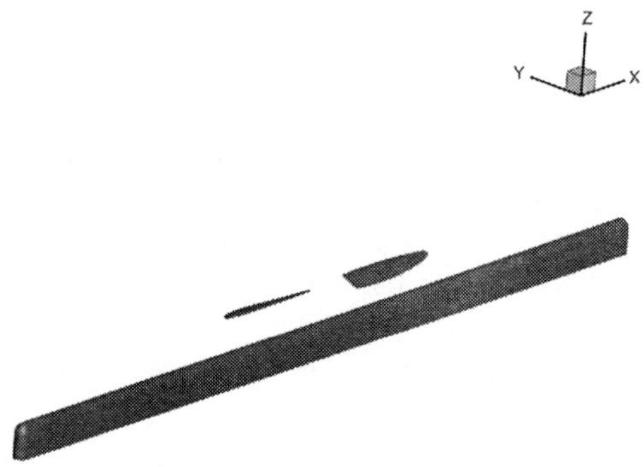

Figure 5. Boom Placement

Table 4. Overview of propeller test cases with nearby boom geometry

Case	Sweep	Type of Boom	Boom Location	Sweep
8	N/A	Full	0.25R	N/A
9	N/A	Full	0.50R	N/A
10	N/A	Full	0.25R	Aft
11	N/A	Semi-Infinite	0.25R	N/A

Finally, Cases 12 and 13 explored the effect of aft and forward sweep on the aeroacoustics without the boom geometry. Cases 14 and 15 included chord variations with rectangular and 2:1 tapered planforms, still with a linear distribution of twist, and Case 16 included transition modeling for the baseline propeller case with a free stream turbulence intensity (FSTI) of 0.10%. These are all outlined in Table 5.

Table 5. Overview of propeller test cases with chord, sweep, and turbulence modeling variations

Case	Planform	Sweep	Model
12	Ellipse	Aft	Fully Turbulent
13	Ellipse	Forward	Fully Turbulent
14	Rectangular	N/A	Fully Turbulent
15	2:1 Taper	N/A	Fully Turbulent
16	Ellipse	N/A	Transition, FSTI 0.10%

All sixteen of these cases were run in GARFIELD and coupled to ACUM to give the resulting Figure of Merits and OASPL to assess aeroacoustic performance.

AERODYNAMIC RESULTS

The aerodynamic results for the sixteen cases will be presented in two sections, I. Isolated Propeller Cases and II. Propeller Cases with Boom Geometry.

I. Isolated Propeller Cases

First, the effect of twist on the aerodynamic performance was evaluated for Cases 1-4. Each case was trimmed to as close to 1,000N as possible and the Figure of Merit was calculated based off of the power prediction from CFD or BEMT. The Figure of Merit is a hover efficiency, varying from 0 to 1, with 1 being 100% efficient. A good propeller design should yield a Figure of Merit in the range of 0.7-0.8. The Figure of Merits for Cases 1-4 are shown in Table 6.

Table 6. Final Aerodynamic Results for Cases 1-4

Case	BEMT Figure of Merit	CFD Figure of Merit
1	0.783	0.679
2	0.846	0.726
3	0.844	0.730
4	0.845	0.724

Case 1 was the elliptic planform with zero twist, which gives the lowest CFD Figure of Merit of 0.679. Once a linear twist distribution was added in Case 2, the CFD Figure of Merit jumped to 0.726. This makes sense since twist helps to minimize the induced power by having more uniform inflow over the blade. Looking at Cases 3 and 4, having an increased twist rate at 90%R or zero twist rate at 90%R has little impact on the aerodynamic performance. The BEMT results show the same increase in Figure of Merit from zero twist to linear twist, however the Figures of Merit are much higher. As stated above, this is to be expected since BEMT is an optimistic, lower fidelity solver and cannot model the more complex 3D effects in the flow. However, the same overall trends are captured when looking at BEMT and CFD. Having a linear twist distribution significantly improves the aerodynamic performance, but modifying the twist affects the Figure of Merit very little. Thus, introducing twist is worthwhile to increase the aerodynamic efficiency of the aircraft, but modifying the twist rate may only be beneficial if it has a significant impact on noise.

Cases 5-7 explored the impact of increasing the number of blades from two to three, and reducing the RPM from 1800 to 1470. For Case 5, when the RPM was reduced, the chord distribution had to be increased by 1.5 times to maintain the same blade loading. For Case 6, when the RPM was reduced and the number of blades was increased, the chord distribution stayed the same as the baseline case since the RPM reduced the blade loading by the same amount that the increased number of blades increased it by. For Case 7, the number of blades was increased only, therefore the chord distribution had to be decreased by 1.5 times to maintain the same blade loading. These variations resulted in different aspect ratios for Cases 5 and 7 and can be seen along with the CFD solutions at 12 revolutions in Figure 6.

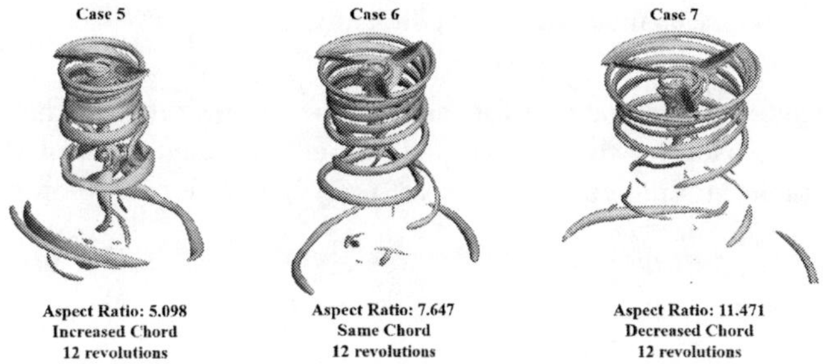

Figure 6. Cases 5-7 solutions

The Figure of Merit results are shown in Table 7.

Table 7. Final Aerodynamic Results for Cases 5-7

Case	BEMT Figure of Merit	CFD Figure of Merit
5	0.856	0.715
6	0.873	0.749
7	0.858	0.748

In order to easier see the effects of number of blades and RPM, the chordwise force was first multiplied by $N_b c/R$ and then was scaled by the local mach number, M to obtain a force analagous to power. The spanwise distribution of power for Cases 5-7 alongside Case 2, the baseline case, are shown in Figure 7. The area under each curve is equivalent to the power.

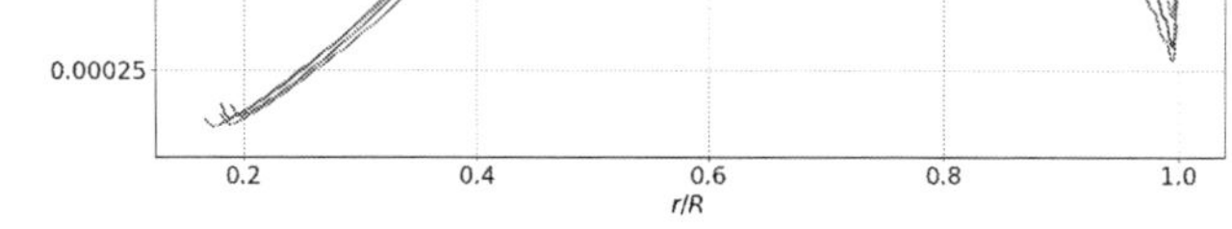

Figure 7. Cases 5-7 Power

Case 6 clearly shows the smallest area and thus the least power consumption, which should correlate to the highest Figure of Merit. This is shown in both the BEMT and CFD results. Since the RPM was reduced while increasing the number of blades, the loading on each blade is reduced causing less power consumption and higher efficiency. Therefore, Case 6 shows the largest Figure of Merit of 0.749. When the RPM was kept at the nominal value of 1800, and the number of blades was simply increased (Case 7), the Figure of Merit also improved from the baseline case, but came just shy of that of Case 6. If the RPM was simply reduced (Case 5), the aerodynamic performance decreased since the propeller was not able to produce as much thrust as efficiently.

The effect of sweep on the isolated propeller was also investigated. The amount of forward sweep was equivalent to that swept back for the aft sweep so the elliptic planform had a straight trailing edge. The swept propeller blades are shown in Figure 8.

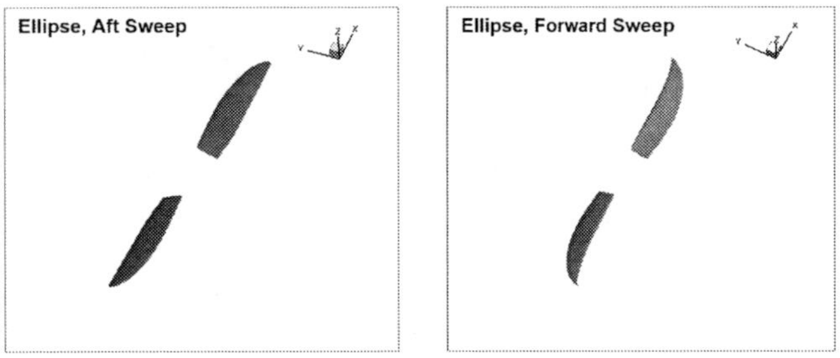

Figure 8. Aft and Forward Swept Elliptic Propeller Blades

In addition, a rectangular blade and 2:1 tapered blade were run, to illustrate the effects of chord variation. Lastly, Case 16 was the baseline case, except no longer run fully turbulent, to see how allowing a laminar region over the blade to occur would decrease drag and thus increase aerodynamic performance. The Figure of Merit results for the remaining isolated propeller cases are outlined in Table 8. The BEMT Figure of Merit is not reported for Cases 12, 13, or 16, since the current BEMT model cannot be used to predict the aerodynamic performance of swept blades and does not have a transition model.

Table 8. Final Aerodynamic Results for Cases 12-16

Case	BEMT Figure of Merit	CFD Figure of Merit
12	N/A	0.730
13	N/A	0.719
14	0.797	0.711
15	0.820	0.718
16	N/A	0.744

For Cases 14 and 15, the BEMT and CFD Figure of Merits follow the same trend, showing a slight benefit in aerodynamic performance by tapering the blade as was first expected when choosing the baseline elliptic planform. Case 12 showed a slight increase in aerodynamic performance by implementing aft sweep to alleviate the loading at the tip of the blade, whereas Case 13 with the forward sweep decreased the Figure of Merit from the baseline case. Case 16 showed significant improvement from the baseline case, as expected with decreased drag from the transition model versus a fully turbulent case.

II. Propeller Cases with Boom Geometry

In reality, propellers do not run in isolation so a boom was placed underneath the propeller to analyze its effect on the aerodynamic performance since it would most likely result in a distortion of the wake being blown down by the propeller. This in fact was the case as shown in Figure 9. The Figure of Merits for Cases 8-11 including the booms are outlined in Table 9.

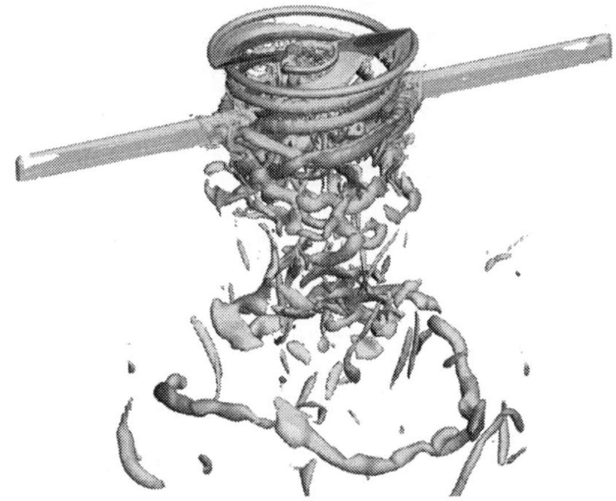

Figure 9. CFD solution of Boom 0.50R below propeller

Table 9. Final Aerodynamic Results for Cases 8-11

Case	CFD Figure of Merit
8	0.701
9	0.719
10	0.727
11	0.713

Case 8, was the baseline case with the boom placed at 0.25R below, causing quite a decrease to the aerodynamic performance from 0.726 to 0.701. When the boom was placed 0.50R below (Case 9), the Figure of Merit still dropped, but by a smaller amount. This makes sense because the closer the boom is located to the propeller the more the wake will be distorted and the more power it will take to provide the same amount of thrust. Case 10, used the aft swept propeller with the boom 0.25R below, and it resulted in a slight increase of Figure of Merit. This is because the aft swept propeller in isolation provided a small increase in aerodynamic performance, so even when the boom is placed underneath it is still more efficient than the baseline case and even more efficient than the baseline case with the boom placed further away. Case 11 contained the semi-infinite boom, which resulted in decreased aerodynamic performance as well, just not as much as the full boom for Case 8, but still more than the full boom placed farther away in Case 9. While the boom presence decreases the aerodynamic performance, it does not effect it by a huge amount. The unsteadiness of the blade was analyzed by looking at the variation of thrust in time, which can be seen in Figure 10. The thrust coefficient varies, but not substantially, however the resulting acoustics were analyzed to see if the boom had a large impact on the noise and if this should be of concern.

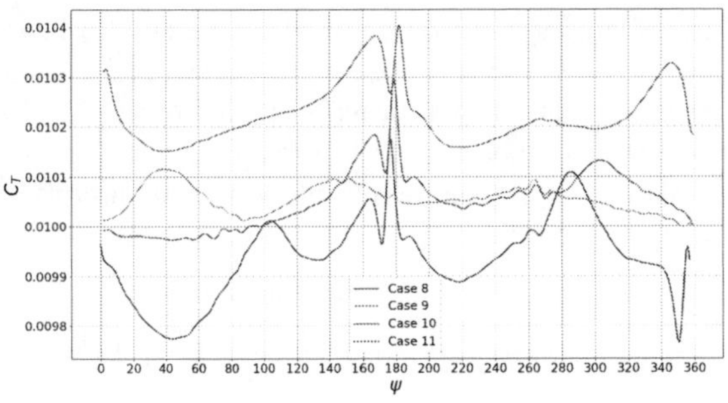

Figure 10. Thrust History of Boom Cases

ACOUSTIC RESULTS

The acoustic results for the sixteen cases will be presented in two sections, I. Isolated Propeller Cases and II. Propeller Cases with Boom Geometry.

I. Isolated Propeller Cases

The normal and chordwise forces, C_nM^2 and C_cM^2, were obtained by GARFIELD at 95 spanwise points along the blade, which were then input into ACUM and ultimately changed to pressure fluctuations via the FWH equation. The SPL in dB as a function of elevation angle was compared to that obtained by BEMT. The results for Cases 1-7 are shown in Figure 11.

The dotted lines represent the BEMT solutions, while the solid lines represent the CFD solutions. There is a consistent 1 dB shift from the BEMT solution to the CFD solutions for all cases shown. Since BEMT is an optimistic, lower fidelity solver, it makes sense that it is constantly predicting about 1 dB lower for all elevations as compared to that of CFD.

For Cases 1-4, it can be seen that adding a linear twist distribution decreases the maximum SPL by only about 1 dB, and further variations in the twist result in negligible SPL differences. This indicates that the design parameter of twisting the blade does not impact the noise. This suggests that twist only be used to help the aerodynamic performance since it should not have a significant negative or positive effect on the noise.

Cases 5-7 show a more promising impact on the noise. Once again BEMT shows a 1 dB optimistic shift from the CFD results. Case 6 is shown to be the quietest, almost about 10 dB quieter from the baseline case. This shows that when the RPM is reduced along with increasing the number of blades, a significant decrease in noise can result. It is no surprise that

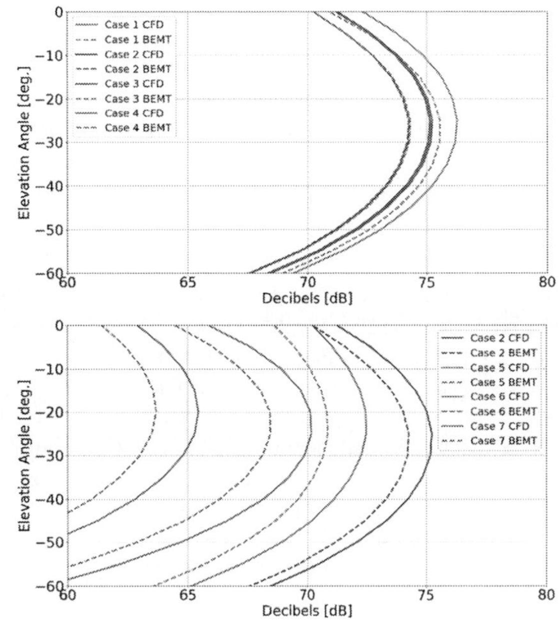

Figure 11. Polar Directivity Plots of Cases 1-4, and Cases 5-7 as compared to the baseline case, Case 2.

the most efficient case aerodynamically, is the most efficient case acoustically. This is due to the fact that the airloads are smaller, which contribute to the smaller power consumption, higher Figure of Merit, and then smaller pressure fluctuations and decreased noise. Furthermore, reduced tip speed has a strong effect on lowering noise. For all of these cases, the maximum SPL, or the OASPL occurred at an elevation 25° below the propeller.

Instead of solely looking at the dB levels, a FFT analysis was used to apply A-weighting to the dB levels at various frequencies. The A-weighted OASPL's were computed to compensate for the relative loudness perceived by the human ear. Some acousticians prefer to look at the dBA levels for this reason. The FFT analysis for the baseline propeller case is shown in Figure 12.

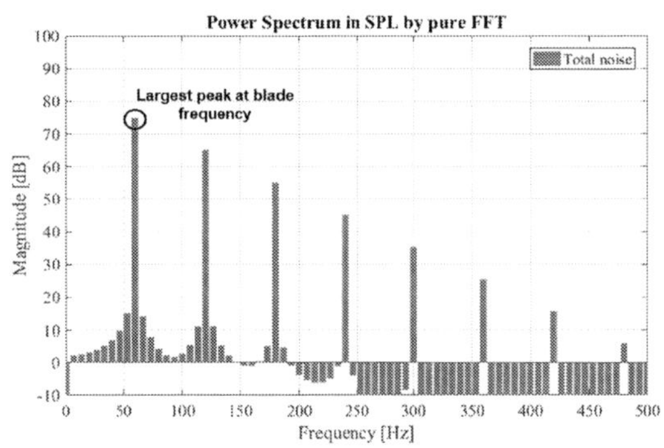

Figure 12. FFT Analysis of the baseline case of dB levels at various frequencies

The largest peak occurs at the blade frequency of about 60 Hz, which is the dominant frequency seen by the propeller. The dB levels slowly diminish at the higher frequencies. When the dB values are summed up, the maximum SPL at the blade frequency ends up becoming the OASPL. This is the noise level that is used to compare the propeller cases to one another, being careful to take into account the effect that higher frequencies may have on the noise levels. Since the noise was found to be mostly coming from lower frequencies for all of these propellers, the A-weighted OASPL's mostly followed the same trend as that without A-weighting.

This is more clearly shown in Figure 13. Case 6 illustrates the largest impact on minimizing the noise of the aircraft by reducing RPM and increasing the number of blades. Cases 1-4 show negligible differences, indicating that twist is not a design variable to help minimize noise. Cases 5 and 7 still show promising results, decreasing the OASPL by about 3 dB and 5 dB respectively. This identifies both RPM and number of blades as main drivers for improving aeroacoustic performance. The ideal minimization in noise seems to be achieved when both of these design parameters are utilized.

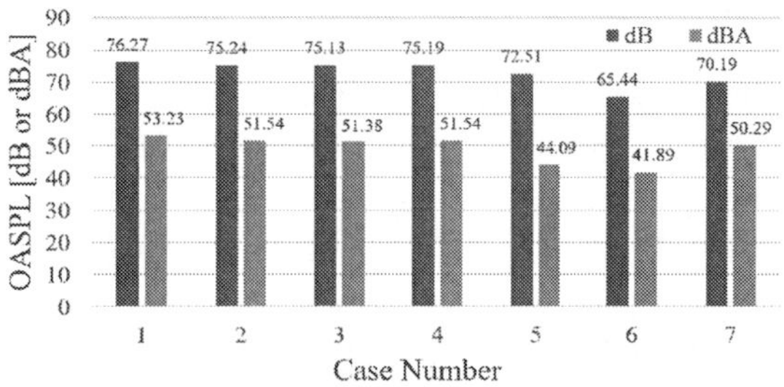

Figure 13. dB and dBA levels at 50 meter radius, 25° below the propeller, for Cases 1-7

A polar directivity plot was also created for isolated propeller Cases 12-16 and is shown in Figure 14. Again, the maximum SPL occured 25° below the propeller. When comparing Cases 12-16 with the baseline case, the forward swept propeller, Case 13, has the best acoustic performance. While the aft swept propeller, Case 12, showed an improvement in aerodynamic performance, the aft sweep resulted in a slight increase in noise (less than 1 dB).

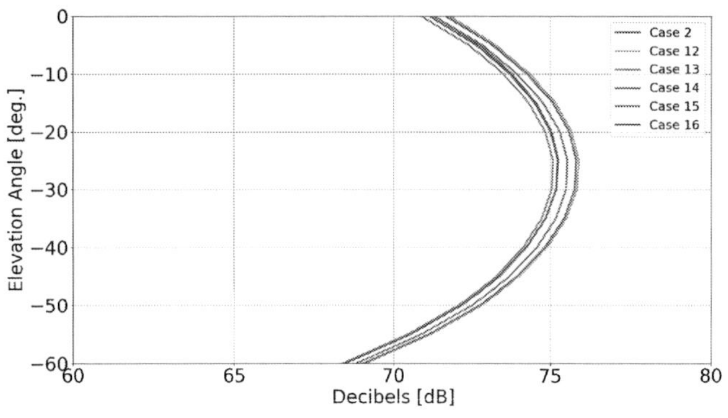

Figure 14. Polar Directivity Plot of Cases 12-16, as compared to the baseline case, Case 2.

This is most likely because in hover the acoustic planform is swept back and the aft sweep may be more aligned with the acoustic planform. This may result in the weaker sources of noise for the aft swept propeller arriving at more nearly the same time at the observer than for the unswept blade. The forward swept propeller had shown a decrease in aerodynamic performance, but indicates a slight benefit to the noise, since the sources of noise all arriving at the observer at the same time may be alleviated due to the forward sweep combating the acoustic planform. However, implementing a forward swept propeller does not impact the noise a significant amount, and arguably an observer won't be able to hear the difference between the forward sweep, aft sweep, or unswept blade since they are all within 1 dB of each other. This indicates that the design parameter of sweep may not be optimal to minimize noise, but can be used to make the aircraft aerodynamically more efficient.

The rectangular and 2:1 taper planforms (Case 14 and 15), show higher OASPL's than the baseline elliptic planform as expected. The elliptic planform decreases the noise less than 1dB from the rectangular planform, therefore, the design parameter of taper is also not promising when trying to minimize noise levels. Finally, Case 16 shows the slightest improvement from the baseline case (difference of about 0.02 dB) when transition modeling was used. The drag was decreased by allowing laminar flow over a region of the blade, but since the sources of noise coming from drag are much smaller than the normal forces, the decrease in drag barely impacted the noise level. This suggests that the majority of the simulations run fully turbulent are adequate when predicting noise levels since transition modeling had no significant effect.

II. Propeller Cases with Boom Geometry

For the propeller cases with the boom geometry, the OASPL was computed as a function of elevation and time. All boom cases contained the maximum OASPL at 25° below the propeller as before, but the maximum occurred at different azimuth locations. The variation in dB levels in time is shown in Figure 15.

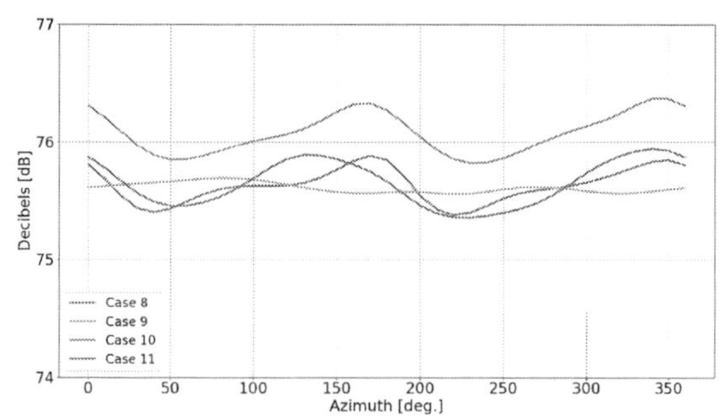

Figure 15. Sound Level Variation in time for Cases 8-11, 25° below the propeller

Case 9 showed the least variation in noise in time since the boom was placed further away. Case 11 showed the largest noise fluctuation due to the impulsiveness of the asymmetry of the semi-infinite boom. The aft swept propeller, Case 10, contained the highest OASPL for all azimuth locations. This agrees with the isolated aft swept propeller analysis above, since we would expect the aft swept propeller to be a detriment to the noise due to the acoustic planform.

The maximum OASPL for each boom case was chosen from Figure 15 and a FFT analysis was performed to obtain the OASPL in dBA as well. These results are shown in Figure 16 for comparison. The boom placement or type of boom did not show a significant variation in the maximum OASPL. Placing the boom further away only decreased the noise by about 0.2 dB, and the semi-infinite boom only increased the noise by about 0.05 dB, both of which would be negligible to an observer.

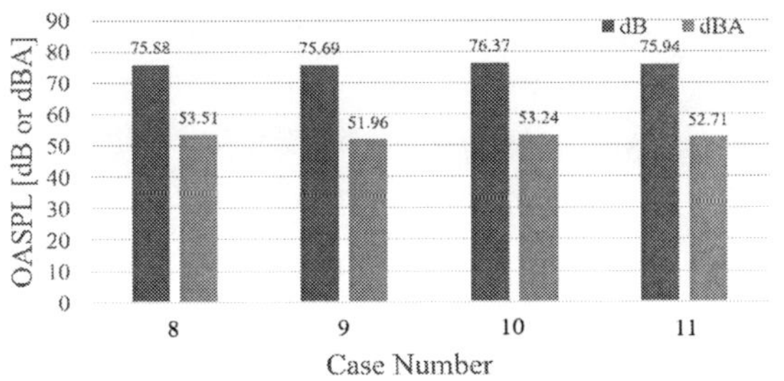

Figure 16. dB and dBA levels for Cases 8-11

The aft-swept propeller contained the highest OASPL of 76.37 dB compared to the boom placed at the same distance with the baseline propeller at 75.88 dB. This suggests that it may be acceptable to use aft-sweep to increase the aerodynamic performance and settle on the slight increase in noise depending on the specific design requirements. The aft swept propeller with the boom is less than 1 dB higher than the baseline case without the boom, which further points out that the boom presence and the swept design parameter do not significantly effect the OASPL. A boom is included in the Aurora Flight Sciences eVTOL aircraft design, and it is found that a boom does not significantly impact the aerodynamic or acoustic performance.

CONCLUSIONS

A successful computational aeroacoustic framework was created and implemented for various propeller configurations. Preliminary test cases were identified using BEMT and selected propeller cases were chosen to be solved using CFD in the in-house, widely tested and trusted GARFIELD solver. Aerodynamic forces were taken from both BEMT and CFD to be evaluated in the Ffwocs Williams and Hawkings (FWH) equation. Observer locations in a 50 meter radius hemisphere below the propeller were chosen and the sources of noise were computed as pressure fluctuations at each of these points. The OASPL was calculated using an FFT analysis to break down the dB levels into various frequencies to then be summed together. An A-weighting analysis was included to attempt to report noise levels that compensate for the relative loudness perceived by the human ear.

Various isolated propeller cases were run varying chord, twist, RPM, number of blades, sweep, and transition modeling. A plausible boom geometry was then placed underneath the propeller at various distances to investigate the effect of nearby geometries and unsteady airloads on the noise. Primary conclusions drawn from this study are:

1. Design parameters such as twist and chord variation did not show a significant impact on the acoustic performance, even though a tapered, twisted blade improved the aerodynamic performance.

2. Design parameters such as RPM and number of blades were identified as the main drivers of improving aeroacoustic performance. When these design parameters are used together the OASPL was decreased by about 10dB.

3. Aft sweep significantly improves the aerodynamic performance, but increases OASPL due to its likely alignment with the acoustic planform. Forward sweep decreases aerodynamic performance, but offers less than 1 dB alleviation in the noise. Therefore, sweep is not a recommended design parameter to improve aeroacoustic performance.

4. A nearby boom geometry did not significantly impact the aerodynamic or acoustic performance. The aerodynamic performance increased the further away the boom was placed as expected. The asymmetry of the semi-infinite boom caused larger pressure fluctuations and was louder than both full boom distances tested. However, the boom's presence, type of boom, and distance had negligible impacts on the acoustics to an observer (less than 1dB).

Author contact: Bernadine Passe bepasse@comcast.net James Baeder baeder@umd.edu

ACKNOWLEDGMENTS

The authors would like to thank Dr. Ashish Bagai and Martin Kearney of Aurora Flight Sciences for their guidance and support in this computational aeroacoustic analysis for eVTOL applications.

REFERENCES

[1] Swartz, I. Kenneth, *"Charging Forward, New eVTOL Concepts Advance"*, Vertiflite, July/August 2017.

[2] Hubbard, H. Harvey, et al. *"Aeroacoustics of Flight Vehicles: Theory and Practice"* Vol. 1 (1991), Chapters 1-2.

[3] Jude, Dylan and Baeder, James *"Extending a Three-Dimensional GPU RANS Solver for Unsteady Grid Motion and Free-Wake Coupling"*, AIAA Scitech, 2016.

[4] Leishman, J. G., *Principles of Helicopter Aerodynamics*, Cambridge University Press, New York, NY, 2000, Chapter 3.

[5] Amiraux, Mathieu *"Numerical Simulation and Validation of Helicopter Blade-Vortex Interaction Using Coupled*

CFD/CSD and Three Levels of Aerodynamic Modeling", University of Maryland, PhD Dissertation, 2014.

[6] A.S. Morgans, S.A Karabasov, A.P. Dowling, and T.P. Hynes, *"Transonic Helicopter Noise"*, AIAA Journal, Vol. 43, No.7 (2005), pp. 1512-1524.

Developing Noise Metrics for New VTOL Aircraft
A Progress Report

David Josephson *dlj@josephson.com*
Josephson Engineering, Inc.
Santa Cruz, California
Uber Technologies, Inc.
San Francisco, California

Existing aircraft noise measurements have been used to estimate effects of changes in distant, large airports but have not been shown to predict the acoustic impact of helicopters or urban VTOL aircraft. The emerging aerial ridesharing market requires a different approach to inform discussions with communities about the potential acoustic impact. An approach to predicting disturbance potential of integrated operations has been defined using partial specific loudness spectra, in the presence of a background soundscape. Time-varying loudness is used to assure that the additional loudness from impulsive sound such as rotor BVI is included in the estimate. Additional corrections are being investigated to approximate the annoyance of the aircraft sound, not only its loudness. Due to the variability of sonic background, no single spectrum is valid as a reference for this method when used for vehicle comparison, so a different metric is needed for that. Research is underway to define a weighting function and method for time-varying loudness estimation applicable to individual vehicles in typical urban background. In the interim, a simplified approach has been identified using components of existing aircraft certification and environmental acoustic metrics.

Introduction

Existing airport noise metrics were designed to identify compatible land uses in the vicinity of airports. Airport changes may reduce the value of property subject to increased noise impact, so prediction methods take on special importance. Annualized daily averages such as DNL are employed, along with tables of land uses considered compatible with certain long-term average levels of noise. While this approach has been successful in identifying properties requiring remediation when airports are expanded, it is of limited utility in evaluating new urban aircraft operations in close proximity to users. A new approach is required to define a scalable acoustic cost, that cities and aircraft operators can balance with the benefits of introducing urban air mobility to a community.

A "noise budget" is based on statistical analysis of people who will actually hear a given aircraft operation. Flight operations with more people, who experience higher likelihood of audibility are associated with potential for increased disruption of activities, and this framework is being used as the basis for listener surveys to calibrate the "noise cost" of each operation in a city. Audibility is specifically chosen over noisiness because in many cases it's the trigger for reaction whether or not the sound is intrusive.

Noise budget considerations drive vehicles to be quieter, flown over routes with fewer people present, higher background noise, and in operating modes that project less noise toward the ground. Route planning will depend on balancing these factors with trip length

Presented at The AHS International/Vertical Flight Society Technical Meeting and eVTOL Symposium.

costs, routing around other air traffic, and demand factors. A consistent requirement exists for vehicles that have the lowest likelihood for disturbance at a given distance. This paper describes methods for evaluating the components of audibility or annoyance based on physiological principles.

Two modes of community encounter with urban VTOL noise are anticipated: enroute and terminal. In the enroute phase, impact of overflights (which will be very quiet) is expected to be related to the number of overflight events and the peak loudness of each event that's noticeable in the ambient background. This supports the definition of an maximum overflight value, presented as an undertrack radius of acoustic impact probability around the vehicle on its flight path. Near a vertiport, however, people will be more accustomed to VTOL operation and an integrated exposure metric is expected to be the most indicative of acoustic impact, in addition to the event count. In each case, quantification of loudness or noisiness during each encounter is required.

Several metrics are being evaluated for use with emerging VTOL aircraft. It won't be possible to make a definitive choice until they have been tested with actual aircraft sounds and compared with human listener responses. While each has theoretical advantages, different questions are being asked and no single metric may be suitable. We need both a vehicle metric, as a design guide and standard of comparison, and an operations metric, that will help inform the discussion between aircraft operators and cities about the tradeoffs to be considered when establishing aerial ridesharing in a community. Such a multiple standard is well established in the transport jet community, with EPNdB used for certification, detailed spectra (by confidential arrangement with the airframe manufacturers) for propagation calculations for airport studies, and A-weighted SPL for discussions with the community.

Noise Certification

Initial evaluation of VTOL noise emissions uses metrics required for noise certification of aircraft according to US FAR 36, Appendix H or ICAO Annex 16, Chapter 8. The aircraft is flown on prescribed takeoff, overflight and approach profiles and three microphones undertrack and sideline to the flight path capture the sound.

Comprehensive corrections for atmospheric variation and flight path variations are built in to the protocol. The instantaneous measurement is tone-corrected effective perceived noise level or EPNL(T) and the event-integrated results are reported in EPNdB. This is an early metric developed by Kryter[1] and others at BBN to correspond to the noisiness of aircraft sound. If vehicles lack strong impulsive components or complex prominent tones, EPNL(T) is probably adequate to establish relative noisiness of the different vehicles proposed to date and known to the author, even if they have somewhat different spectral characteristics. EPNL(T) incorporates a fairly steep high frequency boost, that correponds with more modern research on sensitivity to noisiness, and a correction that tracks prominent tones. Future work will identify the expected errors when vehicles with unique sound characteristics are compared using EPNL(T).

Helicopter noise certification in FAR 36 sets the approach flight profile at 6 degrees, intended to capture the noisiest operating mode of a conventional helicopter being flown manually. Emerging VTOL aircraft are likely to be flown with a high degree of automated trajectory control, allowing, for example, precise selection of approach angles for minimum noise. When this is employed, the vehicle should be tested using the "variable noise reduction system" procedure, as described in the current ICAO Environmental Technical Manual[2]. This will permit the vehicle designer to specify the optimum operating modes for noise, and if necessary to require their use.

Sound Pressure and Exposure Metrics

Traditional urban noise studies have mostly been conducted using frequency-weighted sound pressure metrics based on studies done in the 1930s with single tones[3]. They were designed to evaluate the circuit gain needed to make speech audible on analog telephone circuits. Due to the relatively uniform spectral and temporal characteristics of most existing aircraft, simplified metrics using A-weighted sound pressure level have been sufficient for comparisons between types, although A-weighting is optimized for lower sound pressure levels than are expected for overflights.

In existing exposure metrics such as DNL, sound pressure level is integrated over the duration of the loudest part of an encounter. A simple power integration

$$L_{AE} = 10\ Log_{10} \frac{1}{T_0} \int_{t_1}^{t_2} \left(\frac{P_A(t)}{P_0} \right)^2 dt\ dB$$

is used to define L_{AE} as the continuous A-weighted sound pressure lasting T_0 seconds containing the same energy as the aircraft noise from t_1 to t_2, where that interval is the period during which the sound level is within 10 dB of the encounter maximum. When normalized to T_0 = 1 s, this is referred to as sound exposure level or SEL. A similar integrating function can be used with other instantaneous metrics.

A-weighted SPL is effective for comparing noises that are otherwise similar in spectral and temporal content. Extensive research on hearing[4] has established physiological bases for large differences in perceived acoustic magnitude between two sounds having identical sound pressure levels. An early experiment by Steudel[5] in Germany describes "the amazing fact that a 1000 Hz tone is about as loud as when only one in twenty half-cycles of it are presented." The latter case represents a sound pressure level 15 dB less than the former, but indeed the two sounds he describes are about equally loud, due to the wider excitation pattern (higher harmonic content) of the latter.

Components of Loudness

Since emerging VTOL aircraft may have very different sonic signatures, a more accurate instantaneous metric than sound pressure level is needed to compare them. Otherwise, a vehicle that is actually louder or more annoying may be evaluated as having a lower acoustic magnitude. We're interested in understanding how humans perceive sound, in order to minimize the effect that new aerial operations have.

Loudness is the most important part of this potential effect, and estimating loudness requires understanding of how the brain – long before it recognizes or reacts to the sound – processes acoustic input. Human hearing functions through the auditory cortex, which receives trains of neural impulses from frequency-selective regions along the basilar membrane of the cochlea. The excitation pattern of impulses over the roughly 30,000 nerve fibers from each ear, forms a spectrum analyzer that informs the brain of relative magnitudes of spectral components received at each interval in time. In order to estimate the acoustic magnitude of sounds, it's helpful to analyze the human hearing function in terms of excitation patterns that produce equal sensations of loudness, rather than simply measure the sound pressure.

A steady sinusoidal pressure variation will excite one group of nerve fibers corresponding to one location on the basilar membrane. The sensation of loudness corresponds to the number of impulses received per unit time. Sounds at a different frequency excite a different group of fibers, and if the two frequencies are far apart, the brain sums the loudness of the two groups as if they were separate sounds. If the sounds are close together in pitch, the stronger sound can raise the detection threshold for nearby spectral regions, known as frequency masking.

Temporal masking also occurs, resulting in different responses depending on length of excitation. Short bursts of sound are perceived as nearly equal to their peak value while the loudness of somewhat longer intervals is perceived as closer to their rms average. Time constants are different for onset and decay.
In order accurately to report physiologically perceived loudness, we must adjust at least for

- Frequency and amplitude of each spectral component (correcting for varying sensitivity with frequency)
- Spectral summation
- Spectral masking (additional loudness due to multiple widely separated frequencies)
- Temporal masking (additional loudness due to rapid onset of sound)

Just as Fletcher and Munson did with single tones in silence, loudness is studied by comparing sounds, asking subjects to adjust levels of sounds until they are of "equal" loudness, or to report when one sound is about double, triple, etc. the magnitude of another. More recent work[6] Patterns have emerged to quantify the components of loudness, leading to two current standardized methods of assigning numeric values to the experience[7]. One is the one-third-octave band method of Zwicker, described in ISO 532-1[8]. It is appropriate for broadband sounds without strong spectral discontinuities. A more recent method was developed by Moore and Glasberg[9] and is described in ISO 532-2[10] and referred to as the Cambridge loudness method[11]. Recently this method has been extended to cover additional loudness due to temporal characteristics, which is now being standardized as ISO 532-3 or TVL, (currently in FDIS stage of standards.) This is the most comprehensive perceptual model for sound perception available today and forms the basis of our noise metric. While it's the best we have, TVL still underestimates the noisiness of some sounds, and further work is required. A loudness method is considered to be more accurate when it corresponds more closely with human listener trials for the sounds being tested.

The result of a loudness estimation is stated in the linear unit sone, or the logarithmic-like unit of loudness level, phon. A sound having a loudness of 1 sone or 40 phon is as loud as a 1 kHz tone of 40 dB SPL. Twice as loud is 2 sone or 50 phon.

A preliminary loudness metric applicable to aircraft comparison follows the flight test method of ICAO Annex 16 but substitutes TVL for EPNL. The result would be reported as maximum long-term loudness in sone for each of takeoff, overflight and approach modes rather than L_{Amax} or $EPNL_{max}$, and using the same microphone locations.

Partial Loudness and Background

Once the momentary loudness of an aircraft is determined, it can be used to replace sound pressure or EPNL(T) in evaluating the difference between aircraft types. Schomer[12] has suggested "loudness level sound exposure level" or LLSEL to replace A-weighting in SEL metrics. But, aircraft don't operate in a silent background, so it's also necessary to evaluate the degree to which ambient sounds mask the audibility of an aircraft. This is particularly important at medium to close range where atmospheric attenuation isn't dominant as it is for jets at distant airports. Studying the loudness exposure of communities near vertiports will require evaluation of this masking effect.

The total loudness a listener experiences can be separated into the contributions of each sound source. Vehicle partial loudness is the contribution of a vehicle operation to the total, when in the presence of the background. We are concerned about background spectrum because it determines how much of the vehicle sound is masked.

A study of eight locations in a major metropolitan area is being completed to derive representative loudness spectra over different levels of road traffic activity. It was begun in late February, 2018 and will continue for 12 months, but clear patterns have already been found. Throughout the day, atmospheric attenuation (lower frequencies travel further before being diminished by the atmosphere) establishes a baseline energy spectrum from natural sources where the sound pressure level in each octave is roughly 6 dB less than the octave below it. There is a broad peak around 1 kHz from tire noise on highways, the magnitude of which can be correlated with average traffic speed, volume and distance from the highway[13]. Additional spectral details are due to patterns of local activity.

The background noise study also demonstrates that a time-varying metric was not necessary to capture loudness increase due to rapid onset of vehicle traffic events. A significant difference between results from stationary loudness and time-varying loudness methods was observed only at monitoring locations within 500 ft from a freeway. This realization makes collection of background loudness much simpler, as slower spectral averaging is sufficient to capture the background noise characteristic.

Because the background varies so much over the course of a day, selecting representative spectra for estimation of vehicle partial loudness is not practical. The noise study is being used to define a predictive model to estimate background noise spectrum and level at any location based on road, rail and air traffic in the vicinity.

Loudness Demonstrations

Loudness has sometimes been described as a psychological or "subjective" quantity, by definition variable from one subject to the next. A few examples of loudness variations due to spectral or temporal complexity can demonstrate that the differences are fairly uniformly perceived by a broad range of people. The slide deck accompanying this paper[14] includes demonstrations of loudness due to wider excitation patterns, and the significantly increased audibility of time-varying sounds either in silence or in the presence of typical urban background. Examples are presented of synthesized and actual helicopter sounds that are clearly audible even though their level is 14 dB less than the background sound in which they are presented. Loudness models aim to predict the loudness perceived by the average listener, and are an objective measure that can be calculated.

Application of Loudness to Vehicle Comparison

There is too much variation both in spectrum and temporal characteristics to pick just one background sound for use in partial loudness estimation, and it is impractical to optimize vehicle design for a variety of background noise conditions. Therefore we are developing a variant of the TVL method, time-varying noisiness or TVN, that captures noisiness while preserving its sensitivity to additional loudness of time-varying events such as rotor BVI noise.

Loudness metrics are validated by asking human subjects to rate the relative loudness of different sounds, or to adjust their amplitude so they seem to be equal. While loudness is the most important part of noisiness or annoyance with a sound, it is not the only variable. Further work is required to capture "equal noisiness" contours among people listening to these sounds. Rizzi, Christian[15] and others have suggested that sound-quality metrics such as sharpness, harshness, roughness and fluctuation strength may be needed to make effective comparisons of noisiness or annoyance between different sounds. Some of these parameters, such as roughness and fluctuation strength, are already in principle covered by the time-varying loudness method we are using, but adjustments of the weighting functions by frequency may be needed to replace an external sharpness correction, for example. TVN is functionally similar to a noisiness correction applied to TVL.

Interim Metrics

While progress is made on a comprehensive noisiness metric, interim procedures are needed to compare emerging aircraft designs. EPNdB is not ideal but given the range of feasible VTOL aircraft, probably good enough to gauge relative noisiness, with the exception of impulsive sounds from blade-vortex interaction in some helicopters. Molino[16] reviewed proposals for an impulsiveness correction and suggested that one was not needed; we draw a different result from review of the same reports. Nordtest NT ACOU 112,[17] which was initially developed for wind-farm measurements, adds a penalty of up to 5 dB (similar to methods suggested in Molino's report) depending on extent of rapidly-varying sound components. This is a simple way to achieve similar results to application of time-varying loudness in accordance with newer loudness metrics.

Aircraft having no impulsive sounds and only a single dominant tone can be compared using unmodified EPNL(T) or EPNdB. When rapid onset events are present, NT ACOU 112 corrections can be employed as suggested in BS 4142[18] to follow the additional loudness from this source.

Conclusions

Initial evaluation of aircraft noisiness can be accomplished using measurements compliant with FAA/ICAO noise certification flight tests, using the special noise reduction system protocol when necessary to allow the use of intended approach and takeoff flight paths. The result is reported in EPNdB for each of three flight procedures and is comparable with results obtained for helicopter noise certification.

A noise budget for integrated exposure in a city can be computed using time-varying loudness in the presence of background sounds for each flight operation. Work is underway on suitable reporting metrics that capture integrated annoyance over the affected population.

When vehicles having widely varying spectral and/or temporal noise characteristics are compared, the EPNL(T) metric used in noise certification may not be appropriate. The flight protocol and its corrections may be used, but with time-varying loudness (TVL) in place of EPNL(T) as the instantaneous magnitude during the test. A modification to TVL, or TVN, may be necessary to capture time-varying noisiness rather than time-varying loudness.

Acknowledgements

This work was begun in 2016 with initial funding from the NASA On-Demand Mobility program, through National Institutes of Aerospace subcontract T16-6500-JEI. Subsequent support has been from Uber Technologies' Elevate project. The assistance of Dr. Josef Schlittenlacher, Cambridge University Neuroscience has been greatly helpful.

References

[1] Kryter, K. D. "Review of research and methods for measuring the loudness and noisiness of complex sounds," NASA CR-422 (1966) http://hdl.handle.net/2060/19660011809

[2] ICAO Doc 9501, Environmental Technical Manual Vol. 1 https://www.icao.int/environmental-protection/Documents/SGAR.2015.ETM.Vol.1.pdf

[3] Fletcher, H. and Munson, W. A. "Loudness, its Definition, Measurement and Calculation" Bell System Technical Journal 12:4 (October 1933)

[4] Zwicker, E. and Fastl, H. Psychoacoustics Facts and Models, Springer (3rd Ed., 2006)

[5] Steudel, U. Über Empfindung und Messung der Lautstarke (Mitteilung aus dem Institut für Schwachstrom Technik, Technische Hochscule Dresden, 1932 (translation at www.josephson.com/aero/pub/Steudel_trans.pdf)

[6] (Levitt, 1971) https://asa.scitation.org/doi/abs/10.1121/1.1912375

[7] Scheuren, J. "ISO 532 – Living and working with alternate loudness standards" INTERNOISE 2014 conference https://www.acoustics.asn.au/conference_proceedings/INTERNOISE2014/papers/p941.pdf

[8] International Standardization Organization, ISO 532-1:2017 Acoustics – Methods for Calculating Loudness Part 1

[9] Moore, B. C. J., Glasberg, B., Baer, T. "A Model for the Prediction of Thresholds, Loudness and Partial Loudness" Journal of the Audio Engineering Society 45:4 (1997)

[10] International Standardization Organization, ISO 532-2:2017 Acoustics – Methods for Calculating Loudness Part 2

[11] Moore, B. C. J. "Development and Current Status of the Cambridge loudness methods," Trends Hea. 2014; 18 https://www.ncbi.nlm.nih.gov/pmc/articles/PMC4227665/

[12] Schomer, P. "Evaluation of loudness-level weightings for assessing the annoyance of environmental noise." J. Acoust Soc Am 110:5 (November 2001)

[13] Can, A., Leclercq, L., Lelong, J., Bottledoren, D. "Traffic noise spectrum analysis: dynamic modeling vs. experimental observations." Applied Acoustics, Elsevier, 2010, vol71, n8, p764-70.

[14] http://www.josephson.com/pub/aero

[15] Rizzi, S. et al, "Annoyance to Noise Produced by a Distributed Electric Propulsion High-Lift System" AIAA Journal 10:2514 (2017)

[16] Molino, J. "Should Helicopter Noise be Measured Differently from Other Aircraft Noise? – A Review of the Psychoacoustic Literature" (1982) NASA Contractor Report 3609 https://ntrs.nasa.gov/archive/nasa/casi.ntrs.nasa.gov/19830005666.pdf

[17] Nordtest, Prominence of impulsive sounds and for adjustment of LAeq (NT ACOU 112)

[18] British Standards Institute, Methods for assessing sound of an industrial or commercial nature (BS4142:2014)

Aerodynamic and Mechanical Design of a Morphing Winglet for a Quadrotor Biplane Tail-sitter

Peter Ryseck
Undergraduate Research Assistant

Derrick Yeo
Assistant Research Scientist

Vikram Hrishikeshavan
Assistant Research Scientist

Inderjit Chopra
Alfred Gessow Professor and Distinguished University Professor
Director of Alfred Gessow Rotorcraft Center

Department of Aerospace Engineering
University of Maryland, College Park, MD 20740

ABSTRACT

Multi-mode micro air vehicles with hover and high cruise speed capabilities can serve a wide range of mission requirements but are limited in their performance in either mode due to the design compromises that come from operating across a wide speed range. This paper describes a morphing winglet for a Quadrotor Biplane Tail-sitter (QBiT) that is proposed to improve the efficiency within a broad flight envelope. The ability of the morphing winglet to increase lift at low speeds and decrease lift at high speeds is quantified in three ways. First, a vortex-lattice method solver (VLM) that accounts for vortical interactions between wing surfaces was used to determine lift characteristics and aerodynamic loads on the vehicle at various dihedral angles and aircraft angles of attack. Through this investigation, it was found that a dihedral angle of 10° generates maximum lift regardless of angle of attack. Second, a VLM analysis was used again to understand the lift distribution along the span of the upper and lower wings at two specific dihedral angles: -90° in box wing mode and 20° in open winglet mode. Third, experimental flight tests showed that open winglet mode provided a 30% power difference at 4 m/s forward flight versus box wing mode. Additionally, for the same 175W power consumption, box wing mode supports a higher cruise airspeed of 7m/s compared to 5.5m/s in open winglet mode. Thus, it is envisaged that a 20° dihedral angle can be useful for low speed cruise mode and -90° dihedral angle for high speed mode when conducting a mission.

INTRODUCTION

Morphing aircraft attempt to expand their operational capabilities by tailoring their configuration to different flight regimes. Today, morphing technologies are being investigated for small rotorcraft unmanned air vehicles (UAV) for military, package delivery, and search and rescue applications. Current advancements in morphing schemes show that aircraft size, flight range, and flight performance envelopes can be further improved over baseline characteristics (Ref. 1).

Organisms such as birds and flying insects often morph their wings in order to execute different flight modes. Inspired by nature, vehicles such as the Variable Gull-Wing Morphing Aircraft mimic the variable wing geometry of a seagull allowing the glide ratio to change from 1 to 11 depending on the geometry of the wings (Ref. 2). Thus, the morphing capability of the vehicle increases the mission capabilities by allowing for a stable descent with both a minimized and maximized horizontal glide distance.

Morphing technologies have also been investigated as a means of providing roll control and lateral and directional stability. Mills and Ajaj studied the performance characteristics of folding wingtips to serve as control effectors for a micro-air vehicle (MAV) and found that varying the winglet angle improved stability and could augment existing ailerons for roll control (Ref. 3).

Figure 1. Current version of the Quadrotor Biplane Tail-sitter (Ref. 4)

In this work we design a morphing winglet system to expand the speed range of a Quadrotor Biplane Tail-sitter QBiT. Figure 1 shows a current version of the QBiT. It uses four rotors for thrust in both hover and forward flight mode. In hover, the trailing edges of the biplane wings point down with the rotor thrust vector pointing up. Once in forward flight, the vehicle uses its biplane wings to provide the necessary lift, while using RPM variation to produce control moments of roll, pitch and yaw. As a result of the lift that is produced by the wings in forward flight, the thrust and power required for steady level flight is significantly reduced compared to a conventional quadrotor.

Presented at the 2019 8th Autonomous VTOL Technical Meeting & 6th Electric VTOL Symposium, Mesa, AZ, USA, Jan. 29-31, 2019.

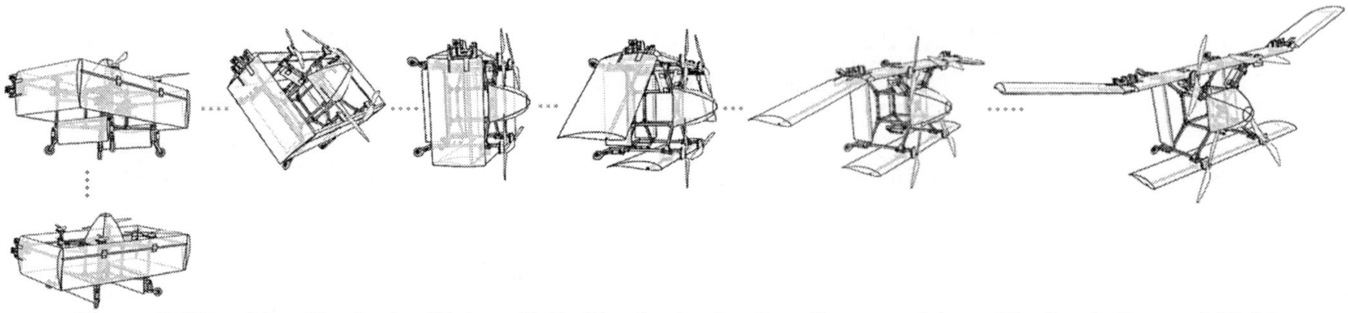

Figure 2. Morphing Quadrotor Biplane Tail-sitter beginning from hover and transitioning to forward flight

Current versions of the QBiT have fixed wing geometries which are inevitably sub-optimal for diverse mission scenarios. Hrishikeshavan, et.al found that the sizing of a QBiT wing and aspect ratio significantly impact flight performance and power efficiency at various operating speeds (Ref. 4). Thus, a morphing winglet applied to the QBiT configuration could broaden the flight capabilities while also consuming less power (Fig. 2).

The goal of this research is to improve and further expand the multi-mission capabilities of the QBiT platform. In this paper, we present the design and fabrication of the morphing QBiT and evaluate aerodynamic advantages through the use of experimental test flights as well as a vortex lattice-method (VLM) solver. The paper first discusses the aerodynamic design of the winglet using VLM and the performance differences between wing geometry with and without winglets. Next, the mechanical design of the morphing winglet and vehicle platform is discussed. Lastly, key findings from experimental flight data are presented.

AERODYNAMIC ANALYSIS AND DESIGN

A Vortex Lattice Method (VLM) 3D solver was used to determine two metrics: First, estimate the winglet dihedral angle (Γ) (Fig. 3) for maximum lift and then second, determine the lift distribution of the upper and lower wings at two fixed dihedral angles as well as a baseline case without winglets. The VLM solver, Tornado (MatLab based), offers a first order analysis based on the theory of an ideal flow (Ref. 5). The solver assumes 1) incompressible, inviscid and irrotational flow, 2) lifting surfaces are thin and 3) angle of attack is small. These assumptions are suitable for the low angle of attack cruise regime.

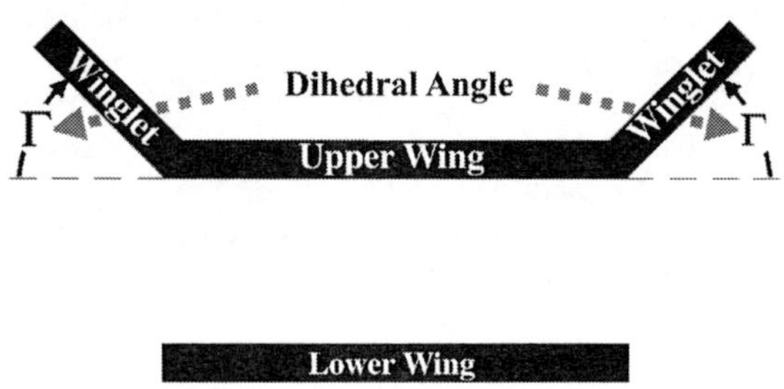

Figure 3. A simplified diagram of the vehicle defining the upper main wing, lower main wing, winglet and dihedral angle

Maximum Lift at Cruise Configuration: Results and Discussion

Our initial objective was to determine the dihedral angle of the winglet for maximum lift. To determine this, linear mesh types with dihedral angles varying from -90 to 90° at angles of attack from 0 to 15° were tested. Figure 4 shows three example geometries.

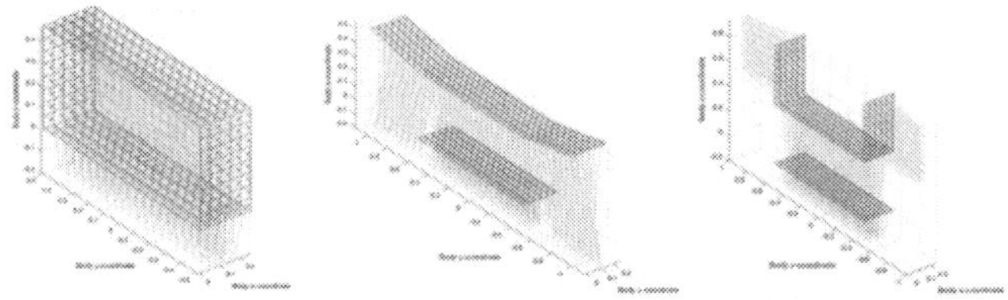

Figure 4. Vortex Lattice Method Geometry Examples. Dihedral angles from -90 to 90° were tested.

The results yielded coefficient of lift (C_L) values of the main wings and winglets. These values were then used to determine the total lift force (L) of the vehicle using Eq (1). A velocity (u) of 15 m/s was chosen based on previous QBiT flights. The surface area (S) of each wing was used to determine lift.

$$C_L \equiv \frac{L}{qS} = \frac{L}{\frac{1}{2}\rho u^2 S} \tag{1}$$

The values of C_L generated from VLM were used to produce Total Lift vs. Dihedral Angle, as shown in Fig. 6. As shown by the dotted line, maximum lift occurs when the dihedral angle is set to 10°. This result was consistent from 0 to 15° angle of attack.

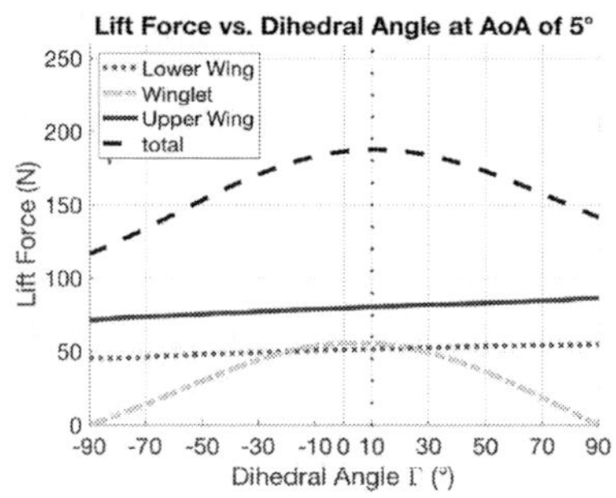

Figure 5. Lift vs. Dihedral Angle at AoA of 5°. Maximum lift obtained at a positive dihedral angle of 10°.

This result is in agreement with Gerontakos and Lee who found that a wing tip extension mounted at a dihedral angle induced additional lifting circulation not present on a planar wing (Ref. 6). At a dihedral angle of 10°, the tip-vortex flow is diffused, reducing the induced drag (Ref. 6). Beyond 10° however, this effect is diminished by the reduction of effective wing span that accompanies a larger tip angle.

Lift Distribution at Cruise Configuration: Results & Discussion

Two dihedral angles were further explored to analyze the lift distributions of the upper and lower wings. First, a dihedral angle of -90° representing box wing mode. Second, a dihedral angle of 20° representing open winglet mode. A third geometry was also created, representing a baseline case with no winglets. The lift sharing between the upper and lower wing changes based on the configuration type, with all three configurations producing more lift on the upper wing than the lower. Figures 6(a), 6(b), 7(a) and 7(b) show the results of these cases at an angle of attack of 0 and 10° of the lower and upper main wings (upper wing refers to the upper main wing without winglets), as is outlined in Fig. 3. Tables 1(a) and 1(b) show the specific C_L of these same upper and lower wings in their respective flight mode and angles of attack.

At both 0 and 10° angle of attack, the baseline configuration has similar lift distributions on the upper and lower wings, with 6% more overall C_L on the upper wing than the lower at both angles of attack. This effect is further illustrated with a delta C_P distribution (Figs. 8-9). As was found in the C_L distribution plots, the pressure distribution (delta C_P) of the baseline case is similar on the upper and lower wings.

This symmetry between the upper and lower wing lift distribution is affected when winglets are added to the baseline wing system. Box wing mode results in approximately 54% more overall C_L on the upper wing than the lower for 0° angle of attack. Open winglet mode produces approximately 33% more overall C_L on the upper wing than the lower for 0° angle of attack. In box wing and open winglet mode for 0° angle of attack, the C_L distribution of the upper wing is mostly consistent at a C_L of 0.7, whereas the lower wing more closely represents the baseline case. This asymmetry is observed again in the delta C_P distribution, which shows that the delta C_P along the span of the upper wing is mostly constant whereas the baseline configuration is not, as shown by the lighter regions on the wing tips.

At 10° angle of attack, the trends change for box and open winglet mode. Overall, box wing mode produces approximately 25% more overall C_L on the upper wing than the lower for 10° angle of attack. Open winglet mode produces approximately 38% more overall C_L on the upper wing than the lower for 10° angle of attack. Open winglet mode's upper wing C_L distribution remains mostly constant as was found at 0° angle of attack, but the box wing's upper wing begins to taper off at the wingtips and instead follows a trend skewed more similarly to the baseline case as shown in Fig. 7(a). On the lower wing, the C_L distribution at 10° angle of attack show that all 3 cases follow the same trend and remain at the same values of C_L. The delta C_P distribution shows the same results. In open winglet mode, the distribution of the upper wing is consistent along the span at both 0 and 10° angle of attack, whereas the distribution in box wing mode and the baseline case is not.

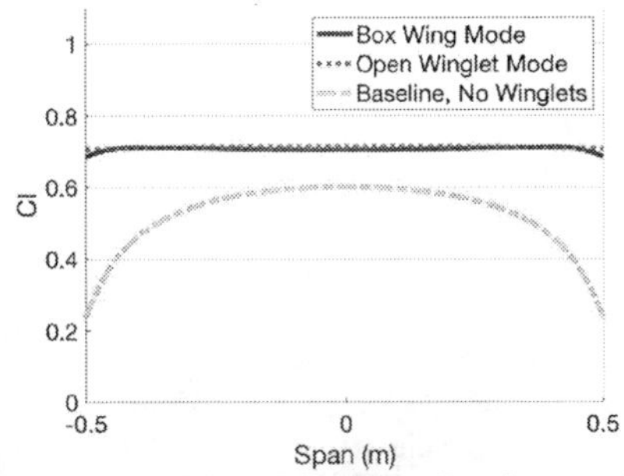

Figure 6a Upper Wing Lift Distribution at AoA of 0°: Open, Box and Baseline Geometries. Box and open winglet mode have greater C_L distributions compared to the baseline configuration.

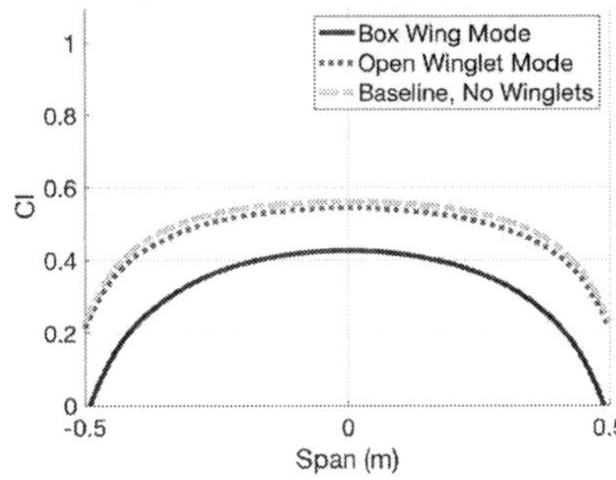

Figure 6b. Lower Wing Lift Distribution at AoA of 0°: Open, Box and Baseline Geometries.

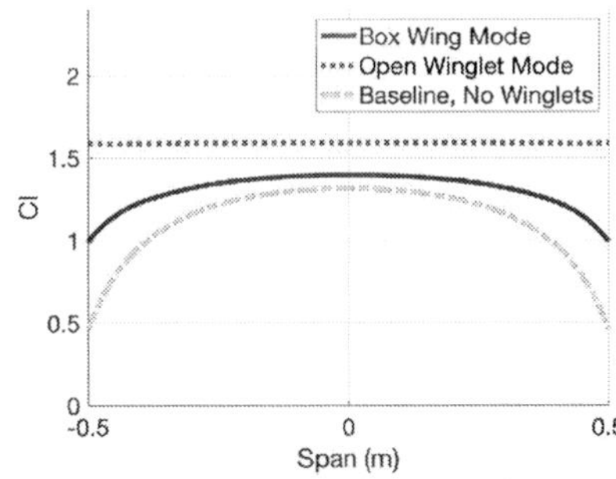

Figure 7a. Upper Wing Lift Distribution at AoA of 10°: Open, Box and Baseline Geometries. Box and open winglet mode have greater C_L distributions compared to the baseline configuration.

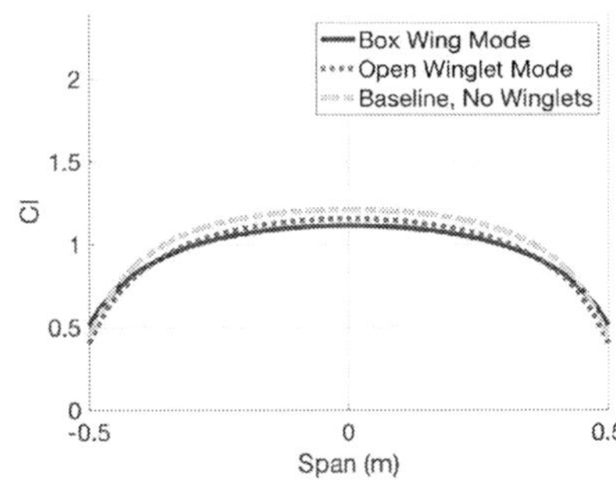

Figure 7b. Lower Wing Lift Distibution at AoA of 10°: Open, Box and Baseline Geometries. All three geometries have similar C_L distributions.

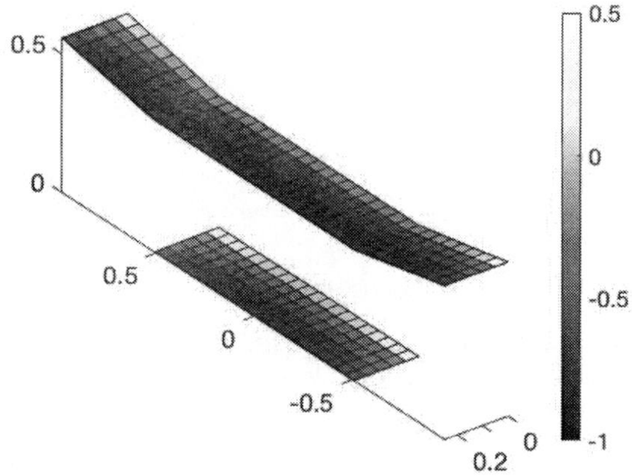

Figure 8a. Delta CP distribution at 0 degrees AoA: Open Winglet. Delta CP is constant along the span of the upper wing.

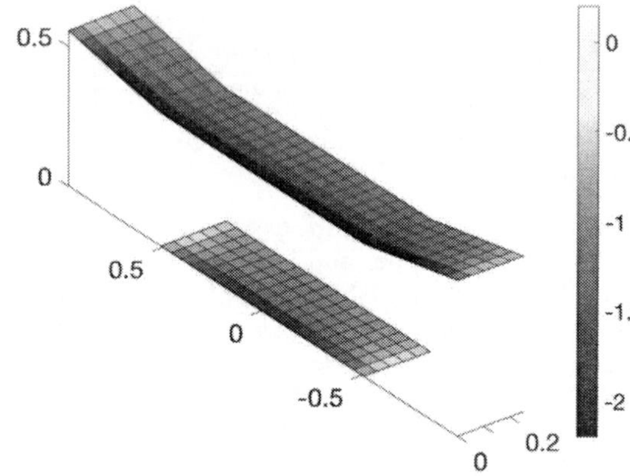

Figure 9a. Delta CP distribution at 10 degrees AoA: Open Winglet. Delta CP is constant along the span of the upper wing.

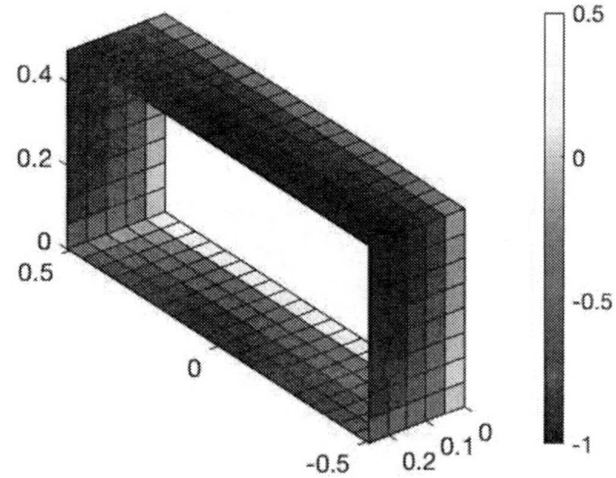

Figure 8b. Delta CP distribution at 0 degrees AoA: Box Wing. Delta CP is constant along the span of the upper wing.

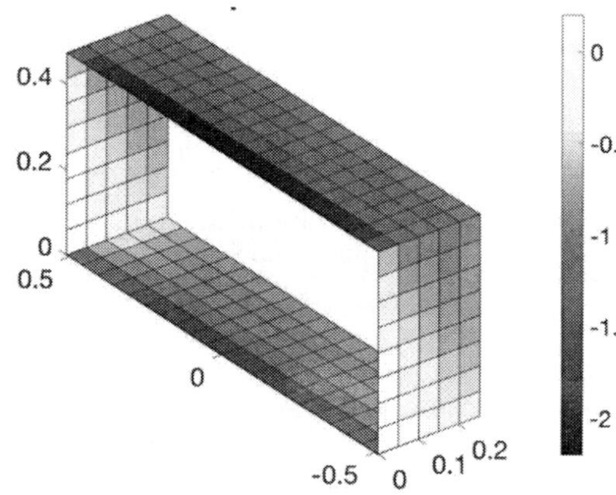

Figure 9b. Delta CP distribution at 10 degrees AoA: Box Wing. Delta CP begins to taper off at wingtips of upper wing.

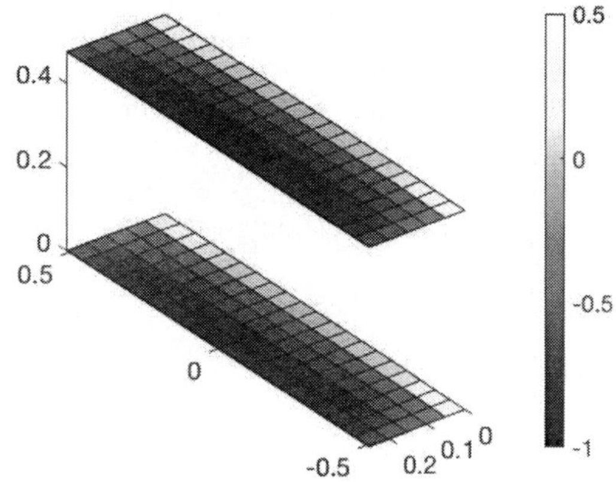

Figure 8c. Delta CP distribution at 0 degrees AoA: Baseline. Delta CP is symmetric on upper and lower wings.

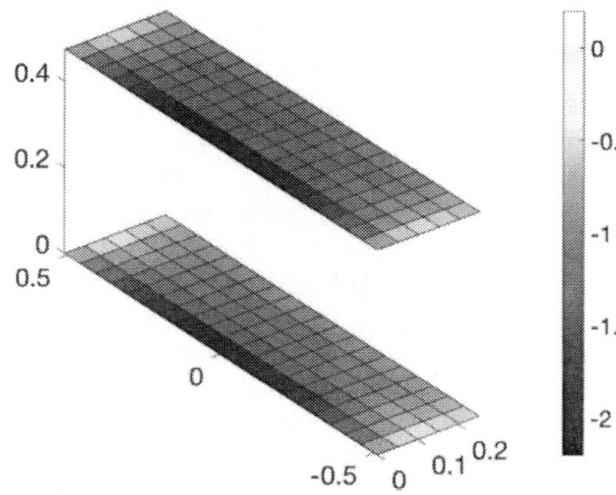

Figure 9c. Delta CP distribution at 10 degrees AoA: Baseline. Delta CP is symmetric on upper and lower wings.

Table 1a. C_L of Upper and Lower Main Wings at 0° AoA

Baseline:	C_L
Upper Wing	0.52
Lower Wing	0.49
Box Wing Mode:	C_L
Upper Wing	0.70
Lower Wing	0.32
Open Winglet Mode:	C_L
Upper Wing	0.71
Lower Wing	0.47

Table 1b. C_L of Upper and Lower Main Wings at 10° AoA

Baseline:	C_L
Upper Wing	1.13
Lower Wing	1.049
Box Wing Mode:	C_L
Upper Wing	1.31
Lower Wing	0.98
Open Winglet Mode:	C_L
Upper Wing	1.59
Lower Wing	0.99

As shown in tables 1(a) and 1(b), the C_L difference between the upper and lower wings for the baseline case is marginal and may be attributed to the wing interactions of the upper and lower wings. However, the difference between the upper and lower wings is much greater for the box and open winglet modes compared to the baseline case. In open winglet mode, the winglets extend the lift distribution of the upper wing, as shown in Figs. 6(a) and 7(a). For box wing mode, the asymmetric lift distribution between the upper and lower wing is in good agreement with the findings of Gall and Smith who noted a similar lift distribution for their box wing study (Ref. 7). It is to be noted that this asymmetry may also be impacted by the choice of winglet airfoil (Ref 8.)

The lift distribution difference between open and box wing mode, as shown in Figs. 6(a), 6(b), 7(a) and 7(b), can also be attributed to the interaction of the winglet with the main wing. The addition of the winglets either in box or open winglet mode significantly impact the lift distribution over the wings. This is also illustrated in the pressure distributions over the wing surfaces as shown in Figs. 8-9. The altered lift distribution due to the presence of winglets in either the open or box wing modes serve to reduce induced drag penalties, and hence, also impact the trailing vorticity distributions (Ref. 6).

MECHANICAL IMPEMENTATION

The mechanical design of the morphing winglet mechanism was influenced by the results found in the aerodynamic analysis and design section. First, the dihedral angle limits for the experimental vehicle were set to cover from -90 to at least 10° in order to achieve box wing mode at -90° and maximum lift at 10°, as was estimated from the VLM analysis. Furthermore, the wing loading requirements of the winglet were estimated from the load predictions in Fig. 5, setting the minimum torque requirement of the winglet at 2.5 Nm. In terms of the geometry, the dimensions of the winglets had to be large enough in order to achieve box wing mode using the pre-existing biplane wings of the vehicle. Lastly, the chord length and airfoil were chosen to be the same as the pre-existing biplane wings for consistency. These parameters are tabulated in Table 2.

Table 2. Winglet Design Parameters

Parameter	Value (units)
Wing Span	0.5 meters
Chord	0.254 meters
Min. Dihedral Angle	-90°
Max. Dihedral Angle	10°
Winglet Min. Torque	2.5 Nm
Airfoil	Wortmann FX 63-137

Worm Gear Transmission

A worm drive is capable of high torque with a 10:1 gear ratio, reducing most of the strain on the servo. The torque requirement of the servo is further reduced since the dynamics of the gears do not allow for back driving of the servo. In the end, the HITEC D955 servo was chosen for its high torque and compact size. This transmission system is able to transfer approximately 25 Nm of torque, setting the mechanism's overall safety factor at 10. With this setup, a full transition from -90 to 20° can be accomplished within 2.5 seconds.

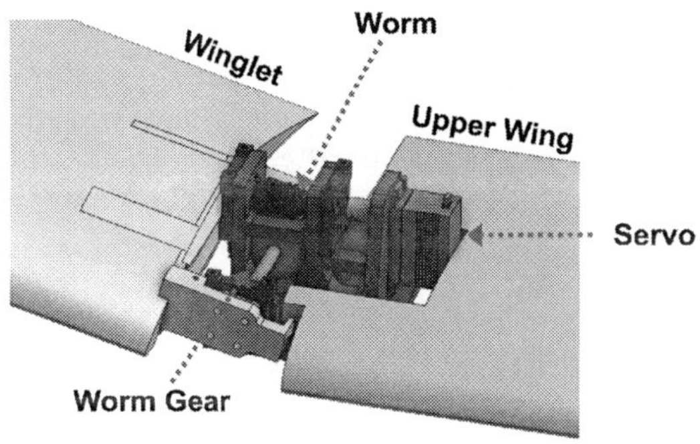

Figure 10. Worm Drive Mechanism labeled

Figure 11. Fabricated Version of the Winglet Mechanism

OVERVIEW OF THE MORPHING QUADROTOR-BIPLANE EXPERIMENTAL VEHICLE

The vehicle uses four brushless motors configured in two pairs above the upper and lower wing (X-configuration) (Fig. 12). The vehicle's main wings span 1.0 meters with a chord of .254 meters. The upper and lower wings have an offset of 0.5 meters from one another (Fig. 13). The vehicle uses four fixed pitch variable RPM rotors with a diameter of .381 meters for propulsion and control in roll, pitch and yaw. The motors are powered by two 6s 3300 mAh LiPo batteries in parallel, providing 6600 total mAh of capacity to the vehicle at 22.2 volts. The morphing quadrotor-biplane weighs approximately 3.93 kg with the winglets and their supporting mechanism contributing up to 25% of the gross take-off weight.

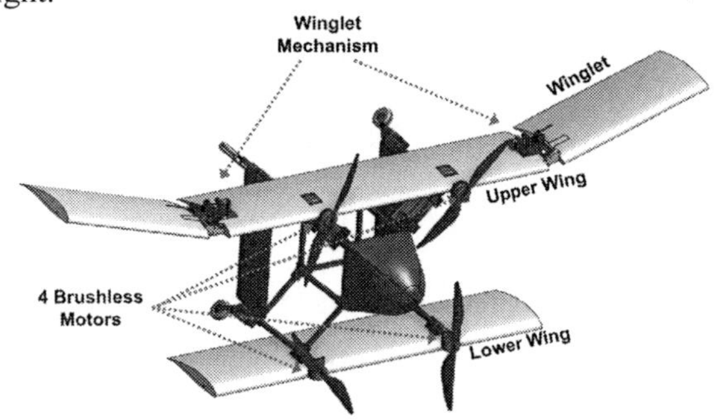

Figure 12. Morphing Winglet Quadrotor Biplane Tail-sitter

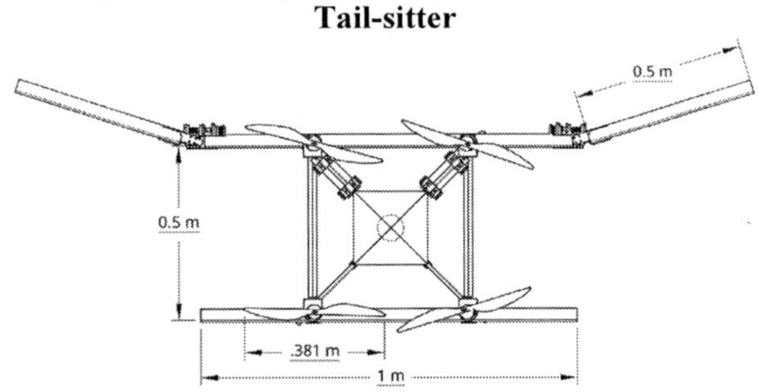

Figure 13. Front view of Morphing QBiT with key dimensions

Table 3 provides a breakdown of the vehicle's weight in greater detail. In order to offset the weight of the winglet mechanisms on the upper wing, the batteries were positioned on the lower structural members, as opposed to inside the fuselage, in order to keep the center of gravity at the center of the vehicle.

Table 3. Weight Distribution of the Morphing QBP

System	Weight (g)	% Total
Batteries	1010	26
Fuselage	903	23
Winglet Mechanism	775	20
Upper and Lower Wings	447	11
Motors	371	9
Landing Gear	197	5
Winglet Wings	192	5
Avionics, Wires	35	<1
Total	**3930 grams**	**100%**

Microcontroller: ELKA-R

Onboard attitude stabilization was implemented using a custom embedded lightweight kinematic autopilot (ELKA-R). The microcontroller is developed in-house at the Alfred Gessow Rotorcraft Center and weighs only 1.7 grams (Ref. 9). The Cortex-M4 microprocessor was chosen for ELKA-R for its large memory and high clock speed. For orientation, the board has an integrated MPU-9150 IMU which includes a 3-axis magnetometer, accelerometer and gyroscope. Due to the large pitch changes throughout flight, a quaternion-based feedback controller was implemented.

Flight Data Collection Electronics

The experimental vehicle was flown with a set of sensors installed to measure power system voltage, current, and vehicle airspeed. The avionics package uses a raspberry Pi Zero for storage on an SD card. In total, two ELKA boards were used. One ELKA board functioned exclusively as the autopilot while the other was used for collecting pilot inputs, vehicle pitch, yaw and roll attitude. A set of sensors was also used to collect basic operating characteristics. First, the RPM of all four motors was measured using an integrated ESC RPM feedback. Second, a flow sensor was used to determine the vehicle's airspeed. Third, a voltage and current sensor was used to measure power consumption. Similar setups have been used on past vehicles (Ref. 9).

FLIGHT TESTING RESULTS

Figure 14. Box Wing Mode (left), Open Wing Mode (right)

Flight Data Collection: Results

The morphing QBiT was flight tested to provide experimental support for the first order VLM analysis. In order to determine power draw characteristics, the vehicle was flown at two separate and fixed dihedral angles: box wing mode, at -90° dihedral angle and in open winglet mode, at 20° dihedral angle (Fig. 14). The vehicle was equipped with onboard data collection to determine power draw and airspeed in the two flight modes.

The results from these flight tests are shown in Figs. 15, 16 and 17. Each data point represents one second of trimmed steady level flight in its respective mode. Figure 16 shows the power savings between both modes (P_{open} - P_{box}), as represented by ΔP. The figure shows that in hover, box wing mode required a similar amount of power compared to open winglet mode. At higher cruise airspeeds, the power draw decreases for both modes, as is expected from previous flight test results (Ref. 9). The cruise power savings in open winglet mode versus box wing mode increases steadily and reaches a local maximum of 100 watts at approximately 4m/s, at which point the ΔP begins to decrease. Thus, open winglet mode required 30% less power than box wing mode at a cruise speed of 4m/s.

Flight test results also show that box wing mode allows for higher cruise airspeeds. At 5.5m/s the power required for trimmed flight in open winglet mode reaches a minimum of ~175W and starts to increase slightly to a maximum measured speed of 6m/s. At the same power setting of 175W, the cruise speed in box mode is 7m/s.

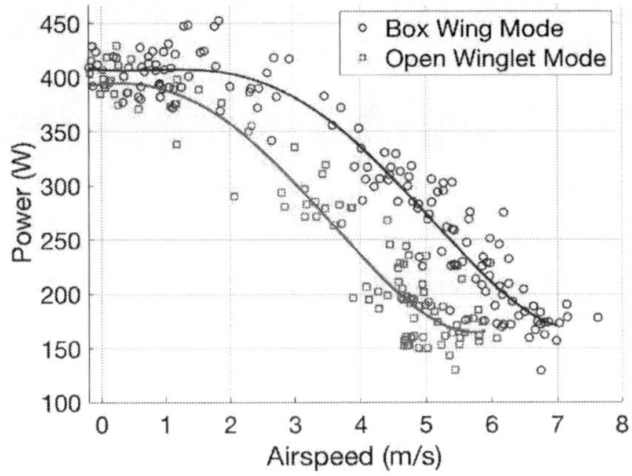

Figure 15. Experimental Power vs. Cruise Airspeed. Open winglet mode requires less power compared to box wing mode.

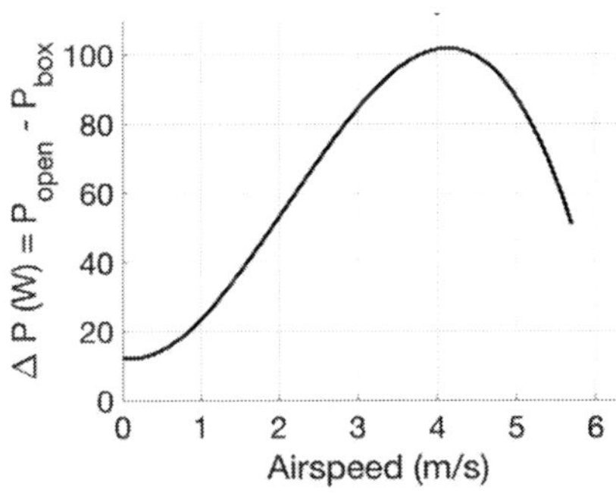

Figure 16. Experimental Δ P (P_{open} - P_{box}) vs. Cruise Airspeed

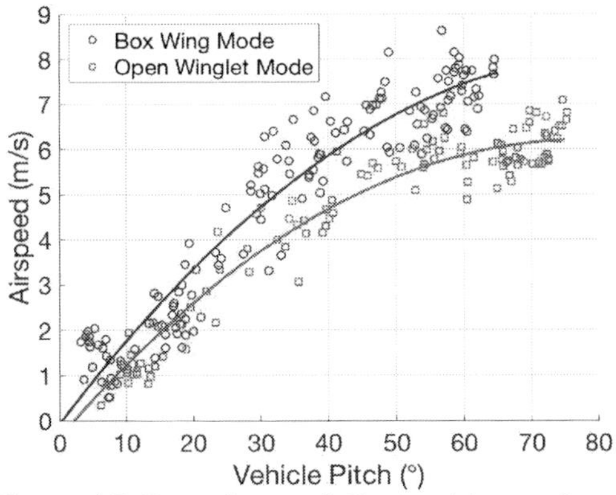

Figure 17. Experimental Cruise Airspeed vs. Pitch. Box wing has a greater speed for the same vehicle pitch.

Flight Data Collection: Discussion

These results indicate that open winglet mode requires less power for low airspeed cruise and is suited for loiter operations. Open winglet mode has a larger lifting surface on the upper wing and a higher aspect ratio compared to box mode, resulting in a larger overall lift coefficient. The VLM analysis predicted that open winglet mode would produce more lift than box mode. The experimental airspeed vs. power data supports this prediction.

Current flight data shows that the box wing mode supports a higher maximum airspeed and is suitable as a high-speed dash configuration. Folding the winglets mean they do not contribute to the total lift of the vehicle, thereby requiring a higher cruise speed and more thrust. Specifically, flight data shows that for the same 175W power consumption, box wing mode supports a cruise airspeed of 7m/s compared to 5m/s with the open wing. These conclusions are consistent with general aerodynamic theory and the vortex lattice method analysis of the wing system.

CONCLUSIONS

Hybrid vehicles such as the Quadrotor Biplane Tail-sitter (QBiT) with hover and high cruise speed capabilities can serve a wide range of mission requirements but are limited due to the design compromises that come from operating across a wide speed range. The focus of this paper was to develop a Morphing Winglet QBiT and investigate the aerodynamic performance gains. Unlike existing QBiTs, the morphing winglet QBiT has two variable winglets capable of rotating 110 degrees, allowing the vehicle to be optimized for power efficiency across a wider flight regime. The following is a summary of some key conclusions drawn from this study:

1. The morphing winglet QBiT weighs 3.93 kg with 4 rotors at a diameter of .381 meters. It has a maximum wing span of 2.0 meters at a dihedral angle of 0° and a minimum wing span of 1.0 meter at a dihedral angle of -90°. The platform successfully demonstrated hover, transition to forward flight mode, and sustained forward flight in box wing (Γ=-90°) and open winglet mode (Γ=20°).
2. The winglet mechanism provided a maximum torque of 25 Nm, 10 times greater than the estimated maximum winglet loads.
3. A first order vortex lattice method study predicted that a winglet dihedral angle of 10° produced maximum lift. VLM also predicted that the upper wing produces more lift than the lower in all three cases: box wing, open winglet, and a standard biplane configuration without winglets. At 0° angle of attack, box wing mode has a C_L difference of 54%, open winglet 33%, and baseline 6%. At 10° angle of attack, box wing mode has a C_L difference of 25%, open winglet 38%, and baseline 6%. These were in agreement with what has been reported in current literature.
4. At increasing steady level cruise airspeeds, both box wing and open winglet mode power draw

requirements were reduced. Open winglet mode consumed less power compared to box wing mode throughout the transition, peaking at a maximum power difference of 100 watts at 4 m/s, which represents 30% less power than box wing mode.

5. At the same pitch, the vehicle's airspeed in box wing mode is consistently higher, supporting the idea that box wing mode is capable of greater speeds than open winglet mode. In flight testing, box wing mode supported a higher max measured cruise airspeed of 7m/s compared to 5.5m/s with the open wing at 175W.

AUTHOR CONTACT INFORMATION

Peter Ryseck – pryseck@umd.edu
Dr. Derrick Yeo – dyeo@umd.edu
Dr. Vikram Hrishikeshavan – vikramh@umd.edu
Dr. Inderjit Chopra – chopra@umd.edu

REFERENCES

1. Min Z., Kien V.K., Richard L.J.Y., "Aircraft morphing wing concepts with radical geometry change," *The IES Journal*, Vol. 3, No. 3, pp. 188–195, 2010.

2. Abdulrahim, M. and Lind, R. 2004. "Flight Testing and Response Characteristics of a Variable Gull-Wing Morphing Aircraft," *Proceedings of AIAA Guidance, Navigation, and Control Conference and Exhibit*, 16-19 August, Providence, RI, *AIAA*, 2004-5113, pp. 1664–1679.

3. Mills, J.; Ajaj, R. "Flight Dynamics and Control Using Folding Wingtips: An Experimental Study." *Aerospace*, Vol. 4, No. 2, 2017.

4. Hrishikeshavan, V., Bogdanowicz, C. and Chopra, I. "Performance and testing of a quad rotor biplane micro air vehicle for multi role missions," *52nd Aerospace sciences meeting,* Maryland, MD, 13–17 Jan 2014, pp.1–17.

5. Melin, T., "Tornado, a Vortex Lattice Program for Conceptual Aircraft Design," Vol. 135, 2010.

6. Gerontakos, P. and Lee, T., "Effects of winglet dihedral on a tip vortex," *Journal of Aircraft*, Vol. 43, No. 1, 2006, pp. 117 – 124.

7. Gall, P. D. and Smith, H.C., "Aerodynamic Characteristics of Biplanes with Winglets," *Journal of Aircraft*, Vol. 24, No. 8, 1987, pp. 518–522.

8. Schiktanz, D. and Scholz, D., "Box Wing Fundamentals-an Aircraft Design Perspective," *AeroAircraft Design and Systems Group*, Univ. of Hamburg, 2011, pp. 601–615.

9. Phillips, P., Hrishikeshavan, V., Yeo, D. and Chopra, I., "Flight Performance of a Package Delivery Quadrotor Biplane," *Proceedings of the 7th AHS Technical Meeting on VTOL Unmanned Air Systems and Autonomy*, Mesa, AZ, Jan 24–26, 2017.

10. Blondeau J., Richeson J., Pines D.J., Norfolk A. "Design, development and testing of a morphing aspect ratio wing using an inflatable telescopic spar". *44th AIAA/ASME/ASCE/AHS Structures, Structural Dynamics, and Materials Conf.*, 7–10 April, Norfolk, VA, 2003.

11. Bourdin, P., Gatto, A. and M.I. Friswell. "Aircraft Control via Variable Cant-Angle Winglets," *Journal of Aircraft*, Vol. 45, No. 2, 2008.

12. Kamlet, Matt. "NASA Tests New Alloy to Fold Wings in Flight." *NASA*, 19 Jan. 2018. [Online] Available: www.nasa.gov/centers/armstrong/feature/nasa-tests-new-alloy-to-fold-wings-in-flight.html.

13. Weierman, J. et al., "Winglet Design and Optimization for UAVs," *28th AIAA Applied Aerodynamics Conference*, 2010.

Stabilization of a Quadrotor in Wind with Flow Sensing: Linear Modeling and Control for Attitude and Position Hold

William Craig
Graduate Research Assistant
University of Maryland
College Park, MD, USA

J.T. Lewis
Undergraduate Research Assistant
University of Maryland
College Park, MD, USA

Derek A. Paley
Willis H. Young Jr. Professor of Aerospace Engineering Education
University of Maryland
College Park, MD, USA

ABSTRACT

As quadrotor missions increase in complexity, we must also seek to increase their capability. This work addresses the ongoing challenge of flight in windy conditions through the use of a custom flow-sensing and feedback-control architecture utilizing two separate probes to measure lateral and longitudinal flow. A linear model is developed by performing system identification using CIFER® software on a 210mm quadrotor that runs customized Cleanflight firmware. A linear controller is then optimized for disturbance rejection using CONDUIT® software to meet desired handling qualities specifications. Experiments performed in an indoor gust-generation facility show improvement in both short and long time-scale gusts through the use of flow feedback.

NOTATION

e_θ	Pitch error, rad
e_q	Pitch rate error, rad/s
e_u	Longitudinal velocity error, m/s
e_x	Longitudinal position error, m
K_θ	Pitch attitude gain
K_q	Pitch rate gain
K_u	Longitudinal velocity gain
K_x	Longitudinal position gain
K_{sat}	Saturation function slope gain
M_{Δ_u}	Angular-acceleration response to wind
q	Pitch rate, rad/s
u	Longitudinal speed, m/s
$\mathbf{V}_\infty$	Wind velocity, m/s
X_{Δ_u}	Linear-acceleration response to wind
Δ_u	Longitudinal wind speed, m/s
Δ'_u	Measured longitudinal wind speed, m/s
θ	Pitch angle, rad
θ_{1s}	Longitudinal control input, %
θ_{d_1}	Outer-loop contribution to desired pitch angle, rad
θ_{d_2}	Inner-loop contribution to desired pitch angle, rad
τ_{1s}	Longitudinal time delay, s
ω_i	Angular velocity of propeller i, rad/s

Presented at the Vertical Flight Society Autonomous VTOL Technical Meeting and eVTOL Symposium 2019, Mesa, Arizona, Jan. 28–Feb. 1, 2019.

INTRODUCTION

In recent years, quadrotors have become highly capable tools with both commercial and military viability. The ability to take off and land vertically makes them attractive compared to fixed-wing Unmanned Aerial Systems (UAS) and their mechanical simplicity is an advantage compared to single-main-rotor helicopters. Current and foreseen missions, including agricultural monitoring, package delivery, and aiding in natural disasters, require stable and reliable flight in a variety of conditions.

Traditionally, stable and reliable flight for quadrotors is achieved with a trial-and-error methodology, where nominal gains are chosen based on experience, then tested in flight and adjusted as necessary. Trial and error may yield an initially unstable platform, but is often successful due to the mechanical simplicity, structural integrity, and low cost of replacement parts for quadrotors. A more rigorous approach may also be taken, using, for example, eigenvalue analysis (Ref. 1), robust control techniques (Ref. 2), or Lyapunov stability criteria (Refs. 3, 4) to provide flight stability based on system dynamics.

More recently, an approach typically used to evaluate full-scale helicopter stability and performance based on handling qualities metrics has also been pursued (Ref. 5), which involves first identifying a linear model of the vehicle, then designing a controller such that the vehicle meets the desired handling qualities specifications. A systematic approach that uses CIFER® software (Ref. 6) for frequency-domain system identification and CONDUIT® software (Ref. 7) for controller gain optimization has been used in Refs. 8, 9. We also rely on CIFER® and CONDUIT® software for model identification and controller optimization, using handling quality specifications informed by Refs. 5, 8, 9.

As quadrotors become increasingly reliable, a variety of work is being performed to extend their mission capability, including aerobatic flight (Ref. 10), sensing and perception of unknown environments (Refs. 2, 11, 12), flying with suspended loads (Refs. 13, 14), and in-flight hardware failures (Ref. 15). This work addresses flight stability of quadrotors in windy conditions, which has also been investigated by Refs. 2, 9, 11, 16–18. Gremillion *et al.* use distributed accelerometers to directly estimate the forces and moments acting on the vehicle (Ref. 2), which are then directly cancelled in control feedback. Berrios *et al.* use the Control-Equivalent-Turbulence-Input (CETI) model from Ref. 19 to design a controller that is inherently robust to disturbances with frequency content typical of turbulent wind by optimizing gains in CONDUIT® using handling qualities criteria (Ref. 9). Yeo *et al.* use flow probes to measure downwash from another quadrotor to estimate its position and avoid the resulting disturbance (Ref. 11).

The feedback control approach described here relies on onboard flow measurements in the lateral and longitudinal directions from multi-hole probes to estimate the aerodynamic forces and moments acting on the quadrotor. By combining flow measurements with inertial sensing, the controller is able to react to wind before the aerodynamic forces and moments have propagated to the rigid-body dynamics, yielding improved performance compared to using inertial sensing alone. Work validating the application of flow feedback for gust rejection using nonlinear feedback-linearization controllers was performed for a one degree-of-freedom (DOF) pitch stand in Yeo *et al.* (Ref. 16), a three DOF attitude stand in Craig *et al.* (Ref. 17), and for a six DOF free flight quadrotor in Craig *et al.* (Ref. 18). Here we use a linear control framework optimized for gust disturbance rejection with CONDUIT® software. The linear control uses flow feedback to cancel the effect of wind on the quadrotor dynamics. The flow sensor package consists of lateral and longitudinal facing probe pairs connected to a microcontroller with flexible tubing (Ref. 11), which measures differential pressure and transmits a digital signal to the onboard flight controller.

The contributions of this paper are (1) the addition of flow feedback to a linear controller optimized for gust rejection using handling qualities criteria and (2) the experimental demonstration of the benefit of adding flow feedback to a linear controller that has been designed for gust rejection. This work parallels the work in Craig *et al.*, in which a nonlinear controller was designed with flow feedback for gust disturbance rejection, but the use of an optimized linear framework is beneficial because it highlights the effect of flow feedback in an optimized, linear controller specifically designed for gust disturbance rejection.

The outline of the paper is as follows. The first section describes the software and hardware used for system identification and experimental testing. The second section describes the system identification process using CIFER® software. The third section uses the identified model to design an optimal controller based on handling qualities criterion in CONDUIT®. The fourth section shows experimental results of the six DOF quadrotor subjected to a series of gusts, and the final section summarizes the paper and discusses ongoing work.

SYSTEM DESCRIPTION

Figure 1 shows the 540g custom-built quadrotor used in this work. The vehicle has a 210mm carbon fiber Shendrones Krieger frame and a Matek F405-STD flight controller paired with a Matek FCHUB-6S power distribution board. The flight controller includes a three-axis accelerometer and three-axis gyroscope for inertial sensing. EMAX Lightning 20A ESCs control EMAX RS-2205 motors, which are paired with Gemfan 5030 five inch diameter propellers. Figure 1 also shows the differential-pressure lateral and longitudinal flow probes highlighted in red.

Fig. 1. Quadrotor used for experimental testing with flow probes highlighted in red

The overall experimental testbed is shown in Figure 2. Outer-loop sensing is performed using an 18 camera, 15' x 10' x 12' motion capture facility that records position and attitude data at 100Hz. The data recorded in motion capture is streamed locally to Matlab, which runs the outer loop position and heading controller. Desired attitude is transmitted using a 2.4

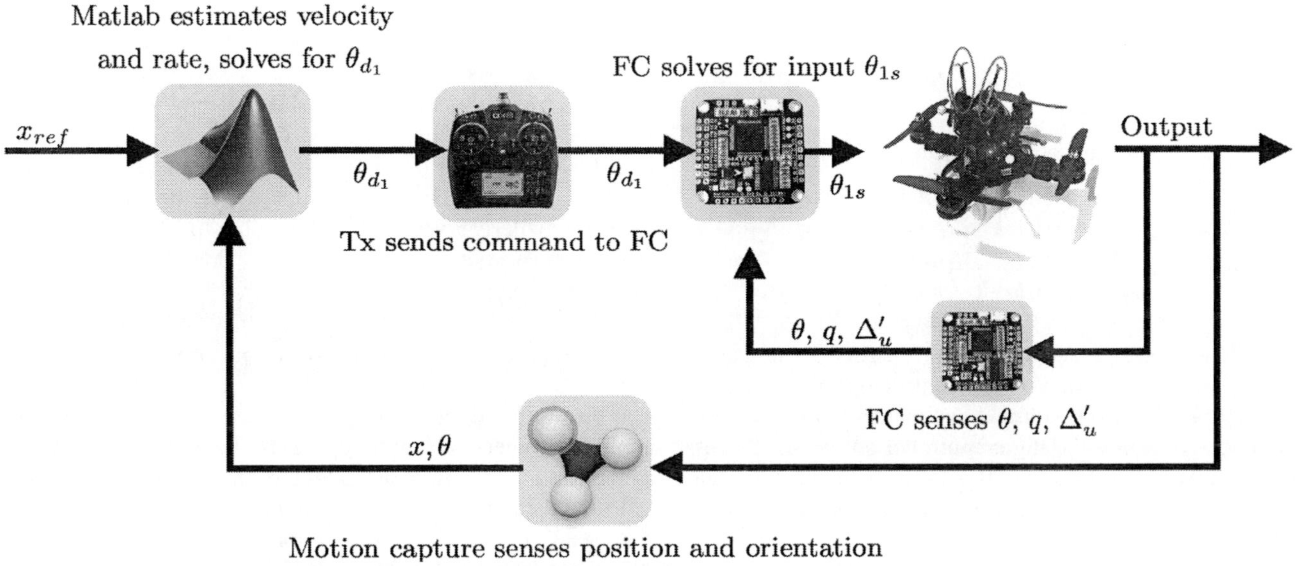

Fig. 2. Inner- and outer-loop experimental control structure for longitudinal direction

GHz Spektrum DX8 radio transmitter to the quadrotor's onboard flight controller, which solves for desired inputs. The onboard flight controller runs open-source Cleanflight software for inner-loop attitude stabilization, which has been customized to run the optimized linear controller and incorporate serial flow-sensor data. Onboard data collection is completed using the built in Blackbox feature that records states and flow data at 250Hz.

Fig. 3. Gust generation system consists of a set of eight Dyson fans behind remotely operated blinds

Gust rejection testing was completed in an indoor gust facility, shown in Figure 3, which consists of eight Dyson fans behind a set of remotely operated blinds, controlled by an Arduino through LabVIEW, that open and close to produce gusts. Two square-wave pattern gusts of different periods were tested to investigate the effects of longer and shorter frequency scales. First, the blinds were opened for five seconds at a time, giving the vehicle a chance to reach steady-state type behavior. Second, the blinds were opened for two seconds at a time, thereby not allowing the quadrotor to to reach steady state during the gust or recover fully after a gust. The quadrotor experienced a wind speed of ten knots in the five-second gust testing, and eight knots in the two-second gust testing.

MODEL IDENTIFICATION

Modeling results are presented for the longitudinal degree of freedom (the process is the same for each of the other degrees of freedom). Define inertial frame $\mathscr{I} \triangleq (O, \mathbf{e}_1, \mathbf{e}_2, \mathbf{e}_3)$ in a north, west, up orientation with O at a known point on the ground, and body frame $\mathscr{B} \triangleq (O', \mathbf{b}_1, \mathbf{b}_2, \mathbf{b}_3)$ in a forward, left, up orientation with O' at the center of mass of the quadrotor, as shown in Figure 4. In the longitudinal degree-of-freedom, u is the velocity along $\mathbf{b}_1$, q is the pitch rate along $\mathbf{b}_2$, θ is the pitch angle along $\mathbf{b}_2$, and Δ_u is the wind velocity in the $\mathbf{b}_1$ direction. Let X_u and M_u be the linear acceleration and angular acceleration response to velocity u, respectively. We also define X_{Δ_u} and M_{Δ_u} as the linear and angular acceleration response to wind Δ_u, respectively. The state-space description of the system is

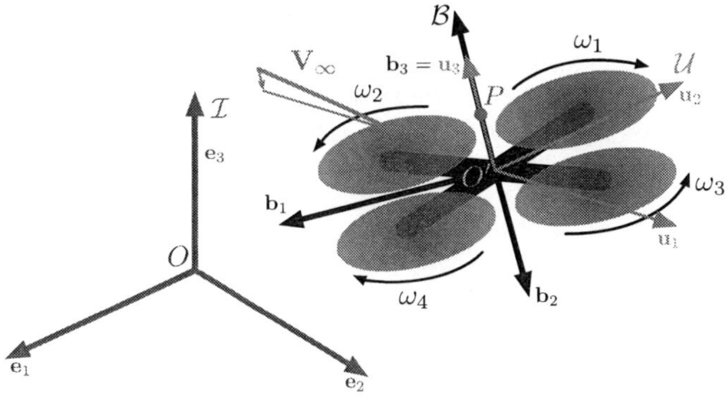

Fig. 4. Quadrotor reference frames and wind velocity

$$\begin{bmatrix}\dot{u}\\ \dot{q}\\ \dot{\theta}\end{bmatrix} = \begin{bmatrix}X_u & 0 & g\\ M_u & 0 & 0\\ 0 & 1 & 0\end{bmatrix}\begin{bmatrix}u\\ q\\ \theta\end{bmatrix} + \begin{bmatrix}0\\ M_{\theta_{1s}}\\ 0\end{bmatrix}\theta_{1s}(t-\tau_{1s}) + \begin{bmatrix}X_{\Delta_u}\\ M_{\Delta_u}\\ 0\end{bmatrix}\Delta_u, \quad (1)$$

$$y = \begin{bmatrix}1 & 0 & 0\\ 0 & 1 & 0\\ 0 & 0 & 1\\ 0 & 0 & 0\end{bmatrix}\begin{bmatrix}u\\ q\\ \theta\end{bmatrix} + \begin{bmatrix}0\\ 0\\ 0\\ 1\end{bmatrix}\Delta_u. \quad (2)$$

System identification is performed with CIFER®, which uses frequency sweeps over each channel to develop a linear model fit. Sweep profiles are performed automatically by inserting a frequency chirp into the flight-controller software. Chirps sweep from 0.62 rad/s to over 100 rad/s; an example is shown in Figure 5. Large spikes in the data are user inputs to limit translational drift.

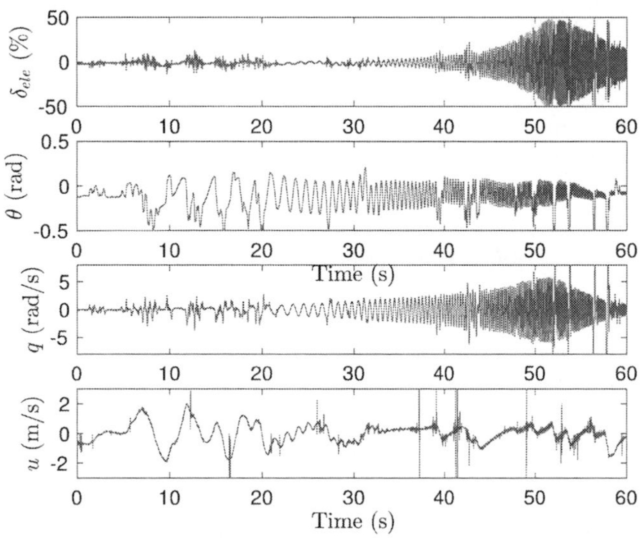

Fig. 5. Chirp data for longitudinal degree of freedom used as input to CIFER®

The resulting model fit to the flight data is shown in Figure 6. Stability derivatives are presented in Table 1, as well as Cramer-Rao bounds and insensitivities showing the level of confidence of the identification (Ref. 6). The value for X_u has been fixed based on insensitivity analysis. The model has also been validated in the time domain against independent data from separate flight tests, showing good agreement in Figure 7.

Table 1. Model parameters (see Eq. 1)

Parameter	Value	C-R %	Insensitivity %
X_u	-0.05	-	-
M_u	-10.26	4.54	1.59
$M_{\theta_{1s}}$	5.79	3.16	1.09
τ_{1s}	0.0196	5.49	2.74

We initially predicted that the effect of wind moving over the quadrotor is equivalent to the effect of the quadrotor moving forward in air, such that $X_{\Delta_u} = -X_u$ and $M_{\Delta_u} = -M_u$. However, after experimental results showed only a small effect of adding flow feedback, the gust stability derivative was adjusted based on prior force-torque measurement experimental results (Ref. 16) to be approximately twice the magnitude of the velocity derivatives. Work is ongoing to identify the gust stability derivative X_{Δ_u} using frequency domain system identification techniques.

CONTROLLER DEVELOPMENT

The quadrotor flight controller structure shown in Figure 8 uses a fast inner-loop running at 2kHz to control the attitude dynamics, and a slower outer-loop running at 100Hz to control the position and heading dynamics. The outer loop applies gains to velocity and position in proportional-integral (PI) control, and also uses a saturation function to limit the maximum desired velocity produced by the position error. The gain-adjusted velocity error is transmitted to the inner loop, where the onboard wind measurements are incorporated to produce the desired attitude. Gains are applied to attitude and rate in a proportional-derivative (PD) controller and combined with flow feedback to produce the final inputs to the motors.

To implement the PID controller, we feed back position and speed to the outer loop; and attitude, attitude rate, and wind speed to the inner loop. This example shows the development in the longitudinal direction (the process for each other degree of freedom is similar). To limit the maximum reference speed, we pass the desired position through a saturation function, i.e.,

$$u_d = \text{sat}(K_{sat}(x - x_d)), \quad (3)$$

where K_{sat} sets the slope of the saturation function. We then set the position and velocity errors

$$e_u = u - u_d, \quad e_x = x - x_d, \quad (4)$$

and apply a PI controller to yield our desired dynamics, i.e.;

$$\dot{u}_d = -K_u e_u - K_x e_x. \quad (5)$$

Inserting Equation (5) into the dynamics (1) yields

$$\dot{u} = X_u u + g\theta + X_{\Delta_u}\Delta_u = -K_u e_u - K_x e_x. \quad (6)$$

Taking θ as our control input to the outer loop system and solving for the desired pitch angle we find

$$\theta_d = \frac{1}{g}(-K_u e_u - K_x e_x - X_u u - X_{\Delta_u}\Delta_u). \quad (7)$$

The nature of onboard versus offboard sensing necessitates splitting the desired pitch angle into separate outer- and inner-loop components, where the outer loop component θ_{d_1} uses the position and velocity data only available through motion

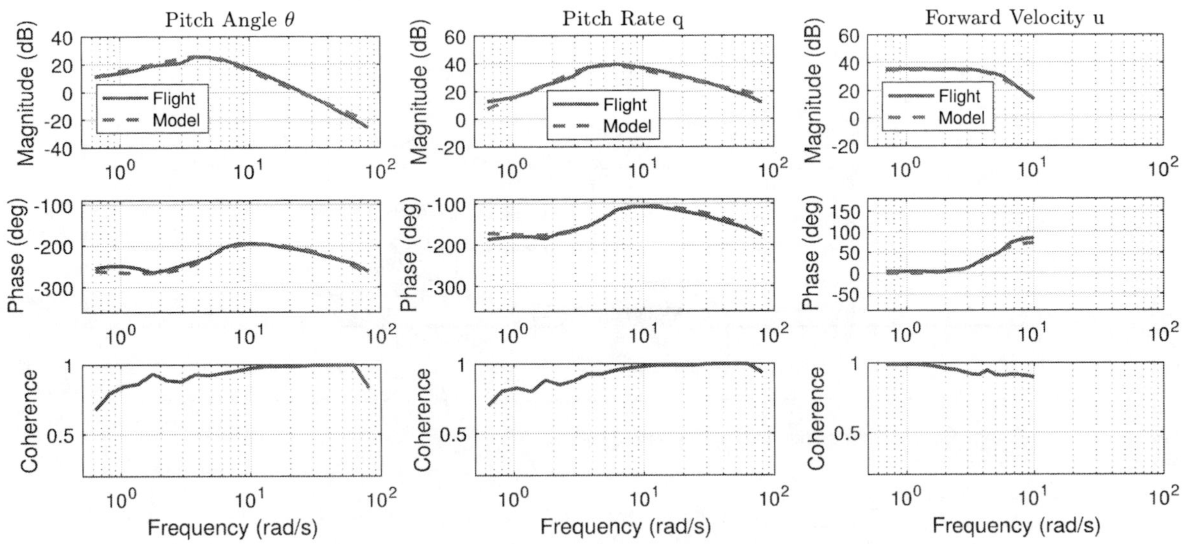

Fig. 6. Model-data agreement using stability derivatives identified in CIFER®

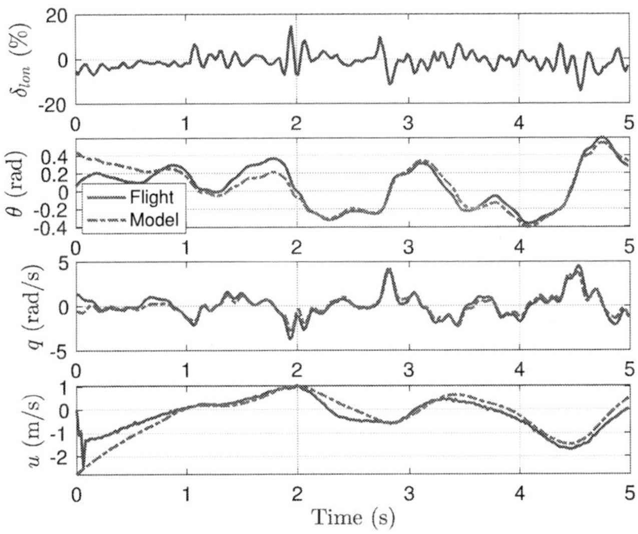

Fig. 7. Model validation against time-series data

capture, and the inner loop component θ_{d_2} uses the wind data only available from the onboard flow sensors. Note, the flow probes measure the combination of wind speed and flight speed, such that the actual measurement $\Delta_u' = \Delta_u - u$. We then calculate the stabilizing portion of the feedback, $1/g(-K_u e_u - K_x e_x)$, in the outer loop, and the dynamic cancellation portion, $1/g(-X_u u - X_{\Delta_u}\Delta_u)$, in the inner loop. Based on the initial assumption that $X_{\Delta_u} \approx -X_u$, we multiplied by the measured flow speed such that $X_{\Delta_u}\Delta_u' = X_{\Delta_u}\Delta_u - X_{\Delta_u}u \approx X_{\Delta_u}\Delta_u + X_u u$ to yield the desired value. Thus, the outer-loop component is

$$\theta_{d_1} = \frac{1}{g}(-K_u e_u - K_x e_x), \tag{8}$$

and the inner-loop component is

$$\theta_{d_2} = \frac{1}{g}(-X_{\Delta_u}\Delta_u'). \tag{9}$$

Once $\theta_d = \theta_{d_1} + \theta_{d_2}$ is identified in the inner loop, we find

$$e_\theta = \theta - \theta_d \tag{10}$$

and

$$e_q = q - \frac{d}{dt}\theta_d. \tag{11}$$

We then establish our desired dynamics:

$$\dot{q} = -K_q e_q - K_\theta e_\theta, \tag{12}$$

and substitute into the system dynamics (1),

$$\dot{q} = M_u u + M_{\theta_{1s}}\theta_{1s} + M_{\Delta_u}\Delta_u = -K_q e_q - K_\theta e_\theta. \tag{13}$$

Rearranging, with θ_{1s} as the input, the value sent to the mixer is

$$\theta_{1s} = \frac{1}{M_{\theta_{1s}}}(-K_q e_q - K_\theta e_\theta - M_u u - M_{\Delta_u}\Delta_u), \tag{14}$$

where we again account for both wind velocity and rigid body velocity using the measurement from the flow sensor to yield

$$\theta_{1s} = \frac{1}{M_{\theta_{1s}}}(-K_q e_q - K_\theta e_\theta - M_{\Delta_u}\Delta_u'). \tag{15}$$

A Simulink model of the controller in combination with the identified linear model yields optimal gains using the CONDUIT® environment. CONDUIT® requires users to choose a number of specifications relating to handling qualities stability and performance, then uses a multiobjective optimization engine to adjust gains such that each of the prescribed handling qualities specifications is achieved. More details on the CONDUIT® software are available in Ref. 7. Handling qualities are characterized as level one, two, or three, with one as the best and three as the worst. The CONDUIT® software is designed to achieve level one handling qualities for all specifications, while also minimizing

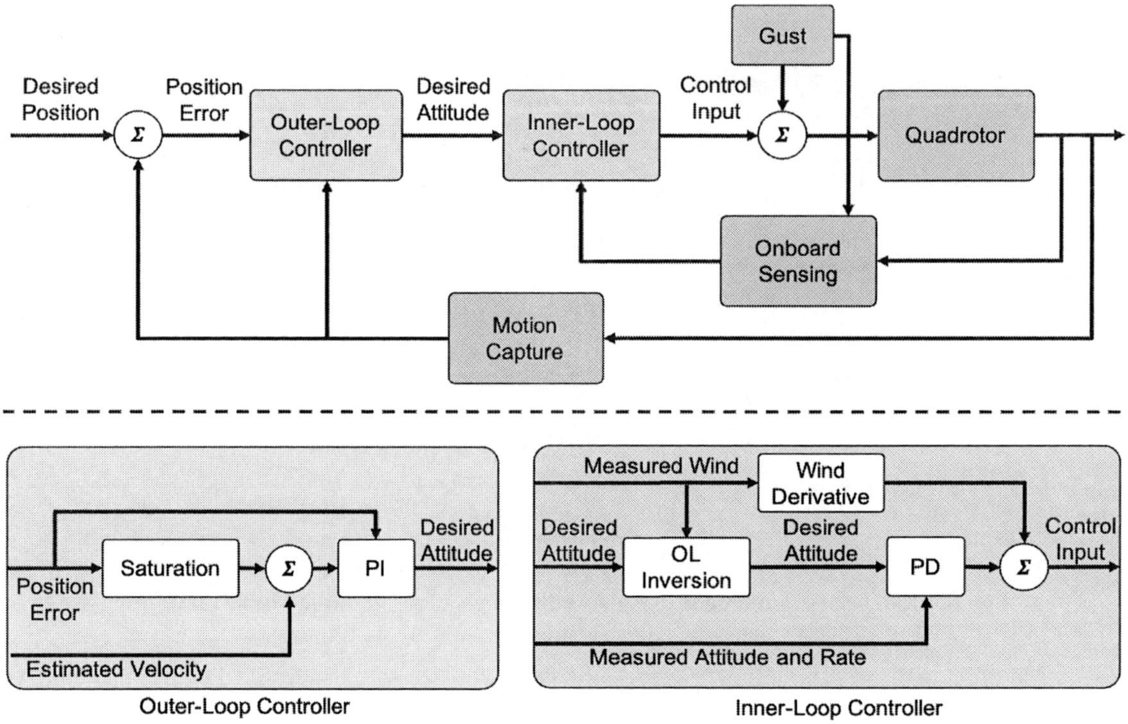

Fig. 8. Quadrotor control architecture

actuator usage. Hard specifications (H) address stability and are required to be met by the optimization, soft specifications (S) address performance based metrics and are desirable but not mandatory, and summed objective (J) specifications are designed to reduce actuator usage and are correspondingly minimized.

Nominal handling qualities specifications in CONDUIT® are designed for full-scale manned aircraft. The small size and unmanned nature of quadrotor UAS lead to different handling qualities requirements, so the CONDUIT® specifications have been adjusted based on previous work using CONDUIT® for small quadrotor UAS (Refs. 5, 8). Additional adjustments were also made based on pilot feedback and flight performance. The specifications to which the gains were optimized for the longitudinal outer loop for position control are shown in Table 3, with comments addressing adjustments to handling qualities boundaries. The resulting handling qualities window is shown in Figure 9, and shows that each of the specifications meets level 1 requirements. The corresponding characteristics of the control law design are shown in Table 3.

EXPERIMENTAL RESULTS

Experimental testing was performed by subjecting the quadrotor to a series of square-wave gusts at different periods in an indoor gust generation facility. Figures 10 and 11 show displacement in the $\mathbf{e}_1$ direction as well as the flow speed experienced by the quadrotor. Figures 12 and 13 show the $\mathbf{e}_1$-$\mathbf{e}_2$ displacement on the left plot and the $\mathbf{e}_1$-$\mathbf{e}_3$ displacement on the right plot over the course of each test.

From Figures 10 and 11 we see improvement in station holding through the use of flow sensing for both tests. The longer time-scale gusts in Figure 10 show a more significant improvement, where both controllers are able to settle before and after the gusts. Without flow feedback, the quadrotor is blown back by the wind where it keeps a very consistent position until the gust ceases, and with flow feedback the quadrotor quickly compensates for the wind, and experiences only a slight movement in the direction of the wind. The shorter time scale results in Figure 11 still show an improvement through the addition of flow feedback, but highlight the benefits of designing the controller for disturbance rejection in CONDUIT® for the controller without flow feedback, and the susceptibility to noise when using flow feedback.

The linear controller produced by CONDUIT® is designed to address higher frequency turbulent wind using the disturbance rejection bandwidth specification (Ref. 20) without the use of flow sensing, and thus shows comparatively better results in the two second gusts versus testing with five second gusts. Adding flow feedback does reduce the $\mathbf{e}_1$ error in the system, however the movement appears slightly more erratic than in the longer time-scale case. The error at the beginning and end of each gust remains consistent between the longer and shorter gusts, however the shorter gusts limit the system's time to stabilize and so we see the quadrotor repeatedly overshooting the desired position as wind conditions change.

The differences in performance for the two time scales are highlighted in Figures 12 and 13, where we see that in the

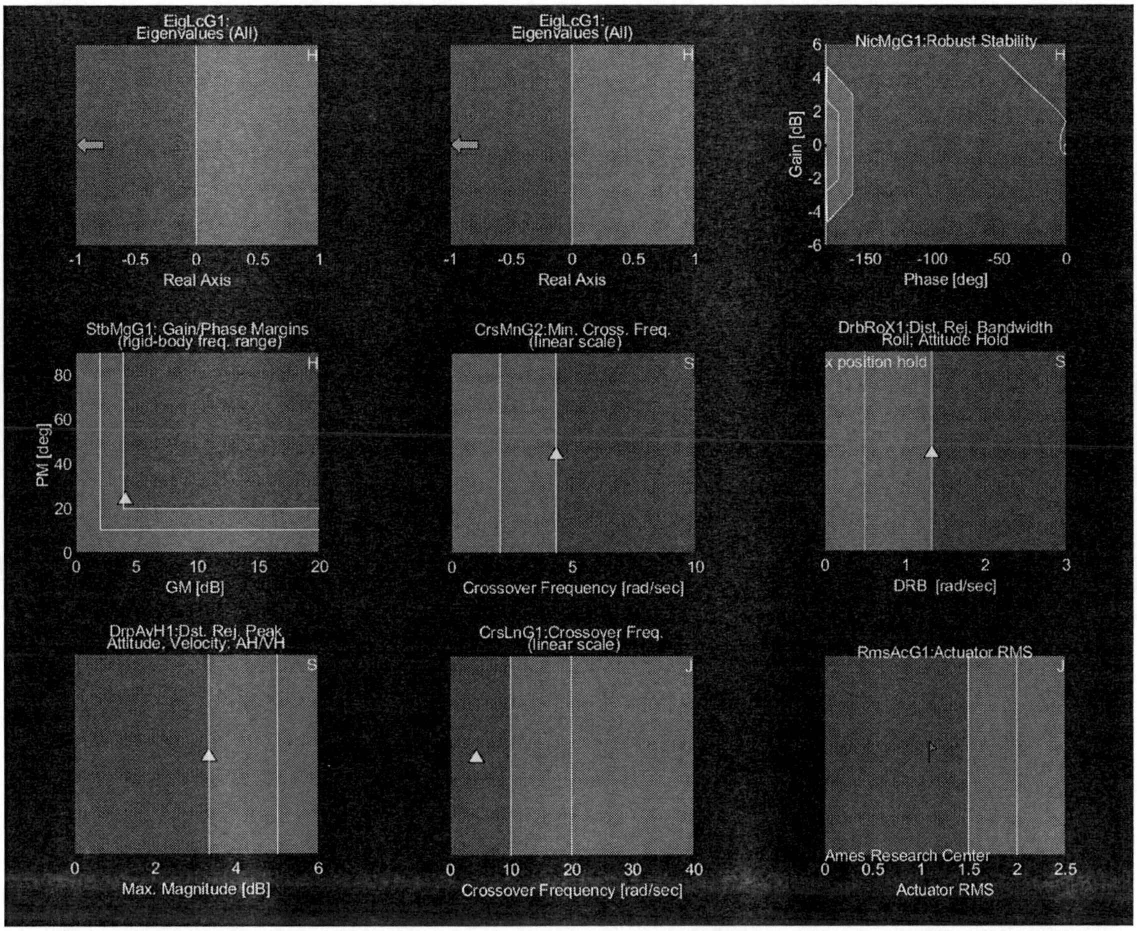

Fig. 9. CONDUIT® handling qualities window. Level one in blue, level two in pink, level three in red. Triangles and flags indicate the value of the handling quality, green arrows indicate an off-screen value, and the line in Nichols robust stability indicates the value across simultaneous changes in gain and phase.

longer time-scale gusts the systems generally maintain a tight position with or without flow sensing apart from the dynamic portion of the gust. Figure 13 shows the quadrotor unable to maintain the tight position in the presence of more rapid gusts for both cases and we also see significant motion in the $\mathbf{e}_2$ direction when flow feedback is utilized, resulting from noise and measurements in the flow signal in the lateral direction. In the longer-time-scale gusts, the $\mathbf{e}_2$ flow error reduces as the vehicle settles. However, the more dynamic wind in the shorter time-scale leads to consistent fluctuation in the $\mathbf{e}_2$ direction in addition to the response to the primary flow in the $\mathbf{e}_1$ direction. Efforts to better address noise in the flow sensors to improve performance is ongoing.

CONCLUSION

This paper describes a linear controller that has been optimized for gust rejection using CONDUIT®, based on system identification performed with CIFER®. Additionally, a custom flow probe package was used to investigate the benefits of flow feedback for gust rejection. Experiments were performed with a 210mm quadrotor system running Cleanflight software, where flow feedback yielded improvements for both short and long time-scale gusts, particularly for the longer time-scale five second gusts when the vehicle had time to settle in the wind. In ongoing work, we are investigating sensor noise to specifically reduce off-axis response and also planning outdoor flight testing.

ACKNOWLEDGMENTS

This work was supported by the University of Maryland Vertical Lift Rotorcraft Center of Excellence Army Grant No. W911W6-17-2-0004.

REFERENCES

[1]Pounds, P., Mahony, R., and Corke, P., "Modelling and control of a large quadrotor robot," *Control Engineering Practice*, Vol. 18, (7), 2010, pp. 691–699.

[2]Gremillion, G. M. and Humbert, J. S., "Disturbance rejection with distributed acceleration sensing for small unmanned aircraft systems," *AIAA Journal*, Vol. 54, (8), 2016, pp. 2233–2246.

Table 2. Design Specifications

Parameter	Description	Type	Std. Bounds?	Comments
EigLcG1	Eigenvalues	Hard	Y	
StbMgG1	Stability Margins	Hard	N	L1/L2 at 4dB, 20 deg
NicMgG1	Nichols Margins	Hard	N	L1/L2 at 160 deg
DrbRoX1	Disturbance Rejection Bandwidth	Soft	N	L1/L2 at 1.33 rad/s
DrpAvH1	Disturbance Rejection Peak	Soft	N	L1/L2 at 3.33 rad/s
CrsMnG2	Minimum Crossover Frequency	Soft	N	L1/L2 at 4.33 rad/s
CrsLnG1	Crossover Frequency	Sum. Obj.	N	L1/L2 at 10 rad/s
RmsAcG1	Actuator RMS	Sum. Obj.	Y	

Table 3. Control Law Design

Controller	Gain Margin [dB]	Phase Margin [deg]	Crossover Freq. [rad/s]	DRB [rad/s]
Pitch Attitude	9.66	21.8	25.8	11.9
Long. Position	4.10	24.8	4.33	1.33

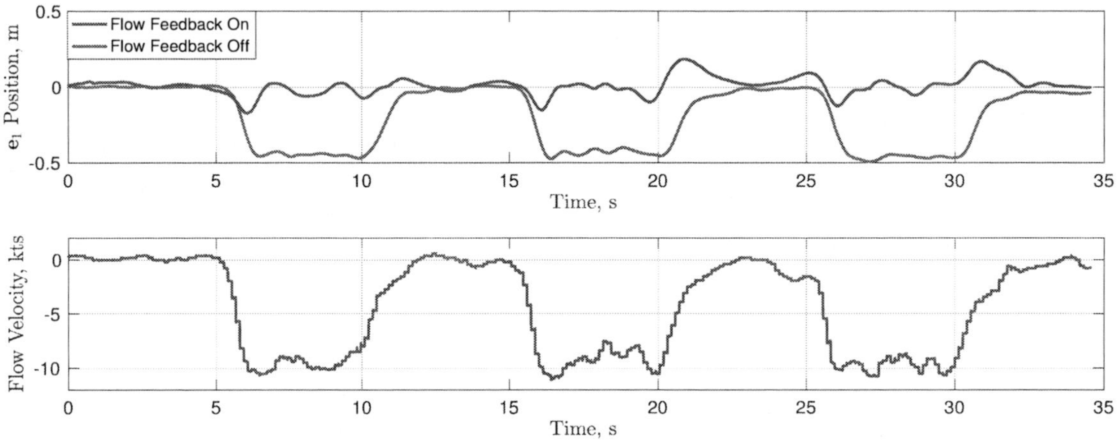

Fig. 10. Experimental quadrotor e_1 position error in response to 5 s duration 10 kt gusts in the $-e_1$ direction. Flow velocity is measured onboard using a custom flow probe.

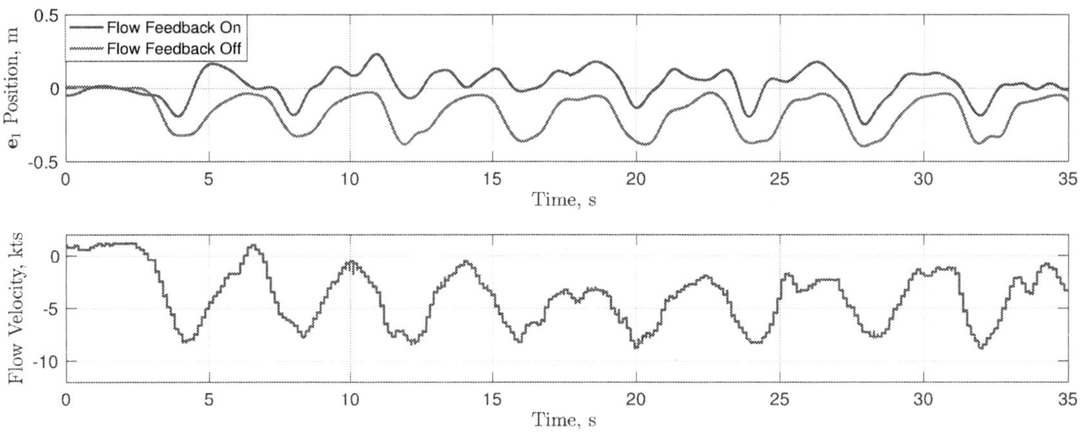

Fig. 11. Experimental quadrotor e_1 position error in response to 2 s duration 8 kt gusts in the $-e_1$ direction. Flow velocity is measured onboard using a custom flow probe.

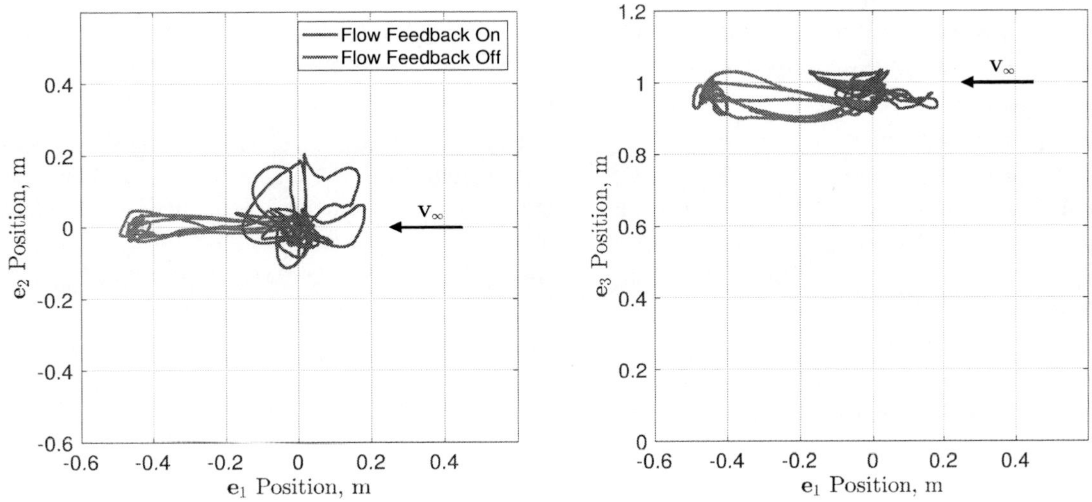

Fig. 12. Experimental quadrotor position response to 5 s duration 10 kt gusts in the $-e_1$ direction.

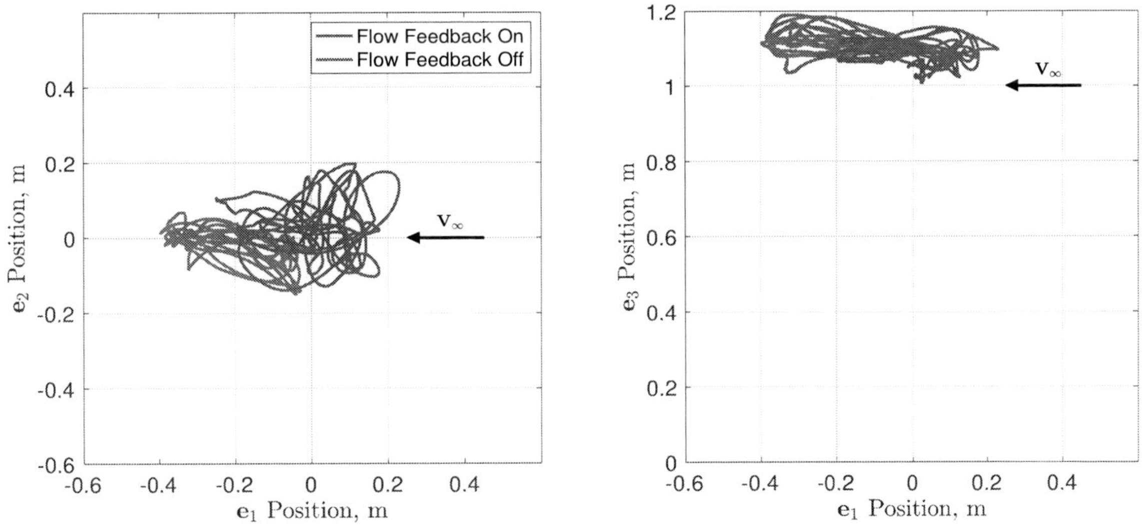

Fig. 13. Experimental quadrotor position response to 2 s duration 8 kt gusts in the $-e_1$ direction.

[3]Cao, N. and Lynch, A. F., "Inner-outer loop control for quadrotor UAVs with input and state constraints," *IEEE Transactions on Control Systems Technology*, Vol. 24, (5), 2016, pp. 1797–1804.

[4]Lee, T., Leok, M., and Mcclamroch, N. H., "Geometric tracking control of a quadrotor UAV on SE(3)," IEEE Conference on Decision and Control, 2010.

[5]Cheung, K. K., Wagster IV, J. A., Tischler, M. B., Ivler, C. M., Berrios, M. G., Berger, T., Juhasz, O., Tobias, E. L., Goerzen, C. L., Barone, P. S., Sanders, F. C., Lopez, M. J. S., and Lehmann, R. M., "An overview of the U.S. Army Aviation Development Directorate quadrotor guidance, navigation, and control project," American Helicopter Society 73rd Annual Forum Proceedings, 2017.

[6]Tischler, M. B. and Remple, R. K., *Aircraft and Rotorcraft System Identification*, American Institute of Aeronautics and Astronautics, Inc., Reston, VA, second edition, 2012.

[7]Tischler, M. B., Berger, T., Ivler, C. M., Mansur, M. H., Cheung, K. K., and Soong, J. Y., *Practical Methods for Aircraft and Rotorcraft Flight Control Design*, American Institute of Aeronautics and Astronautics, Inc., Reston, VA, first edition, 2017.

[8]Wei, W., Cohen, K., and Tischler, M. B., "System identification and controller optimization of a quadrotor UAV," American Helicopter Society 71st Annual Forum Proceedings, 2015.

[9]Berrios, M. G., Berger, T., Tischler, M. B., Juhasz, O., and Sanders, F. C., "Hover flight control design for UAS using performance-based disturbance rejection requirements," AHS Int. 73nd Annual Forum, 2017.

[10]Lupashin, S., Schöllig, A., Sherback, M., and D'Andrea, R., "A simple learning strategy for high-speed quadrocopter multi-flips," IEEE International Conference Robotics and Automation, 2010.

[11]Yeo, D., Sydney, N., Paley, D., and Sofge, D., "Onboard flow sensing for downwash detection and avoidance on small quadrotor helicopters," AIAA Guidance, Navigation, and Control Conference, AIAA SciTech Forum, Paper AIAA 2015-1769, 2015.

[12]Sanket, N. J., Singh, C. D., Ganguly, K., Fermüller, C., and Aloimonos, Y., "GapFlyt: Active vision based minimalist structure-less gap detection for quadrotor flight," *IEEE Robotics and Automation Letters*, Vol. 3, (4), 2018, pp. 2799–2806.

[13]Lee, T., "Geometric control of quadrotor UAVs transporting a cable-suspended rigid body," *IEEE Transactions on Control Systems Technology*, Vol. 26, (1), 2018, pp. 255–264.

[14]Pereira, P. O. and Dimarogonas, D. V., "Control framework for slung load transportation with two aerial vehicles," IEEE Conference on Decision and Control, 2018.

[15]Lanzon, A., Freddi, A., and Longhi, S., "Flight control of a quadrotor vehicle subsequent to a rotor failure," *Journal of Guidance, Control, and Dynamics*, Vol. 37, (2), 2014, pp. 580–591.

[16]Yeo, D. W., Sydney, N., and Paley, D. A., "Onboard flow sensing for multi-rotor pitch control in wind," *Journal of Guidance, Control, and Dynamics*, Vol. 41, (5), 2018, pp. 1193–1198.

[17]Craig, W., Passe, B. E., Yeo, D., and Paley, D. A., "Geometric attitude control of a quadrotor in wind with flow sensing and thrust constraints," Submitted to *IEEE Transactions on Control Systems Technology*, 2018.

[18]Craig, W., Yeo, D., and Paley, D. A., "Geometric control of a quadrotor in wind with flow sensing and thrust constraints: Attitude and position control," AIAA SciTech Forum, Paper AIAA 2019-1192, 2019.

[19]Juhasz, O., Lopez, M. J., Berrios, M. G., Berger, T., and Tischler, M. B., "Turbulence modeling of a small quadrotor UAS using system identification from flight data," Paper January, AHS Technical Meeting on VTOL Unmanned Aircraft Systems, 2017.

[20]Berger, T., Ivler, C. M., Berrios, M. G., Tischler, M. B., and Miller, D. G., "Disturbance rejection handling qualities criteria for rotorcraft," AHS Int. 72nd Annual Forum, 2016.

An Assessment of Heave Response Dynamics for Electrically Driven Rotors of Increasing Diameter

Ariel Walter PhD Student

Michael McKay PhD Student

Robert Niemiec Research Scientist

Farhan Gandhi Redfern Chair in Aerospace Engineering

Colin Hamilton Undergraduate Research Assistant

Center for Mobility with Vertical Lift (MOVE)
Rensselaer Polytechnic Institute
Troy, NY United States

Chris Jaran
Chief Executive Officer
Terrafugia
Woburn, MA, United States

ABSTRACT

An examination is conducted of the effects of increasing rotor size on system heave response time and power consumption, due to the increasing rotor blade inertia. The study aims to establish a rotor size at which using an electric motor and fixed-pitch, variable-RPM rotor may no longer be feasible due to the degradation in system climb rate response time and handling qualities ratings. Several isolated rotors ranging from 1 to 8 feet in diameter are simulated, and the heave responses are examined based on Froude scaled ADS-33E-PRF requirements. A model validation is performed using data from a 28-inch diameter rotor tested on a rotor test stand apparatus. Various feedback controller designs are then considered, including proportional and proportional-derivative control laws. Pure proportional control fails to reach scaled Level 1 requirements due to the underutilization of the motor. Using proportional-derivative control, rotors 6 feet or less in diameter can meet the scaled ADS-33E-PRF Level 1 heave response requirements with variable-RPM control alone; rotors larger than this may not be responsive enough for effective operation.

NOTATION

F	=	Froude Scaling Factor
w	=	Heave Rate (ft/s)
K	=	Final Value of Step Response (ft/s)
t	=	Time (s)
τ	=	Time Delay (s)
T	=	Time Constant (s)
I	=	Rotor Rotational Inertia $(slug * ft^2)$
P	=	Motor Power (hp)
τ_{aero}	=	Aerodynamic Torque
Ω	=	Rotor Rotational Speed (RPM)
T_{rotor}	=	Rotor Thrust
D	=	Rotor Diameter (ft)

Presented at the 2019 Autonomous VTOL Technical Meeting and eVTOL Symposium, Mesa, AZ, Jan 29-31, 2019

INTRODUCTION

The Urban Air Mobility Grand Challenge put forward by NASA (Ref. 1) reflects a growth in interest and development of larger-scale electric VTOL (eVTOL) aircraft, which are currently being considered for a variety of future applications as air transport vehicles for both cargo and passengers. Additionally, Uber issued a call for participation in the design and development of VTOL air taxis (Ref. 2) for passenger transport between cities and suburbs. The scale of these transport aircraft would be significantly larger than the multicopters that are currently commercially available (typically less than 50 lb gross weight).

Many challenges to scaling up eVTOL from their current size exist, some of which are highlighted by Johnson et al. (Ref. 3), who identified several key research areas necessary for the development of air taxis, including operational effectiveness, safety, and performance. Related to these areas are the aircraft handling qualities, which are effectively a quantification of the responsiveness and maneuverability of the aircraft.

Most eVTOL aircraft utilize fixed-pitch, variable-RPM rotors to regulate the thrust and moments on the aircraft, which ties the maneuverability of the aircraft to the rate at which the rotors' speeds can be changed. The increasing rotational inertia of larger rotor blades will cause a greater delay between a commanded and actual change in rotor speed, reducing the agility of multicopters that rely on few, large rotors.

Though well-established for traditional helicopters, handling qualities metrics for evaluating multicopters are still in the process of being refined. Several different groups have attempted to define handling qualities metrics for multi-rotor aircraft. In the current literature, there are studies that provide simple scaling of existing military specifications (Refs. 4, 5), seek to define/develop new handling qualities criteria and evaluation methodology (Refs. 6, 7), as well as provide a redefinition of the existing Cooper-Harper rating scale (Ref. 8). Also, Klyde et al. (Ref. 9) suggest a process to define UAV handling qualities guided by existing military specifications that are altered depending on the weight and associated mission tasks or use cases for the aircraft.

Ivler et al. (Ref. 4) utilized a scaled version of the ADS-33E-PRF (Ref. 10) Mission Task Elements (MTEs) to evaluate the performance of a 3.75 lb quadcopter, and concluded that kinematically scaled (based on the maximum speed of the aircraft) ADS-33E-PRF requirements are an effective way to quantitatively evaluate handling qualities for the quadcopter.

Kinematic scaling cannot be utilized if the maximum flight speed of the aircraft is unknown. However, Froude-scaled MTEs based on the length scaling of the aircraft (Ref. 5) provide another option. The Froude-scaling described by Alverenga et al. (Ref. 5) uses relative length scales to change the upper bounds of level 1 and level 2 handling qualities. The time constant and delay are multiplied by F, which is the square root of the diagonal hub-to-hub distance of the quadcopter divided by the UH-60 rotor diameter (Eq. 1).

$$F = \sqrt{\frac{\text{Hub-to-Hub Distance}}{\text{UH-60 Rotor Diameter}}} \quad (1)$$

Consider the hover and low speed requirement of height response characteristics to a collective control input in the ADS-33E-PRF. For this MTE, the response of the aircraft to a step change in commanded climb rate is fit by a first-order transfer function, in the form of Eq. 2. The two parameters T and τ represent the time constant and delay of the first-order response, and are used to parameterize the fit, while the value of K is fixed by the size of the step change in climb rate. The T and τ which result in the best fit to the simulation (or experimental data) is compared to the level 1 and level 2 bounds given in Table 1.

$$w_{est}(t) = K[1 - exp(-\frac{t-\tau}{T})] \quad (2)$$

Table 1. Heave Rate Response Requirements

Level	Time Constant T (s)	Time Delay τ (s)
1	5.0	0.20
2	∞	0.30

MODELING

In this study, isolated rotors ranging from 1 to 8 feet in diameter are simulated. The rotor response to a collective control input is examined through the simulated heave rate response of the system. This response is then compared to the Froude-scaled (based on length) ADS-33E-PRF boundaries.

Scaling of Heave Handling Qualities Metrics

As described, the method developed in Ref. 5 with the Froude scaling factor in Eq. 1 will be utilized to scale ADS-33E-PRF handling qualities metrics for the multi-rotor aircraft. Based on the geometry of the AeroQuad Cyclone (Fig. 1), a representative 2 kg quadcopter with 12 inch diameter rotors and a hub-to-hub distance of 24 inches, twice the rotor diameter is used to approximate the hub-to-hub distance for a rotor of arbitrary size. Thus, Eq. 1 becomes Eq. 3.

$$F_{\text{isolated}} = \sqrt{\frac{2(\text{Quadcopter Rotor Diameter})}{\text{UH-60 Rotor Diameter}}} \quad (3)$$

Rotor Parameters

The geometry of the simulated rotor is based on the AeroQuad Cyclone quadcopter (Fig. 1). The nominal specifications of the Cyclone rotors are given in Table 2. The blades are assumed to have a linear twist and linear taper. As the rotor diameter is scaled up in simulation, the chord length is appropriately scaled to maintain the solidity and taper ratio.

Fig. 1. AeroQuad Cyclone Quadcopter

Table 2. Nominal Rotor Parameters

Parameter	Value
Rotor Diameter	12 in
Root Airfoil	NACA 4412
Tip Airfoil	Clark Y
Root Chord	0.866 in
Tip Chord	0.343 in
Root Pitch	21.5°
Tip Pitch	11.1°
Rotor Solidity	0.09
Rotor Mass	0.049 lbm

Rotor & Motor Inertia Calculations

CAD models of several multicopter rotors were used to compute the rotational inertia, assuming the rotors were made out of carbon fiber reinforced polymer. A curve was then fit to these data points (Fig. 2) in order to find an approximation of rotor inertia as a function of rotor diameter (Eq. 4), where D is measured in feet, and the inertia is in slug ft^2.

$$I_{rotor} = 4.8 \times 10^{-5} D^5 \tag{4}$$

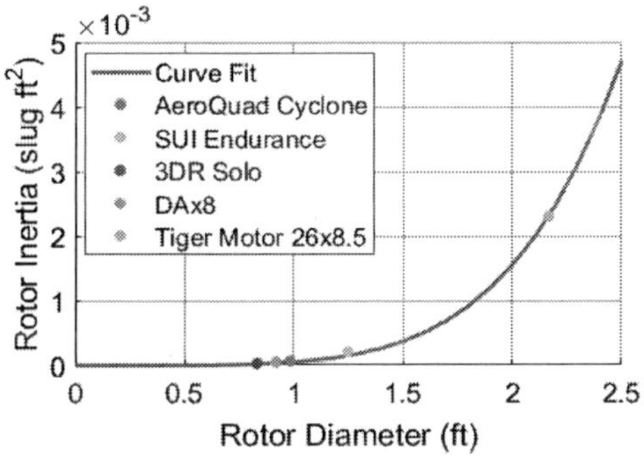

Fig. 2. Estimation of Rotor Rotational Inertia

Motor inertia was computed for several electric motors by taking measurements of the mass, thickness, outer diameter, and height of the rotating part of the motor. The motor inertia was then calculated assuming it was a hollow steel cylinder. Equation 6 was used to map the rated power of the motor to an appropriate rotor diameter and generate the data points in Fig. 3. Fitting a curve to these points yields Eq. 5, where diameter and inertia are measured in ft and slug ft^2, respectively.

$$I_{motor} = 6.8 \times 10^{-6} D^2 \tag{5}$$

In the rotor-motor system, the rotational inertia of the rotor is dominant. Even for the smallest rotor being considered (1 ft diameter) the predicted rotor rotational inertia is around 90% of the total system inertia, shown in Fig. 4. Therefore, the inertia of the motor is neglected, and the rotational inertia of the rotor-motor system is assumed to be equal to the rotational inertia of the rotor.

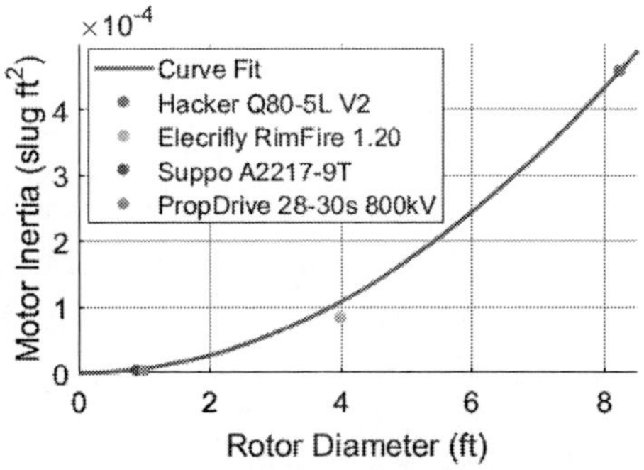

Fig. 3. Estimation of Motor Rotational Inertia

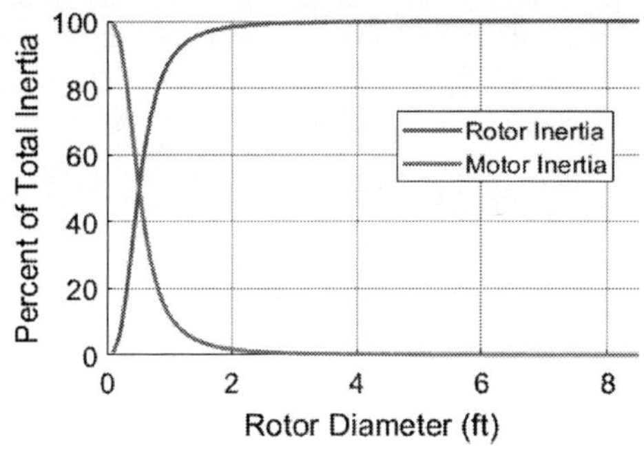

Fig. 4. Estimated Rotor and Motor Inertia as Percent of Total

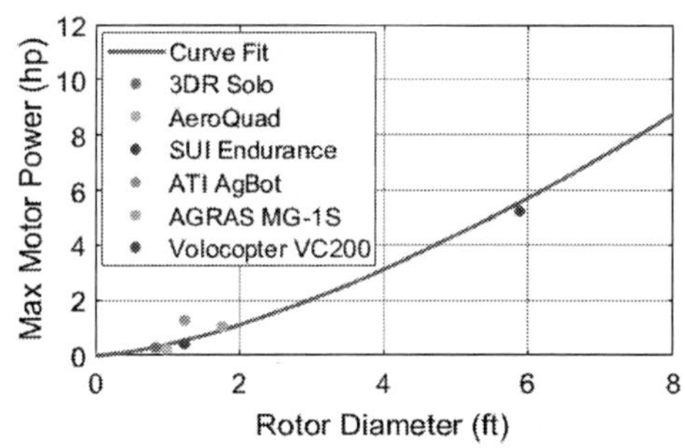

Fig. 5. Estimation of Installed Power from Rotor Diameter

Estimating Installed Power

Naturally, the angular acceleration of the rotor is limited by the power the electric motor can provide. To pair each simulated rotor with an appropriately sized motor, a benchmarking study was conducted based on existing multicopters (Fig. 5). Fitting a curve to this data yields Eq. 6, which suggests that scaling the installed power with the 3/2 power of the rotor diameter (D) is an adequate estimation.

$$P_{max} = 0.39 D^{1.5} \tag{6}$$

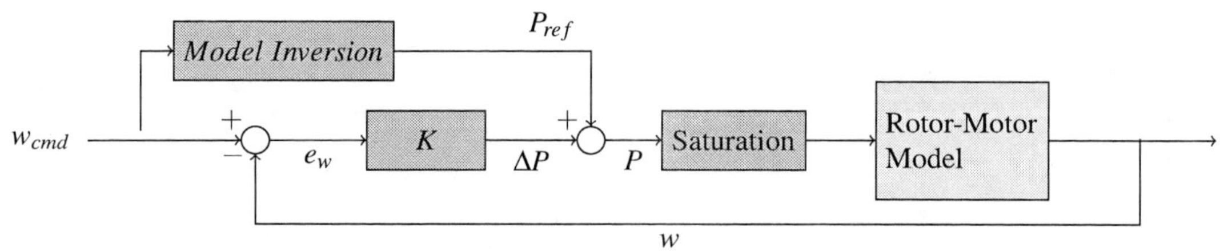

Fig. 6. Control Architecture

Rotor Dynamics Model

The rotational dynamics of the rotor are modeled using conservation of angular momentum (Eq. 7). On the left-hand side is the rate of change of angular momentum, and on the right is the difference between the torque supplied by the motor (modeled as power input divided by rotor speed), and the aerodynamic torque, calculated by the Rensselaer Multicopter Analysis Code (RMAC) (Ref. 11). RMAC uses blade element theory, coupled to a 3x4 Peters-He dynamic wake model, to calculate the phase-averaged rotor forces and moments.

$$I\dot{\Omega} = \frac{P}{\Omega} - \tau_{aero} \tag{7}$$

Rotor heave dynamics are isolated by constraining all but the heave degree of freedom for the system. This is justified because, in hover, the heave dynamics are independent of the longitudinal, lateral, and directional dynamics. Consequently, the heave equation of motion takes the form of Eq. 8, where w is the heave rate of the rotor and T_{rotor} is the thrust produced by the rotor. A mass (m) corresponding to a 2lb/ft^2 disk loading (based on the AeroQuad Cyclone) is used in Eq. 8.

$$m\dot{w} = T_{\text{rotor}} - mg \tag{8}$$

Controller Architecture

This dynamic model constitutes the bare airframe dynamics of the system, and a feedback controller is designed around it to shape the system response, illustrated in Fig. 6. Linear control design is performed such that the closed-loop dynamics are similar on linear approximations of each system, regardless of rotor diameter. This is accomplished via two methods: proportional (state-space pole placement) and proportional-derivative (frequency domain tuning) control of a SISO system. These control laws form the block denoted by K in Fig. 6. In addition to the feedback controller, a feed forward model inversion is included to improve tracking performance of the motor-rotor system. Finally, a saturation block is included to model the physical limitations of the motor considered for each simulation case (estimated using Eq. 6).

Model Validation

The simulation of the isolated rotor was validated using experimental data from the rotor test apparatus at Rensselaer Polytechnic Institute (Fig. 7), which was designed to measure the forces and moments produced by a rotor up to 3 feet in diameter. For the model validation, the open loop response of a 28 inch diameter rotor to a step input in power was recorded.

Fig. 7. Rotor Test Apparatus

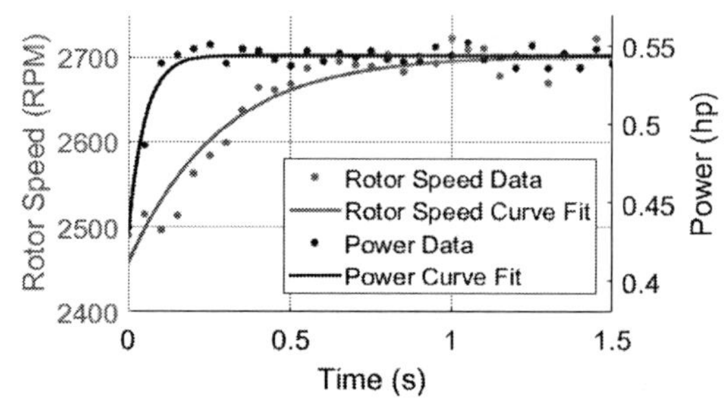

Fig. 8. Experimental Rotor Speed and Power with First Order Curve Fits

Table 3. Rotor Experimental Data Curve Fit Parameters

Response	Time Constant (s)	Time Delay (s)
Motor Power	0.05	0
Rotor Speed	0.28	0.04

A step change in the throttle signal from 23% to 26% received by the ESC at $t = 0.15$ resulted in a rapid change in the mechanical power supplied by the motor from 0.43 hp to 0.54 hp (Fig. 8). Consequently, the rotor speed increased from 2480 RPM to 2700 RPM. Fitting a first-order response (of the form Eq. 2) to both sets of data (Fig. 8) yields the time constant and time delay parameters in Table 3. Based on the relatively low time constant associated with the motor power (18% that of the RPM), it is modeled as an instantaneous change at

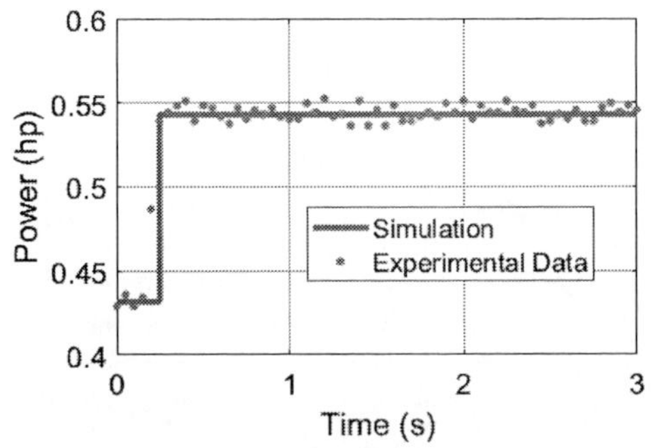

Fig. 9. Power Step Input for Model Validation

$t = 0.15$, as shown in blue on Fig. 9. Using this input, RMAC predicts the change in RPM shown in blue on Fig. 10.

RMAC under-predicts the final rotor speed by around 3.7%, an error consistent with the known behavior of RMAC (Ref. 11), which, similar to other comprehensive analysis codes, tends to over-predict aerodynamic torque. Shifting the simulated response by this constant error (shown by the dotted line in Fig. 10) illustrates the model's ability to capture the transient response, as well as the overall change in rotor speed.

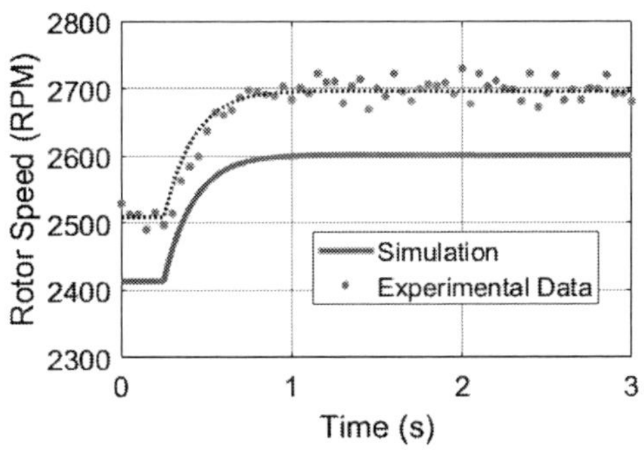

Fig. 10. Open Loop RPM Response for Model Validation

RESULTS

Bare Airframe Response

The bare airframe response was first simulated in order to examine the natural behavior of the system, as well as the effects of increasing rotor rotational inertia. Starting from a steady state hover condition, the power input to the motor was stepped to the value required for a 10 ft/s climb rate at the beginning of the simulation. This power is given as a fraction of the predicted installed power (Eq. 6).

The heave rate responses of the various rotor diameters being considered are shown in Fig. 11. All rotors reach the desired heave rate without any oscillation or overshoot. The change in curvature of the heave rate response for the larger rotors is a result of the increasing rotor rotational inertia.

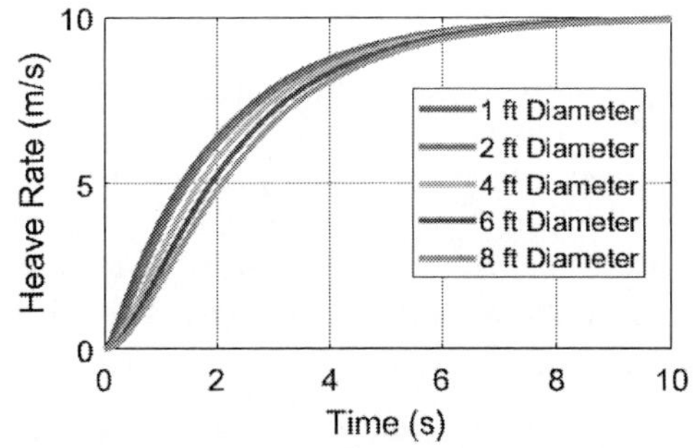

Fig. 11. Open Loop Heave Rate Response

Table 4. Time Constant and Time Delay of Open Loop Responses

Rotor Diameter (ft)	Time Constant (s)	Time Delay (s)
1	1.925	0.063
2	1.986	0.116
4	2.167	0.191
6	2.387	0.247
8	2.626	0.296

A first-order curve is fit to each of the heave responses (as per the ADS-33E-PRF) and the values of the time delay and time constant are calculated using Eq. 2, given in Table. 4. The values are scaled using Eq. 3 and compared to the ADS-33E requirements in Fig. 12. The degradation of handling qualities that occurs with increasing rotor size is a result of the increasing rotor inertia. None of the rotors meet even the scaled Level 2 requirements for the heave rate response, suggesting that open-loop speed control is insufficient for handling qualities purposes, even on small rotors.

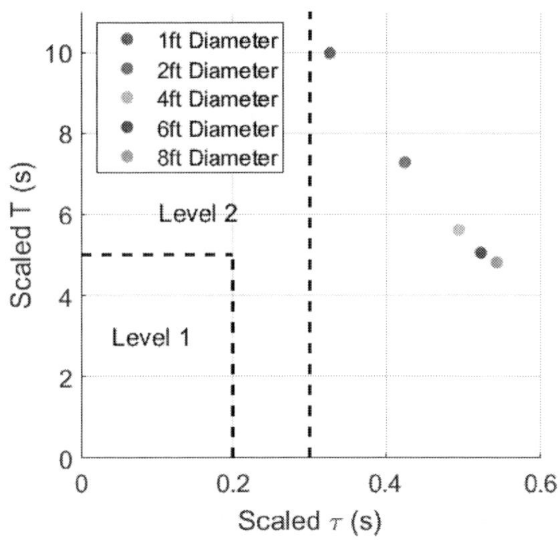

Fig. 12. Open Loop Heave Response Metrics Compared to ADS-33E Requirements

Proportional Control

Because the bare airframe heave dynamics are stable, a proportional gain alone is sufficient to tune the heave response. Due to the presence of the rotational dynamics, the system is second order with two real poles. One pole corresponds to the heave mode while the other corresponds to the rotational dynamics of the motor-rotor system, shown in Fig. 13.

The proportional gain is chosen using pole placement such that the system is critically damped. This design strategy provides the fastest settling time without overshoot, while also providing a predominantly first-order response, per the ADS-33E-PRF standards.

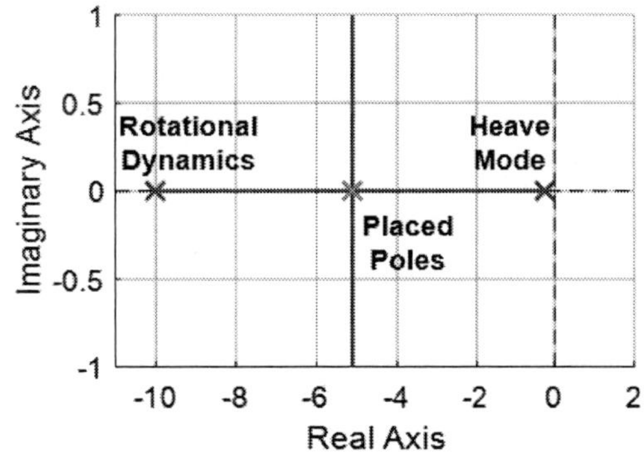

Fig. 13. Root Locus of the 1 ft Rotor with Proportional Control

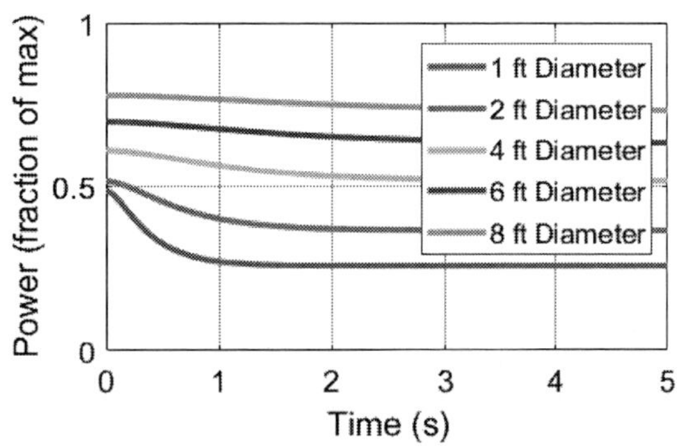

Fig. 14. Input Power with Proportional Control

Starting from a steady hover, a climb rate of 10 ft/s is commanded at $t = 0$. The input power (as a fraction of the installed power) and heave rate response with the proportional controller are shown in Fig. 14 and Fig. 15 respectively. The scaled time constant and delay are compared to the ADS-33E-PRF requirements in Fig. 16. With proportional control alone, only the 1 foot diameter rotor meets the Level 1 requirements. The degradation of handling qualities that occurs with increasing rotor size is a result of the increasing rotor inertia. However, from Fig. 14, it is clear that the use of a proportional gain alone does not utilize the full power capabilities of the motors, indicating that the closed-loop control system can be

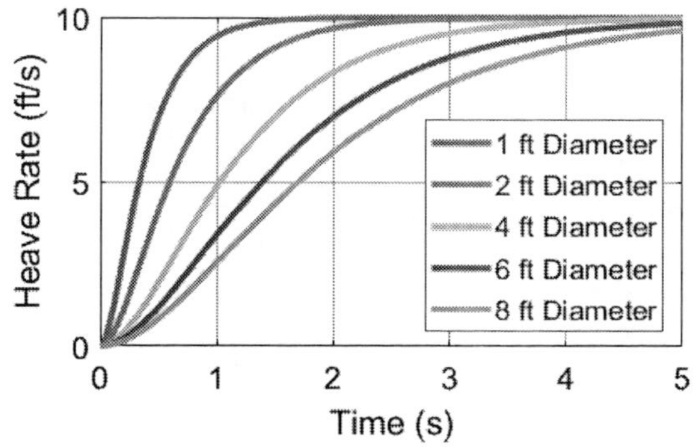

Fig. 15. Heave Rate Response with Proportional Control

redesigned in order to provide a faster response and better satisfy handling qualities metrics. This can be done through the addition of derivative gain.

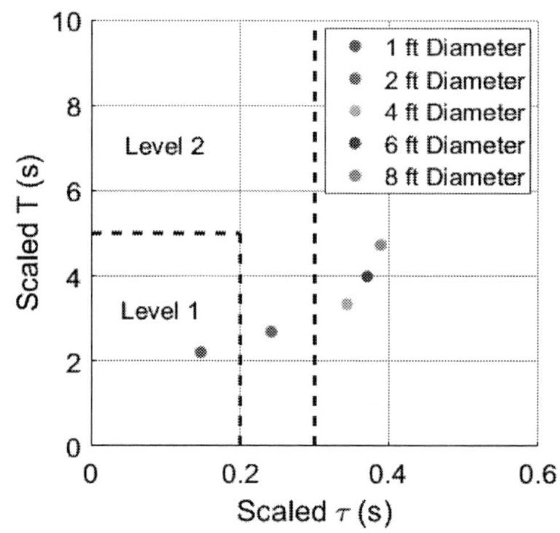

Fig. 16. Heave Response Metrics with Proportional Control Compared to ADS-33E Requirements

Proportional-Derivative Control

The use of a proportional-derivative (PD) controller allows for the further shaping of the heave rate response. If the power constraint is ignored, the rotor responses can be made to match in the time domain. A PD controller is first designed for the 1 foot rotor, such that the closed-loop response is characteristically first-order (no overshoot or oscillation) and the maximum commanded power (at t=0) is equal to the estimated installed power. The PD controllers of the other rotor sizes are then designed such that the closed-loop heave rate response is identical (on the linear models) for all diameters, assuming unlimited power. Comparing the heave response using the nonlinear model (Fig. 17, still ignoring the motor constraint) shows that the linear controllers are equally effective. The power required to achieve this response is shown in Fig. 18 with the exact values of the maximum power given in Table 5. The maximum commanded power for the large rotors greatly exceed the power capacity of their motors, with the largest rotor commanding over 60 times the estimated installed power.

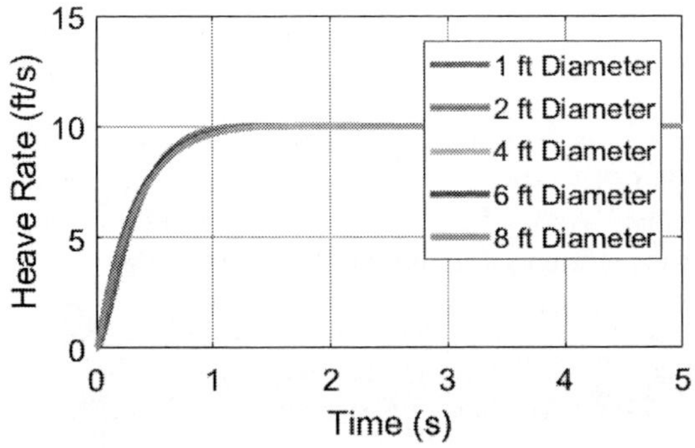

Fig. 17. Heave Rate Response with PD Control, No Power Saturation Limit

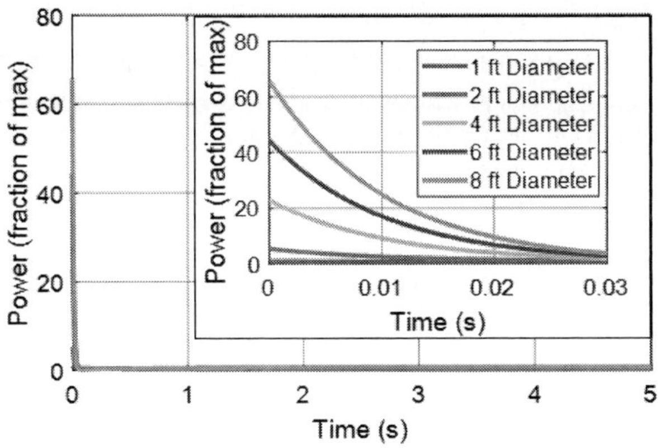

Fig. 18. Input Power with PD Control, No Power Saturation Limit

Table 5. Maximum Commanded Power for Ideal Response

Rotor Diameter (ft)	Maximum Commanded Power (hp)
1	0.4
2	6.2
4	71
6	250
8	590

The simulated heave rate responses presented in Fig. 19 utilize the same controllers used to create the step responses in Fig. 17, but with the inclusion of the power limit. Predictably, restricting the power input to realistic levels slows the system response, with greater lag as the rotor diameter increases. The input power with the saturation limit included is shown in Fig. 20. All of the rotors provide their maximum power at $t = 0$, with larger rotors maintaining this maximum for increasing lengths of time.

The heave rate responses of all the rotor diameters considered meet the ADS-33E-PRF Level 1 requirements (Fig. 21), except for the 8 foot diameter rotor. The limiting factor for the responsiveness of the 8 foot rotor is the time delay (τ) of the heave rate response. This time delay is associated with the initial response of the rotor to a step change in commanded heave rate, while the motor is saturated. Because the motor is

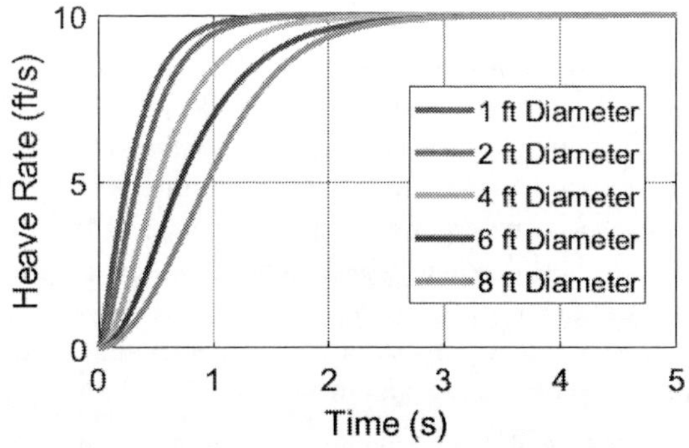

Fig. 19. Heave Rate Response with PD Control

already at its limit, there can be no improvement in the time delay via gain tuning. Thus, the only way an 8 foot diameter rotor can meet the scaled level 1 requirements would be by reducing the inertia of the rotor (via a more complex geometry or less dense material) or by increasing the motor power (and weight).

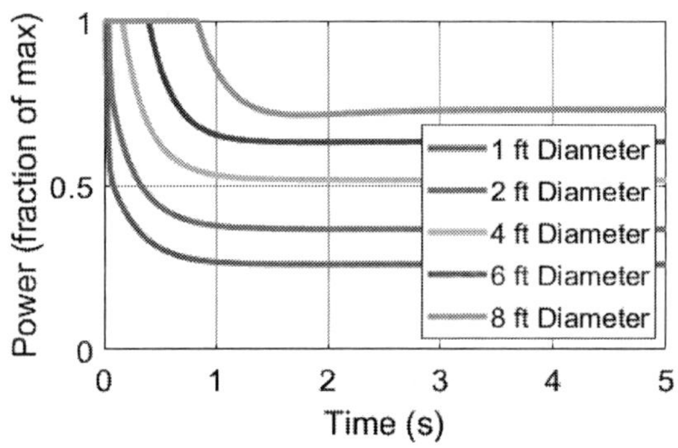

Fig. 20. Input Power with PD Control

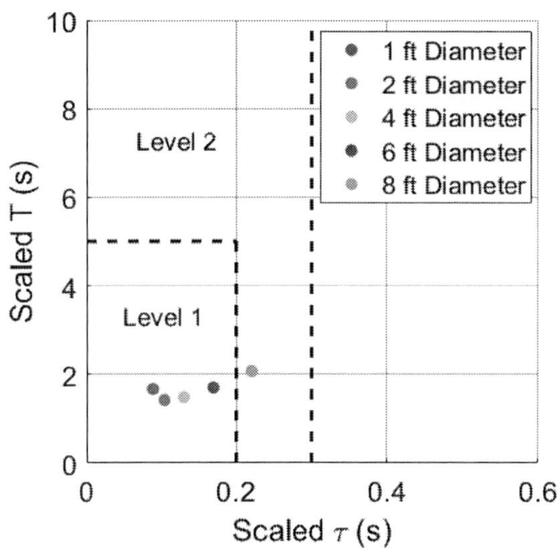

Fig. 21. Heave Response Metrics with PD Control Compared to ADS-33E Requirements

CONCLUSIONS

The effect of rotor radius on the responsiveness of fixed-pitch, variable RPM rotors in heave was explored through simulation. A simple approach based on conservation of angular momentum was found to be sufficient to capture the dynamic response of a rotor in response to a step change in the commanded throttle. Additionally, based on experimental results on a 28 inch diameter rotor, the mechanical power supplied by the motor changed nearly instantaneously, justifying its use as a control input in dynamic simulations.

Due to the natural stability of the heave dynamics, simple proportional control was capable of shaping the response of the heave dynamics, to a degree. However, this was only sufficient for the 1 foot diameter rotor to achieve scaled level 1 handling qualities standards. Including derivative gain allowed greater control and full use of the installed power, allowing rotors up to 6 feet in diameter to meet level 1 requirements. The relative lack of power on the 8 foot diameter rotor caused its time delay to exceed the maximum allowance of the ADS-33E-PRF heave standards, which can only be remedied by changing the design of the rotor or augmenting the motor's capabilities at the expense of weight.

While handling qualities standards are in need of further refinement, preliminary conclusions about the performance of fixed-pitch rotors can be made. Based on the heave dynamics alone, rotors with a diameter of 8 feet or greater will need to reduce the time delay associated with changing the rotor speed, possibly by either increasing the size of the motors (thereby increasing weight), or reducing the rotational inertia (with a more complex blade structure). Alternate configurations could also be considered, such as variable collective pitch which can circumvent the rotational dynamics of the rotor entirely without the need for a swashplate (as no cyclic pitch is required).

REFERENCES

[1]Gipson, L., Dunbar, B., "Urban Air Mobility Grand Challenge," NASA, September 2018, Available: https://www.nasa.gov/uamgc/. [Accessed: December 2018].

[2]"Fast-Forwarding to a Future of On-Demand Urban Air Transportation," UBER Elevate Summit, October 2016.

[3]Johnson, W., Silva, C., and Solis, E., "Concept Vehicles for VTOL Air Taxi Operations," AHS Technical Conference on Aeromechanics Design for Transformative Flight, San Francisco, CA, January 2018.

[4]Ivler, C., Goerzen, C., Wagster, J., Sanders, F., Cheung, K., and Tischler, M., "Control Design for Tracking of Scaled MTE Trajectories on an IRIS+ Quadcopter," AHS International 74th Annual Forum and Technology Display, Phoenix, AZ, May 2018.

[5]Alvarenga, J., Vitzilaios, N., Rutherford, M., Valavanis, K., "Scaled Control Performance Benchmarks and Maneuvers for Small-Scale Unmanned Helicopters," IEEE 54th Annual Conference on Decision and Control, Osaka, Japan, December 2015.

[6]Holmberg, J.A., King, D.J., Leonard, J.R., and Cotting, M.C., "Flying Qualities Specifications and Design Standards for Unmanned Aerial Vehicles," AIAA Atmospheric Flight Mechanics Conference, Honlulu, HI, August 2008.

[7]Cotting, M.C., "Applicability of Human Flying Qualities Requirements for UAVs, Finding a Way Forward," AIAA Atmospheric Flight Mechanics Conference, Chicago, IL, August 2009.

[8]Cotting, M.C., "UAV Performance Rating Scale Based on the Cooper-Harper Piloted Rating Scale," 49th AIAA Aerospace Sciences Meeting, Orlando, FL, January 2011.

[9]Klyde, D.H., Schulze, P.C., Mitchell, D.G., and Alexandrov, N., "Development of a Process to Define Unmanned Aircraft Handling Qualities," AIAA Atmospheric Flight Mechanics Conference, Kissimmee, FL, January 2018.

[10]Anonymous, "Aeronautical Design Standard Performance Specification Handling Qualities Requirements for Military Rotorcraft," Tech Rep., United States Army Aviation and Missile Command, ADS-33E-PRF, March 2000.

[11]Niemiec, R., "Development and Application of A Medium-Fidelity Analysis Code for Multicopter Aerodynamics and Flight Mechanics," Ph. D. Dissertation, Rensselaer Polytechnic Institute, August 2018.

Kinematic Velocity Commander for Obstacle Field Navigation

Chad Goerzen
San Jose State University
Moffett Field, CA

Marc Takahashi
Matt Whalley
US Army Aviation Development Directorate
Aviation & Missile Research, Development & Engineering Center
Moffett Field, CA

MAJ Mark Cleary
Jeffrey Cox
US Army Aviation Development Directorate
Aviation & Missile Research, Development & Engineering Center
Fort Eustis, VA

ABSTRACT

Obstacle Field Navigation is that task of flying toward a destination while avoiding terrain which is sensed through an on-board rangefinder. In this paper, the RiskMinOFN software module that generates a velocity command stream has been updated to maintain groundspeed during turns. It does this while improving safety by explicitly computing the reachable set given dynamic vehicle model constraints. This new method is based on modeling the reachable set during each replan. The kinematic velocity command controller was tested using a simulation study that was validated by a limited number of flight tests on a JUH-60 Black Hawk helicopter. The new velocity command controller is shown to result in more flight trajectories with more consistent airspeed, improved smoothness, and improved safety.

INTRODUCTION

Obstacle Field Navigation (OFN) is essential to operating autonomous aircraft in close proximity to the ground. Using GPS localization to fly along pre-mapped routes is possible for operations high above ground level. However, localization is not accurate or consistent enough to ensure consistent safety close to the ground. Localization tends to be even more degraded in mountainous or urban regions, where higher accuracy is needed. On-board sensing is used to provide the ultimate level of safety. Localization uncertainty has been considered an important aspect of pilot-guided flight from the first days of aviation, resulting in runways and helipads with clear visual markings and radio beacons. The pilot uses these beacons, along with air traffic control communications, to decrease uncertainty of the aircraft location down to the level of a few meters at the end of the flight. In the case of landing in an unprepared site, the pilot may not even know the location of the landing zone, but uses visual feedback to accomplish a safe landing. Here we use on-board sensors to provide safe flight and landing even in the presence of significant localization uncertainty.

The US Army Degraded Visual Environment Mitigation (DVE-M) program is exploring the use of autonomy for use in controlling a helicopter in environments such as darkness, fog, clouds, rain, snow, or dust, where the pilot's vision is impeded. In this program sensors, such as RADAR and LIDAR, are used to build up a current view of the environment and fused to form a terrain representation. The terrain representation is given to the pilot and to autonomous control software, which includes OFN and Safe Landing Area Detection (SLAD). While the DVE-M program is broad in scope and includes visual aids for piloted flight, SLAD, and other technologies, this paper is focused on fully autonomous flight guided by OFN. Since the guiding software, RiskMinOFN v2, has been broadly described in previous publications, this paper is specifically focused on the velocity command control law, which converts the risk-to-go map into a stream of velocity commands given to the inner loop control law.

Prior work

The US Army Aviation Development Directorate (ADD) has been developing OFN algorithms for several years. Early efforts, references [1] and [2], developed a 2-dimensional planner based on Voronoi triangulation. This was then transformed into a 3-dimensional planning using plane slicing, described in references [3] and [4]. The next approach was a quasi-3D A* algorithm capable of handling overhangs, described in [4] and [5]. This has been released as a library called libOFN. Previous AFDD approaches were purely geometric, but the most recent approach uses a full 3D A* algorithm that minimizes risk to the aircraft [6]. In theory, all of these approaches are capable of long distance global planning, but in practice, computational limits make them local planners. Since 2011, development focused on implementing libOFN on an unmanned Black Hawk helicopter at operationally relevant speeds. In [7] libOFN was first demonstrated at speeds up to 20 m/s, and in [8] and [9] the speed was increased to a maximum of 57 m/s. The increase of speed was made possible primarily by increased acceleration limits and sensor range. [10] describes the tight integration of libOFN into the cockpit, giving the pilot the ability to tightly control the level of autonomy and modify the mission, and gives measurements of pilot workload. [11]

Presented at the Vertical Flight Society Autonomous VTOL Technical Meeting and Electric VTOL Symposium, Mesa, Arizona, UAS, Jan. 29-31, 2019.

describes the autonomous system with a focus on inner loop control, including flight test results.

Significant work has been accomplished in planning with a high-fidelity dynamic model, but until recently it has been too computationally expensive to update in real time. Direct optimization has been the standard method for optimal planning for many years, but has high computational complexity. Sections 4.4 and 4.5 of [12] discusses an overview of direct optimization methods. The maneuver automaton (MA), first published in [13], is a method designed to provide control laws that link one state in a roadmap to another, thus bridging the gap between the discrete and continuous world. The maneuver automaton has a fixed library of possible directions, connecting cells of the roadmap using these direction change maneuvers. Sampling-based techniques such as the rapidly-planning random tree (RRT) [14] and probabilistic roadmap (PRM) [15] are complete, but are known not to converge to an optimal solution. More recently the PRM* and RRT* algorithms [16] are modified versions of the original RRT and PRM that have the benefit of having proofs of optimal convergence.

OVERALL SOFTWARE ARCHITECTURE

This paper concerns a component of the RiskMinOFN algorithm. RiskMinOFN has the purpose of keeping the helicopter safe from sensed obstacles while guiding it toward a goal location. This problem is made difficult due to the following uncertainties: execution uncertainty, or the inability to exactly follow a commanded path; sensor uncertainty, or the lack of confidence in a sensor reading; and localization uncertainty, or the lack of confidence in knowing the exact position. RiskMinOFN combines these uncertainties and risks in a unified framework.

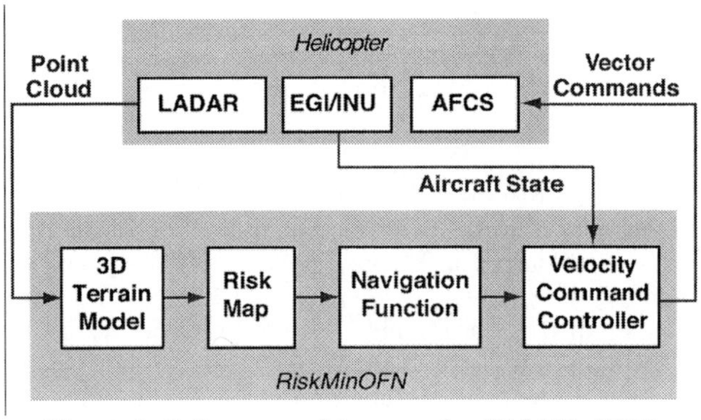

Figure 1. Software architecture for RiskMinOFN.

The RiskMinOFN pipeline (Figure Figure) starts with sensor beams, traces them into a 3D terrain model, calculates a risk map, determines the risk-to-go navigation function, and determines velocity commands. These steps are described below:

3D Terrain Model: Terrain is represented with a three-dimensional evidence grid, a sensor fusion method developed by [17]. The terrain is sized so that it just covers the largest possible sensed area to achieve the highest resolution while still computing the solution in time. The sensor model has an estimate of false positive and false negative laser returns over the length of the beam, and treats the interaction of each beam with each cell as an intersection of a narrow cylindrical beam with a spherical cell, with the radius of the beam increasing as the range increases.

Risk Map: The risk map is a spatial representation of estimated risk that unifies all modeled risk factors. Some risk factors modeled include execution uncertainty, represented as a region of increased risk that decays exponentially with distance from the obstacle, an altitude penalty, a constant risk value that applies to all cells, increased cost for unknown space, and user-programmed airspace restrictions. These risk factors are unified into a global risk map using probabilistic combination with assumed independence between cells. For normal low risk levels, probabilistic combination can be linearized so that the planning problem can be solved efficiently using standard dynamic programming algorithms, such as A*.

Navigation Function: The A* algorithm [18] is used to solve the optimization problem of the minimal risk path, using the risk map to produce the weighted graph. The result of this algorithm is a risk-to-go function, with a minimum value at the goal. When the goal position is a long distance outside the risk map the algorithm places a temporary goal just outside the risk map, and the entire boundary of the array is used as a goal, with distance to the temporary goal being used to initialize the priority queue values for the A* algorithm.

Velocity Command Controller: The navigation function has the property that if a purely local minimization algorithm starts at any point in the navigation function and continues to iterate, it is guaranteed to arrive at the global minimum. The velocity command module takes this role. It searches a local neighborhood for a minimum risk-to-go value and commands the flight path vector in that direction.

Legacy Velocity Command Controller

Currently libOFN uses a heuristic velocity command controller to bridge the gap between a desired direction and a velocity command. This module uses a simple double-integrator with acceleration limits and speed-scheduled vertical speed limits. In flight tests it was found that this module sometimes produces undesirable results. Specifically, the legacy velocity command controller executed an iterated search for a local minimum until the desired look-ahead time, then generates a vector command in that direction with a speed determined by the straight-line distance in clear space. Figure Figure shows an example of an representative error condition that can result from this approach. In this figure, the search jumps over the obstacle and ends up commanding a vector toward the obstacle.

To prevent the velocity command controller from entering an unsafe velocity command, the number of steps in the path is compared to the straight-line distance between the ownship position and the desired location, and once the values differ beyond a threshold parameter, the local minimum search is cut short. In the example shown in Figure 2 the result of this heuristic would be to cut the search short just after the local search passes the obstacle. While this heuristic check improved the algorithm's performance, it occasionally lead to erratic behavior, including large jumps of the local minimum between replans. More importantly, this legacy velocity command controller can not be easily shown in analysis to maintain soundness of the overall software.

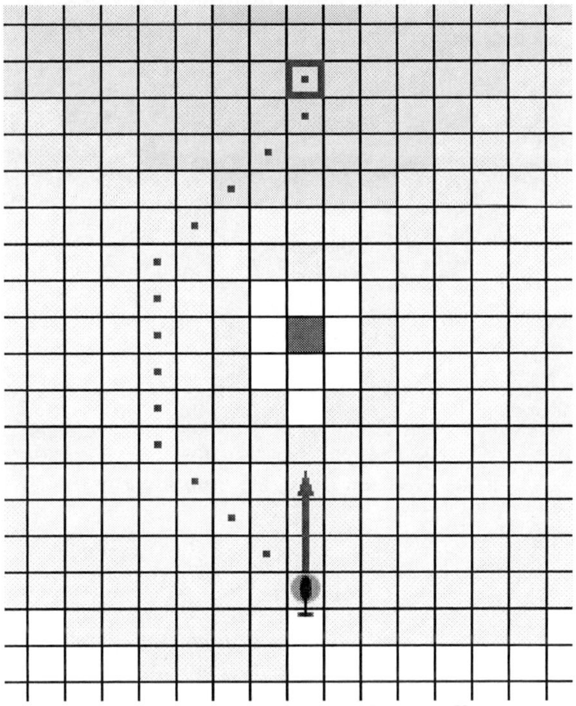

Figure 2: Legacy velocity command controller: example of local minimum search resulting in unsafe velocity command: the local minimum search jumps the obstacle, resulting in a velocity vector pointing toward the obstacle.

One thing the pilots found objectionable to the legacy velocity command controller was in turning about: during drastic course changes, the legacy velocity command controller would slow down the helicopter to keep within safe space. However, in operation it is sometimes not possible to slow down to hover due to a heavy load or high altitude, and pilots are trained to accomplish all turns at a minimum air speed of 60 knots. This requires large radius turns and a more sophisticated velocity command controller.

New Velocity Command Controller

The new velocity command controller, generates many possible simulated paths in a short time horizon, and checks them for safety. This can be considered analogous to the state space sampling of RRT*, although using a structured sampling and branching strategy. Here is the algorithm outline:

1. Generate reachable set: A set of reachable points is estimated by using constant turn-radius paths at constant speed. Each of these constant-speed trajectories is calculated for a different lateral acceleration and vertical rate, and a sufficient number of trajectories are calculated to fill the grid. See Figure Figure for an example in 2 dimensions.

2. Calculate risk along trajectories: Once the trajectories are generated, risk values for positions sampled along each trajectory are summed using probabilistic combination to arrive at the risk along each trajectory. The pseudocode for steps 1 and 2 is shown in the box below labeled KinematicSearch. This generates a three-dimensional array of risk estimates accumulating outward from the current position and limited to the reachable set.

3. Check trajectories for safety: The remainder of the trajectories up to the hover position is checked against the known empty space map. If any point in a given trajectory falls outside the known empty space map, the remainder of the trajectory is eliminated from the candidates.

```
KinematicSearch ( RiskArrayIn, CostArrayOut, currPos, stopPos )
{
  if speed is 0: set CostArrayOut equal to RiskArrayIn and return
  vel = stopPos - currPos // determine velocity vector
  vHat = vel / Magnitude(vel)

  // sample the reachable space with the vector (slope, dTheta, rho)
  for slope in [min_slope, max_slope]
    for dTheta in [min_turn, max_turn]
      theta = atan2(vel0[0], vel0[1])
      pos = currPos
      for rho in [0, distStop]
        pos = pos + STEP*[vHat[1]*sin(theta), vHat[0]*cos(theta), slope)
        theta = theta + dTheta
        accumRiskK = ProbAorB( RiskArrayIn(pos)*STEP, accumRiskK )
        CostArrayOut(pos) = accumRiskK
}
```

4. Search for optimal trajectory: The risk-to-go estimate at each transition point is summed with the risk of the trajectory to produce a total risk estimate for the current plan. The transition point with the lowest summed risk estimate is selected as the optimal plan. The following cost functional uses risk-to-to as calculated by A*, accumulated risk as calculated from step 2, and a distance term to encourage selection of longer time horizons:

$$\text{cost} = \left(1.0 + A\,\text{RTG} + Bk\right)\rho_{\max} / \left(\rho_{\max} + \rho\right) \qquad 1)$$

where cost is the combined cost, RTG is the risk-to-go as computed by A*, k is the accumulated risk as calculated from step 2 above, $\rho_{\max}$ is the maximum radius calculated and ρ is the radius from the starting point. The weighting parameters A and B can be tuned for performance: here A was set to 0.001 and B was set to 0.5.

5. Fly the trajectory and replan: The trajectory chosen in step 4 is flown. Step 3 guarantees that if no future plan is made, the helicopter will stop inside known empty space, which is important to ensure soundness. Repeat this algorithm from step 1 at a time sufficiently early before the transition point to ensure the helicopter will not begin slowing needlessly. The new plan should be completed before the helicopter arrives at the transition point.

Figure 3: 2-dimensional example of a kinematic reachable set. The own-ship position is at the bottom center of the image with current velocity toward the top of the image. Each pixel was given a random value for this example to show how risk accumulates as distance from current position increases. Dark blue pixels are considered to be not in the reachable set for this time horizon.

This algorithm addresses several problems with the existing RiskMinOFN, and represents a tighter integration of the vehicle model into the planning framework. This method may be able to use some of the ideas from RRT* to prove optimality and soundness, and is designed to integrate closely with a vehicle model. Figure Figure illustrates the visualization of the velocity command controller integrated into RiskMinOFN. In this example, the final map combining Risk-to-Go with the kinematic reachable set generated in step 5) of the algorithm is shown as a horizontal slice of the 3-dimensional array, with highest cost shown as red and lowest as green. In addition, the backtrace is shown as a series of white dots, and the safe corridor used for visualization is shown as a series of three gates. It is evident that the lowest cost points are blocked by the obstacle, so the situation visible in Figure Figure will not occur here.

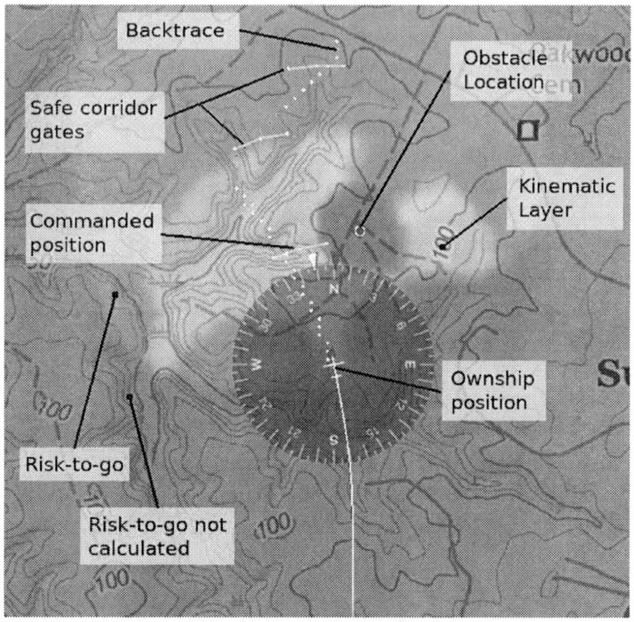

Figure 4: Kinematic layer 2-dimensional slice in context of the planner, from a simulation run.

RESULTS

The new velocity command controller was simulated in comparison to the legacy velocity command controller. It was also compared with piloted flight tests.

Autonomous Flight Control System (AFCS) and Simulation Environment

The AFCS provides the guidance capabilities for the autonomy components in the mission software and consists of the waypoint, tracking, and inner-loop control [19]. The waypoint control accepts a stream of velocity commands in the form of the velocity magnitude, heading, and glide slope that are each filtered through second-order limiting filters, which prevent rapid changes of the velocity vector from entering the tracking control. Once filtered, the velocity is transformed into inertial coordinates and it is used to calculate position and accelerations, which are used to generate a trajectory sent to the tracking control. The tracking control acts as the autopilot around the inner-loop control and forces the system to track an inertially referenced path signal generated from the waypoint control block by passing aircraft-referenced command signals to the inner-loop control.

The inner-loop control is comprised of two parts, the flight-path command and the attitude command systems. The flight-path command system uses the aircraft-referenced outputs from the tracking control and generates collective stick inputs to the aircraft and commanded roll and pitch attitudes and commanded yaw rate to the attitude command system. The attitude command system provides primary three-axis attitude stabilization of the system and generates the lateral, longitudinal, and pedal inputs to the aircraft. In forward flight, turn coordination is enabled on the inner-loop yaw axis. The attitude command system has feedback gains in a model-following architecture based on scheduled linearized design models generated from a UH-60, blade element, simulation model [20] at hover, 40, 80, and 120 knots.

The AFCS can be run in closed-loop in a standalone desktop simulation model consisting of the previously mentioned linear model's stability derivatives interpolated across the 0 to 120 knot speed regime and integrated with full non-linear kinematics. Bounding aircraft trim points are set to preserve stability derivatives that depend on velocity. The desktop simulation has the AFCS and other autonomy components (OFN, SLAD, etc.) fully integrated as would be run in flight test except for hardware specific interfaces, such as 1553 bus and aircraft pilot inceptors.

The sensor was simulated as in previous studies, and the sensor simulator is described in more depth in Ref. [Goerzen 2015]. The sensor simulator uses depth buffers from the OpenSceneGraph 3D graphics library [21] to simulate a scanning LADAR. The simulated ranges are subjected to pseudo-random range noise, pseudo-random walk added to the sensor position estimate, and pseudo-random dropouts based on range.

Validation of Simulation Against Flight Tests

To validate the simulation environment, flight tests were carried out in the following manner. RiskMinOFN is run given inputs from the INU, Radar altimeter, and LADAR. It produces a plan and outputs a commanded velocity to an evaluation pilot through symbology on a flat screen display. The ICE symbology set was used, based on the software described in [22]. This flat screen display also shows a 3-dimensional rendering of the terrain that shows LADAR points. The evaluation pilot, who is flying heads-down, looks at the flat panal display and attempts to match the measured state to the desired state (guidance). The important point is that OFN is producing the guidance signal for the display, and this is the same signal as is given to an autonomous control law in fully autonomous mode.

Figures Figure, Figure, and Figure show an instant when RiskMinOFN is guiding the pilot to avoid a tower. Figure Figure shows a frame from each of the flight video recorders, Figure Figure shows the 3-dimensional synthetic view with rangefinder data and enroute guidance symbology, and Figure Figure shows the moving map display with RiskMinOFN data overlaid. Although the tower appears too small to be clearly visible in the video or synthetic view, it is evident as a red high-risk area in the moving map display.

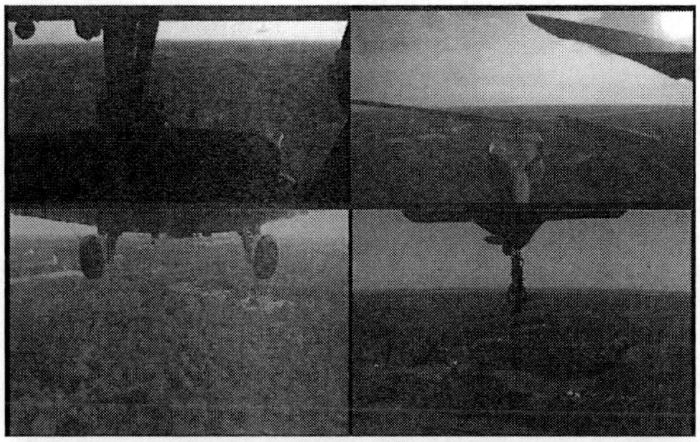

Figure 5: Flight video frames.

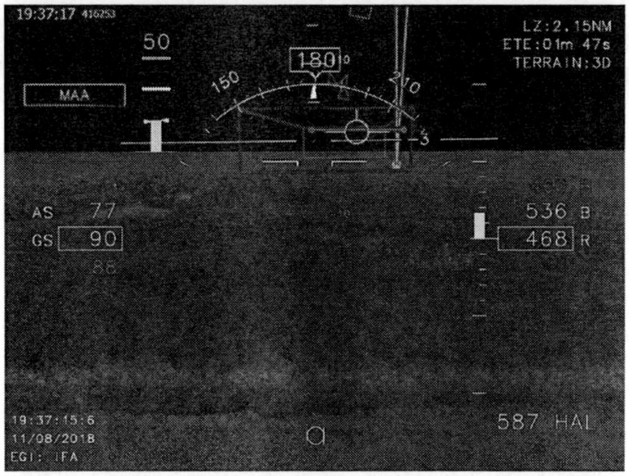

Figure 6: Pilot display A: 3-dimensional synthetic view, with enroute guidance symbology.

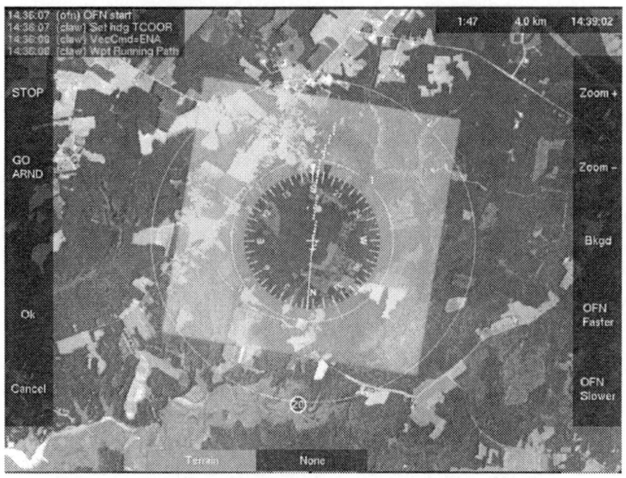

Figure 7: Pilot display B: moving map display.

Two guidance symbology methods were used for the data presented here. The first is a flight director, where the pilot is tasked to move the inceptors to zero out error signals. The second is called the safe corridor, when a flight path vector is rendered in 3-dimensional space, and the pilots is tasked to keep the flight path vector inside a safety box. Tests to compare the two methods are beyond the scope of this paper, but planned for future flights. The flights described here were system integration checks used to verify the software and hardware was properly integrated, and have a secondary purpose of evaluating the simulation framework.

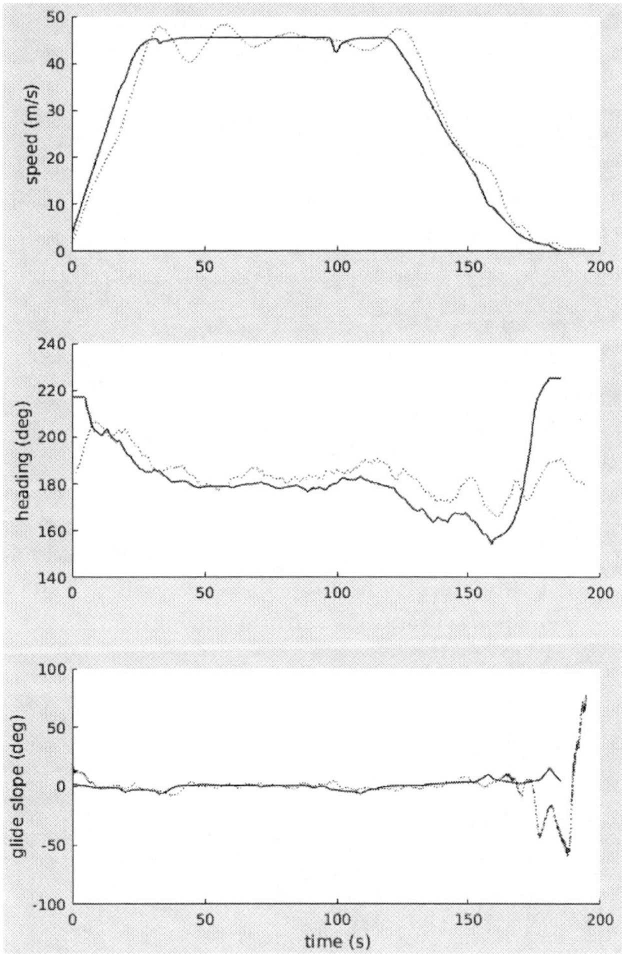

Figure 8: Tracking of human pilot (Pilot A) following an OFN-commanded trajectory. Commanded vector components are plotted with a solid line, and the flown components are plotted as a dotted line.

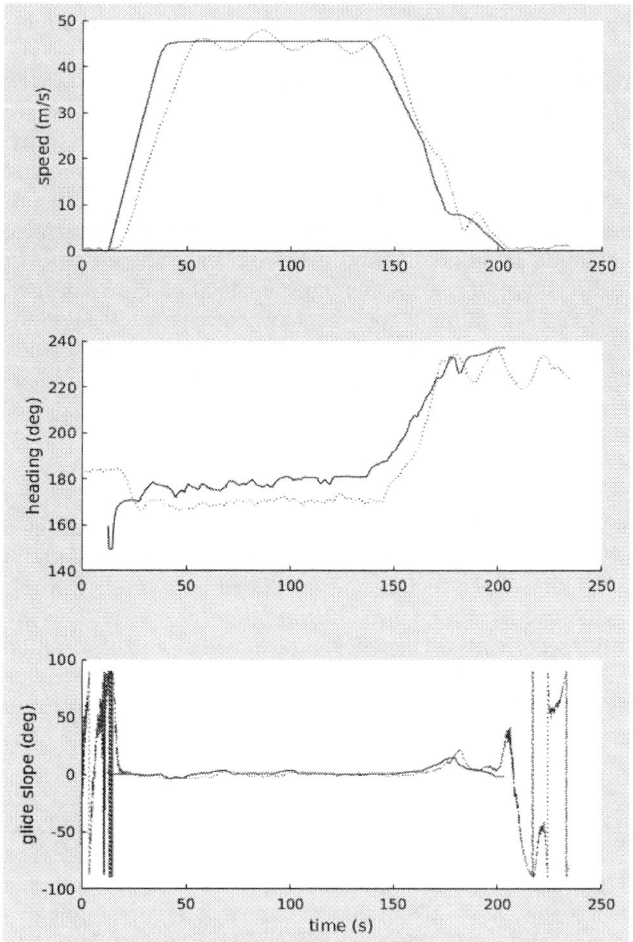

Figure 9: Tracking of human pilot (Pilot B) following an OFN-commanded trajectory. Commanded vector components are plotted with a solid line, and the flown components are plotted as a dotted line.

It is necessary to show that the human pilots have adequate performance in following the guidance symbology. Figures Figure, Figure, and Figure show two pilots flying the same pole avoidance task. Pilot A is following the flight director while Pilot B is following the safe corridor. Both pilots exhibited roughly similar tracking behaviors. They both lagged the desired speed during the acceleration and deceleration phases by several seconds and both oscillated 5 meters per second above and below the desired speed. Both show a bias in lateral tracking: this may be due to the initial direction vector not being oriented exactly towards the destination and the pilots attempt to correct this by flying off cue. Finally, both tracked well in the heave axis.

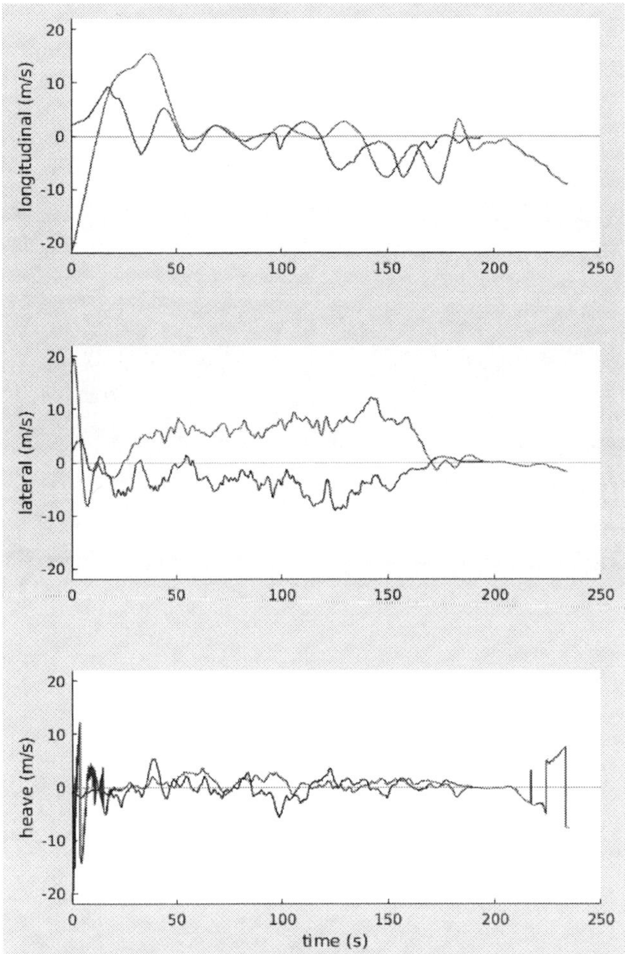

Figure 10: Tracking error in the principal axes for the two pilots. *Black*: pilot A. *Blue*: pilot B.

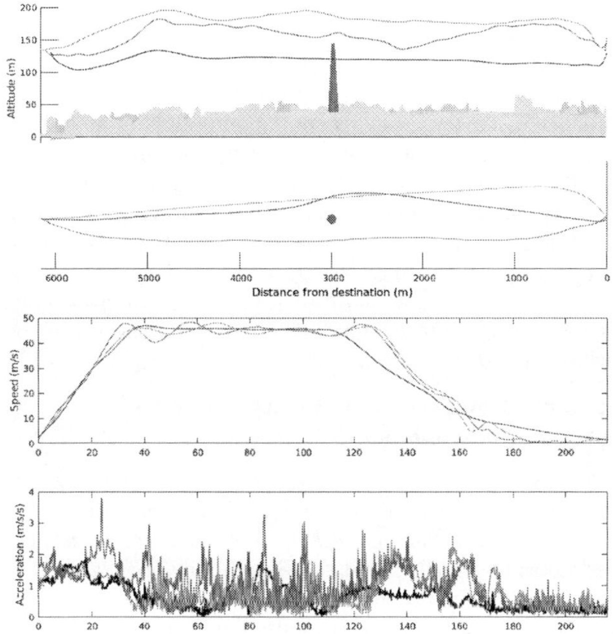

Figure 11: *Terrain mode*. Trajectory plots of piloted flights (red: Pilot A, orange: Pilot B) and simulated AFCS flight (black). *Top.* plan and elevation views; *Middle.* time history of speed; *Bottom.* time history of acceleration magnitude. Radio tower obstacle (magenta) has a height of 113 m above ground level.

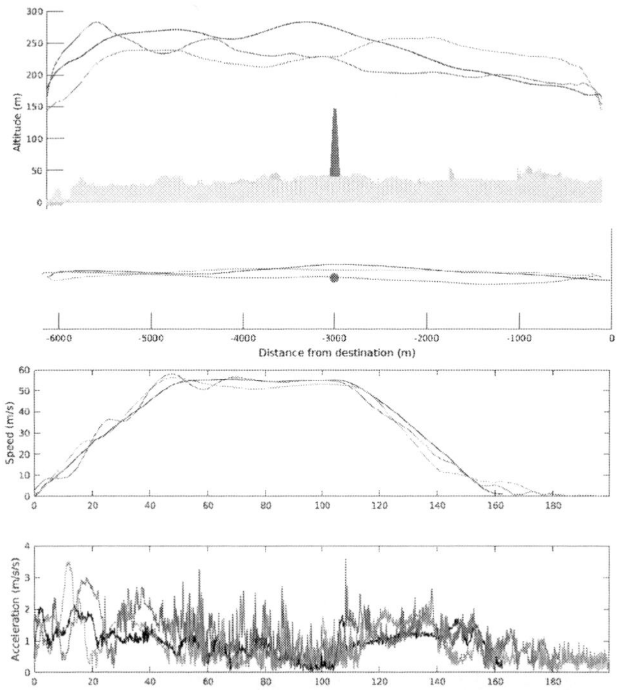

Figure 12: *Cruise mode*. Trajectory plots of piloted flights (red: Pilot A, orange: Pilot B) and simulated AFCS flight (black). *Top.* plan and elevation views; *Middle.* time history of speed; *Bottom.* time history of acceleration magnitude. Radio tower obstacle (magenta) has a height of 113 m above ground level.

The human-piloted flights are used to validate the simulation, keeping in mind that Autonomous Flight Control System in simulation has somewhat better tracking performance. Figures Figure and Figure show trajectories of two piloted flights co-plotted against a trajectory from the simulated automated flight. Figure Figure shows trajectories executed in "Terrain" mode, where the speed is limited to 46 m/s (90 knots) and the desired altitude is 100 m above ground level (AGL). Figure Figure shows trajectories executed in "Cruise" mode, where the speed is limited to 57 m/s (110 knots) and the desired altitude is 330 m above ground level (AGL). The primary difference is that in Cruise mode the helicopter overflies the tower, while in Terrain mode it avoids the tower laterally.

There are several points of consistency between the simulation and flight cases. In the Terrain mode cases, the helicopter avoids the tower laterally, while in the Cruise mode cases, the helicopter over-flies the tower. The avoidance distance is roughly similar, although in the piloted flights the pilots don't avoid as aggressively as in the simulation. Speed and acceleration profiles are roughly in line between simulation in flight – the difference is that in the piloted flights, the speed tends to oscillate with a 20 second period. This may be due to improper gains in the visual pilot interface, and will be looked at in future flights. Overall, the flight tests show that system integration to be nearly complete, and a good match with simulation.

Qualitative Comparison of Velocity Command Controllers in Simulation

To illustrate the difference between planners during a typical maneuver, consider a turn-about. One of the motivations for creating the new planner was to keep a more consistent speed during turns. Figure Figure demonstrates the difference between the two planner. With the legacy velocity command planner it was difficult to execute sharp turns without losing speed while still remaining a safe distance from terrain in all cases, as a result at high speeds sharp turns require a reduction in speed. The new planner is better able to maintain speed, since the wide turn is explicitly tested against the terrain. Figure Figure shows several characteristic differences: the new planner takes a wider turn, but only loses 3 m/s of speed compared to the legacy planner which loses 10 m/s. The new planner executes the turn with a lower peak acceleration, roll, and pitch. Finally, the new planner holds a more consistent collective position.

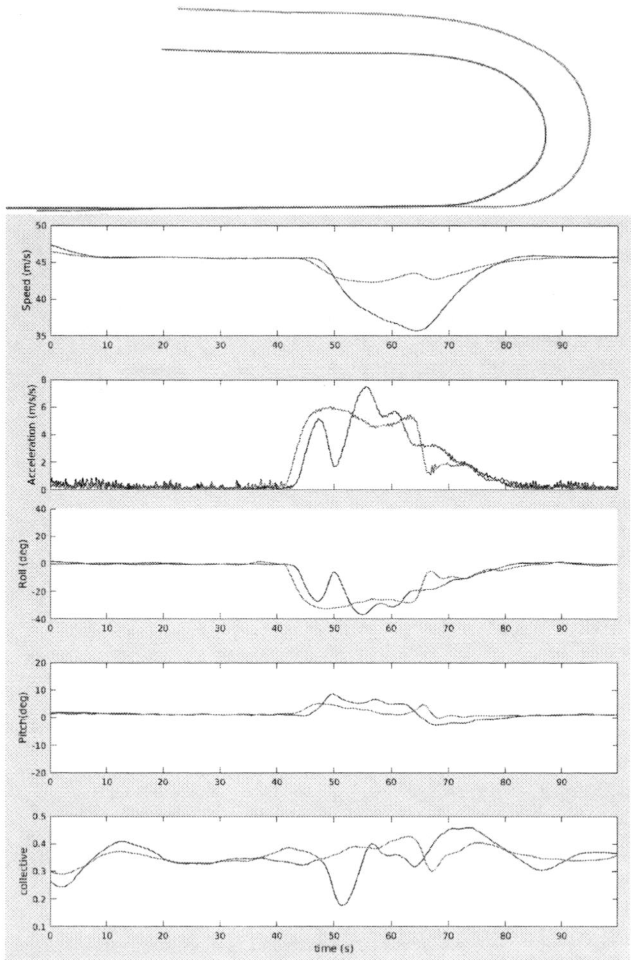

Figure 13: Example run showing a turn-about maneuver, where the commanded direction is instantaneously reversed. The legacy planner (black) takes a tighter turn at lower speed, while the new planner (red) turns wider without losing much speed. Time histories for magnitude of speed in inertial space, magnitude of acceleration, roll, pitch, and collective are shown for both trajectories.

Table 1. Data values for multiple simulations.

Legacy Controller	#1	#1	#1	#2	#2	#2	#3	#3	#3	#4	#4	#4	#5	#5	#5	Mean	Std. Dev.
Duration (s)	668	639	645	231	240	227	424	433	439	626	639	650	574	586	597	507.9245	163.8423
Acceleration (m/s/s)	1.43	1.36	1.2	1.26	1.46	1.18	1.24	1.34	1.3	1.5	1.76	1.63	0.99	0.99	1.07	1.315044	0.221106
Ang. Rate (deg/s)	3.7	3.89	3.66	3.46	3.59	2.88	3.02	3.17	3.61	4.07	5.04	4.18	3.31	3.3	3.54	3.628128	0.533496
1% Terrain Dist. (m)	67.5	57.3	62.3	61.2	54.9	54.1	63.8	56.7	72.1	76.3	73.1	74.3	76.9	61.3	70.4	65.474	7.960355
New Controller	**#1**	**#1**	**#1**	**#2**	**#2**	**#2**	**#3**	**#3**	**#3**	**#4**	**#4**	**#4**	**#5**	**#5**	**#5**	**Mean**	**Std. Dev.**
Duration (s)	617	625	606	228	240	224	425	436	421	612	615	607	580	578	583	493.2019	153.7947
Acceleration (m/s/s)	1.21	1.29	1.54	0.91	0.92	1.12	0.9	0.77	0.81	1.02	1.02	0.84	0.77	0.96	0.82	0.994743	0.217308
Ang. Rate (deg/s)	4.29	3.87	4.45	2.95	2.72	3.48	2.45	1.98	2.08	4.14	3.39	2.79	2.47	3.36	2.67	3.141019	0.790802
1% Terrain Dist. (m)	74.8	67	61.2	69.3	65.1	59.9	72.4	67.7	74.5	79.8	76.6	78.1	79.7	74.3	77.4	71.83867	6.498402
Relative Duration	0.92	0.98	0.94	0.99	1	0.98	1	1.01	0.96	0.98	0.96	0.94	1.01	0.99	0.98	0.975147	0.02657

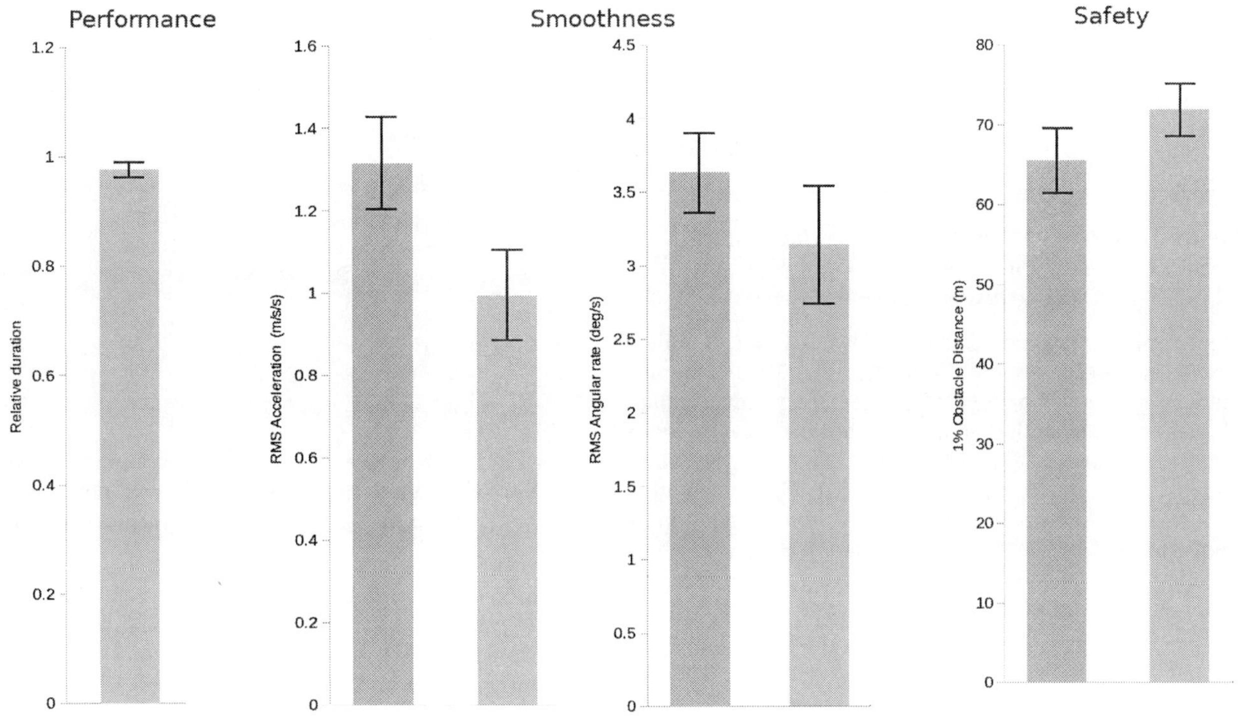

Figure 14: Performance metrics simulated flights in the area of Mount Hamilton for the legacy controller (*black*) and the new controller (*red*).

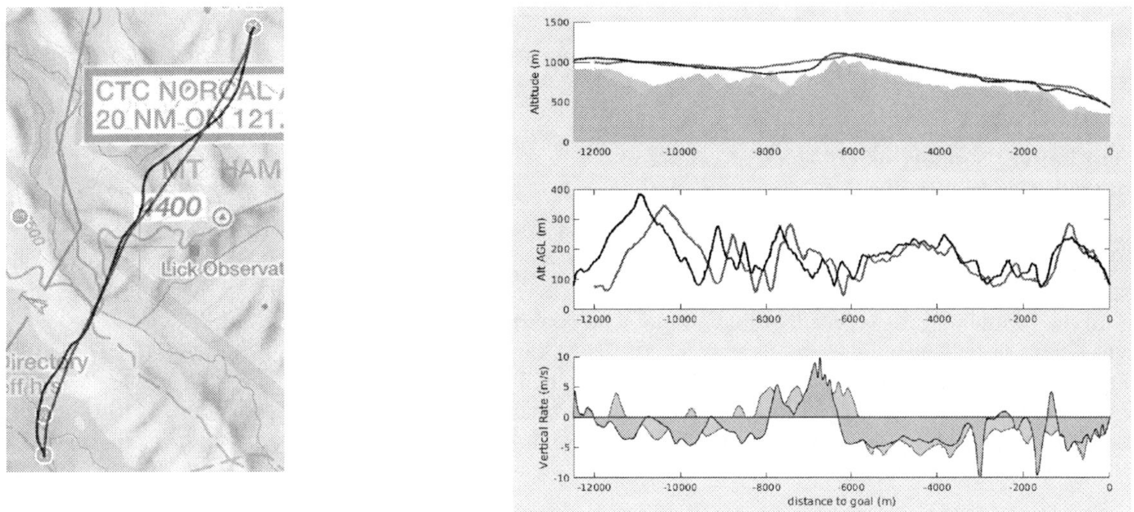

Figure 15: Example run showing resulting trajectories for both the legacy controller (black) and new controller (red). Flight paths flown (left), altitude profile (right top), height above ground level (right middle), and vertical rate (right bottom).

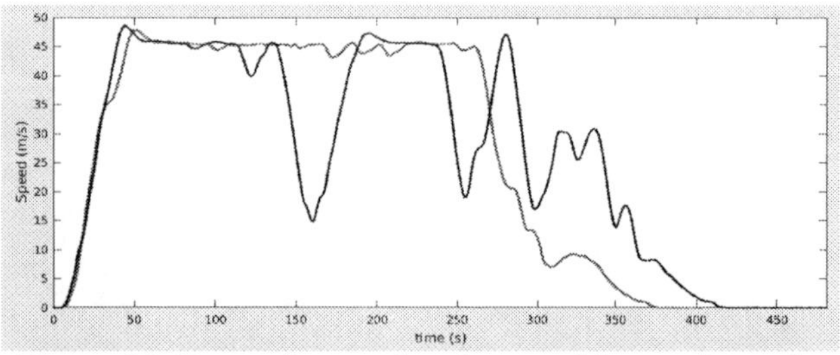

Figure 16: Example run showing resulting trajectories for both the legacy controller (black) and new controller (red). Time history of ground speed

A representative mission simulated near Mount Hamilton shows even more extreme differences. The 12 km mission includes flying over a ridge and descending into a valley, and was run for both the legacy and the new velocity command controller. Figures Figure and Figure show simulated variables for each co-plotted. In Figure Figure it is evident that the legacy controller tends to command more abrupt turns, climbs, and descents. However, both controllers keep a similar profile above ground level. In Figure Figure it is evident that the new controller is better able to keep a constant airspeed, while the legacy controller commands significant deceleration when it discovers unexpected obstacles.

Table 2. Commanded points

Path ID	Start	Command
1	21	30 28
2	28	23
3	23	24
4	24	32 22
5	22	21

Quantitative Comparison of Velocity Command Controllers in Simulation

The terrain around Mount Hamilton, CA, was used for statistical characterization because it was flown in previous tests and has significant vertical relief. The waypoints were run in the following order given in Table Table, where start gives the initial waypoint, and command lists the commanded points. When multiple points are commanded, the intermediate points are fly-through points and the final point is the hover point. All waypoints are located at around 100 meters above ground level:

Figure Figure shows example trajectories for the two planners. There were three runs simulated for all of these five segments, for a total of fifteen runs in all for each planner. Figure Figure shows one example of the three runs.

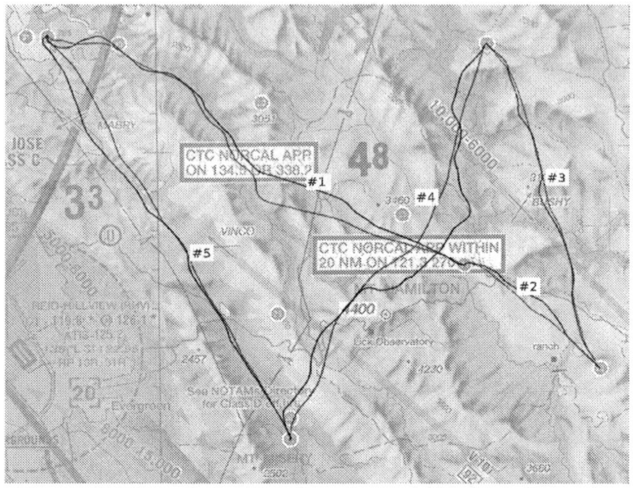

Figure 17: Example run showing resulting trajectories for both the legacy controller (black) and new controller (red). The path IDs are shown in white rectangles.

As with the previous example, the new velocity command controller tends to take a more direct, smoother path. The actual path differs from run to run, but the performance metrics remain the same. These metrics are listed for three categories: performance, smoothness, and safety. Performance uses task duration, smoothness uses acceleration and angular rate, and safety uses 1% terrain clearance.

Duration: Time from hover to hover.

Acceleration: The root mean square (RMS) value of the magnitude of the acceleration with the gravity vector removed.

Angular rate: The RMS value of the magnitude of the angular rate vector.

Terrain clearance: The one percent value: the radar altimeter measures a value higher then this for 99 percent of the flight.

These simulation results are listed in Table Table and summarized in Figure Figure. Each path ID appears three times for each of the three runs. Although the duration for each task differs, the tasks were all given equal weight when averaging. Paths of different length have different durations, thus they cannot be meaningfully averaged like the other metrics can. Instead, the relative duration, or the ratio of the duration with the new controller to the duration with the legacy controller, is used. Each of the metrics is shown with its 95% confidence interval. As can be seen, the new controller has marginally shorter duration, possibly due to the more consistent speed. Both accelerations and angular rates were lower for the new velocity command controller. Additionally, the new velocity command controller was better able to maintain a minimum distance from the terrain.

CONCLUSIONS

A new kinematic velocity command controller was developed based on explicit simulation of the reachable set:

1. A new velocity command controller based on a reachable set search was shown to provide a more reliable way to convert from the discrete solution provided by A* on a grid to a flyable velocity command. It allows arbitrary flight dynamics to be simulated, although even a simple flight dynamics model produces significantly improved results over the previous controller.

2. The simulation environment was validated by comparing with flight tests, and the new velocity command controller was used in these tests. In these flight tests the pilot followed symbology generated by the new velocity command controller.

3. The new velocity command controller was shown in simulation to have improved smoothness and safety as compared to the legacy controller, as well as accomplished the goal of maintaining a more consistent ground speed during turns.

ACKNOWLEDGMENTS

Thanks to MAJ Zachariah Morford and CW4 Joseph Fay who served as evaluation pilots, as well as LTC Carl Ott and MAJ Mark Cleary who served as safety pilots for the test flights outlined here. Thanks also to Jeffrey Cox, John Archdeacon, and Stuart Summerfield who served as flight system operators, and to SFC Maurice Nanton who served as crew chief. Additional thanks to Zoltan Szoboszlay, John Archdeacon, and Nathan Mielcarek for their support in developing the ICE pilot cueing software, and to Greg Schulein who developed application and hardware support software.

REFERENCES

1. Howlett, J., Schulein, G., Mansur, H., "A Practical Approach to Obstacle Field Route Planing for Unmannned Rotorcraft" American Helicopter Society 60th Annual Forum Proceedings, Baltimore, MD, June 2004.

2. Whalley, M., Freed, M., Harris, R., Takahashi, M., Schulein, G., and Howlett, J., "Design, Integration, and Flight Test Results for an Autonomous Surveillance Helicopter," American Helicopter Society International Specialists Meeting on Unmanned Rotorcraft Proceedings, Mesa, AZ, Jan. 2005.

3. Howlett, J., Tsenkov, P., Whalley, M., Schulein, G., Takahashi, M., "Flight Evaluation of a System for Unmanned Rotorcraft Reactive Navigation in Uncertain Urban Environments" American Helicopter Society 63rd Annual Forum Proceedings, Virginia Beach, VA, May 2007.

4. Tsenkov, P., Howlett, J., Whalley, M., Schulein, G., Takahashi, M., Rhinehart, M., and Mettler, B., "A System for 3D Autonomous Rotrcraft Navigation in Urban Environments" AIAA Guidance, Navigation and Control Conference Proceedings, Honolulu, HI, August 2008.

5. Whalley, M., Tsenkov, P., Takahashi, M., Schulein, G., Goerzen, G., "Field-Testing of a Helicopter UAV Obstacle Field Navigation and Landing System" American Helicopter Society 65th Annual Forum Proceedings, Grapevine, TX, May 2009.

6. Goerzen, C. and Whalley, M., "Minimal risk motion planning: a new planner for autonomous UAVs in uncertain environments," American Helicopter Society International Specialists Meeting on Unmanned Rotorcraft Proceedings, Tempe, AZ, Jan. 2011.

7. Whalley, M., Takahashi, M., Goerzen, C., Schulein, G., Fletcher, J., Moralez, E., Ott, C., Olmstead, M., Savage, J., Burns, H., and Conrad, B., "Flight Test Results for Autonomous Obstacle Field Navigation and Landing Site Selection on the RASCAL JUH-60A," American Helicopter Society 69th Annual Forum Proceedings, Phoenix, AZ, May 2013.

8. Whalley, M., Takahashi, M., Goerzen, C., Schulein, G., Fletcher, J., Moralez, E., Ott, C., Olmstead, M., Savage, J., Burns, H., and Conrad, B., "Autonomous Black Hawk in Flight: Obstacle Field Navigation and Landing-site Selection on the RASCAL JUH-60A," *Journal of Field Robotics*, Vol. 31, Issue 4, pages 591-616, Jul/Aug 2014.

9. Goerzen, C. Whalley, M., Takahashi, M., Schulein, G., Ott, C., Minor, J., Savage, M., Burns, H., and Conrad, B., "High Speed Canyon Flight: Obstacle Field Navigation Envelope Expansion on the RASCAL JUH-60A Black Hawk," Sixth American Helicopter Society International Specialists Meeting on Unmanned Rotorcraft Proceedings, Chandler, AZ, January 2015.

10. Whalley, M., Takahashi, M., Berrios, M., Mansur, H., Minor, J., Morford, Z., Ott, C., Goerzen, C., and Schulein, G., "Flight test results for a new mission-adaptive autonomy system on the RASCAL JUH-60A Black Hawk," American Helicopter Society 72nd Annual Forum Proceedings, West Palm Beach, FL, May 2016.

11. Takahashi, M., Whalley, M., Goerzen, C., Schulein, G., Mansur, H., Ott, C., Minor, J., and Morford, Z., "Autonomous Rotorcraft Flight Control with Multi-Level Pilot Interaction in Hover and Forward Flight," American Helicopter Society 72nd Annual Forum Proceedings, West Palm Beach, FL, May 2016.

12. Goerzen, C., Kong, Z., Mettler, B., "A Survey of Motion Planning Algorithms from the Perspective of Autonomous UAV Guidance." *Journal of Intelligent and Robotic Systems*, Nov. 2009.

13. Schouwenaars, T., Mettler, B., Feron, E., How, J., "Robust Motion Planning Using a Maneuver Automaton with Built-in Uncertainties," in proceedings of the American Control Conference (ACC), June 2003.

14. LaValle, S. And Kuffner, J., "Randomized kinodynamic planning." *International Journal of Robotics Research*, 20(5), 378-400, May 2001.

15. Kavraki, L.E., Svestka, P., Latombe, J.-C., Overmars, M.H., "Probabilistic roadmaps for path planning in high-dimensional configuration spaces," in IEEE Transactions on Robotics and Automation, 12:4, 566-580, Aug. 1996.

16. Karaman, S., Frazzoli, E., "Incremental Sampling-based Algorithms for Optimal Motion Planning," in Proceedings of the Robotics: Science and Systems Conference (RSS), 2006.

17. Martin, M. C., & Moravec, H. P. (1996). Robot evidence grids (CMU publication CMU-RI-TR-96-06), retrieved from http://www.frc.ri.cmu.edu/~hpm/project.archive/robot.papers/1996/RobotEvidenceGrids.pdf

18. Hart, P. E.; Nilsson, N. J.; Raphael, B. (1968). "A Formal Basis for the Heuristic Determination of Minimum Cost Paths". *IEEE Transactions on Systems Science and Cybernetics* SSC4. 4 (2): 100–107.

19. Takahashi, M., Whalley, M., Goerzen, C., Mansur, H., Ott, C., Minor, J., Morford, Z., and Schulein, G., "Autonomous Rotorcraft Flight Control with Multi-Level Pilot Interaction in Hover and Forward Flight," *Journal of the American Helicopter Society*, Vol. 62, No. 3, July, 2017.

20. Kim, F. D., Celi, R., and Tischler, M. B., "High Order State Space Simulation Model of Helicopter Flight Mechanics," *Journal of the American Helicopter Society*, Vol. 38, No. 4, Oct., 1993.

21. OpenSceneGraph (Version 3.0.1) [Computer software]. Retrieved from http://www.openscenegraph.org/

22. Szoboszlay, Z., Davis, B., Fujizawa, B., Minor, J., Osmon, M., Morford, Z., "Degraded Visual Environment Mitigation (DVE-M) Program, Yuma 2016 Flight Trials in Brownout," American Helicopter Society 73rd Annual Forum Proceedings, Fort Worth, TX, May 2017.

Improved Gust Tolerance of a Cyclocopter UAV using Onboard Flow Sensing

Elena Shrestha
Graduate Research Assistant
University of Maryland
College Park, MD, USA

Derrick Yeo
Assistant Research Scientist
University of Maryland
College Park, MD, USA

Vikram Hrishikeshavan
Assistant Research Scientist
University of Maryland
College Park, MD, USA

Inderjit Chopra
Alfred Gessow Professor and Distinguished University Professor
University of Maryland
College Park, MD, USA

ABSTRACT

This paper describes the gust rejection study of a twin-cyclocopter micro air vehicle (MAV) with two cyclorotors and an anti-torque nose rotor. A gust rejection controller relying on velocity feedback and onboard flow sensing was implemented in a closed-loop feedback system. Tethered experiments were conducted with the vehicle mounted on a 6-DOF stand in front of a synthetic gust generation device capable of providing up to 4 m/s step gust input. Planar gusts along the longitudinal and lateral axis of the cyclocopter and crosswinds were systematically studied. Both pitch control (varying nose rotor rpm) and thrust vectoring control strategies were evaluated to counteract perturbation along the longitudinal axis while the cyclocopter used differential rotational speed of the cyclorotors for lateral gusts. Results showed that thrust vectoring was more effective than pitch control in reducing displacement from gust. Pitch control using the position feedback controller resulted in a maximum gust tolerance of 2.8 m/s with a duration of 3 seconds while thrust vector control using the combined flow feedback and position feedback controller was able to withstand 4 m/s step gust input with one second duration with only 0.01 m displacement. In addition, flow feedback was more effective than position feedback in minimizing position error. The cyclocopter was also able to mitigate step gusts of 2.8 m/s in magnitude and 3 seconds duration with crosswind components at 30 degrees from the longitudinal axis. The difference in performance can be attributed to the cyclocopter's higher control authority using thrust vectoring.

INTRODUCTION

Micro air vehicles (MAVs) have gained increased popularity in the recent decade due to their applications in both military and civilian settings. Because of their low inertia and operational speeds, MAVs are particularly susceptible to environmental disturbances such as gusts. Flow fields in urban settings are especially prone to spatial and temporal variations arising from scattered layouts of buildings and trees. These complex and unpredictable gusts can lead to critical perturbations in vehicle attitude, lifting-surface stall, and ultimate loss of control. When subjected to an edgewise gust, MAVs experience both drag force in the direction of the flow and pitching moment that rotates the vehicle away from the source.

Zarovy et al. (Ref. 1) and Hrishikeshavan et al. (Ref. 2) evaluated the disturbance rejection capability of rotary-winged MAVs subjected to artificial gust profiles from rotary fans up to 3 m/s. These studies largely focused on the single-main-rotor and shroud configurations. An alternative rotary-winged configuration is the cyclocopter (Fig.1), which relies on the unconventional cycloidal rotors (cyclorotor). The cyclorotor rotates about a horizontal axis where the blades span parallel to the axis and perpendicular to the direction of flight (Fig.2). Using a four-bar linkage pitching mechanism, the blade pitch angle cyclically varies at each point along the circular trajectory (Ref. 3). The resulting blade kinematic produces a net thrust after a full revolution. The magnitude of the thrust vector is controlled by the rotational speed of the cyclorotor while the direction of the thrust vector is varied by phasing the cyclic blade pitch.

This thrust vectoring capability offers unique advantages for the cyclocopter configuration such as a power efficient forward flight by maintaining level attitude, effectively reducing parasitic drag (Ref. 4). The cyclocopter does not undergo any mechanical changes (e.g tilting rotor) in order to transition from hover to forward flight since the cyclorotors provide both lift and propulsive thrust. A previous study by the University of Maryland demonstrated level forward flight (zero pitch attitude) up to 5 m/s (Ref. 5). The ability to rapidly vector main cyclorotor thrust could also prove beneficial for disturbance rejection along the longitudinal axis. While there has not been any untethered experimental studies investigating the response of a cyclocopter MAV to gust perturbation,

Presented at the Vertical Flight Society Autonomous VTOL Technical Meeting and Electric VTOL Symposium, Mesa, Arizona, US, Jan. 29–31, 2019.

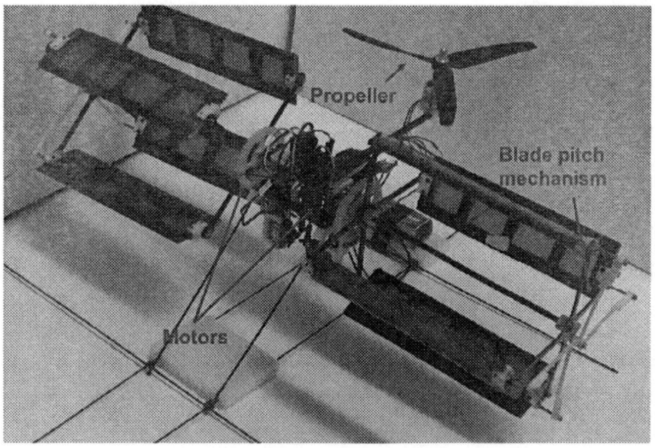

Fig. 1. Cyclocopter MAV.

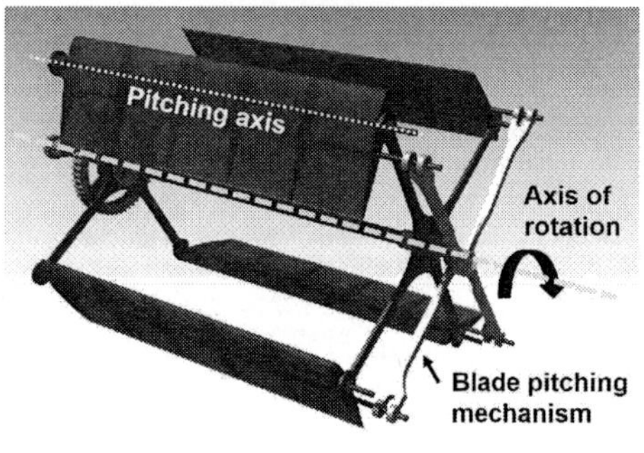

Fig. 2. Cyclorotor.

Hrishikeshavan et al. conducted a maneuverability and disturbance rejection analysis using a linear 6-DOF flight dynamics model of the cyclocopter in hover and estimated that a 550g cyclocopter had longitudinal and lateral gust tolerances of 7.9 m/s and 17 m/s, respectively (Ref. 6).

In this paper, the disturbance rejection capability of the cyclocopter is investigated through series of experiments using a synthetic gust generation device. Since the cyclorotor has the unique capability of providing both lift and propulsive thrust by phasing the blade pitch cyclic, it is possible to reject gust disturbance isolated along the longitudinal axis using thrust vectoring. In addition, feedback control schemes of position feedback using a motion capture system and flow feedback using onboard flow sensing are also evaluated. Results of the comparative study of gust rejection strategies are provided for the 6-DOF experiments in front of a synthetic gust generation device for step gust inputs up to 4 m/s and 6 seconds in duration. Results also include performance for crosswinds up to 30 degrees from the longitudinal axis.

CYCLOCOPTER VEHICLE SPECIFICATION

A 550 gram twin-rotor cyclocopter was developed with a hybrid configuration utilizing both the cyclorotor and a conventional rotor (Refs. 5, 7). Since both the cyclorotors spin in the same direction, a large pitching reaction moment needed to be counter-balanced by the forward-mounted propeller. The vehicle has a lateral dimension of 0.381 meters (1.25 feet), longitudinal dimension of 0.457 meters (1.5 feet), and a height of 0.305 meters (1 foot) (Fig. 1). Each cyclorotor has a diameter of 0.152 meters (6 inches), blade span of 0.171 meters (6.75 inches), a chord of 0.051 meters (2 inches), and is independently driven by a 75 watts out-runner motor through a 7.5:1 single-stage transmission. The horizontal nose rotor is directly driven by a 920 kV brushless outrunner motor and is fitted with a 3-bladed GWS 9050 propeller. All three rotors are powered by a single 3S 11.1 volt 850 mAh Lithium-Polymer battery.

GUST REJECTION STRATEGIES

Possible methods to alleviate gust effects include increasing the vehicle's control authority and improving the closed-loop feedback system. Both were experimentally evaluated for the cyclocopter through series of tethered experiments in front of a custom gust generation device. The following sections provide an overview of the control scheme and sensor measurements used by the gust controller.

Control Input

Conventional rotary-winged MAVs like quadcopters rely on differential rpm to recover from gust perturbation. Through this control input, the vehicle achieves a non-zero pitch attitude angle, thereby increasing its effective surface area and potentially reducing its gust rejection capability. For the cyclocopter configuration, there are two possible gust rejection control strategies for disturbance along the longitudinal axis: (1) pitch control by actuating nose rotor rotational speed and (2) thrust vectoring the cyclorotors (Fig. 3).

The cyclocopter has the unique capability of thrust vectoring by phasing the blade pitch cyclic. The resulting thrust vector has both a vertical lift component and a horizontal propulsive thrust component. By vectoring the resultant thrust to increase its propulsive thrust component, the cyclocopter is able to maintain a zero pitch attitude angle in forward flight and independently translate forward and backward without adjusting the vehicle's pitch angle.

Each cyclorotor is independently controlled by a 75 watts brushless outrunner motor through a 7.5:1 single-state transmission while the nose rotor is controlled by a 920kV brushless outrunner motor. In addition, the blade pitching mechanism on each rotor is independently guided by a rotary BMS-373MG servo through a 2:1 gear ratio which allows up to 120° phasing. As a result, the control strategy for the cyclocopter relies on a unique combination of rpm modulation and thrust vectoring to achieve roll, pitch, yaw, and forward flight.

Onboard Flow Sensing

A number of approaches were considered to measure freestream velocity for low operational range. A hot-wire or ultra-sonic anemometer offers high resolution and 3D measurement of velocity, but has a high weight penalty making

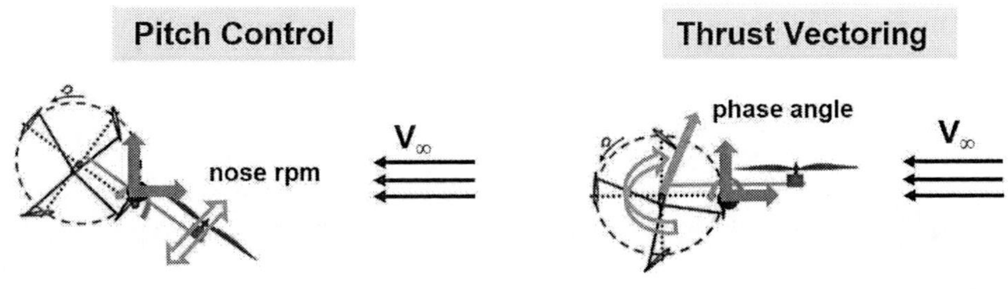

Fig. 3. Pitch control vs. thrust vectoring control.

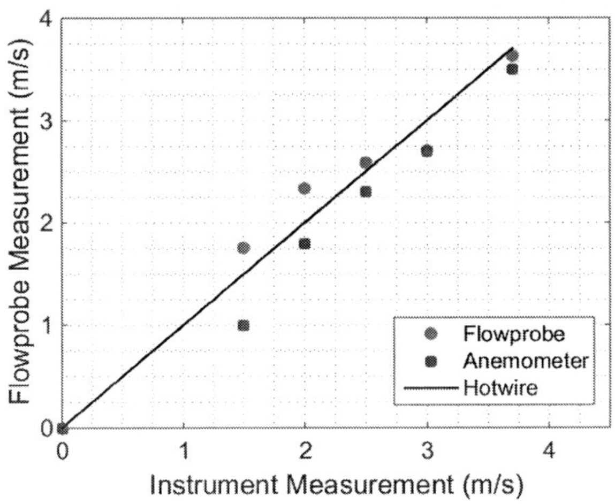

Fig. 4. Flowprobe calibration compared to hot-wire and rotary anemometer.

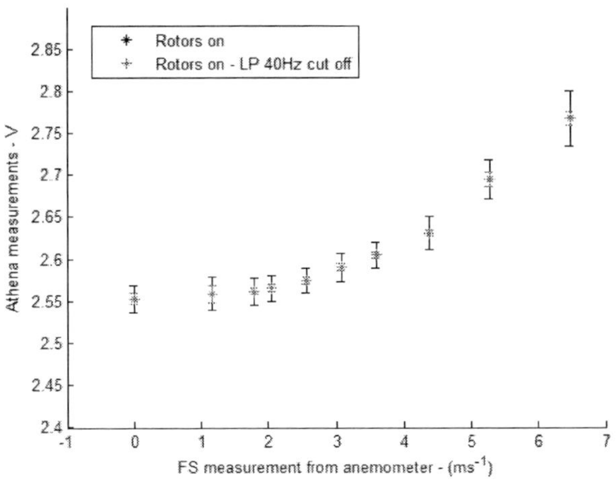

Fig. 5. Filtered pressure signal statistics.

them unsuitable for MAVs. A mechanical approach that uses miniature rotating vanes or a turbine would provide adequate resolution at low speeds, but will be mechanically complex and sensitive to vibrations from the cyclorotors. Pressure-based instrumentation is inherently compact and suited to the MAV-scale but traditional pitot-static probes are prone to poor signal-to-noise ratios at low airspeeds.

The pressure-probe design used in this work relies on measuring differential pressure. Each measurement axis of the two-component flowprobe uses one opening facing towards the freestream and other oriented with the flow in order to provide a differential pressure measurement (Refs. 9, 10). Any forward velocity causes a degree of pressure loss on the aft-facing port, enabling the probe to provide an amplified measurement of dynamic pressure with higher sensitivity at low speeds. Unlike a traditional pitot-static probe that uses a forward-facing port for total pressure and a ring of interconnected static ports perpendicular to the flow, the measurement in this system is inherently directional, allowing airspeed components to be measured.

Freestream velocity was continuously measured throughout the experiment using an analog Honeywell HSCDRR series differential pressure sensor. The sensor measures pressure range of $\pm 249.08 Pa$ and has an input voltage of 3.33 V. The analog voltage measurements from the differential pressure sensor are converted to ADC (analog-to-digital) values with a 32-bit resolution. Pressure taps on the sensor are individually connected to ports on the flowprobe. The pressure measurements from the sensor are converted to velocity using a scaled version of the pitot static equations (Ref. 11). A multi-probe setup with flowprobes along X-Y axes was used to enable 2D velocity measurement. The components of velocity are calculated using calibration gains (K_u, K_v, K_b), pressure measurements (X_1, X_3, Y_2, Y_4), and density (ρ):

$$u = K_u \sqrt{\frac{2K_b(X_1 - X_3)}{\rho}} \tag{1}$$

$$v = K_v \sqrt{\frac{2K_b(Y_2 - Y_4)}{\rho}} \tag{2}$$

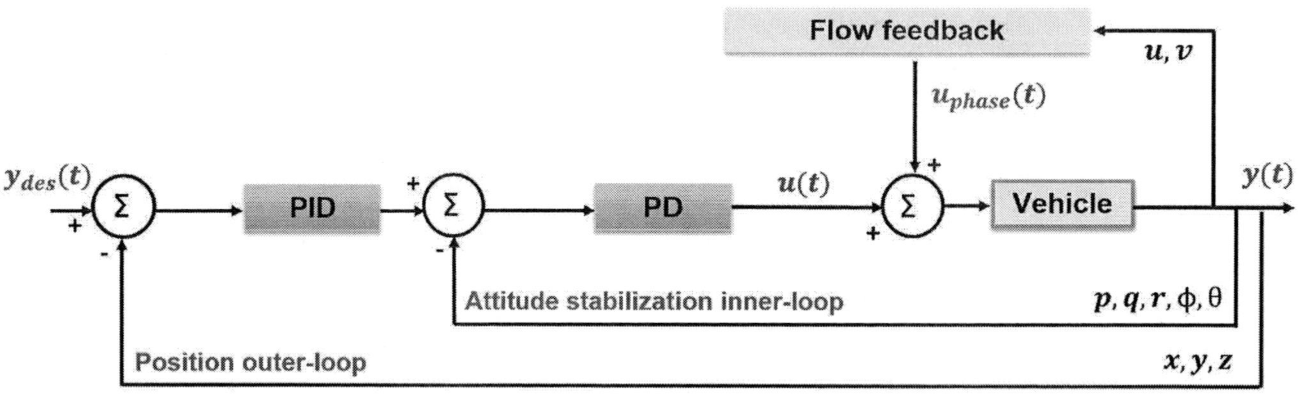

Fig. 6. Closed-loop feedback system with gust rejection controller.

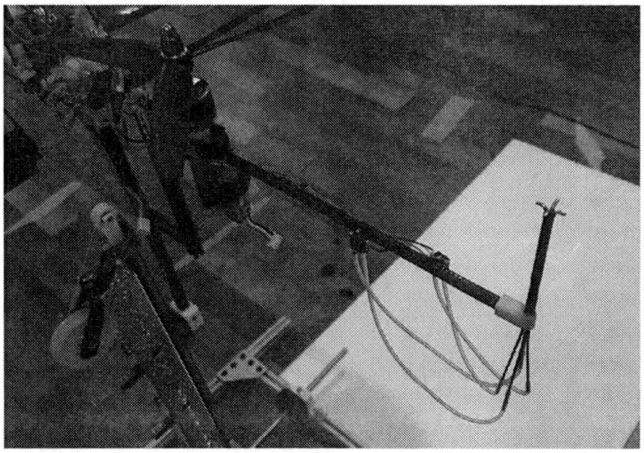

Fig. 7. Onboard flow sensing mounted on the cyclocopter.

Symmetry of the bi-directional flow probe resulted in equivalent calibration gains ($K_u = K_v$). With this setup, the total velocity magnitude (V) and side-slip angle (β) can be calculated using:

$$V = \sqrt{u^2 + v^2} \tag{3}$$

$$\beta = tan^{-1}\frac{v}{V} \tag{4}$$

Flowprobe measurements were calibrated against measurements from both a hot-wire anemometer and rotary anemometer in the wind tunnel between 1 m/s and 8 m/s. Figure 4 shows velocity measurements of all three sensors. Additional calibration was experimentally performed with the probes mounted on the vehicle to account for the flow field disturbances specific to a cyclocopter. In order to filter high frequency noise, the wind tunnel pressure data is first characterized by its power spectral density (PSD). By comparing the power spectrum with the cyclorotor powered down and with the rotors spinning, contributions from the flow field and the vehicle can be independently identified. Harmonics were noted at 135 Hz and 200 Hz, which suggests that either the pressure transducer measurements are being affected more by probe/vehicle structural response than actual flow or by the forward mounted propeller rotor. In addition, a harmonic near 35 Hz was noted with an incident free stream, suggesting that this might be flow field dependent. As a result, a low pass filter with a cut-off freq of 40 Hz was chosen. The filtered measurements are compared with raw measurements in Fig. 5.

CLOSED-LOOP FEEDBACK SYSTEM

The closed-loop feedback system on the cyclocopter contains an attitude stabilization inner-loop and a position control outer-loop augmented with a gust controller (Fig. 6). The inner-loop is implemented onboard through ELKA, the 1.4g custom onboard flight controller. ELKA houses a STM-32 ARM Cortex M3 microprocessor, integrated MPU-9150 IMU, a 2.4GHz RF transceiver communicating with the pilot's commercial transmitter, and a nRF24L01 wireless transceiver that transmits data to the ground control station (Ref. 12). The outer-loop relies on the VICON motion capture system and communicates with the ground control station to relay position measurements to ELKA. The following sections provide a detailed summary of the controllers implemented on the cyclocopter.

Attitude Stabilization

The inner-loop attitude stabilization uses a proportional-derivative (PD) controller and the feedback states include attitude rates (p, q, r) and pitch and roll Euler angles (θ,ϕ). To account for gyroscope drift, a high pass complementary filter is used with a 4 Hz cut-off frequency for the sensor measurements (Refs. 2, 13, 14). Vibrations from the rotary systems affected accelerometer measurements, which are processed through a low pass complementary filter with a 6 Hz cut-off to reject disturbances from the rotor vibrations occurring at significantly higher frequencies than the body dynamics.

Position Control

The closed-loop feedback system was modified with a proportional-integral-derivative (PID) controller for autonomous position control. Position data from the VICON motion capture system were transmitted to the onboard autopilot via a LabVIEW interface on the ground control system. The position trim setpoints $(x_{des}, y_{des}, z_{des})$ were initialized at the start of each experiment. The position controller outputs desired attitude angles (ϕ, θ, ψ) and are the setpoints for the attitude stabilization inner-loop controller.

The ground control station communicates with ELKA through a wireless IEEE 802.15.4 data link at 2.4 GHz frequency. Onboard IMU data is streamed to a LabVIEW interface, which permits data logging, controls the gust setup, and transmits settings for the closed-loop feedback system. All the data processing and feedback control calculations are performed onboard by ELKA. A separate connection between a commercial receiver and transmitter ensures a constant and fail-safe connection and enables manual control of the vehicle.

Gust Controller

A basic approach to mitigating gust effects is to rely on inertial measurements coupled with closed-loop position feedback. The drag force and pitching moment experienced by the cyclocopter would affect body angular rates and displacement. Inertial measurements are then used to calculate control inputs required to recover initial position and attitude. With position feedback, the gust controller only comes into effect after the vehicle has been perturbed. Onboard instrumentation provides real-time flow measurements that can be used to account for flow disturbances. The flow feedback controller would then predict impending vehicle response and provide control inputs to negate gust effects in advance of their occurrence (Ref. 15). Unlike position feedback, flow feedback could potentially enable the vehicle to maintain its initial attitude throughout the gust perturbation. However, the control scheme is sensitive to the choice of onboard miniature flow sensor and requires a priori knowledge of control inputs suitable for measured disturbances.

A gust controller with a flow feedback scheme was added to the feedback system and only provided a collective thrust vector input to the vehicle. Unlike differential thrust vectoring, collective thrust vectoring results in forward and backward translation of the cyclocopter instead of a change in orientation. Inputs to the gust controller includes X-Y velocity components (u, v) from a miniature dual-probe flow sensor (Fig. 7). The appropriate phase angle control input for a given velocity was determined a priori in the wind tunnel and from forward flight experiments:

$$\Phi = K_{\Phi} V + \Phi_0 \tag{5}$$

The nominal phase angle command (Φ) is equivalent to measured velocity (V) multiplied by a gain (K_{Φ}) that was fine-tuned during these experiments. Since the experiments were conducted in hover, the initial phase angle is zero $(\Phi_0 = 0)$. However, the controller can be adapted to forward flight if the cyclocopter's flight speed is subtracted from measured velocity.

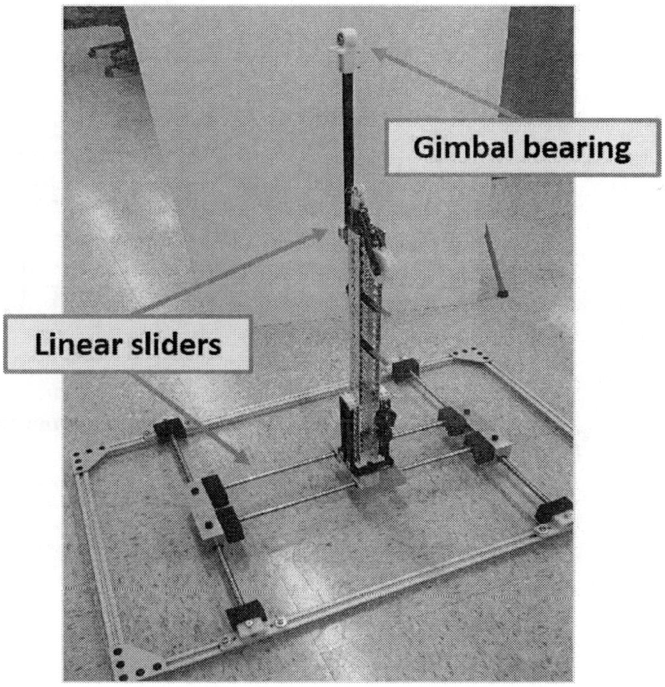

Fig. 8. 6-DOF experimental stand.

TETHERED EXPERIMENTS (6-DOF)

Previous wind tunnel experiments showed that a gust perturbation along the longitudinal axis leads to a pitch up moment, which consequentially results in backwards translation (Ref. 16). Pitch control enabled the cyclocopter to recover level attitude for ramp gust profiles up to 10 m/s with a settling time of 12 seconds. However, the 1-DOF experiments did not account for translation or secondary effects on roll and yaw body rates. In addition, a slow ramp gust profile enabled the controller to respond earlier at a lower gust magnitude instead of the selected magnitude.

Tethered experiments using a custom gust generation device and 6-DOF test stand were conducted to evaluate the effectiveness of pitch control and thrust vectoring control for a gust along the longitudinal axis. Both position feedback and flow feedback control schemes were tested for step gust profiles with peak magnitudes of 4 m/s and durations of 1, 3, and 6 seconds. The following sections describe the 6-DOF test stand, custom gust generation device, and the gust rejection controller.

Experimental Stand

In order to emulate free-flight conditions, the cyclocopter was mounted on a 6-DOF stand (Fig. 8) assembled using three linear ball-bearing sliders and a gimbal bearing. The gimbal mount enables up to ±16° in pitch, ±20° in roll, and ±16° in yaw while the linear sliders permit translation of 0.53 m across x-axis, 0.27 m across y-axis, and 0.15 m across z-axis. The vehicle's main mount is attached to a pulley system and a

Fig. 9. Synthetic gust generation device with shutter system.

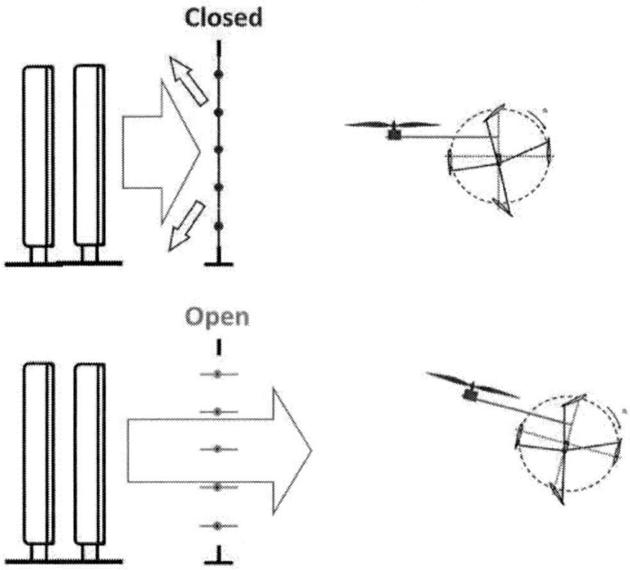

Fig. 10. Shutter system mechanism provides step gust inputs.

counterweight of 300 grams to offset weight of the mount. As a result, the cyclorotors are able to operate at a nominal 1400 rpm cruise settings during these tests. The coefficient of friction for the ball-bearing sliders is 0.001. However, as the setup does have a considerable amount of inertia due to weight of the aluminum carriages these are best suited to comparing the effectiveness between different gust rejection strategies rather than as predictions for free-flight performance.

Synthetic Gust Generation Device

To produce repeatable step gust perturbations, a custom gust generation setup was developed in this study (Fig. 9). These bladeless fans output a stream of air from small ducts outlining its boundaries. A set of three fans were grouped together and placed 0.089 meters apart. Another set of fans were stacked directly behind the first row but with a 0.10 meters horizontal offset in order to augment the airflow. The setup was placed inside the testing volume (3.0 m x 2.4 m x 1.8 m) enclosed by the VICON Motion Capture System.

The fans are arranged behind a shutter system that is actuated using two Hitec HS-1000SGT rotary servos. The shutter mechanism is controlled through the ground control station and is automated to open and close for the desired duration (Fig. 10). By controlling flow speed using shutters, the Dyson fans are always operating in order to achieve a step gust profile instead of the ramp profile produced in the wind tunnel. Variable parameters of the gust setup include fan power, distance between adjacent fans, distance between front and back rows of fans, location of shutter system, and distance to the test area. These parameters were adjusted in order to achieve steady uniform flow throughout the test region.

An important characteristic of the gust generation setup is that the flow field within the test region must have a steady unidirectional profile so that the vehicle is not subjected to additional aerodynamic effects from turbulence. A flow field survey of the test region, 1.32 m from the fans, was conducted using a differential pressure sensor mounted on an automated carriage mechanism (Fig. 11). The survey measured the average freestream velocity at 11 points evenly distributed along the cross-stream location (0-80cm) and at 6 different flow probe heights (0-16.8cm). A sweep of the flow field shows that the mean freestream velocity is uniform across the 80 cm by 16.8 cm test region. Two rows of fans generated a maximum freestream velocity of 4 m/s with variance of 0.15 (Fig. 12).

$$V_{rms} = \sqrt{\sum_{i=1}^{n} \frac{V_i - \bar{V}}{n-1}} \tag{6}$$

$$T = \frac{V_{rms}}{\bar{V}} \times 100 \tag{7}$$

Using equations 6 and 7, the turbulence at the maximum generated freestream velocity is approximately 11.4%. Overall, the synthetic gust generation device is capable of produc-

Fig. 11. Custom mechanism to survey flow field at 66 discrete points in test cross-section.

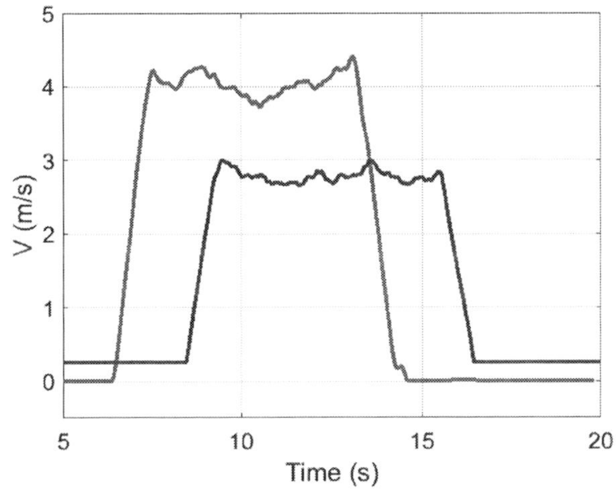

Fig. 12. Magnitude of step gust input for one vs. two rows of Dyson fans.

ing step gust inputs of 2.8 m/s and 4 m/s at various step durations. The setup can also generate sinusoidal gust profiles by actuating the shutters. For the experiments, the duration of gust input was set at 1, 3, or 6 seconds (Fig. 13).

RESULTS AND DISCUSSION

While the 1-DOF wind tunnel studies explored both longitudinal and lateral gust response, 6-DOF experiments focused on longitudinal gust response in order to evaluate effectiveness of thrust vectoring control. Initial experiments assessed pitch control and thrust vectoring control against a step gust perturbation. These experiments relied only on position feedback to detect disturbances. At the beginning of each experiment, the vehicle maintained a steady hover at approximately 0.1 m longitudinal offset from the slider bounds while centered laterally. Afterwards, a step gust input was provided with magnitudes of 2.8 m/s and 4 m/s and durations of 1, 3, and 6 seconds. The chosen performance metric was displacement from the desired hover position. After selecting the control input, additional experiments were conducted to specifically evaluate performance of position feedback and flow feedback controllers.

Pitch Control vs. Thrust Vectoring

Effectiveness of position feedback using pitch control was investigated because preliminary wind tunnel results showed that the cyclocopter is able to maintain pitch control authority in a 10 m/s gust perturbation. However, the 1-DOF experiment only assessed the vehicle's ability to recover pitch attitude and not the hover trim. For this experiment, the attitude stabilization for pitch used a PD controller and position feedback used a PID controller. Therefore, the cyclocopter will not maintain a level attitude ($\theta = 0^\circ$) during a gust perturbation but instead use the non-zero pitch attitude to recover the trim position on the horizontal slider (thrust = drag). This control strategy is similar to how typical rotary MAVs such as quadcopters recover from gust perturbation.

Figure 14 shows the 6-DOF response to a 2.8 m/s gust perturbation using pitch control to recover trim. The plots display displacement along both lateral and longitudinal axis, step gust input, and the controller's response in terms of resulting pitch angle. The dashed line indicates maximum achievable pitch angle due to constraints on the gimbal stand. As evident from the figure, a step gust input causes the cyclocopter to pitch-up and subsequently translate backwards by 0.34 meters. In addition, gust along the longitudinal axis does not have a significant impact on lateral translation. The feedback controller is ultimately able to recover desired position on the slider with a 16° pitch down attitude when operating at 1400 rpm.

Because of its thrust vectoring capability, the cyclocopter is able to provide both lift and propulsive thrust by phasing the blade pitch cyclic. Position feedback using thrust vector-

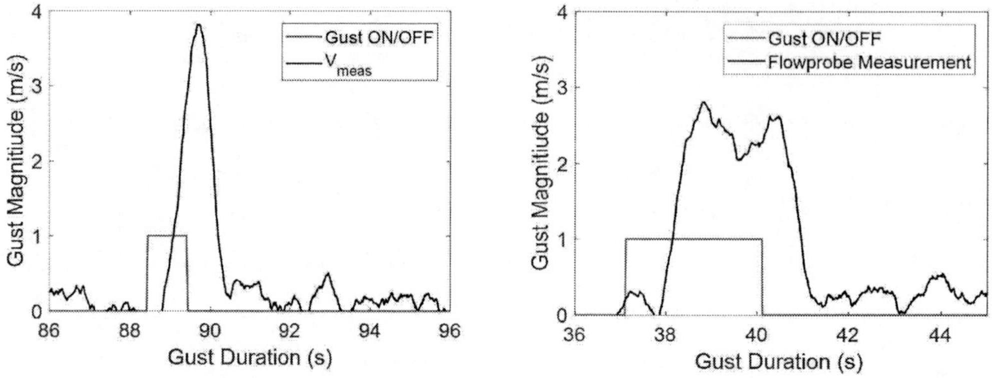

Fig. 13. Variations in magnitude and duration of gust profiles using the shutter mechanism.

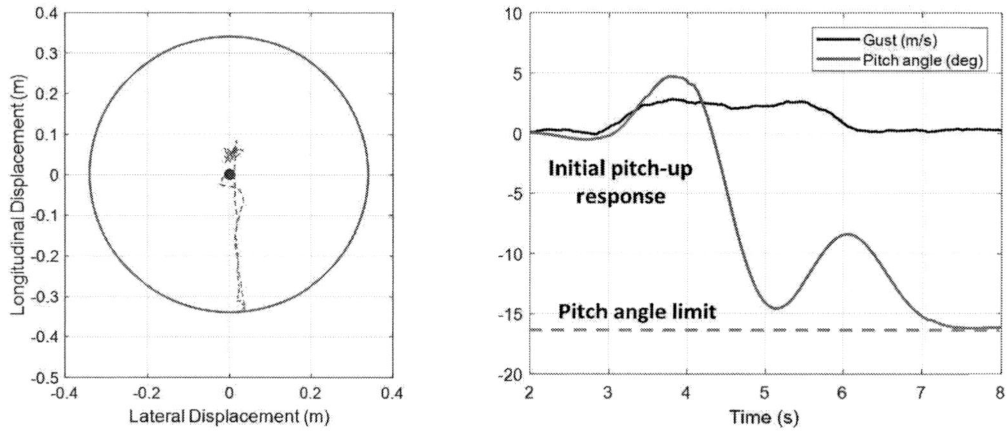

Fig. 14. Displacement and pitch control input for 2.8 m/s step gust.

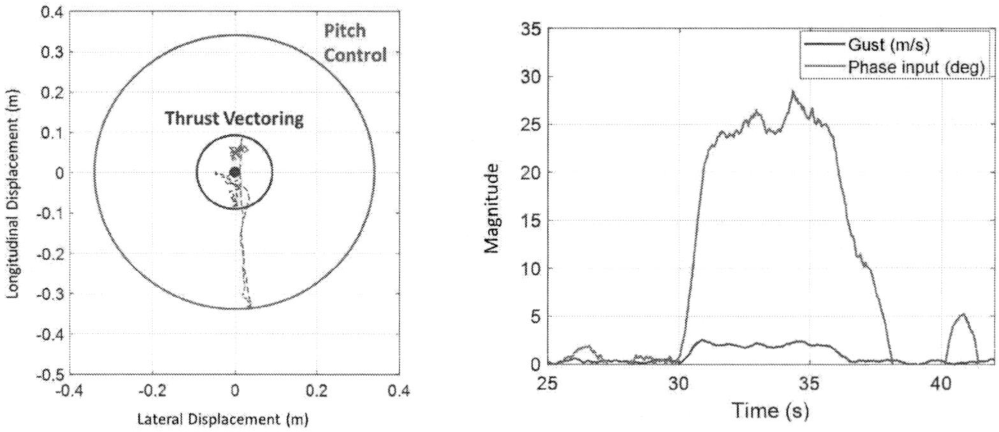

Fig. 15. Displacement and thrust vector input for 2.8 m/s step gust.

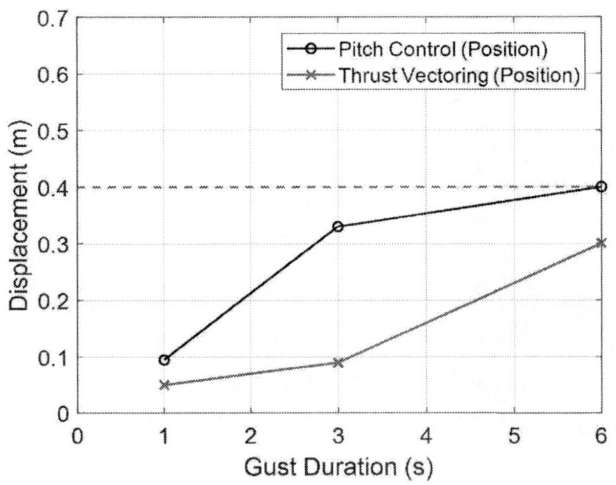

Fig. 16. Displacement vs. duration of gust for pitch control and thrust vectoring inputs.

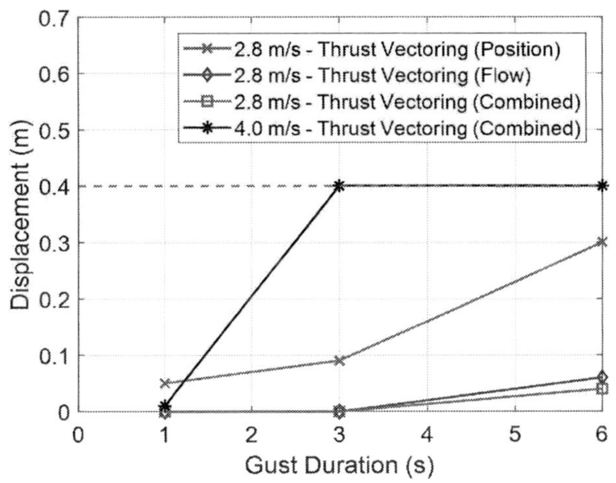

Fig. 17. Displacement vs. gust durations for 2.8 m/s and 4 m/s.

ing resembles the forward flight control strategy previously implemented on the vehicle. Therefore, it is entirely possible for the cyclocopter to maintain both a level attitude (θ = 0°) and hover at the setpoint during a gust perturbation.

Figure 15 shows a comparison of the resulting displacement between pitch control and thrust vectoring. A phasing input of 26° counteracts the drag force from the 2.8 m/s gust perturbation and follows a similar step profile. Even with a phasing input, the cyclocopter still experiences a pitch-up moment, but it is stabilized by the inner-loop PD controller. The cyclocopter translates only 0.1 m along the longitudinal axis since the propulsive thrust was increased by instantaneously rotating the thrust vector instead of incrementally modulating the nose rotor's rotational speed.

Figure 16 summarizes the displacement results for both pitch control and thrust vectoring for 2.8 m/s at 1, 3, and 6 seconds durations. During the experiments, the linear slider enabled 0.53 m of translation along the perturbed axis. Accounting for the vehicle's initial position, displacements of 0.4 m or above were categorized as a failed controller (i.e vehicle was not able to tolerate gust). Each data point on the graph was an averaged value of multiple trials after selecting the final gains of the position feedback controller.

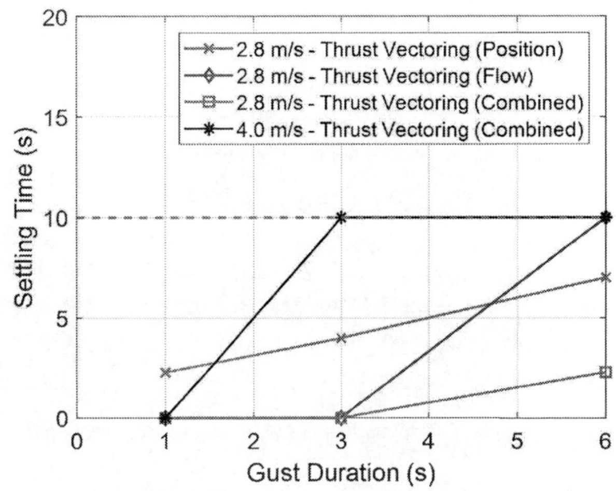

Fig. 18. Settling time vs. gust durations for 2.8 m/s and 4 m/s.

Pitching forward to counteract a longitudinal gust perturbation resulted in a maximum gust tolerance of 2.8 m/s with a duration of 3 seconds. The maximum pitch attitude (θ = 16°) at 1400 rpm on the 6-DOF stand did not produce adequate propulsive thrust to counteract the drag on the vehicle. While the vehicle was able to output a pitching moment to overcome 10 m/s gust perturbation in pitch attitude, it was not able to recover the longitudinal translation trim using the same control strategy. Thrust vectoring resulted in reduced displacement for each gust profile and was therefore selected as the control input for the gust controller.

Position Feedback vs. Flow Feedback

Figure 17 shows resulting displacement of a gust controller using thrust vectoring control coupled with (1) position feedback, (2) flow feedback, and (3) combined position and flow feedback at both 2.8 m/s and 4 m/s step gust inputs. The results show that flow feedback is more effective than position feedback in minimizing the vehicle's response to gust. In flow feedback, the controller provides an input to negate predicted response of vehicle based on measured velocity instead of correcting after the vehicle responds to the disturbance. The combined flow feedback and position feedback controller was able to withstand 4 m/s step gust input with one second duration with only 0.01 m displacement. For 4 m/s magnitude and durations of 3 s and higher, the combined controller was unable to counteract the pitching moment due to gust and failed to recover position.

The importance of position feedback is evident when considering settling time. Settling time is defined as the time to recover desired position. Since the position feedback controller has a 3 cm buffer from the setpoint, settling time is only

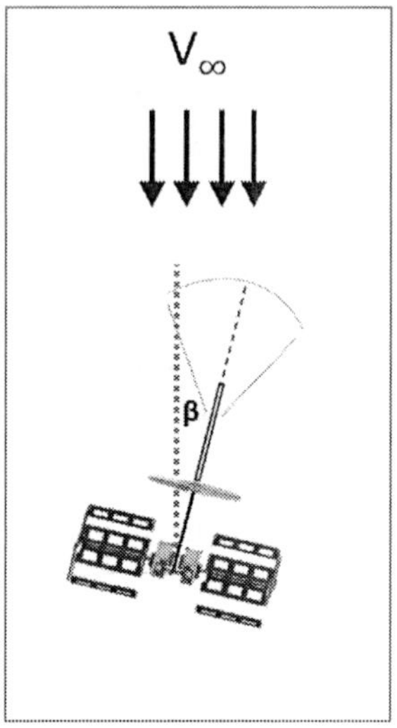

Fig. 19. Setup for crosswind gust scenarios ($\beta > 0$).

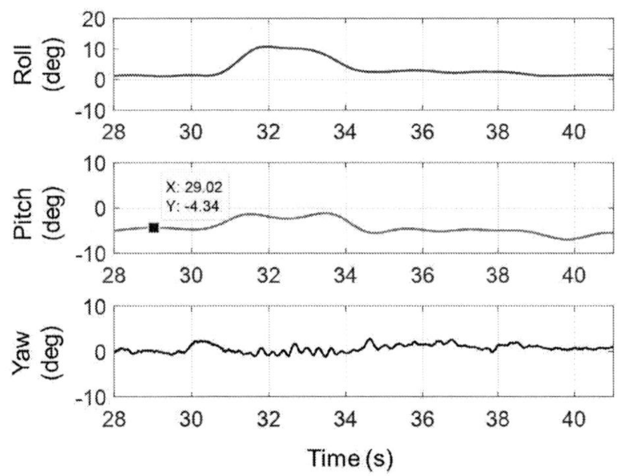

Fig. 20. Attitude response for step gust input at 30 degrees azimuth from longitudinal axis.

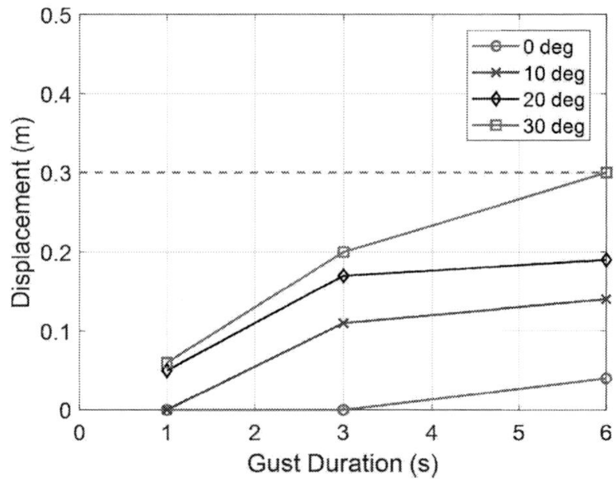

Fig. 22. Displacement vs. gust duration for crosswinds using combined controller.

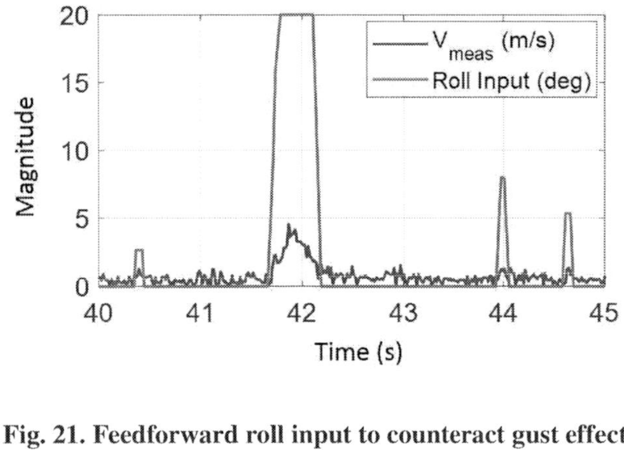

Fig. 21. Feedforward roll input to counteract gust effects.

measured after the vehicle passes this point. The buffer or dead zone prevents the gust controller from oscillating about the setpoint since the vehicle is resting on a ball bearing carriage. In addition, settling time of 10 seconds or greater was set as the time limit for the controller to recover trim. It should be noted that settling time is dependent on control system gains and is being evaluated in these experiments to highlight significance of position feedback.

Figure 18 shows that the combination of flow feedback and position feedback results in a lower settling time. Intuitively, the settling time result is unsurprising because only the position feedback controller uses displacement error in its control scheme. The flow feedback controller does not measure position and is intended only to mitigate changes in the vehicle's inertial state based on flow measurements.

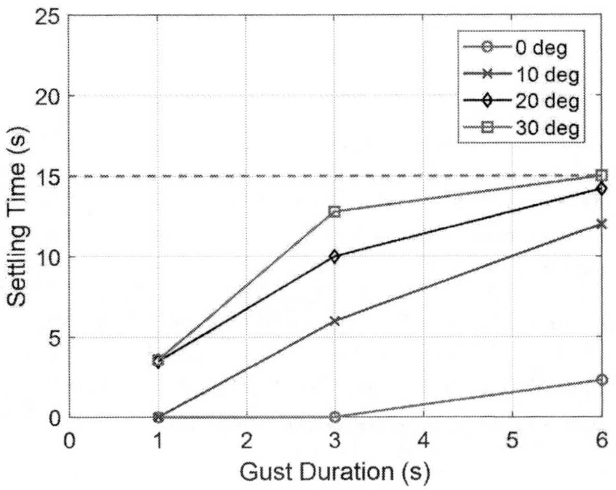

Fig. 23. Settling time vs. gust duration for crosswinds using combined controller.

Crosswind Scenario

In a realistic scenario, MAVs can encounter gust perturbation from any direction in flight. To capture some of these effects additional experiments were conducted with a incoming gusts directed at 10, 20, and 30 degrees azimuthal variation from the cyclocopter's longitudinal axis (Fig. 19). The gust controller relied on thrust vectoring, combined flow feedback and position feedback in order to reduce settling time. The acceptable maximum displacement was lowered to 0.3 m to keep the vehicle within the Dyson fan test cross-section and the maximum allowable settling time was increased to 15 seconds.

Unlike the longitudinal gust experiments where only a pitch response was excited, crosswind tests affect both pitch and roll angles (Fig. 20). The cyclocopter responds with a roll right and pitch up reaction to a step gust perturbation. In addition to the inner-loop attitude stabilization, an additional feedforward input was supplemented to the gust controller. The roll input depends on the y-component flow measurement (v_y) and is calculated similarly to the thrust vectoring input. Because the magnitude of the y-component flow measurement (v_y) is lower than the resultant velocity (V), the measurements are sensitive to noise from vibrations. Therefore the feedforward roll inputs were non-zero above a set saturation point, paralleling the step gust profiles (Fig. 21).

Figures 22 and 23 show that the cyclocopter was able to tolerate step gusts of 2.8 m/s in magnitude and 3 seconds duration with crosswind components at 30 degrees from the longitudinal axis. The decline in performance at higher azimuth angle can be attributed to the cyclocopter's lower control authority in roll. Similar results were observed in the previous 1-DOF wind tunnel experiments (Ref. 16).

CONCLUSION

The gust tolerance capability of the cyclocopter was investigated through series of experiments using a synthetic gust generation device. The gust rejection controller relies on the cyclorotor's unique thrust vectoring capability to reject perturbation isolated along the longitudinal axis. Feedback control schemes of position feedback using a motion capture system and flow feedback using onboard flow sensing were also compared. The following are key conclusions from this study:

1. A gust controller with a flow feedback scheme was successfully added to the closed-loop feedback system to augment the outer-loop PID position controller. Inputs to the gust controller included position measurements from the VICON motion capture system and airspeed measurements from a miniature dual-axis flow probe.

2. For a step gust input along the longitudinal axis, thrust vectoring was more effective than pitch control in reducing displacement from gust. Pitch control using the position feedback controller resulted in a maximum gust tolerance of 2.8 m/s with a duration of 3 seconds while thrust vector control using the combined flow feedback and position feedback controller was able to withstand 4 m/s step gust input with one second duration with only 0.01 m displacement.

3. Flow feedback was more effective than position feedback in minimizing the vehicle's response to gust. While the settling time results measured effectiveness of the gains used in the controller, the displacement results demonstrated that thrust vectoring coupled with onboard flow sensing improved the ability of the cyclocopter to mitigate gusts.

4. The cyclocopter was able to recover from step gusts of 2.8 m/s in magnitude and 3 seconds duration with crosswind components at 30 degrees from the longitudinal axis. The decline in performance at higher azimuth angle can be attributed to the cyclocopter's lower control authority in roll.

5. A synthetic gust generation device was assembled using six Dyson fans. A thorough sweep of the flow field showed that the mean freestream velocity was uniform across the 80 cm by 16.8 cm test region. The setup generated 2.8 m/s and 4 m/s gust with variance of 0.15 and a step gust profile using an automated shutter system.

ACKNOWLEDGMENT

This research was supported by the Army's MAST CTA Center for Microsystem Mechanics with Dr. Brett Piekarski (ARL) and Mr. Chris Kroninger (ARL-VTD) as Technical Monitors and by the National Science Foundation (NSF) Graduate Research Fellowships Program (GRFP) under Grant No. DGE 1322106.

REFERENCES

[1]Zarovy, S., Costello, M., and Mehta, A., "Experimental method for studying gust effects on micro rotorcraft," *Journal*

of Aerospace Engineering, Vol. 227, No. 4, April 2013, pp. 703-703.

[2]Hrishikeshavan V., and Chopra I., "Performance, Flight Testing of Shrouded Rotor Micro Air Vehicle in Edgewise Gusts," *Journal of Aircraft*. Vol. 49, pp.193–205

[3]Benedict, M., "Fundamental Understanding of the Cycloidal-Rotor Concept for Micro Air Vehicle Applications," Ph.D Thesis, Department of Aerospace Engineering, University of Maryland College Park, December 2010.

[4]Jarugumilli, T., Benedict, M. and Chopra, I., "Wind Tunnel Studies on a Micro Air Vehicle-Scale Cycloidal Rotor," *Journal of the American Helicopter Society*, Vol. 59, No. 2, April 2014, pp. 02208 (1–10).

[5]Shrestha, E., Hrishikeshavan, V., Benedict, M., Yeo, D., and Chopra, I.,"Development of Control Strategies for a Twin-Cyclocopter in Forward Flight," *Journal of the American Helicopter Society*, Vol. 61, No. 4, October 2016, pp. 1–9.

[6]Hrishikeshavan, V., Benedict, M., and Chopra, I., "Identification of Flight Dynamics of a Cylcocopter Micro Air Vehicle in Hover," *Journal of Aircraft*, Vol. 52, No. 1 (2015), pp. 116-129.

7

[8]Shrestha, E., Hrishikeshavan, V., Yeo, D., Benedict, M., and Chopra, I., "System Identification of a Cyclocopter Micro Air Vehicle in Forward Flight," Proceedings of the American Helicopter Society 71st Annual Forum, Virginia Beach, VA, May 5-7, 2015.

[9]Yeo, D., Sydney, N., and Paley, D., "Onboard Flow Sensing for Rotary-Wing UAV Pitch Control in Wind," AIAA Guidance, Navigation, and Control Conference, AIAA SciTech, Jan. 2016.

[10]Yeo, D., Shrestha, E., Paley, D., and Atkins, E., "Experimental Development of a Rotorcraft UAV Downwash Model for Real-Time Disturbance Localization and Avoidance," Proceedings of the AIAA SciTech Forum, Kissimmee, FL, January 5–9, 2015.

[11]Yeo, D., Sydney, N., Paley, D., and Sofge, D. "Onboard Flow Sensing for Downwash Detection and Avoidance with a Small Quadrotor Helicopter," AIAA Guidance, Navigation, and Control Conference, AIAA SciTech, Jan. 2015.

[12]Hrishikeshavan, V., and Chopra, I., "Refined Lightweight Inertial Navigation System for Micro Air Vehicle Applications," AHS Specialists' Meeting on Unmanned Rotorcraft and Network-centric Operations, Chandler, AZ, Jan 20–22, 2015.

[13]Georgy, J., Noureldin, A., Korenberg, M., and Bayoumi, M., "Modeling the Stochastic Drift of a MEMS-Based Gyroscope in Gyro/Odometer/GPS Integrated Navigation," IEEE Transactions on Intelligent Transportation Systems, Vol. 11, (4), Dec 2010, pp. 856–872.

[14]Y. K. Thong, M. S. Woolfson, J. A. Crowe, B. R. Hayes-Gill, and R. E. Challis, "Dependence of Inertial Measurements of Distance on Accelerometer Noise," Meas. Sci. Technol., Vol. 13 , (8), pp.1163–1172 , 2002.

[15]Watkins, S., Mohamed, A., Prudden, Sam., Tennent, D., Marino, M., and Fisher, A., "Upstream Wind Sensing for Small Unmanned Aerial Vehicles," Proceedings of the5th IEEE International Workshop on Metrology for AeroSpace, June, 2018.

[16]Shrestha, E., Yeo, D., Hrishikeshavan, V., Benedict, M., and Chopra, I., "Gust Disturbance Rejection Study of a Cyclocopter Micro Air Vehicle," Proceedings of the 72nd Annual National Forum of the American Helicopter Society, West Palm Beach, Florida, May 17–19, 2016.

Development of a Reconfigurable Multicopter Flight Dynamics Model from Flight Data Using System Identifcation

Mark J. S. Lopez **Mark B. Tischler** **Ondrej Juhasz**
Aerospace Engineer Senior Technologist Aerospace Engineer
Aviation Development Directorate
U.S. Army RDECOM, Aviation & Missile Center
Moffett Field, CA
Anthony Gong **Frank C. Sanders**
Research Associate Senior Research Associate
San Jose State University
Ames Research Center
Moffett Field, CA

ABSTRACT

Multirotor unmanned aerial systems (UAS) are prone to tubulent wind conditions and gust disturbances. Improving gust rejection performance is a critical technology to enable multirotor UAS operations in highly turbulent conditions and has been a recent topic of interest for study. This work is focused on developing flight dynamics models to better understand the flight characteristics of these multirotor aircraft which directly affect their ability to reject gusts and other external disturbances. This paper uses flight data and system identification to develop and validate a reconfigurable multicopter model. A Joint Input-Output system identification method is used to determine contributions from individual motors. The reconfigurable model is used to gain physical insight on the effects of rotor number and spacing on flight dynamic characteristics and gust rejection capability.

NOTATION

Symbols

$\boldsymbol{M}$	Mixing matrix from summed actuator signals to individual actuator signals
$\boldsymbol{r}$	Vector of reference signals in JIO method
$\boldsymbol{y}$	Vector of output signals
$\boldsymbol{\delta}_A$	Vector of actuator signals in JIO method
$\boldsymbol{\delta}_F$	Vector of individual actuator command signals
$\boldsymbol{\delta}_{Sum}$	Vector of summed actuator command signals

INTRODUCTION

Multirotor unmanned aerial systems (UAS) are especially prone to gust disturbances. The Aviation Development Directorate (ADD) at Ames Research Center has been working with the Defense Advanced Research Projects Agency (DARPA) under an effort to explore methods for improving gust rejection capabilities in UAS. One part of this effort is developing flight dynamics models to better understand the flight dynamics characteristics of these multirotor aircraft which directly affect their ability to reject gusts and other external disturbances.

Flight dynamics models for multirotor UAS are not sufficiently mature. While physics-based models exist and are well validated and understood for traditional manned-size full-scale rotorcraft, for UAS the model paramters are rarely known accurately (e.g. aerodynamic coefficients) and assumptions that go into these physics-based models may not apply at the UAS scale and have not been well validated. Furthermore, while a physics-based model approach may be appropriate for full-scale rotorcraft, the development and validation cycle of physics-based modeling is currently unable to keep up with the the rapid fly-crash-fix development cycles of UAS.

There is currently much ongoing work to make physics-based models appropriate for multirotor UAS. Bristeau studied the effects of propeller aerodynamics on flight dynamics for a quadrotor UAS (Ref. 1). Russel performed a study on the level of modeling fidelity needed for comprehensive analysis of multirotor UAS (Ref. 2). Most recently Niemiec has taken steps toward a reconfigurable multicopter flight dynamics model based on first principles (Ref. 3).

While the physics-based model approach is continually improving, system identification (Ref. 4) is an effective approach for obtaining accurate multirotor flight dynamics models from flight data that has been used with great success. Wei first used system identification on a quadrotor UAS to obtain a

---

Presented at the 8th Biennial Autonomous VTOL Technical Meeting, Mesa, AZ, January 29-31, 2019.
Distribution Statement A: Approved for public release; distribution is unlimited.
Disclaimer: thoughts, opinions, and conclusions are those of the authors and do not specifically represent those of the Department of Defense.

bare-airframe model from closed-loop flight-test data (Ref. 5). Juhasz used system identification to obtain a bare-airframe and turbulence models of an Iris+ quadrotor (Ref. 6). Berrios used the identification results from Juhasz in a control system optimization effort that greatly improved the gust rejection capabilities (Ref. 7).

The work presented herein is an approach for a generic, reconfigurable multicopter flight dynamics model, which is developed based on system identified results from flight data.

SYSTEM IDENTIFICATION

The reconfigurable model is developed based on flight identified models of quad and octocopter vehicles. A hexacopter vehicle was also built and identified. However, the hexacopter was not used in development of the reconfigurable model, it was used only to verify the reconfigurable model. The quad, hexa, and octocopter vehicles were built using common parts. The primary difference among the configurations is the number and location of arms and rotors. Each vehicle has a hub-to-hub diagonal distance of 1.27 meters, with 0.46 meter diameter rotors. The quad, hexa, and octocopter have masses of 6.1, 7.1, and 8.2 kg respectively, and pictures of each vehicle are shown in Fig. 1.

To develop the reconfigurable model with user selectable number and location of rotors, contributions from individual rotors are needed. Due to the way that forces and moments are usually allocated for multicopter control (all motors simultaneously increase or decrease RPM appropriately to obtain the total desired force or moment command), all motor inputs are inhereintly highly correlated. This property of highly correlated inputs is difficult to deal with for traditional system identification methods (Ref. 4). The work presented herein uses a frequency domain system identification approach of Ref. 4, with an extra post-processing step to address the issue of highly correlated inputs: the Joint Input-Output (JIO) methdology with a careful choice of input signals and transformations.

The JIO method was first developed by Akaike (Ref. 8) to address noise. More recently, the JIO method has been used extensively by Gennaretti et al (Ref. 9) and Hersey et al (Ref. 10) to identify inflow models with highly correlated inputs. Knapp et al (Ref. 11) and Berger et al (Refs. 12, 13) also used the JIO method to identify flight dynamics models with highly correlated inputs. The JIO method is used here to obtain frequency responses and state space models where the inputs are the commands for each individual motor. The method used herein is similar to the method demonstrated by Berger (Ref. 13), however there are additional considerations for the purposes of the reconfigurable model.

System Identification Signals

At this point, it is useful to define the nomenclature that will be used for describing the system identification and JIO methodology. There are 3 general signals that are used in the JIO methodology: the reference, actuator, and output signals.

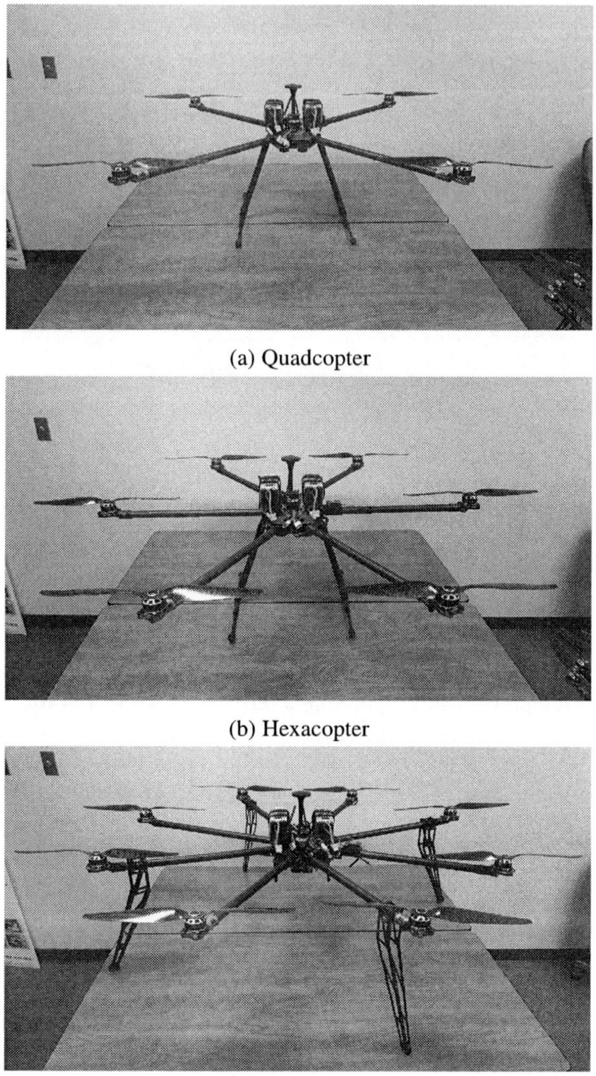

(a) Quadcopter

(b) Hexacopter

(c) Octocopter

Fig. 1: Pictures of each multirotor vehicle

The output signals $\boldsymbol{y}$ are all rigid body responses of interest and include the standard attitudes and rates in roll, pitch, and yaw $(\phi, \theta, \psi, p, q, r)$ in addition to the standard body velocity derivatives and accelerations $(\dot{u}, \dot{v}, \dot{w}, a_x, a_y, a_z)$.

The reference signals $\boldsymbol{r}$ are signals used for intermediate frequency response calculations, and are chosen here as the external vehicle commands. For example, during a frequency sweep, the external sweep command is considered as the reference signal; for a vehicle doublet, the external doublet command is considered the reference signal.

The actuator signals $\boldsymbol{\delta}_A$ correspond to the vehicle control effectors, which are the inputs of interest for flight dynamics, but are highly correlated and difficult to use as inputs with traditional system identification methods (Ref. 4). Actuator signals can be individual control effectors (e.g. individual motor commands $\boldsymbol{\delta}_A = \boldsymbol{\delta}_F = [\delta_{F_1}, \delta_{F_2}, ..., \delta_{F_N}]^T$), or alternatively actuator signals can be chosen as pseudo-control effectors (e.g.

sums and differences of motor combinations $\boldsymbol{\delta}_A = \boldsymbol{\delta}_{Sum}$).

Both individual control effectors and pseudo-control effectors are valid choices for actuator signals, and both choices have independent uses. For comparison, a generic block diagram is shown in Fig. 2. In the diagram, $\boldsymbol{P}$ is the bare airframe plant, $\boldsymbol{F}$, $\boldsymbol{C}$, and $\boldsymbol{H}$ are generic flight control blocks, and $\boldsymbol{M}$ is the mixing matrix to convert from summed actuator commands to individual actuator commands. Fig. 2a shows the actuator signals selected as individual motor commands $\boldsymbol{\delta}_A = \boldsymbol{\delta}_F$. Fig. 2b shows the actuator signals selected as summed motor commands $\boldsymbol{\delta}_A = \boldsymbol{\delta}_{Sum}$.

Pseudo-control effectors here are chosen as sums and differences of motor combinations referred to as "summed" actuator commands $\boldsymbol{\delta}_{Sum}$. For a pair of two motors δ_{F_i} and δ_{F_j}, the summed commands would be symmetric $\delta_{symm} = 1/2(\delta_{F_i} + \delta_{F_j})$ and differential commands $\delta_{diff} = 1/2(\delta_{F_i} - \delta_{F_j})$. For a group of four motors (tetrad), the "summed" actuator commands are chosen as standard col, lon, lat and ped commands which correspond to the four standard decoupled heave, pitch, roll, and yaw commands normally used in flight dynamics and control. The mixing from summed and individual actuator commands is chosen such that mixing matrix is square (number of summed actuator commands is equal to the number of individual actuator commands) and has mutually orthogonal rows and columns; these choices allow the mixing matrix to be invertible so that actuator signals can be easily converted between summed and individual actuator commands.

For a quadcopter with four individual motors ($\boldsymbol{\delta}_F = [\delta_{F_1}, \delta_{F_2}, \delta_{F_3}, \delta_{F_4}]^T$), the four summed actuators are straightforward ($\boldsymbol{\delta}_{Sum} = [col, lon, lat, ped]^T$) and correspond to control of decoupled axes (heave, pitch, roll, yaw). Thus, standard system identifcation methods (Ref. 4) can be used to obtain frequency responses with respect to the four summed actuators with no issues (Ref. 6). The inputs can then be transformed from summed actuators to individual motors, using the mixing matrix $\boldsymbol{M}$, as follows: $\boldsymbol{\delta}_F = [\boldsymbol{M}]\boldsymbol{\delta}_{Sum}$.

For an octocopter with 8 individual motors ($\boldsymbol{\delta}_F = [\delta_{F_1}, \delta_{F_2}, \delta_{F_3}, \delta_{F_4}, \delta_{F_5}, \delta_{F_6}, \delta_{F_7}, \delta_{F_8}]^T$), the 8 motors are divided into 2 groups of 4 (tetrads). As shown in Fig. 3, Tetrad 1 consists of motors 1, 2, 5, and 6, while Tetrad 2 consists of motors 3, 4, 7, and 8. For each octocopter tetrad, the four motors are combined into summed commands col, lon, lat and ped. Summed commands for Tetrad 1 are col_1 (motors 1, 2, 5, and 6 generate net positive thrust), lon_1 (motors 1 and 2 generate more thrust, while 5 and 6 generate less thrust, i.e. net positive pitching moment), lat_1 (motors 1 and 6 generate more thrust, while 2 and 5 generate less thrust, i.e. net positive rolling moment), and ped_1 (motors 1 and 5 generate more thrust, while 2 and 6 generate less thrust, i.e. net positive yawing moment). Similarly, summed commands for Tetrad 2 are col_2 (motors 3, 4, 7, and 8 generate net positive thrust), lon_2 (motors 3 and 8 generate more thrust, while 4 and 7 generate less thrust, i.e. net positive pitching moment), lat_2 (motors 7 and 8 generate more thrust, while 3 and 4 generate less thrust, i.e. net positive rolling moment), and ped_2 (motors 3 and 7 generate more thrust, while 4 and 8 generate less thrust, i.e. net positive yawing moment.). The 8 octocopter summed actuator commands are $\boldsymbol{\delta}_{Sum} = [col_1, lon_1, lat_1, ped_1, col_2, lon_2, lat_2, ped_2]^T$, which are summarized in Table 1 and the summed-to-individual actuator mixing equation given in equation 1.

Table 1: Summed octocopter actuator commands.

Summed Actuator Command	Description
col_1	Tetrad 1 creates net positive heave force
lon_1	Tetrad 1 creates net positive pitch moment
lat_1	Tetrad 1 creates net positive roll moment
ped_1	Tetrad 1 creates net positive yaw moment
col_2	Tetrad 2 creates net positive heave force
lon_2	Tetrad 2 creates net positive pitch moment
lat_2	Tetrad 2 creates net positive roll moment
ped_2	Tetrad 2 creates net positive yaw moment

$$\underbrace{\begin{bmatrix} \delta_{F_1} \\ \delta_{F_2} \\ \delta_{F_3} \\ \delta_{F_4} \\ \delta_{F_5} \\ \delta_{F_6} \\ \delta_{F_7} \\ \delta_{F_8} \end{bmatrix}}_{\boldsymbol{\delta}_F} = \underbrace{\begin{bmatrix} 1 & 1 & 1 & 1 & 0 & 0 & 0 & 0 \\ 1 & 1 & -1 & -1 & 0 & 0 & 0 & 0 \\ 0 & 0 & 0 & 0 & 1 & 1 & -1 & 1 \\ 0 & 0 & 0 & 0 & 1 & -1 & -1 & -1 \\ 1 & -1 & -1 & 1 & 0 & 0 & 0 & 0 \\ 1 & -1 & 1 & -1 & 0 & 0 & 0 & 0 \\ 0 & 0 & 0 & 0 & 1 & -1 & 1 & 1 \\ 0 & 0 & 0 & 0 & 1 & 1 & 1 & -1 \end{bmatrix}}_{\boldsymbol{M}} \underbrace{\begin{bmatrix} col_1 \\ lon_1 \\ lat_1 \\ ped_1 \\ col_2 \\ lon_2 \\ lat_2 \\ ped_2 \end{bmatrix}}_{\boldsymbol{\delta}_{Sum}} \tag{1}$$

It is important to note that for an octocopter with 4 control channels (heave, pitch, roll ,yaw), the redundancy in summed actuators and control allocation for each channel leads to redundant actuators being fully correlated. Specifically when the vehicle is flown closed loop, pitch channel feedback will be allocated to both lon_1 and lon_2, meaning that lon_1 and lon_2 will always be fully correlated. Similarly, roll channel feedback will result in lat_1 and lat_2 being fully correlated, and yaw channel feedback will result in ped_1 and ped_2 being fully correlated. Often the heave axis is flown open loop meaning that col_1 and col_2 can be uncorrelated. The high correlation between redundant summed actuators is addressed using the JIO method.

Joint Input-Output Method

The system identification goal herein is to determine a bare airframe frequency response matrix $[\boldsymbol{y}/\boldsymbol{\delta}_A]$, where $\boldsymbol{\delta}_A$ are the desired bare airframe inputs or motor commands (actuator signals) and $\boldsymbol{y}$ is the vector of vehicle outputs. Due to the high correlation between bare-airframe inputs, direct identification of the bare airframe frequency response matrix $[\boldsymbol{y}/\boldsymbol{\delta}_A]$ is not possible and the JIO method (Refs. 12, 13) is used instead. The JIO method computes the bare airframe frequency response matrix indirectly by first computing responses with

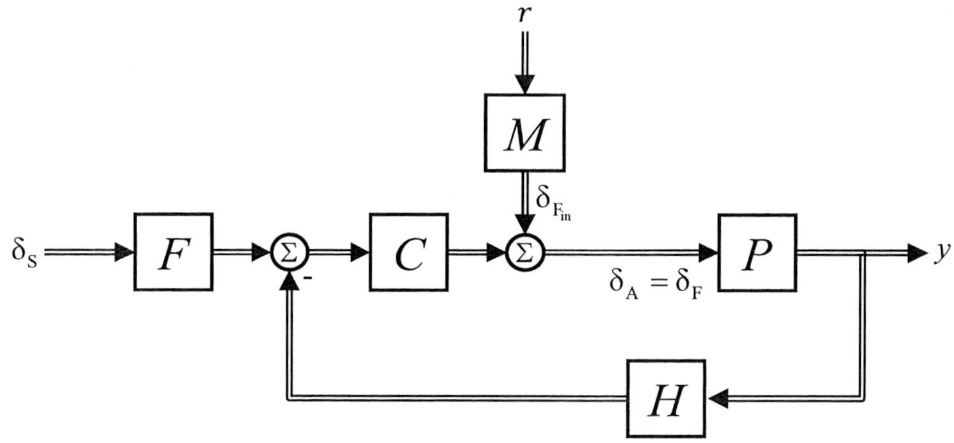

(a) Individual Motor Actuator Signals

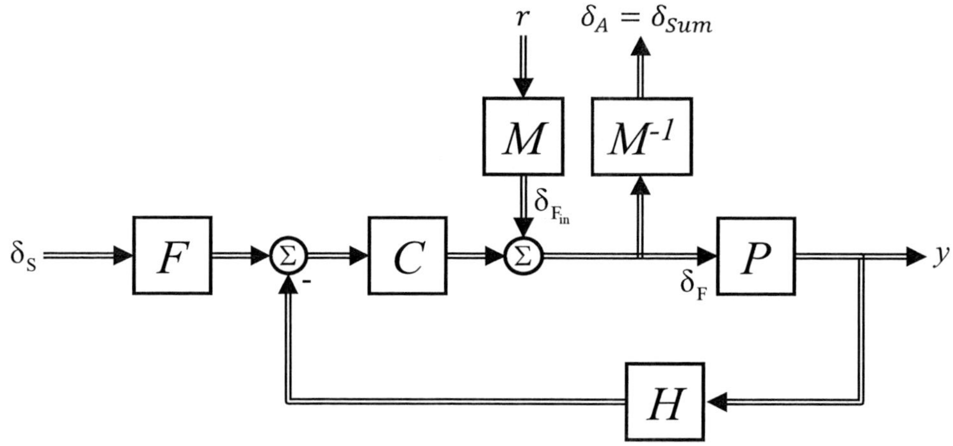

(b) Summed Actuator Signals

Fig. 2: Generic closed-loop block diagram

respect to reference signals $\boldsymbol{r}$, which herein are the external sweep commands. The actuator-to-reference frequency response matrix $[\boldsymbol{\delta}_A/\boldsymbol{r}]$ and output-to-reference frequency response matrix $[\boldsymbol{y}/\boldsymbol{r}]$ are first computed, then at each frequency ω, the bare airframe response matrix is simply the product of the actuator-to-reference frequency response matrix inverse with the output-to-reference frequency response matrix:

$$\left[\frac{\boldsymbol{y}}{\boldsymbol{\delta}_{\mathrm{A}}}(j\omega)\right] = \left[\frac{\boldsymbol{y}}{\boldsymbol{r}}(j\omega)\right]\left[\frac{\boldsymbol{\delta}_{\mathrm{A}}}{\boldsymbol{r}}(j\omega)\right]^{-1} \quad (2)$$

In scalar form, equation 2 can be thought of a simply a chain rule calculation of y/δ_A.

It should be noted that the reference signals are chosen such that they can be independently actuated (and therefore decorrelated) and also that the resulting actuator signals $\boldsymbol{\delta}_A$ are all linearly independent (i.e. that the acutator-to-reference response matrix $[\boldsymbol{\delta}_A/\boldsymbol{r}]$ is square and invertible at all desired frequencies).

For the purposes of obtaining models with individual motor command inputs, there are several considerations when selecting reference signals and actuator signals to process through the JIO method.

The first consideration is with respect to desired responses from external vehicle commands. While one method would be to command frequency exitations to a single individual motor doing so can cause the vehicle to response to be completely coupled between all axes. This can result in responses which have poor signal quality due to the lack of control authority from a single motor. Instead, the vehicle excitations are chosen such that the responses remain decoupled and on-axis as much as possible. Specifically, external vehicle commands are chosen to align closely with the traditional heave, roll, pitch, and yaw commands. For the quadcopter with 4 motors, only 4 vehicle commands are needed and are chosen as the traditional heave, pitch, roll, and yaw commands.

For the octocopter with 8 motors shown in Fig. 3, 8 linearly independent external vehicle commands are needed and are chosen as heave, roll, pitch, and yaw for each tetrad of motors: $\boldsymbol{r} = [col_1, lon_1, lat_1, ped_1, col_2, lon_2, lat_2, ped_2]^T$ as shown in

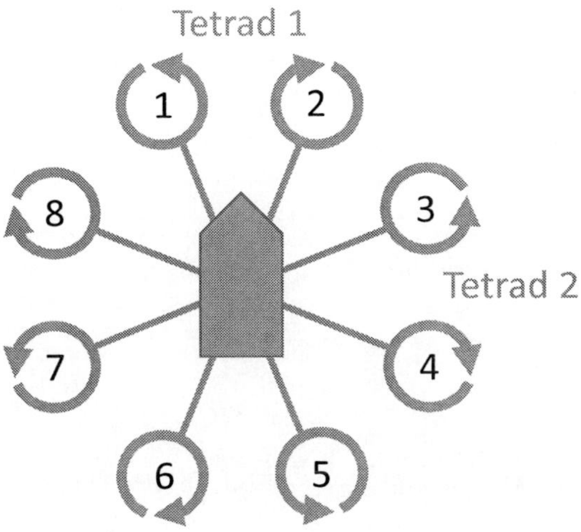

Fig. 3: Octocopter rotor orientations.

Table 1. This choice of vehicle commands means that for any particular command, only 2 summed actuator commands are highly correlated for any given frequency sweep excitation. For example, for a pitch sweep command of lon_1, only the summed actuator signals lon_1 and lon_2 will be correlated, with all other summed actuator signals being uncorrelated due to the on-axis nature of the commands. Selection of the vehicle sweep commands and summed actuator signals in this manner means that for any given sweep, only 2 summed actuators need to be considered for the JIO calculation in equation 2. For example, for an octocopter pitch sweep command, one only needs to consider summed actuators $\boldsymbol{\delta}_A = [lon_1, lon_2]^T$; for an octocopter roll sweep command, one only needs to consider summed actuators $\boldsymbol{\delta}_A = [lat_1, lat_2]^T$.

The second consideration is the choice of actuator signals to use in the JIO method. The JIO method can be used to directly obtain frequency responses to individual motors (e.g. $\boldsymbol{\delta}_A = \boldsymbol{\delta}_F$) as illustrated in Fig. 2a. Results for pitch rate response to individual motors is shown in Fig. 4. δ_{F_2} and δ_{F_6} have higher control power compared to δ_{F_4} and δ_{F_8} as indicated by the higher magnitudes for δ_{F_2} and δ_{F_6} above 3 rad/s. The difference in control power in Fig. 4 matches the expected differences based on the rotor placements and differences in moment arms with respect to the vehicle center of gravity.

While Fig. 4 shows that directly obtaining frequency responses to individual motors is possible (e.g. $\boldsymbol{\delta}_A = \boldsymbol{\delta}_F$), there are also variations between the smoothness of responses and therefore overall data quality (e.g., pitch rate q/δ_{F_2} and q/δ_{F_6} have much smoother magnitude and phase responses compared to q/δ_{F_4} and $/q\delta_{F_8}$). It was found that obtaining frequency responses with the summed actuator signals as inputs (e.g. $\boldsymbol{\delta}_A = \boldsymbol{\delta}_{Sum}$) yielded the best coherence and frequency response quality. The reason for this is that vehicle excitation commands were chosen to be heave, pitch, roll, or yaw commands, which are all directly correlated to summed actuator commands in each axis. For example, for a pitch sweep command, individual motors $\boldsymbol{\delta}_F$ will be correlated with the pitch sweep command, but will also be correlated with any

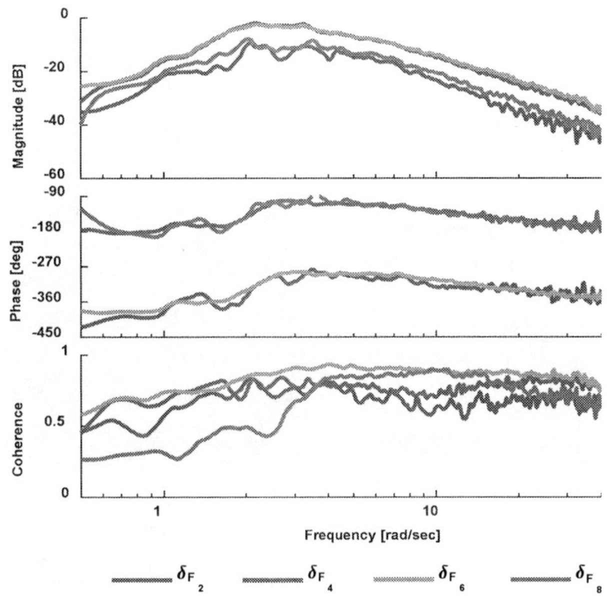

Fig. 4: Octocopter pitch rate response to individual motors q/δ_{F_i}, extracted directly with $\boldsymbol{\delta}_A = \boldsymbol{\delta}_F$.

off axis responses due to disturbances (e.g. roll feedback to a roll disturbance will be allocated to all individual motors); in contrast, summed actuator commands lon_1 and lon_2 will be correlated with the pitch sweep command, but will be uncorrelated with any off-axis disturbances (e.g. roll feedback and yaw feedback to disturbances are not allocated to lon_1 and lon_2) due to the decoupled nature of summed actuator inputs.

System Identification Process

The full process is for obtaining models with individual actuators is shown in Fig. 5. The description of each step is shown on the left, and an example for the octocopter in pitch is shown on the right.

The first step is to sweep each reference command $\boldsymbol{r}_i$. In this case, the reference command sweeps $\boldsymbol{r}_i$ are mixed into sweeps of all individual actuators $\boldsymbol{\delta}_{F_{in}}$ as shown in Fig. 2b. In the diagram, $\boldsymbol{M}$ is the mixing matrix to convert from summed actuator commands to individual actuator commands. The individual actuator commands (i.e. commands to each motor $\boldsymbol{\delta}_F = [\delta_{F_1}, \delta_{F_2}, ..., \delta_{F_N}]^T$ are fully correlated. The summed actuator commands $\boldsymbol{\delta}_{Sum}$ are computed by multiplying total individual actuator commands $\boldsymbol{\delta}_F$ with the mixing matrix inverse. Any redundant summed actuator commands will also be fully correlated; for example, an octocopter r_1=lon_1 vehicle sweep command will excite lon_1, lon_2, and q, with lon_1 and lon_2 being redundant control inputs and therefore completely correlated.

The second step is to extract the bare airframe frequency responses using the JIO method. In this case, the sweep commands are the references $\boldsymbol{r}$, the actuator signals in the JIO method are the summed actuators $\boldsymbol{\delta}_A = \boldsymbol{\delta}_{Sum}$ as shown in Fig. 2b, and the outputs any desired vehicle responses $\boldsymbol{y}$. For the

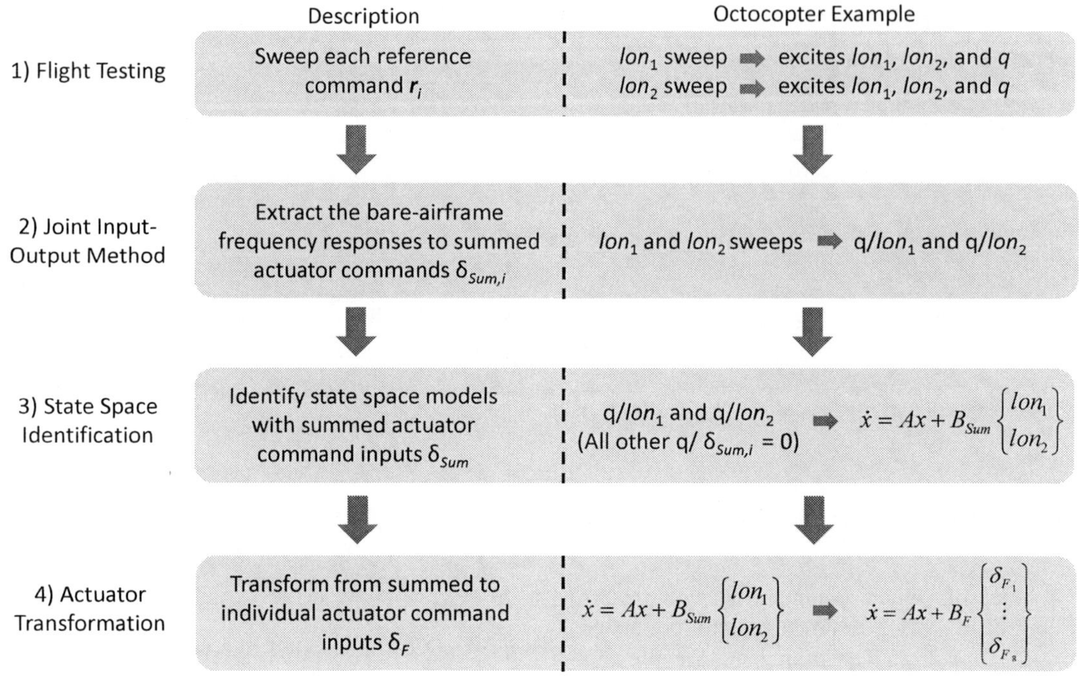

Fig. 5: Process for obtaining models with individual actuator commands.

octocopter pitch example, $r_1 = lon_1$ and $r_2 = lon_2$ sweeps are used to extract frequency responses q/lon_1 and q/lon_2. It should be noted that the choice of references and summed actuator commands results in all axes (heave, roll, pitch, and yaw) being mutually decoupled (e.g. for the octocopter, only lon_1 and lon_2 will cause any pitch rate response.

The third step is then to identify state space models with respect to the summed actuator command inputs. Once this appropriate frequency responses have been obtained throught the Joint-Input Output method, this state space identification is straightforward and follows the standard identification procedure (Ref. 4). Guidelines from Tischler are used for parameter reliability and model acceptability (Ref. 4). Cramér-Rao bounds are within 20% and insensitivities are within 10% for all identified parameters indicating good reliaibility. Table 2 shows the average and maximum cost functions for each configuration. All models have average cost functions $J_{ave} < 50$ (Ref. 4), indicating that overall each individual model is an excellent fit to its corresponding flight data. Maximum cost functions are also displayed for each configuration and are all below 100, indicating good levels of fit for corresponding frequency responses from flight data.

Table 2: Cost Functions for Identified State Space Models.

Configuration	Average Cost Function	Maximum Cost Function
Quadcopter	40.8	89.9 (q/lon)
Hexacopter	45.5	98.4 ($\dot{u}/lon$)
Octocopter	27.5	61.2 ($\dot{u}/lon$)

It should be noted that there are 3 caveats for the state space identification, which are specific to the goal of the reconfigurable multicopter model. 1) All longitudinal and lateral derivatives are constrained to be equal after accounting for differences in inertia. 2) All actuator dynamics and delays are constrained to be equal. 3) Quad, hexa, and octocopter state space models all have the same structure. Each of the constraints are assumptions based on the symmetry of the vehicles and common parts used for each rotor and each vehicle. One consequence of these caveats is that the state space identification must be performed with all axes simultaneously; even though the vehicle dynamics are decoupled, the way to enforce symmetries among the axes and motors is by simultaenously identifying all axes with the inter-axis constsraints.

The final step is to transform the state space inputs from summed actuator commands to individual actuator commands. This is straightforward as the mixing matrix $\boldsymbol{M}$ was chosen to be invertible. A similar process is done to the states corresponding to actuator dynamics (motor lags), which is possible due to the constraint that all actuators are the same.

This process for identifying state space models with individual actuator command inputs is repeated for the quad, hexa, and octocopter. The final state space form is shown in Fig. 6 with individual actuator command inputs. The mass and inertia terms are explicity separated from aerodynamic terms in the stability and control matrices for the purposes of the reconfigurable model. All 3 configurations share the same state space form. The stability derivatives $X_w, Y_r, Z_u, Z_q, L_r, M_w, N_v$, and N_p are all fixed to 0 due to their corresponding frequency responses being dropped as a result of the decoupled responses (Ref. 4). Additionally, the stability derivatives X_q

Mass and Inertias — Rigid Body Stability Derivatives — Control Derivatives — Motor Lags

$$
\begin{bmatrix}
m & 0 & 0 & 0 & 0 & 0 & 0 & 0 & 0 & 0 & 0 & \cdots & 0 \\
0 & m & 0 & 0 & 0 & 0 & 0 & 0 & 0 & 0 & 0 & \cdots & 0 \\
0 & 0 & m & 0 & 0 & 0 & 0 & 0 & 0 & 0 & 0 & \cdots & 0 \\
0 & 0 & 0 & I_{xx} & 0 & 0 & 0 & 0 & 0 & 0 & 0 & \cdots & 0 \\
0 & 0 & 0 & 0 & I_{yy} & 0 & 0 & 0 & 0 & 0 & 0 & \cdots & 0 \\
0 & 0 & 0 & 0 & 0 & I_{zz} & 0 & 0 & 0 & 0 & 0 & \cdots & 0 \\
0 & 0 & 0 & 0 & 0 & 0 & 1 & 0 & 0 & 0 & 0 & \cdots & 0 \\
0 & 0 & 0 & 0 & 0 & 0 & 0 & 1 & 0 & 0 & 0 & \cdots & 0 \\
0 & 0 & 0 & 0 & 0 & 0 & 0 & 0 & 1 & 0 & 0 & \cdots & 0 \\
0 & 0 & 0 & 0 & 0 & 0 & 0 & 0 & 0 & 1 & 0 & \cdots & 0 \\
0 & 0 & 0 & 0 & 0 & 0 & 0 & 0 & 0 & 0 & 1 & \cdots & 0 \\
\vdots & \vdots & \vdots & \vdots & \vdots & \vdots & \vdots & \vdots & \vdots & \vdots & \vdots & \ddots & \vdots \\
0 & 0 & 0 & 0 & 0 & 0 & 0 & 0 & 0 & 0 & 0 & \cdots & 1
\end{bmatrix}
\begin{bmatrix} \dot{u} \\ \dot{v} \\ \dot{w} \\ \dot{p} \\ \dot{q} \\ \dot{r} \\ \dot{\phi} \\ \dot{\theta} \\ \dot{\psi} \\ \dot{m}_1 \\ \dot{m}_2 \\ \vdots \\ \dot{m}_n \end{bmatrix}
=
\begin{bmatrix}
X_u & 0 & X_w & 0 & X_q & 0 & 0 & -mg\cos(\theta_e) & 0 & 0 & 0 & \cdots & 0 \\
0 & Y_v & 0 & Y_p & 0 & Y_r & mg\cos(\theta_e) & 0 & 0 & 0 & 0 & \cdots & 0 \\
Z_u & 0 & Z_w & 0 & Z_q & 0 & 0 & -mg\sin(\theta_e) & 0 & Z_{m_1} & Z_{m_2} & \cdots & Z_{m_n} \\
0 & L_v & 0 & L_p & 0 & L_r & 0 & 0 & 0 & L_{m_1} & L_{m_2} & \cdots & L_{m_n} \\
M_u & 0 & M_w & 0 & M_q & 0 & 0 & 0 & 0 & M_{m_1} & M_{m_2} & \cdots & M_{m_n} \\
0 & N_v & 0 & N_p & 0 & N_r & 0 & 0 & 0 & N_{m_1} & N_{m_2} & \cdots & N_{m_n} \\
0 & 0 & 0 & 1 & 0 & 0 & 0 & 0 & 0 & 0 & 0 & \cdots & 0 \\
0 & 0 & 0 & 0 & 1 & 0 & 0 & 0 & 0 & 0 & 0 & \cdots & 0 \\
0 & 0 & 0 & 0 & 0 & \sec(\gamma_e) & 0 & 0 & 0 & 0 & 0 & \cdots & 0 \\
0 & 0 & 0 & 0 & 0 & 0 & 0 & 0 & 0 & -\omega_{m_1} & 0 & \cdots & 0 \\
0 & 0 & 0 & 0 & 0 & 0 & 0 & 0 & 0 & 0 & -\omega_{m_2} & \cdots & 0 \\
\vdots & \vdots & \vdots & \vdots & \vdots & \vdots & \vdots & \vdots & \vdots & \vdots & \vdots & \ddots & \vdots \\
0 & 0 & 0 & 0 & 0 & 0 & 0 & 0 & 0 & 0 & 0 & \cdots & -\omega_{m_n}
\end{bmatrix}
\begin{bmatrix} u \\ v \\ w \\ p \\ q \\ r \\ \phi \\ \theta \\ \psi \\ m_1 \\ m_2 \\ \vdots \\ m_n \end{bmatrix}
+
\begin{bmatrix}
0 & 0 & \cdots & 0 \\
0 & 0 & \cdots & 0 \\
0 & 0 & \cdots & 0 \\
0 & 0 & \cdots & 0 \\
0 & 0 & \cdots & 0 \\
N_{f_1} & N_{f_2} & \cdots & N_{f_n} \\
0 & 0 & \cdots & 0 \\
0 & 0 & \cdots & 0 \\
0 & 0 & \cdots & 0 \\
\omega_{m_1} & 0 & \cdots & 0 \\
0 & \omega_{m_2} & \cdots & 0 \\
\vdots & \vdots & \ddots & \vdots \\
0 & 0 & \cdots & \omega_{m_n}
\end{bmatrix}
\begin{bmatrix} f_1(t-\tau_1) \\ f_2(t-\tau_2) \\ \vdots \\ f_n(t-\tau_n) \end{bmatrix}
$$

Fig. 6: State space model with individual actuator commands.

and Y_p are fixed to 0 as they were found to be insenstive for all 3 vehicles. Additionally, the stability derivative N_r is fixed to 0 for the quad and hexacopter configurations, as it was insensitve for those configurations. A final note is that there are motor lags present to account for motor dynamics; this is similar to previous work that found actuator dynamics were needed for other multicopter configurations (Ref. 6).

RECONFIGURABLE MULTICOPTER MODEL

The reconfigurable multicopter model is derived from quad and octocopter flight identified models. The hexacopter configuration was also flight tested and a model was extracted, however this data was only used in validation of the reconfigurable model, not in the model development. The reconfigurable model is developed by producing a fit between each individual state space term of the quad and octocopter identified models.

Linear least squares fits are used for terms which are expected to vary linearly with the number of rotors or rotor placement. For example, mass varies linearly with the number of rotors, while inertias vary linearly with the rotor placement. It should be noted that inertias from identified models are computed from a spreasdsheet buildup of individual components and are fixed in the actual state space identification process. Figure 7 shows the pitch inertia for the reconfigurable model along with the values used in system identified models. The dashed line indicates the fit used in the reconfigurable model (RM Fit). The red "x", black upward facing triangle, and blue circle indicate the pitch inertia for the quad, hexa, and octocopter respectively based on values used in the system identified models (based on a spreadsheet buildup). The downward facing magenta triangle indicates the pitch inertia for the hexacopter based on the reconfigurable model (Hexa RM). The reconfigurable model precisely predicts the hexacopter pitch inertia computed from a spreadsheet buildup as the Hexa RM and Hexa ID markers lie directly on top of one another.

The reconfigurable multicopter model is based on fits of the "basic derivatives" as defined by McRuer (Ref. 14). The

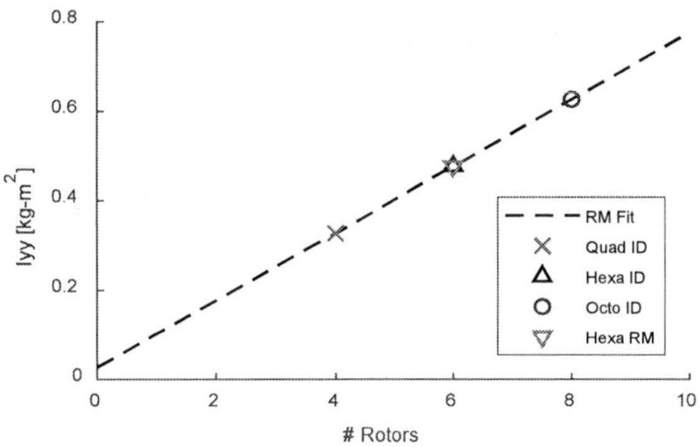

Fig. 7: Pitch inertia for reconfigurable model.

basic stability and control derivatives are in terms of force or moment, for example the basic pitch damping derivative has units of $[(Nm)/(rad/s)]$. This is in contrast to standard dimensional derivatives, which are in terms of acceleration, for example the standard dimensional pitch damping derivative has units of $[(rad/s^2)/(rad/s)] = [1/s]$. For clarity, the standard dimensional derivatives can be obtained by dividing the basic derivatives with the appropriate mass or inertia term. For example, standard dimensional pitch damping is obtained by dividing basic pitch damping by pitch inertia: $M_{q,dimensional} = M_{q,basic}/I_{yy}$. The standard dimensional derivatives are normally used in state space flight dynamic model identification. The basic derivatives are independent of mass and inertia, and are used in the reconfigurable model in order to separate the effects of inertia from aerodynamic effects.

Aerodynamic terms such as heave, roll, and pitch damping vary linearly with the number of rotors or rotor placement and are fit using linear least squares fits. Figure 8 shows the basic pitch damping derivative for the reconfigurable model along with values from system identified models. Error bars are shown indicating a 1-standard deviation confidence based on Cramér-Rao bounds. The reconfigurable model has an overall linear trend with the number of rotors as depicted in Fig. 8. The reconfigurable model accurately predicts the hexacopter

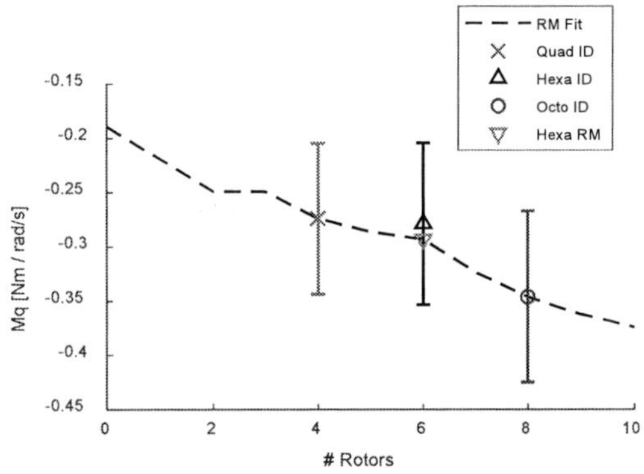

Fig. 8: Pitch damping for reconfigurable model (basic deriative).

basic pitch damping from the flight identified value, as the Hexa RM marker is very close to the Hexa ID marker, and well within the Hexa ID error bounds.

Step function fits are used for terms which do not vary linearly with the number of rotors, for example control derivatives are greatly affected by interference effects, which are a nonlinear function of rotor spacing. Figure 9 shows the basic pitch control derivative for an individual motor, relocated to be 1 meter from the vehicle center of gravity. Basic control power for an individual motor is shown in terms of moment [Nm] per motor input command [PWM]. For configurations where the number of rotors is less than 8, each motor has the same control power as an individual motor from the quadcopter. For configurations where the number of rotors is 8 or more, each motor has the same control power as an individual motor from the octocopter. Figure 9 shows a reduction in inidivdual motor control power between when the number of rotors is increased from 6 to 8. This reduction in individual motor control power is due to the decrease in rotor spacing resulting in rotors becoming physically close enough that significant aerodynamic interference occurs.

After accounting for the step function in individual motor control power and also accouting for changes in rotor placement and inertia, the total dimensional pitch control power for the reconfigurable model is shown in Fig. 10. Total dimensional control power is shown in terms of vehicle angular acceleration [rad/s^2] per mixer input command [PWM]. Due to all factors combined (aerodynamic interference and changes in rotor placement and inertia), the quadcopter vehicle acutally has the largest total dimensional control power in pitch. Compared to the quadcopter, the hexacopter has significantly less total pitch control power. The octocopter has more total pitch control power compared to the hexacopter, but not as much as the quadcopter. Thus, adding more rotors can result in a net increase or decrease in total control power, which may contradict initial intuition.

Similar validations for all individual state space terms (Fig. 6) were performed between the hexacopter directly identified model and the hexacopter configuration from the reconfigurable model. All terms from the reconfigurable model based hexacopter were within the 1-standard deviation error bounds from the directly identified hexacopter model, validating that the reconfigurable model is able to predict individual terms within the accuracy of the directly identified hexacopter model.

FLIGHT DATA VALIDATION

After the reconfigurable model has been validated at the hexacopter flight identified configuration for all state space model terms, it is next checked against the identified hexacopter in comparison to the actual flight data using frequency responses.

Figure 11 shows the pitch rate response due to a longitudinal mixer input for flight data, state space model directly identified from flight data (Hexacopter ID), and state space model from the reconfigurable model (Hexacopter RM). The directly identified and reconfigurable model responses lie directly on top of each other, both with nearly identical costs (J = 73 vs 75). This shows that the reconfigurable model is able to predict the pitch rate response just as well as the directly identified hexacopter model.

The costs for all frequency responses are compared for the reconfigurable model and the directly identified hexacopter model in Table 3. The Hexa ID column is the cost of the directly identified state space model compared to the flight data. The Hexa RM column is the cost of the reconfigurable model compared to the hexacopter flight data. For all frequency responses, the reconfigurable model has a cost similar to the directly identified model. On average, the reconfigurable model has a cost of 46.0 which closely compares to the average directly identified model cost of 45.5, meaning that the reconfigurable model is able to predict flight data frequency responses just as well as the directly identified hexacopter model.

Table 3: Cost Functions for Hexacopter Identified and Reconfigurable Models.

Response	Hexa ID	Hexa RM
a_z/col_1	13.6	13.6
a_z/col_2	60.7	56.6
a_z/col_3	79.4	83.9
$\dot{u}/lon$	98.4	97.8
q/lon	73.3	75.1
a_x/lon	45.3	49.3
$\dot{v}/lat$	27.2	27.2
p/lat	23.9	23.8
a_y/lat	28.3	26.8
r/ped	5.4	5.4
Average	45.5	46.0

The reconfigurable model and directly identified hexacopter model are checked in the time domain with doublet response flight data. Figure 12 shows the pitch rate response due to a longitudinal doublet for flight data, state space model directly identified from flight data (Hexacopter ID), and state space

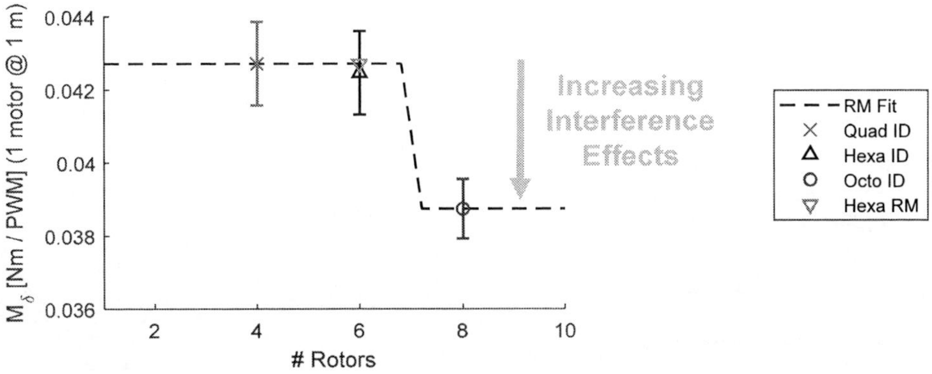

Fig. 9: Pitch control power for an individual motor (basic derivative).

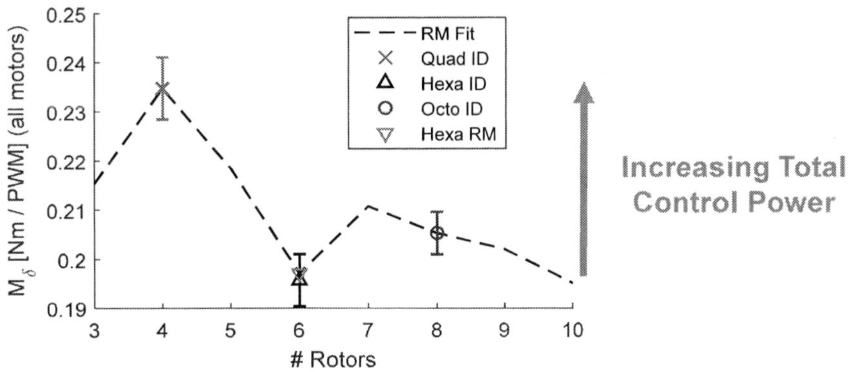

Fig. 10: Total pitch control power for reconfigurable model (dimensional derivative).

model from the reconfigurable model (Hexacopter RM). The directly identified and reconfigurable model responses lie directly on top of each other.

The RMS cost of the time response data can be Froude scaled based on multicopter hub-to-hub distance scaled to UH-60 rotor diameter to so that existing RMS cost guidelines can be applied (Ref. 4). Both models have the same scaled RMS cost $J_{RMS,Scaled} = 0.78 < 2$, and both models have a Theil Inequality Coefficient $TIC = 0.08 < 0.3$. This shows that the reconfigurable model is able to predict the pitch rate response in the time domain just as well as the directly identified hexacopter model, and both are extremely accurate (Ref. 4).

The $J_{RMS,Scaled}$ and TIC values for all axes are shown in Table 4. The ID columns are the scaled RMS cost and TIC of the directly identified state space model compared to the flight data. The RM columns are the scaled RMS cost and TIC of the reconfigurable model compared to the hexacopter flight data. For all doublet responses, the reconfigurable model has a cost and TIC similar to the directly identified model. On average, the reconfigurable model has an RMS cost $J_{RMS,Scaled} = 0.73$ and $TIC = 0.13$ which closely compares to the average directly identified model $J_{RMS,Scaled} = 0.70$ and $TIC = 0.11$, meaning that the reconfigurable model is able to predict flight data frequency responses just as well as the directly identified hexacopter model.

Table 4: Scaled RMS Cost and Thiel Inequality Coefficients for doublet verification.

Input	ID Cost	RM Cost	ID TIC	RM TIC
col_{mixer}	0.43	0.55	0.07	0.09
lon_{mixer}	0.79	0.79	0.08	0.08
lat_{mixer}	1.28	1.28	0.16	0.16
ped_{mixer}	0.31	0.31	0.17	0.18
Average	0.70	0.73	0.11	0.13

CONCLUSIONS

To study the inherent effects of the number of rotors on gust rejection capability, a quad, hexa, and octocopter were built and used to obtain system identified flight dynamics models. The quad and octocopter configurations were then used to develop a reconfigurable multicopter model, which was validated with hexacopter flight data and a directly identified model. This work supports the following conclusions:

1. Frequency response system identification augmented with the Joint Input-Output method can be used to extract stability and control contributions from individual rotors of a multicopter vehicle.

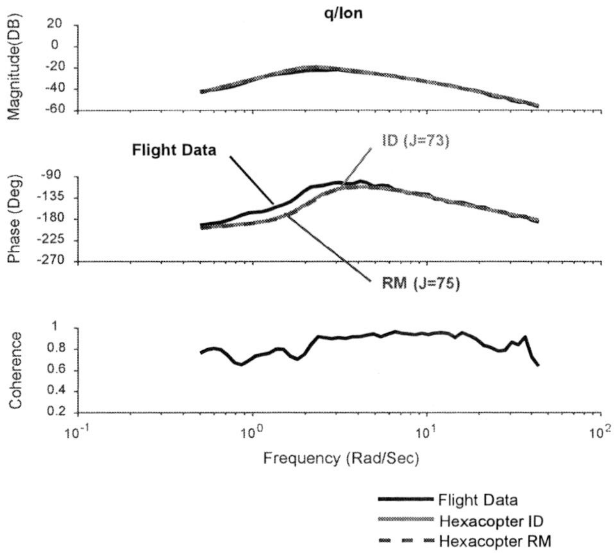

Fig. 11: Pitch rate response to longitudinal mixer input.

2. A reconfigurable multicopter model based on identified quad and octocopter configurations can be used to predict hexacopter flight dynamics parameters and responses just as well as a directly identified model.

3. Individual rotor control power may be reduced when the number of rotors is increased. This is due to decreased rotor spacing which results in significant aerodynamic interference effects between rotors.

4. Increasing the number of rotors on a multicopter may not inherently add control power or gust rejection capability. Adding rotors affects inertia, rotor placement, and individual rotor control power and all 3 effects combined can result in an increase or decrease in total control power.

ACKNOWLEDGMENTS

This work was sponsored by the Defense Advanced Research Project Agency, Tactical Technology Office (DARPA TTO) under the Gust Rejection effort at the Aviation Development Directorate (ADD) - Ames, managed by Dr. Alexander M. G. Walan. The authors would like to thank the Gust Rejection team at DARPA TTO for their support. The authors would also like to thank Dr. Mark F. Costello for his support of the gust rejection work at ADD.

REFERENCES

[1]Bristeau, P., Martin P., Salaun, E., and Petit, N., "The Role of Propeller Aerodyanmics in the Model of a Quadrotor UAV", Proceeding of the European Control Conference 2009, Budapest, Hungary, August 23-26, 2009.

[2]Russel, C. R. and Sekula, M. K., "Comprehensive Analysis Modeling of Small-Scale UAS Rotors", presented at the AHS International 73rd Annual Forum & Technology Display, Fort Worth, TX, May 9-11, 2017.

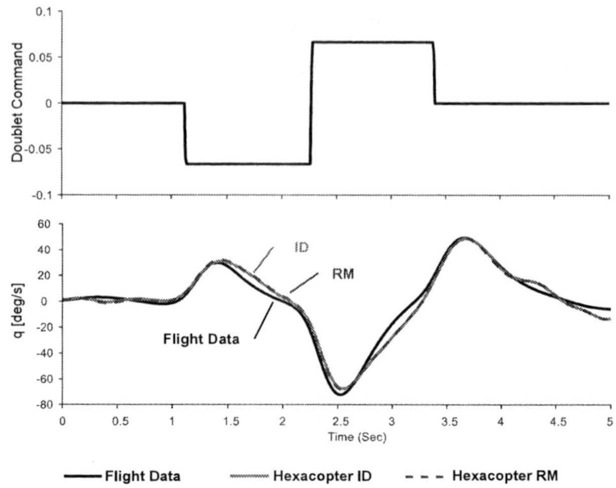

Fig. 12: Pitch rate response to doublet command.

[3]Niemiec, R., Gandhi, F., and Singh, R., "Control and Performance Analysis of a Reconfigurable Multi-Copter", presented at the AHS International 73rd Annual Forum & Technology Display, Fort Worth, TX, May 9-11, 2017.

[4]Tischler, M. B. and Remple, R. K., *Aircraft and Rotorcraft System Identification: Engineering Methods with Flight Test Examples*, 2nd Edition, American Institute of Aeronautics and Astronautics, Inc., Reston, VA, 2012.

[5]Wei, W., Cohen, K., and Tischler, M. B., "System Identification and Controller Optimization of a Quadrotor UAV", presented at the AHS 71st Annual Forum, virginia beach, VA, May 5-7, 2015.

[6]Juhasz, O., Lopez, M. J. S., Berrios, M. G., Berger, T., and Tischler, M. B., "Turbulence Modeling of a Small Quadrotor UAS Using System Identification from Flight Data", presented at the Seventh AHS Technical Meeting on VTOL Unmanned Aircraft Systems, Mesa, AZ, January 24-26, 2017.

[7]Berrios, M. G., Berger, T., Tischler, M. B., Juhasz, O., and Sanders, F. C., "Hover Flight Control Design for UAS Using Performance-based Disturbance Rejection Requirements", presented at the AHS International 73rd Annual Forum & Technology Display, Fort Worth, TX, May 9-11, 2017.

[8]Akaike, H., "Some Problems in the Application of the Cross-Spectral Method," Spectral Analysis of Time Series, edited by B. Harris, Wiley, New York, 1967, pp.81-107.

[9]Gennaretti, M., Gori, R., Serafini, J., Cardito, F., and Bernardini, G., "Identification of Rotor Wake Inflow Finite-State Models for Flight Dynamics Simulations," CEAS Aeronautical Journal, Vol. 8, No. 1, 2017, pp. 209230.

[10]Hersey, S., Celi, R., Juhasz, O., Tischler, M. B., Rand, O., and Khromov, V., "State-Space Inflow Model Identification and Flight Dynamics Coupling for an Advanced Coaxial Rotorcraft Configuration" , presented at the AHS International

73rd Annual Forum & Technology Display, Fort Worth, TX, May 9-11, 2017.

[11]Knapp, M. E., Berger, T., Tischler, M. B., and Cotting, M. C., "Development of a Full Envelope Flight Identified F-16 Simulation Model", AIAA Paper 2018-0525, 2018.

[12]Berger, T., Tischler, M. B., Knapp, M. E., and Lopez, M. J. S., "Identification of Multi-Input System in the Prescence of Highly Correlated Input", AIAA Journal of Guidance, Control, and Dynamics, June 27, 2018.

[13]Berger, T., Tischler, M. B., Knapp, M. E., and Lopez, M. J. S., "Identification of Multi-Input System in the Prescence of Highly Correlated Inputs", presented at the Atmospheric Flight Mechanics Conference at AIAA SciTech, San Diego, CA, January 7-11, 2019.

[14]McRuer, D. T., Graham, D., and Ashkenas, I., *Aircraft Dynamics and Automatic Control*, Princeton University Press, Princeton, NJ, 1973.

An Integrated Simulation Platform to Validate Safety of Autonomous VTOL Vehicles

Bernard DION
Chief Technical Officer
ANSYS Systems
Canonsburg, PA, USA

Nicolas DALMASSO
Innovation Director
ANSYS Optis
Sophia-Antipolis, France

ABSTRACT

Autonomous vehicles systems rely upon sensors and embedded software for localization, perception, motion planning and execution. Autonomous vehicle systems can only be released after developers have demonstrated their ability to achieve the required level of safety. Today's hands-off autonomous vehicle systems are largely built with deep learning algorithms that can be trained to make the right decision for nearly every situation. These systems, however, lack the detailed requirements and architecture that have been used until now to validate safety-critical software, such as that which controls commercial aircrafts. Physical testing is clearly an essential part of the development process, but billions of miles of road testing or flight hours would be required to validate the safety of autonomous systems and software. The autonomous vehicle open simulation platform that is presented in this paper addresses this challenge by integrating physics, electronics, and embedded systems and software to accurately simulate complete autonomous vehicle systems in a fraction of the time and cost required for physical testing.

INTRODUCTION

The ANSYS autonomous vehicle open simulation platform addresses the challenge of validating autonomous vehicle safety and performance by integrating physics, electronics, and embedded systems and software to accurately simulate complete autonomous vehicle systems in a fraction of the time and cost required for physical testing.

ANSYS capabilities span the simulation of all sensors, including lidars, cameras, radars, and ultrasonic sensors; the multiphysics simulation of physical and electronic components; the analysis of system functional safety; as well as the design and automatic code generation of safety-certified embedded software. Sensor simulation can be integrated into a closed-loop simulation environment that interacts with road traffic or flight simulators, enabling several hundred thousand scenarios to be executed virtually.

ANSYS medini analyze helps to manage the safety validation process by implementing key safety analysis methods such as failure modes and effects analysis (FMEA) or Fault Tree Analysis (FTA) while supporting safety analysis and design according to the required safety standards. This integrated solution will achieve end-to-end safety in deep-learning-based and other autonomous vehicle systems, while integrating complementary safety analyses such as the ones proposed on in the Safety of the Intended functionality (SOTIF) standard preview [1].

Presented at the Vertical Flight Society Autonomous VTOL Technical Meeting and Electric VTOL Symposium, Mesa, Arizona, UAS, Jan. 29-31, 2019.

CHALLENGES TO LARGE SCALE AUTONOMOUS SYSTEMS ADOPTION

The greatest obstacle that needs to be overcome prior to large-scale rollout of hands-off fully autonomous vehicles is developing software capable of safely handling every possible scenario without intervention from the operator. For example, consider the challenge of identifying any imaginable pedestrian, animal, vehicle, or other obstacle that might suddenly appear on a road or in the sky ahead, along with the endless number of scenarios that might affect the information received by the automated system's sensors such as weather conditions, lighting, haze, obstructed views, etc.

Today's autonomous systems typically address this challenge with **machine learning/deep learning software** that can be trained to recognize patterns without having to be specifically programmed for every possible situation that could arise [2].

The SAE Industry Survey about "Insight into the Evolution of Autonomous Vehicles" [3] found that "public confidence and adoption" is the number one barrier to widespread adoption of fully autonomous vehicles. Autonomous vehicle systems based on deep learning can only be released to the public after manufacturers have proven that they meet today's high safety standards. How should this software be validated?

Traditional safety validation methods are no longer valid

Safety-critical software that has been used until now, such as the software controlling advanced driver assistance systems (ADAS) or flight-control systems (FCS), are validated using the system and embedded software lifecycle models defined in the ISO 26262 or ARP 4754/DO-178 safety standards.

As described in Figure 1 below, with this approach, engineers specify the detailed requirements of the systems, define a software architecture, and create subsystems in parallel to address each of these requirements.

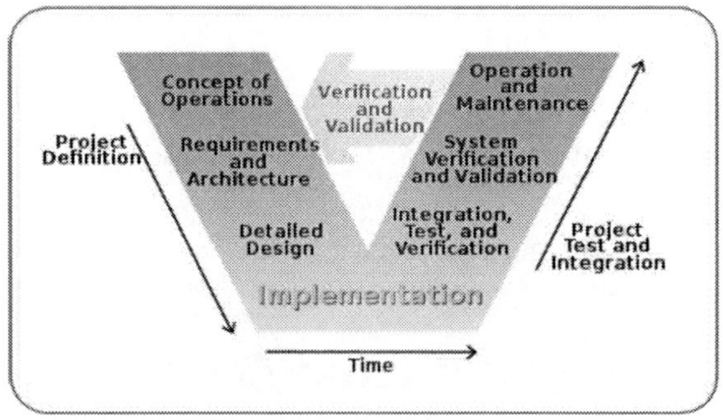

Figure 1. Lifecycle model as defined in ISO 26262 (automotive) or ARP 4754/DO-178 (aeronautics)

The left side of the V shows how the system is broken down into smaller and smaller subsystems. The right side of the V shows the process of validating the system, starting with the smallest subsystems and climbing the V to a system-level validation.

Traditional safety validation methods are no longer valid

Physical testing such as road testing or flight testing is clearly an essential part of development but it does not provide a viable solution to the question of how to validate autonomous systems. The problem is that autonomous system testing primarily consists of routine situations that will not be sufficient to identify all potential situations that may lead to failures. Akio Toyoda, President of Toyota Motors, said that fully autonomous vehicles would have to be driven 8.8 billion miles to ensure their safety.[1] Furthermore, this enormous test driving program would only serve to validate a single autonomous driving system. If you change your code or experience a failure, you must reset your mileage count to zero and start over again.

The autonomous systems industry needs closed-loop simulation in which a virtual vehicle is driven in a realistic virtual world by real autonomous vehicle software. Simulating every detail of autonomous driving is an enormous challenge that requires accurate simulation of a wide range of elements:

- Obstacles such as pedestrians and other vehicles

[1] https://www.forbes.com/sites/alanohnsman/2016/10/03/toyotas-robot-car-line-in-the-sand-8-8-billion-test-miles-to-ensure-safety/#20144c716f01

- The road or the sky and its state, such as being drenched in rain
- The vehicle including its mechanical characteristics, state, location and behavior
- The environment: level of visibility, material properties, weather conditions etc.

THE REGULATORY CONTEXT

Figure 2 below describes the regulatory context for Autonomous cars, which we will now consider.

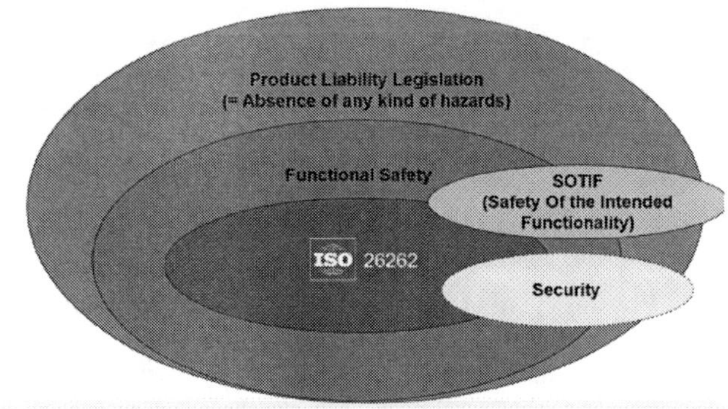

Figure 2. The regulatory context for automotive

The ISO 26262 [4] automotive safety standard is taken as the basic safety standard that relates to functional safety, meaning prevention of accidents caused by E/E control systems. This means prevention of failures and malfunctions, or detection at runtime and taking the system into a safe state. Functional safety analysis tracks malfunctions down to either defective hardware parts or design errors.

In addition, the SOTIF standard (PAS 21448) [1] is currently being created. The scope of SOTIF goes beyond the E/E systems and considers classes of errors due to limitations of performance linked to the relationship between sensors and perception algorithms that can be based on deep learning and neural networks, and also errors due to mismatch between the autonomy system and a human operator for systems that are not fully autonomous.

As discussed above, machine learning algorithms are hard to implement and verify according to safety standards such as ISO 26262 (or DO-178C) [5]. As illustrated by Figure 3 below, accidents of autonomous vehicles often cannot be tracked down to a single root cause but are caused by misfit of environment and system assumptions.

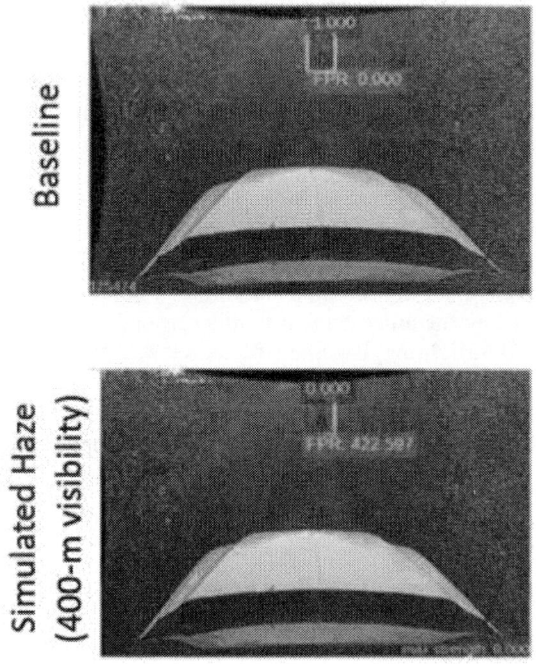

Images: Carnegie Mellon University

Figure 3. Pedestrian detection is not performing after introduction of a low level of haze

PAS 21448 (SOTIF) only covers the lower levels of autonomy, L1 and L2. It should be later extended to cover the upper levels of autonomy, L3 to L5.

Similar efforts now must be undertaken by the aeronautics industry and the aeronautics certification authorities (FAA, EASA, etc.) to cover similar problems for autonomous VTOL vehicles. This safety standard could typically complement the ARP 4754, ARP 4761, DO-254 and DO-178 regulatory framework.

It should be noted that this is a very difficult topic when the goal is to achieve full autonomy and it may take several years to find a solution to it and reach consensus about it.

THE ANSYS INTEGRATED SIMULATION PLATFORM

As illustrated by Figure 4 below, ANSYS provides the first complete simulation environment which includes virtual world models and operating scenarios, physically accurate sensor models, AV embedded software development tools and vehicle dynamics (flight dynamics for airplanes) and functional safety and cybersecurity analysis.

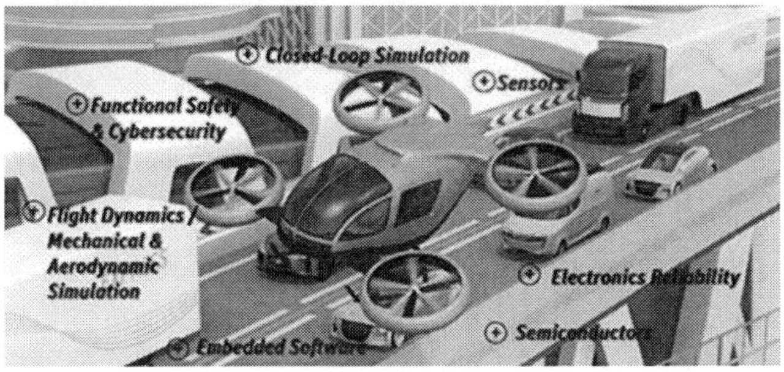

Figure 4. ANSYS is building the only comprehensive solution for simulating autonomous vehicle miles driven or flight hours, flown or maneuvered that can accurately test millions of scenarios

Figure 5 below, illustrates both initial open loop simulation and closed loop simulation.

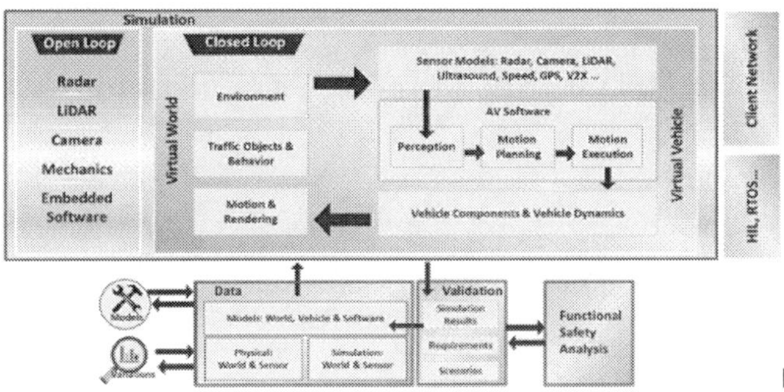

Figure 5. The ANSYS autonomous vehicle simulation platform

Open-loop simulation is used to prepare for Closed-loop simulation of the sensors (see next section on sensors).

Once open-loop simulation has been achieved, closed-loop simulation can be used to run physics-based simulation of the vehicle within its environment (flight dynamics), the sensors (radars, camera, lidars, etc.) and the AV embedded software. Regarding flight dynamics, a particularly difficult Fluid Dynamics problem has to be handled for the Vertical Take-Off and Landing phases; this may be tackled by specific uses of ANSYS Fluent. Closed loop simulation allows virtual testing with Software-in-the-Loop (SIL) simulation and Hardware-in-the-Loop (HIL) testing.

A Data layer has been introduced to manage all the data related to simulation (simulation models, simulation scenarios, simulation results, sensor data coming from real-world, etc.).

Functional safety analysis is acting as a cockpit for the whole simulation environment.

Figure 6 below illustrates simulation of a drone in a urban virtual environment that includes physical effects such as reflexion of light on glass buildings for camera sensors detections and effects of wind on flight dynamics.

Figure 6. ANSYS VRXPERIENCE closed-loop simulation of a drone in a urban environment, including flight dynamics and wind.

Figure 7 below illustrates the vertical takeoff of a drone in a urban environment. Vertical takeoff is driven by the output of camera sensors installed on the drone.

Figure 7. Drone vertical takeoff

The ANSYS interactive autonomous systems simulator considers all the interactions of the autonomous vehicle with the environment, the obstacles, and provides guidance on how to engineer the vehicle to ensure its compliance with safety regulations and customer expectations. Based on realistic operating conditions, this autonomous vehicle simulator makes the same reliable decisions as the future real-world connected vehicle.

Simulating the sensors

By providing a realistic physics-based sensor response in real time for cameras, lidars, radars and ultrasonic sensors, the ANSYS simulation platform provides engineers with all the information needed to create the safest autonomous systems. Autonomous vehicles require increasing numbers of optical sensors, including cameras for visible and infrared detection and lidars for a 360-degree 3D view of the operating environment.

As an example, ANSYS SPEOS optical simulation software validates the performance of the optical sensors through physics-based simulation that accurately represents their real-world performance. SPEOS enables engineers to perform a detailed physics simulation of optical cameras and lidars, taking optical lenses, mechanics, sensors, materials and light properties into account and merging images obtained by multiple cameras.

Next, SPEOS simulates the in-vehicle installation to determine the impact of the installation on images produced by the cameras and lidars. Finally, a reduced order model (ROM) of the camera and lidar models are integrated into an operating scenario that provides the performance needed for real-time simulation. The simulation accurately duplicates the images generated by the cameras and lidar, tracing rays of light as they enter the sensors to identify factors such as sun glare and reflections on (glass) buildings while validating the perception of the sensors.

This is illustrated by the Figure 8 below.

Figure 8. From left: 1) Detailed camera component simulation with SPEOS; 2) Simulation of in-vehicle installation of the cameras with SPEOS; 3) Simulation of camera sensor performance with VRXPERIENCE

A similar process applies with ANSYS HFSS for radars.

Managing the safety validation process

ARP 4761 [6] requires that aeronautics safety engineers evaluate and assess the safety of aircraft systems by identifying relevant hazards and failure modes for electronic components.

As described above, the question of how to develop autonomous VTOL vehicles shall be more comprehensively addressed in an additional standard, playing a similar role to SOTIF to provide guidelines for documenting operating scenarios, safety analysis of those scenarios, the verification of both safety scenarios and triggering events, i.e.explicit set of conditions that initiate a failure condition, and the validation of the vehicle in its environment.

The complete system safety analysis is described in the Figure 9 below which proposes model-based safety analyses, all organized around a model of the system architecture.

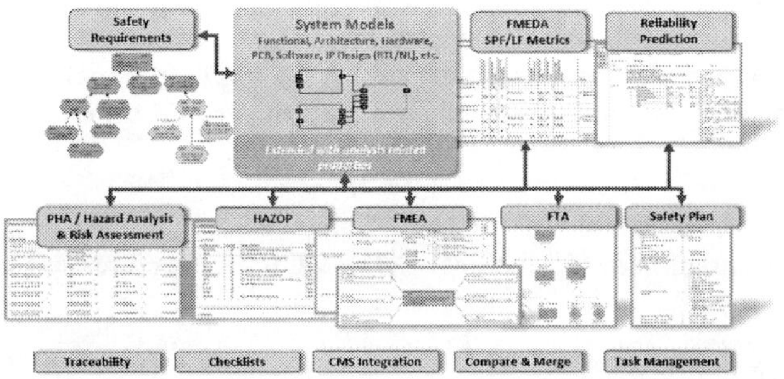

Figure 9. The ANSYS medini analyze model-based functional safety analysis environment. Just as the ISO 26262 version of medini is currently augmented with SOTIF analyses, the ARP 4761 version of the tool will be augmented with the specific autonomous VTOL safety regulation, when available

Overall, we propose the following organization of the Product Lifecycle:

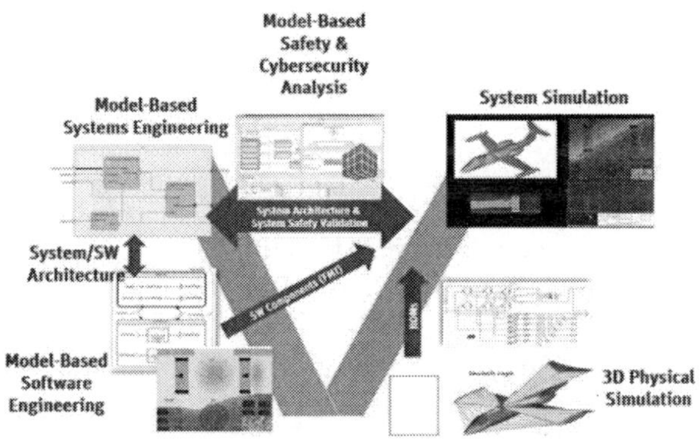

Figure 10. The proposed V cycle, which includes system architecture and embedded software development, detailed 3 Physics simulation of the sensor models and production of reduced order models (ROMs) for system-level simulation. Functional safety analysis with medini sits in the middle of the V cycle

Model-based design of the AV embedded software

– Overhauling the control loop

The Autonomous flight control loop is described in Figure 11 below. It includes perception taking as input data coming from sensors and performing fusion, planning to determine a flight plan and execution to drive the actuators.

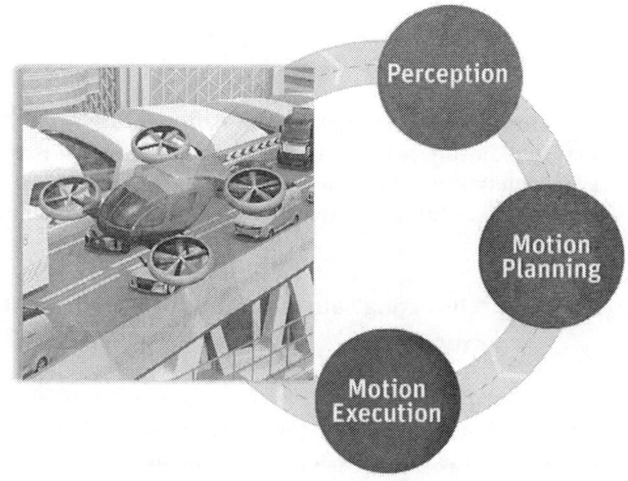

Figure 11. The automated flight control loop

A combination of machine learning / deep learning and classical control logic is being used to implement a fully autonomous vehicle control loop. This control loop is executed in cyclic fashion so that the vehicle can respond to constant environmental changes.

– Defining the control logic

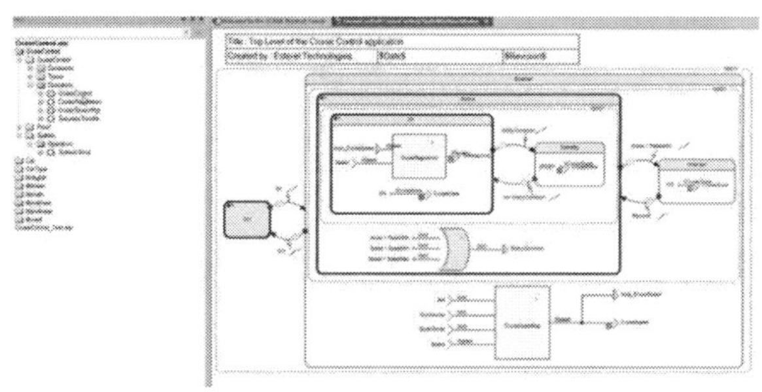

Figure 12. Designing control logic with ANSYS SCADE model-based software engineering tools

In order to design and generate code for the control logic, ANSYS developed a comprehensive solution capable of automatically generating code from software models for autonomous vehicles. The ANSYS SCADE Suite KCG code generator has been qualified for DO-178/DO-330 at TQL 1 (for developing Level A applications).

The SCADE Suite KCG code generator takes as input a SCADE model made of a combination of state machines and data flows and produces as output an equivalent C program.

The KCG code generator, used together with SCADE Test, will ensure that the embedded software meets the DO-178C safety standards objectives in a very efficient way [7]. It will also help to drastically reduce the development and certification time.

– Architecting safe autonomous vehicle control software

To verify the safety of the complex algorithms used for perception, motion planning and execution, engineers must break down the overall autonomous vehicle architecture into a meaningful set of components.

A Carnegie Mellon University research team proposed an architecture that will guarantee safety for each of these components based on a DOER-CHECKER principle [8], [9].

The detailed architecture is composed of a primary algorithm (DOER) that may be extremely complex, undergo frequent updates and be difficult to verify. This primary algorithm is paired with a corresponding safing gate (CHECKER) that verifies that the outputs of the primary algorithm are correct. If the safing gate detects a problem, it transfers control to a safing channel algorithm that produces a short-duration mission, such as pulling the car to the side of the road.

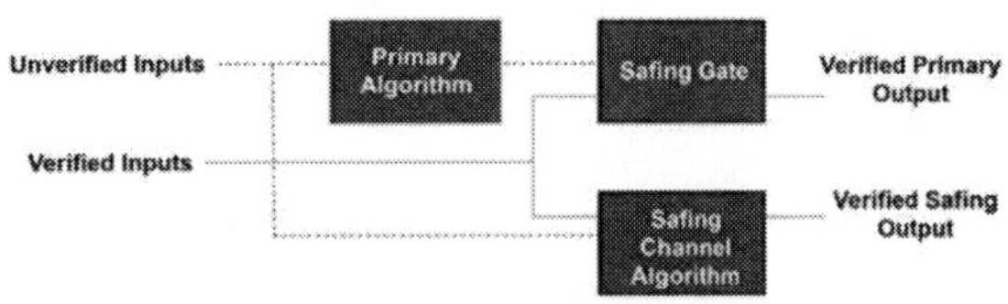

Figure 13. The primary channel produces a long-duration mission with no defined end state, while the safing channel produces a short-duration mission that ends in a safe state.

The detailed safety requirements of the safing gates can be established so that their implementation satisfies the highest-level safety requirements of ASIL D in ISO 26262 or DAL A in DO-178C, avoiding the need for the DOER to meet these requirements. The DOER can fail arbitrarily, do wrong things in the worst possible way, because the CHECKER turns the DOER into a fail silent component that shuts down whenever it does not produce correct data. For example, if the safing gate (CHECKER) detects a planning failure such as -- the planned path intersects with an obstacle -- it shuts down the planning algorithm (DOER).

As illustrated by Figure 14 below, this can be seen as an instantiation of the classical COM-MON architecture.

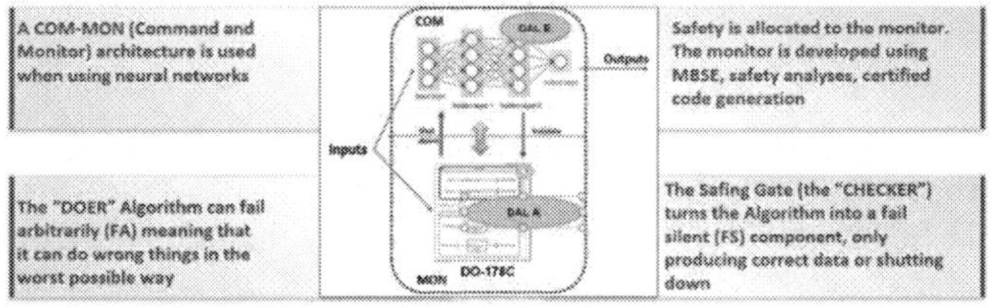

Figure 14. The safing architecture implemented through COM-MON.

CONCLUSIONS

The ANSYS autonomous vehicle simulation platform can verify the safety of complex autonomous systems by integrating physics, electronics, embedded systems and software simulation. The operating scenario model animates the motion of the test vehicle and other vehicles and objects in a test run. Sensor models observe the surroundings in the virtual world and output sensor signals. Signal processing models and deep learning identify objects, autonomous vehicle position and operating conditions from sensor data. Control algorithms make control decisions, generate actuator inputs and display information and decisions to the passenger/operator. Vehicle component models use actuator inputs such as flight controls and throttle commands. The flight dynamics model computes the position, velocity and orientation of the test vehicle.

This integrated platform makes it possible to quickly and economically simulate any vehicle, with any combination of sensors, with any control system in any operating scenario, thus drastically reducing the time that will be needed to achieve both the performance and safety goals for these applications.

Authors contact:

Bernard Dion bernard.dion@ansys.com, Nicolas Dalmasso nicolas.dalmasso@ansys.com

REFERENCES

[1] ISO/PAS 21448:2019 Preview, Road vehicles -- Safety of the intended functionality, SAE, 2019.

[2] Ian Goodfellow, Yoshua Bengio, Aaron Courville, *Deep Learning*, MIT Press, 2016.

[3] SAE Survey: "Insight into the Evolution of Autonomous Vehicles", SAE, 2018, www.ansys.com/campaigns/autonomous-engineering/sae-survey.

[4] ISO 26262, ed. 2, "Road vehicles – Functional safety", SAE, 2018.

[5] DO-178C, DO-178C, Software Considerations in Airborne Systems and Equipment Certification, RTCA, 2012.

[6] ARP 4761, Guidelines and methods for conducting the safety assessment process on civil airborne systems and equipment, SAE, 1996.

[7] Bernard Dion, "A Cost-Effective Model-Based Approach for Developing ISO 26262 Compliant Automotive Safety Related Applications", SAE World Congress, SAE, 2016.

[8] Kane, Chowdhury, Datta and Koopman, "A Case Study on Runtime Monitoring of an Autonomous Research Vehicle (ARV) System," RV 2015.

[9] Koopman, P., Wagner, M., "Toward a Framework for Highly Automated Vehicle Safety Validation," SAE World Congress, SAE 2018.

Reinforcement Learning Control for Quadrotors using Snapdragon Flight

Eric Solomon
Graduate Research Assistant

Abhishek Shastry
Graduate Research Assistant

Vikram Hrishikeshavan
Assistant Research Scientist

Inderjit Chopra
Alfred Gessow Professor and
Director of the Alfred Gessow Rotorcraft Center

Alfred Gessow Rotorcraft Center
University of Maryland
College Park, Maryland, U.S.A

ABSTRACT

As unmanned aerial systems(UAS) begin to operate in more dynamic and stochastic environments as presented by urban setting, new problems for tracking and control are introduced that must be addressed. Reinforcement-Learning(RL) techniques for control combined with deep-learning are promising methods for aiding UAS in such environments. This paper is an exploration of use of some of the popular Reinforcement Learning algorithms in aerospace, that have already been successfully used in non-aerospace applications such as self-driving cars. The vehicle chosen for this study is a quadrotor. The simulation results demonstrates that a RL controller performs slightly better than PID (Proportional-Integral-Derivative) in well-controlled deterministic environments and hence shows potential for more rigorous testing in highly uncertain environments. Also, some estimations of the processing capability required for these algorithms is presented and is used to demonstrate that these can be executed on some of the commercially available flight computers for drones such as Qualcomm Snapdragon Flight™.

*

Nomenclature

ACKTR	Actor-Critic using Kronecker-Factored Trust Region
API	Application Programming Interface
DQN	Deep Q-network
ELKA	Enhanced Lightweight Kinematic Autopilot
FFNN	Feed Forward Neural Network
LSTM	Long Short-Term Memory network
MDP	Markov Decision Process
NLP	Natural language Processing
PID	Proportional Integral Derivative controller
PPO	Proximal Policy Optimization
RL	Reinforcement Learning
ROS	Robotic Operating System
TRPO	Trust Region Policy Optimization
VREP	Virtual Robot Experimentation Platform

INTRODUCTION

Commercial drone is one of the fastest growing areas and as per the Technavio's market research and analysis report (Ref. 1), it is forecasted to reach $15 billion in value by 2022. The vehicle's applications include geospatial mapping, inspection of civil structures, crowd security and surveillance, aerial photography, search and rescue, and cargo delivery. It is evident that most of these tasks will require safe autonomous operations in an urban setting. The urban environment is very diverse, rapidly changing and largely unpredictable(stochastic), unlike the high skies where most aerospace vehicles have been operating for more than a century. Hence there is a requirement for the development and validation of robust adaptive navigation and control tools, more complex than the conventional PID controller.

Quadrotors have proven to be an important benchmark for testing flight controls using complex control laws. It is mechanically simpler than a conventional helicopter while its dynamics is relatively more straightforward and intuitive than other types of drones. This greatly simplifies the process of controller development for this vehicle. Because many of the popular simulation techniques for adaptive control often require a large amount of memory and processing capability, these techniques must be adapted with the computationally limited real-world hardwares in mind such as Qualcomm Snapdragon Flight™ (Ref. 2). This paper begins with the first portion of the controller design process: simulation.

We explore two types of reinforcement learning(RL) control in this paper, one which requires action (or control) to be a small discrete set and the other which requires it to be continuous. We find that it is best to use continuous action control policies and that neural training is most feasibly carried out by separating the quadrotor's translational and rotational degrees of freedom. The results are merely an exploration of the usage of reinforcement learning for quadrotor control.

Presented at the Vertical Flight Society Autonomous VTOL Technical Meeting and Electric VTOL Symposium, Mesa, Arizona, USA, Jan. 29-31, 2019.

BACKGROUND

Neural networks as a tool have been popular recently for both model prediction and control. Punjani and Abbeel (Ref. 3) showed the ability of feed-forward neural networks to predict quadrotor's trajectory over a short horizon successfully. Ogunmolu and Olalekan (Ref. 4) has shown that deep neural networks are effective model predictors from input-output data. Neural networks with layers originally proven useful for time-series prediction and Natural Language Processing (NLP), namely recurrent and Long Short-term memory networks (LSTM), have shown improved performance in model predction (Ref. 5). Recently, Lou and Guo (Ref. 6) have demonstrated in simulations, that reinforcement learning algorithms can provide good trajectory tracking even in the presence of strong gusts. RL has a few proven real world applications such as the program AlphaGo Zero by Google Deepmind that mastered and defeated world champions in the game of GO (Ref. 7) and a robot that learned how to flip pancakes from human demonstrations (Ref. 8).

The basis of all Reinforcement Learning (RL) algorithms employing neural networks for control is Markov Decision Process (MDP), which is represented by the following figure:

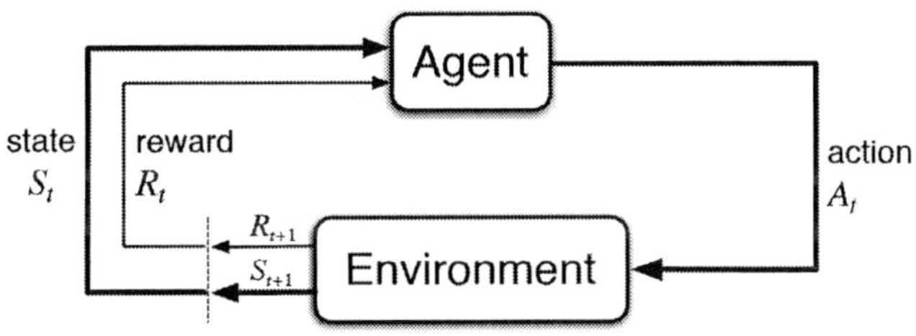

Fig. 1. Markov Decision Process (Ref. 9)

The environment in MDP is the external system to be controlled. The agent is the program executing actions on environment and receiving rewards and the current state of the environment in return. Action by an agent is same as control input to a system in classical control terminology. A policy in MDP is any function which outputs an action for a given state of the environment. It is to be noted that an agent can either follow a policy or act randomly, which is common during training phase of RL algorithms. All RL algorithms are focussed on finding optimum policy functions for any given environment.

RL controllers have primarily two benefits over traditional analytical controllers namely model-free algorithm and emergent behaviour.

Model-free algorithm: RL controller design does not require the knowledge of system's dynamics apriori. It learns the dynamics during training phase. Hence by nature, it can adapt to changes in plant's dynamics if the training is continued online. This adaptive nature of RL controllers, also make it more robust than PID controllers. It can by design tackle problems like gust, turbulence, ground effect (on one or multiple rotors), partial actuator failure, etc.

Emergent behaviour: For RL controllers, system behaviour emerge during training so as to perform the task as decided by the reward function optimally. For PID these are by design limited to the linear regime and are fixed throughout the system operation. The behaviour of an RL system can be nonlinear and change as the environment changes. This comes with the downside that since the behaviour is not known apriori, it could be undesirable in environments the algorithm has not been trained before.

Guided-policy learning has been a stepping stone towards fully model-free real-time neural control. An offboard trained controller has shown performance benefits at test time. This type of controller has performed faster at test time and maintained good performance in stochastic and unstructured environments (Ref. 10). Simulation results have shown the simple DQN (Deep Q-networks) reinforcement learning algorithm to perform on par with and occasionally better than human pilots (Ref. 12).

Recent reinforcement learning (RL) algorithms have been focused on speeding up the training phase. Techniques such as neural network freezing and experience replay have shown to be beneficial (Ref. 13) for controller performance over time. TRPO (Ref. 14) and PPO (Ref. 15) present better bounds on the approximation performance function versus the actual performance function, which has led to significant improvements in controller performance. Hierarchical learning techniques show the benefit of simplifying control actions into known solutions (Ref. 16). Recently, the gradient noise scale of a training dataset has shown to be directly related to the difficulty of training with a particular dataset batch size (Ref. 17).

METHODOLOGY

This section describes the designs for the reinforcement learning quadrotor controllers present in this study. First is a description of the controller structure as a connectionist system. Following that is a description of the adaptive algorithm.

Controller structure

This simulation experiment presents a comparison between two popular deep reinforcement learning algorithms: Deep Q-Network (DQN) and Proximal Policy Optimization (PPO). DQN is a value iteration method of RL, that is to say it learns to approximate the value of each (state, action) pair in the environment as represented by a neural network. The agent in the algorithm then typically follows the greedy policy of taking the action with the highest value (Ref. 18). DQN has been successfully used for solving many atari-style games and even to simulate the optimum move of a car stuck in traffic (Ref. 19). However, DQN requires the action set to be discrete, which has significant limitations for continuous real-time systems such as quadrotor. This is overcome by another type of RL algorithms known as policy gradient algorithms in which the policy is represented as another neural network. This type of RL algorithms has been used for autonomous driving in cars (Ref. 20), optimization of chemical reactions (Ref. 21), etc. This paper uses one such policy gradient algorithms known as Proximal Policy Optimization (PPO). PPO is presented here in actor-critic style and hence requires two neural nets, one for the policy (the actor) and another for the value (the critic) of each state.

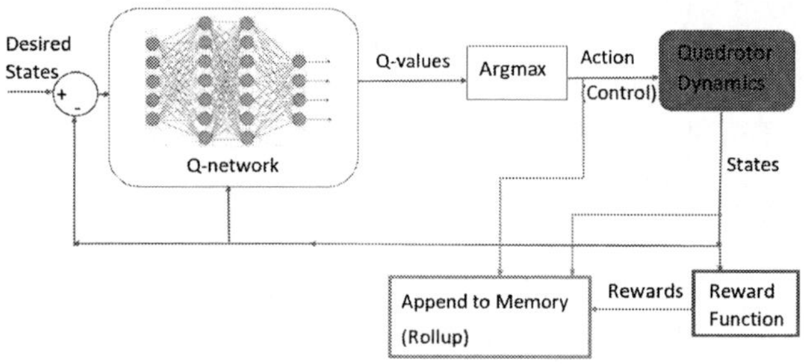

Fig. 2. Deep Q-Network structure. The neural network takes the vehicle state as input and calculates the Q-value of all actions. Under greedy policy, the action with max Q-value is executed.

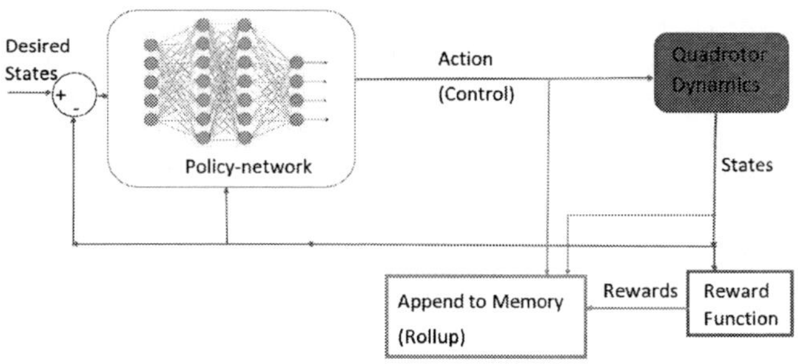

Fig. 3. Proximal Policy Optimization structure. In addition to the value network(as in DQN), policy is also represented by a neural net. The output of policy network is used for control and while the value network is used to train the policy network.

The simulation experiment is designed to test the ability of RL algorithms to control a hovering quadrotor. Neural networks used anywhere in the controller are simple feed forward neural networks. The input data set being used by both the controllers is the 6 Degree-of-Freedom (DoF) pose representing the current state, concatenated with the 6 DoF pose error (Fig. 2). The output of DQN network is the value for all the actions. The output of PPO's critc network is the same as DQN network while the output of its actor or policy network is a probability distribution over the continuous actions space. Since DQN requires discrete action space, the continuous action space is discretized for this purpose into 'bins' for every action variable. The bins in this work for every variable is to increase or decrease or no change of motor rpm of quadrotor by a fixed value. This set of bins is chosen because it requires a coarser discretization of the control action space, yet it allows for a high granularity of control actions to be achieved.

Controllers design

This subsection describes the adaptive control algorithm for DQN and PPO. High variance in results is an important problem in reinforcement learning(or deep learning in general)algorithms, because it leads to non-repeatability of results. Hence different techniques used to minimize the effect of variance on controller training is also presented here.

Network Freezing and Data Rollout The first methods of minimizing the effect of variance are called neural network freezing or controller freezing and data rollout. Freezing is the process of keeping an adaptive controller at a static state. At the same time, data is being collected in the form of the next 'rollout'. A rollout is conceptually the same as a line search in the reinforcement learning sense. It is a collection of data representing the previous test sample. In our case, the rollout is constructed of 1000 data points or a full test to the failure/success point, depending upon which occurs first. By not calculating an adaptive step after each state update, and instead calculating the adaptive step based on the expected value of a batch, the effect of noisy data points is filtered out. An alternative for rollout construction that is not explored in this paper is to keep a sliding filter representing the current rollout. This allows the adaptive step to be calculated at a quicker rate.

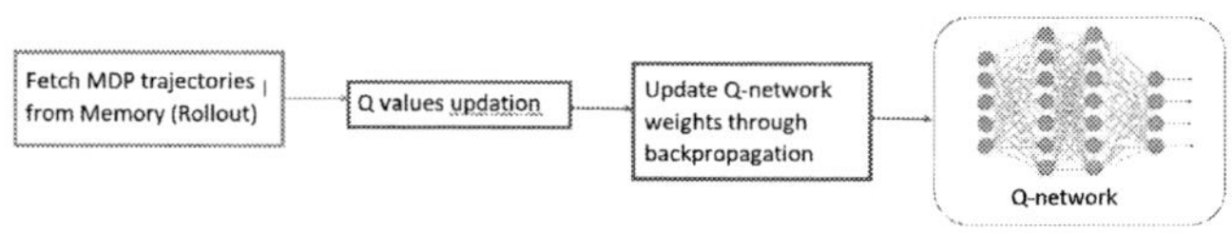

Fig. 4. DQN weight update. Training data for the neural nets is generated by executing the current policy for N(=1000) iterations. This is then used to to train the neural nets in the form of data rollout from memory.

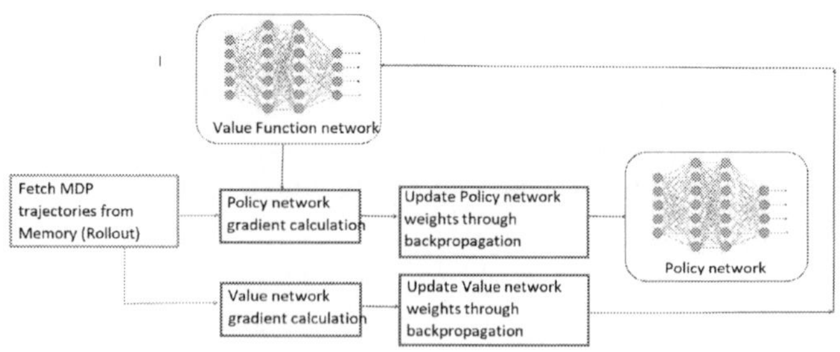

Fig. 5. PPO weight update. Training data for the neural nets is generated by executing the current policy for N(=1000) iterations. This is then used to to train the neural nets in the form of data rollout from memory.

```
function CONTROLANDROLLUP ▷ Control freezing and data roll-up
    while in flight do
        N ← 1000
        v ← []                              ▷ Memory of experiences
        for 1 to N do
            s ← quadstate        ▷ control is via DQN or PPO
            r, a, s_next ← control(s)   ▷ Reward and Action from Control
            v ← v :: [r, a, s, s_next]      ▷ Append to memory
        end for
        NORMALIZE(v)
        REINFORCE(v) in parallel   ▷ reinforce is also via DQN or PPO
    end while
end function
```

Parameter freezing The next method used in this paper to minimize the effect of variance is called parameter freezing. Thus, the controller neural networks are trained to integrate more degrees of freedom into control as time goes on. For this study, non-integrated degrees of freedom are controlled by a simple linear gain controller. Ideally, the controller will be fully model-free. However, this was infeasible for a preliminary study. The reinforcement learning controller integrates degrees of freedom in the order shown below.

Parameter freezing/training in steps

1. Planar (horizontal) position
2. Vertical position
3. Roll/Pitch
4. Yaw

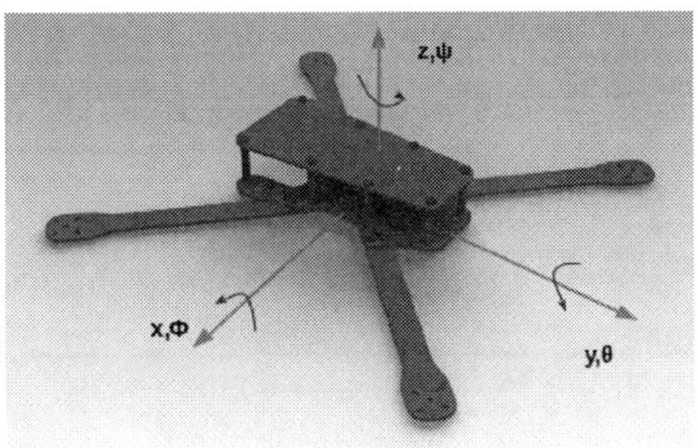

Fig. 6. Motion convention used for the quadrotor in V-REP simulation environment. For DQN's restricted DOF simulations, all linear motions are restricted, and rotational motion in z is restricted for 2-DOF simulation and additionally motion in x is restricted for 1-DOF simulation.

Data normalization Data and parameter normalization has shown to be an extremely effective and very simple method of dealing with data variance. The form of normalization known as 'batch normalization' is shown below (Ref. 22). It is performed using the data from the previous data rollout.

Input: A set of values of x $(x_1, x_2, ..., x_m)$
$$\text{BatchNormalize}(x_i) = \frac{\gamma_b(x_i - \mu)}{\sqrt{\sigma^2 + \varepsilon}} + \beta$$
where, $\mu = \frac{1}{m}\sum_i x_i$, $\sigma^2 = \frac{1}{m}\sum_i (x_i - \mu)^2$ and γ_b, β are 'learnable' parameters

DQN parameter update The DQN parameter update method relies on the principles of Q-Learning, which is a popular reinforcement learning method that is guaranteed to find the optimal parametrization of a function(the policy) in a metric space with a given state evaluation function or value function. Value functions typically have the following form.

$$V_\pi(s_i) = \sum_{j=i}^{\infty} \gamma^{j-i} r(s_j, a_k = \pi(s_j))$$

where $r(s_j, a_k)$ is the reward obtained for taking action a_k in state s_j and $\gamma < 1$ is a discount factor (a hyperparameter) to indicate rewards in near future have preference over long-term rewards. Q-learning optimizes the policy for a given value function using another function known as Q-value function of a state-action pair and is defined as,

$$Q(s_i, a_k) = r(s_i, a_k) + \gamma V_\pi(s_{i+1})$$

It is evident that under the optimal policy(π^*), which maximizes the value function, the following relation holds:

$$V_{\pi^*}(s_i) = \max_{a_k} Q^*(s_i, a_k) = \max_{a_k}(r(s_i, a_k) + \gamma V_{\pi^*}(s_{i+1}))$$

Henceforth, we get the following:

$$Q^*(s_i, a_k) = r(s_i, a_k) + \gamma \max_{a_k} Q^*(s_{i+1}, a_k)$$

This is the Bellman equation of optimality (Ref. 23) written in terms of Q-value function. The update law for the Q-value function is derived from the bellman equation using fixed-point iteration method

$$Q_{new}(s_i, a_k) = (1 - \alpha) Q_{old}(s_i, a_k) + \alpha(r(s_i, a_k) + \gamma \max_{a_k} Q_{old}(s_{i+1}, a_k))$$

where α is the relaxation factor also known as 'learning rate' in reinforcement learning literatures.

Using this framework, the process of finding the optimal policy is simply reduced to process of finding a good reward function to optimize. Simple reward functions have recently overtaken expertly crafted reward functions in popularity due to their tendency to produce more robust control solutions (Ref. 24). Here, the reward function is the 2-norm of normalized state versus the normalized desired state.

$$n, n' = BatchNormalize(s, s')$$

```
function DQNREINFORCE(rollout vector)
    γ ← 0.95                                        ▷ Discount factor
    α ← 0.7                                         ▷ Learning rate
    N ← size(rollout vector)
    for 1 to N do
        r, a, s, s_next ← rollout_vector.next
        q_target ← r + γ Q_Horizon(s_next)
        UPDATE WEIGHTS(q_target, Q(s), α)            ▷ Weight
update of Q-network
    end for
end function
```

$$R(n,n') = |n-n'|_2$$

The gradient of the Q-value function with respect to the parametrized neural controller is then taken and is used to update with the standard Q-Learning formula.

$$\begin{aligned}\delta w =& \alpha[r+\gamma \max_a Q(s_{i+1},a,w) - Q(s_i,a,w)]\nabla_w Q(s,a,w)w \\ =& w+\alpha(q_{target}-q)\delta wq\end{aligned}$$

PPO parameter update The PPO parameter update method simultaneously updates the controller (actor) and control action evaluator (critic). The update of the evaluator is essentially the same as the update process for the DQN network. In fact, DQN can be thought of as using only the control action evaluator as a controller itself. PPO parameter update makes use of the advantage function instead of the value function for determining controller performance. The important difference is that the advantage function generates a dataset with much lower variance.

```
function PPOREINFORCE(rollout vector)
    ε ← 0.2
    δ ← 0.1
    A ← AdvantageRollup(rollout vector, γ)
    UPDATE WEIGHTS(A)                               ▷
end function
```

```
function ADVANTAGEROLLUP(rollout vector,γ)
    v ← rollout vector
    T ← length(rollout vector)
    A ← []
    for t in T do
        d ← reward(v,t) + γV(s_{t+1}) − V(s_t)    ▷ Advantage
    end for
    for t in T do
        A ← A :: Σ_t^T γ^t d_t                      ▷ Rollup
    end for
end function
```

The policy performance function is the standard PPO clipping function. It is shown below for reference. PPO and its predecessor TRPO rely on the principles of statistical information divergence in order to bound the parameter update step. The bounded region, after simple tuning, has the properties that it takes an adaptive step in the performance maximization direction that is relatively large compared to other algorithms but not too large.

$$L_t(\theta) = min(R_t(\theta)A_t, clip(R_t(\theta)), 1-\varepsilon, 1+\varepsilon)A_t$$

SIMULATION RESULTS

Simulation validation is performed in two stages. The first is on a simple set of reduced degree of freedom non-linear quadrotor rotational dynamics model using only DQN. The second is on a more robust nonlinear quadrotor model described in Virtual Robot Experimentation Platform (V-REP) (Ref. 25). The simplified dynamics is useful as a validation for reinforcement learning on quadrotors in general. The 1-DOF model tests the ability of DQN to control pitch only. All other degrees of freedom have been constrained. The 2-DOF model incorporates control of pitch in addition to roll.

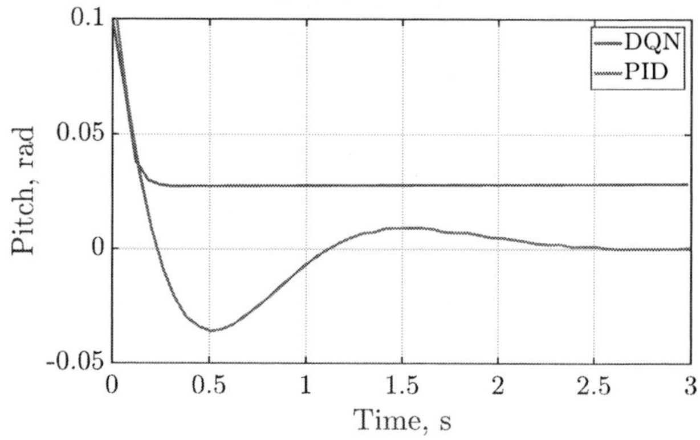

Fig. 7. Validation of Deep Q-Learning on a quadrotor to perform pitch control. The steady state error in DQN is believed to be the result of coarse discretization of continuous controls for quadrotor

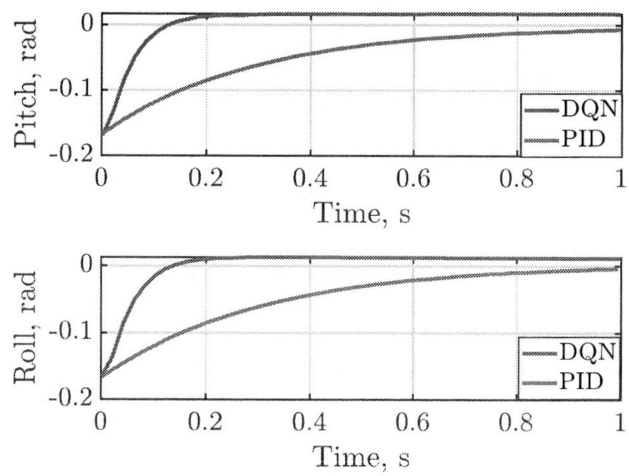

Fig. 8. Validation of Deep Q-learning on a quadrotor for 2d-attitude control. DQN performs only slightly better than linear PID.

As observed from the figures, DQN has atleast similar or slightly better performances to PID controller. The DQN controller though has a steady state error, the source of which is believed to be the discretization of the continuous action set for the controller. In the light of the above results, it is important here to point out that the benefits of RL controllers over traditional analytical controllers as model-free algorithm, being adaptive to changing environments and free from extensive online manual tuning, makes it worthwhile to implement it on flight hardware for validation in future.

Though DQN algorithm demonstrates good performance in the above simulation environment, it cannot be used in real-world environments primarily because it requires the action space to be discrete and finite. To overcome this limitation, continuous action reinforcement learning algorithms are required. Hence, it was decided to incorporate V-REP and the popular OpenAI implementations (Ref. 26) of modern reinforcement learning algorithms. Two continuous action reinforcement learning algorithms, PPO and ACKTR are tested (ACKTR algorithm is not described here, but is accessible through OpenAI baselines (Ref. 29)). For this set of simulation tests, the controller type is modified. A baseline PID controller is created for position and attitude control. This is so that the RL controller model can be easily adapted to the real-world, where avoiding catastrophic failure is of much greater importance.

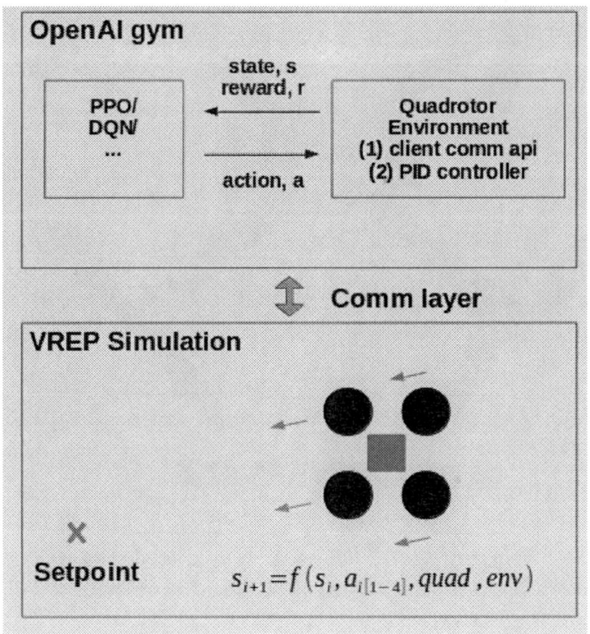

Fig. 9. V-REP + OpenAI quadrotor environment.

Although consistent improvement over PID is not yet shown, initial results suggest that the chosen experimentation method is promising. RL algorithms are run for approximately one million timesteps with time-stepping parameter of 0.001. For both PPO and ACKTR, only about 10 percent of the simulation rollouts ended in a stable quadrotor, however the successful rollouts are spread relatively evenly over epochs with the exception that success does not happen during the first 50 or so rollouts. No pattern for learning beneficial behaviors is immediately visible by taking a selection of the successful flight tests over training epochs. It is suggested that future simulations, experiment with hyperparameter tuning and different reward functions to achieve better results. The present results are not conclusive of successful continuous action RL control, but are nonetheless important. They demonstrate the repeatable and simple process by which a quadrotor with full-state, nonlinear dynamics is able to be tested in simulation. This process is useful for controller and general process design for RL-based quadrotor flight. For example, one may use the V-REP in conjunction with OpenAI simulation environment to examine RL controller learning processes that avoid catastrophic failure by building up control authority incrementally.

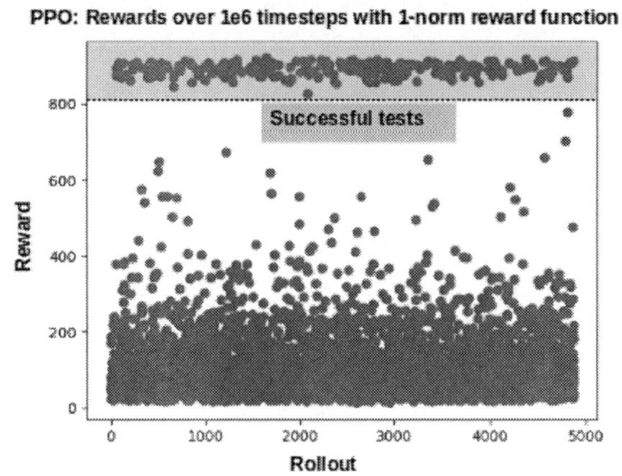

Fig. 10. PPO cumulative rewards over 10^6 timesteps and 5000 rollouts. Rewards over 800 represent tests that end due to reaching a time limit. These results do not demonstrate successful RL control. Hyperparameters tuning(such as learning rate) and different rewards functions are avenues of exploration for better results.

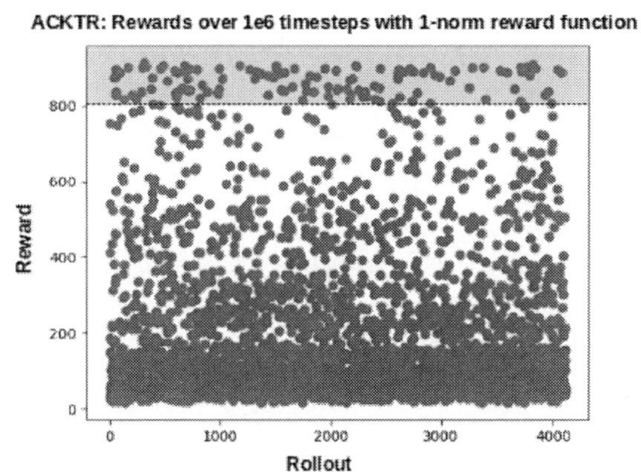

Fig. 11. ACKTR cumulative rewards over 1e6 timesteps and 5000 rollouts. Rewards over 800 represent tests that end due to reaching a time limit.

HARDWARE VALIDATION

A quadrotor platform (Fig.15) utilizing in-house developed ELKA and Qualcomm Snapdragon Flight ™ (Ref. 2) (Ref. 34) has been developed at University of Maryland. Localization using Visual-Inertial Odometry(VIO) has been implemented on Snapdragon Flight, the hardware architecture for which is presented in Figure.12. The position estimation results from VIO are shown in Figure13. A baseline PID controller has been implemented on the hardware and its results are presented in Figure14.

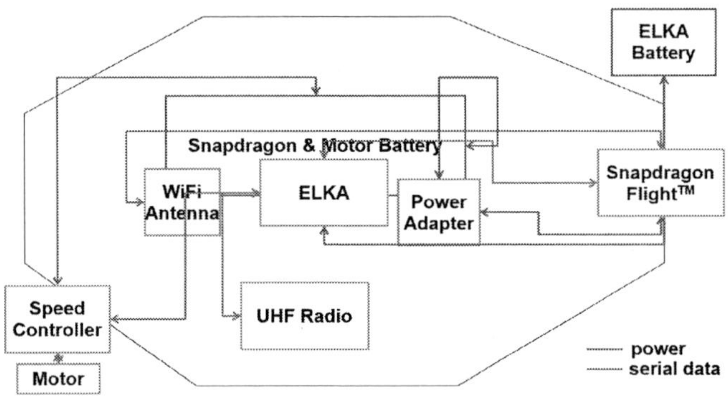

Fig. 12. Hardware architecture for Visual-Inertial Odometry. All the primary hardware components and communication between them is shown here.

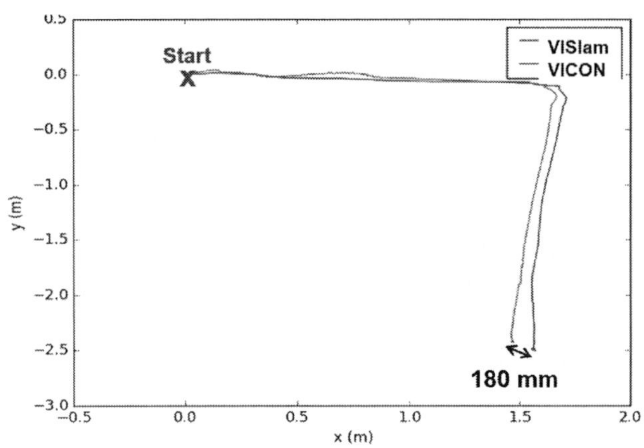

Fig. 13. Results of localization using Visual-Inertial Odometry(VISlam). The Vicon data is assumed as ground truth here. The drift in the VISlam position estimation small enough for it to be used reliably in position control.

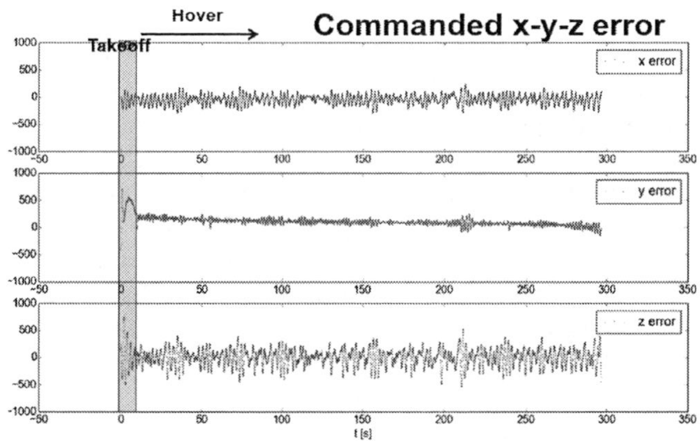

Fig. 14. Results of our naive PID controller for hover position control using VISlam. In order to circumvent tedious gain tuning process, we plan to implement an RL controller within our hardware framework. This will also help with the vehicle's operation in stochastic environments such as in presence of gust, unintended ground effect on one or multiple rotors, partial motor failure, etc.

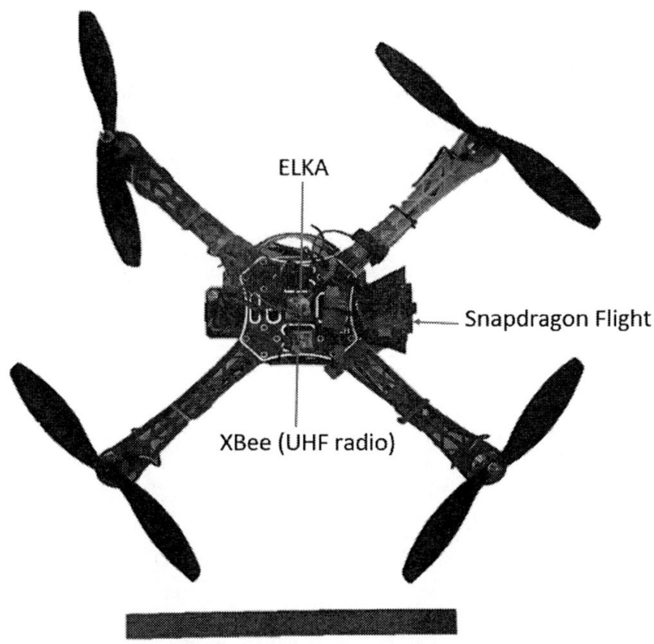

Fig. 15. Snapdragon™-ELKA (Enhanced Lightweight Kinematic Autopilot) controlled quadrotor with a ruler shown for size reference

Future hardware testing will utilize our developed quadrotor hardware using Qualcomm Snapdragon Flight™ coupled with Robotic Operating System (ROS) and Apache MXNet C++ API for tensor processing. Preliminary hardware benchmark tests for the feasibility of a system of symbolic graph computations for neural/tensor processing on an Intel i5-3320M show a PPO algorithm will require maximum operation frequency of 250Hz (Fig. 16). This is sufficient compared to the necessary position control rate of 20Hz and would allow for 100,000 updates in 7 minutes, which is slightly shorter than the battery life of the tested quadrotor in hover. Also, the processor onboard the snapdragon flight runs at 2.26 Ghz which is similar to base frequency of 2.6Ghz of Intel i5-3320M. This suggests that it is possible to run the neural controller at a similar rate of 250Hz snapdragon. The onboard memory is limited to 2GB which might at worst limit the number of parameters of neural nets and the training batch size to be used for online training.

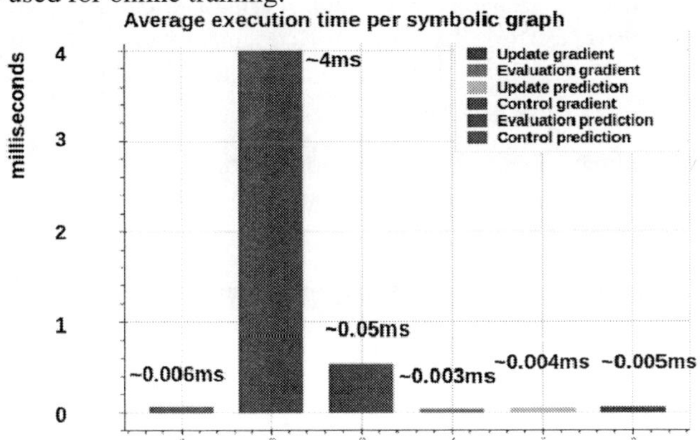

Fig. 16. MXNET PPO implementation time to execute each of the necessary symbolic graph computations. These benchmark tests were performed on a computer with Inteli5-3320M processor, that has a base frequency of 2.6 GHz. This benchmarking result shows that the algorithms can be executed on the Snapdragon at acceptable update rates.

FUTURE WORK

In continuation of this work, the first step is to validate a policy gradient based RL control on quadrotor's attitude dynamics in well-controlled deterministic simulation environment(i.e. without any real world uncertainities such as gust, ground effect, motor failure, etc.). Following this, the method is to be validated on position control of quadrotor in the same environment with tracking performance compared to PID or traditional linear compensators. After that, the controller's performance in stochastic simulation environment such as in the presence of gust, partial motor failure, etc will be tested. All the simulation results will then be verified with experiments on a real quadrotor. It is to be noted here that unlike classical controls, RL control has no proof of guaranteed stability, hence rigorous simulation testing, before hardware testing is even more important to have any kind of confidence on its ability.

CONCLUSION

This paper introduces reinforcement learning algorithms as a viable option for quadrotor control. It discusses their potential benefits over conventional PID controller and presents ways in which RL controllers can tackle problems inherently present in applications of drone in civilian airspace. The paper also describes two popular reinforcement learning algorithms (DQN and PPO) and demonstrate their ability to control a simplified model of a quadrotor in hover. The basis for more robust testing is then established in a repeatable and useful way. Some benchmark results on a conventional computer has been shown, which demonstrates that the amount of processing required for neural control is within the available limit of some of the flight computing boards available in market. This makes it feasible to apply these algorithms in real-time environment on drones. In addition, the paper also discusses methods to create a hybrid controller using both reinforcement learning and conventional PID to mitigate the effects of catastrophic failure on the quadrotor system.

Author contact:Eric Solomon, esolomon1221@gmail.com
Abhishek Shastry, shastry@umd.edu
Vikram Hrishikeshavan, vikram.h1@gmail.com
Inderjit Chopra, chopra@umd.edu

ACKNOWLEDGEMENTS

This work was supported by NAVAIR under the program under the Air Vehicle Test and Evaluation Engineering contract N00421-13-2-0001 with navy lead Mr. James Bumbaugh.

REFERENCES

[1]Technavio, "Global Consumer Camera Drones Market 2018-2022", March 2018, Infiniti Research Limited.

[2]Solomon, E., Hrishikeshavan, V., Chopra, I., (2018). Autonomous Quadrotor Control and Navigation with Snapdragon Flight. 74th Annual American Helicopter Society Forum: Phoenix, AZ.

[3]Punjani A. and Abbeel, P., "Deep learning helicopter dynamics models.", Robotics and Automation (ICRA), 2015 IEEE International Conference on. IEEE, 2015, pp 3223-3230.

[4]Ogunmolu, Olalekan, Xuejun Gu, Steve Jiang, and Nicholas Gans. "Nonlinear systems identification using deep dynamic neural networks." arXiv preprint arXiv:1610.01439 (2016).

[5]Mohajerin, N. and Waslander, S., "State initialization for recurrent neural network modeling of time-series data", International Joint Conference on Neural Networks (IJCNN), Anchorage, AK, 2017, pp. 2330-2337.

[6]Lou, W., and Guo, X. "Adaptive Trajectory Tracking Control using Reinforcement Learning for Quadrotor. International Journal of Advanced Robotic Systems". https://doi.org/10.5772/62128

[7]David Silver, Julian Schrittwieser, Karen Simonyan, Ioannis Antonoglou, Aja Huang, Arthur Guez, Thomas Hubert,

Lucas Baker, Matthew Lai, Adrian Bolton, Yutian Chen, Timothy Lillicrap, Fan Hui, Laurent Sifre, George van den Driessche, Thore Graepel and Demis Hassabis, "Mastering the game of Go without human knowledge", Nature volume 550,pages 354359, 19 October 2017.

[8]Kormushev, P., Sylvain C., and Darwin G. Caldwell. "Robot motor skill coordination with EM-based reinforcement learning." In Intelligent Robots and Systems (IROS), 2010 IEEE/RSJ International Conference on, pp. 3232-3237. IEEE, 2010.

[9]https://towardsdatascience.com/introduction-to-various-reinforcement-learning-algorithms-i-q-learning-sarsa-dqn-ddpg-72a5e0cb6287

[10]Zhang, T., Kahn, G., Levine, S. and Abbeel, P., "Learning Deep Control Policies for Autonomous Vehicles with MPC-Guided Policy Search", 2016 IEEE International Conference on Robotics and Automation (ICRA).

[11]Hwangbo, Jemin, Inkyu Sa, Roland Siegwart, and Marco Hutter. "Control of a quadrotor with reinforcement learning." IEEE Robotics and Automation Letters 2, no. 4 (2017): 2096-2103.

[12]Polvara, R., Patacchiola, M., Sharma, S.K., Wan, J., Manning, A., Sutton, R. and Cangelosi, A. "Autonomous Quadrotor Landing using Deep Reinforcement Lerarning", CoRR, 2017.

[13]Hessel, M., Modayil, J., van Hasselt, H., Schaul, T., Ostrovski, G., Dabney, W., Horgan, D., Piot, B., Azar, M., Silver, D., "Rainbow: Combining Improvements in Deep Reinforcement Learning", 2017.

[14]Schulman, John, Sergey Levine, Pieter Abbeel, Michael Jordan, and Philipp Moritz. "Trust region policy optimization." In International Conference on Machine Learning, pp. 1889-1897. 2015.

[15]Schulman, J., Wolski, F., Dhariwal, P., Radford, A., Klimov, O. (2017). Proximal Policy Optimization Algorithms. arXiv:1707.06347v2.

[16]Frans, K., Ho, J., Chen, X., Abbeel, P., Schulman, S., "Meta Learning Shared Hierarchies", 2017.

[17]McCandlish, S., Kaplan, J., Amondei, D., OpenAI Dota Team, "An Empirical Model of Large-Batch Training, 2018.

[18]Mnih, V. et al, "Human-level control through deep reinforcement learning", Nature, doi:10.1038/nature14236, 2015.

[19]Fridman,Lex et al, "DeepTraffic: Crowdsourced Hyperparameter Tuning of Deep Reinforcement Learning Systems for Multi-Agent Dense Traffic Navigation", NIPS 2018, Deep Reinforcement Learning Workshop, doi:10.5281/zenodo.2530457 .

[20]Yu, April, Raphael Palefsky-Smith, and Rishi Bedi. "Deep reinforcement learning for simulated autonomous vehicle control." Course Project Reports: Winter 2016 (CS231n: Convolutional Neural Networks for Visual Recognition) (2016): 1-7,

[21]Zhou, Zhenpeng et al, "Optimizing Chemical Reactions with Deep Reinforcement Learning", ACS Central Science 2017 3 (12), 1337-1344 , doi:10.1021/acscentsci.7b00492 .

[22]Ioffe, S., Szegedy, C., "Batch Normalization: Accelerating Deep Network Training by Reducing Internal Covariate Shift", Proceedings of the 32nd International Conference on Machine Learning, PMLR 37:448-456, 2015.

[23]Bellman, R.E., "Dynamic Programming", Princeton University Press, Princeton, NJ. Republished 2003: Dover, ISBN 0-486-42809-5

[24]Heess, N., et al, "Emergence of Locomotion Behaviours in Rich Environments", arXiv:2707.02286v2, 2017.

[25]Coppelia Robotics, "Virtual robot experimentation platform", 2018.

[26]OpenAI, "Gym: A toolkit for developing and comparing reinforcement learning algorithms", 2018.

[27]OpenAI, "Baselines: high-quality implementations of reinforcement learning algorithms", 2018.

[28]Klopf, A.H., "Brain Function and Adaptive Systems - A Heterostatic Theory", Air Force Cambridge Research Laboratories, 1972.

[29]Wu, Y., et al., "Scalable trust-region method for deep reinforcement learning using Kronecker-factored approximation", arXiv:1708.05144v2, 2017.

[30]Annaswamy, A.B., Aziz, A. (1992). An Adaptive Controller for Aerospace Vehicles. Guidance, Navigation and Control Conference: Hilton Head Island, SC.

[31]Berrios, M., Berger, T., Tischler, M., Juhasz, O., Sanders, F., (2017). Hover Flight Control Design for UAS Using Performance-based Disturbance Rejection Requirements. 73rd Annual American Helicopter Society Forum: Fort Worth, TX.

[32]Lewis, F. (2005). Neural Networks in Feedback Control Systems. Mechanical Engineers Handbook, John Wiley: NY

[33]Heess, N. et al. (2017). Emergence of Locomotion Behaviors in Rich Environments. arXiv:1707.02286v2.

[34]Solomon, E., Vorwald, C., Hrishikeshavan, V., Chopra, I.. (2017). Visual Odometry Onboard a Micro Air Vehicle Using Snapdragon Flight. 7th AHS Technical Meeting on VTOL Unmanned Aircraft Systems and Autonomy: Mesa, AZ.

[35]Gremillion, G., Humbert, J. (2010). System Identification of a Quadrotor Micro Air Vehicle. AIAA Atmospheric Flight Mechanics Conference.

[36]Basri, M., Husain, A., Danapalasingam, K. (2015). GSA-based optimal backstepping controller with a fuzzy compensator for robust control of an autonomous quadrotor UAV. Aircraft Engineering and Aerospace Technology: An International Journal. Vol 87 Iss 5 pp. 493-505.

[37]Weiss, S., Achtelik, M., Lynen, S., Achtelik, M., Kneip, L., Chli, M., Siegwart, R. (2013). Monocular Vision for Long-term Micro Aerial Vehicle State Estimation: A Compendium. Journal of Field Robotics. Vol 30 Iss 5.

[38]Wang, J., Olson, E. (2016). AprilTag 2: Efficient and robust fiducial detection. Proceedings of the IEEE/RSJ International Conference on Intelligent Robots and Systems (IROS).

Horizon Estimation from Thermal Cameras for Unmanned Aerial Vehicles

Keith Sponsler
kspons@tamu.edu

Srikanth Saripalli
ssaripalli@tamu.edu

J. Mike Walker '66 Department of Mechinical Engineering
Texas A&M University
College Station, TX

***Abstract*— Obstacle avoidance is essential for the integration of autonomous Unmanned Aerial Vehicles (UAVs) into the National Airspace. Thermal imaging provides visual information in situations that a typical camera can not. Sky-ground segmentation is useful for detecting obstacles. In this paper, we propose a technique of sky-ground segmentation and obstacle detection using thermal imaging, and its use for control in quadrotor UAV systems. Sky-ground segmentation is performed by using deep neural networks and image processing techniques. We present results from flight trials in the field which demonstrate that our detection algorithms identify obstacles within the UAV's field of view.**

I. INTRODUCTION

Obstacle avoidance is essential for Unmanned Aerial Vehicles (UAVs) to be integrated into the National Airspace. Small UAVs can only carry a limited set of sensors because of size and weight constraints. Typical cameras are unusable at night, in bright sunlight, and other extreme conditions. Thermal imaging is an alternative methodology for UAV guidance that succeeds where a normal camera cannot. Thermal radiation and light are both electromagnetic radiation, but they differ in wavelength. Visible light has a wavelength of 0.4 to 1 micrometer whereas thermal radiation has a wavelength of 8 to 12 micrometers.

Sky-ground segmentation is useful for classifying obstacles. Objects that appear in the ground segment will generally not interfere with UAV operations. Altitude can be increased to avoid ground-segment objects. Objects in the sky segment, however, may collide with the UAV. Sky-segment objects may be other aircraft, manned or unmanned, and avoidance may not be as trivial as increasing altitude. The avoidance of obstacles in the sky-segment should be prioritized. Categorizing an object based on the image segment it occupies allows for obstacle prioritization and it streamlines avoidance procedures.

In this paper, we propose a method for sky-ground segmentation and obstacle detection using thermal images. Previous work on sky-ground segmentation and horizon detection has been performed on visible light images [1] [2] [3]. Visible light cameras have been developed widely, and as a result they have high resolution, fast frame rates, small size, and are relatively inexpensive. However, there are occurrences where visible light imaging does not perform well i.e. dark or nighttime scenes, transitions from bright to dark scenes (and visa versa), scenes with bright lights and lens flare, and . Thermal imaging on the other hand is not as widely developed. As such, thermal image detectors are more costly, have lower resolution, and slow frame rates relative to visible light detectors. However, thermal cameras can provide useful image data in environments where visible cameras fail. Thermal imaging is especially useful in gathering visual data for night time and normal vision-impaired situations.

Horizon detection has been approached in literature for visual images, but not yet in thermal. Various machine learning methods have been used to successfully identify a horizon line in an image [4] [5]. In the technique proposed in this paper, deep machine learning is used in the first step of horizon detection. The results of the deep model's prediction is then analysed and an obstacle detection heuristic is applied to the prediction.

II. MAIN BODY

Experimental Testbed

Data was collected using a 3DR Solo quadrotor [6]. A Global position system receiver (GPS) and inertial measurement unit sensor (IMU) are built into the 3DR Solo system. Two cameras were mounted to the front underside of the UAV. Thermal images were captured at 8hz with a ThermApp camera [7]. Monocular images were captured at 30hz with a Firefly MV camera from Point Grey [8]. To record data and process image information, an onboard computer was attached to the UAV. The onboard computer was a Raspberry Pi 3 b (RPI) [9]. The RPI is both low power and lightweight, traits that are desirable for an on-board system. The RPI runs the Robot Operating System (ROS) [10] to organize the flight information and communicate between the pi and the UAV. ROS is software that facilitates the coordination of every element of autonomous flight. With ROS, test flights can be flown, and all the data (camera streams, location information, attitude data, etc) can be stored in a ROSbag (re-playable file format) [11] for post processing off board. The RPI connected to both cameras via usb and to the UAV by WIFI. Their information is transmitted to the RPI via WIFI through MAVROS [12]. MAVROS enables the RPI to send flight commands, such as takeoff and land,

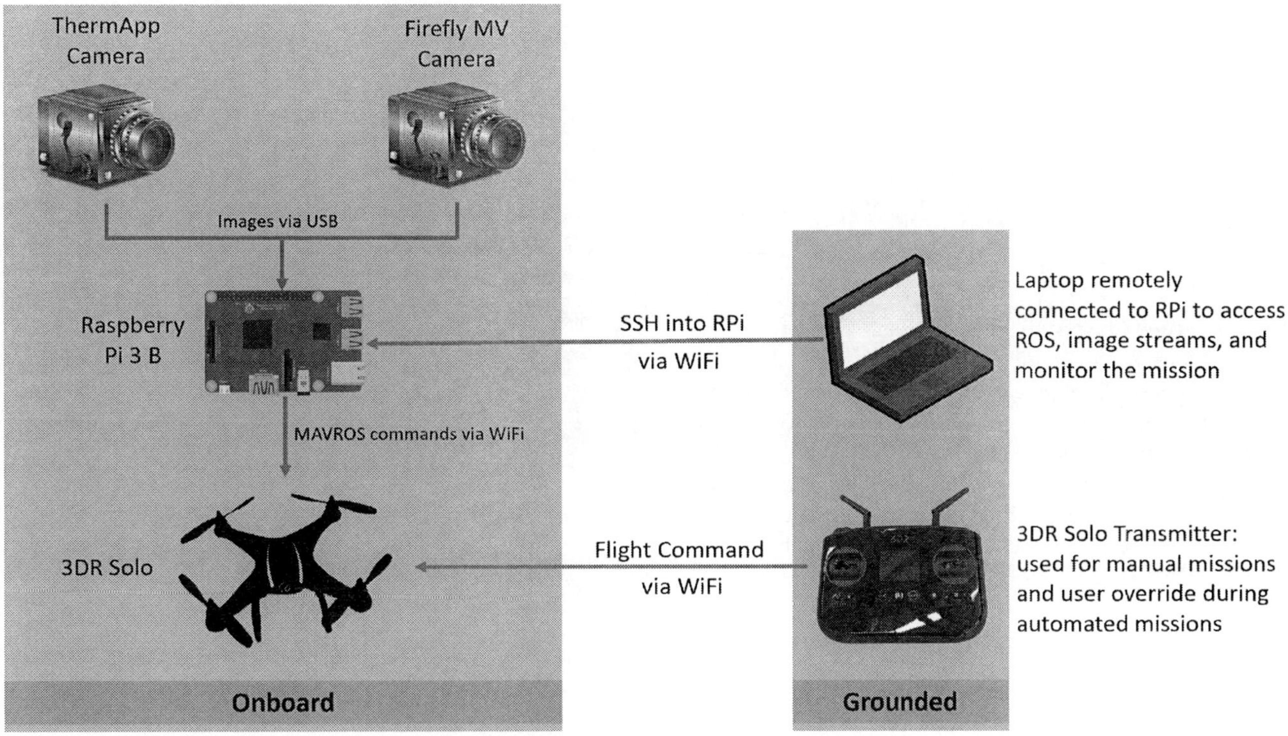

Fig. 1. A flowchart of the equipment used for data collection and the methods of communication between devices.

to the UAV. With this configuration, the pilot can monitor all flight information on the ground in real time through a remote connection to the RPI. It also is set up such that full autonomy can be implemented with no additional hardware. The pilot can fly manually to collect data. During an autonomous flight, the ground pilot can take control of the UAV by moving any joystick or pressing any button on the transmitter. Figure 1 visually represents the hardware and communication standards used for collecting data. Figure 2 is a photograph of the experiment's UAV with equipment on board.

The thermal camera has a 384 x 288, 17 μ detector. A long wave infrared, 6.8 mm lens was used for collecting data. Long wave infrared is the electromagnetic spectrum from 8 to 12 micrometers [13]. The frame rate of the thermal camera is 8.7 Hertz. The field of view of the setup is 55 horizontal and 41 vertical. The sky in thermal images appears colder than the ground. That is, sky-pixels on the whole have a lower average pixel intensity than the ground-pixels. This is the principal on which we build the segmentation heuristic. There are additional considerations, however. As the sky meets the ground, the pixel intensity of sky-pixels increases. Additionally, clouds belong to the sky-class but have higher pixel intensities than the sky itself. By training a deep neural network on labeled images we may obtain a model that has built-in accommodation for the slight gradient above the horizon and the inclusion of clouds as sky-pixels. The speed of the UAV on an automated mission is constrained by the camera and the speed at which images are processed. A detection could be declared when a single pixel in the sky-segment is significantly brighter than its neighbors. However due to noise, a single pixel is an untrustworthy measurement. A square block of two by two pixels is the smallest positive detection size. Now let us assume that this obstacle detection technique is aimed at avoiding obstacles that are larger than 2 square centimeters. To avoid collision with an object of this size, we must fly at a safe speed according to the object detector's abilities. For safety calculations, the ThermApp thermal camera's resolution of 284 by 384, field of view of 41 by 55, and frame rate of 8.7 frames per second are used. The camera can detect an obstacle measuring two square centimeters at a distance of 0.069 meters. Given the frame rate, the UAV should travel at a velocity of no more than 0.6 meters per second directly forward. The speed limit can be raised by increasing the frame rate or resolution. In this

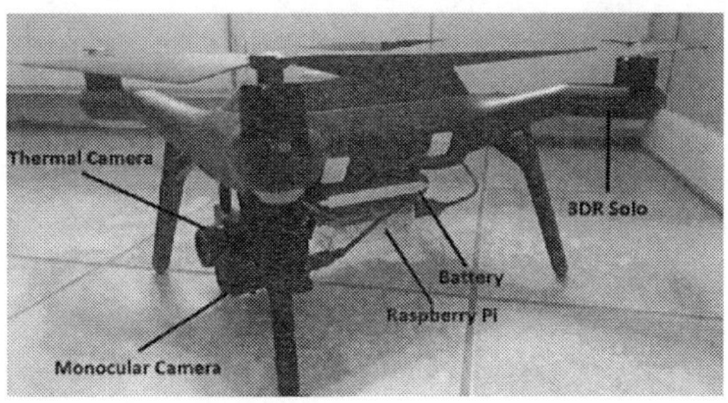

Fig. 2. A picture of the 3DR Solo with on board ThermApp camera, Point Grey Firefly MV camera, Raspberry Pi 3 B, and external battery

setup, the main limitation to speed is the camera frame rate. The image processing process can run at 30 frames per second. It takes an additional 0.03-0.05 seconds from the moment the image is read to send a command to the UAV.

The datasets were organized such that the flights summarize a variety of lighting environments. The datasets are categorized by the time in which the flight took place: dawn, daylight, dusk, and nighttime, and over water. Over water refers to an additional dataset containing flights near a large body of water at dusk, where reflections of the setting sun were of special interest. Each dataset consists of at least four ROSbag files, 30 seconds each, that have MAVROS topics, thermal camera data, and monochrome camera data. Additionally, each dataset contains flights from at least two separate days. Data from different days has varying conditions from sunny to cloudy. The datasets each contain more than 1000 thermal images. The entirety was flown and recorded in central Texas in hilly to flat terrain.

Horizon Estimate

Sky-ground segmentation begins with a horizon estimate given by a single shot detector. The You Only Look Once (YOLO) detector was used to identify where the sky and ground appear in each image [14]. YOLO is a neural network that is used to detect objects in images. The advantage of YOLO over other neural networks is the speed at which detections are made. Different frameworks, such as GoogLeNet or Resnet may provide slightly better detection, but speed is of great importance when dealing with robotics in real time. The detector needs to operate quickly so that the information for navigation can be inserted into the controls scheme of the vehicle.

The detector is trained such that it should estimates the largest contiguous regions of each class. The horizon line can be interpolated within the region separating the bounding boxes. YOLOv3 was trained on 51 labeled thermal images. From each ROSbag recording in the datasets, at least one thermal frame was copied and labeled. Each image was labeled with two bounding boxes, one for sky and one for ground. The labels were applied such that they encompass the largest contiguous portion of their class. Care was taken to ensure mutual exclusivity of the labeled regions. No pixels were labeled as belonging to both sky and ground. A rule of thumb for supervised machine learning is that more data generally gets better results. Given a relatively small training set of only 51 examples of each class, the results are fairly good. Figure 4 shows four examples of successful detections. The detector tends to fail when once class is not visible in the image. This can be prevented by taking less agressive movements with the UAV. The yolo detector also fails when an obstacle obstructs a majority of the the sky. This becomes a large issue when flying close to large, tall objects, especially within urban environments. Figure 5 shows four examples of unsuccessful detections. After bounding boxes on the sky and ground are generated, the average pixel value of each region is calculated. First, the average value of pixels within the sky bounding box calculated. Pixels of this value or lower make up a majority of the sky in each image, usually the upper portion. Every pixel that has a value lower than this sky-average, is assumed to be sky. As the sky meets the ground the average pixel intensity of the sky increases. This may be caused by radiation scattering in the atmosphere. To pick up these brighter pixels that belong in

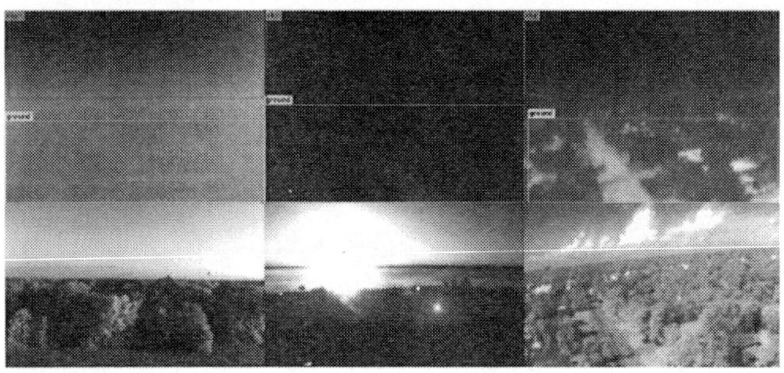

Fig. 3. Top Row: Thermal image with YOLOv3 predicted bounding box. Bottom Row: Respective visual image

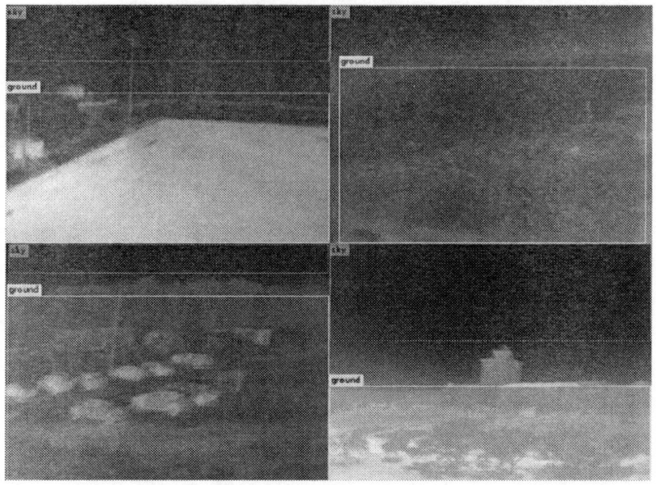

Fig. 4. Four examples of successful detections

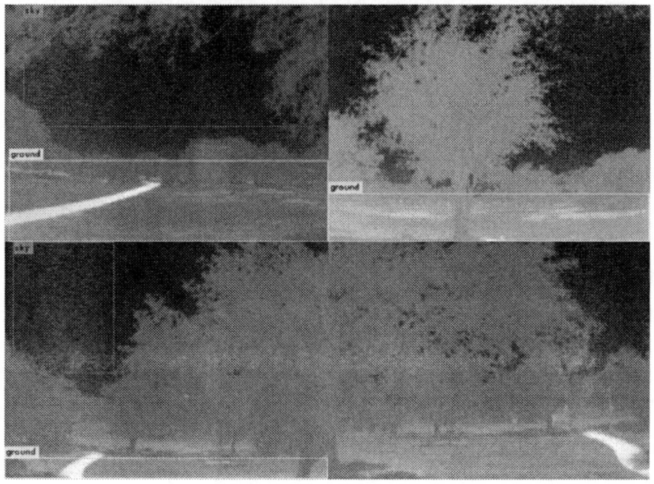

Fig. 5. Four examples of unsuccessful detections

the sky segment, we use a higher threshold based off the pixel values within the area between the ground bounding box and the sky bounding box. The pixel intensity of the region sandwiched by the two predictions is sorted into a histogram with 16 bins. There tends to be a peak or plateau to the histogram plot. Take the pixel value at the max or plateau and threshold the image at that value in the region above the ground prediction bounding box. The average values of the sky bounding box and the peak value within the interim histogram are approximations that return good results. For best results, begin with these parameters and manually tune. Selected parameters will work for an entire flight or until the weather and temperature changes significantly.

Finding Obstacles

To detect and avoid obstacles, let us assume that the UAV is traveling at a constant velocity with its camera facing forward. Using the sky-ground segmentation thresholded image, we can predict if there are obstacles in our path that need to be avoided. If there are no obstacles in front of the forward traveling UAV, assume that the horizon line is approximately linear. An obstacle within the path of the forward moving UAV will appear as a deformation of the horizon line obtained in the above methods. If an object is in front of the UAV, the sky-ground segmentation will no longer be approximately flat. Buildings, poles, and trees are common obstacles that will deform the straight-line horizon with this method. This gives us the opportunity to use sky-ground segmentation to avoid these obstacles.

Given the black and white thresholded image of sky-ground, as seen in Threshold 1 in Figure 6, the horizon line is the interface between black and white pixels. For the purposes of finding irregularities in the horizon line, the information in the image can be condensed into an array of one dimension. Let H be the approximated horizon line, where each element is given by the number of sky-segment pixels present before the first ground segment pixel in each column. Threshold 2 in Figure 6 corresponds to the data contained in R. In an environment without obstacles, there would be no portion of the plot above noise level (the horizontal line in Figure 6, Object Detection). Obstacles in the sky-segment become apparent in this binary-image form. Let A be the same size as R and defined such that

$$A_n = \frac{\Sigma_{j=-k}^{k} H_{i-j}}{2k} \tag{1}$$

$$\forall i \ni k < i < (\text{image width}) - k,$$

$$A_n = 0 \text{ otherwise}$$

k is the size window which we are averaging H to get A. For our experiment, $k = 5$ pixels. Now define Ω as

$$\Omega_n = \frac{|A_n - H_n|}{H_n} \tag{2}$$

A plot of Ω is given in Figure 6. A higher value of Ω implies a greater disruption of a straight line horizon. A threshold value, α, can be used to discriminate noise from a detection.

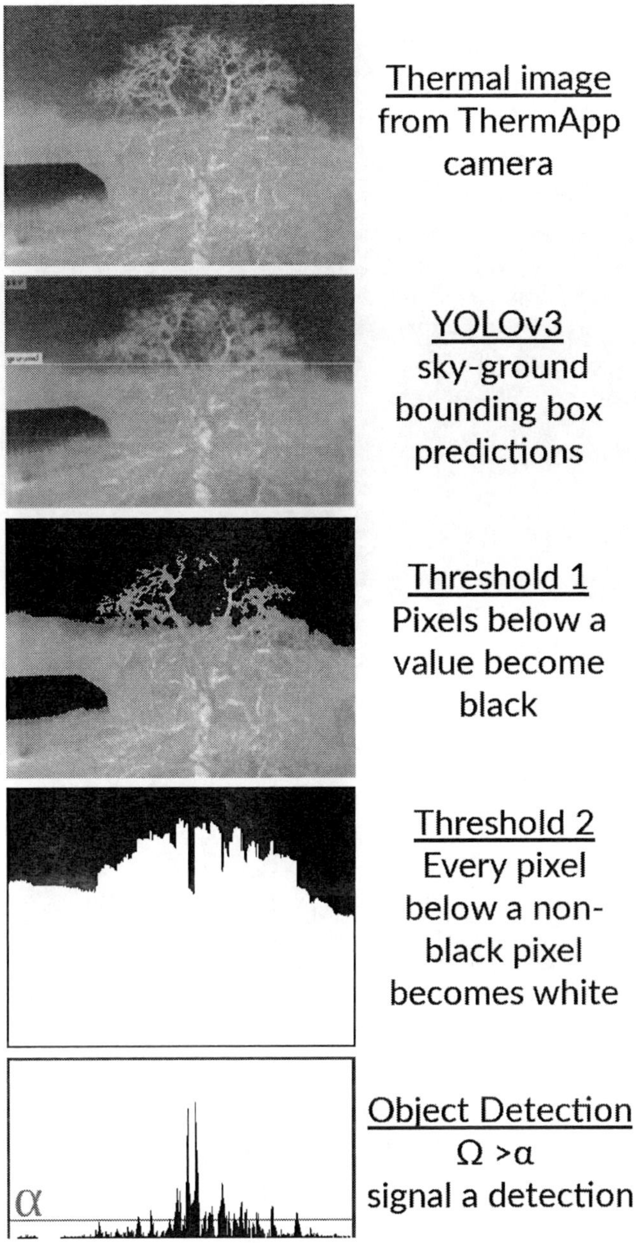

Fig. 6. Obstacle Detection Pipeline

If $\Omega_n > \alpha$, a detection has taken place in column n. An α value of 0.05 was determined to effectively filter out noise in this experimental setup. Detections near the center of the image raise safety concerns for the UAV. If the UAV continues straight, a collision is likely to occur. Detections that occur on the periphery of vision do not pose immediate threats if the UAV is traveling in a straight line. The UAV may add a horizontal velocity vector to its path to avoid a collision.

Control Architecture

The UAV flights are automated through MAVROS. MAVROS is a an adaptation of Micro Air Vehicle Link (MAVLink) for use in ROS [15]. The flight controller on the

Fig. 7. Another example of the object detection pipline. From top to bottom: Visual spectrum gray scale image, thermal image, Threshold 1, Object Detection

3DR Solo manages the low level control. High level control is dictated by MAVROS. In this experiment, we assume that the action being performed is forward facing movement only. The mission given to the UAV for testing begins by arming at performing a simple takeoff to 30 feet above ground level. If the obstacle detection algorithm identifies an object, it will interrupt the mission with a defensive flight maneuver. The "MAVROS mavsetp global vel $<v_x><v_y><v_z>$" changes the UAV's current velocity to match the input velocity. Upon detect an obstacle, the RPI will use the velocity command to add a horizontal component to the UAV's velocity such that the magnitude of the velocity vector remains constant. The sign and order of the horizontal velocity is described in detail in obstacle detection. Once the UAV's path is clear of obstacles, the autopilot takes over, orients the UAV towards the next way-point, and begins forward movement towards the way-point.

III. Concluding Remarks

A technique for obstacle detection from thermal imaging has been proposed. The results have demonstrated that thermal images provide useful data for autonomous flights on UAVs. Thermal imaging can be used complimentary to visual cameras, and provide much more visual information than visual images in certain conditions. The procedure for sky-ground segmentation will become more robustness by collecting and labeling diverse environments. Once we have validated our algorithm against manual flight data, we will begin autonomous flight testing. Full autonomy is possible with the hardware already present.

References

[1] S. Todorovic, M. C. Nechyba and P. G. Ifju, "Sky/ground modeling for autonomous MAV flight," 2003 IEEE International Conference on Robotics and Automation (Cat. No.03CH37422), Taipei, Taiwan, 2003, pp. 1422-1427 vol.1.

[2] Hong-Zhao Yuan, Xiu-Qiong Zhang and Zi-Liang Feng, "Horizon detection in foggy aerial image," 2010 International Conference on Image Analysis and Signal Processing, Zhejiang, 2010, pp. 191-194. doi: 10.1109/IASP.2010.5476135

[3] Gui-Qiu Bao, Shen-Shu Xiong and Zhao-Ying Zhou, "Vision-based horizon extraction for micro air vehicle flight control," in IEEE Transactions on Instrumentation and Measurement, vol. 54, no. 3, pp. 1067-1072, June 2005. doi: 10.1109/TIM.2005.847234

[4] S. Fefilatyev, V. Smarodzinava, L. O. Hall and D. B. Goldgof, "Horizon Detection Using Machine Learning Techniques," 2006 5th International Conference on Machine Learning and Applications (ICMLA'06), Orlando, FL, 2006, pp. 17-21.

[5] T. G. McGee, R. Sengupta and K. Hedrick, "Obstacle Detection for Small Autonomous Aircraft Using Sky Segmentation," Proceedings of the 2005 IEEE International Conference on Robotics and Automation, Barcelona, Spain, 2005, pp. 4679-4684.

[6] 3DR Solo Quadrotor https://3dr.com/solo-drone/:.

[7] Therm-App Camera https://therm-app.com/therm-app/:.

[8] Firefly MV Camera https://www.ptgrey.com/firefly-mv-03-mp-mono-usb-20-micron-mt9v022:.

[9] RaspberryPi Computer https://www.raspberrypi.org/products/raspberry-pi-3-model-b/:.

[10] Robot Operating System http://wiki.ros.org/kinetic:.

[11] Rosbag http://wiki.ros.org/rosbag

[12] MAVROS http://wiki.ros.org/mavros:.

[13] LWIR wide angle lens https://therm-app.com/product/6-8mm-lens/

[14] Redmon, Josep and Farhadi, Ali, YOLOv3: An Incremental Improvement, arXiv, 2018

[15] https://mavlink.io

Amphibious Quadcopter with Smooth Powered Air-Water Transition

Ilya Semenov
Graduate Research Assistant

Vikram Hrishikeshavan
Assistant Research Scientist

Inderjit Chopra
Alfred Gessow Professor and
Distinguished University Professor

Alfred Gessow Rotorcraft Center
Department of Aerospace Engineering
University of Maryland
College Park, MD, USA

ABSTRACT

Hybrid amphibious and aerial vehicles face the significant challenge of transitioning between and operating within two very different environments. In this study a quadrotor configuration is described that uses no additional propulsion systems or specialized sensing equipment to complete a powered air-water transition. Central to its operation is the ability to sense the amount of water the rotor is encountering using basic telemetry of motor state and the subsequent ability to predict thrust at any point through the transition. The assembled vehicle is described and experimental measurements were conducted with conventional fixed pitch rotor blades and their performance in air, water and transition media is determined. While transitioning from air to water thrust and power consumption increases and rotor speed decreases rapidly. Empirical metrics were then extracted to inform the controller to maintain constant thrust through transition between air and water. A reduced order blade element momentum theory (BEMT) analysis is introduced and resulting correlation with experimental results is discussed. An empirically derived metric is defined and used to demonstrate sensing of water encountered by the rotor. Additionally, an empirical algorithm for predicting thrust through the transition is introduced and demonstrated against experimental results of lowering a rotor into water at constant throttle. These methodologies are incorporated to achieve a smooth transition from air to water or water to air using a PID controller.

INTRODUCTION

Hybrid Unmanned Aerial/Underwater Vehicles (HUA/UVs) have the challenge of robustly operating within and transitioning between two vastly different environments. This research is primarily focused on overcoming this issue.

Many bioinspired designs have been put forward. Some examples emulate diving birds like the gannet (Ref. 1) or birds that land on the water and then submerge like a duck (Ref. 2) or like flying fish capable of propelling out of the water and gliding (Ref. 3). An example of a gannet style design is shown in Figure 1. These designs have had some success, but are limited by the inability to hover and face many challenges in the air-water transition. The present research investigates a quadrotor for air-water transition.

The benefits of a quadrotor configuration include increased agility and controllability underwater compared to traditional submarine or fixed wing designs. The increased maneuverability allows for better obstacle avoidance and the ability to approach a target from multiple angles (Ref. 4).

The Autonomous Quadcopter With Underwater Ability (AQWUA) is an assembled demonstration vehicle that is designed to be capable of diving underwater using a standard quadcopter configuration with no additional propulsion or buoyancy systems and with active rotors during the transition states. Similar research efforts are underway at Rutgers University and Oakland University, but their approach differs from the AQWUA. The Naviator from Rutgers uses 8 propellers on 4 arms, mounted coaxially. They perform the transition by shutting-off the motor set going through the surface of the water. As a result, the vehicle is not properly equipped for transitioning in water where the rotor plane angle relative to the water surface exceeds a physically defined limit (Refs. 5, 6). The Oakland LoonCopter utilizes a buoyancy tank and ballast system for transition and underwater performance (Ref.7). This results in a significant payload penalty and increases complexity of the control strategy due to center of gravity offsets. The advantages of our AQWUA revolve around the lack of an additional propulsion system, which is compensated by advanced control of the motors through the transition. Additionally the AQWUA is nearly neutrally buoyant and submerges under its inertia, the force of gravity, and rotor thrust only.

In order to modulate motor controls during the transition, multi-medium propeller dynamics need to be investigated. Previous work has been done with CFD techniques to model air-water interfaces. Some examples of this are the exploration of 2-D and 3-D bodies impacting the water surface performed in Reference 8 as well as wave impact forces on floating bodies (Ref. 9). These techniques are

Presented at the Vertical Flight Society Autonomous VTOL Technical Meeting and Electric VTOL Symposium, Mesa, Arizona, UAS, Jan. 29-31, 2019.

computationally intensive and therefore are not optimal for use in control-theoretic analysis. This study focuses on reduced order modeling and correlation with experimental data, expanding on the work started by Crane and Faruque (Ref. 10).

A single propulsion system of four rotors may allow for use in variable environments, larger payloads, and faster transitions. The present research is focused around the multi-medium propeller dynamics and their application to control of HUA/UVs, which is unique and innovative. There are many potential applications of such systems. The technology in the AQWUA can be equipped with specific payloads for use in: naval mine detection, marine biology research, hydrogeology research, cave exploration, marine pollution mitigation and inspection of underwater structures such as bridge posts, pipes, cables, and offshore energy facilities.

The paper is structured as follows: the vehicle construction and component description is discussed followed by results from rotor air/water performance measurements. Subsequently using these measurements, a Transition index (TI) is developed to predict the thrust as the vehicle transitions through the media. Lastly, a control strategy implementing the transition index is proposed and preliminary results are presented.

Figure 1: Gannet-like bioinspired design performing water-air transition maneuver (Ref. 11).

VEHICLE DESCRIPTION

The AQWUA, shown in Figure 2, is an X configuration quadrotor that is built around a waterproof enclosure for the main control electronics, shown in Figure 3. The control electronics package features a Navio 2 shield onboard a Raspberry Pi 3 as the main autopilot running ArduCopter. ArduCopter is a full featured open source autopilot that is modifiable for custom applications and can support autonomous operation. Besides the core autopilot the other key component is a motor control package that intercepts the pulse width modulation (PWM) motor throttle signals output by the autopilot. The signal is processed by an Arduino Micro designated as the Scaler Board that implements the control loop for stable transition behavior. This Scaler Board is attached to a Signal Relay board that transfers sensor signals to a separate Arduino 101 designated as the Telemetry Processor. This is necessary as the Arduino Micro does not support enough interrupt functions to process all of the sensor data and input throttle signals simultaneously. The Telemetry Processor logs the data to an SD card and processes it to estimate the amount of water the rotor is encountering using the Transition Index (TI). This is passed over a serial link to the Scaling Board for use in implementation of control. The flowchart of these components is shown as Figure 4. Not represented in the schematic is a Corona rP8D1 receiver and PPM converter that allow for radio communication on a 35MHz frequency processed by the Navio autopilot. This specific frequency was chosen for its ability to penetrate water at a moderate depth.

Figure 2: A Quad With Underwater Ability (AQWUA), featuring cylindrical waterproof enclosure attached to 550mm frame.

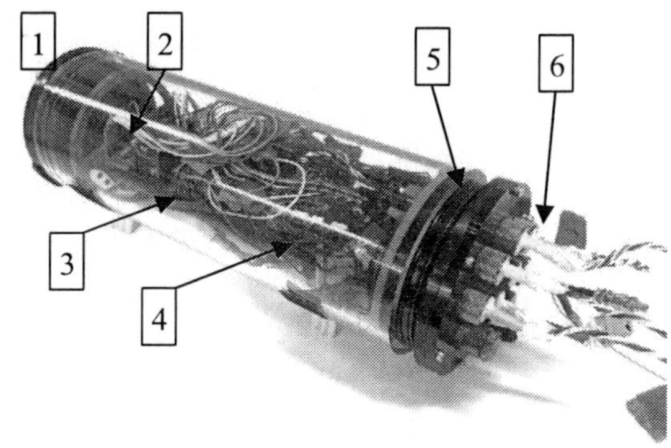

Figure 3: The waterproof enclosure, labeled from left to right are: (1.) the front endcap, (2.) the telemetry processor, (3.) the scaler-relay board, (4.) the Navio2, (5.) the rear endcap opened to show double O-rings, and (6.) cable penetrators.

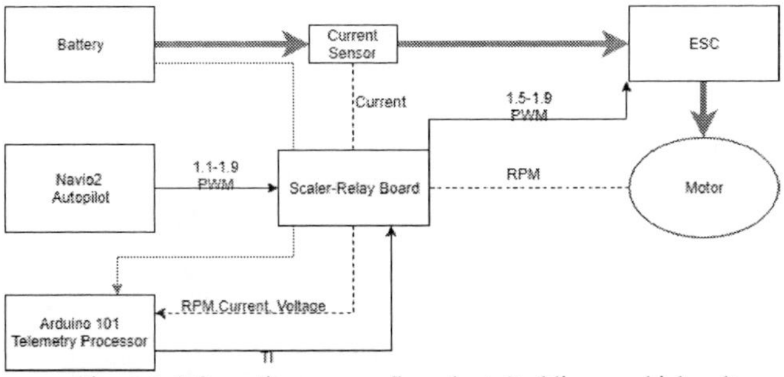

Figure 4: Information process flow chart. Red lines are high voltage lines, dotted lines are raw sensor data, and solid lines are processed data.

The hardware includes the main body composed of a Blue Robotics acrylic/aluminum waterproof enclosure with redundant O-rings and cable penetrators mounted by a 3d printed part and hose clamps to a X525 fiberglass and aluminum frame. The motor mounts are 3d printed supporting a propulsion system composed of carbon composite 15 inch propellers on GARTT ML4108 620KV motors run by Castle Creations Micro Mamba Crawler edition electronic speed controllers (ESCs), all powered by a MaxAmps 3S 4550mah 120C battery switched by a IXYS power MOSFET.

Waterproofing is an everlasting challenge in the assembly of these types of vehicles. The AQWUA main electronic components: the Navio, Arduinos, Battery, SD card, MOSFET, and radio hardware are all sealed in a Blue Robotics 4'' waterproof enclosure. This enclosure features double O-rings for the main flanges supporting endplates with a single O ring. One endplate has holes in which Blue Robotics cable penetrators and vent are installed. Each cable penetrator is attached to the endplate with an O ring. Each cable has additional marine sealant or liquid electrical table on the other end to seal the cable jacket. This design ensures good protection even at depths exceeding 50 meters. Some components require individual waterproofing as they are assembled outside of the enclosure. The current sensors are wrapped entirely in heat shrink in which silicone was applied during the heat treatment process. This procedure places the entire components in a gapless silicone layer. Due to heat production and the necessary exposure of the heat sink the ESCs are waterproofed with a different strategy. The ESCs were disassembled and the main board coated in multiple coats of high temperature conformal coating, the leads were additionally coated in small amounts of silicone. The motors can operate in water without waterproofing but will deteriorate due to corrosion quickly. Each motor is disassembled and CorrosionX HD, a dielectric water repellent grease and lubricant, is applied.

The AQWUA is designed to be slightly positively buoyant. This is important for underwater operations for minimal energy consumption in hold position idle modes, but still allows for easy recovery as a failsafe.

EXPERIMENTAL RESULTS

In order to quantify rotor performance in the transition region between air and water, several experiments were carried out.

Experimental data was obtained via test setup, schematic shown in Figure 5. As shown, a variable height platform supports an arm assembly with the motor and rotor at the end of it, which allows the lowering of a rotor into the water gradually. The motor support mounts to an S-shape DYLY-103 10KG strain gauge force sensor connected to a Sparkfun HX711 force gauge amplifier module for data collection via Arduino Micro. Current data is obtained via Allegro ACS712 20A Current Sensor, voltage data is passed through a resistor voltage divider and read by the Arduino Micro directly, then scaled appropriately. RPM is measured using A3114 Hall Effect sensor and motor mounted neodymium magnets. The ESC used is a Castle Creations Micro Mamba Crawler Edition, which is controlling a Turnigy MultiStar 4114 320Kv 22 Pole brushless motor. The container of water used has a maximum water volume diameter of 34.5 inches, tapers down to 24 inches, and has a height of 14 inches. It is important to note the possible presence of ground effect and overall swirl of the water due to the size of the container; the effects will be more prevalent for larger diameter rotors and include increased thrust due to ground effect, as well as increased power due to swirl and churning of the water volume.

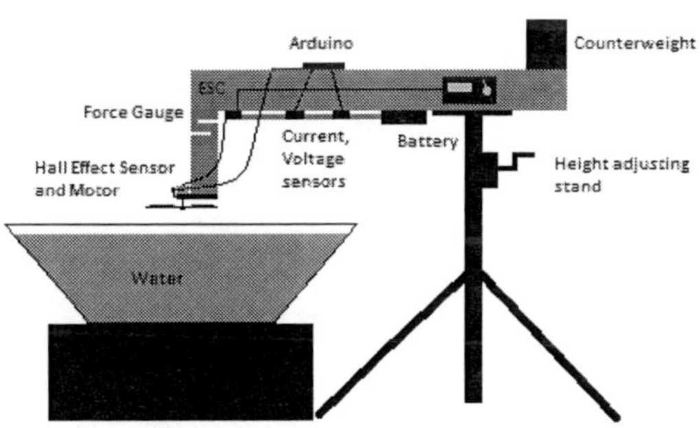

Figure 5: Air/water performance measurement setup.

This test stand was used with three different rotors: a carbon composite 15 inch diameter propeller, carbon composite 9 inch diameter propeller, and a plastic 3-D 8 inch diameter propeller pictured in Figure 6. These will be henceforth referred to as CC15, CC9, and 3D8 respectively.

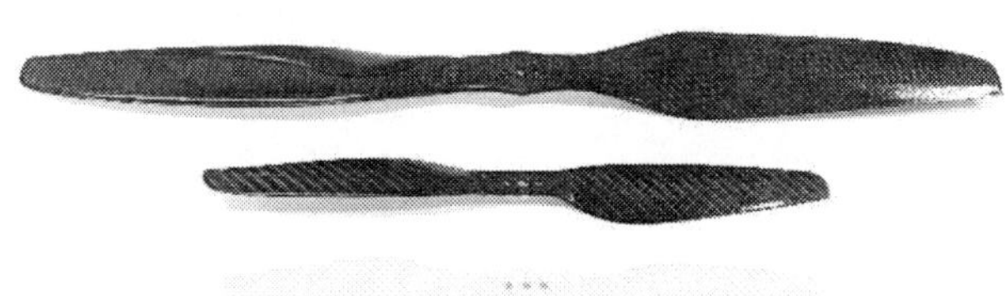
Figure 6: Propellers tested for air/water performance measurements. From top to bottom: CC15, CC9, 3D8.

The differences in performance of completely in water and in air conditions were considered first, and then applied to analysis of the transient behavior.

Variation of Thrust with RPM for air and water conditions for the different rotors are shown in Figures 7, 8, 9. Each rotor was throttled up to maximum steady behavior, which is lower in water than air due to the high forces involved. Data near 1500 RPM exhibited test stand resonance and were omitted.

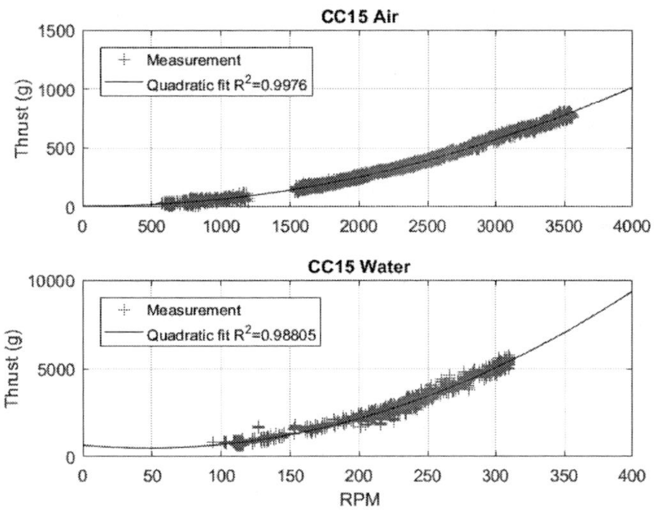

Figure 7: Rotor dia. 15" (CC15) thrust measured data with quadratic fit.

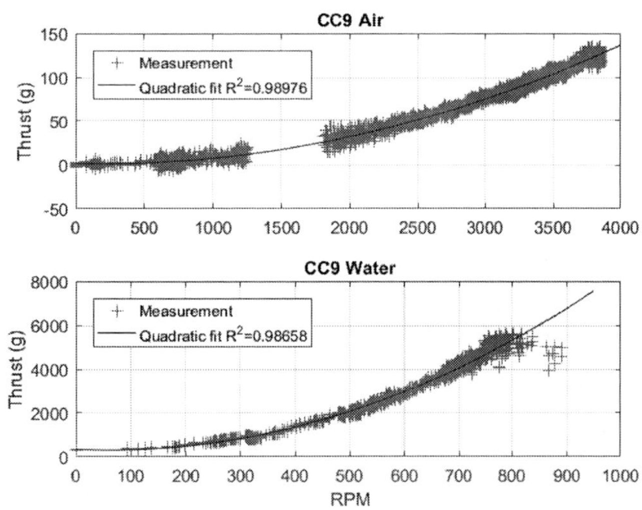

Figure 8: Rotor dia. 9" (CC9) thrust measured data with quadratic fit.

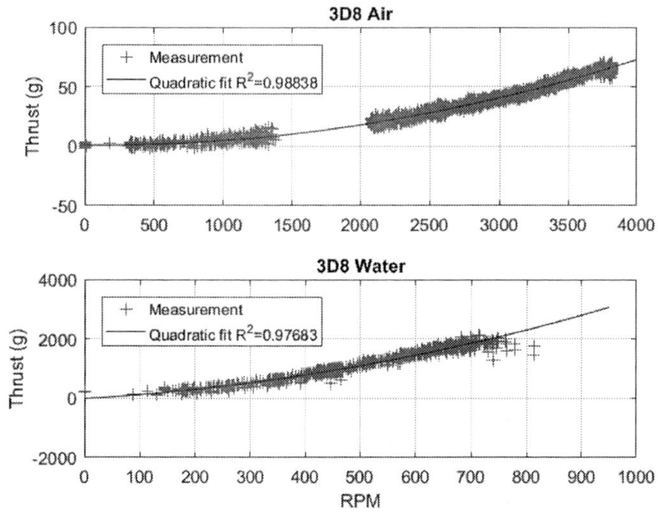

Figure 9: Rotor dia. 8" (3D8) thrust measured data with quadratic fit.

A major challenge of operating a rotor in both air and water conditions is exemplified in the thrust plots. The range of thrust achieved can be orders of magnitude larger in water than the design conditions in air. Additionally, the rotor speed range can be orders of magnitude lower in water than in air for the same motor. Because brushless DC rotors are optimized for specific speed and torque conditions, operating in two extremes can limit efficiency. Gearing is a possible solution, such as the concept proposed by Tan et al. (Ref. 11). However, it was not implemented on the AQWUA for mechanical simplicity.

This challenge is represented in the electrical power consumption of the motor with these rotors in both mediums, plotted in Figures 10, 11, 12.

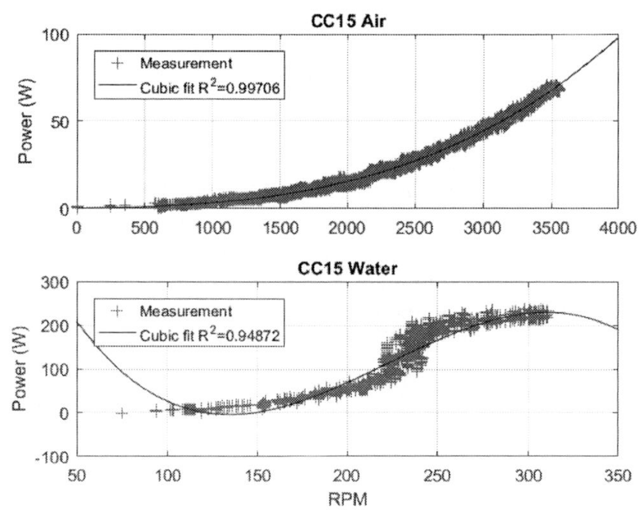

Figure 10: Rotor dia. 15" (CC15) power experimental data.

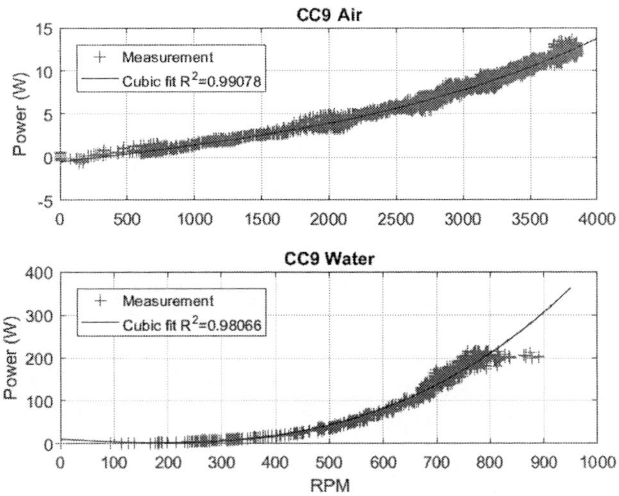

Figure 11: Rotor dia. 9" (CC9) power experimental data.

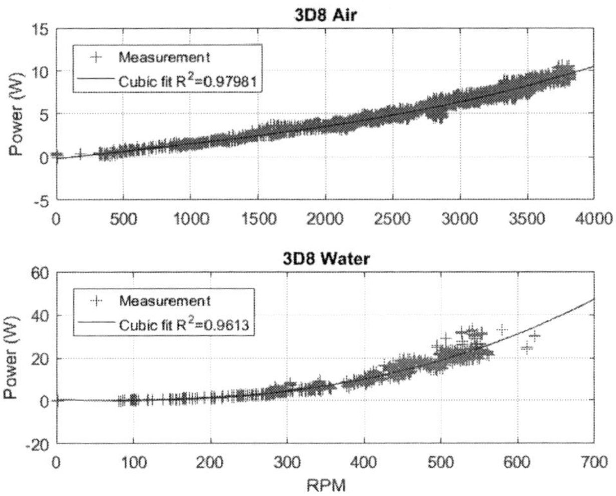

Figure 12: Rotor dia. 8" (3D8) power experimental data.

It can be seen that power consumption in water is much larger than that in air. Additionally, the calculated coefficient of power (C_P) values in water are 2-3 times larger than that in air. This can be attributed to the effect of cavitation that occurs only in water, or the increased effect of swirl, as well as sub-optimal motor performance is extreme off-design conditions.

Additional experiments were conducted to analyze the behavior of the rotor in the transition mode between air and water. A series of tests lowering the rotors at zero angle into the water at various constant throttles were conducted as well as a series of tests throttling each rotor up at two constant depths.

When lowering the rotor into the water at constant throttle, the rotor speed fell and thrust increased as expected, this is plotted in Figure 13. The thrust increases steadily as the rotor is submerged further underwater, however, rotor speed decreases very rapidly when the water is first encountered and continues to decrease steadily as rotor depth increases.

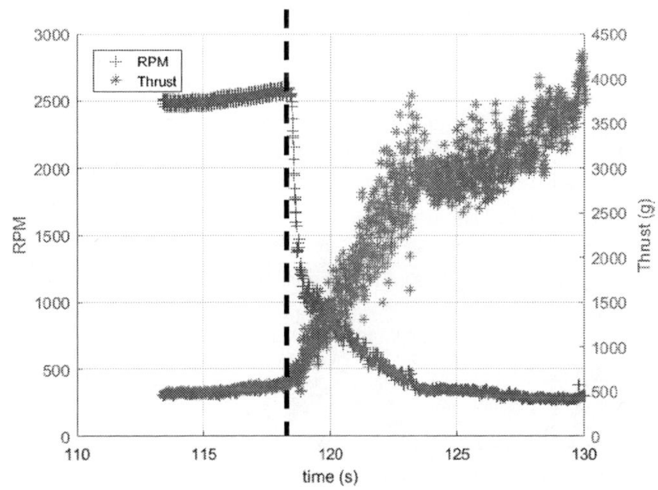

Figure 13: Thrust and RPM as CC15 is lowered into water at 60% throttle, black dotted line indicates when the rotor first encounters water surface. RPM can be observed to decrease rapidly, whereas Thrust increases more slowly.

When the rotor is held at a constant depth of 1.5 inches below water surface ($0.1\frac{z}{R}$) and RPM is varied two observations can be made; shown in Figure 14. First, at high RPM, an oscillatory behavior of water movement occurs. At high enough RPMs the rotor is able to displace a volume of water and operate in a mostly air condition. Water is then re-engulfed and the rotor encounters it again, displacing another volume of water. This motion occurs at a frequency on the order of 1 Hz. Second, a local maxima in thrust is seen at 4 seconds, circled in black. This occurs at the maximum rotor speed before a significant amount of water is displaced allowing the rotor to continue to operate in a pure underwater condition and therefore generate comparatively more thrust.

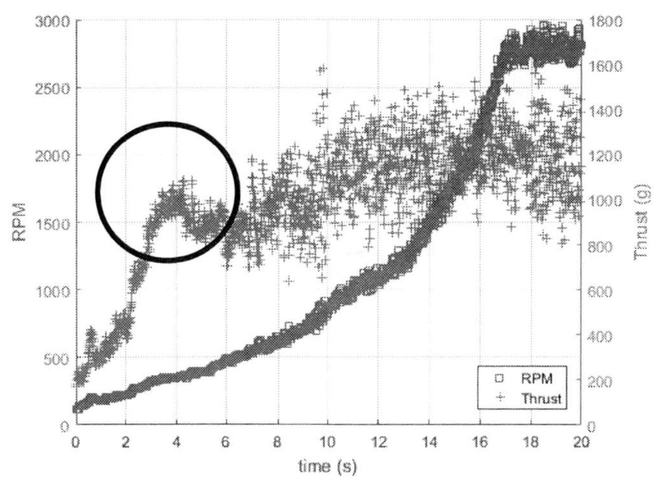

Figure 14: CC9 throttling up at a depth of 1.5 inches, circled is a peak in thrust due to operation in a fully water condition. As RPM increases futher, water is displaced and the rotor transitions to an air operating condition, producing less thrust.

The relationship between rotor speed and thrust produced can be used to derive thrust in unsteady multi-medium conditions.

ANALYSIS AND MODELING

In order to predict rotor performance in various mediums modeling was performed using BEMT analysis with the methods in Reference 12. The process flowchart is shown in Figure 15. The equations used in the analysis are as follows:

$$\lambda = \frac{\sigma * Cl_\alpha}{16 * F}\left(\sqrt{1 + \frac{32F}{\sigma Cl_\alpha}\theta r} - 1\right) \quad (1)$$

$$F = \left(\frac{2}{\pi}\right)\cos^{-1} e^{-f} \quad (2)$$

$$f = \frac{N_b}{2}\left(\frac{1-r}{\lambda}\right) \quad (3)$$

$$dC_T = \frac{\sigma}{2} c_l(\alpha) r^2 dr \quad (4)$$

$$dC_P = \frac{1}{2}\lambda\sigma c_l(\alpha) r^2 dr + \frac{1}{2}(C_{d0} + d_1\alpha + d_2\alpha^2) r^3 dr \quad (5)$$

Where λ is inflow, F and f are tip loss correction factors, σ is local solidity, Cl_α is lift coefficient curve slope of the airfoil, $c_l(\alpha)$ is lift coefficient at angle of attack $\alpha = \theta - \frac{\lambda}{r}$ where θ is local blade pitch angle, N_b is number of blades, C_{d0}, d_1, d_2 are drag coefficients and r is non-dimensional spanwise location along blade.

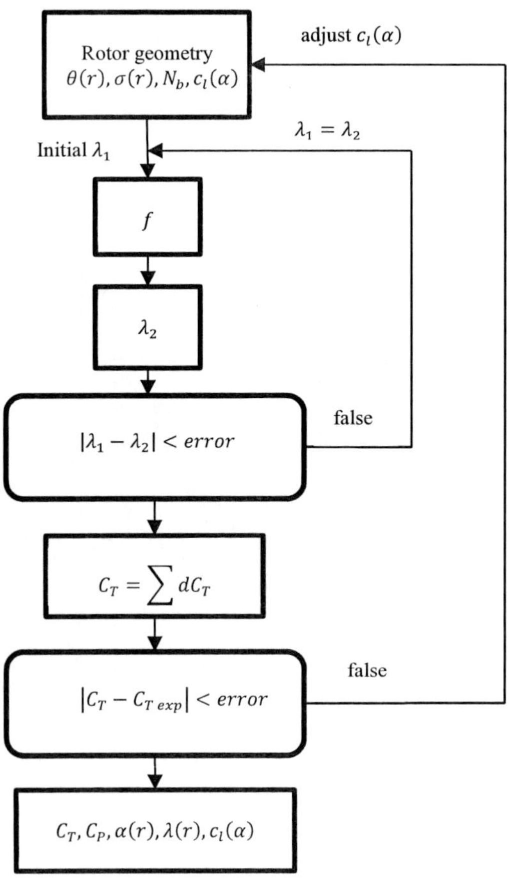

Figure 15: BEMT pseudocode flowchart.

Rotor parameters were measured and shown in Figure 16, a constant $C_{l\alpha}$ of 5.7 was assumed for the carbon composite rotors.

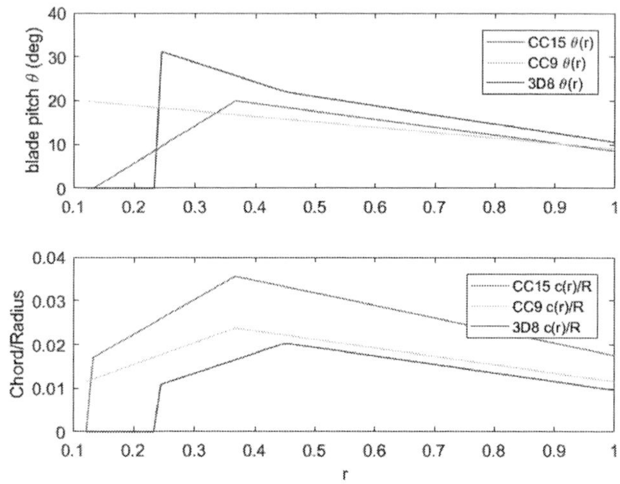

Figure 16: Blade pitches and non-dimensional chords.

Table 1: Comparison of C_T and Reynolds number for air and water conditions of each rotor.

	CC15	CC9	3D8
C_T Air	.0110	.0112	.00985
% Standard Deviation	3.8%	4.2%	5.3%
C_T Water	.0117	.0123	.00999
% Standard Deviation	3.8%	7.9%	5.4%
% Difference C_T	6.9%	9.7%	1.5%
¾ span Re at 75% max thrust in air	71300	31700	24600
¾ span Re at 75% max thrust in water	104000	112000	66500

From measurements, the coefficient of thrust (C_T) can be found for completely in air and in water isolated conditions, and are represented in Table 1.

The measured C_T was found to be about 4-6% higher in water than air for all the measured rotors. This can be possibly due to effect of increased Reynolds number in water. The kinematic viscosity of water is about 1/18th that of air, whereas the rotational speeds are typically about 5 times less in water than in air depending on the propeller. As a result, the Reynolds number is consistently larger in water and the lift coefficient increases (Ref. 13). The effect has been shown to be significant, with changes in lift coefficient of up to .3 reported by Winslow (Ref. 14) for a NACA0012 airfoil between Reynolds numbers of 10,000 and 100,000. However, the quantitative influence of Reynolds number on the lift curve slope of a given airfoil is currently only found via computationally intensive CFD techniques or targeted experimental testing.

The airfoil of each carbon composite rotor most closely resembles a cambered plate. A maximum angle of attack of 8.80° and 7.27° is seen for the 15 inch and 9 inch rotors respectively, justifying a linear lift curve model with C_{la}=5.7/rad. The cambered plate lift curve offset values found via methodology shown in Figure 15 are presented in Table 2 for air and water conditions. In contrast, the 3D8 rotor has a flat plate airfoil and exceeds stall conditions as shown in Figure 17, therefore a bilinear lift curve model was used with a corner angle of 7 degrees, and a constant post-stall slope of 1.15/rad as is consistent with Reference 14. For this rotor the pre-stall lift curve slope was adjusted and found to be 5.24/rad in air and 5.33/rad in water.

Table 2: Lift curve characteristics based on BEMT modeling and experimental C_T.

	CC15	CC9
Offset in air (degrees)	-1.85	-2.02
Offset in water (degrees)	-2.48	-2.87

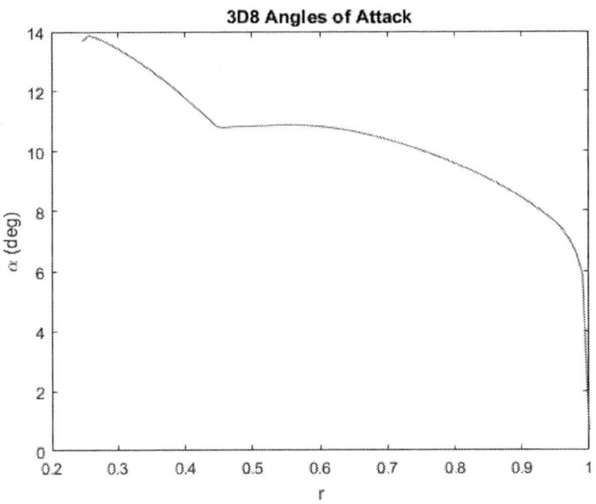

Figure 17: Spanwise angles of attack of the 3D8 rotor.

Using these airfoil models, thrust is predicted well for each rotor in air and water conditions, as shown in Figure 18 for the CC15 blade. The limitations of measuring electrical power and the off-design motor conditions result in an unknown ESC-motor efficiency, however in air a nominal value of 65% shows a good comparison with test data as illustrated in Figure 19 for CC15. In water the motor efficiency drops, a value of 20% (a non-optimal value at low RPM) shows a good correlation of predictions with test data in Figure 20 for CC9. These values of motor efficiency are nominal as supported by analysis in Reference 11. For a more robust analysis measurements such as torque are necessary for accurate motor efficiency changes and additional effects such as cavitation and swirl also need to be developed in the model.

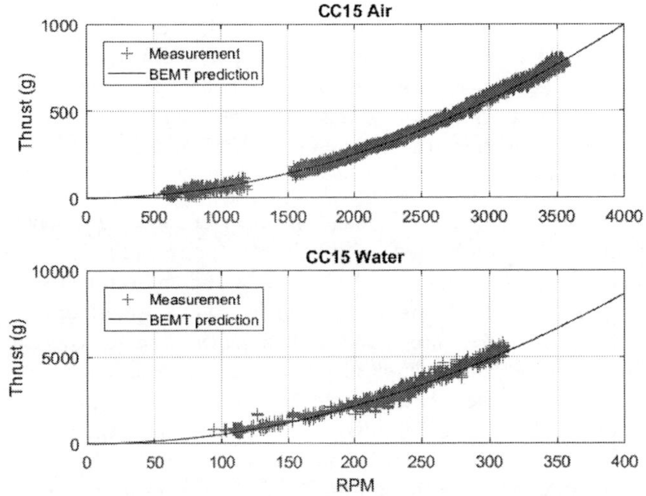

Figure 18: Correlation of BEMT thrust prediction for CC15 rotor.

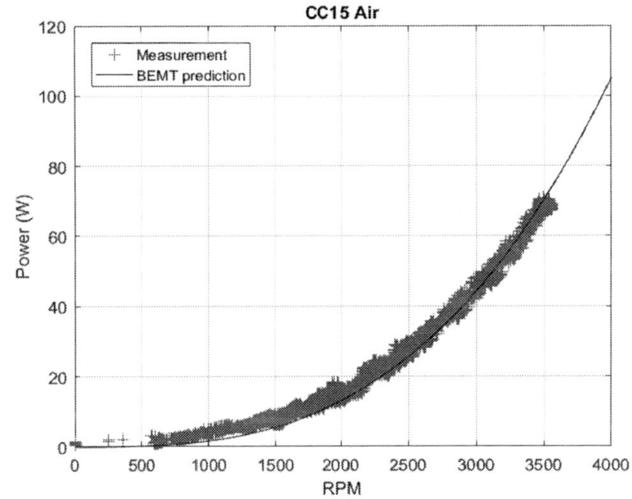

Figure 19: Correlation of power prediction with test data in air exemplified by CC15 with assumed motor-ESC efficiency of 65%.

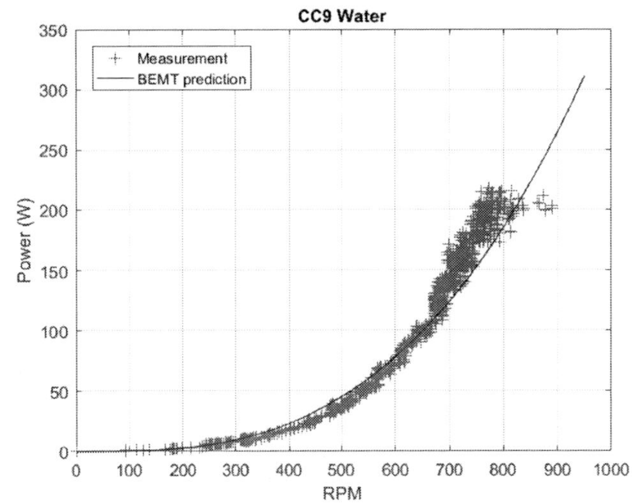

Figure 20: Correlation of power prediction with test data in air exemplified by CC9 with assumed motor-ESC efficiency of 20%.

Transition analysis

For smooth powered transition, it is necessary to develop low order models for rotor thrust prediction as it encounters the air-water transition media. Crane et al. (Ref. 10) attempted this by introducing the concept of effective density. However, it was defined only in terms of rotor height above water. As will be shown, the rotor speed also should be taken into account and therefore an alternate index is presented.

In order to predict the amount of water encountered, a new metric is introduced: Transition Index (TI). Using experimental data, the RPM of a given rotor motor combination in air and water can be found for all throttle levels. These data are found to be smooth and a polynomial fitted curve can be applied to both conditions. These two curves are labeled RPM_a and RPM_w for the resulting RPM at a given throttle for air and water conditions respectively. Throttle is represented as a value between 0 and .5 due to the reversible nature of the motor. An alternate approach may be to store the RPM data in a table and use lookup and extrapolations procedure to find the same information. The Transition Index TI is determined as:

$$TI = 1 - \frac{RPM - RPM_w}{RPM_a - RPM_w} \tag{6}$$

Where RPM is the measured RPM at the current throttle, and RPM_a and RPM_w are the rotor speeds at the same throttle in air and water respectively. The methodology for this is described graphically in Figure 21; at a certain throttle setting (solid green vertical line), TI is determined linearly based on the corresponding RPM_a and RPM_w values using Eq. (6). TI returns 0 when the present RPM matches expectation for fully in air conditions and 1 for fully in water conditions with a linear interpolation for conditions in transition.

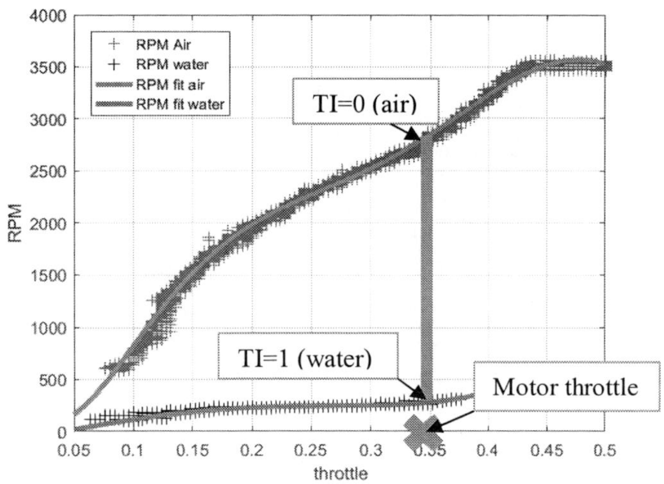

Figure 21: A graphical display of how TI is calculated using RPM vs throttle data for CC15.

Using empirically derived data, this metric allows for a real-time detection of rotor operating media: air, water, or air-water transition.

TI is useful for determining the amount of water the rotor is encountering with no additional sensors such as those onboard the Naviator (Ref. 6). An example of this is observed when the rotor is operating at a fixed height just below the water surface such as in Figure 22. As the RPM increases, the water is observed to be displaced and the rotor encounters less resistance as it operates in more air conditions. TI changes accordingly as more water is displaced. Labelled: (1.) the rotor speed is minimal and thus operates in pure water condition, (2.) rotor speed increases, water begins to be displaced, and the rotor begins to transition to a more in-air condition, (3.) rotor speed reaches a maximum and the closest to air condition is reached for this height.

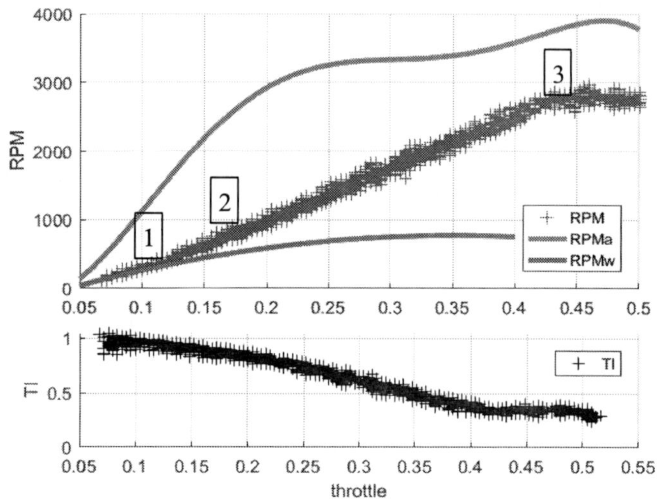

Figure 22: CC9 rotor operating 1.5 inches below water surface, TI computed inline based on measured RPM data. Labelled: (1.) rotor in water mode, (2.) rotor begins transition, (3.) rotor at maximum air condition.

TI is applicable at constant throttle, varying height conditions as well. The rotor was lowered into and then raised out of the water, when the rotor encounters the water the thrust increases and the RPM decreases. The Transition Index reflects that water is being encountered in this case as well, shown in Figure 23.

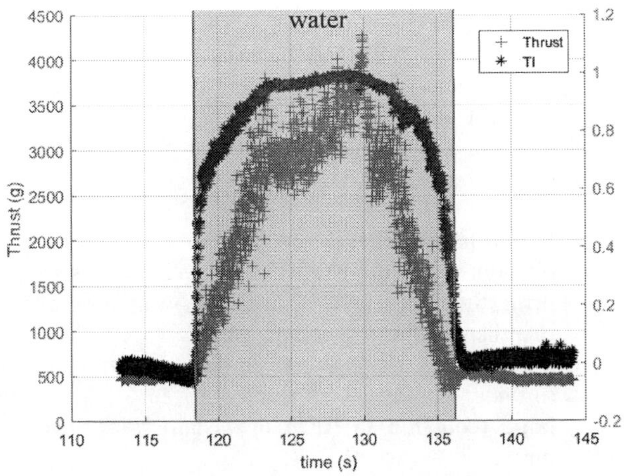

Figure 23: Variation of thrust and TI as CC15 rotor is lowered from air into water and raised back into air. TI becomes 1 when the CC15 rotor is submerged and 0 when it is not, seen as an increase in thrust.

In addition to being useful as a sensing metric, TI can also be used to predict thrust through the transition. This is extremely valuable to developing a control strategy for a stable air-water transition onboard the vehicle control using only readily available data. The methodology is analogous to the derivation of TI. The thrust for pure air and pure water conditions as a function of throttle is known from experiment. A curve fitting polynomial is applied to simplify the data, however a lookup table is also valid. The resulting thrust in air or in water for a given throttle is labeled T_a and T_w respectively and the current throttle setting is labeled thr. The resulting predictive equation is:

$$T_{predicted} = (T_w - T_a)|TI|^{A+B*thr} + T_a \tag{7}$$

The constants A and B are determined from performance measurement data. Applying this relationship shows a good prediction, some examples with the 15 inch rotor are shown in Figure 24.

The ability to predict thrust through the transition onboard a quadrotor amphibious vehicle is valuable for developing a control strategy for a smooth powered transition. This also allows for transition at various vehicle angles relative to the water surface since it is possible for one vehicle rotor to achieve the same performance as another in two different mediums. This predictive ability is also useful for unsteady environments.

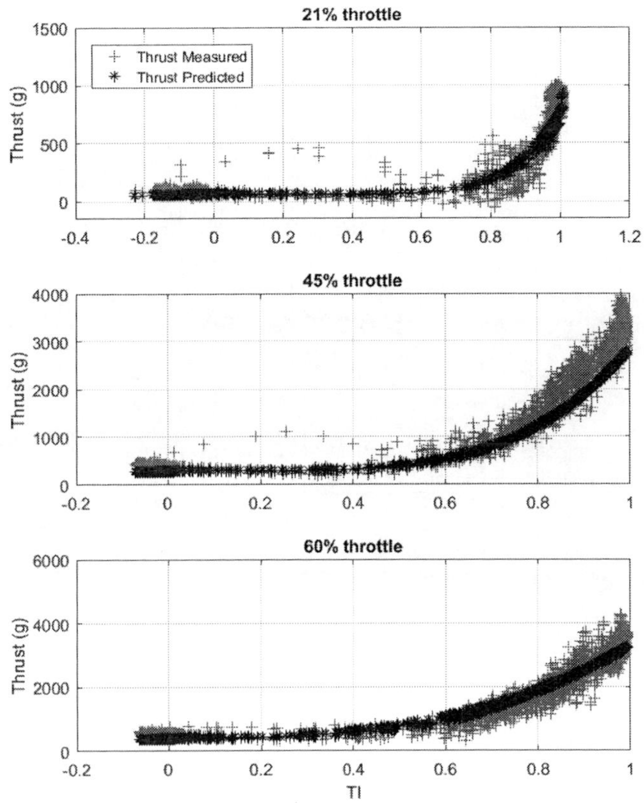

Figure 24: Prediction of thrust through the transition versus experimental data is shown for three throttle levels for CC15 rotor.

A control strategy is outlined for keeping constant thrust through the transition. This goal allows for stable vehicle operation through the transition with minimal propulsion or sensing configuration changes for a quadrotor. More advanced thrust control is also possible for another configuration. In order to achieve this a proportional integral derivative (PID) control is applied to the motor throttle in series with the autopilot's own controls. This architecture allows for minimal modification of the autopilot, and applies to any existing vehicle control. The process flow chart is shown in Figure 25. The system shown is designed so that the thrust in any condition is as close to the thrust expected in a pure air condition as possible.

This control strategy was applied to a condition where the 15 inch blade is lowered into water. The thrust results are shown in Figure 26 compared against a constant throttle with no control case. It can be seen that in the control-ON case the thrust remains relatively constant at all values of transition index as opposed to the control-OFF case. Additionally the initial peak in thrust during first contact as attenuated by the control by up around 60%.

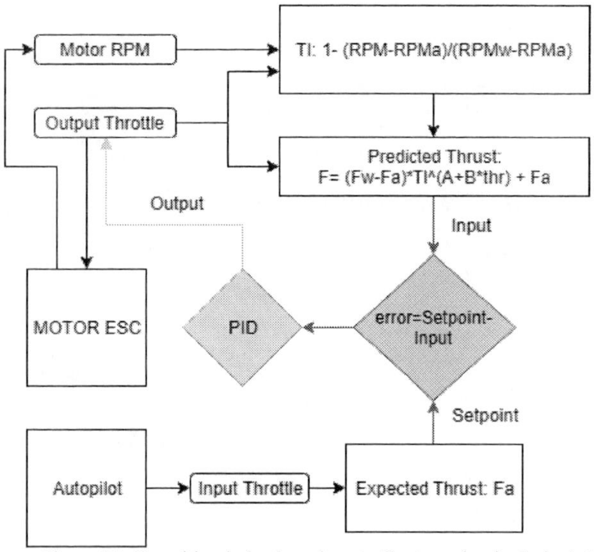

Figure 25: Transition index based controller to maintain desired thrust through air-water transition.

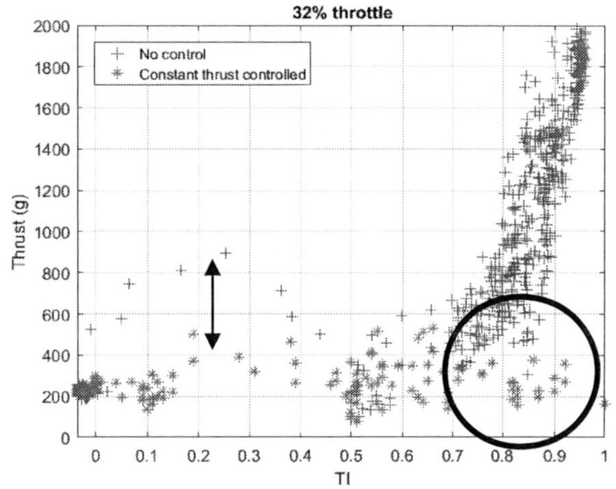

Figure 26: Control strategy shows two improvements. The thrust shock of entering the water is attenuated significantly (double arrow line). The thrust when submerged is held roughly equal to air condition (circled).

The success of this PID controller is a strong indicator to the potential of using the Transition Index for stable multi-rotor vehicle performance in multi-medium conditions. The set point thrust level can be modulated in more sophisticated control schemes according to vehicle characteristics such as buoyancy, weight, rotor geometry, motor characteristics or mission to achieve the desired performance.

CONCLUSIONS

Amphibious hybrid quadrotors provide a solution to performing a mission in both air and water with superior maneuverability underwater and the ability to hover in air. A Quad With Underwater Ability (AQWUA) is a demonstration vehicle with no additional propulsion systems or specialized sensing equipment capable of performing a powered rotor air-water transition. This allows for performing transitions at arbitrary angles relative to the water surface with mission specific control strategies.

Experiment and analysis of rotors was performed for air, water, and transition conditions. Based on this, the Transition Index (TI) is defined, indicating the amount of water present, returning 0 in air and 1 in water. TI can be subsequently used to predict thrust through the transition, this prediction is applicable to amphibious quadrotor control strategies and are being tested on the AQWUA.

1.) Experimental results are compared against blade element momentum theory (BEMT) analysis, and using the coefficient of thrust (C_T) thrust produced in either medium is well predicted.
2.) Motor efficiency is shown to decrease dramatically in water conditions from 65% to 20% because of large reduction in RPM to sustain same level of thrust.
3.) The Transition Index is introduced and shown to predict level of water encountered without specialized sensors, representing 0.0 and 1.0 in in air and water conditions respectively.
4.) The Transition Index is also applied for an empirical methodology for successfully determining thrust in air, water, and transition regions.
5.) Using the predicted thrust within a simple thrust controller is shown to be effective at keeping a constant thrust through the transition.

Future Work

Future work includes refining the analysis for more accurate prediction of power in water by incorporating a refined model of motor efficiency in conjunction with torque measurements. Additional work will also include applying the thrust predictive algorithm into a control strategy onboard the demonstrator vehicle for flight testing comparisons of transitions with and without the control strategy.

Author contact:
Ilya Semenov isemenov@umd.edu
Vikram Hrishikeshavan vikramh@umd.edu
Inderjit Chopra chopra@umd.edu

REFERENCES

[1] Fabian, A., Feng, Y., Swartz, E., Thurmer, D., and Wang, R., "Hybrid Aerial Underwater Vehicle (MIT Lincoln Lab)," 2012 SCOPE Projects, Paper 8 (2012)

[2] Aigeldinger, T., and Fish, F., "Hydroplaning by ducklings: overcoming limitations to swimming at the water surface," *Journal of Experimental Biology*, vol. 198, (7), 1995, pp. 1567-1574.

[3] Desbiens, A. L., Pope, M. T., Christensen, D. L., Hawkes, E. W., and Cutkosky, M. R. "Design principles for efficient, repeated jumpgliding." *Bioinspiration & biomimetics*, vol. 9, (2), 2014, pp. 1- 12.
doi: 10.1088/1748-3182/9/2/025009

[4] Alzu'bi, H, Akinsanya, O., Kaja, N., Mansour, I., and Rawashdeh, O., "Evaluation of an aerial quadcopter power-plant for underwater operation." IEEE 10th International Symposium on Mechatronics and its Applications (ISMA), Sharjah, United Arab Emirates, Dec. 8-10, 2015.
doi: 10.1109/ISMA.2015.7373488

[5] Maia, M M., Soni, P, and Diez, F. J., "Demonstration of an aerial and submersible vehicle capable of flight and underwater navigation with seamless air-water transition." *CORR arXiv:*1507.01932 (2015).

[6] Maia, M M., Mercado, D. A., and Diez, F. J., "Design and implementation of multirotor aerial-underwater vehicles with experimental results." Paper IEEE 961-966, IEEE/RSJ International Conference on Intelligent Robots and Systems (IROS) Vancouver, BC, Canada, September 24-28, 2017.
doi: 10.1109/IROS.2017.8202261

[7] Alzu'bi, H., Mansour, I., and Rawashdeh, O., "Loon Copter: Implementation of a hybrid unmanned aquatic–aerial quadcopter with active buoyancy control." *Journal of Field Robotics,* Vol. 35, (5), 2018 pp 764-778.
doi: 10.1002/rob.21777

[8] Yang, Q., and Qiu, W., "Numerical simulation of water impact for 2D and 3D bodies." *Ocean Engineering*, Vol. 43, 2012, pp. 82-89
doi: 10.1016/j.oceaneng.2012.01.008

[9] Yamasaki, J., Miyata, H., and Kanai, A., "Finite-difference simulation of green water impact on fixed and moving bodies." *Journal of Marine Science and Technology*, Vol. 10, (1), pp. 1-10
doi: 10.1007/s00773-005-0194-1

[10] Crane J, Faruque I. "Testing and characterization of hybrid unmanned aerial/underwater vehicle at the air-water interface". American Helicopter Society International Annual Forum and Technology Display Annual forum proceedings, Fort Worth, Texas, May 2017, pp.2986-2993

[11] Tan, Y. H., Siddall, R., and Kovac, M., "Efficient aerial–aquatic locomotion with a single propulsion system." *IEEE Robotics and Automation Letters*, Vol. 2, (3) Jul. 2017, pp. 1304-1311.
doi: 10.1109/LRA.2017.2665689

[12] Leishman, J. G., *Principles of Helicopter Aerodynamics*, Cambridge University Press, New York, NY, 2000, Chapter 3.

[13] Winslow, J., Otsuka H., Govindarajan, B., and Chopra, I., "Basic Understanding of Airfoil Characteristics at Low Reynolds Numbers (10^4–10^5)." *Journal of Aircraft,* Vol. 55, (3), Dec. 2017, pp. 1050-1061.
doi: 10.2514/1.C034415

[14] Winslow, J. M., "Understanding of Low Reynolds Number Aerodynamics and Design of Micro Rotary-Wing Air Vehicles." Master's thesis, University of Maryland, College Park, Maryland, 2016.
doi: 10.13016/M29Z9H

Disruptive Advance in P.M. Motor Technology Provides Higher Top Speed and Extended Range

Lawrence Zepp
Chief Technology Officer
Dura-Trac Motors, Inc.
Fort Wayne, IN, USA

ABSTRACT

The electric vertical takeoff and landing (eVTOL) industry can benefit from a disruptive advance in permanent-magnet motor technology that increases top speed and extends range. The Dura-Trac Motors field-weakening technology is mechanical, fail safe and well proven in the commercial-vehicle hybrid industry for 15 years. This technology is being introduced to the eVTOL market. The technology can be part of a ground-up design motor or added to existing brushless motor designs as a motor-manufacturer option, requiring few modifications.

DuraTrac mechanical field-weakening technology is far superior to electronic field-weakening for achieving higher motor speeds. During cruise flight, DuraTrac acts as a transmission, enabling motor top speed to be increased up to 40% on command. Because DuraTrac technology preserves high motor efficiency, range is also extended. The Dura-Trac technology allows greater flexibility for the user to select greater speed and range to suit the mission while continuing to provide the high torque required for hover and lift performance.

INTRODUCTION

The advance of multi-rotor drones and work by NASA and others in distributed electric propulsion has spurred the expanding capabilities of eVTOL craft. These advancements focus on multiple electric-lift motors with simple fixed-pitch props and redundant fault-tolerant controls. Brushless permanent-magnet (PM) electric motors provide a power dense, light weight and fault tolerant propulsion method. Nonetheless, further advancement in motors and controls is needed to optimize performance, range and efficiency with existing and future battery and energy storage technologies.

This paper introduces a method that is applicable to all brushless PM motors by extending the dynamic speed range beyond the normal limits without sacrificing efficiency. DuraTrac field weakening provides the advantages of a transmission or an adjustable-pitch propeller without the complexity of these alternatives.

RISE OF EVTOL AIRCRAFT

In the past eight years, we have seen the convergence of electric motor, battery, GPS and electronic control technologies to enable the rapid growth of the drone market. This development of quad, hex and octo drones using redundant high-output brushless PM and simple fixed-pitch props has formed the justification for distributed electric propulsion (DEP).

Early proponents of DEP have been Dr. Mark Moore of NASA, JoeBen Bevirt of Joby Aviation and others. The NASA X-57 research plane used (18) electric motors to energize wing lift in the LEAPTech Project. (Ref.1) (Ref. 2) The use of (4) or more brushless PM motors with a simple fixed-pitch prop provides distributed propulsion, simplicity, and redundancy for the rapid expansion of eVTOL designs emerging during the last few years.

Kenneth I. Swartz summarized this expansive growth of eVTOL craft in his article "Charging Forward – New eVTOL Concepts Advance." (Ref. 3) This excellent article chronicled the Uber Elevate Summit 2017 with over 500 stakeholders in attendance.

UNDERSTANDING THE ELECTRIC MOTOR – THEY'RE NOT ALL THE SAME

We are surrounded with an increasing number of highly efficient motors that use both permanent magnets and electric commutation – commonly called brushless PM motors. These designs can have the magnetic rotor on the inside of the coils (internal rotor) or on the outside of the coils (external rotor or outrunner). External-rotor designs are favored as spindle motors for your hard-disk drive and computer cooling fan due to their higher torque for a given size. External rotor PM motors are often a popular choice for drone and eVTOL lift motors.

Understanding the operating characteristics of PM brushless motors can improve our applications in the eVTOL field. For example, the same rules apply to internal and external rotor PM motors. The performance of these motors is generally

Presented at the Vertical Flight Society Autonomous VTOL Technical Meeting and Electric VTOL Symposium, Mesa, Arizona, UAS, Jan. 29-31, 2019.

very linear, which helps us understand, estimate and compare performance.

For a given motor design, motor speed (RPM) is linearly proportional to voltage. With increasing load (torque), the speed curve usually has a constant slope. Output torque is linearly proportional to the current (Amps) and the current curve usually maintains a constant slope. This produces a characteristic graph that is shown in Fig. 1. Note that for a given torque, you can read the speed on the solid line and amps on the dashed line. Compared to other electric motor types, these curves in a brushless PM motor are amazingly linear and constant slope. This makes predicting PM motor performance much easier. For a more detailed explanation, see (Ref. 4).

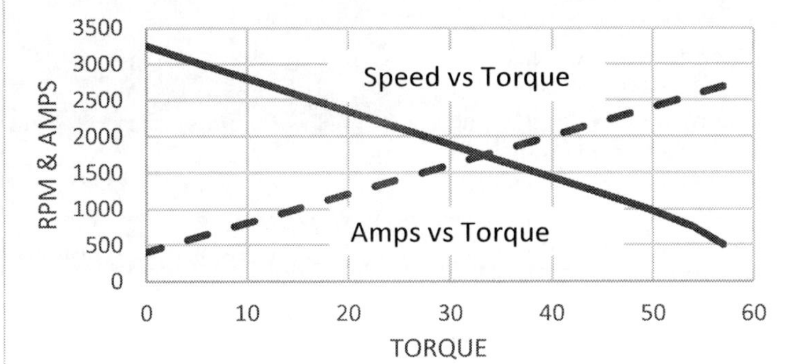

Fig. 1 – Typical relationship between speed, torque and amps in a brushless PM motor

The key motor/prop performance specification for eVTOL applications is vertical lift thrust. Lift thrust is often set at 125%-150% of the static hover thrust required to enable vertical acceleration and to allow a margin for high-altitude and/or high-temperature conditions causing reduced performance. This lift thrust point is often near the maximum power or maximum efficiency points on the motor curve or the motor is oversized for the application.

What Brushless Motors Can/Cannot Do

Brushless motors that utilize the new high-energy rare-earth magnets have especially high torque and power outputs for their weight. This makes these magnets well suited for eVTOL and drone service. One characteristic of permanent-magnet motors that can limit performance is a narrow dynamic-speed range. Because the strong permanent magnets create their own field, at increasing motor rpm the magnets induce an increasing voltage in the stator coils, called the "Back Electromotive Force," or Back EMF.

When this back EMF builds up to the level of the supply voltage of the batteries, then increases in motor rpm hit a wall and the motor is at maximum rpm. With the high torque, lower-speed motors suited to eVTOL and prop-drive applications, this means that when a motor is unloaded by factors like aerodynamic lift of the vehicle in cruise, the increase in rpm is modest at best.

This paper illustrates how the brushless motor rpm performance can be extended by using a mechanical field-weakening technique that achieves significantly extended cruise speed and range while preserving high efficiency. Mechanical field weakening acts like a transmission or torque converter, but it is a light and fail-safe option that can be added to existing motor designs (Ref. 4).

Configuration of a Typical eVTOL Air Vehicle

The Surefly eVTOL air vehicle, a product of Workhorse Group, Inc., is representative of an (8) electric motor vehicle where lift motors provide both lift and forward thrust. When the eVTOL vehicle transitions from vertical hover to cruise mode, the horizontal centerline is tilted to provide forward thrust just as a helicopter does. Increasing forward speed generates aerodynamic lift and lift reduces required thrust to maintain level flight. As we see in Fig. 1 of a typical PM performance curve, reduced torque allows only a modest increase in motor and prop speed. This is due to the narrow dynamic speed range of the high torque/lower speed motor.

How Does the PM Motor Work in an eVTOL Application? Surefly as an Example.

In the Surefly vehicle by Workhorse Group (Fig 2), maximum cruise speed is listed as 75 mph (121 km/hr) using the motor/prop technology available. Even though the body and arm shape may generate significant amounts of aerodynamic lift at cruise, the back EMF and characteristic PM motor curve restrict the motor-rpm dynamic-speed range. If a transmission could be added between the motor and the prop, motor rpm could be increased during cruise.

Fig. 2 – Surefly – Courtesy Workhorse Group, Inc.

Unfortunately, weight and complexity make this hard to justify. If a controllable pitch prop were available, then changes in prop pitch would extend the cruise speed range. The added complexity of controlling and synchronizing pitch of (8) props provides several additional failure modes to consider. Most eVTOL vehicles rely on the simplicity and redundancy of a fixed pitch prop that is directly driven by a brushless PM motor.

Field Weakening – Mechanical vs. Electronic

Dura-Trac technology can be added to an existing brushless-motor design or used in a new motor design. By weakening the magnetic effect between the rotor and stator coils on command, cruise speed can be extended by up to 40%. The mechanical field weakening effect acts like a transmission, but it is a simple and fail safe way to extend motor rpm that is inside the motor case. This mechanical method of field weakening preserves the motor high efficiency, unlike electronic field weakening methods. Electronic methods use additional energy to "buck" the rotor magnetic field, resulting in a major efficiency drop. Fig. 3 compares extended motor speed using Dura-Trac compared to electronic field-weakening methods (Ref. 5).

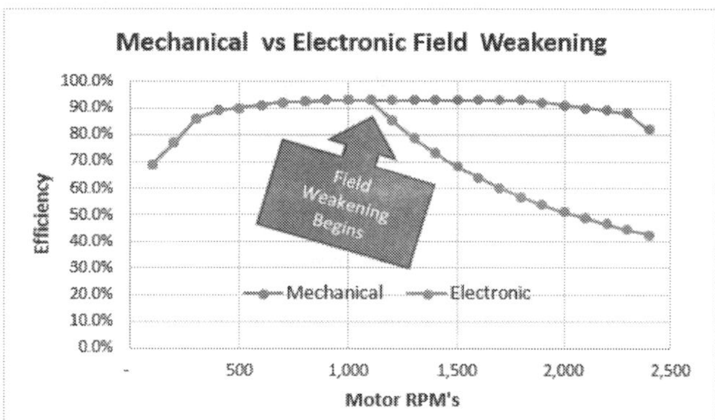

Fig. 3 – Efficiency difference between Mechanical and Electronic Field Weakening. Courtesy Duratracmotors.com

Does Dura-Trac Really Work?

Commercial shuttle buses and delivery trucks require very high launch torques due to their weight. For a commercial vehicle hybrid, a conventional brushless motor could produce this massive torque, but the top speed of the motor was limited to 10 mph (16 km/hr)! The application required a top speed of 35 mph (56 km/hr) for urban and suburban applications. The Dura-Trac motor achieved the required expanded speed range by applying field weakening to this motor.

The Dura-Trac hybrid accelerated the bus from a stop with full-launch torque. As speed approached 10 mph, the motor controller moved the magnet rotor to field weaken as needed, functioning as a magnetic transmission. During deceleration, the field weakening effect was reversed to optimize regen braking of the brushless PM motor. In instrumented long-term testing, the Dura-Trac system increased shuttle bus miles per gallon (MPG) (L/km) by 25%. This hybrid system has been in successful operation since 2005. This technology was torture tested and Cummins Engine approved for heavy-duty commercial service. The eVTOL industry can gain similar performance advantages by adding this disruptive technology to existing or new motor designs.

Fig. 4 – Crosspoint Kinetics hybrid using Dura-Trac field weakening technology – Photo courtesy Crosspoint Kinetics

HOW TO IMPLEMENT DURA-TRAC FIELD WEAKENING IN EVTOL VEHICLES

Most eVTOL motor applications do not need the massive launch torque or wide dynamic speed range required by commercial vehicles. Field weakening, on a reduced scale, can provide a very beneficial increase in cruise speed and range in many eVTOL applications. (Ref. 6) The enabling factor in the eVTOL operating cycle is the benefit of generated aerodynamic lift due to body shape and increased cruise speed. Any lift generated can proportionally reduce motor torque required.

By incorporating limited magnetic field-weakening shift to a typical external rotor PM motor, the resulting rpm increase can enable significant increases in eVTOL cruise speed and range performance. Let's look at how this disruptive PM brushless technology can be applied to a typical eVTOL craft.

Dura-Trac Field Weakening Application Example

Baseline performance of a two seat eVTOL vehicle using (8) conventional external rotor brushless motors with fixed pitch props is used as an example. This system enables hover and vertical climb capability for a maximum gross weight of 1,800 lbs. (816 kg). In cruise mode, the fixed motors achieve a cruise speed of 75 mph (121 km/hr).

If the body and strut profile is optimized for aerodynamic lift generation, then 20-30% generated lift may be achieved at 75 mph (121 km/hr). With this additional lift reducing the required motor torque, partial magnet field weakening of the rotor can increase motor / prop rpm to achieve a higher cruise speed. In this example motor, a 15% field weakening shift increases rpm by 17% and a 30% shift increases rpm by 40%. Fig. 5 shows the typical torque/speed/efficiency of this method.

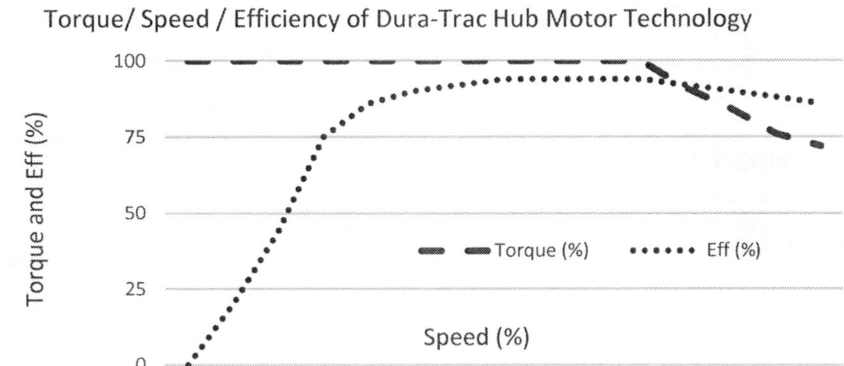

	0	10	20	30	40	50	60	70	80	90	100	110	120	130	140
Torque (%)	100	100	100	100	100	100	100	100	100	100	100	91	84	76	72
Eff (%)	0	20	44	75	86	90	92	94	94	94	94	92	90	88	86

Fig. 5 – Typical torque/speed/efficiency using DuraTrac mechanical field weakening

This method enables important cruise-speed extension that can be added to existing motor designs. Because the majority of fully engaged motor efficiency is preserved by this small amount of magnet decoupling, this also enables important range extension benefits using existing energy storage technologies, as shown in Fig. 5. This method is superior to electronic field weakening, which drops efficiency significantly – creating a severe penalty for increased speed that spoils range, as shown in Fig. 3.

Adding Field Weakening Technology to a Typical Motor

A cross section of a typical external rotor (or out-runner) brushless motor for eVTOL is shown in Fig. 6. The domed rotor housing carries the high energy magnets on the inside diameter and this structure rotates on the shaft. The prop is bolted to the hub, allowing a simple and rugged connection. The motor stator contains the drive coils and is stationary and concentric inside the hub.

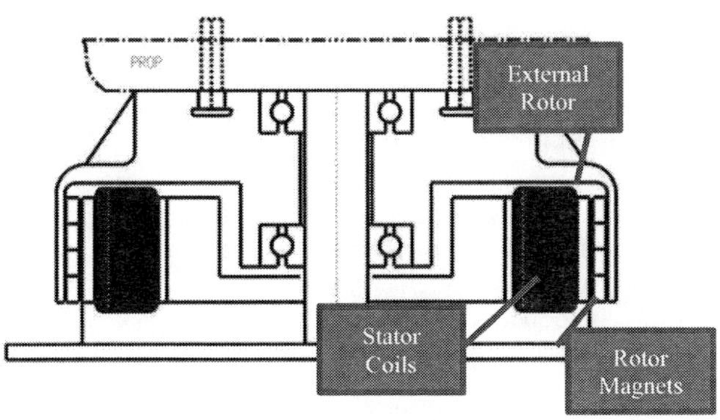

Fig. 6 – Cross section view of a typical external rotor PM motor for eVTOL applications

A cross section of a DuraTrac field weakening brushless motor for eVTOL is shown in Fig. 7. Note that in this fully engaged position, the rotor magnets and stator coils are common with the conventional motor, which makes conversion to a field weakening design much easier and more cost effective.

The motor performance is identical to the conventional motor shown in Fig. 6 and full-design torque and speed for hover is achieved. The domed rotor housing contains an internal hub with a splined connection to the motor shaft.

In this diagram, a ball spline connection is shown. The shaft rotates with the external rotor structure and contains a hub that is bolted to the prop. A shift actuator is located between the rotor hub and the stator coils. Several technologies are available to provide the electrically controlled axial shift of the external rotor.

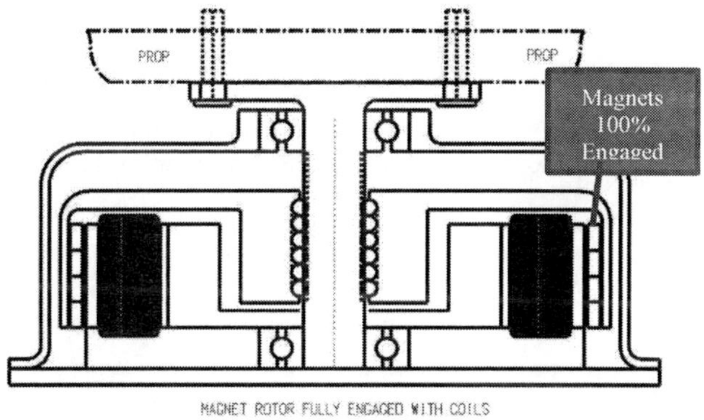

Fig. 7 - Cross section view of a typical DuraTrac PM motor with rotor magnets fully engaged with the stator coils

Effect of Dura-Trac on Flight Performance

When the air vehicle reaches cruise speed and aerodynamic lift is generated, a portion of the maximum design torque can be traded for additional motor rpm by the use of field weakening. In many eVTOL applications, two-position axial field-weakening may be sufficient for simple and reliable control of extended speed. (In other cases, multiple shift positions are available using variable shift actuator technology.)

In Fig. 8, the spinning external rotor has been extended 15% by the shift actuator. Now 85% of the magnets are active with the stator coils and back EMF is reduced. The motor now follows a higher speed/lower maximum torque performance curve that enables a typical 17% increase in motor rpm. If the shift actuator is turned off or disabled, then the magnets automatically retract to the 100% engaged full torque position for hover.

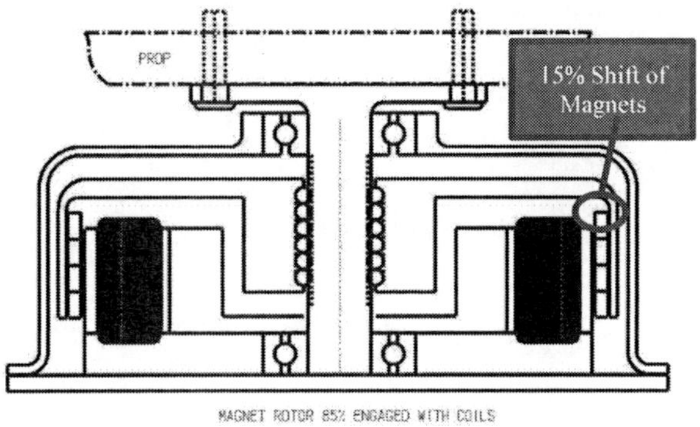

Fig. 8 – A typical DuraTrac PM motor with rotor magnets 85% engaged (15% shifted) with the stator coils to achieve a 17% rpm increase

If additional aerodynamic lift is generated, a greater portion of the maximum design torque can be traded to additional motor rpm by the use of field weakening. In Fig. 9, the spinning external rotor has been extended 30% by the shift actuator. Now 70% of the magnets are active with the stator coils and back EMF is reduced. The motor now follows a higher speed/lower maximum-torque performance curve that enables a typical 40% increase in motor rpm. If the shift actuator is turned off or disabled, then the magnets automatically retract to the 100% engaged full torque position for hover.

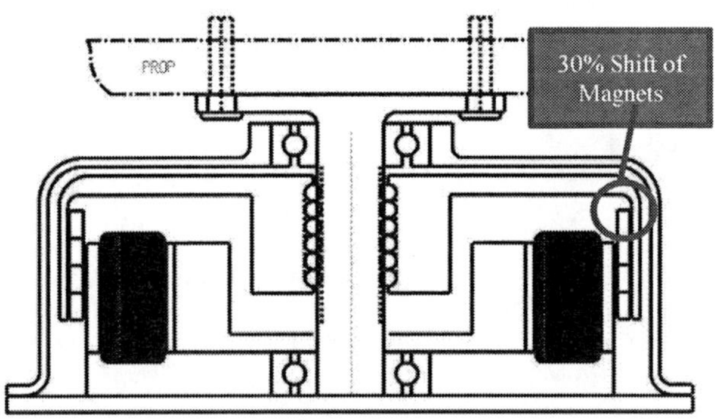

Fig. 9 – A typical DuraTrac PM motor with rotor magnets 70% engaged (30% shifted) with the stator coils to achieve a 40% rpm increase

The Physics (and Math) of Field Weakening a PM Motor

For a detailed discussion of the physics of both DuraTrac mechanical field weakening and electronic field weakening, see the white paper on the web site (Ref. 5). Have just a little time for some math? The following paragraphs explain the physics behind calculating the efficiency and torque of permanent magnet motors. Take time to read and understand. We promise there's no test.

As illustrated in Fig. 1 and Fig. 5, permanent magnet motors are linear with respect to their torque/speed performance and the ratio changes created by field weakening. While the following example is for just one motor configuration, the data indicate how mechanical field weakening can be applied to any PM brushless motor design.

Extended speed = (1 ÷decimal engagement) x Base Speed

- 100% rotor engagement, max speed 1,310 rpm's, at 96 V.
- 75% rotor engagement (25% decoupling) the speed increases to 1,746 rpm's. In this condition, 75% of the rotor is exposed to the stator coils (DuraTrac) or the magnets are bucked to reduce their attraction to 75% (Electronic methods). Because of the linearity of PM motors, this new maximum speed at 75% will be speed at 100% multiplied by 1÷ the decimal engagement 1 ÷ 0.75 = 1.33 *x* 1,310 = 1,746 max rpm's. The 1,746 rpm maximum agrees well with the dyno data for 75% rotor engagement.
- 50% rotor engagement (50% decoupling) the speed increases to 2,620 rpm's. In this condition, 50% of the rotor is exposed to the stator coils (DuraTrac) or the magnets are bucked to reduce their attraction to 50% (Electronic methods). Using the same method as above, we calculate: 1 ÷ 0.50 = 2.00*x*1,310 = 2,620 max rpm's. The 2,620 maximum agrees well with the dyno data for 50% rotor engagement.

If the original fully engaged motor torque is referenced from dyno data or a published performance curve, torque estimation is even easier. Multiply the fully engaged torque by the percentage of rotor engagement to get an estimate of resulting torque at the extended speed.

Adding Field Weakening Tech to Existing PM Motors

Existing motor designs used in the eVTOL field have a significant advantage when DuraTrac mechanical field weakening technology is added. The motor is already designed with appropriate performance for direct drive of the prop; weight, size and efficiency factors also have been optimized. The two most expensive sub-assemblies of the motor can be used with little or no modification: (i) laminated stator with wound coils; (ii) high-energy magnets. The stack length and air gap between the rotor and stator can remain the same.

Because many eVTOL applications require only a small percentage of shift, length change is small (Ref. 6). A rotor hub and shaft with a splined connection and an actuator does need to be added with one longer end bell for the axial stroke required. In many applications, a simplified two position actuator may be used that is electrically actuated from the motor controller. These features make testing the DuraTrac mechanical field weakening much easier and quicker, because many existing components can be applied.

Reliability, Efficiency and Dangers of Electronic Field Weakening

More than 20 years of reliability and efficiency history of on-road electric vehicles using electronic field weakening with PM motors can be applied to similar conditions in the eVTOL market. A DuraTrac white paper discusses the methods and impacts of additional energy used to buck the PM rotor magnets to simulate field weakening for extended speed (Ref. 5). Very few references reveal the high input of current needed and the major drop in motor efficiency when these currents are taken into account. The axial decoupling used by the DuraTrac system removes a portion of the magnets from the proximity of the stator, preserving high efficiency (Fig. 3).

The other danger presented by electronic field weakening is that the system is not fail safe if a controller malfunction occurs. Several on-road electric vehicles using electronic field weakening have experienced sudden motor-drive failures when the motor was operating beyond the normal maximum speed and using electronic field weakening.

When electronic-bucking field collapses, a large voltage spike can be generated, potentially failing the entire system. If a single motor controller shuts down in an eVTOL due to software, overheat or other conditions, a sudden dynamic braking of the motor causing a voltage surge can fail ALL motor-drive circuitry without rigorous voltage-spike safeguards. With typical interconnected battery packs, motor-controller power supplies and control circuits, electronic field weakening presents a serious risk to air vehicle safety and reliability.

CONCLUSIONS

1) The eVTOL industry can benefit from Dura-Trac Motors disruptive technology of PM motor axial field weakening to extend speed and range during cruise by 20-40%.
2) Aerodynamic lift generated during cruise can enable extension of the PM motor top speed through mechanical field weakening without the negative effects of electronic field weakening.
3) Many eVTOL PM motor applications require only a 15%-30% field weakening for performance enhancement, while retaining compact size of the PM motors.
4) DuraTrac field weakening technology can be added to existing motor designs, retaining many of the high cost components like the stator assembly and high strength magnets.
5) This technology was torture tested and Cummins Engine approved for heavy-duty bus commercial service. The eVTOL industry can gain similar performance advantages by adding this disruptive technology to existing or new motor designs.
6) DuraTrac mechanical field weakening retains high-motor efficiency and is fail safe, reliable and robust, unlike methods of electronic PM motor field weakening to extend rpm.

Author Contact: Lawrence Zepp; LarryZepp2@gmail.com

ACKNOWLEDGMENTS

Many thanks are due to the following individuals and companies for advancing the state-of-the-art of PM motor design and demonstrating the effectiveness of the mechanical PM motor field weakening concept.

- Jerry W. Medlin; Dura-Trac Motors principal investor, co-inventor of the field weakening PM motor patents.
- James R. Hendershot; Brushless motor design mentor, world renowned brushless motor expert and co-inventor of separated stator tooth patents.
- John R. Dabels; Electric vehicle marketing leader from EV-1 to present, Dura-Trac Motors President and technical presentation advisor.
- John McNichols; President of Energetic Drive Systems and Dura-Trac Motors patent licensee.

REFERENCES

1 Stoll, Alex M. et al, "Drag Reduction Through Distributed Electric Propulsion", Aviation Technology, Integration, and Operations Conference, 16-20 June 2014, Atlanta, Georgia, https://ntrs.nasa.gov/archive/nasa/casi.ntrs.nasa.gov/20140011507.pdf

2 Moore, Mark D., "Distributed Electric Propulsion", NASA Presentation, https://aero.larc.nasa.gov/files/2012/11/Distributed-Electric-Propulsion-Aircraft.pdf

3 Swartz, Kenneth I., "Charging Forward – New eVTOL Concepts Advance", Vertiflight, July / August 2017, http://evtol.news/2017/06/22/charging-forward-new-evtol-concepts-advance/

4 Zepp, Lawrence P., "Extended Speed and Range Accelerates eVTOL Development", Nov. 2018, Dura-Trac Motors, https://duratracmotors.files.wordpress.com/2018/11/2018-q4-duratrac-motors-inc-pdf-evtol-overview.pdf

5 Zepp, Lawrence P., "Advantages of Mechanical vs. Electronic-Field Weakening"., Nov. 2018, Dura-Trac Motors, https://duratracmotors.files.wordpress.com/2018/12/18-12-28-Advantages-of-Mechanical-vs-Electronic-Field-Weakening-Zepp-Paper.pdf

6 Dabels, John R., "DuraTrac Motor Technology Retrofit Space Guidelines", Dec. 2018, Dura-Trac Motors, https://duratracmotors.files.wordpress.com/2018/12/DuraTrac-Retrofit-Space-Requirement-Guidelines.pdf

Complexities of Lightning Certification of VTOL Aircraft

G.K. Tian*, J.P. McKennon †

NTS Lightning Technologies, Pittsfield, MA 01201, 413-499-2135
**George.Tian@nts.com, Project Engineer*
†Justin.McKennon@nts.com Chief Engineer, Department Manager – Engineering and Simulation

Abstract

Any conductive object flying through the air is at risk of being struck by lightning. Transport authorities such as the FAA require aircraft to be certified for lightning safety. This a difficult process, and is particularly challenging for VTOL rotorcraft for two main reasons. The first is the lightning zoning for VTOL rotorcraft, where the zoning must be done for both vertical and horizontal flight modes for shapes of rotorcraft with no set standard. The second reason is the extensive use of carbon composite and fiber glass in constructing the rotorcraft. The poor electrical conductivities can cause puncturing or EMI and conducted currents on electrical components when lightning strikes the rotorcraft. This can be mitigated by incorporating expanded metal foils or diverter strips to protect against puncturing, or metallic shields or in-line circuit protection elements for EMI or conducted currents. Multiphysics modeling can be utilized to help evaluate lightning risks and the effectiveness of lightning protection methods.

Introduction

VTOL rotorcraft has seen expanding development for both commercial and military use. While metrics such as efficiency and flight range are optimized for during the design stage, all aircraft, including VTOL rotorcraft, must be certified for safety for flight by various safety regulations such as the FAA. One often overlooked safety certification is lightning safety certification. Given that regular aircraft is struck by lightning about once per thousand flight hours [1], if they are not properly protected from lightning it can have devastating consequences. These can range from Electromagnetic Interference (EMI) causing functional upset [2] in electrical components to fuel tank explosions.

Certifying VTOL rotorcraft for lightning safety presents unique challenges that are not usually present for typical aircraft. Firstly, the lightning zoning of VTOL rotorcraft is not straightforward due to the multiple rotors it has along with the different flight modes potentially changing the positions of the rotors. This can cause portions of the rotorcraft to experience lightning strikes of greater severity in areas where it would normally not if it were a normal aircraft. Secondly, many VTOL rotorcraft are made mostly out of fiber glass or carbon composites, with conductive electrical components such as antennas inside the rotorcraft near the exterior surface. Lightning strikes could either directly strike the electrical components, or conduct on the outer surface and cause EMI effects on the components.

Necessary protection methods such as shielding, diverter strips or expanded metal meshes should be included in the design in order to successfully pass certification. These protection methods should be implemented as early in the development process as possible to evaluate their effectiveness and effects on flight performance. Computer simulations can be used to aid this development process.

Lightning Zoning of Aircraft

Certification of typical aircraft for lightning safety starts by determining the lightning attachment zones. These in turn define the threat levels each portion of the aircraft is likely to see in the event of a lightning strike. The detailed process of lightning zoning is explained in ARP5412B [3]. A brief overview will be presented here.

Lightning Zones

Lightning zones can be broken up into three general zones: Zones 1, 2 and 3. Zone 1 indicates areas of the aircraft where initial lightning attachment, known as the first return stroke, is likely. These areas are generally parts of the aircraft that are protruding or have sharp edges, such as rotors, tailfins, landing gears etc. Zone 1 is further broken up into Zones 1A, 1B and 1C, with A having a low chance of hang on, B having a high chance of hang on, and C having a low chance of hang on with reduced amplitude. Hang on refers to the duration the lightning is attached to the particular location. Areas that have a high chance of hang on are typically not moving relative to the aircraft, such as landing gears of tail fins. Areas that have low chance of hang on are typically areas that are moving relative to the aircraft, such as rotors.

Zone 2 are areas where it is unlikely to experience and initial lightning attachment, but is likely to experience subsequent return strokes. Lightning typically consists of an initial return stroke and multiple subsequent return strokes, where the subsequent return strokes are of lower amplitude. Zone 2 is also known as the swept stroke zone as it indicates areas where the lightning will reattach when the aircraft is in motion. Similar to Zone 1, Zone 2 is also broken up into zone 2A and 2B, where A and B have low and high chance of hang on, respectively.

Zone 3 indicates areas that are not in Zones 1 or 2 but are in the conduction path of the entry and exit points of the return strokes. These areas are unlikely to experience any lightning attachments.

Lightning Zoning - Helicopter

VTOL rotorcraft in the vertical flight mode is similar to that of helicopters. The figure below shows an example of lightning zoning of a helicopter from ARP5412B [2].

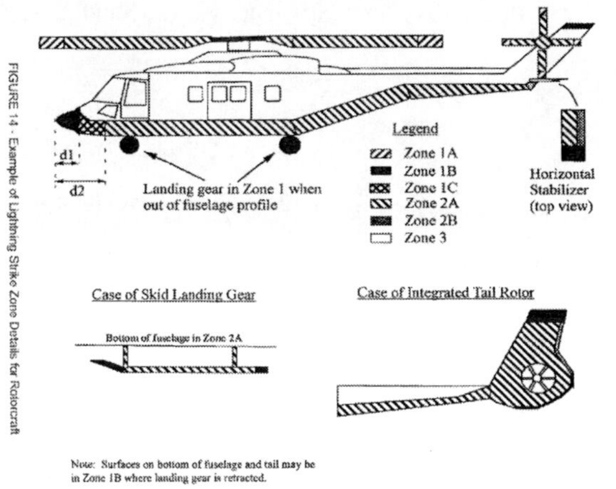

ARP5412B Figure 14 showing Lightning Zoning of a Helicopter

From the above example, it can be seen that the outer most edges with sharp protrusions of the helicopter are in Zone 1, with the rotating parts being in Zone 1A and the non-moving parts being in Zone 1B. Areas such as the nose of the helicopter would also overlap with a Zone 2B area, however since Zone 1B is considered more worst-case Zone 1B takes precedence.

Lightning return strokes last on the order of one to two seconds. Since the helicopter is assumed to be capable of ascending more than its height within this time duration, the under surface of the main rotor blade is highlighted as Zone 2A due to the swept stroke jumping from the rotor and onto the under surface of the main body.

Lightning Zoning - Propeller Aircraft

VTOL rotorcraft in horizontal flight mode is similar to that of propeller aircraft. The following figure shows an example of lightning zoning of a dual propeller plane from ARP5412B [2].

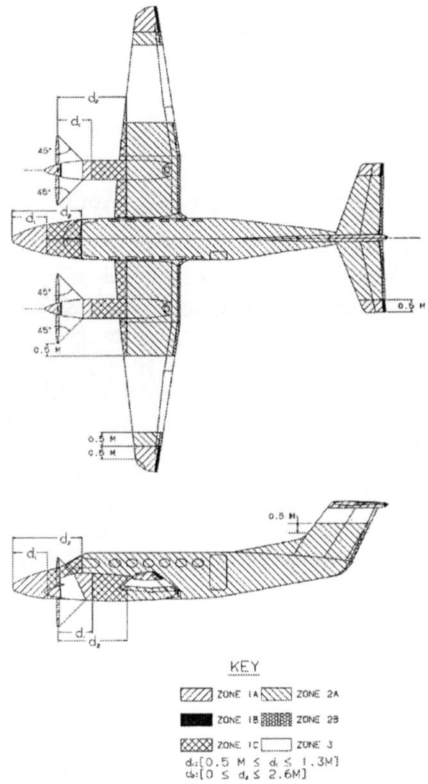

ARP5412B Figure 13 showing Lightning Zoning of a Dual Propeller Plane

Similar to the helicopter example, sharp protrusions on the aircraft are in Zone 1. Since the plane is assumed to be moving forward horizontally, much of the main fuselage is in Zone 2A, with the trailing wing tips and tail fin tips in Zone 1B.

Since the airplane is moving horizontally, the entirety of the fuselage is either in Zone 1 or 2. Compare this to the helicopter case previously where only the under surface of the main body was considered Zone 2A, with the majority in Zone 3. Additionally, the nose of the plane is considered Zone 1A, whilst in the helicopter case it is considered Zone 1B, which is a more severe case.

Lightning Zoning of VTOL Rotorcraft

Lightning zoning typical aircraft such as helicopters and airplanes is a relatively straightforward process as they only have one mode of flight, fixed geometries, and typically have standardized shapes. VTOL rotorcraft on the other hand have two modes of flight, vertical and horizontal. Not only do both modes of flight have to be certified for both flight modes, but rotation of engines further complicates both the lightning zoning and protection of the rotorcraft.

In addition to the above, VTOL rotorcraft designs are far from standardized, with many variations between each design. This includes but is not limited to: the number and position of the rotors, whether the rotors are enclosed in a metallic casing or not, the spacing of rotors from the main body and from each other, the angle and amount of rotation the rotors rotate. Zoning non-standard aircraft such as VTOL rotorcraft must be done carefully due to these factors.

Fiber Glass and Composite vs. Metal

In order to boost flight performance, many aircraft designers, including VTOL rotorcraft designers, opt to use fiber glass or carbon composite materials instead of aluminum when constructing the main body of their vehicles. While fiber glass or carbon composite can offer better mechanical properties at a lower weight cost than aluminum, the electrical conductivity is orders of magnitude lower. For insulative materials such as fiber glass, the possibility of puncturing and striking sensitive electronics becomes a problem. For slightly more conductive materials such as carbon composites, EMI effects or conducted currents could potentially disrupt the electronics.

Lightning Strikes and Puncturing

When lightning attaches to an aircraft or rotorcraft, it may potentially puncture the material. This generally occurs when conductive objects inside the rotorcraft are near the exterior surface near an attachment location, and the exterior surface has poor conductivity. The likelihood of puncturing is determined by various factors, including whether or not the surface is painted, how uniform the paint is, the conductivity of the material, and the homogeneity of the material [4]. Carbon composites are highly inhomogeneous, and fiber glass is insulative. Both of these increase the likelihood of puncturing.

The damage from puncturing is threefold. The first is that the lightning attaches to the conductive object inside the rotorcraft instead of the exterior surface. These conductive objects tend to be sensitive electrical components such as antennas which will fail if subjected to these lightning currents. The second is that since the lightning has attached to inside the rotorcraft, it exposes the interior components to large EMI effects. The third is that the physical hole left from the puncture may cause mechanical failures depending on the size of the hole.

Protection from Puncturing

Protection from puncturing can be done by modifying the exterior surface of the rotorcraft. This can be done by either adding an expanded metal mesh into the material of the main body, or by adding diverter strips to ensure that the lightning attaches to the exterior surface rather than internal components. Adding an expanded metal mesh into the main body prevents puncturing by increasing its effective conductivity, thereby avoiding the possibility of lightning attaching onto an interior conductive object. This has the added benefit of shielding the interior of the rotorcraft. This protection method does add a significant amount of weight to the rotorcraft. Attaching diverter strips to the exterior surface can prevent punctures by having the lightning attach onto the diverter strips rather than the interior conductive object. This is a lower weight option than the conductive mesh, but does not provide the additional shielding benefit.

A simple Finite Element Method (FEM) model was constructed in COMSOL Multiphysics to help demonstrate the effects of diverter strips. An antenna is placed in a fiber glass radome and grounded on the bottom. It is subjected to a DC electric field from above. The figure below shows a surface plot of the electric field intensity with the diverter strips inactive (left) and active (right).

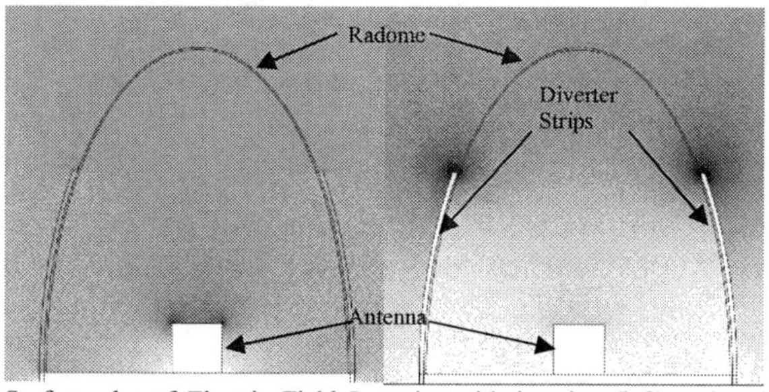

Surface plot of Electric Field Intensity with inactive (left) and active (right) diverter strips of an antenna inside a radome. Darker areas indicate higher intensity electric fields. Both images are plotted on the same scale. Note that the conductive objects (antenna, active diverter strips, ground) are white inside as the electric fields inside conductors are zero.

From the figure above, it can be seen that with the diverter strips inactive, the antenna's two top corners have high electric field intensity. This encourages streamer development at these corners which will encourage lightning attachment at these areas. With the diverter active, the antenna corners have lower electric field intensities, and the diverter strips have even higher intensities than seen on the antenna corners with inactive diverter strips. This shows that not only do the diverter strips lower field intensity around the antenna corners, but they are more likely to develop streamers and have lightning attach to them.

Care must be taken when implementing these protection methods. Simply adding expanded metal meshes or diverter strips is not enough to protect from puncturing. The expanded metal meshes must fully encapsulate the main body, and must be connected to relevant ground points on the rotorcraft. Diverter strips only work if they are grounded to conductive objects able to divert the lightning current away.

Effects of Low Electrical Conductivity

There are two major consequences of the lower electrical conductivity of carbon composite compared to aluminum. The first is the reduced shielding effects carbon composites give against electromagnetic waves. Large amounts of sensitive electrical components are used in many VTOL rotorcraft, which if not properly protected can be susceptible to EMI effects. When lightning attaches to a rotorcraft, large amounts of fast changing currents flow from the entry to the exit points. This current path is generally across the exterior surface of the rotorcraft and encapsulates any electrical components. For aluminum rotorcraft, this creates a coaxial configuration which cancels out most of the magnetic fields generated from the lightning currents, thus providing shielding for sensitive electronics. However, for carbon composite rotorcraft, the reduced electrical conductivity results in this shielding effect being reduced. This is exacerbated if the carbon composite has irregular conductivities, resulting in non-uniform current flows across the exterior surfaces.

The second consequence of lower conductivity of carbon composites is that the outer surface of the rotorcraft has a higher impedance than if it were made with aluminum. This means that any electrical components that are grounded to the exterior surface of the rotorcraft will experience higher currents than if the rotorcraft were made of aluminum.

Protection from EMI and Currents

The consequences of carbon composite's lower conductivity can be mitigated through various means. Components that are grounded to the exterior surface can be protected by shielding them in metallic boxes, grounding the shield in two locations, or by implementing in-line circuit protection mechanisms. Shielding will protect the components from EMI effects and will reduce the amount of conducted current that goes through it from the exterior surface. Although effective, this solution adds weight through the added aluminum used to shield the components. Using in-line circuit protection mechanisms such as metal oxide varistors or transient voltage suppression diodes (transorb) can mitigate the EMI and conducted current effects, but can also potentially affect the signal quality of the circuit.

For components that are floating relative to the exterior surface, shielding and grounding the shield can also protect it from EMI effects. The shielding can also be forgone and the component can simply be grounded to the exterior surface. While this saves on weight, it means that the component will be a parallel path for the current on the exterior surface of the rotorcraft.

Utilizing Multiphysics Modeling to Aid Design and Certification

Traditional design and certification processes involves testing physical test articles, either the entire aircraft/rotorcraft or sections of it. Engineering testing may have also been performed to evaluate either the need for lightning protection in certain areas, or to evaluate their effectiveness. This can be a costly process as many tests are destructive due to the nature of lightning, which necessitates multiple test articles. Evaluating multiple lightning protection method further compounds the costs. Furthermore, if the lightning protection method fails at the certification stage, a complete redesign may be necessary.

Testing is necessary in the design and certification process, but it also has many limitations. In order to measure quantities such as current, physical instruments such as current transformers or Rogowski coils must be placed around the object in question. For test that involve the entire rotorcraft, certain areas may be inaccessible and thus not able to be measured. In addition, these measurement devices may also pick up noise from other sources.

All of the above problems are especially true for VTOL rotorcraft. The multiple modes of flight of VTOL rotorcraft necessitate non-fixed geometries. This combined with the fact that this is an emerging technology with drastically varying designs and very few certified for lightning safety, which means that there is no set standard for lightning zoning the rotorcrafts. There is a high amount of risk if the traditional test-only process is used for VTOL rotorcraft lightning safety since there is a dearth of data for lightning protection and zoning for them.

To reduce the costs and risks associated with the test-only process, computer simulation techniques have been used to aid in the design and certification process. In the past, the use of SPICE-type circuit models were used to evaluate lightning protection methods. These types of models have a significant drawback. If not explicitly included, SPICE-type circuit models would not account for various physical phenomena such as capacitance and inductance of objects, field coupling effects, thermal effects etc. This is especially problematic in the lightning environment as lightning currents involve high amounts of energy in short durations. For example, one part of a Zone 1A strike is Current Component A, which is a 200 kA peak, 6.4 µs rise time and 69 µs decay to 50% waveform [5]. This large, fast changing current creates high intensity magnetic and electric fields that can induce currents and voltages in other conductors. These effects must be taken into account to accurate evaluate lightning protection methods.

Multiphysics modeling, such as COMSOL Multiphysics used in the radome with diverter strips example, can be used to take into account all of these effects. By building the geometry with the correct sizes and spacings, and by applying the correct material properties and boundary conditions, all of the above-mentioned phenomena are implicitly calculated. While it is possible to include these effects in a SPICE-type circuit model, for large, complex geometries such as an entire VTOL rotorcraft, this becomes exceedingly complex. These complexities include the number of elements that can influence a given object's capacitance or inductance, and the number of field coupling sources for a given conductor.

By utilizing Multiphysics modeling, lightning protection methods can be evaluated without having to manufacture any test articles. While there are limitations to the model's accuracy, which is based on the number of approximations made to make the model solvable, the model can provide good estimations of the currents and voltages throughout the system. These values can then be used to eliminate protection methods that do not meet the lightning safety criteria. The methods that pass the criteria in the simulations can be tested with less risk than before. Multiphysics modeling can be used before or in parallel with the design process. This allows designers to have engineering level data early in the development life cycle that can be used to evaluate risks before certification testing.

The below figures show an example of a Multiphysics model made in COMSOL Multiphysics of a turboprop engine shaft with a sensor. In this example, there was concern that a lightning strike to the engine would cause enough voltage to develop between a sensor and the beta ring such that a spark would form.

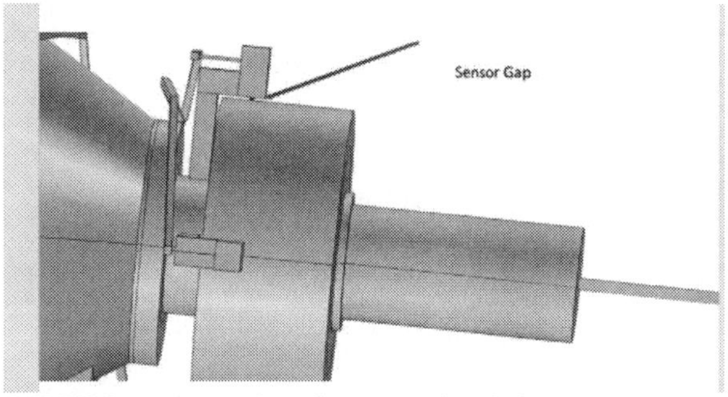

COMSOL geometry of a turboprop engine shaft

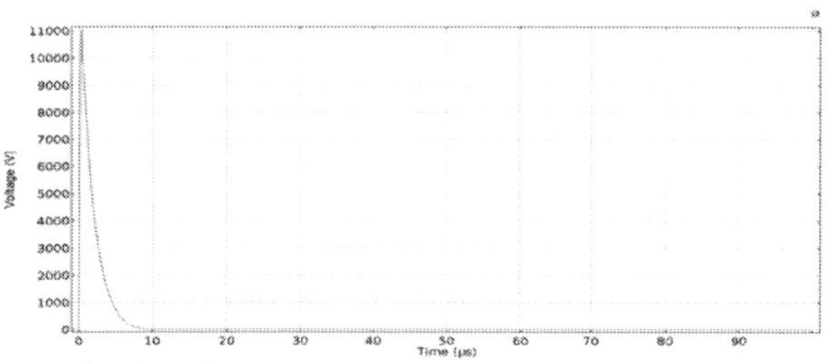

Plot of Voltage between Beta Ring and Sensor

From the plot above it can be seen that the voltage between the beta ring and sensor peaks at 11 kV. With the gap being less than 1 mm in length, a voltage of 11 kV would indicate the possibility of a spark. This example shows how Multiphysics modeling can be used to evaluate risk

Conclusions

Certifying VTOL rotorcraft for lightning safety presents unique challenges in terms of lightning zoning and threat assessment, especially due to the materials used, dual flight mode and non-standard geometries. The dual flight nature of VTOL rotorcraft means they must be tested and certified for both modes to ensure they can withstand the worst-case lightning zones of both helicopters and airplanes. The high variation of shapes of rotorcraft means that there is no standard lightning zoning method for VTOL rotorcraft compared to aircraft such as business jets. Optimization of efficiency and flight range lead to the usage of fiber glass and carbon composite materials, which have poor electrical conductivities. These materials' poor conductivities lead to either puncturing of the material, or EMI or currents adversely affecting sensitive electronics inside the rotorcraft. Protection methods such as expanded metal meshes, diverter strips, shielded enclosures or in-line circuit protection mechanisms can be implemented to prevent these effects. All of these protection methods must be evaluated for their effectiveness. Multiphysics modeling can be used to aid in this process by evaluating both the risk of lightning and the lightning protection effectiveness before physical testing is performed. By using Multiphysics modeling, engineers can obtain data before any testing is performed, provoding confidence that their rotorcraft is adequately protected from the effects of lightning strikes.

References

[1] C. E. Rash, Flight Safety Foundation, When Lightning Strikes, Aircraft Design incorporate system to protect against direct and indirect damage, July 9 2010

[2] U.S. Department of Transportation, Federal Aviation Administration, General Aviation Lightning Strike Report and Protection Level Study, August 2004

[3] Aerospace Recommended Practice, ARP5414B, Aircraft Lightning Zoning, Feb 1st 2012

[4] Franklin A. Fisher, J. Anderson Plumer, Rodney A. Perala, Lightning Protection of Aircraft 2nd Edition, 2004, pp. 175-176

[5] Aerospace Recommended Practice, ARP5412B, Aircraft Lightning Environment and Related Test Waveforms, Jan 2013, pp 26-27

Usability of a Steering Wheel Control Concept for a Highly Augmented Personal Helicopter

Bianca I. Schuchardt
Research Engineer
DLR German Aerospace Center
Braunschweig, Germany

ABSTRACT

Recent developments in the sector of urban air mobility have also increased the interest in personal helicopters. Helicopters are typically controlled by center stick, collective lever and pedals, while sidesticks offer alternative arrangements. This paper reports on a third option: the control of a helicopter by steering wheel. The car-inspired steering wheel control concept promises to be intuitively usable, especially for novice pilots with driving experience. The steering wheel replaces the center stick and the original yaw pedals are used for speed or acceleration commands. For altitude commands two different inceptors are proposed, either a conventional collective lever or rocker switches behind the steering wheel. Horizontal movements are commanded by an 8-way-switch on the wheel. In the paper both steering wheel configurations are compared to conventional helicopter controls in ground-based simulation. All investigated configurations received high usability ratings from the test group of helicopter pilots, while two groups of test candidates without flying experience gave better ratings for the control configurations with steering wheel compared to conventional helicopter controls.

NOTATION

AC	Attitude Command
AcC	Acceleration Command
ACT/FHS	Active Control Technology / Flying Helicopter Simulator
ATPL(H)	Air Transport Pilot License (Helicopter)
AVES	Air VEhicle Simulator
CPL(H)	Commercial Pilot License (Helicopter)
DLR	German Aerospace Center
GLM	General Linear Model
PAV	Personal Aerial Vehicle
PPL(H)	Private Pilot License (Helicopter)
RC	Rate Command
SUS	System Usability Scale
TC	Turn Coordination
TRC	Translational Rate Command
βC	Sideslip Angle Command
γC	Flight Path Angle Command

INTRODUCTION

Urban air mobility and the dream of personal air vehicles have recently experienced high attention in popular media but also made it into research (e.g. Ref. 1) and industry (e.g. Refs. 2 and 3). On the way to fully-autonomous air vehicles with human payload, easy-to-fly air vehicles are the logical next step for opening the skies to a wider range of potential users. Also for optionally piloted air vehicles, intuitive human-system integration is essential for maximum user acceptance. This paper reports on the development and usability assessment of a car inspired control concept for highly augmented rotorcraft.

Already in 1987 Drees (Ref. 4) proposed an "easy-to-fly" rotorcraft that would feature a car-like control and advanced control laws. In the 1940s Gazda (Refs. 5 and 6) had tried to build a helicopter with steering wheel control but could not successfully finish his project. Today's development of fly-by-wire architectures has removed the constraints of control devices that have to be mechanically linked to the rotorcraft's actuators (Ref. 7). This offers new possibilities of freely redesigning the conventional pilot controls for more comfort and efficiency. Sidesticks are the most popular competitors to conventional controls and constantly under research (Ref. 8). Car-like controls on the other hand promise an intuitive pilot interface that can be comfortably used by professional pilots as well as by future pilots who can profit from their – often extensive – experience in car driving.

The research documented in this paper, has its origins in the European project myCopter (Ref. 9). The development of the used steering wheel control concept for rotorcraft has been published in Refs. 10 and 11. Jones et al. (Ref. 12) also conducted investigations on car-like control concepts for PAVs, but they focused on thrust vectoring systems.

In a first piloted study in DLR's Air Vehicle Simulator (AVES, Ref. 31), the steering wheel control concept has demonstrated its viability and led to a reduction in workload

Presented at the Vertical Flight Society Autonomous VTOL Technical Meeting and Electric VTOL Symposium, Mesa, Arizona, USA, Jan. 29-31, 2019.

and improved handling qualities compared to conventional helicopter controls (Refs. 11, 13). In a second ground-based simulator study, the objective was to investigate the usability of the steering wheel control concept for different user groups. Helicopter pilots, car drivers and novice candidates with no experience in flying or driving had the chance to compare the steering wheel control concept to conventional helicopter controls. The achieved workload reduction has already been published in Refs. 11 and 14, while this paper focuses on the usability of the developed control concepts.

PERSONAL HELICOPTER

The idea of personal aerial vehicles (PAV), or a combination of aircraft and automobile has been there since the first flying machines took off from the ground (Ref. 15). Modern companies are still exploring these possibilities, today with modern technology (e.g. Ref. 16). Personal helicopters offer the advantage of vertical take-off and landing, especially useful in urban environments. Growing interest on personal helicopters was already prognosticated by Macauley in 1944 (Ref. 17). Today, companies like Volocopter (Ref. 18), Lilium (Ref. 19), Urban Aeronautics (Ref. 20), Joby Aviation (Ref. 21), Kitty Hawk (Ref. 22) or Ehang (Ref. 23) are constantly presenting new designs for personal rotorcraft. Uber is exploring future mobility concepts in cooperation with Bell (Refs. 2 and 24). Other established aircraft manufacturers are also exploring the PAV market with concepts like the CityAirbus (Refs. 3 and 25) and Vahana (Ref. 26) or Aurora Flight Sciences, recently acquired by Boeing (Ref. 27).

The European project myCopter (Ref. 9) had its start before the recent boom of PAV development evolved. Basic research was conducted on the technologies necessary for enabling future personal aerial transportation systems. The control concepts enrolled in this paper were inspired by the myCopter vision of potential future personal rotorcraft, available to a wide range of users. Nevertheless, the control concepts can also be applied to highly augmented helicopters already existing today, such as DLR's research helicopter ACT/FHS (Ref. 28). Highly augmented means in this case that the helicopter is equipped with fly-by-wire control and a model-following control environment that allows for the integration of advanced stabilization and control laws, see also Ref. 29.

STEERING WHEEL CONTROL CONCEPT

The idea of the proposed control concept is to use a steering wheel as primary control interface that takes the place of the conventional helicopter's cyclic stick. Figure 1 and 2 show two different set ups that were investigated. Turning the wheel results in flying coordinated turns in forward flight. In hover condition, the wheel commands a pure yaw rate. Conventional yaw pedals are repurposed for horizontal acceleration and deceleration. Finally, the height is controlled either by a conventional collective lever (see Figure 1, Configuration Wheel + Collective) or by rocker switches that are located behind the wheel (see Figure 2, Configuration Wheel + Switches) The right switch is used for ascending and the left for descending. An additional 8-way-switch in the center of the wheel allows precision manoeuvers, such as flying horizontally forward, backward, sideward, or diagonally at small speeds.

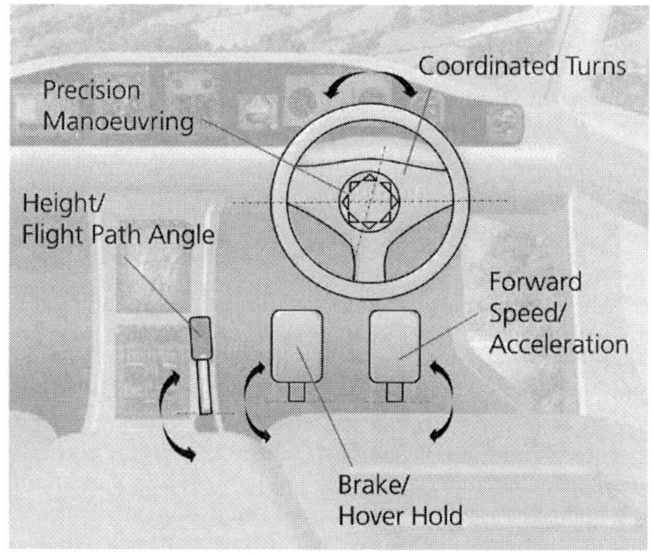

Figure 1. Steering wheel control concept with collective lever for height control (Wheel + Collective).

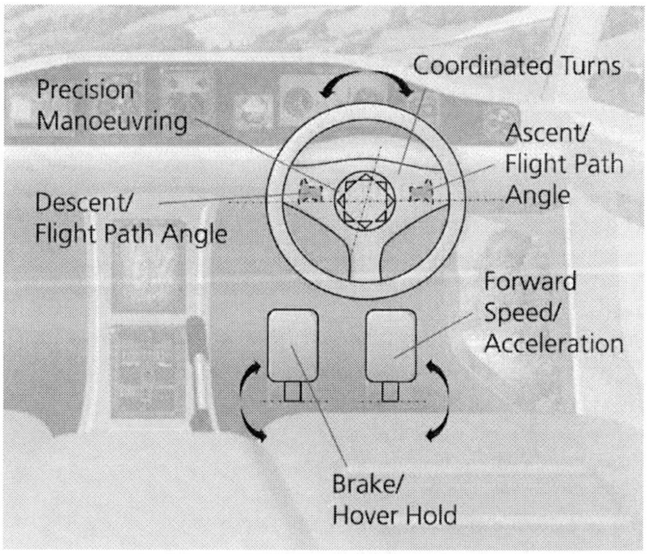

Figure 2. Steering wheel control concept with lever switches for height control (Wheel + Switches).

The underlying control laws were originally developed for usage with simulated PAVs during the myCopter project and represent the behavior of a highly-augmented personal helicopter. They offer a very high level of stabilization and augmentation such that minimal training is required even for flight-naïve users (Ref. 30). These control laws have successfully been flight tested on DLR's research helicopter and in-flight simulator ACT/FHS in combination with conventional controls (Refs. 11 and 29).

Figure 3 shows the control laws that were used in combination with conventional helicopter controls. In the longitudinal and lateral axes a Translational Rate Command

(TRC) is implemented for the low speed flight regime. This response type relates the control deflections of the pilot's center stick (cyclic) linearly to commands for the longitudinal and lateral ground speed. When the stick is returned to the neutral position, the rotorcraft returns to hover condition. Above 15 kn, linear blending starts towards the forward flight control mode which is fully active at 25 kn forward speed. The forward flight mode is of Acceleration Command (AcC) type in the longitudinal axis which means that the rotorcraft's acceleration command is proportional to the pilot's inceptor deflection. Thus, the current airspeed is held when the inceptor is returned to neutral. In the lateral axis an Attitude Command (AC) with attitude hold is implemented for speeds above 25 kn. Therefore, a lateral control input results in a proportional roll angle command.

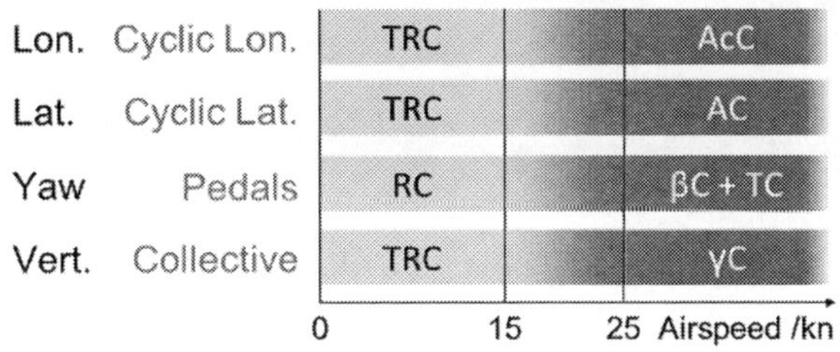

Figure 3. Control laws for conventional controls.

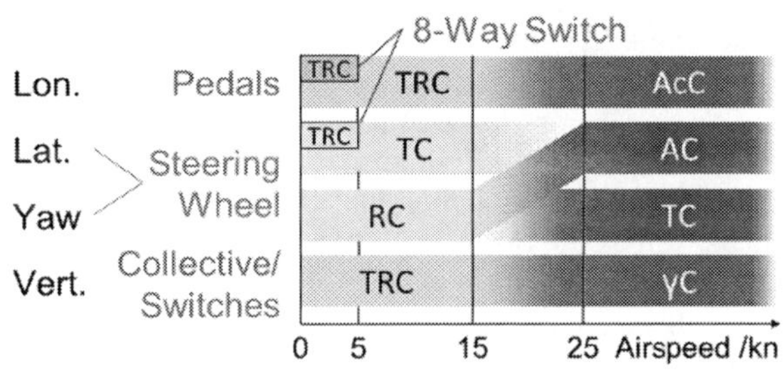

Figure 4. Modified control laws for steering wheel control.

The yaw control is designed as Rate Command (RC) response type in hover condition and for low speed. The yaw rate command is proportional to the pilot's pedal input. In fast forward flight the response type changes to a Sideslip Angle Command (βC) with Turn Coordination (TC). This allows flying coordinated turns (free of sideslip) in forward flight without additional pilot inputs.

Altitude is controlled with the collective lever via a vertical TRC response type in hover mode and changes to a Flight Path Angle Command (γC) in forward flight. Interaxis coupling is not present in the selected response type configuration apart from the turn coordination.

The mode change is designed one-directional such that the blending is active between 15 and 25 kn when accelerating. In order to return from the fast forward mode to the slow mode, the pilot must decelerate the rotorcraft back to hover while AcC is active until hover is achieved. Then, in hover, TRC is activated. The respective hysteresis prevents unintended mode changes while being in the transition phase.

Figure 4 shows the changes in the control laws for usage with the steering wheel. The dynamics of the longitudinal axis remain unchanged but are now controlled via pedals. The steering wheel controls both lateral and yaw axes with a turn coordination being implemented over the complete speed range. For low speeds the wheel is primarily used for yaw commands. These are linearly blended over into roll attitude commands between 15 and 25 kn. The vertical axis response remains the same as for the conventional inceptor arrangement but can now be controlled either via collective lever or switches. The precision maneuvers that can be commanded via the 8-way switch are implemented as additional TRC in the longitudinal and lateral direction for speeds up to 5 kn.

EXPERIMENTAL SETUP

The study was conducted at DLR's AVES (Air VEhicle Simulator, see Figure 5 and Ref. 31). The helicopter cockpit (resembling DLR's EC135 ACT/FHS) was used on the fixed-base simulation platform.

Figure 5. DLR's Air Vehicle Simulator Center AVES.

Figure 6. AVES helicopter cockpit featuring conventional and steering wheel inceptors.

Figure 6 shows the inside of the helicopter cockpit with the steering wheel prototype installed. This prototype was not attached to the airframe but rested on the pilot's knees. This allowed for easy interchangeability between the different control concepts. When the steering wheel was used, the conventional center stick was deactivated and trimmed to the most forward position.

MATLAB/Simulink was used for the development of the control laws. Details can be found in Refs. 11, 13 and 32. The integration into the AVES simulation environment was conducted with the in-house developed 2Simulate software (Ref. 33).

The following three control configurations were investigated in the study:

- Conventional helicopter controls as reference,
- Steering wheel control and car-like pedals combined with a conventional collective lever, and
- Steering wheel control and car-like pedals combined with rocker switches for height control.

For the usability testing the same simulated scenarios and the same group of test candidates were used as for the workload investigations published in Refs. 11 and 14. After a concise training session in the simulator with all of the control configurations, the flight tasks were assigned. The first scenario was adapted from the slalom mission task element defined in ADS-33E-PRF (Ref. 34). The test candidates had to ascend from a runway and accelerate the simulated rotorcraft up to 40 kn. Then they had to follow the actual slalom. A head-down tunnel-in-the-sky display (Ref. 35) showed the desired flightpath. An urban environment was selected as location for the second scenario. Here, the test candidates had to fly from a parking area through a street canyon with several turns until they landed on a rooftop helipad. The desired flightpath was again shown on the tunnel display.

During the evaluation runs flight performance and workload were measured as already published in Ref. 11 and 14. The study was completed by the evaluation of the control concepts' usability. Brooke's System Usability Scale (SUS, Ref. 36) was used. This scale was originally proposed for the evaluation of office software. Due to its universal character it can be used to evaluate any kind of system; in this case the three different control concepts. It is a questionnaire with ten statements that are to be agreed or declined on a five-point scale. All answers are then weighed and summed for an overall usability score lying between 0 and 100. The lower the score, the lower the overall usability is. In the appendix, the used scale can be found.

After rating the usability, the test candidates were asked how hard or easy the flight tasks were with the different control concepts and what features of the control concepts they liked best. The answers where again given on a five-point scale.

The group of test candidates consisted of six helicopter pilots (PPL(H), CPL(H), ATPL(H) or military), six car drivers and five inexperienced candidates with neither driving nor flying license. The test candidates' experience is summarized in Table 1. The helicopter pilots were between 34 and 58 years old with 60 to 5,400 flight hours. The drivers' age ranged from 27 to 52. They drove 2 to 20 hours as weekly average. The inexperienced candidates' age ranged from 16 to 33 years. Four women and thirteen men participated in the study. All of the test candidates were right handed.

Table 1. Experience of test groups (*M* – mean, *SD* – standard deviation).

	Pilots		Drivers		Inexperienced	
	M	*SD*	*M*	*SD*	*M*	*SD*
Age	40.2	8.5	36.8	9.3	26.4	6.3
Years with driver's licence	22.2	8.6	19.0	9.1	-	-
Weekly driving time	3.7	1.6	10.3	3.7	-	-
Years with helicopter licence	14.1	11.1	-	-	-	-
Total helicopter flight hours	1843	1856	-	-	-	-
Number	6		6		5	

RESULTS

Mean usability ratings (*M*) and standard deviations (*SD*) are given in Table 2. The descriptive statistics show that the helicopter pilots gave better averaged usability ratings for the conventional controls compared to the steering wheel configurations. The configuration wheel + collective was second and wheel + switches last. In contrast to that, the drivers as well as the inexperienced test candidates gave best ratings to the configuration wheel + switches. For the drivers wheel + collective was second and the conventional controls last. Conventional controls were second for the inexperienced users and the configuration wheel + collective last.

Table 2. Descriptive statistics of the usability ratings.

	Pilots		Drivers		Inexperienced	
	M	*SD*	*M*	*SD*	*M*	*SD*
Conventional	81.8	10.4	47.0	12.2	64.0	16.4
Wheel + Collective	74.0	25.8	62.2	15.8	58.8	28.5
Wheel + Switches	77.3	19.4	70.2	6.1	86.6	12.0

In order to assess if the found differences in the ratings are statistically significant, an analysis of variance was conducted. The analysis was performed using the statistics software SPSS (Ref. 37). The implemented GLM (General Linear Models) repeated measures procedure with mixed design was used as an overall statistical test. A correction of the degrees of freedom according to Greenhouse-Geisser was conducted whenever the Mauchly test of sphericity was significant. Pairwise comparisons after Bonferroni were conducted afterwards and post hoc t-tests showed pairwise differences for significant interferences between configurations and groups of test candidates (see also Ref. 38).

The error probability α was initially set to 0.05. When the level of significance p is smaller than α, results are statistically significant. Trends were taken into account up to values of 0.1. For completeness not only the level of significance p is given for the results but also the test sizes $F(df1, df2)$ (for the analysis of variance) and $T(df)$ (for t-tests) in dependence of the degrees of freedom df. The effect size η^2 can take values between 0 and 1 with higher values representing higher strength of the investigated effect.

For the usability ratings, the analysis of variance shows a trend in the interference between configuration and user group (F [2.88, 20.15] = 2.33, p = 0.107, η^2 = 0.25). This confirms that the groups statistically differ in their ratings. Post hoc t-tests show that the descriptive differences in the helicopter pilots' ratings are not statistically significant (p > 0.1). Figure 7 illustrates this finding. The usability ratings are shown in boxplots, separated by user group and configuration. A bold horizontal line shows the median; the colored box contains the upper and lower quartiles of the data; minimum and maximum ratings are shown by vertical antennas, outliers by dots accompanied by the participant's number.

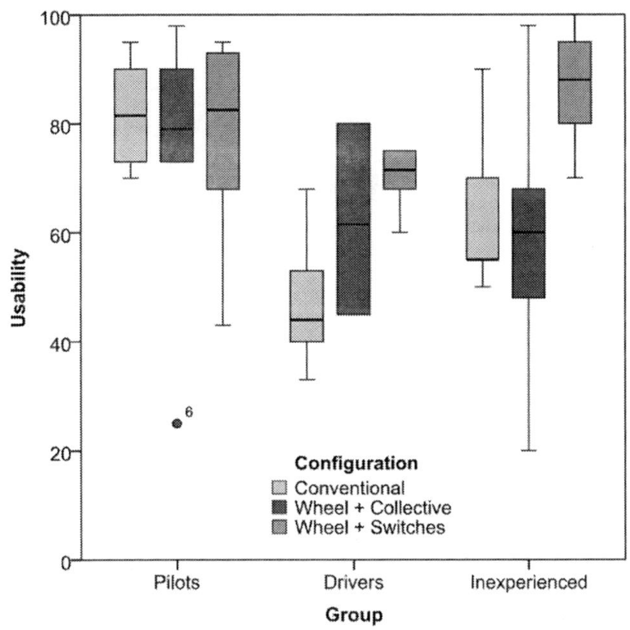

Figure 7. Usability rated by different user groups with steering wheel or conventional controls.

For the drivers the differences in the ratings are clearer. They gave statistically lower usability ratings for the conventional controls compared to the configuration wheel + collective (T [5] = -2.71, p = 0.042) and wheel + switches (T [5] = 4.66, p = 0.006). The difference between both steering wheel configurations is not significant (p > 0.1).

For the inexperienced users the trend is the same as for the car drivers when conventional controls and wheel + switches are compared (T [4] = -2.39, p = 0.075). As the spreading of ratings is high for the configuration wheel + collective, the differences in the ratings compared to the other configurations are not significant (p > 0.1).

When the helicopter pilots were confronted with the idea of steering wheel control for rotorcraft for the first time in this or previous studies, several of them reacted rather skeptical. Nevertheless, after flying the first turns with the steering wheel in the simulator, all of them could be convinced that the wheel is actually usable in flight. One pilot commented that each of the tested control configurations had their individual advantages and that he could hardly decide for a favorite. Another pilot mentioned that flying turns was especially easy with the wheel.

The drivers in general reacted very positive towards the steering wheel. One of them commented, after having switched from conventional controls to steering wheel, that he felt "much more comfortable now". Another driver said he could "feel" the curves better when flying with the steering wheel.

Even without prior driving experience, the inexperienced test candidates could easily adapt to the steering wheel control. Two of them explicitly appreciated that they had to handle fewer individual inceptors when flying the configuration wheel + switches.

The results of the final comparative questionnaire are summarized in Figure 8. A significant interference between group and configuration resulted on the question how hard or easy the control of the flight direction was (F [4, 28] = 5.05, p = 0.003, η^2 = 0.42). For the helicopter pilots a trend can be seen for better ratings for the steering wheel + switches compared to conventional controls (T [5] = -2.24, p = 0.076). Their other ratings do not differ statistically (p > 0.1). For both groups without flying experience direction control was easier with the steering wheel than with the conventional controls. (Drivers: wheel + collective vs. conventional (T [5] = -5.40, p = 0.003), wheel + switches vs. conventional (T [5] = -5.53, p = 0.003); inexperienced: wheel + collective vs. conventional T [4] = -3.67, p = 0.021), wheel + switches vs. conventional (T [4] = -3.09, p = 0.037). Both steering wheel configurations do not differ statistically (p > 0.1).

For the question how hard or easy controlling the airspeed was, the averaged answers of the three groups differ. Nevertheless, a significant result for the factor configuration was found (F [2, 28] = 5.46, p = 0.010, η^2 = 0.28). Overall, the configuration wheel + switches was rated as easier than the conventional controls (p = 0.045), while the other ratings do not differ statistically (p > 0.1).

A significant interference between configuration and group was found for the last question regarding the altitude control (F [4, 28] = 3.90, p = 0.012, η^2 = 0.36). The helicopter pilots rated the altitude control to be easiest with conventional controls compared to wheel + collective (p = 0.025) or wheel + switches (p = 0.025). The other test candidates had individual preferences. Therefore no significant differences could be investigated for these groups (p > 0.1).

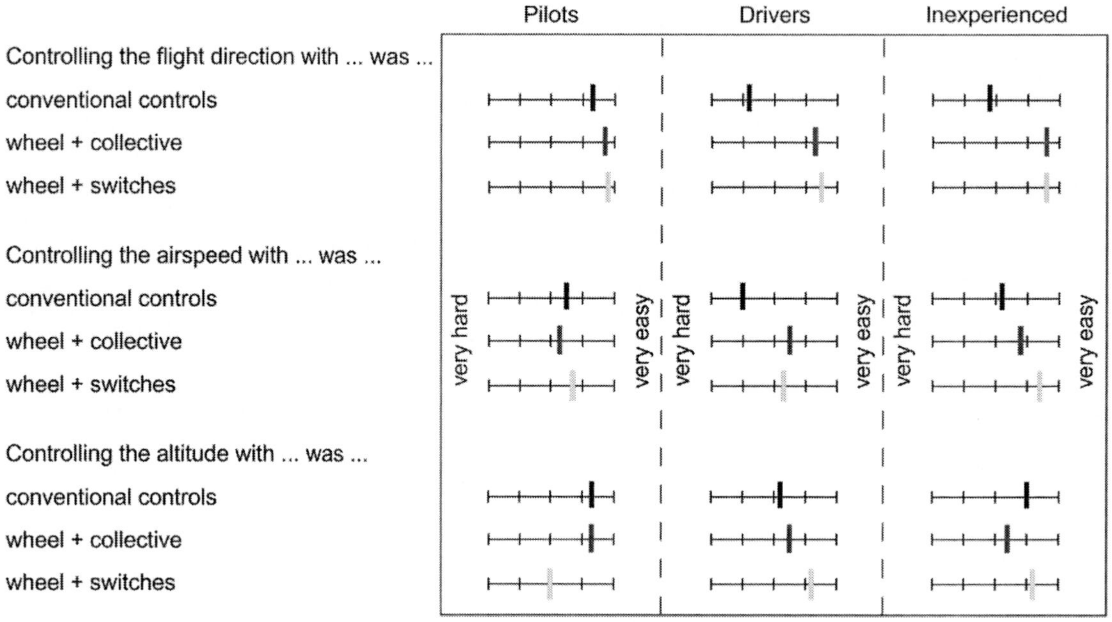

Figure 8. Averaged comparative ratings of the control concepts.

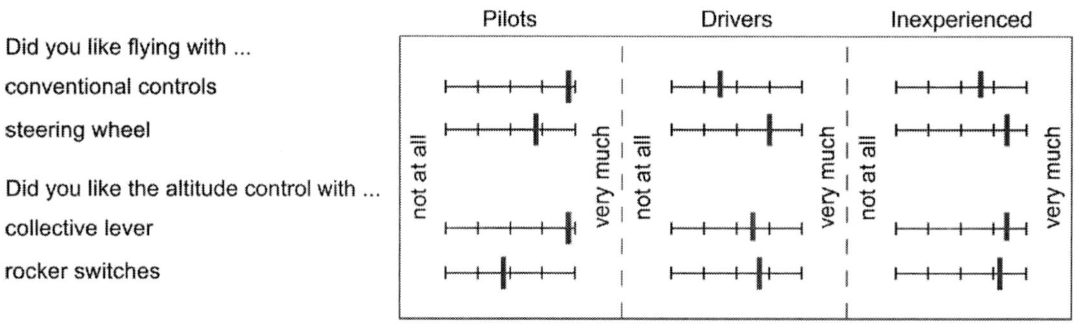

Figure 9. Averaged preference ratings for the control concept features.

The last two questions explicitly address the personal preference for steering wheel or conventional controls and for collective lever or rocker switches. Figure 9 shows the averaged results. Again, a significant interference between configuration and group was found for the first question (F [2, 14] = 3;82, p = 0.047, η^2 = 0.25). The helicopter pilots tend to prefer the conventional controls (T [5] = 1.94, p = 0.111). Descriptively, the other two groups prefer the steering wheel, although a significant difference could only be found for the drivers (T[5] = -2.24, p = 0.076).

A significant interference between configuration and group was also found for the altitude control question (F [2, 14] = 6.31, p = 0.011, η^2 = 0.47). The helicopter pilots' preference was clearly on the collective lever in comparison to the switches (T [5] = 5.48, p = 0.003). For the other groups no favorite was found (p > 0.1).

DISCUSSION

This usability study has shown that the helicopter pilots gave rather high usability scores to all three investigated control concepts. Nevertheless, when directly being asked, they clearly preferred the conventional controls. This is a sign of the rather conservative attitude of the helicopter pilots towards changes in their traditional working environment. For the other test groups without flying experience, or potential future pilots of personal helicopters, the steering wheel control concept was much more attractive and generally received better usability, comparative and preference ratings. Due to the limited availability of the simulator, the training prior to this study was kept to a minimum. In future studies the training effect could be interesting to be further investigated.

For the altitude control a clear favorite inceptor could not be identified. While the helicopter pilots preferred the collective lever, the other test candidates favored either the collective or the switches. The advantage of the switches is that the user can keep both hands on the steering wheel during altitude changes. The disadvantage is that the direction of movement is not as intuitive as for the collective lever (right/left switch for up/down). Another disadvantage that was identified in the study was that the travel range of the switches is too small (only few millimeters). This even led to a digital feeling for some of the test candidates. For a future steering wheel design, the switches' travel range and input sensitivity should be extended.

For controlling the speed or acceleration, the direction of movement is more intuitive for the center stick than for the pedals. Nevertheless, only very few misused pedal inputs were observed during the training phase and none during the evaluation runs. Critical comments were given on the design of the automatic speed blending function between low speed and high speed regime. During the mode change, individual speed control was hardly possible. The automatic function should be improved for future testing.

CONCLUSIONS

This paper focuses on the usability rating for a previously developed steering wheel control concept for rotorcraft. The idea is to offer an intuitive control concept to helicopter pilots as well as to novice users who can profit from their previous experience in car driving. The following findings were made:

- The conventional center stick could be replaced by a steering wheel prototype. The helicopter simulator's yaw pedals were repurposed for speed control. Altitude commands were introduced either by a conventional collective lever or by rocker switches located behind the wheel. An additional 8-way-switch in the center of the wheel allowed for horizontal maneuvers in the low speed regime.
- The helicopter pilots participating in the study gave similarly high usability ratings to both steering wheel configurations as well as to the reference configuration with conventional helicopter inceptors.
- A group of test candidates with driving experience but without pilot license and a group of inexperienced users with neither driving nor flying license gave best usability ratings for the steering wheel configuration in combination with switches for altitude control.
- When directly being asked which control configuration was easiest to use, the helicopter pilots opted for the conventional controls. The other test candidates favored either the configuration steering wheel + collective lever or steering wheel + switches.

Several of the pilots were initially very skeptical regarding the integration of a steering wheel into a helicopter. Nevertheless, the reported study shows that this control concept is suitable for a wide range of users with different levels of experience.

ACKNOWLEDGMENTS

The author would like to acknowledge the participating test candidates, the members of the FT-HUB flight control group as well as the AVES team for their support in the completion of this study. This research activity has received funding from the European Commission's Seventh Framework Program for the project myCopter – Enabling Technologies for Personal Aerial Transportation Systems under grant agreement no. 266470.

Author contact:
Bianca I. Schuchardt, bianca.schuchardt@dlr.de

REFERENCES

[1] Anon., "Commission Welcomes European Cities Joining the Urban Air Mobility Initiative", https://ec.europa.eu/transport/media/news/news/2018-05-30-commission-welcomes-european-cities-joining-urban-air-mobility-initiative_en, May 30, 2018.

[2] Holden, J. and Goel, N., "Fast-Forwarding to a Future of On-Demand Urban Air Transportation", White Paper, Uber Elevate, October 27, 2016, https://www.uber.com/info/elevate/.

[3] Steuer, G., von Kursell, G., "Iron Bird Power On: CityAirbus Reaches Next Milestone", Airbus Helicopters, December 19, 2017, https://www.airbus.com/newsroom/press-releases/en/2017/120/iron-bird-power-on--cityairbus-reaches-next-milestone.html.

[4] Drees, J.M., "Prepare for the 21st Century - The 1987 Alexander A. Nikolsky Lecture", Proceedings of the 43rd American Helicopter Society Annual Forum, St. Louis, MO, USA, May 1987.

[5] Anon., "Personal Perspective on Helicopter History", Rotor Magazine, Helicopter Association, International, p. 43, autumn 1999.

[6] LaChance, D., "The Rise and Fall of the Helicospeeder", Hemmings Motor News, December 2006, https://www.hemmings.com/magazine/hmn/2006/12/The-Rise-and-Fall-of-the-Helicospeeder/1382419.html.

[7] Stiles, L., Mayo, J., Freisner, A., Landis, K., and Kothmann, B., "Impossible to Resist: The Development of Rotorcraft Fly-by-Wire Technology", Proceedings of the 60th American Helicopter Society Annual Forum, Virginia Beach, VA, USA, June 2004.

[8] Rakotomamonjy, T., Binety, L, and Müllhäuser, M., "ONERA-DLR Joint Research on Tactile Cueing for Reactive Obstacle Avoidance Dedicated to Low Speed Helicopter Manuevers", Proceedings of the 42nd European Rotorcraft Forum, Lille, France, September 2016.

[9] Nieuwenhuizen, F.M., Jump, M., Perfect, P., White, M.D., Padfield, G.D., Floreano, D., Schill, F., Zufferey, J.C., Fua, P., Bouabdallah, S., Siegwart, R., Meyer, S., Schippl, J., Decker, M., Gursky, B., Höfinger, M., and Bülthoff, H.H., "myCopter - Enabling Technologies for Personal Aerial Transportation Systems", 3rd International HELI World Conference, Frankfurt/Main, November 2011.

[10] Gursky, B.I. and Müller, D., "Novel Steering Concepts for Personal Aerial Vehicles", 62. Deutscher Luft- und Raumfahrtkongress, Stuttgart, Germany, September 2013.

[11] Schuchardt, B.I., "PKW-ähnliche Steuerungskonzepte für den Einsatz im Hubschrauber", DLR Ger-

man Aerospace Center, Institute of Flight Systems, FB 2017-65 / TU Braunschweig, FB 2017-10, doctoral thesis, 2017.

[12] Jones, M., Perfect, P., Jump, M., and White, M., "Investigation of Novel Concepts for Control of a Personal Aerial Vehicle," Proceedings of the 70th American Helicopter Society Annual Forum, Montréal, Canada, May 2014.

[13] Schuchardt, B.I., "Towards Handling Qualities Evaluations of a Personal Aerial Vehicle in Ground-Based and In-Flight Simulation", Proceedings of the 71st American Helicopter Society Annual Forum, Virginia Beach, Virginia, USA, May 2015.

[14] Schuchardt, B.I., "Workload Reduction through Steering Wheel Control for Rotorcraft", Proceedings of the 43rd European Rotorcraft Forum, Milan, Italy, September 2017.

[15] Orna, B., "Modest Experimenter – Vuia's Powered Flights: the Successes of a Little-Known Pioneer", *Flight*, 69(2462):365-366, March 1956.

[16] Anon., "Terrafugia, the Future of Personal Transportation", 2018. http://www.terrafugia.com/, last access 12/05/2018.

[17] Macauley, C.B.F., *The Helicopters are Coming*, Whittlesey House, 1944.

[18] Anon., "Volocopter. Reinventing Urban Mobility", 2017, https://www.volocopter.com/, last access 12/05/2018.

[19] Wiegand, D, "The Lilium Jet. The world's first electric vertical take-off and landing jet", Lilium GmbH, 2018, http://lilium-aviation.com/, last access 12/05/2018.

[20] Anon., "Urban Aeronautics, Leading the Fancraft Revolution", http://www.urbanaero.com/, last access 12/05/2018.

[21] Anon., "Take the high road", http://www.jobyaviation.com/, last access 12/05/2018.

[22] Anon., "Kittyhawk. Things are looking up", 2018, https://kittyhawk.aero/, last access 12/05/2018.

[23] Anon., "Ehang184, Autnomous Aerial Vehicle", 2018, http://www.ehang.com/ehang184/, last access 12/05/2018.

[24] Anon., "Uber Elevate", Bell Helicopter Textron Inc., 2018, https://www.bellflight.com/company/innovation/uber-elevate-summit, last access 12/05/2018.

[25] Anon., "CityAirbus Media Backgrounder", Airbus Helicopters, June 2017, https://www.airbus.com/content/dam/corporate-topics/publications/backgrounders/City-Airbus-Backgrounder-June2017.pdf, last access 12/05/2018.

[26] Anon., "Vahana. The Next Technological Breakthrough in Urban Air Mobility", Airbus, 2018, https://www.airbus-sv.com/projects/1, last access 12/05/2018.

[27] Anon., "Aurora Flight Sciences, a Boeing Company", 2018, https://www.aurora.aero/, last access 12/05/2018.

[28] Kaletka, J., Kurscheid, H., and Butter, U., "FHS, the New Research Helicopter: Ready for Service", Proceedings of the 29th European Rotorcraft Forum, Friedrichshafen, Germany, September 2003.

[29] Schuchardt, B.I., and Greiser, S., "Preparing In-Flight Simulation of a Personal Aerial Vehicle on DLR's Research Helicopter ACT/FHS", 65. Deutscher Luft- und Raumfahrtkongress, Braunschweig, Germany, September 2016.

[30] Perfect, P., White, M.D., and Jump, M., "Development of Pilot Training Requirements for Personal Aerial Vehicles", Proceedings of the 40th European Rotorcraft Forum, Southampton, UK, September 2014.

[31] Duda, H., Gerlach, T., Advani, S., and Potter, M., "Design of the DLR AVES Research Flight Simulator". Proceedings of the AIAA Modeling and Simulation Technologies (MST) Conference, Boston, MA, USA, August 2013.

[32] Perfect, P., Jump, M., and White, M., "Methods to Assess the Handling Qualities Requirements for Personal Aerial Vehicles", *Journal of Guidance, Control, and Dynamics*, Vol. 38, No. 11, 2015, pp. 2161-2172, DOI: 10.2514/1.G000862.

[33] Gotschlich, J., Gerlach, T., and Durak, U., "2Simulate: A Distributed Real-Time Simulation Framework", ASIM STS/GMMS Workshop, Reutlingen, Germany, February 2014.

[34] Anon., "ADS-33E-PRF, Aeronautical Design Standard, Performance Specification, Handling Qualities Requirements for Military Rotorcraft", USAAMC, March 2000.

[35] Gursky, B.I., Olsman, W.F.J. and Peinecke, N., "Development of a Tunnel-in-the-Sky Display for Helicopter Noise Abatement Procedures", *CEAS Aeronautical Journal*, 5(2):199-208, June 2014.

[36] Brooke, J., "SUS - A Quick and Dirty Usability Scale", In Thomas, B., Weerdmeester, B.A., and McClelland, A.L. (editors): *Usability Evaluation in Industry*, Taylor and Francis, London, 1996.

[37] Anon., "IBM SPSS Advanced Statistics 20", IBM Corporation, 2011.

[38] Bühner, M. and Ziegler, M., *Statistik für Psychologen und Sozialwissenschaftler*, Pearson Studium, Munich, 2009.

APPENDIX

	1 Strongly disagree	2	3	4	5 Strongly agree
I think that I would like to use this system frequently.	○	○	○	○	○
I found the system unnecessarily complex.	○	○	○	○	○
I found that the system was easy to use.	○	○	○	○	○
I think that I would need the support of a technical person to be able to use this system.	○	○	○	○	○
I found the various functions in this system were well integrated.	○	○	○	○	○
I thought there was too much inconsistency in this system.	○	○	○	○	○
I would imagine that most people would learn to use this system very quickly.	○	○	○	○	○
I found the system very cumbersome to use.	○	○	○	○	○
I felt very confident using the system.	○	○	○	○	○
I needed to learn a lot of things before I could get going with this system.	○	○	○	○	○

Figure 10. System Usability Scale (SUS). "To calculate the SUS score, first sum the score contributions from each item. Each item's score contribution will range from 0 to 4. For items 1, 3, 5, 7, and 9 the score contribution is the scale position minus 1. For items 2, 4, 6, 8 and 10, the contribution is 5 minus the scale position. Multiply the sum of the scores by 2.5 to obtain the overall value of SU" (Ref. 36).

Integrated Occupant Safety for Urban Air Transport Emergency Landing Applications

Gerardo Olivares
gerardo.olivares@wichita.edu
Director Crashworthiness
National Institute for Aviation Research
Wichita, KS, USA

Fernando de Caralt
fdecaralt@brsaerospace.com
CEO
BRS Aerospace
Saint Paul, MN, USA

Adrian Vina
avina@brsaerospace.com
Head of Engineering
BRS Aerospace
Saint Paul, MN, USA

ABSTRACT

Occupant Safety must be an integral part of the overall technical and management processes associated with the design, development and operation of Urban Air Transport systems. Nowadays the crashworthiness design for aerospace applications under 14 CFR *.561 and *.562 only address the dynamic response of the seat and restraint system during emergency landing conditions. In order to improve the survivability rate of occupants an integrated safety approach is required. This paper will provide an outline on the different elements that constitute the integrated safety concept; Pre-crash sensing technologies, Parachute Ballistic Recovery Systems, Landing Gear-Airframe Crashworthy Structural concepts, high-energy absorbing seats, and advanced restraints.

INTRODUCTION

Structural design for airplane safety combines airworthiness and crashworthiness design objectives to varying degrees. Airworthiness design objectives pertain to the ability of the airframe to withstand design loads, or to maintain safety of flight of the airplane relative to the operational environment. Crashworthiness design objectives pertain to safety of the occupants relative to the airplane. The development of current aircraft crash dynamics standards dates back to the 1970's during which product liability grew for small aircraft's manufacturers. To address the crashworthiness characteristics of transport category aircraft, small general aviation aircraft, and rotorcraft, Federal Aviation Administration (FAA) and NASA initiated a wide range of research and development programs. These programs represented a concentrated effort to analyze the aircraft behavior and the occupant characteristics through interrelated studies of accident data, dynamic analyses of crash events; full-scale aircraft impact tests, and aircraft seat tests. A panel named GASP (General Aviation Safety Panel) was also formed in 1978 to make recommendations on crashworthiness requirements. They considered survivable accidents for which the floor remained intact. The results of these studies formed the basis for the development of crashworthiness design standards for civil aircraft. These requirements are defined in the Federal Aviation Regulations (FARs) Parts 23, 25, and 27 for general aviation aircraft, transport aircraft, and rotorcraft respectively. These regulations were first proposed in 1982 and they became effective in 1988. Nevertheless, the crashworthiness requirements only applied to the dynamic performance of the seats and occupants, and no crashworthiness requirements were defined for the structure. In the evolution of the regulations, there is at present no specific dynamic regulatory requirement for airplane level crashworthiness. However, the FAA requires an assessment of each new model airplane to ensure that the airplane will not significantly deviate from typical dynamic characteristics found in previous designs (Ref. 1 and 2).

Presented at 8th Biennial Autonomous VTOL Technical Meeting & 6th Annual Electric VTOL Symposium. Jan 28th, 2019, Mesa, USA.

Occupant Safety must be an integral part of the overall technical and management processes associated with the design, development and operation of Urban Air Transport systems. Nowadays the crashworthiness design for aerospace applications under 14 CFR *.561 and *.562 only addresses the dynamic response of the seat and restraint system during emergency landing conditions. In order to improve the survivability rate of occupants an integrated safety approach is required. This paper will provide an outline on the different elements that constitute the integrated safety concept: Pre-crash sensing technologies, Parachute Ballistic Recovery Systems, Landing Gear-Airframe Crashworthy Structural concepts, high-energy absorbing seats, and advanced restraints.

AEROSPACE CRASHWORTHINESS

The FAA promulgates standards for occupant protection at the seat installation level, with the presumption that the airframe provides an acceptable level of crashworthiness. Thus when an applicant proposes to use unconventional fuselage structure (materials, design, or both), the FAA has written special conditions to ensure the level of crash protection is equivalent to that provided by a traditionally-configured metallic airplane. These special conditions have been comparative in nature, and do not establish performance standards that are independent of traditional metallic skin-stringer-frame architecture for airframe crashworthiness.

Many factors influence the crashworthiness of an airframe, including materials used for construction, geometry, structural design philosophy, and fuselage size. The key elements of crashworthy airframe design are managing energy absorption and maintaining structural integrity. For the most part, energy absorption is managed through plastic deformations and controlled failures of the landing gear system and lower fuselage structure. As shown in Figure 1, maintaining the integrity of the structure is a balance between keeping loads within human tolerance levels, retaining items of mass, preserving a survivable volume and maintaining access to emergency exits. Existing airworthiness requirements mainly focus on the safety of flight, and not crashworthiness, consequently when deviating from the traditional aircraft architectures an adequate level of safety cannot be assured. This is specially a problem for eVTOL platforms were the vehicle architecture, materials (extensive use of composites and advanced materials), and systems (batteries, electric motors…etc) are non-conventional and there is not enough historical data to assess the level of safety for new designs.

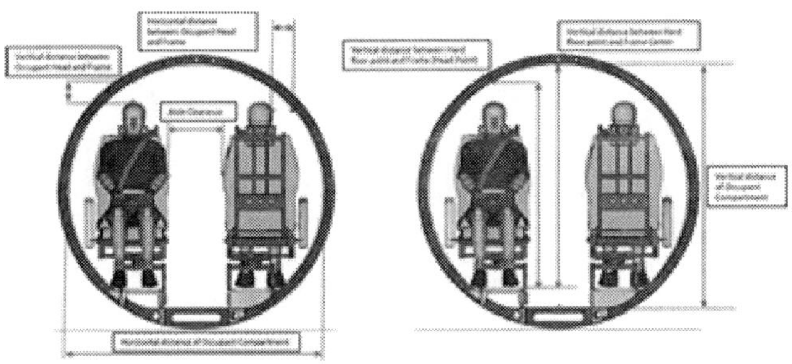

Figure 1. Survivable Space Evaluation (Ref. 2).

General Aviation Safety and Crashworthiness Standards

Due to their size and operational envelop eVTOL vehicles will fall in a category similar to General Aviation Aircraft. General Aviation fatalities have remained constant since 2005 as shown in figure 2. Crashworthiness in general aviation aircraft is a topic that requires substantial consideration due to the number of fatalities, yet due to overly prescriptive regulatory requirements over the past several decades, has not advanced to the same level as 14 CFR Part 25 Aircraft or other modes of transportation. The recent reorganization of the small airplane airworthiness rules (14 CFR Part 23) and the use of globally accepted consensus standards as a means of compliance (ASTM, EUROCAE, SAE, RTCA, etc.) allows for greater innovation and flexibility in the way survivability and crashworthiness are approached. It is because of these recent developments that there is now an opportunity to define crashworthiness design guidelines and standards for general aviation and eVTOL aircraft.

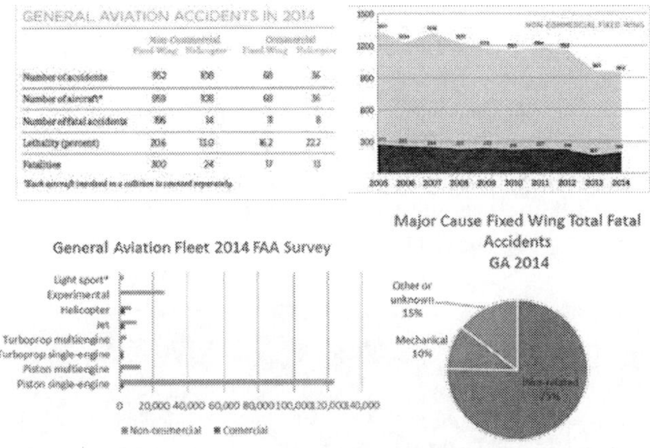

Figure 2. Survivable Space Evaluation (Ref. 2).

Crashworthiness Design Requirements

The aircraft must be designed to provide for adequate occupant survivability during survivable crash events. In order to demonstrate adequate occupant survivability, compliance must be shown with the following criteria:

- *Maintain a survivable volume:* as shown in Figure 1, the fuselage structural deformation will not result in infringement of the occupant's normal living space so that passenger survivability will not be significantly affected. Avoid occupant to aircraft interior structure contacts during the dynamic impact event.

- *Maintain Survivable Deceleration Loads to Occupants:* During the crash event, the occupant injury criteria thresholds specified in 14 CFR *.562 must not be exceeded for the load levels experienced by the occupants.

- *Retention of Items of Mass:* All occupants must be protected during the crash impact event from the release of seats, overhead bins, and other items of mass due to the impact loads and resultant structural deformations of the supporting airframe and floor structures.

- *Maintain Egress Paths:* Post crash, the fuselage structure must provide suitable egress paths to evacuate the occupants.

The National Institute for Aviation Research has conducted a 15-year R&D program studying the crashworthiness of 14 CFR Part 25 composite and metallic aircraft structures. The major conclusion of the research is that in order to provide real world safety a design must meet all the aforementioned pass/fail criteria. The research has shown that for a 14 CFR part 25 business jet aircraft equipped with compliant 25.561 and 25.562 seats is by itself not capable of meeting real world impact scenarios, in this case for a 30-ft\sec vertical impact velocity. As shown in Figure 3, the aircraft meets the requirements for survivable volume, retention of items of mass, and egress path; but due to the lack of energy absorbing capabilities of the airframe subfloor, the seat fails to maintain survivable deceleration levels to the occupants that result in fatal lumbar injuries (lumbar load values in excess of 2500 lb.) (Ref 2.).

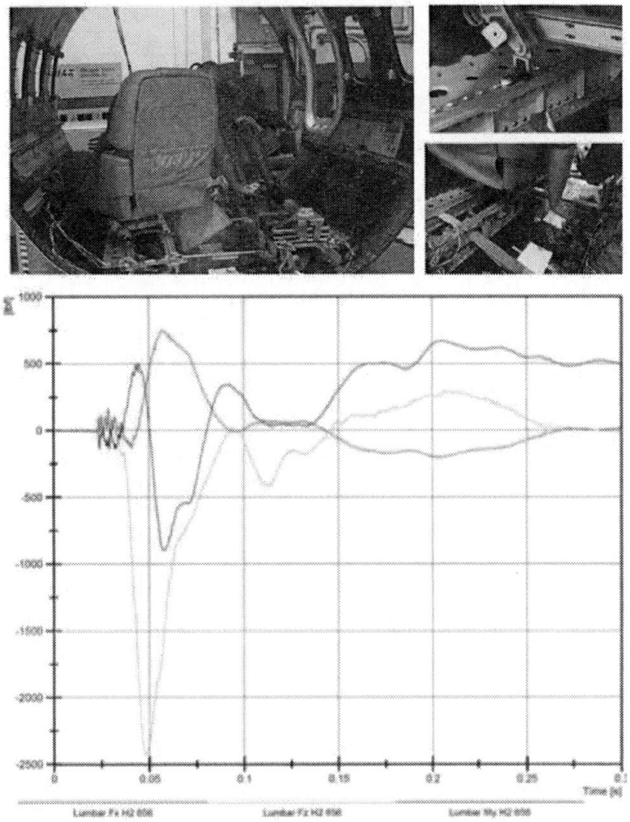

Figure 3. NIAR Composite Fuselage Vertical Drop Test – Post Impact Evaluation and Occupant Lumbar Load (Ref. 2).

INTEGRATED SAFETY

Occupant safety must be an integral part of the overall technical and management processes associated with the design, development and operation of eVTOL Urban Air Transport systems. In order to provide real word safety an integrated safety design approach is required. As shown in figure 4 the different elements that constitute the integrated safety concept are; Pre-crash sensing technologies, Parachute Ballistic Recovery Systems, and a combination of Landing Gear-Airframe crashworthy structures, high-energy absorbing seats, and advanced restraints. Two strategies can be used to mitigate the outcome of an emergency landing condition. If the aircraft has enough height with respect to the ground when failure occurs a Ballistic Recovery System can be used. One of the advantages of ballistic recovery systems is that the aircraft impact velocity and attitude can be controlled prior to impact hence allowing to optimize the airframe, seat and advanced restraints to provide an optimum level of safety to occupants. If the BRS system cannot be deployed then it is more challenging to design a crashworthy aircraft due to the unknown attitude and velocity prior to impacting on the ground. Other area that will need to be addressed by eVTOL aircraft manufacturers will be the integrity of the battery systems during impact and the risk of post-impact fires that could prevent the safe egress of the passengers.

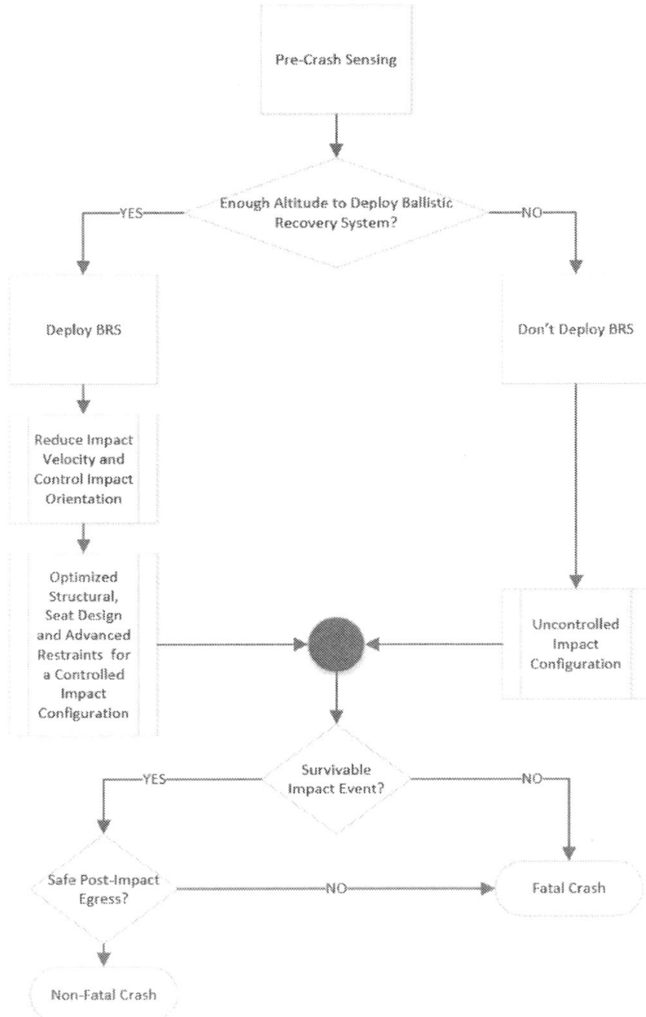

Figure 4. Integrated Safety Crashworthiness.

Active Safety Pre-Crash Sensing

The implementation of active safety systems can prevent or mitigate the outcome of eVTOL crashes. The autonomous nature of eVTOL Urban Air Transport systems will have a significant effect in the reduction of fatalities caused by human error. As shown in figure 2 the fatalities caused by human error account for more than 75 % of the cases for GA fixed wing aircraft.

Ballistic Recovery Systems Description

The Ballistic Recovery System (BRS) recovers the aircraft from a non-nominal situation and sets the aircraft into a previously defined landing configuration. When the event is detected, a signal is sent for the activation of the BRS.

As shown in Figure 5, a typical BRS consists of different components that can be grouped into the following subsystems:

- *Parachute:* A non-steerable round parachute used for the recovering of the aircraft slowing the speed down. It is the intrinsic simplicity of such parachute what brings reliability to the system. An important element added to this parachute is the slider. The slider controls the opening rate and gives control over the maximum loads reached during the opening process.

- *Rocket:* A solid fuel rocket extracts and stretches the parachute.

- *Igniter:* The element that receives the input signal and initiates the combustion in the rocket.

- *Frame:* A mechanical support device that works as a compartment and is used to store the system inside the aircraft.

- *Egress panel:* A breakable surface where the rocket followed by the parachute break through to extract the recovery system from inside the aircraft.

- *Harness:* A three or four point harness assembly acts as a joining element between the parachute and the aircraft.

- *Attachment Points (AP):* The harness is attached to the aircraft with front attachment points and rear attachment points.

A BRS is initiated by manual or autonomous activation of the igniter by one of the occupants or the on-board system. The igniter, by means of redundant, dual firing sequence, activates the solid-fuel rocket that after developing significant internal pressure fractures its aluminum tension restraints and breaks through the aircraft window. The rocket continues straight out from its launch base, taking up slack from the parachute attachment lanyard and pulls the packed parachute from its storage container. The rocket and its payload continue their outwards trajectory until the parachute, lines and riser are pulled to their full lengths and the parachute has begun to deploy, developing high drag and tearing the aircraft attachment harnesses from their respective fairings (harness covers).

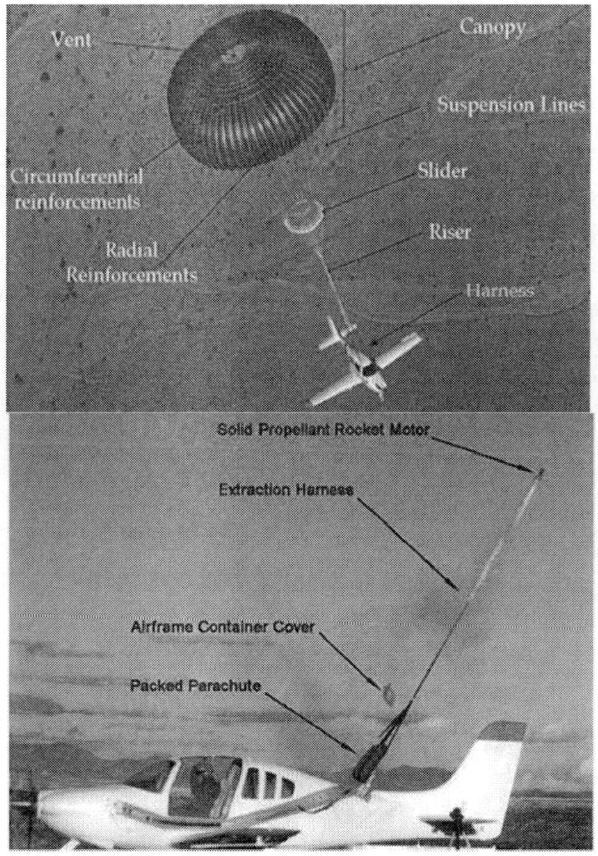

Figure 5. Recovery System and Parachute Elements.

The airplane dynamics play an important role in the deceleration of the aircraft. The inflation loads are transferred to the airplane through the attachment's harness. Since the forward harness straps are attached to the front carry-thru structure, located above of the airplane's CG, the inflation loads apply a pitching moment to the airplane, which results in a nose up rotation. Some pitch rotation is desirable since the airplane's angle of attack is increased well beyond stall and its effective drag area increases dramatically. This helps to decelerate the aircraft and relieves some of the loading on the parachute. Moreover, the force of the parachute is transmitted to the aircraft as the sum of a linear deceleration and a rotation of the system. Therefore, the pitching moment also helps to absorb some of the energy in the disreef process. This concept has been proved over different General Aviation certification programs when flight tests were conducted using an aircraft attachment harness that was configured such that the initial loads were applied to the aft airplane attachment points located behind the airplane CG. No pitch up moment was induced in these tests and the disreef inflation loads were considerably higher as a result. However, excessive rotation can result in unstable airplane dynamics and should be avoided. The magnitude of the pitch rotation is controlled by the harness geometry and the location of the harness/airframe attachment points relative to the airplane's center of gravity. The angle of the parachute riser relative to the airplane's longitudinal axis projected on a vertical plane, is referred as the elevation angle, ϑ, in degrees (°). Although the flight conditions when the activation occurs have an effect, in the first stage of the parachute inflation ϑ is usually between 5° and 15°. Then, a stable ϑ occurs when the inflation load vector intersects the airplane's center of gravity. The lengths of the aircraft harness straps are configured to provide an optimum stable ϑ based on the empirical data. The airplane will continue to pitch up until the stable ϑ is reached. Once in this position, all three harnesses work, and the airplane will start pitching down until the parachute force is vertical. At any point along this nose-down rotation, the airspeed and parachute drag have dropped considerably, and the slider can move down the suspension lines, allowing the parachute to open fully.

Once the parachute is completely inflated, the loads have dropped and the parachute is vertical, all harness straps are tight and allow the airplane to hang at the desired touchdown attitude.

Integrated Structural Crashworthiness Design: Landing Gear System, Airframe and High Energy Seat

In order to optimize a typical eVTOL (see Fig. 6) crashworthiness characteristics it is recommended that an integrated safety approach be used concurrently during the airworthiness design process. Currently the seats are designed to meet emergency landing dynamic impact conditions (see Fig. 7) and the airframe is designed to meet airworthiness design requirements.

Figure 6. Quadrotor eVTOL Architecture (Ref 3.).

To highlight the importance of the integrated safety approach lets evaluate the survivability of a typical Quadrotor eVTOL using an energy based analytical method. For this example, the following design parameters are evaluated:

- Vertical Impact Velocity: 30 ft.\sec
- MTOW: 3500 lbs.
- Energy Absorbing Seat Stroke: 5 inches
- Maximum Subfloor Crushable Space: 10 inches
- Skid Structure Maximum Crushable Space: 30 inches
- 50 % Energy Absorbing Structural Efficiency
- 50th Percentile Occupant

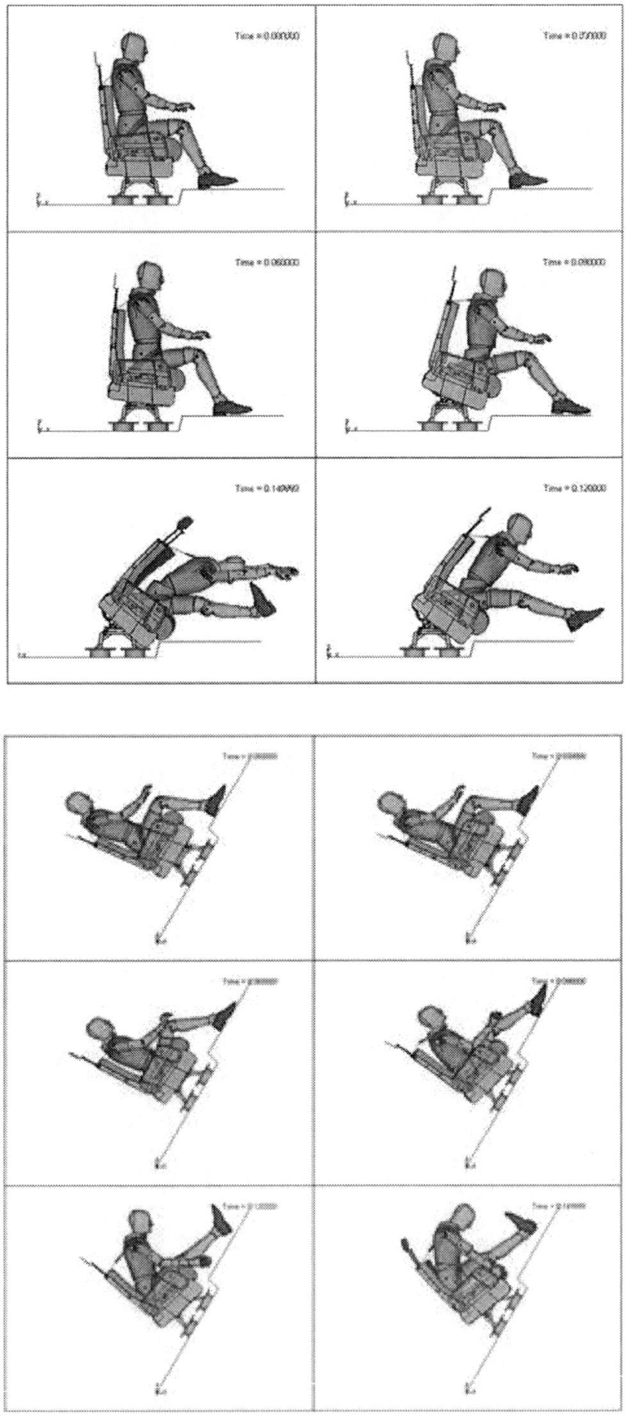

Figure 7. Typical 14 CFR *.562 Type I and II Dynamic Seat Test Conditions (Ref. 2).

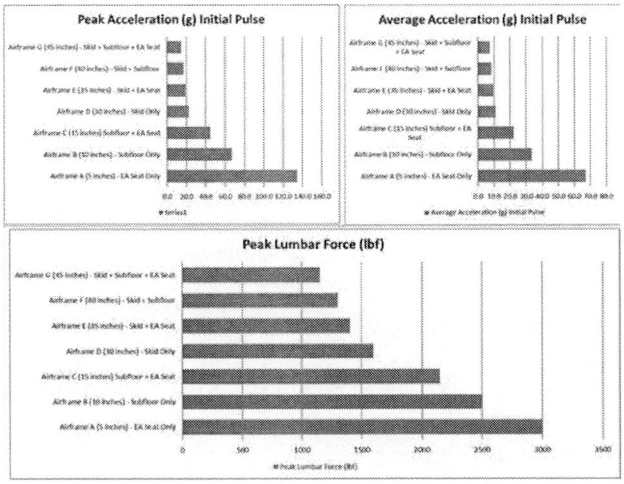

Figure 8. eVTOL Energy Based Method Design Evaluation

As summarized in figure 8 the real world survabability rate improves with the integration of the various energy absorbing components that define the eVTOL vehicle. The following conclusions can be observed::

- The highest risk of fatality corresponds to the current industry approach of crashworthiness design driven by dynamic seat certification requirements per 14 CFR *.562. The lumbar load for this configuration is 3000 lbs.

- The optimum solution corresponds to the proposed integration of the skid/landing gear system, the subfloor space and the energy absorbing seat. The lumbar load for the passenger is in the 1100 lb. range.

This preliminary sizing exercise highlights the importance of the proposed integrated safety approach. This approach for crashworthiness design needs to be implemented from the conceptual design stage of the vehicle, since the crashworthiness optimization of the various structural elements cannot be implemented once the design has been driven only by airworthiness requirements.

The implementation of preliminary sizing simplified analytical methods and detail design finite element models can be used to develop the next generation of crashworthy eVTOL vehicles. Extensive research has been conducted through the FAA JAMS program and internal NIAR research to develop and validate computational techniques that can support the virtual development and certification of future crashworthy aircraft designs (See Fig. 9 for integrated safety R&D Examples).

Figure 9. Integrated Safety Methods Verification and Validation Examples (Ref. 2).

CONCLUSIONS

Structural design for airplane safety combines airworthiness and crashworthiness design objectives to varying degrees. Airworthiness design objectives pertain to the ability of the airframe to withstand design loads, or to maintain safety of flight of the airplane relative to the operational environment. Crashworthiness design objectives pertain to safety of the occupants relative to the airplane. Occupant safety must be an integral part of the overall technical and management processes associated with the design, development and operation of eVTOL Urban Air Transport systems. In order to provide real word safety an integrated safety design approach is required.

The use of Ballistic Recovery Systems enables the designers to control the impact velocity and attitude of the aircraft during emergency landing conditions. The preliminary sizing exercise presented in this paper highlights the importance of the proposed integrated safety approach. This approach for crashworthiness design needs to be implemented from the conceptual design stage of the vehicle, since the crashworthiness optimization of the various structural elements cannot be implemented once the design has been driven only by airworthiness requirements.

REFERENCES

[1]Federal Register: June 11, 2007 (Volume 72, Number 111), Proposed Rules, Page 32021-32023

[2]Olivares G., Airframe "Crashworthiness – Certification by Analysis", NIAR Technical Report FAA003, 2019.

[3]Jonshon W., *Designing eVTOL for the Mission, From VTOL to eVTOL Workshop*, 2018

.

A Review of Current Technology and Research in Urban On-Demand Air Mobility Applications

Nicholas Polaczyk
BSc. Student Aerospace Engineering
Iowa State University
Ames, Iowa, United States

Enzo Trombino
BSc. Student Aerospace Engineering
Iowa State University
Ames, Iowa, United States

Dr. Peng Wei, Ph. D
Assistant Prof. Aerospace Engineering
Iowa State University
Ames, Iowa, United States

Dr. Mihaela Mitici, Ph. D
Assistant Prof. Aerospace Engineering
Delft University of Technology
Delft, the Netherlands

ABSTRACT

The purpose of this report is to summarize the current technology and challenges facing the market for electric vertical takeoff and landing (eVTOL) aircraft for urban on-demand air mobility applications. This report has two parts. The first part reviews existing electric drones for on-demand air mobility services for human transportation and package delivery drones. In this section, eVTOL projects are classified and compared by design parameters, specifications, and performance. While electric aircraft are the primary focus, some hybrid-electric and gas-powered aircraft will be used for comparison purposes. The second component reviews research publications discussing optimization and urban strategies for these aircraft. Challenges currently facing the industry are also discussed. These challenges include battery technology, efficiency, and safety. In this section, applications of eVTOL technology and future development are also reviewed.

INTRODUCTION

With a rising population and more cars on the road, travelling across metropolitan areas is slow and wastes energy, fuel, and productivity. Holden and Nikhil (Ref. 1) anticipate people and products moving more quickly and efficiently over medium-range distances with electric-powered aircraft. Through design characteristics and specifications, the performance details of each aircraft vary drastically. In this report we review these characteristics and performance specifications.

This report consists of two main parts. The first discusses 44 eVTOL projects currently implemented or being tested and designed. Their parameters are discussed in four subsections; design and configuration; motor and battery specifications; takeoff weight and payload capacity; and performance. Relevant data are presented at the beginning of each section. Designs are compared, and the trends shown in the data are reviewed. The second part reviews existing research with three subsections; eVTOL design, battery technology, and on-demand mobility with eVTOL aircraft. These publications discuss current and future trends and challenges for urban on-demand air mobility.

Presented at the Vertical Flight Society Autonomous VTOL Technical Meeting and Electric VTOL Symposium, Mesa, Arizona, UAS, Jan. 29-31, 2019.

PART I — CURRENT TECHNOLOGY

Aircraft Design and Configuration

Table 1. Aircraft Design and Configuration Specifications

Aircraft	Payload Type[1]	Number of Rotors	Tilt-Wing, Ducted[2]	Electric, Hybrid, Other[3]
Advanced Tactics Panther	C	6	-	E
AeroMobil 5.0	P	2	-	H
Airbus Helicopters CityAirbus	P	8	D	E
Airbus Skyways	C	8	-	E
Airbus Vahana A³	P	8	T	E
Aquinea Volta	O	2	-	E
Astro Aerospace Elroy (Passenger Drone)	P	16	T	E
ASX MOBi	P	6	T	E
Aurora eVTOL	P	9	-	E
Aurora LightningStrike	O	24	T D	H
Bartini Flying Car	P	8	D	E
Boeing Cargo Drone	C	8	-	E

Table 1. (Continued)

Aircraft	Payload Type[1]	Number of Rotors	Tilt-Wing, Ducted[2]	Electric, Hybrid, Other[3]
Carter Air Taxi	P	2	-	E
Davinci ZeroG	P	8	-	E
Digi Droxi	C	6	-	E
eHang 184	P	8	-	E
Joby S2	P	16	-	E
Joby S4	P	6	-	E
KARI PAV	P	8	-	E
Kármán XK-1	P	8	D	E
Kitty Hawk Cora	P	13	-	E
Kitty Hawk Flyer	P	10	-	E
La Poste Drone	C	6	-	E
Lilium Jet	P	36	D	E
MMIST Snowgoose	C	2	-	O
Moeller Skycar	P	4	D	O
NASA Puffin	P	2	-	E
Opener BlackFly	P	8	T	E
Pop.Up Next	P	8	-	E
Rolls-Royce eVTOL	P	6	T	H
Sikorsky Firefly	P	2	-	E
Solution F	P	1	-	E
Swiss Postal Matternet M2	C	4	-	E
Terrafugia TF-2 Lift+Push	P	5	-	E
Terrafugia TF-2 Tiltrotor	P	2	-	E
Terrafugia TF-X	P	2	-	H
Trek FlyKart 2	P	1	D	E
Urban Aeronautics CityHawk	P	2	D	O
Vimana	P	8	T	H
Volocopter 2X	P	18	-	E
Volocopter VC200	P	18	-	E
Workhorse Horsefly	C	8	-	E
Workhorse Surefly	P	8	-	O
XTI TriFan	P	3	D	H

[1] Passenger (P), Cargo (C), Other (O)

[2] Tilt-Wing (T), Ducted (D)

[3] Electric (E), Hybrid-Electric (H), Other (O)

The designs for the aircraft listed in Table 1 can vary dramatically depending on the project. Designs can be divided into two main categories. The first group uses open-mounted, fixed motors as the primary source of lift. The second group vectors thrust to different configurations while in takeoff and landing and during cruise.

In the first category, the most common configuration uses only horizontally-mounted fans to generate lift, like a common drone. This type of design is simple but can limit the range and cruising speed of the aircraft, to be discussed in further detail in a later section. This group can be further simplified into three subgroups. In this report, these three subgroups will be labelled as multicopters, rotorcraft, and winged multicopters. This report reviews 27 fixed motor designs as shown in Table 1.

Out of the 27 fixed motor designs reviewed, 16 projects fit the multicopter architecture. For the purposes of this report, this is defined as an aircraft that has only vertically-mounted props for vertical thrust. These props can be configured in several ways. For example, as noted by the World eVTOL Aircraft Directory, Ullman et al, and eHang (Ref. 2, 3, 4), the eHang 184 use a quadcopter-style design with eight motors mounted X-style on four corner supports. Similarly, according to the World eVTOL Aircraft Directory, Matternet Inc., and SureFly, (Ref. 2, 5, 6), three other designs use an X architecture; the Workhorse Surefly, Swiss Postal Matternet M2, and the Pop.Up Next. According to the World eVTOL Aircraft Directory, Advanced Tactics Inc., MediaRoom, and Passenger Drone, (Ref. 2, 7, 8, 9), the Advanced Tactics Panther, Boeing Cargo Drone, Astro Aerospace Elroy Passenger Drone, and Kitty Hawk Flyer fix their props along the fuselage. According to the World eVTOL Aircraft Directory and Ullman et al (Ref. 2, 3), two Volocopter aircraft, the 2X and VC200, mount their props above the fuselage using a branching architecture surrounded by an outer circular framework. According to Airbus, Workhorse, Davinci, and DPDgroup (Ref. 10, 11, 12, 13), the Airbus Skyways and Workhorse Horsefly delivery drones have traditional octocopter layouts, while the Davinci ZeroG and La Poste DPD Group Drone use traditional hexacopter designs; each having twelve and six open-mounted props, respectively. Uniquely, the Kármán XK-1 uses a ducted hexacopter architecture for lift and uses two additional vertically-mounted ducted fans for forward thrust as shown on the World eVTOL Aircraft Directory (Ref. 2).

There are five rotorcraft designs discussed in this report; the Carter Aviation SR/C Air Taxi; MMIST Snowgoose CQ-10B; Sikorsky Firefly; the Solution F helicopter; and the Aquinea Volta (Table 1). According to Carter (Ref. 14), the SR/C Air Taxi is a winged autogyro, using a central rotor for lift during takeoff, landing, and hover, while the high aspect ratio wing plus the autorotating rotor provide the lift during cruise conditions. This is made possible with the use of a rotating, horizontally-mounted propeller in the rear providing forward thrust during cruise and preventing fuselage rotation during takeoff, hover, and landing. According to MMIST (Ref. 15), the Snowgoose CQ-10B does not use wings for lift, however the design uses a fixed horizontally-mounted propeller in the empennage for forward thrust. According to the World eVTOL Aircraft Directory (Ref. 2), the Sikorsky

Firefly and Volta aircraft follow a traditional helicopter design with single lift rotor and tail rotor systems, while the Solution F uses two counter-rotating rotors for lift and a rudder at the tail for yaw control vectoring the rotor downwash.

Per the World eVTOL Aircraft Directory (Ref. 2), six aircraft follow a winged multicopter design; the AeroMobil 5.0, Aurora eVTOL, Kitty Hawk Cora, NASA Puffin, Terrafugia TF-2 Lift + Push, and the Opener BlackFly. The AeroMobil is a winged flying car design with two vertically-fixed propellers at the wingtips. The Aurora eVTOL, Kitty Hawk Cora, and Terrafugia TF-2 Lift + Push use horizontally-mounted motors for vertical takeoff and landing. During cruise, the aircraft use wings and a vertically-mounted thrust motor in the rear for added efficiency. The NASA Puffin and Opener BlackFly have fixed motors on fixed wings and rotate the chassis to generate thrust either vertically for takeoff and landing or near horizontally for cruising flight.

The second group of aircraft studied uses thrust-vectoring in the form of rotating wings, ducted fans, or both (Table 1). Of the aircraft discussed in this report, 16 aircraft fit this description. Different designs use one or both of these characteristics, which generally offer a longer range and faster cruise speed relative to the fixed-motor configurations. As Olcott discusses in Reference 16, ducted fans increase thrust for a motor with the same output and reduce noise. However, as noted by Finger et al in Reference 17, these ducts can add weight and drag to the design. As Duffy et al discuss in Reference 18, the advantage of tilt-wing designs is that they allow for reduced drag during cruise compared to fixed-rotor designs. The mechanisms used to vector the thrust increase the weight of the propulsors. A disadvantage however is the increased complexity and therefore lower reliability.

As referenced by the World eVTOL Aircraft Directory, Bartini, Lilium, Moorman, XTI Aircraft Company, and Yoeli (Ref. 2, 19, 20, 21, 22, 23), and shown in Table 1, six of the studied designs incorporate ducted fans; the Bartini Flying Car, Lilium Jet, Aurora Lightning Strike, Urban Aeronautics CityHawk, XTI Aircraft TriFan 600, and the Moller Skycar M400. Five of these aircraft have ducts that rotate to different configurations for takeoff, hover, and landing, and cruise conditions. These aircraft are the Bartini Flying Car, Lilium Jet, Lightning Strike, TriFan 600, and Skycar M400. The proof of concept CityHawk has fixed ducted fans for lift and uses vanes in the ducts to direct the thrust.

Fourteen aircraft use wings for lift according to the World eVTOL Aircraft Directory, Lilium, XTI Aircraft Company, AirSpaceX, Stoll et al, Lovering, and Vimana Global (Ref. 2, 20, 22, 24, 25, 26, 27). These aircraft are the MOBi AirSpaceX, Airbus A[3] Vahana, Joby S2, Joby S4, Lilium Jet, Digi Robotics Droxi, KARI PAV, Terrafugia TF-2 Tiltrotor, Terrafugia TF-X, Vimana Global AAV, Aurora Lightning Strike, Rolls-Royce eVTOL, Moller Skycar M400 and XTI TriFan 600 (Table 1). The AirSpaceX, A[3] Vahana, Rolls-Royce, and Vimana designs have open-mounted motors on rotating wings. These wings are rotated vertically during takeoff, landing, and hover, and are rotated horizontally during cruise. The Joby S2, S4, Digi Robotics Droxi, Moller Skycar, Lilium Jet, TriFan 600, Terrafugia TF-2 Tiltrotor and TF-X have fixed wings and use mechanisms to rotate the motors. The XTI TriFan 600, Moller Skycar M400, and the Lilium Jet use both ducted fans and rotating motors over fixed wings. The Aurora Lightning Strike uses ducted fans mounted below rotating wings for thrust-vectoring during vertical takeoff and landing.

Aircraft Motor and Battery

Table 2. Aircraft Motor and Battery Specifications

Aircraft	Motor Output[1] (kW)	Total Output[2]	Battery Density[3]	Battery Capacity[4]
Advanced Tactics Panther	-	-	-	-
AeroMobil 5.0	-	75	-	-
Airbus Helicopters CityAirbus	100	800	-	110
Airbus Skyways	-	-	-	-
Airbus Vahana A[3]	45	360	200	-
Aquinea Volta	-	90	-	22
Astro Aerospace Elroy (Passenger Drone)	-	-	-	-
ASX MOBi	-	-	-	-
Aurora eVTOL	-	-	-	-
Aurora LightningStrike	125 (18x), 90 (6x)	279 0	-	-
Bartini Flying Car	40	320	200	64
Boeing Cargo Drone	-	-	-	-
Carter Air Taxi	-	-	300	-
Davinci ZeroG	-	-	-	52.8
Digi Droxi	0.8	4.8	-	-
eHang 184	19	152	140	17
Joby S2	-	-	-	-
Joby S4	-	-	-	-
KARI PAV	-	-	-	-
Kármán XK-1	30	240	-	-
Kitty Hawk Cora	-	-	-	-
Kitty Hawk Flyer	-	-	-	-
La Poste Drone	-	-	-	-
Lilium Jet	-	-	-	320

Table 2. (Continued)

Aircraft	Motor Output[1] (kW)	Total Output[2]	Battery Density[3]	Battery Capacity[4]
MMIST Snowgoose	-	-	-	-
Moeller Skycar	-	535	-	-
NASA Puffin	44	88	-	-
Opener BlackFly	-	-	-	12
Pop.Up Next	20	160	-	70
Rolls-Royce eVTOL	-	500	-	-
Sikorsky Firefly	-	142	-	-
Solution F	20	40	-	-
Swiss Postal Matternet M2	-	-	-	-
Terrafugia TF-2 Lift+Push	-	-	-	-
Terrafugia TF-2 Tiltrotor	-	-	-	-
Terrafugia TF-X	-	-	-	-
Trek FlyKart 2	-	-	-	9.6
Urban Aeronautics CityHawk	-	-	-	-
Vimana	60	480	-	-
Volocopter 2X	2	32	-	-
Volocopter VC200	3.9	70.2	-	-
Workhorse Horsefly	-	-	-	-
Workhorse Surefly	-	150	-	15
XTI TriFan	-	750	-	-

[1] Per Motor, kW

[2] Overall Power, kW

[3] Battery Energy Density, Wh/kg

[4] Battery capacity, kWh

Battery technology poses a challenge for eVTOL aircraft. As Luo et al discusses in Reference 28, the theoretical limit for the lithium ion batteries currently used in these aircraft is 250 Wh/kg. As the data in the above table show, all but one aircraft has an energy density below this value (according to Carter in Reference 14, the Carter SR/C Air Taxi has only been proposed and has not been prototype tested nor commercially developed).

Different design configurations have differing requirements for power output of their motors. As mentioned by the World eVTOL Aircraft Directory, Ullman et al, Bartini, Vimana Global, Pradeep and Wei, and Volocopter (Ref. 2, 3, 19, 27, 29, 30) and shown in Table 2, the eHang 184, Kármán XK-1, Pop.Up Next, Volocopter 2X, and VC200, with their vertically-mounted open props, require far less power than their thrust-vectoring counterparts. This can be seen in Table 2; while passenger-carrying multicopters have an average total output of 113 kW, the thrust-vectoring aircraft use an average of 490.8 kW of total output. However, of the aircraft studied in this report, passenger-carrying thrust-vectoring aircraft have a higher passenger capacity than passenger-carrying multicopters. When taking passenger capacity into account, multicopter aircraft have an average of 79.7 kW output per passenger, while thrust-vectoring aircraft have an average output of 123.1 kW per passenger. Helicopters, similarly to multicopters, have an average total output of 90.5 kW and 66.9 kW per passenger. An important parameter to keep in mind is how the motor thrust is being applied in each aircraft. With non-winged multicopter designs, nearly all available thrust is being forced in the downward direction to keep the aircraft flying. Therefore, a higher percentage of the energy capacity is dedicated to vertical lift and not horizontal thrust as there are no wings to relieve the motors of this required vertical force generation. As mentioned earlier and shown in Table 1, there are 16 such multicopter designs discussed in this report. For example, in the case of the Vimana Global AAV the wings generate lift, however, the motor output is higher to account for the necessary thrust to keep the aircraft from stalling. This report discusses 19 multicopter and thrust-vectoring aircraft that incorporate wings, shown in Table 1. Although most thrust-vectoring aircraft incorporate wings, this is not always the case. For example, the Bartini Flying Car has no wings; only eight rotating ducted fans. In this case, the thrust vectors increase to account for the lift and thrust simultaneously, which can be done with less power output using a ducted fan rather than open motors. Bartini in Reference 19 claims a lift-to-drag ratio of 0.8, allowing more of the thrust force to be dedicated to horizontal flight during cruise.

For different gross weights of aircraft, different amounts of power are required to get the aircraft off the ground and flying, which in turn depends on the design of the aircraft. With many projects still in development, data on battery capacity and energy density are limited. These capacities currently range from 140 to 300 Wh/kg (Table 2). According to Ullman et al (Ref. 3), it can be generally inferred eVTOL aircraft with maximum gross takeoff weights of 1500 kg will require 120 horsepower (89 kW) of output for a range of 90 miles (145 kilometers). As aircraft approach a max gross takeoff weight of 1800 kg, the required battery mass increases more significantly.

The battery capacity of the aircraft is dependent on the mass, volume, and type of battery used on the aircraft. The data shown in Table 2 gives the total capacity for the respective aircraft. The requirement for battery capacity is mainly determined by the motor power, and subsequently, the maximum takeoff weight.

Takeoff Weight and Payload Capacity

Table 3. Takeoff and Payload Specifications

Aircraft	Passenger Capacity	Payload Capacity (kg)	Max Gross Weight (kg)
Advanced Tactics Panther	-	3.2-10	-
AeroMobil 5.0	4	-	-
Airbus Helicopters CityAirbus	4	-	-
Airbus Skyways	-	4	-
Airbus Vahana A^3	2	200	815
Aquinea Volta	1	10	520
Astro Aerospace Elroy (Passenger Drone)	2	120	260
ASX MOBi	4	499	-
Aurora eVTOL	2	225	800
Aurora LightningStrike	-	737	5900
Bartini Flying Car	4	400	1100
Boeing Cargo Drone	-	225	-
Carter Air Taxi	4	499	2494.8
Davinci ZeroG	1	150	240
Digi Droxi	-	5	14
eHang 184	1	100	260
Joby S2	2	177	907
Joby S4	4	-	-
KARI PAV	1	-	500
Kármán XK-1	2-8	200	-
Kitty Hawk Cora	2	181	-
Kitty Hawk Flyer	1	-	-
La Poste Drone	-	4	-
Lilium Jet	5	-	2000
MMIST Snowgoose	-	260.8	-
Moeller Skycar	4	-	-
NASA Puffin	1	91	-
Opener BlackFly	1	113	-
Pop.Up Next	2	-	-
Rolls-Royce eVTOL	5	-	-
Sikorsky Firefly	2	-	929.9
Solution F	1	75	245
Swiss Postal Matternet M2	-	2	11.5
Terrafugia TF-2 Lift+Push	4	500	4080
Terrafugia TF-2 Tiltrotor	4	500	3630
Terrafugia TF-X	4	-	-
Trek FlyKart 2	-	-	-
Urban Aeronautics CityHawk	2	-	1930
Vimana	4	400	1050
Volocopter 2X	2	160	450
Volocopter VC200	2	150	450
Workhorse Horsefly	-	4.5	-
Workhorse Surefly	2	181	680
XTI TriFan	6	816	2404

The main parameters that affect the payload of eVTOL aircraft are battery weight and capacity, the motor output, and the configuration of the design. The higher the energy density of the batteries, the lower the mass for an equal amount of energy, and therefore a higher mass can be dedicated to the payload of the aircraft. The higher the motor output, the more thrust can be generated. The configuration determines how much motor output (and therefore energy stored in the batteries) is required to fly the aircraft, which can be increased or decreased based on aerodynamic optimization for each type of design.

Table 3 shows the XTI TriFan 600 and Vimana Global AAV hybrid aircraft have some of the highest payloads out of the sample of aircraft. This is due to their higher thrust power output due to their hybrid designs. In terms of fully electric aircraft, the winged, ducted fan, and thrust-vectoring aircraft have much higher payload capacities than their horizontal-mounted motor counterparts. For example, the ducted fan Bartini Flying Car has a 400 kg capacity versus the eHang 184's 100 kg capacity. Note, from Ullman et al and Bartini (Ref. 3, 19) and shown in Table 2, both aircraft have 64 kWh and 17 kWh battery capacities, respectively. This shows a clear trend; higher battery capacities lead to higher payload capacities. This result is quite trivial. Since the Bartini Flying Car has a four-passenger capacity compared to the single passenger eHang, the larger size requires a larger battery to carry the necessary payload.

The strongest fully electric design comes from the Lilium Jet. The design combines all three of the aforementioned parameters giving the aircraft an estimated 5 passenger payload, the highest of the sampled projects. This further supports the trend of these parameters leading to a higher payload. As additionally noted in Reference 3, its 320 kWh battery capacity (Table 2) is high enough to facilitate such a heavy payload.

Aircraft Performance

Table 4. Aircraft Performance Specifications

Aircraft	Cruise Speed (kph)	Flight Time (min)	Range (km)
Advanced Tactics Panther	113*	5-29*	-
AeroMobil 5.0	-	-	700
Airbus Helicopters CityAirbus	120	15	30
Airbus Skyways	-	-	-
Airbus Vahana A^3	230	26	100
Aquinea Volta	-	40	-
Astro Aerospace Elroy (Passenger Drone)	70*	25	29
ASX MOBi	241	25	104
Aurora eVTOL	180	-	-
Aurora LightinigStrike	555	-	-
Bartini Flying Car	300	30	150
Boeing Cargo Drone	-	-	-
Carter Air Taxi	282	-	256
Davinci ZeroG	70*	25*	-
Digi Droxi	100 (150*)	90	-
eHang 184	60	25	38
Joby S2	322	60	322
Joby S4	225	-	246
KARI PAV	200	15	50
Kármán XK-1	450*	-	-
Kitty Hawk Cora	180	19	100
Kitty Hawk Flyer	-	12-20	-
La Poste Drone	30*	-	25
Lilium Jet	300	60	300
MMIST Snowgoose	-	-	-
Moeller Skycar	533*	350	1300
NASA Puffin	241*	19	80*
Opener BlackFly	128*	30	64
Pop.Up Next	150*	-	50
Rolls-Royce eVTOL	402.3	-	804.7
Sikorsky Firefly	159.3*	15	-
Solution F	-	10	-
Swiss Postal Matternet M2	36	-	20
Terrafugia TF-2 Lift+Push	240*	-	400
Terrafugia TF-2 Tiltrotor	333*	-	555
Terrafugia TF-X	320	-	800
Trek FlyKart 2	102*	30	-
Urban Aeronautics CityHawk	270	60	360*
Vimana	244	-	900
Volocopter 2X	100	27*	27*
Volocopter VC200	100	-	-
Workhorse Horsefly	80*	30	-
Workhorse Surefly	113*	60	113
XTI TriFan	555*	-	2222

* Indicates maximum value

The data presented in Table 4 show most aircraft have similar flight times (18 of which have 25-30-plus minute flight times), but significantly different ranges and cruise speeds. For example, the Volocopter 2X travels at a top speed of 70 kph with 32 kW of power output, a whole 10 kph faster than the eHang 184, despite the 184 outputting nearly five times the power (Tables 2 and 4), both with multicopter architectures. Similarly, the Bartini flying car outperforms the Vimana AAV with a top speed of 300 kph with a 320 kW output, which is 56 kph faster than the 480 kW AAV, despite its design incorporating hybrid technology and a higher motor power output.

The maximum flight time, despite the marginal differences, is mostly dependent on the battery capacity and motor output of the aircraft. For example, the Lilium Jet has a battery capacity of 320 kWh and a flight time of 60 minutes (calculated using given data from range and cruising speed), while the Bartini Flying Car has a 64 kWh battery capacity for half the flight time (Tables 2 and 4). However, this is a general trend and not always a definite quantifiable difference, as seen when comparing the eHang 184 and the Davinci ZeroG. Both aircraft have the same maximum flight time despite the ZeroG having a 52.8 kWh battery and the 184 having 17 kWh, each with a similar 240 and 260 kg maximum takeoff weight, respectively.

The longest-range aircraft are those with higher cruising speeds. Some also use longer flight times; the hybrid AeroMobil, TriFan 600, Terrafugia TF-X, Rolls-Royce, and AAV use their high energy capacities to increase their flight time and therefore their range (Tables 2 and 4). Both the Joby S2 and S4 combine high velocities and long flight times resulting in impressive potentials. The Lilium Jet follows closely behind while boasting a five-passenger payload to the S2's two passenger cockpit. On the other end of the spectrum, the eHang 184 and Volocopter 2X are limited in their ranges by their low flight time and maximum cruise speed due to their designs. For the aircraft with low flight times (25 minutes or less), their ranges are limited by their cruise speeds.

PART II — EXISTING RESEARCH

Aircraft Design

Moore and Goodrich at NASA in Reference 31 have put out goals for increased investment in on-demand aviation, so-called "thin-haul" applications. The primary challenges for this emerging market are in costs of operation, reliability, safety, and emissions, which can be negated by integrating electric propulsion and autonomy to designs. For increased efficiency, NASA sees significant contributions to hybrid-electric propulsion systems, eventually integrating technologies such as high aspect ratio wings and fully electric propulsion systems, as a long-term goal (Ref. 31). Safety through autonomy can be increased by simplifying piloting skills for currently piloted aircraft in order to decrease pilot mistakes, which can make self-preserving autopilot systems less complex and more reliable and eventually making full autonomy simpler.

For urban on-demand air mobility, electric power sources are more advantageous than gas-powered propulsion. According to Moore and Fredericks in Reference 32, electric motors have a higher thrust-to-weight ratio than gas-powered turbine engines due to having fewer high technology/high complexity components. Electric motors are also much easier to scale up or down due to their relative simplicity. This opens a number of possibilities to overcome aerodynamic and structural challenges posed by traditional aircraft. For example, motors can be placed on the chassis where they make the most significant change to the attitude of the aircraft rather than being limited to where they can be structurally supported. Ullman notes in Reference 3 that for winged aircraft, incorporating the propulsor with the upper boundary layer is simpler with more compact and less sophisticated propulsors. This increases lift as pressure on the upper boundary of the wing is lowered.

Hybrid aircraft are also common for urban applications, as discussed in Part I. As Patterson, M. D. et al discuss in Reference 33 the use of fuel in hybrid aircraft gives the aircraft a higher range compared to fully electric aircraft, as fuel has a higher energy storage than current batteries. Since batteries have this limitation, the higher battery mass can decrease efficiency at a point as range increases. This maximum value can be increased by increasing the specific energy density of the batteries, which comes as a result of increased development in battery technology. Additionally, according to Moore (Ref. 34), the current industry for flying taxi services puts costs for travel under \$1.50 per passenger mile. This number is expected to decrease as the aircraft become more common, and the technology is further developed.

Currently, according to Duffy et al in Reference 18, eVTOL design takes into account three factors; cost, safety, and noise pollution. In Reference 35, Fredericks et al note low cost comes from an increase in efficiency, and electric motors offer a solution with higher thrust to weight ratios than their gas-powered counterparts. However, the battery weight is not burned in flight like gas. Because of this, eVTOL aircraft require a significant amount of power during takeoff and hover. This power requirement can be mitigated if the aircraft incorporates wings into the design, making this high required power limited in time, and therefore allowing for a smaller battery with less power exertion required during cruise. Additionally, according to McDonald in Reference 36, the use of ducted electric motors as opposed to open motors increases efficiency during high altitude cruise, which, for urban applications, is not a likely scenario. The performance of these motors is reduced at lower altitudes which are more common in these applications. Otherwise, noted by Xu (Ref. 37), the use of wings can increase the value of the L/D by nearly a factor of 2; increasing the value from a reasonable 3 in horizontally-fixed rotor aircraft to up to 5.5 as seen in winged helicopters. Ultimately, optimization for electric aircraft needs to mitigate high battery weights by incorporating high lift over drag ratios. According to Pornet and Isikveren (Ref. 38), These higher ratios can be obtained via larger propellers, more efficient weight distribution among the structure, and increasing the flow over control surfaces to increase responsiveness to controller inputs. As Riboldi and Gualdoni discuss in Reference 39, previous design iterations can be scaled to improve designs. During the sizing optimization process, the use of data from previous projects is beneficial as battery and motor weights, power outputs, and their relations can be interpolated from these trends. As battery technology progresses to higher energy densities per unit mass or volume, previous designs can be iterated until a new optimal efficiency is obtained for a given mission. More specifically, from Reference 40, the process follows a trend of preliminary sizing, incorporating the battery weights and motor parameters, and the process is iterated until said optimum value is obtained. Generally, this is accurate for scaling within 10% accuracy of the parameters of the aircraft.

Reference 25 notes safety is increased in eVTOL aircraft in part by using distributed electric propulsion in order to add redundancy. Despite the definition of distributed propulsion being broad, as noted by Gohardani in Reference 41, an increased safety factor implies redundancy in the propulsion system, therefore making the aircraft less susceptible to an emergency scenario in the event of a single propeller failure. For example, according to Reference 17, in octocopter designs like the aforementioned eHang 184, having two motors on each branch of the X configuration adds redundancy to the lift propulsion system and therefore increases the aircraft's factor of safety. The X configuration also has higher lateral stability compared to a + design in turbulent conditions, according to Bershadsky in Reference 42. Challenges in implementing redundancy in electric propulsion come from the limitations of modern electric generators. As mentioned in Reference 43, generators that have a higher power output are volumetrically larger and heavier, making them difficult to implement in aircraft. This is a more significant challenge for commercial flight, but with a flying taxi application, less power is required to fly the aircraft and therefore smaller generators can be used. Additionally, as noted in Reference 44, the Auxiliary Power Units (APUs) used in these aircraft are likely required to be specialized for these designs individually. The maximum thrust value required for the aircraft is especially important to know early in the design process, as a higher required output will require a different APU entirely, possibly with different dimensions and mass than another unit of a different size. Therefore, it is important to note the maximum required output of the motors and cap the system at said value to avoid APU or motor failure, overheating, or structural damage.

Additionally, the noise produced by the aircraft can limit its applications geographically based on codes in the urban area of application. This is discussed in Reference 45. For example, if the aircraft is too loud to fly over an urban area, its flight plan may be required to fly over a local body of water or take a non-direct flight plan in order to comply with said ordinances, which can lead to longer flight times and less efficient use of energy. This is most effectively combated with a complete redesign of the aircraft if noise pollution is too significant. Replacing the motors with units focused on limiting noise reduces pollution by about 1 dBA. Increasing the aspect ratio of the fan blade can be effective, however the structural integrity and rotational speed of the fan must be taken into account, so the safety is not compromised, and the increased rotational speed does not increase noise pollution if the diameter of the rotor is increased.

Batteries

As mentioned in Part I, current battery technology is a limiting factor in electric aviation. Batteries with higher energy densities are still being developed. With current battery capacities around 200 Wh/kg, the applications are limited to small one or two passenger aircraft. This will likely increase with the increase in capacities per unit mass and will likely become more useful in aircraft propulsion in the next 20-40 years, according to Reference 46. In the future, eVTOL aircraft will incorporate batteries that are currently in development that have a higher potential energy density, such as the 2500 Wh/kg potential lithium-sulfur batteries or 3000 Wh/kg potential hydrogen fuel cells. Currently, according to Reference 28, the battery mass accounts for about 45% of the total mass of these eVTOL aircraft. In Reference 31, NASA challenged the industry to build "structural batteries" in order to more effectively and efficiently use this weight while keeping the safety standards set by the companies' respective governing bodies.

On-Demand Mobility

Fleet management in urban on-demand air mobility differs significantly from fleet management currently used in industrial aircraft. As noted in Reference 47, a significant challenge for these aircraft is full autonomy as pilots add cost, increase the likelihood of an accident, and cut the already low payload of these aircraft significantly. This however puts the responsibility on the manufacturer to design a safe system where the human error is now in the hands of a designer rather than a pilot. This also adds costs for maintenance and may require certification with currently unwritten regulations. Reference 48 adds the autonomous system must also be able to stick to a flight plan consistently and precisely in order to avoid collisions with surrounding hazards, maintain enough battery power to complete the flight, and for the overall management of the fleet. As discussed in Reference 49, fleets need to be managed via a remote operations center, or ROC, that can gather data about weather, fleet designated airports (so-called vertiports), and the aircraft within the fleet. This would act similarly to a current airport but with fewer personnel that could provide all-around support to ensure the proper distribution of personnel for all different roles required for the ports to operate. Research into the most efficient personnel support structure has yet to be performed. However, according to Reference 50, the airspace capacity of a geographical area will be highly constrained with limited air traffic control (ATC) personnel, therefore limiting scalability at lower altitudes, in which these aircraft would operate. Reference 51 suggests as these aircraft fleets are incorporated into the urban landscape, the ATC system will become less realistic as a management system as there will be too many aircraft to be managed in such a small airspace. Therefore, the system design and safety will be the responsibility of the aircraft manufacturers. Regulations will need to change, for managing an entire fleet of on-demand aircraft will be no simple task for flight planners and air traffic management personnel. With this lack of human communication between ATC and pilots, there must be a communication system between ROCs and the autonomous aircraft for warnings about weather, terrain, airspace, nearby aircraft, and the urban landscape, according to Reference 52. These systems account for and project the demand of the fleet's services and make sure the service is operated smoothly.

As discussed in Reference 29, using an efficient flight trajectory can lower flight times and decrease energy consumption, adding efficiency to the fleet. Current eVTOL aircraft have a higher energy consumption, limiting their efficiency. Not only this, but the power required for the aircraft to climb to higher altitudes would drastically reduce battery charge, making a higher altitude cruise less viable.

Additionally, a descent with a smaller angle and maintaining constant cruise altitude for less time results increased efficiency. Kleinbekman et al in Reference 53 perform a computational model for the optimization of aircraft arrivals over the span of one hour. Assuming the use of the eHang 184 aircraft and 40 incoming aircraft per hour, an aircraft can arrive optimally approximately every 80 seconds. This allows the aircraft to use an energy efficient descent angle to limit battery consumption. Wei and Pradeep (Ref. 54) use an algorithm to optimize the arrival sequencing of winged and wingless eVTOL aircraft. According to their models, winged and wingless eVTOL aircraft should use separate airways for final approach. This is because winged aircraft have higher cruising speeds which allow them to overtake the wingless aircraft, making the landing process more efficient.

According to Kohlman in Reference 55, the most economically efficient fuel system for VTOL aircraft is using natural gas for VTOL applications. This is due to low energy costs and rapid turn-around time compared to all-electric designs. However, this cost advantage comes with added complexity and negative traits such as increased noise and increased emissions. Similarly, hybrid-electric systems are favorable in turn-around times, range, and costs, but come with higher complexity, noise pollution, and emissions.

German et al in Reference 56 produced a case study of delivery aircraft in the San Francisco Bay area. A significant issue discussed was the necessary density for vertiports and their distribution throughout the area. Based on the population around those vertiports, the estimated demand must also be taken into account. If the demand in one area is particularly high due to population density, the traffic can be mitigated through higher vertiport density or larger vertiports in these areas. The report also discusses basic aircraft parameters for the sake of speed of delivery and range, which have been discussed in earlier sections of this report. Ultimately, the project compares two designs for unmanned aerial vehicles (UAVs) for package delivery and sees which one can deliver more mass and volume of packaging per 12 hour "shift", acting as a well-organized initial model for such technology in a potential market. The effectiveness of the system comes down to two main factors; the effectiveness of the infrastructure and the organization of the fleet.

CONCLUSIONS

The future of on-demand transportation lies in eVTOL UAVs. Current projections from the American Institute of Aeronautics and Astronauts (AIAA) in Reference 57 projects more than 660 million people will use UAVs each year by 2035. These projections also anticipate the cost per passenger mile to drop to $0.20 in that same time frame. This will account for a significant portion of mid-range transportation. Through a sample of case studies, Uber in Reference 1 estimates the average commuting time to be reduced by about 88%, and for prices to drop close to current ride sharing fares in the long term. The talking points of this report have focused on current designs and research into the future of urban on-demand air mobility services. With further development in aircraft performance and fleet management, urban on-demand air mobility will likely become more prevalent in modern society.

Author contact: Nicholas Polaczyk polaczyk@iastate.edu
Enzo Trombino enzo@iastate.edu
Dr. Peng Wei, Ph. D pwei@iastate.edu
Dr. Mihaela Mitici, Ph. D M.A.Mitici@tudelft.nl

ACKNOWLEDGMENTS

This project was supported by the resources made available to students and staff members at Iowa State University in Ames, IA and Delft University of Technology in Delft, the Netherlands. The engineering exchange program offices at both Iowa State University and Delft University of Technology made the authors' collaboration possible.

REFERENCES

[1] Holden, Jeff, and Goel, Nikhil. "Fast-Forward to a Future of On-Demand Urban Air Transportation." Uber. 27 October, 2016. https://www.uber.com/info/elevate/

[2] "EVTOL Aircraft Directory." *Electric VTOL News™*, Electric VTOL News™, 4 Jan. 2019, evtol.news/aircraft/.

[3] David G. Ullman, Vincent Homer, Patrick Horgan, Richard Ouellette. *Comparing Electric Sky Taxi Visions*. June 2017.

[4] *EHANG|Official Site-EHANG 184 Autonomous Aerial Vehicle Specs*.

[5] Matternet, Inc. *In Action for Public Health: MATTERNET M2*. Mar. 2017.

[6] *SureFly | Workhorse*.

[7] *Autonomous AT Panther Air/Ground Robot – On Sale | Advanced Tactics, Inc.*

[8] "Boeing Unveils New Unmanned Cargo Air Vehicle Prototype." *MediaRoom*, 10 Jan. 2018.

[9] "Specs - Passenger Drone." *Passenger Drone*.

[10] 8th, February. "Airbus Completes First Flight Demonstration for Its Commercial Parcel Delivery Drone 'Skyways.'" Airbus.

[11] *Aerospace | Workhorse.*

[12] "Davinci." *Davinci ZeroG.*

[13] DPDgroup. *The DPDgroup Drone Parcel Delivery 2.0.*

[14] Carter, Jay. *Carter SR/C Air Taxi.* Uber Elevate Summit, Dallas, Texas.

[15] *CQ-10.*

[16] Olcott, Dennis D. "Tri-Fan VTOL Conceptual Design." *16th AIAA Aviation Technology, Integration, and Operations Conference*, American Institute of Aeronautics and Astronautics, 2016.

[17] Finger, D. F., et al. "The Impact of Electric Propulsion on the Performance of VTOL UAVs." *66. Deutscher Luft- Und Raumfahrtkongress DLRK 2017*, 2017.

[18] Duffy, Michael J., et al. "A Study in Reducing the Cost of Vertical Flight with Electric Propulsion." *17th AIAA Aviation Technology, Integration, and Operations Conference*, American Institute of Aeronautics and Astronautics, 2017.

[19] "Bartini.aero — the Flying Car." *Bartini.aero — the Flying Car.*

[20] *Lilium.*

[21] Moorman, Robert W. "XTI Aircraft Refines Its TriFan 600 BizJet." *Verti-Flite*, March/April 2018.

[22] "TriFan 600 | XTI Aircraft Company." *XTI Aircraft Company.*

[23] Yoeli, Rafi. "Ducted Fan Utility Vehicles and Other Flying Cars." *2002 Biennial International Powered Lift Conference and Exhibit*, American Institute of Aeronautics and Astronautics, 2002.

[24] *ASX – ASX.*

[25] Alex M. Stoll, JoeBen Bervirt, Percy P. Pei, Edward V. Stilson. *Conceptual Design of the Joby S2 Electric VTOL PAV.* 2014.

[26] Lovering, Zach. "Vahana's First Flight a Success – Vahana." *Vahana*, Vahana, 1 Feb. 2018.

[27] *Vimana Global.*

[28] Luo, Jiajie, et al. *PRELIMINARY DESIGN OF VTOL PERSONAL FLYING MACHINE – DEREAMWINGS.* Oct. 2016.

[29] Pradeep, Priyank, and Peng Wei. "Energy Efficient Arrival with RTA Constraint for Urban eVTOL Operations." *2018 AIAA Aerospace Sciences Meeting*, American Institute of Aeronautics and Astronautics, 2018.

[30] *Volocopter - Product.*

[31] Moore, Mark, and Ken Goodrich. *On-Demand Mobility: Goals, Technical Challenges, and Roadmaps.* ntrs.nasa.gov, 2015.

[32] Moore, M. D., and W. J. Fredericks. "Misconceptions of Electric Propulsion Aircraft and Their Emergent Aviation Markets, AIAA 2014-0535." *52nd AIAA Aerospace Sciences Meeting*, 2014.

[33] Patterson, M. D., et al. "Performance Analysis and Design of On-Demand Electric Aircraft Concepts." *12th AIAA Aviation Technology, Integration, and Operations (ATIO) Conference and 14th AIAA/ISSMO Multidisciplinary Analysis and Optimization Conference*, American Institute of Aeronautics and Astronautics, 2012.

[34] Moore, Mark D. *The Third Wave of Aeronautics: On-Demand Mobility.* SAE Technical Paper, 2006.

[35] Fredericks, William J., et al. "Benefits of Hybrid-Electric Propulsion to Achieve 4x Cruise Efficiency for a VTOL UAV." *2013 International Powered Lift Conference*, American Institute of Aeronautics and Astronautics, 2013.

[36] McDonald, Robert A. "Electric Propulsion Modeling for Conceptual Aircraft Design." *52nd Aerospace Sciences Meeting*, American Institute of Aeronautics and Astronautics, 2014.

[37] Xu, Jia. *Design Perspectives on Delivery Drones.* RAND, 2017.

[38] Pornet, C., and A. T. Isikveren. "Conceptual Design of Hybrid-Electric Transport Aircraft." *Progress in Aerospace Sciences*, vol. 79, Elsevier, Nov. 2015, pp. 114–35.

[39] Riboldi, C. E. D., and F. Gualdoni. "An Integrated Approach to the Preliminary Weight Sizing of Small Electric Aircraft." *Aerospace Science and Technology*, vol. 58, Elsevier, Nov. 2016, pp. 134–49.

[40] Tyan, Maxim, et al. "Comprehensive Preliminary Sizing/resizing Method for a Fixed Wing--VTOL Electric UAV." *Aerospace Science and Technology*, vol. 71, Elsevier, 2017, pp. 30–41.

[41] Gohardani, Amir S. "A Synergistic Glance at the Prospects of Distributed Propulsion Technology and the

Electric Aircraft Concept for Future Unmanned Air Vehicles and Commercial/military Aviation." *Progress in Aerospace Sciences*, vol. 57, Elsevier, Feb. 2013, pp. 25–70.

[42] Bershadsky, Dmitry. *Electric Multirotor Design and Optimization*. Georgia Institute of Technology, 2017,

[43] ---. "Challenges of Future Aircraft Propulsion: A Review of Distributed Propulsion Technology and Its Potential Application for the All Electric Commercial Aircraft." *Progress in Aerospace Sciences*, vol. 47, no. 5, Elsevier, July 2011, pp. 369–91.

[44] De Tenorio, Cyril. *Methods for Collaborative Conceptual Design of Aircraft Power Architectures*. Georgia Institute of Technology, 2010.

[45] Brown, Arthur, and Wesley Harris. "A Vehicle Design and Optimization Model for On-Demand Aviation." *2018 AIAA/ASCE/AHS/ASC Structures, Structural Dynamics, and Materials Conference*, American Institute of Aeronautics and Astronautics, 2018, doi:10.2514/6.2018-0105.

[46] Hepperle, Martin. *Electric Flight - Potential and Limitations*. elib.dlr.de, 2012.

[47] Moore, Mark D., and Kenneth H. Goodrich. "High Speed Mobility through On-Demand Aviation." *2013 Aviation Technology, Integration, and Operations Conference*, American Institute of Aeronautics and Astronautics, 2013.

[48] Alnaqeb, Abdullah H., et al. "Online Prediction of Battery Discharge and Flight Mission Assessment for Electrical Rotorcraft." *2018 AIAA Aerospace Sciences Meeting*, American Institute of Aeronautics and Astronautics, 2018, doi:10.2514/6.2018-2005.

[49] Nneji, Victoria C., et al. "Functional Requirements for Remotely Managing Fleets of On-Demand Passenger Aircraft." *2018 AIAA Aerospace Sciences Meeting*, American Institute of Aeronautics and Astronautics, 2018.

[50] Vascik, P. D. and Hansman, R. J., "Scaling Constraints for Urban Air Mobility Operations: Air Traffic Control, Ground Infrastructure, and Noise," Paper AIAA 2018-3849, 2018 Aviation Technology, Integration, and Operations Conference, Atlanta, GA, June 25-29, 2018.

[51] Hansman, R. John, and Parker Vascik. *Constraint Identification In On-Demand Mobility For Aviation Through An Exploratory Case Study Of Los Angeles*. dspace.mit.edu, May 2018.

[52] Gawdiak, Yuri, et al. "Air Transportation Strategic Trade Space Modeling and Assessment Through Analysis of On-Demand Air Mobility with Electric Aircraft." *12th AIAA Aviation Technology, Integration, and Operations (ATIO) Conference and 14th AIAA/ISSMO Multidisciplinary Analysis and Optimization Conference*, American Institute of Aeronautics and Astronautics, 2012, doi:10.2514/6.2012-5594.

[53] Kleinbekman, I. C., Mitici, M. A. and Wei, P., "eVTOL Arrival Sequencing and Scheduling for On-Demand Urban Air Mobility," Paper AIAA/EIEE Digital Avionics Systems Conference (DASC), London, England, UK, September 2018.

[54] Pradeep, P. and Wei, P., "Heuristic Approach for Arrival Sequencing and Scheduling for eVTOL Aircraft in On-Demand Urban Air Mobility," AIAA/IEEE Digital Avionics Systems Conference (DASC), London, England, UK, September, 2018.

[55] Kohlman, Lee W. and Patterson, Michael D., "System-Level Urban Air Mobility Transportation Modeling and Determination of Energy-Related Constraints," Paper AIAA 2018-3677, 2018 Aviation Technology, Integration, and Operations Conference Proceedings, Atlanta, GA, June 25-29, 2018.

[56] German, Brian, et al. "Cargo Delivery in by Passenger eVTOL Aircraft: A Case Study in the San Francisco Bay Area." *2018 AIAA Aerospace Sciences Meeting*, American Institute of Aeronautics and Astronautics, 2018.

[57] "Projected Demand and Potential Impacts to the National Airspace System of Autonomous, Electric, On-Demand Small Aircraft." *12th AIAA Aviation Technology, Integration, and Operations (ATIO) Conference and 14th AIAA/ISSMO Multidisciplinary Analysis and Optimization Conference*, American Institute of Aeronautics and Astronautics, 2012, doi:10.2514/6.2012-5595.